WILEY PLUS
www.wileyplus.com

ALL THE HELP, **RESOURCES**, AND PERSONAL **SUPPORT** YOU AND YOUR STUDENTS NEED!

www.wileyplus.com/resources

2-Minute Tutorials and all of the resources you & your students need to get started.

Student support from an experienced student user.

Collaborate with your colleagues, find a mentor, attend virtual and live events, and view resources.
www.WhereFacultyConnect.com

Pre-loaded, ready-to-use assignments and presentations. Created by subject matter experts.

Technical Support 24/7 FAQs, online chat, and phone support.
www.wileyplus.com/support

Your *WileyPLUS* Account Manager. Personal training and implementation support.

DISCOVERING PHYSICAL GEOGRAPHY

Second Edition

DISCOVERING
PHYSICAL GEOGRAPHY

Second Edition

ALAN F. ARBOGAST | MICHIGAN STATE UNIVERSITY

WILEY

JOHN WILEY & SONS, INC.

VICE PRESIDENT AND EXECUTIVE PUBLISHER   Jay O'Callaghan
EXECUTIVE EDITOR   Ryan Flahive
ASSOCIATE EDITOR   Veronica Armour
MARKETING MANAGER   Margaret Barrett
MEDIA EDITOR   Lynn Pearlman
EDITORIAL ASSISTANT   Darnell Sessoms
MARKETING ASSISTANT   Susan Matulewicz
CREATIVE DIRECTOR   Harry Nolan
INTERIOR DESIGN   Wendy Lai
COVER DESIGN   Howard Grossman
FRONT COVER PHOTO   Bill Hatcher/National Geographic/Getty Images, Inc.
BACK COVER PHOTO   Courtesy of Alan Arbogast
SENIOR PRODUCTION EDITOR   Sujin Hong
SENIOR ILLUSTRATION EDITOR   Sandra Rigby
SENIOR PHOTO EDITOR   Jennifer MacMillan

This book was set by Prepare and printed and bound by Quad/Graphics. The cover was printed by
Quad/Graphics.

This book is printed on acid free paper. ∞

Founded in 1807, John Wiley & Sons, Inc. has been a valued source of knowledge and understanding for
more than 200 years, helping people around the world meet their needs and fulfill their aspirations. Our
company is built on a foundation of principles that include responsibility to the communities we serve and
where we live and work. In 2008, we launched a Corporate Citizenship Initiative, a global effort to address
the environmental, social, economic, and ethical challenges we face in our business. Among the issues we
are addressing are carbon impact, paper specifications and procurement, ethical conduct within our business and among our vendors, and community and charitable support. For more information, please visit our
website: www.wiley.com/go/citizenship.

Evaluation copies are provided to qualified academics and professionals for review purposes only, for
use in their courses during the next academic year. These copies are licensed and may not be sold or
transferred to a third party. Upon completion of the review period, please return the evaluation
copy to Wiley. Return instructions and a free of charge return shipping label are available
at www.wiley.com/go/returnlabel. Outside of the United States, please contact your local representative.

ISBN 978-0-470-52852-5
ISBN: 978-0-470-91778-7 (BRV)

Printed in the United States of America

10 9 8 7 6 5 4 3 2 1

For Jenn, Hannah, and Rosie

ALAN F. ARBOGAST is a Professor of Geography at Michigan State University in East Lansing, Michigan. He is an award-winning instructor at MSU where he teaches a variety of classes at the undergraduate and graduate levels. Most of these courses are related to physical geography, geomorphology, and human/environment interactions. Alan frequently teaches large lecture classes and is known for being an enthusiastic instructor who uses innovative approaches. He has also taught extensively in Australia and New Zealand in association with MSU's Study Abroad program.

Alan's research focuses on Holocene landscape evolution of eolian, coastal, and fluvial environments in the Great Lakes region. Most of his research focuses on the age and formation of coastal sand dunes along Michigan's Great Lakes. Alan has published over 30 combined scientific papers, book chapters, and monographs related to this work. His research has been funded by NASA, The National Science Foundation, and the State of Michigan. He is a member of the Association of American Geographers, Geological Society of America, and American Quaternary Association. Alan is married to Jennifer and has two daughters, Hannah and Rosie.

Introduction to Physical Geography is a high-enrollment course at most universities. The usual goal of this course is to help students understand the Earth as a natural system and how various processes on the planet operate over time and space. Given the interactions that occur among these natural processes, physical geography requires integration of many different topics. For example, students are expected to understand how seasonal Earth–Sun relationships affect atmospheric circulation, which in turn influences the distribution of vegetation. In addition, physical geography is an applied discipline that can inform decisions about environmental issues such as global warming, earthquake hazards, coastal erosion in populated areas, soil degradation, and deforestation, to name a few.

Discovering Physical Geography: A Visually-Oriented, Interactive Approach

Everyone associated with geography is fully aware that geographical literacy in the United States is very poor. This poor comprehension exists not only with respect to the fundamental issue of locating places, but also in understanding the age and processes associated with physical landscapes. Because many students enroll in a physical geography course only to fulfill a general education requirement in natural science, they frequently have little enthusiasm for the subject or fail to see the relevance of the class in their lives. In addition, a high percentage of students are simply afraid of science and are thus intimidated by the course. As a result, they believe that their chances of success in the class are low.

Discovering Physical Geography confronts these barriers to learning and classroom success in a number of unique ways. At a fundamental level, the text is written in a conversational style that is easily understandable to the average non-major. This writing style was not chosen at the expense of science as the topics covered are dealt with in the scope and breadth as they are elsewhere. In fact many topics, such as Earth/Sun geometry, reconstructing past climates, ongoing climate change, geologic time, tides, and formation of sand dunes, to name a few, are covered in a more comprehensive way in this text than in others. The discussions are just presented in a more accessible way to help the non-majors and those intimidated by science as they grapple with the concepts. My goal was to help them make that connection a bit more readily.

Another way that this text confronts learning barriers is by offering students rich graphics and striking photos that depict physical processes and the natural variability of the landscape in memorable ways. The quality and breadth of the illustrations are designed to spark students' interest and help them see the relevance of physical geography to their daily lives. The illustrations are accompanied by a dynamic tool, *GeoDiscoveries,* which is an interactive, web-based multimedia resource. *GeoDiscoveries* consists of a variety of animations and simulations that allow students to visualize and manipulate many of the factors associated with geographical processes and see the results over time and space. The multimedia will enhance students' learning as they participate more closely with geographical processes and will reinforce the integrative nature of the discipline by showing related variables in motion. This form of active learning will, in turn, help promote long-term retention of the material. The multimedia is fully integrated within the chapter text in distinct sections that direct students to the related modules on the website and explain to students what they should expect to learn by interacting with it.

Each *GeoDiscoveries* module on the website also includes a variety self-assessment questions for students. Students can use these questions to test their understanding of topics, or instructors can assign them as homework. Such questions allow both students and instructors to assess learning. They also provide the foundation for exam questions that are independent of class lectures. The *GeoDiscoveries* modules should motivate more students to interact with the textbook and media because they will more readily see their connection with the course. They contain a variety of self-assessment tools that will engage the interest of *all* students enrolled in physical geography, not just those who are scientifically inclined or have a background in Earth science.

Changes for the Second Edition

Students and faculty often wonder why it is necessary to produce new editions of textbooks. The need for new editions in geography is particularly necessary because events frequently occur that change the nature of the cultural and physical landscape. Similarly, scientists learn more about Earth processes between editions that advance the state of knowledge. Since the first edition of this text was written, a number of important events have occurred on Earth that require coverage in a second edition. For example, the extent of the annual Antarctic ozone hole shows signs of changing. A new system of tornado classification, the *Enhanced Fujita Scale*, is now used to estimate the strength of tornadoes. Using this system, the monstrous Greensburg tornado in 2007 was the very first EF5 tornado every recorded, one that wiped out 95% of this small town in Kansas. An intense tsunami hit the islands of American Samoa, Samoa, and Tonga in September 2009. This tsunami devastated coastlines, killed over 100 people, and further demonstrated the need for systematic warning programs. In 2010 alone, a massive earthquake struck Haiti and extensive flooding occurred in Pakistan. The earthquake killed over 200,000 people and made another 1,000,000 homeless. Flooding in Pakistan was related to the summer monsoon and displaced about 15 million people.

On the climate front, concentrations of atmospheric carbon dioxide on Earth have risen still further, from about 384 ppm in 2007 to 390 ppm in 2010. This continued increase is causing the vast majority of climate scientists to become even more concerned about the effects of climate change and potential future warming of Earth. As a result, the melting of the Greenland ice cap apparently continues to accelerate, including the break off of a chunk of ice in summer 2010 that was the size of Manhattan Island in New York. In a similar vein, marine corals in Indonesia are now bleaching at a rapid rate due to warming ocean waters. In this context, the global community continues to struggle with how to deal with climate, including the tepid discussions about tradable carbon credits in the U.S. Although these are not the only changes in the physical geography of Earth, they illustrate why a new edition is warranted.

Aside from the new coverage of topics in this edition, another reason to create a new one was simply to improve upon the first and make it a better resource for students. The first edition was well received, and a number of reviewers and users of it offered excellent suggestions to strengthen the text. In this context, a variety of new figures was constructed to better illustrate concepts, such as those associated with tides and eolian processes, as well as enhanced discussions of certain topics. Given the strong emphasis and integration of technology, twenty-two new *GeoDiscoveries* modules were included in this text. Fourteen of these modules are brand new, with the rest absorbed from existing sources that were previously untapped. These new modules were chosen to strengthen topical discussions and to increase comprehensive coverage of concepts. A new global climates module was created, for example, that focuses on the specific Köppen climate regions. This module enabled the text discussion of global climates to become much more streamlined and thus more readable. In addition, a climate change module was constructed that nicely illustrates the fundamentals of climate change science. This module includes a great animation of a potential future warming scenario on Earth from the National Center for Atmospheric Research. A fantastic new module illustrating the concept of El Niño is also included, as is a module that focuses on the glacial mass budget. This latter module includes links to great time-lapse studies of glaciers in Greenland and Alaska. A new movie from the United States Geological Survey is also included that demonstrates the rate of land loss in southern Louisiana due to rising sea levels and subsidence. A variety of similar new modules are sprinkled throughout the text.

I am particularly excited about a theme that has been significantly enhanced in this edition of the text. This theme is entitled *Human Interactions* and focuses on some of the many ways that physical geography is relevant to human/environment interactions such as global climate change, agriculture, and energy production. Although the first edition contained some discussion of these issues, such as the evidence for human-induced global climate change, deforestation, and human impacts on streams, it was not a consistent theme of the book. Given the growing relevance of human/environmental issues in the world, and the important role physical geographers play in understanding and solving them, this theme was systematically strengthened in this edition by adding the following *Human Interactions* discussions:

1. *Solar energy*—In the context of our stated goal to increase sustainable energy resources, a discussion of this energy source was included in Chapter 4 (*The Global Energy System*). This discussion focuses on the methods used to obtain this energy, the pros and cons of using it, and the geography of solar energy in the U.S.

2. *Wind energy*—Given the growing desire to develop sustainable energy resources, a discussion of this energy source was included in Chapter 6 (*Atmospheric Pressure, Wind, and Global Circulation*). This discussion focuses on the methods used to obtain this energy, the pros and cons of using it, and the geography of wind energy in the U.S.

3. *Humans and soils*—The relevance of soils to people was more thoroughly discussed in Chapter 11 (*The Global Distribution and Character of Soils*) by including a section showing how soils are important

engineering considerations. In addition, a discussion of the National Resource Conservation Soil Survey and its relevance to agriculture is now included.

4. *Humans and petroleum*—Given the significance of petroleum in the modern economy, Chapter 12 (*Earth's Internal Structure, Rock Cycle, and Geologic Time*) now includes a discussion of petroleum geography. This discussion includes a brief history of oil production, the geography of known reserves, current consumption patterns, and the concept of peak oil.

5. *A new chapter*—This text includes a new chapter, Chapter 20 (*Relevance of Physical Geography to Environmental Issues*) that focuses entirely on the significance of physical geography to human/environment interactions. The chapter begins with a discussion of the history of human population on Earth, future projections, and why these concepts are important regarding resource consumption. Subsequently, the chapter focuses on three case studies that clearly demonstrate the relevance of physical geography. The first such study examines the growth of Las Vegas, Nevada, and how it relates to the geography of water in the Southwest U.S. This discussion includes description of water rights, the history of Colorado River Compact, and future concerns. The second case study investigates the process of soil salinization and how it is a growing problem in California's San Joaquin Valley and in Australia. The final case study examines the geography of panda habitat in China. This discussion describes the history of habitat loss, current trends, and how geographers have helped to potentially stabilize the situation.

Special Features of the Text

To help students navigate their way through the book and better appreciate the nature and scope of physical geography, the chapters include a number of special and innovative features:

- **Discover...**—This feature presents an opportunity for students to discover the patterns and causes of particularly interesting geographic phenomena, such as rainbows, wildfires, and unusual clouds (to name a few) on Earth. Each chapter has one such feature, which includes a photo and explanation of how geographers interpret this aspect of the Earth system. The goal of this feature is to make students realize that there is more to the physical landscape than meets the eye, which will hopefully spark their interest in what they see around them.

- **GeoDiscoveries Multimedia**—Each chapter contains several multimedia modules that explain to students what they can expect to see and learn as they interact with the *GeoDiscoveries* simulations and animations on the text's website. The website media also include a variety of self-assessment questions for students.

- **Visual Concept Check**—To provide students with a means of self-testing within the flow of chapter content, this feature offers a scenario with an illustration and questions to test students' understanding of key chapter concepts. Answers to the visual concept checks appear at the end of the chapter.

- **Key Concepts to Remember**—This feature is an interim summary that appears after specific sections of the chapter to help students check their comprehension of the key concepts covered.

- **Locator Maps with Photographs**—Photographs of non-U.S. sites are accompanied by a small map indicating the location of the site shown.

- **Marginal Glossary**—Key terms are set in boldface type in the text and defined at the foot of the page for easy recognition and reference.

- **The Big Picture**—This concluding section uses a photograph to explain how the concepts learned in the chapter relate to the overall scope of physical geography, using an illustration to demonstrate the relationship.

- **Summary of Key Concepts**—The main points of the chapter are summarized.

- **Check Your Understanding**—Self-assessment questions at the end of the chapter allow students to test their comprehension.

Acknowledgments

A project of this scope naturally required the help and support of a number of people. I would first like to thank the friendly people at John Wiley & Sons, Inc., for their faith in my vision and for giving me the freedom and resources to see it through. In this context, I have to give special thanks to a group of four who, in reality, are good friends who happened to help me create a great book. Ryan Flahive deserves special kudos for recruiting me, backing me up many times, and making a number of good things happen along the way. Veronica Armour was an organizational force who effectively guided the process and just made it a whole lot of fun, even when she was about to get married. For the striking photographs, I am especially indebted to Jennifer MacMillen for her excellent judgment and kind disposition. Through two editions she has consistently made me feel special and I really appreciate that. Suzanne Thibodeau recreated her important role as my guru to help me properly organize the new Chapter 20. I will always be indebted to Suzanne for the help she has given me through two editions.

A variety of other people associated with Wiley have routinely demonstrated their incredible creativity, talent, and good nature. Lynn Pearlman and Josh Barinstein did an outstanding job with media development and integration. My thanks also go to Sandra Rigby for the rich illustration program; and for the beautiful design that supports all our efforts, to Harry Nolan. For the expert coordination of all the players involved in producing the book, I am extremely grateful to Sujin Hong. Major thanks also to Margaret Barrett, who oversaw the evolution of a creative marketing strategy, and to the many Wiley sales reps who enthusiastically carried it out.

I also want to extend my thanks to the many geographers and scientists who are doing the research that needs to be done in this complex time of Earth history. Their research is crucial to a better understanding of our world and for the development of sustainable land-use practices. Many members of this team served as reviewers for this book and I want to thank them for taking the time out of their very busy schedules to help out. In particular I wish to thank Dr. Patrick Hesp at Louisiana State University for his thorough and helpful review of Chapters 18 and 19. I also wish to thank my good friend Dr. Randall Schaetzl at Michigan State University for his advice and help with photographs. Other reviewers include:

John All, *Western Kentucky University*

John Anderton, *Northern Michigan University*

Jake Armour, *University of North Carolina, Charlotte*

Barbara Batterson-Rossi, *Cuyamaca College (California)*

Kevin Baumann, *Indiana State University*

Sheryl Beach, *George Mason University*

Jason Blackburn, *Baton Rouge Community College*

Greg Bohr, *California Polytechnic State University, San Luis Obispo*

Margaret Boorstein, *Long Island University, C.W. Post*

William Budke, *Ventura College (California)*

Michaele Ann Buell, *Northwest Arkansas Community College*

David Cairns, *Texas A&M University*

Tom Carlson, *University of Washington, Tacoma*

Nicole Cerveny, *Mesa Community College*

Philip Chaney, *Auburn University*

Richard Cooker, *Kutztown University*

Ron Crawford, *University of Alaska, Anchorage*

James Davis, *University of Utah*

Lisa DeChano, *Western Michigan University*

Carol DeLong, *Victor Valley College*

Mike DeVivo, *Grand Rapids Community College*

Lynn Fielding, *El Camino College*

Donald Friend, *Minnesota State University, Mankato*

Colleen Garrity, *SUNY Geneseo*

Alan Gaugert, *Glendale Community College*

David Goldblum, *University of Wisconsin, Whitewater*

John Greene, *University of Oklahoma, Norman*

Duane Griffin, *Bucknell University*

Chris Groves, *Western Kentucky University*

Patrick Hesp, *Louisiana State University*

Miriam Helen Hill, *Jacksonville State University*

Curt Holder, *University of Colorado, Colorado Springs*

David Harms Holt, *Miami University of Ohio*

Michael Holtzclaw, *Central Oregon Community College*

Robert Hordon, *Rutgers University*

Hixiong (Shawn) Hu, *East Stroudsburg University*

Peter Jacobs, *University of Wisconsin*

Scott Jeffrey, *Community College of Baltimore County, Catonsville*

Kris Jones, *Long Beach City College*

Stacy Jorgenson, *University of Hawaii, Manoa*

Theron Josephson, *Brigham Young University, Idaho*

Trudy Kavanagh, *University of Wisconsin, Oshkosh*

Ryan Kelly, *Lexington Community College*

Joseph Kerski, *U.S. Geological Survey*

John Keyantash, *California State University, Dominguez Hills*

Eric Keys, *Arizona State University*

Marti Klein, *Saddleback College*

Dafna Kohn, *Mt San Antonio College*

Jean Kowal, *University of Wisconsin, Parkside*

Jack Kranz, *California State University, Northridge*

Barry Kronenfeld, *George Mason University*

Steve LaDochy, *California State University, Los Angeles*

Charles Lafon, *Texas A&M University*

Jeff Lee, *Texas Tech University*

Elena Lioubimtseva, *Grand Valley State University*

James Lowry, *Stephen F Austin State University*

David Lyons, *Century College*

Michael Madsen, *Brigham Young University, Idaho*

Emmanuel Mbobi, *Kent State University, Stark*

Christine McMichael, *Morehead State University*

Beverly Meyer, *Oklahoma Panhandle State University*

Armando V. Mendoza, *Cypress College*

Peter Mires, *Eastern Shore Community College*

Laurie Molina, *Florida State University*

Christopher Murphy, *Community College of Philadelphia*

Steven Namikas, *Louisiana State University*

Andrew Oliphant, *San Francisco State University*

Darren Parnell, *Kutztown University*

Charlie Parson, *Bemidji State University*

Brooks Pearson, *University of West Georgia*

Robert Pinker, *Johnson County Community College*

Rhea Presiado, *Pasadena City College*

Kevin Price, *University of Kansas*

David Privette, *Central Piedmont Community College*

Carl Reese, *University of Southern Mississippi*

David Sallee, *University of North Texas*

Justin Scheidt, *Ferris State University*

Peter Scull, *Colgate University*

Glenn Sebastian, *University of Southern Alabama*

John Sharp, *SUNY New Paltz*

Wendy Shaw, *Southern Illinois University, Edwardsville*

Andrew Shears, *Kent State University*

Binita Sinha, *Diablo Vallege College*

Brent Skeeter, *Salisbury University*

Thomas Small, *Frostburg State University*

Lee Stocks, *Kent State University*

Sam Stutsman, *University of South Alabama*

Aondover Tarhule, *University of Oklahoma*

Roosmarijn Tarhule-Lips, *University of Oklahoma, Norman*

Thomas Terich, *Western Washington University*

Donald Thieme, *Georgia Perimeter College*

U. Sunday Tim, *Iowa State University*

Paul E. Todhunter, *University of North Dakota*

Erika Trigoso, *University of Arizona*

Alice Turkington, *University of Kentucky, Lexington*

Lensyl Urbano, *University of Memphis*

Michael Walegur, *Moorpark College*

David Welk, *State Center Community College District, Clovis Center*

Forrest Wilkerson, *Minnesota State University, Mankato*

Dennis Williams, *Southern Nazarene University*

Joy Wolf, *University of Wisconsin, Parkside*

Lin Wu, *California State Polytechnic University, Pomona*

Ken Yanow, *Southwestern College, Chula Vista*

Brent Yarnal, *Pennsylvania State University*

Hengchun Ye, *California State University, Los Angeles*

Zhongbo Yu, *University of Nevada, Las Vegas*

Guoqing Zhou, *Old Dominion University*

Yu Zhou, *Bowling Green State University*

Tongxin Zhu, *University of Minnesota, Duluth*

Charles Zinser, *SUNY Plattsburgh*

Craig ZumBrunnen, *University of Washington*

A thank you also goes to the following professors and companies for allowing us to use select videos and animations in the GeoDiscoveries modules:

AccuWeather, Inc.

Allianz SE

Bill Dietrich, *University of California, Berkeley*

Paul Heller, *University of Wyoming*

Glacier National Park

NOAA Center for Tsunami Research

Andrew Oberhardt

Chris Paola, *University of Minnesota*

Taichiro Sakagami, *Duke University*

Doug Smith, *California State University, Monterey Bay*

The Times-Picayune Publishing Company

University Corporation for Atmospheric Research

U.S. Geological Survey

Last, and most important, I wish to express my heartfelt thanks to my family for their support. My wife, Jennifer, is a wonderful and beautiful person who has enriched my life in ways that I cannot describe. She urged me to write the first edition and has patiently lived with the many ups and downs along the way. I simply could not have done it without her support. My children, Hannah and Rosie, have consistently given me their affection throughout this process and I love them dearly. I hope this effort inspires them someday to stretch the limits of their potential.

Alan F. Arbogast
Professor of Geography
Michigan State University

The Teaching and Learning Package

Discovering Physical Geography, Second Edition is supported by a comprehensive supplements package that includes an extensive selection of print, visual, and electronic materials. Additional information, including prices and ISBNs for ordering, can be obtained by contacting John Wiley & Sons. You can find your local representative and their contact information using our "rep locator" website: www.wiley.com/college/rep.

WileyPLUS

WileyPLUS is an innovative, research-based, online environment for effective teaching and learning.

What Do Students Receive with *WileyPLUS*?

A RESEARCH-BASED DESIGN

WileyPLUS provides an online environment that integrates relevant resources, including the entire digital textbook, in an easy-to-navigate framework that helps students study more effectively.

- *WileyPLUS* adds structure by organizing textbook content into smaller, more manageable "chunks."
- Related media, examples, and sample practice items reinforce the learning objectives.
- Innovative features such as calendars, visual progress tracking, and self-evaluation tools improve time management and strengthen areas of weakness.

ONE-ON-ONE ENGAGEMENT

With *WileyPLUS* for *Discovering Physical Geography, Second Edition*, students receive 24/7 access to resources that promote positive learning outcomes. Students engage with related examples (in various media) and sample practice items, including:

- Practice Questions provide immediate feedback to true/false, multiple choice, and short answer questions.
- www.ConceptCaching.com: caches linked directly to the e-book give additional examples of the concepts.
- GeoDiscoveries Modules and questions.
- Interactive Drag-and-Drop Exercises challenge students to correctly label important illustrations from the textbook.
- Google Earth™ Tours.

MEASURABLE OUTCOMES

Throughout each study session, students can assess their progress and gain immediate feedback. *WileyPLUS* provides precise reporting of strengths and weaknesses, as well as individualized quizzes, so that students are confident they are spending their time on the right things. With *WileyPLUS*, students always know the exact outcome of their efforts.

What Do Instructors Receive with *WileyPLUS*?

WileyPLUS provides reliable, customizable resources that reinforce course goals inside and outside of the classroom as well as visibility into individual student progress. Pre-created materials and activities help instructors optimize their time.

CUSTOMIZABLE COURSE PLAN

WileyPLUS comes with a pre-created Course Plan designed by a subject matter expert uniquely for this course. Simple drag-and-drop tools make it easy to assign the course plan as-is or modify it to reflect your course syllabus.

PRE-CREATED ACTIVITY TYPES INCLUDE

- Questions
- Readings and resources
- Presentation
- Print Tests
- Concept Mastery
- Project

COURSE MATERIALS AND ASSESSMENT CONTENT

- Lecture Notes PowerPoint Slides
- Classroom Response System (Clicker) Questions
- Image Gallery
- Instructor's Manual
- Gradable Reading Assignment Questions (embedded with online text)
- GeoDiscoveries Media Library
- Testbank

GRADEBOOK

WileyPLUS provides instant access to reports on trends in class performance, student use of course materials, and progress toward learning objectives, helping inform decisions and drive classroom discussions.

WileyPLUS. Learn More. www.wileyplus.com.
Powered by proven technology and built on a foundation of cognitive research, *WileyPLUS* has enriched the education of millions of students, in over 20 countries around the world.

Discovering Physical Geography Instructor's Site (www.wiley.com/college/arbogast)

This comprehensive website includes numerous resources to help you enhance your current presentations, create new presentations, and employ our pre-made PowerPoint presentations. These resources include:

- **Image Gallery.** We provide online electronic files for the line illustrations and maps in the text, which the instructor can customize for presenting in class (for example, in handouts, overhead transparencies, or PowerPoints).

- A complete collection of **PowerPoint presentations** available in beautifully rendered, 4-color format, and have been resized and edited for maximum effectiveness in large lecture halls.
- A comprehensive **Test Bank** with multiple-choice, fill-in, matching, and essay questions that is distributed via the secure Instructor's website as electronic files, which can be saved into all major word processing programs.
- **GeoDiscoveries Media Library.** In addition to the modules from the book, this easy-to-use website offers lecture launchers that help reinforce and illustrate key concepts from the text through the use of animations, videos, and interactive exercises. Students can use the resources for tutorials as well as self-quizzing to complement the textbook and enhance understanding of geography. Easy integration of this content into course management systems and homework assignments gives instructors the opportunity to integrate multimedia with their syllabi and with more traditional reading and writing assignments. Resources include:
  - *Animations:* Key diagrams and drawing from our rich signature art program have been animated to provide a virtual experience of difficult concepts. These animations have proven influential to the understanding of this content for visual learners.
  - *Videos:* Brief video clips provide real-world examples of geographic features, and put these examples into context with the concepts covered in the text.
  - *Simulations:* Computer-based models of geographic processes allow students to manipulate data and variables to explore and interact with virtual environments.
  - *Interactive Exercises:* Learning activities and games built off our presentation material. They give students an opportunity to test their understanding of key concepts and explore additional visual resources.
- **Google Earth™ Tours.** Photos from the *WileyPlus* e-book are linked from the text to their actual location on the Earth using Google Earth™.

- An online database of photographs, **www.ConceptCaching.com,** allows professors and students to explore what a physical feature looks like. Photographs and GPS coordinates are "cached" and categorized along core concepts of geography. Professors can access the images or submit their own by visiting www.ConceptCaching.com.
- **Instructor's Manual.** This manual includes chapter overviews, lecture suggestions, and classroom activities.

WILEY FACULTY NETWORK

This peer-to-peer network of faculty is ready to support your use of online course management tools and discipline-specific software/learning systems in the classroom. The WFN will help you apply innovative classroom techniques, implement software packages, tailor the technology experience to the needs of each individual class, and provide you with virtual training sessions led by faculty for faculty.

Student Companion Website (www.wiley.com/college/arbogast)

This easy to use and student-focused website helps reinforce and illustrate key concepts from the text. It also provides interactive media content that helps students prepare for tests and improve their grades. This website provides additional resources that complement the textbook and enhance your students' understanding of Physical Geography:

- **Flashcards** offer an excellent way to drill and practice key concepts, ideas, and terms from the text.
- **GeoDiscoveries Modules** allow students to explore key concepts in greater depth using videos, animations, and interactive exercises.
- **Chapter Review Quizzes** provide immediate feedback to true/false, multiple choice, and short answer questions.
- **Annotated Web Links** put useful electronic resources into context.

A GUIDE TO THE FEATURES

DISCOVER... —This feature presents a photo and demonstrates how visual clues contained within it can be used to "discover" the character of the landscape or environment. The goal of this feature is to make students realize that there is more to the physical landscape than meets the eye, which will hopefully spark their interest in what they see around them.

GEODISCOVERIES MULTIMEDIA — Multimedia modules in every chapter explain to students what they can expect to see and learn as they interact with the *GeoDiscoveries* simulations, animations, and videos on the text's website. The modules in *WileyPLUS* also include a variety of self-assessment questions for students.

• DISCOVER...

MARINE TERRACES ON SAN CLEMENTE ISLAND, CALIFORNIA

The elevation of marine terraces can provide a great deal of information about past ocean levels and tectonic uplift. This image shows marine terraces on San Clemente Island, which is 115 km (~72 mi) west of the southern coast of California. Each of these terraces formed during a prehistoric high sea level associated with an interglacial period, with the highest surface representing the oldest high-level sea stand. Sea level subsequently dropped during glacial periods when enormous volumes of water were stored as ice. At the same time, the old wave-cut bluff was slowly uplifted due to tectonic activity. With each successive interglacial, sea level rose again, creating new wave-cut bluffs that were progressively lower than the previous ones. Given that the rate of uplift is known, it is possible to reconstruct the height of the former sea-level high stands.

 **Geo Discoveries**   www.wiley.com/college/arbogast   **WILEY PLUS**

Migration of Hurricane Katrina

An excellent example of the need for monitoring can be seen in videos showing the path of Hurricane Katrina in 2005. To see this migration, go to the *GeoDiscoveries* website and access the module *Migration of Hurricane Katrina.* The first part of this module is an animated cartoon illustrating the storm's path. Subsequently, a pair of videos follow the path of this catastrophic hurricane as it moved across the Gulf of Mexico. Watch for several things. First, look for the overall rotation of the system and how its path changed over time. Also note the rapid intensification of the system as it migrated across the Gulf. This intensification is obvious because the eye of the storm suddenly becomes very prominent. At this point, the storm was at its most powerful, specifically a Category 5 hurricane. Next, observe how the size of the eye fluctuated as it approached the coast; this occurred because the storm's strength varied somewhat before it made landfall. Finally, notice how the eye suddenly disappeared after the storm reached land. As you watch the video, think about why satellite images such as these are essential for good monitoring and how they save lives. Once you finish the video, be sure to answer the questions at the end of the module to test your understanding of this concept.

KEY CONCEPTS TO REMEMBER—This feature is an interim summary that appears after specific sections of the chapter to help students check their comprehension of the key concepts covered.

VISUAL CONCEPT CHECK—To provide students with a means of self-testing within the flow of chapter content, this feature offers a scenario with an illustration and questions to test students' understanding of key chapter concepts. Answers to the visual concept checks appear at the end of the chapter.

VISUAL CONCEPT CHECK 7.1

The hydrologic cycle has many different components. This image shows a typical farm pond in the Midwest. Explain how this pond is part of the hydrologic cycle. What are two ways in which water could fill the pond? What happens to the water once it becomes stored within the pond? How does w

VISUAL CONCEPT CHECK 7.2

Water vapor is produced in the exhaust of jet engines, such as those on large airliners. As the vapor cools in the air behind the plane, it often forms condensation trails, or contrails. What conditions must be present

VISUAL CONCEPT CHECK 7.3

This satellite image shows the clouds over the Florida peninsula in January 2002. East of Florida is the Atlantic Ocean; west is the Gulf of Mexico. Notice that a distinct pattern of clouds is specifically associated with the Florida peninsula. Given the pattern that you see, which one of the following statements makes the most sense?

a) No convection is occurring over the Florida landmass.
b) Lots of condensation is occurring over the Gulf of Mexico.
c) Strong convection is occurring over the Atlantic Ocean.
d) The temperature of the Florida landmass is warmer than that of the surrounding oceans.

ANSWERS TO VISUAL CONCEPT CHECKS

VISUAL CONCEPT CHECK 7.1

The pond is part of the hydrologic cycle because it is a place where water is temporarily stored. Water can enter the pond either through precipitation or from within the ground itself. Once the water is stored in the pond, it can either flow back into the ground or evaporate into the atmosphere.

VISUAL CONCEPT CHECK 7.2

The answer is *d*, "all of the above." For contrails to form, the air must be close to saturation. Condensation nuclei must be present for water droplets to form, and the plane must be above the level of condensation.

VISUAL CONCEPT CHECK 7.3

The answer is *d*. Given that the cloud pattern is distinctly associated with the peninsula, strong convection must be occurring over the land. For this to happen, the Florida landmass must be warmer than the surrounding water.

The Dry Adiabatic Lapse Rate

In most cases that involve a parcel of surface air, the air is not saturated and the relative humidity is thus less than 100%. Imagine that this parcel begins to rise and expand because it is warmer than the surrounding air. When these conditions occur, the air temperature decreases at what is called the **dry adiabatic lapse rate (DAR)** as the air starts to rise. This rate is not to be confused with the *environmental lapse rate* that was discussed in Chapter 5. The environmental lapse rate refers to the average change in temperature of *still* air with altitude, which can

Dry adiabatic lapse rate (DAR) *The rate at which an unsaturated body of air cools while lifting or warms while descending. This rate is 10°C/1000 m (5.5°F/1000 ft).*

GLOSSARY—Key terms are set in boldface type in the text and defined at the foot of the page for easy recognition and reference.

■ THE BIG PICTURE

In this chapter, you learned about the nature of the hydrologic cycle and humidity. Through the course of this discussion, a model of adiabatic processes described how precipitation occurs due to orographic and convectional uplift, which are easy concepts to visualize. Adiabatic processes are also associated with more complex weather systems, specifically midlatitude cyclones and hurricanes, which can influence large regions in many different ways. These systems integrate the atmospheric circulatory processes discussed in Chapter 6 with the adiabatic processes presented in Chapter 7. The interaction of these various processes can result in powerful weather-makers that release tremendous amounts of energy, as this great image of lightning demonstrates.

THE BIG PICTURE—This concluding section uses a photograph to explain how the concepts learned in the chapter relate to the overall scope of physical geography, using an illustration to demonstrate the relationship.

SUMMARY OF KEY CONCEPTS—The main points of the chapter are summarized.

■ SUMMARY OF KEY CONCEPTS

1. A coastline is a narrow zone where the hydrosphere, lithosphere, and atmosphere interact on a very large scale. These interactions result in very distinctive processes and landforms. Most coastlines are associated with the world's oceans, seas, and gulfs, but some prominent coastlines occur along very large lakes such as the North American Great Lakes.

2. The world's oceans collectively constitute the largest component of the hydrological cycle. Smaller bodies of water associated with oceans are seas and gulfs.

3. The most important agents of coastal change include water fluctuations, tides, waves, and littoral processes. The most extensive water fluctuations occurred when continental glaciers advanced and retreated during the Pleistocene. Tides are daily fluctuations caused by the gravitational pull of the Moon (mostly), and to a lesser degree the Sun. Waves form when wind blows across the water, causing energy within the water to roll forward due to friction. Waves often approach coastlines obliquely, resulting in the littoral processes of (1) longshore current in the water along the shore and (2) swash/backwash of water moving up and down, respectively, on the beach.

4. Erosional coastlines evolve when sediment is removed by coastal processes. Prominent erosional landforms along coastlines are headlands, bluffs, sea stacks, and marine terraces. Depositional coastlines develop where sediment accumulates. These landforms include beaches, spits, baymouth bars, tombolos, and barrier islands.

5. A slight majority of people on Earth live near coastlines. Given this population distribution, coastlines are impacted greatly by human behavior. Most of this impact occurs in the form of engineered structures, such as seawalls and jetties, that protect the shore and maintain the location of shipping channels.

6. It appears that global warming is having a major impact on coastlines around the world. These impacts include bleaching of coral reefs and rising sea levels, which are threatening coastal communities.

CHECK YOUR UNDERSTANDING

1. Define the concept of an air mass.

2. What are the specific characteristics of an mT air mass and how do they differ from the characteristics of a cP air mass?

3. Which air mass is most likely to be associated with precipitation—an mT air mass or a cP air mass? Why?

4. Discuss the evolution and migration of a midlatitude cyclone.

5. Why are midlatitude cyclones a mechanism through which contrasting air masses are mixed?

6. How does the formation of an upper air trough at the 500-mb level result in the development of a midlatitude cyclone?

7. What is the basic difference between a warm front and a cold front? Why is the term *front* used in association with these concepts?

8. Precipitation along a warm front is gradual and long-lasting, whereas it is short-lived and often violent along a cold front. Why does this difference exist?

9. Describe the evolution of a thunderstorm and the various stages it goes through during its life cycle.

10. What is a downdraft and why is it the first step in the dissipation of a thunderstorm?

11. Discuss the evolution of a hurricane in the Northern Hemisphere, including its various stages, movement, and relationship with ocean temperature.

CHECK YOUR UNDERSTANDING— Self-assessment questions at the end of the chapter allow students to test their comprehension.

GOOGLE EARTH™ LINKS WITH PHOTOGRAPHS—Photographs in the *WileyPLUS* e-book are accompanied by a link to Google Earth™ to view a satellite image of the location shown.

BRIEF CONTENTS

1 ■ Introduction to Physical Geography   2

2 ■ The Geographer's Tools   14

3 ■ Earth–Sun Geometry and the Seasons   44

4 ■ The Global Energy System   62

5 ■ Global Temperature Patterns   90

6 ■ Atmospheric Pressure, Wind, and Global Circulation   112

7 ■ Atmospheric Moisture and Precipitation   146

8 ■ Air Masses and Cyclonic Weather Systems   178

9 ■ Global Climates and Global Climate Change   210

10 ■ Plant Geography   248

11 ■ The Global Distribution and Character of Soils   278

12 ■ Earth's Internal Structure, Rock Cycle, and Geologic Time   320

13 ■ Tectonic Processes and Landforms   352

14 ■ Weathering and Mass Movement   392

15 ■ Groundwater and Karst Landscapes   412

16 ■ Fluvial Systems and Landforms   430

17 ■ Glacial Geomorphology: Processes and Landforms   468

18 ■ Arid Landscapes and Eolian Processes   502

19 ■ Coastal Processes and Landforms   528

20 ■ Relevance of Physical Geography to Environmental Issues   560

CONTENTS

1 ■ Introduction to Physical Geography 2

The Scope of Geography 4

Defining Physical Geography 5

The Earth's Four Spheres 8

Organization of This Book 9

Exploring Cause-and-Effect Relationships Holistically 10

Emphasis on Human Interactions with the Environment 10

GeoDiscoveries: An Interactive Tool 11

Focus on Geographical Literacy 11

GeoDiscoveries: *Stream Meandering* 11

Physical Geography Is Interesting, Exciting, and Very Relevant to Your Life 12

The Big Picture 12

2 ■ The Geographers' Tools 14

The Geographic Grid 16

Latitude 17

Longitude 20

Using the Geographic Grid 21

GeoDiscoveries: *Using the Geographic Grid* 21

Maps—The Basic Tool of Geographers 22

Map Projections 22

Map Scale 28

Isolines 28

GeoDiscoveries: *Using Maps* 32

Digital Technology in Geography 33

Remote Sensing 33

Global Positioning Systems 37

Discover... *Infrared Images* 38

Geographic Information Systems 40

GeoDiscoveries: *Using a Geographic Information System* 41

The Big Picture 42

3 ■ Earth–Sun Geometry and the Seasons 44

Our Place in Space 46

The Shape of Earth 46

Earth's Orbit Around the Sun 47

The Earth's Rotation and Axial Tilt 48

The Seasons 51

Solstice and Equinox 52

GeoDiscoveries: *Orbital View and Earth as Viewed from the Sun* 55

Human Interactions: How We See Earth–Sun Geometry on Earth 56

Day and Night 56

Seasonal Changes in Sun Position (Angle) and Length of Day 56

GeoDiscoveries: *Celestial Dome* 58

GeoDiscoveries: *Sun Angle and Length of Day* 58

GeoDiscoveries: *Earth–Sun Geometry and Ancient Humans* 58

The Big Picture 59

Discover... *Sunrise in the Southern Hemisphere* 60

4 ■ The Global Energy System   62

The Electromagnetic Spectrum
and Solar Energy   64

The Electromagnetic Spectrum   64

Solar Energy and the Solar Constant   65

GeoDiscoveries: *The Electromagnetic
Spectrum*   66

Composition of the Atmosphere   66

Constant Gases   67

Variable Gases   67

Discover... *The Formation of Rainbows*   70

Particulates   73

The Flow of Solar Radiation
on Earth   74

Heat Transfer   74

Flow of Solar Radiation in
the Atmosphere   76

Interaction of Solar Radiation and
the Earth's Surface   78

GeoDiscoveries: *The Angle
of Incidence*   80

The Global Radiation Budget   81

GeoDiscoveries: *The Global
Energy Budget*   82

Human Interactions: Solar
Energy Production   86

The Big Picture   88

5 ■ Global Temperature Patterns   90

Layered Structure of the
Atmosphere   92

The Troposphere   92

The Stratosphere   93

The Mesosphere   94

The Thermosphere   94

Surface and Air Temperatures   95

Discover... *Auroras*   97

Human Interactions: Calculating the
Heat Index and Wind Chill   98

Large-Scale Geographic Factors That
Influence Air Temperature   99

Local Factors That Influence
Air Temperature   101

GeoDiscoveries: *Surface Temperature*   101

GeoDiscoveries: *Maritime vs. Continental
Effect*   104

The Annual Range of Surface Temperature
(A Holistic Assessment)   106

GeoDiscoveries: *Global Temperature
Patterns*   108

GeoDiscoveries: *Temperature
and Location*   109

The Big Picture   110

6 ■ Atmospheric Pressure, Wind,
and Global Circulation   112

Atmospheric Pressure   114

Factors That Influence Air Pressure   114

Measuring and Mapping Air Pressure   115

Atmospheric Pressure Systems   116

Low-Pressure Systems   116

High-Pressure Systems   116

The Direction of Airflow   119

Unequal Heating of Land Surfaces   119

Pressure Gradient Force   120

Discover... *Migrating Pressure Systems*   121

Coriolis Force   122

GeoDiscoveries: *Fluctuations in the
Pressure Gradient*   122

Frictional Forces   123

GeoDiscoveries: *The Coriolis Force*   125

Global Pressure and Atmospheric
Circulation   125

Tropical Circulation   126

Midlatitude Circulation   129

Polar Circulation   132

Seasonal Migration of Pressure Systems   132

GeoDiscoveries: *Global Atmospheric
Circulation*   133

GeoDiscoveries: *The Asian Monsoon*   134

GeoDiscoveries: *Global Atmospheric
Circulation and Water Vapor Movement*   135

Human Interactions: Harnessing
Wind Energy   135

Local Wind Systems   137

Land–Sea Breezes   137

Topographic Winds   138

Oceanic Circulation   139

Gyres and Thermohaline Circulation   139

El Niño   141

GeoDiscoveries: *El Niño*   143

The Big Picture   143

7 ■ Atmospheric Moisture and Precipitation   146

Physical Properties of Water   148

Hydrogen Bonding   148

Thermal Properties of Water and Its Physical States   148

GeoDiscoveries: *Latent Heat*   150

The Hydrosphere and the Hydrologic Cycle   151

Humidity   152

Maximum, Specific, and Relative Humidity   153

Dew-Point Temperature   156

GeoDiscoveries: *Atmospheric Humidity*   158

Evaporation   159

Adiabatic Processes   160

The Dry Adiabatic Lapse Rate   160

The Wet Adiabatic Lapse Rate   161

GeoDiscoveries: *Adiabatic Processes*   162

Cloud Formation and Classification   163

Cloud Classification   164

Fog   164

Discover... *Unusual Clouds*   165

Precipitation   165

Types of Precipitation   165

Precipitation Processes   169

GeoDiscoveries: *Orographic Processes*   172

GeoDiscoveries: *Convectional Precipitation*   175

The Big Picture   176

8 ■ Air Masses and Cyclonic Weather Systems   178

Air Masses and Fronts   180

Air Masses   180

Fronts   181

Evolution and Character of Midlatitude Cyclones   184

Upper Airflow and the 500-mb Pressure Surface   184

Interaction of Upper Airflow and Surface Airflow   186

Cyclogenesis   188

GeoDiscoveries: *Formation of a Midlatitude Cyclone*   190

GeoDiscoveries: *Migration of a Midlatitude Cyclone*   191

Thunderstorms   191

Evolution of Thunderstorms   191

Severe Thunderstorms   193

Tornadoes   195

GeoDiscoveries: *Formation of Thunderstorms*   196

Discover... *Fascinating Clouds Associated with Thunderstorms*   198

GeoDiscoveries: *Tornadoes*   200

Tropical Cyclones   201

Hurricanes   202

GeoDiscoveries: *Migration of Hurricane Katrina*   205

The Big Picture   207

9 ■ Global Climates and Global Climate Change   210

Climate and the Factors That Affect It   212

Köppen Climate Classification   213

Geography of Köppen Climates   214

Tropical (*A*) Climates   214

Discover... *Visualizing Climates with Climographs*   218

Arid and Semi-Arid (*B*) Climates   219

GeoDiscoveries: *Tropical Savanna Climate (Aw)*   219

Mesothermal (*C*) Climates   222

GeoDiscoveries: *Humid Subtropical Hot-Summer Climate (Cfa, Cwa)*   224

GeoDiscoveries: *Marine West-Coast Climates (Cfb, Cfc)*   225

Microthermal (*D*) Climates   226

Polar (*E*) Climates   227

Highland (*H*) Climates   228

GeoDiscoveries: *Global Climates* 228

GeoDiscoveries: *Remote Sensing and Climate* 229

Reconstructing Past Climates 229

Pollen Records 230

Tree Ring Patterns 231

Ice Core Analysis 233

GeoDiscoveries: *Reconstructing Past Climates Using Oxygen Isotopes* 236

Causes of Past Climate Change 236

GeoDiscoveries: *The Milankovitch Theory* 238

Human Interactions and Future Climate Change 239

The Carbon Cycle 239

Is Anthropogenic Climate Change Really Occurring? 241

Predicting Future Climate Change 242

GeoDisocoveries: *The Greenhouse Effect and Global Climate Change* 245

The Big Picture 245

10 ■ Plant Geography 248

Ecosystems and Biogeography 250

The Process of Photosynthesis 251

GeoDiscoveries: *Photosynthesis and Respiration* 254

The Relationship of Climate and Vegetation: The Character and Distribution of Global Biomes 255

Forest Biomes 255

Grassland Biomes 261

Discover... *Wildfires* 262

Desert Biomes 264

Tundra Biome 264

Local and Regional Factors That Influence the Geographic Distribution of Vegetation 266

Slope and Aspect 266

Vertical Zonation 267

Plant Succession 268

GeoDiscoveries: *Plant Succession* 269

Riparian Zones 269

Human Interactions: Human Influence on Vegetation Patterns 270

Deforestation and Its Consequences 270

Agriculture in the Midlatitude Grassland Biome 272

Overgrazing 273

GeoDiscoveries: *Deforestation* 273

GeoDiscoveries: *Remote Sensing and the Biosphere* 275

The Big Picture 275

11 ■ The Global Distribution and Character of Soils 278

What Is Soil? 280

Basic Soil Properties 280

Soil-Forming (Pedogenic) Processes 283

Soil-Forming Factors 285

Measurable Soil Characteristics 289

Discover... *Soil Conservation on Steep Slopes* 290

Soil Chemistry 293

Soil pH 293

Colloids and Cation Exchange 294

GeoDiscoveries: *Soil Colloids and pH* 295

Soil Profiles (Reading the Soil) 295

Time and Soil Evolution 297

GeoDiscoveries: *Soil Horizon Development* 298

Soil Science and Classification 298

The Twelve Soil Orders 299

GeoDiscoveries: *African Climate, Vegetation, and Soils* 305

GeoDiscoveries: *Regional Pedogenic Processes* 310

GeoDiscoveries: *North American Climate, Vegetation, and Soils* 314

Human Interactions with Soils 316

The Big Picture 318

12 ■ **Earth's Internal Structure, Rock Cycle, and Geologic Time** 320

Earth's Inner Structure 322

The Major Layers 322

Rocks and Minerals in the Earth's Crust 326

Igneous Rocks 327

Discover... *Exposed Igneous Intrusions* 328

GeoDiscoveries: *Identification of Igneous Environments* 329

Sedimentary Rocks 330

GeoDiscoveries: *Clastic Rocks* 334

The Rock Cycle 336

GeoDiscoveries: *The Rock Cycle* 338

Geologic Time 338

"Telling" Geologic Time 340

GeoDiscoveries: *Geologic Time* 341

Putting Geologic Time in Perspective 341

A Holistic View of Geologic Time and the Rock Cycle: The Grand Canyon and the Spanish Peaks of Colorado 342

Human Interactions with the Rock Cycle: The Case of Petroleum 345

The Big Picture 349

13 ■ **Tectonic Processes and Landforms** 352

Plate Tectonics 354

The Lithospheric Plates 354

Plate Movement 355

GeoDiscoveries: *Continental Drift* 357

Types of Plate Movements 357

Passive Margins 357

Transform Plate Margins 358

Plate Divergence 360

Plate Convergence 360

GeoDiscoveries: *Folding* 368

GeoDiscoveries: *Plate Tectonics* 368

Earthquakes 369

Seismic Processes 369

Human Interactions with Earthquakes as Natural Hazards 374

GeoDiscoveries: *Types of Faults* 379

Volcanoes 379

Explosive Volcanoes 379

GeoDiscoveries: *Walk the Pacific Ring of Fire* 383

Discover... *Crater Lake* 384

Fluid Volcanoes 385

Hotspots 386

GeoDiscoveries: *Volcanoes* 389

The Big Picture 390

14 ■ **Weathering and Mass Movement** 392

Weathering 394

Mechanical Weathering 395

Discover... *The Long-Term Effects of Weathering and Erosion on Mountains* 396

Chemical Weathering 398

Human Interactions: Acid Rain 401

GeoDiscoveries: *Weathering* 402

Mass Wasting 403

Rockfall 403

Soil Creep 405

Landslides 406

Flows 407

Avalanches 408

GeoDiscoveries: *Weathering and Mass Movements* 410

The Big Picture 410

15 ■ **Groundwater and Karst Landscapes** 412

Movement and Storage of Groundwater 414

Discover... *Natural Lakes* 416

Human Interactions with Groundwater 417

The High Plains Aquifer 417

Subsidence 422

Groundwater Contamination 423

GeoDiscoveries: *Hydrologic Cycle*
   and Groundwater   424

Karst Landforms and
   Landscapes   424

Caves and Caverns   425

Karst Topography   426

The Big Picture   428

16 ■ **Fluvial Systems and
Landforms   430**

Overland Flow and Drainage
   Basins   432

Origin of Streams   432

Drainage Basins   433

Drainage Patterns, Density, and
   Stream Ordering   435

GeoDiscoveries: *The Rhine River*   437

Hydraulic Geometry and
   Channel Flow   438

Fluctuations in Stream Discharge   438

Fluvial Processes and
   Landforms   442

Erosion and Deposition   442

Stream Gradation   444

GeoDiscoveries: *Bedload Transport
   and Braided Streams*   445

GeoDiscoveries: *The Graded Stream*   449

Evolution of Stream Valleys and
   Floodplains   449

GeoDiscoveries: *Stream Meandering*   451

Discover... *Mississippi River Meanders*   452

GeoDiscoveries: *Fluvial Geomorphology
   and Stream Processes*   453

Human Interactions with
   Streams   461

Urbanization   461

Artificial Levees   462

Dams and Reservoirs   463

The Big Picture   466

17 ■ **Glacial Geomorphology: Processes
and Landforms   468**

Development of a Glacier   470

The Metamorphosis of Snow to
   Glacial Ice   470

The Glacial Mass Budget   471

GeoDiscoveries: *The Glacial
   Mass Budget*   472

Glacial Movement   472

Types of Glaciers   474

Glaciers in Mountainous Regions   474

GeoDiscoveries: *Glaciers in the
   Cascade Mountains*   476

Continental Glaciers   478

Glacial Landforms   479

Landforms Made by
   Glacial Erosion   479

Deposition of Glacial Drift and
   Resulting Landforms   483

GeoDiscoveries: *Depositional
   Glacial Landforms*   487

History of Glaciation
   on Earth   488

The Wisconsin Glaciation and the Evolution
   of the Great Lakes   489

Discover... *The Channeled Scablands in
   Eastern Washington*   490

Probable Human Impact
   on Glaciers   493

GeoDiscoveries: *Glaciers and
   Climate Change*   494

Periglacial Processes and
   Landscapes   496

Permafrost   496

Ground Ice and Associated
   Landforms   497

The Big Picture   499

18 ■ **Arid Landscapes and Eolian Processes** 502

Arid Landscapes 504
Desert Geomorphology 504
Eolian Erosion and Transport 509
The Fluid Behavior of Wind and Sediment Transport 509
Eolian Erosional Landforms 510
Eolian Deposition and Landforms 513
Airflow and the Formation of Sand Dunes 513
Classification of Sand Dunes and Related Landforms 515
Loess 516
Discover... *Sand Dunes on Mars* 520
GeoDiscoveries: *Eolian Processes and Landforms* 521
Human Interactions with Eolian Processes 521
Desertification in the African Sahel 521
Desertification in the Great Plains of the United States: The Dust Bowl 523
The Big Picture 526

19 ■ **Coastal Processes and Landforms** 528

Oceans and Seas on Earth 530
The Nature of Coastlines: Intersection of Earth's Spheres 531
Processes That Shape the Coastline 531
GeoDiscoveries: *Tides* 537
Coastal Landforms 539
Erosional Coastlines 539
Discover... *Marine Terraces on San Clemente Island, California* 543
GeoDiscoveries: *Waves and Coastal Erosion* 544
Depositional Coastlines 544
GeoDiscoveries: *Longshore Processes and Depositional Coastlines* 549

Human Interactions with Coastlines 553
Coastal Engineering 553
Global Climate Change and the Impact on Coastlines 556
GeoDiscoveries: *Evolution of the Louisiana Coastline* 557
The Big Picture 558

20 ■ **Relevance of Physical Geography to Environmental Issues** 560

A Short History of Human Population 562
Technological Development and Population Growth 562
The Impact of Growing Human Population on the Natural Environment 563
Discover... *Urban Sprawl* 565
Case 1: Water Issues in the Arid American Southwest 566
Establishing Water Rights 568
Will Water Supplies Disappear? 569
A Drought-Prone Future? 571
Case 2: Soil Salinization in Arid and Semi-Arid Lands 572
Ancient Sumeria 574
The Current Global Extent of Salinization 576
The San Joaquin Valley in California 576
Australia 580
The Challenge of Sustainable Agriculture in Semi-Arid and Arid Regions 583
Case 3: Giant Panda Conservation in China 583
The Wolong Nature Reserve 586
The Future Picture 589

■ **Glossary** 591

■ **Photo Credits** 606

■ **Index** 611

Special Features

Discover....

Auroras   97

The Channeled Scablands in
  Eastern Washington   490

Crater Lake   384

Exposed Igneous Intrusions   328

Fascinating Clouds Associated
  with Thunderstorms   198

The Formation of Rainbows   70

Infrared Images   38

The Long-Term Effects of Weathering
  and Erosion on Mountains   396

Marine Terraces on San Clemente Island,
  California   543

Migrating Pressure Systems   121

Mississippi River Meanders   452

Natural Lakes   416

Sand Dunes on Mars   520

Soil Conservation on Steep Slopes   290

Sunrise in the Southern Hemisphere   60

Unusual Clouds   165

Urban Sprawl   565

Visualizing Climates with
  Climographs   218

Wildfires   262

GeoDiscoveries Modules

Adiabatic Processes   162

African Climate, Vegetation,
  and Soils   305

The Angle of Incidence   80

The Asian Monsoon   134

Atmospheric Humidity   158

Bedload Transport and Braided
  Streams   445

Celestial Dome   58

Clastic Rocks   334

Continental Drift   357

Convectional Precipitation   175

The Coriolis Force   125

Deforestation   273

Depositional Glacial Landforms   487

Earth–Sun Geometry and
  Ancient Humans   58

The Electromagnetic Spectrum   66

El Niño   143

Eolian Processes and Landforms   521

Evolution of the Louisiana Coastline   557

Fluctuations in the Pressure Gradient   122

Fluvial Geomorphology and Stream
  Processes   453

Folding   368

Formation of a Midlatitude Cyclone   190

Formation of Thunderstorms   196

Geologic Time   341

The Glacial Mass Budget   472

Glaciers and Climate Change   494

Glaciers in the Cascade Mountains   476

Global Atmospheric Circulation   133

Global Atmospheric Circulation and
  Water Vapor Movement   135

Global Climates   228

The Global Energy Budget   82

Global Temperature Patterns   108

The Graded Stream   449

The Greenhouse Effect and Global
  Climate Change   245

Humid Subtropical Hot-Summer
  Climate (*Cfa*, *Cwa*)   224

Hydrologic Cycle and Groundwater   424

Identification of Igneous
  Environments   329

Latent Heat   150

Longshore Processes and Depositional
  Coastlines   549

Marine West-Coast Climates
  (*Cfb*, *Cfc*)   225

Maritime vs. Continental Effect   104

Migration of a Midlatitude Cyclone   191

Migration of Hurricane Katrina   205

The Milankovitch Theory   238

North American Climate, Vegetation,
  and Soils   314

Orbital View and Earth as Viewed
  from the Sun   55

Orographic Processes   172

Photosynthesis and Respiration   254

Plant Succession   269

Plate Tectonics   368

Reconstructing Past Climates Using Oxygen Isotopes   236

Regional Pedogenic Processes   310

Remote Sensing and Climate   229

Remote Sensing and the Biosphere   275

The Rhine River   437

The Rock Cycle   338

Soil Colloids and pH   295

Soil Horizon Development   298

Stream Meandering   11

Stream Meandering   451

Sun Angle and Length of Day   58

Surface Temperature   101

Temperature and Location   109

Tides   537

Tornadoes   200

Tropical Savanna Climate (*Aw*)   219

Types of Faults   379

Using a Geographic Information System   41

Using Maps   32

Using the Geographic Grid   21

Volcanoes   389

Walk the Pacific Ring of Fire   383

Waves and Coastal Erosion   544

Weathering   402

Weathering and Mass Movements   410

CHAPTER ONE

INTRODUCTION TO PHYSICAL GEOGRAPHY

I want to welcome you to this introductory textbook about physical geography. Physical geography is an exciting scientific discipline that is central to our understanding of Earth and how it functions. Every region of the planet is part of a delicately balanced natural system that has distinctive physical characteristics, including the type of climate, weather, vegetation, soils, and landforms. These characteristics are the result of processes that, in some cases, have been operating for millions of years. These processes often cause incremental changes that are barely noticeable in a human lifetime. In other cases, noticeable changes can occur over the course of a single day. As you read through the text, your first goal should be to examine these physical processes and resulting geographical patterns on Earth. A second goal is to examine some ways that Earth systems and associated processes affect human lives and, in turn, how people impact them. This opening chapter outlines the topics discussed in this book and places them in the context of the overall discipline of geography. Then we discuss the various components and features of the book and how they will assist with your learning.

This view of Wilson Peak in Colorado reflects many geographical processes discussed in this text, including the character of the atmosphere, how water is stored and seen in the hydrosphere, the role of climate and its impact on vegetation, and the way that landscapes evolve over time.

CHAPTER PREVIEW

The Scope of Geography

Defining Physical Geography

Organization of This Book
   On the Web: *GeoDiscoveries* Stream
   Meandering

The Big Picture

LEARNING OBJECTIVES

1. Comprehend the character and scope of geography as a scientific discipline.

2. Discuss the concept of spatial analysis and how it relates to geography.

3. Define the subdisciplines of physical geography.

4. Explain the concept of a natural system.

5. Define the four Earth spheres.

6. Describe how the scientific method is used in physical geography.

7. Discuss why physical geography is relevant to many human/environment issues.

The Scope of Geography

When most people are asked to define the nature of geography, the usual response is it focuses on the locations of countries, capital cities, rivers, and oceans. Although the location of places is certainly a part of geography, the field encompasses *much* more. In fact, you may be surprised to learn that whatever you like to do and wherever you live, geography is very much a part of your everyday life.

Geography is an ancient discipline that examines the spatial attributes of the Earth's surface and how they differ from one place to another. Derived from the Greek words for "Earth description," the concept of geography has likely been a central part of the human experience for tens of thousands of years. It is easy to imagine, for example, that prehistoric hunters and gatherers were intimately aware of their surroundings, including the location and character of forests, streams, lakes, berry patches, animal herds, and competing groups of people. In short, an understanding of local geography would have been absolutely essential to sustain life. It would also have been necessary to be able to describe those patterns for future generations so they, in turn, could thrive.

So, for thousands of years at least, geography was a descriptive discipline that focused on the generalized location and character of places and things. Slowly, however, geography became an academic discipline with numerous specialized subfields. Scientists became experts in areas such as geology, meteorology, ecology, and human cultural differences. Interest in geography grew especially between the 15th and 19th centuries when explorers like Christopher Columbus, Ferdinand Magellan, James Cook, Charles Darwin, and Lewis and Clark began to investigate parts of the world that were previously unknown to people of European descent (including Americans). These explorers, as well as many others, brought detailed descriptions of exotic places and animals to a keenly interested public. The new knowledge and perspectives gained from this time were a major driving force for the development of the modern world.

The trend toward specialization in geography has continued to the present time. Most geographers consider themselves first of all to be either physical or human specialists. Within these two broad fields are a range of geographical subdisciplines, as shown in Figure 1.1a. Although each of these subfields has a unique focus, such as soils or agricultural land use, geographers draw from many of these subfields when they analyze any particular spatial pattern. For example, to fully understand human settlement patterns in Africa, it is important to consider the interaction of subfields such as climatology, soils, and vegetation, to name a few. In turn, to understand the nature of soils in any given place, you must consider the effects of climatology, vegetation, geomorphology, and perhaps even regional cultural practices (Figure 1.1b).

Although the focus of this book is physical geography, it is important to know that all subfields of geography are based on the same five themes of location, movement, place, human/environment interaction, and region. Location refers to the exact position where something is found

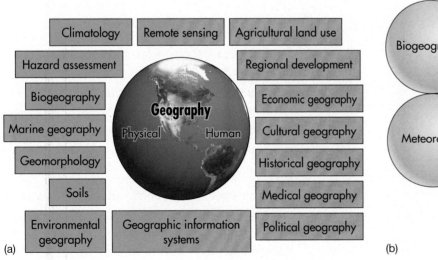

(a)

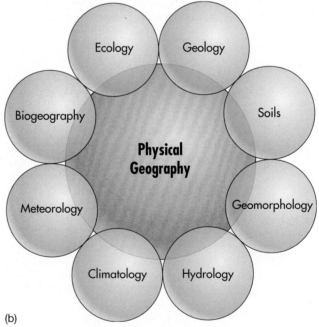

(b)

Figure 1.1 Subdisciplines of geography. (a) The field of geography can be broadly subdivided into physical and human geography. Many subfields occur within these two broad categories, with many that overlap. (b) Physical geography itself overlaps with several other areas of science. Many of these areas require a good understanding of chemistry and physics as well.

on Earth. As the name implies, movement is related to the way that geographical variables flow from one locality to another. Human/environment interaction acknowledges the complex bond between people and nature. The concept of place is an important theme in geography because it reflects the characteristics that make a certain location distinct. Finally, a region is an area that shares one or more common characteristics.

In the context of the five themes, all geographers share a common methodology that makes them part of the overall discipline. Geographers use a method known as **spatial analysis**, which attempts to explain patterns or distributions of specific variables over physical space. Most geographers are fundamentally interested in knowing two things: *where* and *why*. In other words, geographers examine why a specific environmental or cultural variable in a specific place or region has one particular set of characteristics, whereas the same variable someplace else has different characteristics.

Consider for a moment an example involving cultural diversity. The Middle East is a region of Earth that is culturally diverse, with many different religious sects, dialects, and tribal identities. Although this diversity has produced a rich heritage, it has also resulted in a great deal of conflict between and among various cultural groups. A cultural geographer can examine the spatial distribution of the many groups in the area (in

---

Spatial analysis *A method of analyzing data that specifically includes information about the location of places and their defining characteristics.*

---

other words, *where are they?*), as well as study why people in one place differ in, say, their political or religious views, from people someplace else. The geographer might look for similarities (such as language) among groups across physical space, which might explain why certain people align themselves politically with others. In the course of this study, the geographer would have integrated several variables into one picture, including language, religion, history, and climate. Such a study might contribute to an understanding of why people differ across this region and why sources of conflict remain.

Defining Physical Geography

Now that the fundamental nature of geography as a broad field has been defined, let's focus on a more comprehensive definition of physical geography. You probably already have some kind of an interest in physical geography, whether you know it or not. For example, have you ever wondered why violent storms occur? You may know that tornadoes frequently occur in the central United States, especially in the springtime (Figure 1.2a). Do you know why? Maybe you wonder why large mountains are found in Washington but not in Texas (Figure 1.2b). Perhaps you have heard about the Sahara Desert and wonder why it is so dry there and why much of it is covered with sand dunes (Figure 1.2c). Like many people, you might enjoy the seashore and wonder why nice beaches form in some places (Figure 1.2d) but not in others. If you have asked yourself questions like these, then you are

(a)

(b)

(c)

(d)

Figure 1.2 Some elements of physical geography. (a) A tornado in the central United States (see Chapter 8). (b) The Olympic Mountains overlooking Puget Sound in Washington (see Chapter 13). (c) Sand dunes in the Sahara Desert (see Chapters 9 and 18). (d) Coastline at Big Sur in northern California (see Chapter 19).

Figure 1.3 Examples of energy flows on Earth. (a) Earth receives its energy from the Sun in the form of solar radiation (see Chapters 3 and 4). (b) The atmosphere circulates energy around Earth, as can be seen in this stream of clouds (see Chapter 7). (c) Some energy is transferred when water flows from the atmosphere to Earth as rain (see Chapter 5). (d) Some of the energy on the surface of Earth is transferred by flowing water (see Chapter 16).

probably interested in physical geography at some level. The fact is most people are; they just do not realize it.

Simply stated, **physical geography** involves the spatial analysis of the various physical components and natural processes of the Earth. Some examples of Earth's physical components are air, water, rocks, vegetation, and soil. The term **process** broadly refers to a series of actions that can be measured and that produce a predictable end result. In physical geography, these processes are fundamentally products of the energy that flows from the Sun to Earth in the form of solar radiation. Once this energy reaches Earth, it then flows from one place to another on the planet in various forms. Some examples of natural processes directly related to the flow of solar radiation (Figure 1.3) are the circulation of the atmosphere (Chapter 7), the distribution of vegetation (Chapter 10), the formation of soils (Chapter 11), and the movement of water in the air, in streams, and collecting in lakes (Chapters 5 and 16).

As you will see throughout the book, many processes behave in an interconnected way within *natural systems* where one environmental variable has a direct impact on another. Given these relationships, physical geographers often invoke **systems theory** in their studies because it is a holistic framework through which someone can analyze

---

Physical geography *Spatial analysis of the physical components and natural processes that combine to form the environment.*

Process *A naturally occurring series of events or reactions that can be measured and that result in predictable outcomes.*

Systems theory *The examination of interactions involving energy inputs and outputs that result in predictable outcomes.*

and/or describe a group of variables that work together to produce some definable result. Another way to look at natural systems is that they are greater than the sum of their parts. Such systems depend on a set of energy inputs that flow in some way to various kinds of predictable outputs. Physical geographers understand that natural systems are complex and that processes within them do not occur in isolation from one another. Instead, they are interconnected in often complicated ways that may at first appear chaotic. Despite this apparent complexity, natural systems are, in fact, self-organizing entities that internally adjust toward an equilibrium condition. These adjustments can be readily explained by understanding how dynamic feedbacks, oscillations, and delays in reaction time influence outcomes.

To see an example of how environmental variables relate to one another and can be viewed holistically within a natural system, imagine you want to explain the spatial distribution of rivers in the United States. One way to see the geographical concentration of rivers in the country is with a map showing the location of gauging stations, which are places where the U.S. Geological Survey continuously monitors the flow of water in the streams (Figure 1.4). Note that the eastern part of the country contains many more gauging stations than the interior west in places like Nevada and Utah. This pattern reflects the fact that the eastern part of the United States includes more streams than the interior west.

The question a geographer would ask about this is: *Why* do more rivers occur in the eastern United States than in the western part of the country? This question would naturally lead you to holistically examine streams as natural systems that reflect the input of water from some kind of source to the output of water actually flowing in rivers and creeks. Understanding these relationships, in turn, helps explain the geographical patterns observed in Figure 1.4. The simple reason for this geographical distribution of streams in the United States is that more precipitation falls in the eastern part of the United States than in the western states. Later you will study why this geographical variability occurs, but for now it is sufficient to say it exists because the atmosphere over the eastern United States typically contains more water than in the interior parts of the western states. Thus, more water flows from the atmosphere to the ground as precipitation in the eastern United States than in the western states. Some of this water flows directly across the Earth's surface as input into streams. A great deal of it slowly absorbs into the ground where it is steadily released into streams. As a result of these interconnected processes, the eastern United States contains more streams than the western part of the country.

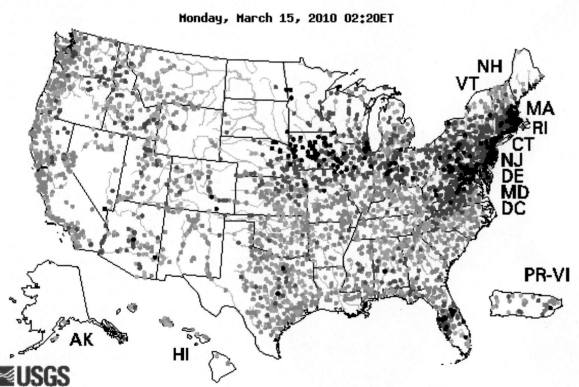

Monday, March 15, 2010 02:20ET

Figure 1.4 Location of stream gauging stations in the United States. The dense concentration of stations in the eastern half of the country reflects the fact that many more streams occur in the eastern United States than in the interior western states. The map is color-coded to illustrate the amount of water in any given stream on March 15, 2010, relative to its average flow on that day calculated over time. For example, red dots reflect very low water levels, whereas dark blue/purple dots mean that stream flow was relatively high on that day. Black dots represent localities on streams that have very high flow and may be close to flooding.

The Earth's Four Spheres

As you can imagine, a huge number of component/process combinations exist for geographical study in a holistic way. In physical geography, these various combinations can be grouped into the four "spheres" on Earth (Figure 1.5):

1. **Atmosphere**—The atmosphere is the gaseous shell that surrounds Earth. This sphere is composed of many critical components essential to life, such as oxygen, carbon, water, and nitrogen, to name a few, that flow around Earth.

2. **Lithosphere**—The lithosphere is the solid part of Earth, including soil and minerals. A good example of a natural system in this sphere is the way in which water, minerals, and organic matter flow in the outermost layer of the Earth to form soil. This sphere provides the habitat and nutrients for many life forms.

3. **Hydrosphere**—The hydrosphere is the part of Earth where water, in all its forms (solid ice, liquid water, and gaseous water vapor), flows and is stored. This sphere is absolutely critical to life and is one with which humans regularly interact—for example, through irrigation and navigation.

4. **Biosphere**—The biosphere is the living portion of Earth and includes all the plants and animals (including humans) on the planet. Various components of this sphere regularly flow from one place to another, both on a seasonal basis and through human intervention. Humans interact with this sphere in a wide variety of ways, with agriculture being a very prominent example.

These four spheres overlap to form the natural environment that makes Earth a unique place within our solar system. Physical geography examines the spatial variation within these spheres, how natural systems work within them, the observable outcomes in each, and the manner in which components flow from one sphere to another.

Physical geography can be a descriptive discipline that simply characterizes the nature of the Earth's spheres in specific regions. A simple example of such a descriptive focus would be to acknowledge that the western part of the United States is mountainous, whereas the central part of the country consists mostly of relatively level plains. Physical geography is also a science because research is conducted within the framework of the scientific method, which is the systematic pursuit of knowledge through the recognition of a problem, the formulation of hypotheses, and the testing of hypotheses through the collection of data by measurement, observation, and experiment. The conclusions derived from the systematic application of the scientific method contribute to the formulation of scientific theories and laws that explain how Earth functions.

A simple example of using the scientific method in physical geography would be to test the hypothesis that water in streams within a certain region is acidic. This hypothesis would best be tested by collecting water samples from a number of streams and conducting chemical analyses on them to determine their acid levels. If high acid levels were indeed obtained, then an effort would be made to explain why that pattern occurred. In addition to understandings produced in these kinds of analyses, physical geographers also test hypotheses about all sorts of natural phenomena by collecting information from the atmosphere, rocks, soils, ice cores, satellite im-

Atmosphere   *The gaseous shell that surrounds Earth.*

Lithosphere   *A layer of solid, brittle rock that comprises the outer 70 km (44 mi) of Earth.*

Figure 1.5 The four Earth spheres. Each sphere encompasses a major component of the Earth's natural environment.

Hydrosphere   *The part of Earth where water, in all its forms, flows and is stored.*

Biosphere   *The portion of Earth and its atmosphere that supports life.*

(a)

(b)

(c)

(d)

Figure 1.6 Examples of collecting scientific data about Earth. (a) Many space shuttle missions are designed to obtain measurements about the atmosphere, oceans, and the distribution and character of plants, among many other things. (b) To learn about the behavior of streams in the past, scientists study the type of sediment deposited by the stream through time. This picture is from one of my class field trips in the Great Plains. (c) One way to learn about past climate changes on Earth is to obtain samples of ancient ice on the Greenland and Antarctic ice caps. (d) New methods of surveying enable scientists to obtain accurate measurements about elevation and location.

ages, the Earth's magnetic field, and even other planets (Figure 1.6). For these investigations to occur, it is essential that scientists understand physical laws and have the ability to mathematically analyze and compare them.

Organization of This Book

The chapters in this book are organized to provide you with a good understanding of the fundamental concepts associated with physical geography. They contain infor-

mation that ranges in scale from global to local, which will allow you to better grasp your place both in the world and even in your neighborhood. Chapter 2 will focus on the various kinds of tools that geographers use in their work, such as maps, remote sensing, and geographical information systems. Chapters 3 through 5 center on our relationship with the Sun (Chapter 3), the way we receive solar radiation (Chapter 4), and how those interactions relate to temperature (Chapter 5). The processes discussed in these chapters will prepare you for the topics that will be covered in the rest of the book. Chapters 6 through 9 revolve around the atmosphere, including the way that air circulates within it (Chapter 6), precipitation processes (Chapter 7), weather systems (Chapter 8), and global climate patterns (Chapter 9). The text then examines the influence of the atmosphere and how it interacts with Earth's other spheres by focusing on plant geography in Chapter 10 and soils in Chapter 11.

Chapters 12 through 19 of the book deal mainly with the lithosphere and hydrosphere. Chapter 12 describes the Earth's internal structure, rock cycle, and geologic time. This discussion leads directly into Chapter 13, which focuses on the lithosphere and tectonic landforms. From there, we turn your attention in Chapter 14 to the way that rocks weather and how sediment moves through mass wasting processes. Chapters 15 and 16 discuss the way that water moves on Earth and how it is stored within it. Chapter 15 focuses specifically on groundwater processes and the formation of landforms such as caves. In Chapter 16 we look at how water flows across the surface in stream systems and the landforms that result. Chapters 17 through 19 are devoted to specific geomorphic processes and the resulting landforms, including glaciers (Chapter 17), eolian (wind) processes and arid landscapes (Chapter 18), and coastal regions (Chapter 19).

Exploring Cause-and-Effect Relationships Holistically

As you work through these chapters, you will constantly see the effects of interactions between the four Earth spheres. To understand how such interactions work, take a quick look at a recent drought in the southeastern United States, which hit Georgia particularly hard between 2005 and early 2009. With this problem in mind consider the following question, one that encompasses elements of the atmosphere and the hydrosphere: How was the quantity of water in rivers in the southeastern United States affected by the drought? A testable hypothesis would be that the quantity of water in rivers decreased. You could test this hypothesis by collecting data about the amount of water in the streams during the drought and comparing those values to normal water levels. In all probability, the amount of water dropped significantly during the drought years. The reason for this decreased water level is that a significant drought would result in less water flowing from the atmosphere (as precipitation) to Earth. As a result, less water would then be available as an input to streams across the Earth's surface. In addition, the quantity of water stored in the ground likely decreased, which would also have reduced river levels because a great deal of water in streams is derived from the ground. Potential further impacts of this drought could have been that some forms of vegetation became less common or that the likelihood of fire increased.

This book systematically explores these kinds of cause-and-effect relationships in a variety of ways. One way that this material is presented is through the traditional use of text accompanied by photographs, diagrams, and tables. Each chapter contains detailed discussions that connect important concepts to events that you may have experienced. Each chapter also ends with a visual preview of the next chapter specifically designed to show how one chapter relates to the next.

Emphasis on Human Interactions with the Environment

Many of the scientific analyses associated with physical geography are driven by the growing impact that human activities have within and among the Earth spheres. Given the nature of human impact on the natural environment, physical geography is at the forefront of research on many environmental issues that face the world today. These issues include (to name a very few):

- **Global climate change**—Human industrial activities are increasing the levels of carbon dioxide in the atmosphere. Abundant evidence suggests that this relationship is contributing significantly to global climate change.
- **Deforestation**—The clearing of the tropical rainforests is occurring at a very rapid rate, leading to soil erosion, loss of wildlife habitat, and species extinctions.
- **Farmland loss**—Due to increasing global population, farmland is being converted to zones of economic development and residential housing. This loss of farmland is resulting in more intensive farming of agricultural soils still in use, which increases the risk of soil erosion and pollution due to the extensive application of fertilizers and pesticides.
- **Natural hazards**—Hazards occur when extreme events result in danger to humans. Examples of natural hazards include hurricanes, tornadoes, flooding, earthquakes, and volcanoes. Natural hazards are a particularly important area of geographical study because as the global population grows, increasing numbers of people are moving into areas that are susceptible to extreme natural events.

A recent natural disaster in the United States vividly illustrates the integrated nature of physical geography and the critical role that geographers play with respect to solving real-world problems. As you probably know, Hurricane Katrina devastated the Gulf Coast in the summer of 2005, causing billions of dollars of damage, much of it occurring when the city of New Orleans was extensively flooded. Before the storm reached land, geographers were at the forefront of the effort to monitor the storm's path and predict where the most significant damage would occur. Once the storm passed, geographers began conducting research on the impact that the hurricane had on a variety of issues, including the shape of the coastline, distribution of wetlands, and the regional economy, to name just a few. These studies have profound implications for future environmental decisions, politics, and economic development in the Gulf region. Several chapters contain sections specifically devoted to such human–environment interactions, including a discussion of solar energy production in Chapter 4, wind energy in Chapter 5, and petroleum in Chapter 12, to name a few. The final chapter (Chapter 20) is entirely devoted to these kinds of issues and demonstrates how physical geography is highly relevant to human–environmental interactions.

GeoDiscoveries: An Interactive Tool

These holistic discussions in the text are accompanied by graphics and photographs, as well as a more dynamic tool—interactive digital modules called *GeoDiscoveries*. These modules consist of a variety of animations and simulations that allow you to visualize and manipulate many of the factors associated with geographical processes and see the results over time and space. The animations and simulations will enhance your learning as you participate more closely with geographical processes and will reinforce the interactive nature of the discipline by showing related variables in motion. The media are integrated entirely within the chapter text as distinct sections that explain what you should expect to learn by interacting with them.

Here is a good example of how the digital modules will enhance your learning. As you may already know, rivers are bodies of water that flow from one place to another. This concept is described in great detail in Chapter 16. This discussion is accompanied by a variety of diagrams that illustrate how water flows, using flow lines and arrows embedded within them. It is also supplemented by several *GeoDiscoveries* modules that are accessed on the

text's accompanying website. One of the modules in that chapter shows how streams snake across the river valley in a process called *meandering*. Through this process, the geographical position of streams actually moves through time. The *GeoDiscoveries* module shows this process in animated form, which will enable you to comprehend it better. Have a look now to see what these modules are like.

Focus on Geographical Literacy

In addition to improving your overall understanding of physical geography and how it relates to human/environmental issues, another goal of this book is to enhance your geographical literacy. It is common knowledge that the overall geographical literacy of average Americans is very poor. How many Americans can identify, for example, the countries of the Middle East, where so much of our national focus is presently centered? This illiteracy not only extends to the world at large, but also applies to the United States. In a poll a few years ago, for example, Americans chose more than 30 different states (of the 50 in the country) as being the state of New York.

Geographical literacy also involves knowing the location of distinct physical regions on Earth, such as the Sahara Desert or the Himalaya Mountains, and why they exist. In this context, you will see that discussions include maps of the places described. I also hope that your *visual* geographical literacy will improve by using this text. In other words, this book is designed to sharpen *your eye* so that when *you see* things on the landscape, which you may have previously ignored, you might better appreciate them and why they occur. Two features—*What a Geographer Sees* and *Visual Concept Check*—are specifically designed to help you improve this aspect of your geographical literacy. The *What a Geographer Sees* feature allows you to see the physical landscape through the eyes of a geographer. In other

www.wiley.com/college/arbogast

Stream Meandering

An excellent way to get a feel for the interactive media in this book is by examining a simple animation showing how streams move across the landscape. Go to the *GeoDiscoveries* website and select the *Stream Meandering* module. This animation allows you to visualize the way that streams migrate horizontally through the process of meandering. Throughout the book, you will frequently encounter modules like this one that will give you a better idea of

how geographical processes function. Some of the modules are simple and require only that you carefully watch them. Others are simulations that allow you to manipulate variables to see how outcomes change. Regardless of the type of media, each module will contain a series of questions that can be used to test your understanding of the concepts after you watch the animation or simulation.

words, you will see the kinds of visual cues that geographers use to interpret why physical patterns exist, which will hopefully train your eye to look for similar visual cues. The *Visual Concept Check* features are placed after key topic discussions so you can test your understanding of those topics immediately after encountering them.

Physical Geography Is Interesting, Exciting, and Very Relevant to Your Life

As with any new endeavor you pursue, you can expect that improving your understanding of physical geography may be difficult at times. Many students initially avoid this subject because they feel that geography is boring, or they are intimidated by science, or they see no relevance of geography to their lives. If you genuinely give this subject a chance, however, you will see that physical geography is indeed relevant to *your everyday life* and, most of all, is interesting and even exciting. At a fundamental level, how else can you explain the popularity of weather and nature programs, national and state parks, the travel industry, mountains, or beautiful coastlines? Why do people go on exotic vacations if not, in part, to enjoy the uniqueness of the physical landscape in new places? With a greater understanding of physical geography, you will appreciate those trips more. You may even appreciate the immediate world around you more.

In addition, with an understanding of physical geography you will be able to make *informed* decisions when you are confronted with important environmental issues in your lifetime. Thus, you will become a better citizen, one who is capable of better protecting the best interests of your family and community. For instance, at some point in the future you may be confronted with a choice of where to place a landfill in your city or town. In order to make the most informed decision, one that perhaps ensures the safety of your drinking water, it will be important to understand the geology of the site, the character of soils, and the way water moves through the ground and is stored within it.

It is also possible that you may even decide after using this text that you want to become a physical geographer, as many people have. A number of excellent, well-paying jobs can be obtained with a specialization in physical geography, including environmental analysts, cultural resource managers, conservation agents, teachers, meteorologists, and landscape architects to name a very few. If you decide to pursue such a career, you will find that collecting, analyzing, and reporting about geographical data are very rewarding. I do that kind of work myself and find it incredibly rewarding. Earth is a beautiful and complex place and it is fun to understand how it works. Regardless of whether you are a geographer or not, you will understand the planet better. Enjoy the ride!

▎ THE BIG PICTURE

It is now time to move more deeply into the nature of physical geography. This discussion begins with an overview of the tools that physical geographers use to obtain measurements about Earth. This is important because it will give you a feel for how scientists came to understand how Earth functions. An excellent example of a tool that physical geographers use is satellite remote sensing. The premise of this data-gathering method is that scientists can view Earth from space with satellites and monitor changes that occur across the planet. The image shown here is a satellite image of the Baltimore, Maryland, area. It is an infrared image, which means that it is measuring energy that you cannot see with your naked eye. The gray areas are urban zones associated with the Baltimore metropolitan area, whereas the red zones are places where vegetation is especially dense, such as in a park or in suburban areas. Black is water (part of Chesapeake Bay). The next chapter will describe the different kinds of tools that geographers use and will provide examples of the data that can be collected.

CHECK YOUR UNDERSTANDING

1. Describe the character of geography as a scientific discipline.

2. Explain the concept of spatial analysis and how it is used in geography.

3. What is the nature of physical geography and why is it a subdiscipline of geography?

4. What is a natural system and why can it be viewed holistically with distinct inputs and predictable outputs? Provide an example.

5. What is the scientific method and how is it applied to research problems in physical geography?

6. Compare and contrast the four spheres of Earth and provide an example of how two of them interact with each other.

7. Why can physical geography be used to understand the relationship of humans with the environment?

CHAPTER TWO

THE GEOGRAPHER'S TOOLS

Before you begin to study the various concepts associated with physical geography, it is important to understand some of the tools that geographers use to locate places and to gather and display spatial information. Some of these tools, such as maps, are probably familiar because you have used them yourself. Other aids, such as remote sensing, Geographic Information Systems (GIS), and Global Positioning Systems (GPS), are relatively new and not well understood by most people. This chapter focuses on these tools, with the goal of providing a better understanding of how geographers do their work. The material in this chapter will also help you understand how much of the information presented in this text was obtained and how it should be interpreted.

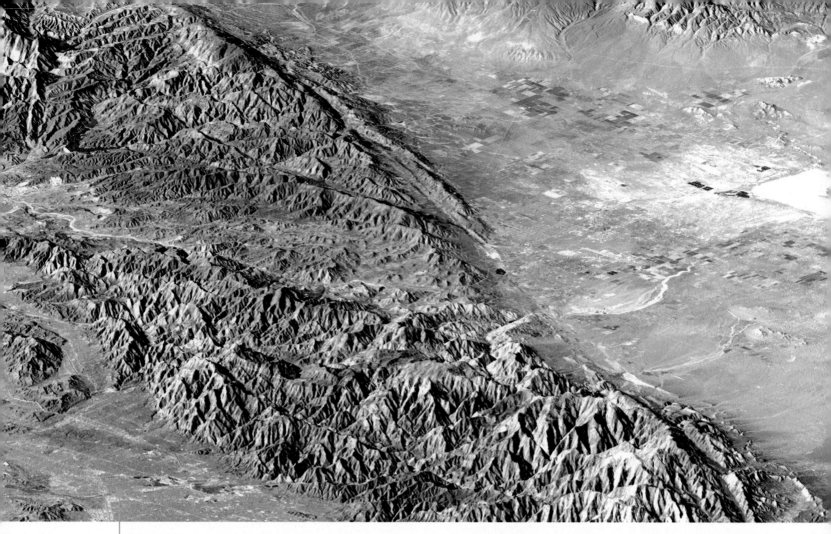

Geographic technology allows us to view landscapes in unique and insightful ways. This photograph combines a satellite image with a digital elevation model to show the relationship of urban areas (in gray shades) and topography near Los Angeles, California. Of particular interest is the nearby location of the San Andreas Fault, which is the linear feature that angles from the upper center of the image to the lower right.

CHAPTER PREVIEW

The Geographic Grid
On the Web: *GeoDiscoveries* Using the Geographic Grid

Maps—The Basic Tool of Geographers
On the Web: *GeoDiscoveries* Using Maps

Digital Technology in Geography
On the Web: *GeoDiscoveries* Using a Geographic Information System

The Big Picture

LEARNING OBJECTIVES

1. Compare and contrast the various components of the Earth's geographic grid.

2. Discuss the concept of map projection and why it is important when designing maps.

3. Explain the concept of map scale and how it is used to represent location.

4. Describe remote sensing and the various ways it is used.

5. Discuss the concept of Global Positioning Systems (GPS) and how location is determined with this method.

6. Understand the nature of Geographical Information Systems (GIS) and how they are used to manage spatial data.

15

The Geographic Grid

As discussed in Chapter 1, one of the five themes of geography is location. One way that location can be determined is by a relative comparison to some other place. For example, a geographer might identify Los Angeles as being south of San Francisco or that Africa is on the eastern side of the Atlantic Ocean. You have no doubt made such a comparison yourself, perhaps by indicating that you live on the west side of town or that your favorite park is in the eastern part of a state.

Although relative location provides an accurate depiction of place compared to some other location, such a statement is very general. In most circumstances, geographers desire to identify the absolute location of a place, which is most effectively done using the geographic grid. If Earth were flat, then it would be easy to construct a grid similar to a sheet of graph paper using the cardinal directions: north, south, east, and west (Figure 2.1). In this imaginary situation, the grid lines would extend from these four directions and intersect at various places on the grid. With this grid in place, it would be easy to choose a place of interest and then determine the distance from this reference point to your location by measuring along the nearest grid lines, using some unit of measurement such as miles or kilometers.

However, Earth is not flat, but instead has a curved surface. In order to account for the Earth's shape when locating places, the grid must consist of a series of intersecting circles, with one set extending north and south and the other set east and west. We can begin by examin-

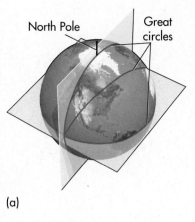

(a)

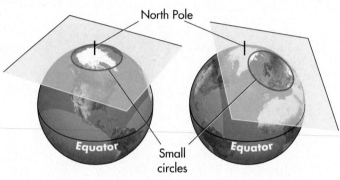

(b)

Figure 2.2 Great and small circles. (a) Examples of great circles on Earth. (b) Examples of small circles on Earth.

ing the circles that are oriented in a north–south direction. These circles converge at the North and South Poles on their way to the other side of Earth. These circles are called **great circles** because they are the largest circles that can be drawn on a sphere (Figure 2.2a). A good analogy of great circles in the Earth's grid system is the set of circles that separate orange segments after the fruit is peeled. If you examine a peeled orange closely, you will see that these great circles all converge on the top and bottom of the fruit and bisect it into many segments.

Great circles can be drawn that bisect Earth in myriad ways (Figure 2.2a). The common feature of each of these circles is they have the center of Earth as their core and they bisect Earth into two equal halves. Great circles are also important because their outlines reflect the shortest distance between any two points on the planet. This concept is particularly relevant to international travel because pilots plot their courses by great circle routes to reach destinations as quickly as possible and save on fuel costs. If you plan to travel from New York City to Moscow, for example,

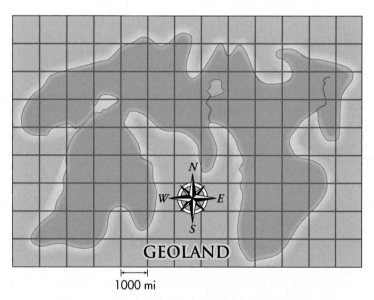

GEOLAND

1000 mi

Figure 2.1 Hypothetical flat land and grid. If Earth were shaped in this way, you could determine your location by measuring the distance from the nearest grid lines.

Great circles *Circles that pass through the center of the Earth and that divide the planet into equal halves.*

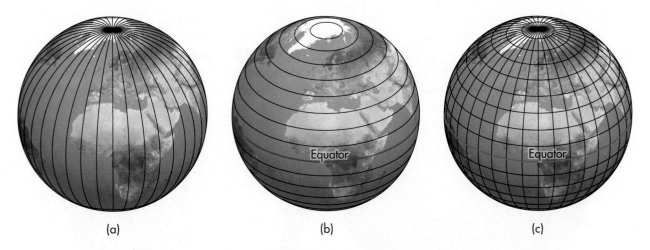

(a)   (b)   (c)

Figure 2.3 Orientation of great and small circles. Together, these circles form Earth's grid system. (a) All the great circles that are oriented north and south intersect at the poles. (b) Except for the great circle at the Equator, all circles oriented east and west on Earth are small circles. (c) Overlay of great and small circles to form the Earth's grid system.

you would probably think that the shortest route is to fly due east across the Atlantic Ocean. But, in fact, the great circle route from New York to Moscow passes north over Greenland.

With respect to the Earth's grid system, the most important group of great circles is those that converge at the North and South Poles. Along with this group of circles, another corresponding set of circles has outlines that extend east and west. These outlines are always parallel to one another and thus never meet. The largest of these parallel circles (the Equator) lies halfway between the two poles and is a great circle because its outline corresponds with the maximum circumference of Earth. North and south of the Equator, however, the parallel circles have smaller diameters and are known as **small circles** (Figure 2.2b).

To visualize this sequence of small circles, imagine what happens when you slice a lemon or pineapple. First, the slices are generally parallel to one another as you work across the fruit. Second, the circumference of each slice is progressively smaller as you move away from the center of the fruit (the great circle of the fruit). As with great circles, small circles can be aligned in any direction on the globe. Geographers use this network of north–south intersecting great circles and east–west parallel small circles to create a grid on the Earth's surface (Figure 2.3) that allows them to determine any location within about 30 m (100 ft). This system is known as the latitude/longitude coordinate system. Let's see how this system works.

Latitude

Latitude is the portion of the grid system that uses the parallel set of circles with outlines extending east and west. Using these circles, latitude measures location north and south of a geographical reference. The natural reference for latitude is the great circle that bisects the middle of Earth at its maximum east–west circumference, otherwise known as the **Equator** (Figure 2.4a). The portion of Earth north of the Equator is called the **Northern Hemisphere**, whereas the area south of the Equator is the **Southern Hemisphere**. Latitude is determined using simple geometry by measuring the angle of any point on the Earth's surface north or south of the Equator. This calculation is accomplished by projecting two lines from the center of Earth to its surface. One of these lines always extends from the center of Earth to the surface along the Equator. The other line extends to the north or south from the center of Earth directly to the location in question, depending on whether it is in the Northern or Southern Hemisphere, respectively

---

Latitude   *The part of the Earth's grid system that determines location north and south of the Equator.*

Equator   *The great circle that lies halfway between the North and South Poles.*

Northern Hemisphere   *The half of Earth that lies north of the Equator.*

Southern Hemisphere   *The half of Earth that lies south of the Equator.*

Small circles   *Circles that intersect the Earth's surface and that do not pass through the center of the planet.*

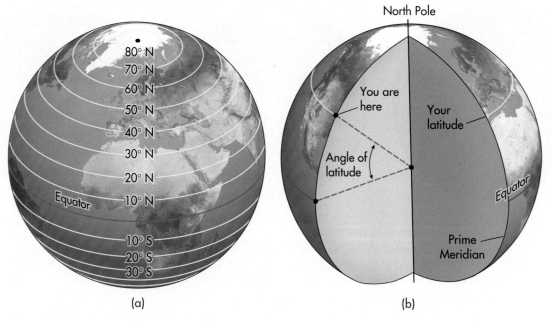

Figure 2.4 Determining latitude. (a) The reference point for latitude is the Equator, which divides Earth into Northern and Southern Hemispheres. Latitude ranges from 0° to 90° N and S. (b) Latitude is calculated by simply measuring the angle of a point relative to the Equator from the center of Earth.

(Figure 2.4b). We can then calculate the latitude of the location from the geometric arc between the two intersecting lines.

In this system the Equator has a value of 0° latitude. Latitudes close to the Equator have small geometric arcs from the Earth's center and, therefore, have relatively low latitude designations such as 5° or 10° N or S (see Figure 2.4a). Localities progressively farther north (N) or south (S) from the Equator have progressively greater angles from the Earth's center and, therefore, have progressively larger latitude designations. The maximum angle is perpendicular (90°) to the center of the equatorial great circle. In the Northern Hemisphere, this angle corresponds to the North Pole (90° N), whereas it correlates with the South Pole (90° S) in the Southern Hemisphere.

As far as distance is concerned, degrees of latitude are separated by about 110 km (69 mi) from each other, regardless of location. In other words, a location on the Earth's surface can be specified only within 110 km north or south at the level of a degree. It is possible to more closely locate places by subdividing degrees of latitude into smaller units. One way to more specifically identify locations is to progressively subdivide degrees of latitude into 60 minutes (′) and each minute into 60 seconds (″). An example of such a latitude designation would be a location in the Northern Hemisphere that has latitude of 39° 52′ 46″ N. At this level of detail, the location of something can be determined within 30.5 m (100 ft) on the Earth's surface.

Another way to designate places with a higher level of precision is simply to convert latitude designations containing minutes and seconds to full decimal notations. This conversion is done with a simple equation that would, for example, change the previous latitude of 39° 52′ 46″ N to 39.8794° N using a full decimal notation. The use of full decimal notations is very common and is most frequently used to designate such important lines of latitude as the Tropic of Cancer (23.5° N), Tropic of Capricorn (23.5° S), Arctic Circle (66.5° N), and the Antarctic Circle (66.5° S) (Figure 2.5). These lines of latitude are related to the Earth/Sun geometric relationship and will be discussed in much more detail in the next chapter.

Lines of latitude are also called **parallels** because they are parallel to the Equator and each other. The value of any given line of latitude remains the same no matter the location, whether it is on one side of Earth (such as in China) or the other (such as in the United States). A common misconception is that lines of latitude must indicate direction east and west because that is the direction that these lines follow. To remember how the system really works, think of degrees of latitude as being numbered sequentially north and south of the Equator.

Although you can certainly think of lines of latitude in individual terms to locate specific places, it is also useful to think of them collectively by region. In that context, lines of latitude can be lumped into nine geo-

---

Parallels *Lines of latitude.*

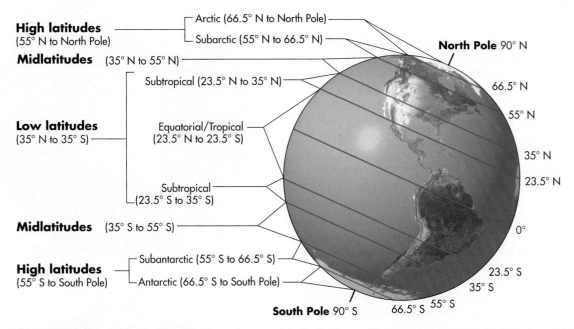

High latitudes (55° N to North Pole)
Arctic (66.5° N to North Pole)
Subarctic (55° N to 66.5° N)

Midlatitudes (35° N to 55° N)

Subtropical (23.5° N to 35° N)

Low latitudes (35° N to 35° S)
Equatorial/Tropical (23.5° N to 23.5° S)

Subtropical (23.5° S to 35° S)

Midlatitudes (35° S to 55° S)

High latitudes (55° S to South Pole)
Subantarctic (55° S to 66.5° S)
Antarctic (66.5° S to South Pole)

North Pole 90° N
66.5° N
55° N
35° N
23.5° N
0°
23.5° S
35° S
66.5° S 55° S
South Pole 90° S

Figure 2.5 Important lines of latitude on Earth and general latitude zones. These zones of latitude often have distinctive climate, vegetation, and soil characteristics.

TABLE 2.1	General Zones of Latitude	
Latitude Zone	Range	General Characteristics
Low latitudes	About 35° N–35° S	Very warm region of Earth with generally consistent weather over the course of the year
Midlatitudes	About 35°–55° in both hemispheres	Seasonal weather with warm/hot summers and cool/cold winters
High latitudes	From about 55° to 90° in both hemispheres	Very cold region of Earth with very short summers

graphic zones (see Figure 2.5). These nine regions can be grouped further into three general zones (Table 2.1) in which many distinct physical processes occur. The **low latitudes** range from about 35° N to 35° S of the Equator and are generally the warmest part of Earth. The **midlatitudes** range from about 35° to 55° latitude, in both hemispheres, and are typically regions of Earth that experience highly variable weather over the course of the year. Finally, the **high latitudes** range from 55° to 90° latitude, again in both hemispheres, and are typically the coldest places on Earth.

Low latitudes   *The zone of latitude that lies between about 35° N and 35° S.*

Midlatitudes   *The zone of latitude that lies between about 35° and 55° in both hemispheres.*

High latitudes   *The zone of latitude that lies from about 55° to 90° in both hemispheres.*

Longitude

The previous discussion demonstrated how latitude is used to identify location north and south of the Equator. It is now time to examine how east and west locations are determined with **longitude**. The foundation of the longitude system is the complex of great circles that pass through the poles. Half of such a great circle is a spherically shaped outline that connects each pole. Each of these half circles is called a meridian of longitude, or simply a **meridian**. In contrast to the latitude system, which has the Equator as the natural reference because it is the only great circle of the parallels, the easterly and westerly longitude calculations require an arbitrary reference meridian because all of them are half outlines of the great circles that pass through the poles. The arbitrary reference meridian for longitude is the **Prime Meridian**, which passes through Greenwich, England, at the Old Royal Observatory (Figure 2.6). This meridian was chosen as the reference meridian of longitude at an 1884 conference in Washington, D.C.

As with the Equator, the Prime Meridian has a value of 0°, and all points north and south along that meridian are at 0° longitude. The geometric basis for determining longitude is that a circle contains 360°. Think of the circle (the Earth) as divided into eastern and western halves by the Prime Meridian. The longitude grid consists of a series of meridians that extend 180° east and west of the Prime Meridian to complete the 360° circle. Using Figure 2.7 as a guide, imagine you are floating in space and looking at a section of the Earth at the Equator. Longitude is calculated by measuring the arc formed by two lines projected from the Earth's center at the Equator. One line extends to the Prime Meridian, the other to the meridian of the

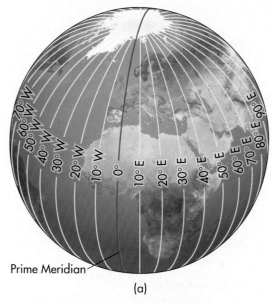

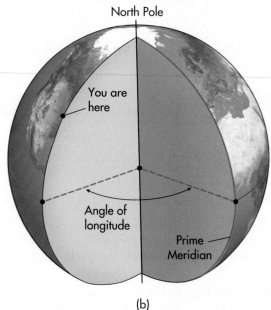

Figure 2.7 Calculating longitude. (a) The longitude grid consists of a series of great circles that converge at the poles and that are measured east and west of the Prime Meridian. (b) Longitude is calculated on the angle of arcs measured from the center of Earth and relative to the Prime Meridian.

location in question. The angle of the arc formed by the two lines determines the longitude. Calculated this way, smaller geometric arcs from the Prime Meridian have lower longitude designations, and greater arcs have higher designations (see Figure 2.7b). As with latitude,

Longitude *The part of the Earth's grid system that determines location east and west of the Prime Meridian.*

Meridians *Lines of longitude.*

Prime Meridian *The arbitrary reference point for longitude that passes through Greenwich, England.*

Figure 2.6 The Prime Meridian. This line of longitude at the Old Royal Observatory in Greenwich, England, is the reference meridian for the longitude system.

1. Lines of longitude run north and south and are also called meridians.

2. Longitude is determined by measuring the geometric arc between two lines projected from the center of Earth to the surface at the Equator and to the meridian in question. The reference point for longitude is the Prime Meridian (or 0°), which divides Earth into eastern and western halves.

3. Meridians are located east and west of the Prime Meridian and extend 180° E and 180° W to complete a full circle.

4. The distance between meridians is greatest at the Equator, but decreases steadily until converging at the poles.

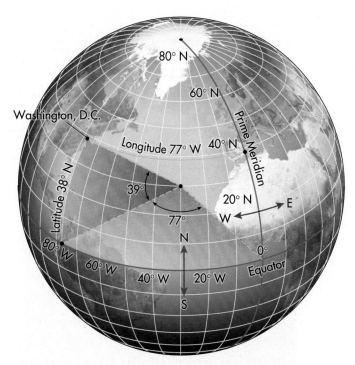

Figure 2.8 Determining latitude and longitude of Washington, D.C. Using the grid network, we can locate Washington, D.C., at approximately 39° N, 77° W.

degrees of longitude can be subdivided into minutes and seconds or be given full decimal notations.

A primary difference between longitude and latitude is that while lines of latitude are always parallel to one another, lines of longitude converge at the poles (Figure 2.7a). Thus, degrees of longitude are about 110 km (69 mi) apart at the Equator and progressively shorten in distance until they meet at the poles. As with latitude, a common misconception about longitude is that meridians are used to determine location north and south because that is the direction in which the semicircle outlines extend. Instead, remember that longitude determines location east and west of the Prime Meridian.

Using the Geographic Grid

After establishing the fundamental north–south and east–west grids through latitude and longitude, the two subsystems can be overlain to form a complete grid network that encompasses Earth (Figure 2.8). Thus, any location can be described in terms of its position north or south of the Equator and east or west of the

Prime Meridian. Imagine, for example, that you wanted to find Washington, D.C., in the grid; Washington is located at 39° N (latitude), 77° W (longitude). To see how you would locate Washington within the grid, examine Figure 2.8 again. Beginning with latitude, find the 38th parallel north of the Equator to identify the location in the Northern Hemisphere. Once this is accomplished, locate the Prime Meridian and arc west to the 77th meridian. The intersection of the 38th parallel and the 77th meridian is the position of Washington, D.C. Notice that you can first find the latitude or you can first find the longitude; the order does not matter, as you wind up at the same place either way. However, remember that by convention the latitude of a location is always specified before the longitude is given.

 www.wiley.com/college/arbogast

Using the Geographic Grid

In order to help you fully comprehend how the complete geographic grid works, go to the *GeoDiscoveries* website and select the module *Using the Geographic Grid*. In this simulation you will be able to select a number of places on Earth and determine their latitude and longitude. This, in

turn, should help you better understand where locations are in physical space. Once you complete the module, be sure to answer the questions at the end of the module to test your understanding of the geographic grid.

Beautiful natural features are scattered around the world on all seven continents. Locate the following places on a globe or map of the world and determine the country in which each one is located.

a) Mont Blanc, 45.50° N, 6.52° E
b) Lake Chad, 13.30° N, 14.00° E
c) Valle de Luna (Valley of the Moon), 23.50° S, 69.00° W

(a)

(b)

(c)

Maps—The Basic Tool of Geographers

Now that Earth's grid system has been discussed, let's investigate the various tools that geographers use to portray spatial information within the grid. Most everyone knows that the primary tool associated with geography is the map. Figure 2.9 shows a typical thematic map (it focuses on a specific theme) and is one with which you are probably familiar. This map shows the United States and includes such geographic information as state boundaries, location of major cities, the course of major rivers, and vegetation patterns. You probably also know that a map such as this one contains a scale, a compass arrow pointing north, and a legend that defines major categories within the map, such as population density, the area covered by cities, the relative size of highways, or the location of mountain ranges. A wide variety of additional information can be portrayed on maps, including the geographic distribution and thickness of rock units, temperature, precipitation patterns, intensity of solar radiation, vegetation, and soils. Much of the information in this text will be presented in map form, so in order to navigate through this book thoroughly, and to be successful in your course, you must first understand the fundamentals of cartography.

Cartography is the subdiscipline of geography that focuses on the many ways to display spatial information

so that it can be used and understood efficiently. The essence of cartography is the manner in which Earth is portrayed in a usable fashion. This task can be accomplished in a variety of ways. The most visually accurate and complete way to illustrate the entire Earth is through the use of a globe (Figure 2.10), which depicts locations on the three-dimensional representation of the planet. Globes can be cumbersome to use, however, and often do not illustrate the level of detail desired. Instead, geographers frequently rely on two-dimensional maps to show various geographic locations and attributes. Many factors must be taken into consideration to represent these features most accurately.

Map Projections

The most important step in the cartographic process is the presentation of three-dimensional information of Earth on a two-dimensional map. In order to create a map, locations on the roughly spherical surface of Earth must somehow be transferred to a flat surface, such as paper or a computer screen. This is done through the process of **map projection**.

A simple way to visualize how projection works is to imagine a light source passing through a curved

---

Cartography *The design and production of maps.*

Map projection *The representation of the three-dimensional Earth on a two-dimensional surface.*

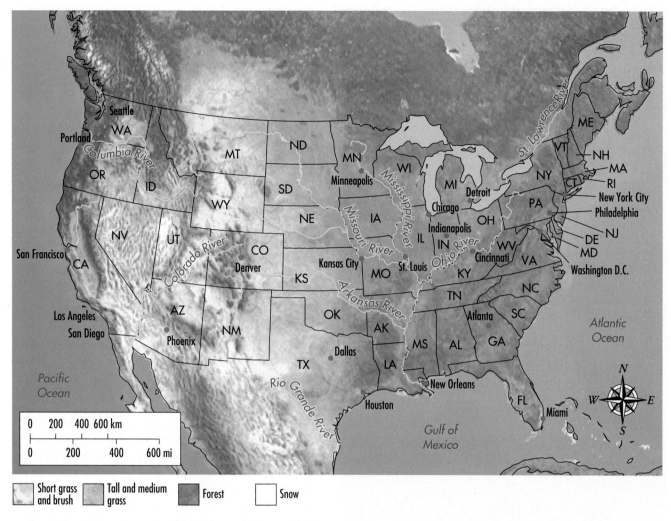

| | Short grass and brush | | Tall and medium grass | | Forest | | Snow |

Figure 2.9 A simple thematic map. This map shows the location of some important features within the United States, including states, major rivers, and cities. Note the legend that reflects general types of vegetation.

translucent surface that has an image on it, projecting the image from the curved surface onto a flat wall (Figure 2.11). Light from the source (the light bulb) streams out toward the wall in straight lines. Before it reaches the wall, however, the light passes through the spherical surface of Earth. As the light beams move through this spherical surface, imagine that they carry with them the collective information from the Earth's

(a)   (b)

Figure 2.10 The globe and planet Earth. (a) A globe portrays geographic information about Earth in a three-dimensional fashion. (b) Compare the image of the globe with this picture of Earth taken from space. This particular Earth view focuses on Africa.

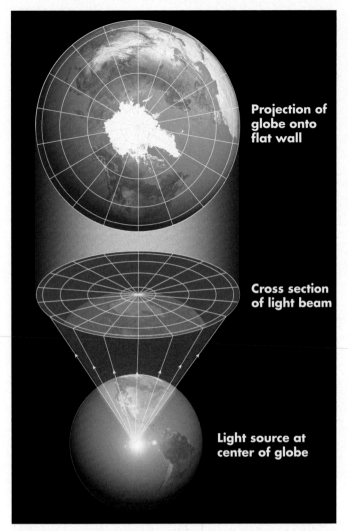

Projection of globe onto flat wall

Cross section of light beam

Light source at center of globe

Figure 2.11 The projection theory. Light from a source projects the information on the curved surface of the Earth onto a plane, forming a flat map.

Figure 2.12 Why map distortion occurs. On the left is the semicircular (undistorted) half of a lime. Once the center of the lime is pressed to the table to make it flat, distortion occurs at the edges. Note that the area covered by the flattened part of the lime is larger than that beneath the uncut part of the lime.

surface through which they just passed. When the light beams reach the wall, this information is then displayed for you to see. Although the image was originally curved on Earth, it becomes flat on the wall. The actual process of map projection theoretically works in the same manner.

Although projection serves the purpose of creating a flat map that is potentially more usable than a globe, a negative by-product results because it is not possible to flatten a sphere (such as Earth) without some distortion of the image. Look at what happens in Figure 2.12 if a lime is sliced in two so that two hemispheres are created. If either of the two hemispheres is flattened on a table, it covers a larger area than it did when it was intact. Also notice that the edges of the flattened hemisphere are severely distorted; in this case, they are split. In fact, the only way to make the hemisphere become perfectly flat is to either stretch or cut the edges.

Maps contain the same kind of distortion because spatial information from the curved surface of Earth is essentially spread over a flat surface. Because of this

distortion, a variety of map projections have been developed, with each kind presenting location and distortion in a different way that can preserve either shape or size but not both. Geographers are intimately aware of the strengths and weaknesses of each kind of projection with respect to spatial analysis and select the type of map they want to use accordingly. Although it is not necessary at this introductory level for you to understand maps as thoroughly as a geographer does, you must know something of the properties that result from the various projections that are used to create them. Such a fundamental understanding can be obtained by examining two simple kinds of projections: conformal and equivalent.

Conformal (or True Shape) A **conformal projection** is a map that maintains the correct shape of features on Earth, but distorts their relative size to one another. A conformal projection is conceptually constructed by placing a translucent cylinder over the globe and shining a light centered within the globe outward so that an image is transferred to the cylinder (Figure 2.13). If the cylinder is then removed from the globe, and cut so that it could be laid flat, a map would result.

One of the most common conformal projections is the Mercator projection (Figure 2.14), which was developed in 1568 by the Flemish geographer and mathematician Gerhardus Mercator. This projection is used frequently because it correctly shows the relative shape of features on the globe. As an example of this accuracy, look at Greenland and the Americas in Figure 2.14 and notice that the actual shapes of these places have been maintained. You

Conformal projection *A map that maintains the correct shape of features on the Earth but distorts their relative size to one another.*

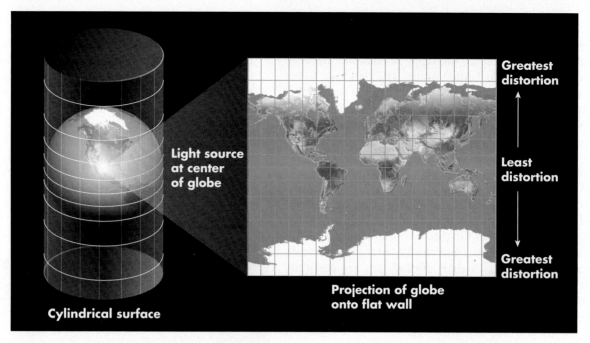

Figure 2.13 A conformal projection. Imagine a cylindrical surface surrounding the globe with a source of light at the center of the globe. The surface is unwound to produce a flat map. The regions of greatest distortion occur at the top and bottom of the map.

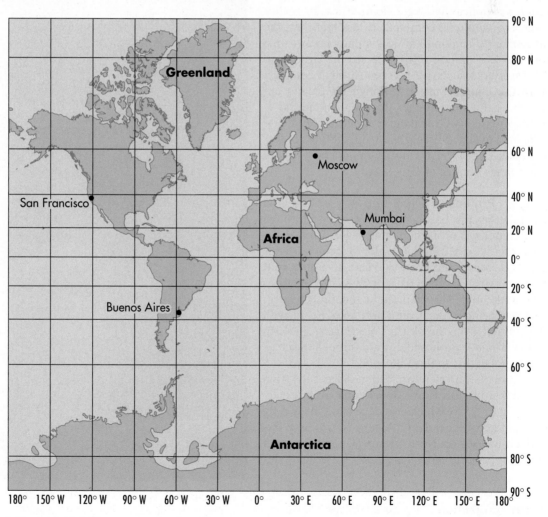

Figure 2.14 A Mercator projection. This map is a conformal projection that maintains shape at the expense of size at high latitudes. Accordingly, scale increases from low to high latitudes.

might ask yourself, why would this be useful? Imagine that you wanted to compare the shape of coastlines between Greenland and the United States. The only way to accomplish this task accurately would be to use a conformal projection such as the Mercator, which would preserve the correct shape of the coastline. Thus, you can make an accurate comparison between these places.

Remember, however, that some spatial attribute must be distorted to keep the shape accurate. In the case of the Mercator projection, the distorted feature is size. Notice how lines of latitude and longitude are at right angles to each other because meridians do not converge at the poles, and that the distance between parallels increases with distance from the Equator. This increase occurs because the higher latitudes are geometrically stretched more during the projection process than the lower latitudes. Another way to look at this type of distortion is to realize the map in Figure 2.14 was stretched much like the lime in Figure 2.12.

You can see this distortion particularly well by looking at Greenland and Antarctica in Figure 2.14. Notice that Antarctica covers almost a quarter of the map and is significantly larger than Africa. In reality, however, the Antarctic continent is less than half the size of Africa (8.7 million km² [3.4 million mi²] vs. 18.8 million km² [7.3 million mi²]). The size of Greenland is also distorted, which can be seen through comparison with the United States. Although Greenland appears to be significantly larger than the United States on the map, in reality Greenland is about 2,175,590 km² (840,000 mi²) in size, whereas the United States covers an area of about 9,582,956 km² (3.7 million mi²)—more than four times the size of Greenland.

Equivalent Projection Once you know how the angular relationships between places compare through use of a conformal projection, you might want to see how their relative size differs. In order to accomplish this visualization, you would need to create an **equivalent projection**, which accurately portrays size features throughout the map. This map can be theoretically constructed by placing a cone over a portion of Earth, with the sharp point over the North Pole (Figure 2.15). In this instance light is projected upward through the cone, which is then laid flat to form the map.

One kind of equivalent projection that results from this projection process is an Albers equal-area projection (see Figure 2.15), which was developed in 1805 by the German geographer H. C. Albers. Notice how this map, which focuses on North America, compares with the Mercator projection in Figure 2.14. One primary dif-

ference is that the lines of latitude are now curved. Given that these particular lines are actually curved on Earth, this kind of map represents a more accurate presentation of the geographic grid than a Mercator projection.

Another difference lies in the respective shapes of landmasses, which can be nicely seen again at Greenland. Remember that in the Mercator projection (Figure 2.14), the shape of Greenland was accurately portrayed but the size was distorted. How does Greenland now look in Figure 2.15? Notice that Greenland looks squatty in the Albers projection compared to the Mercator projection. Why does this difference exist? Remember

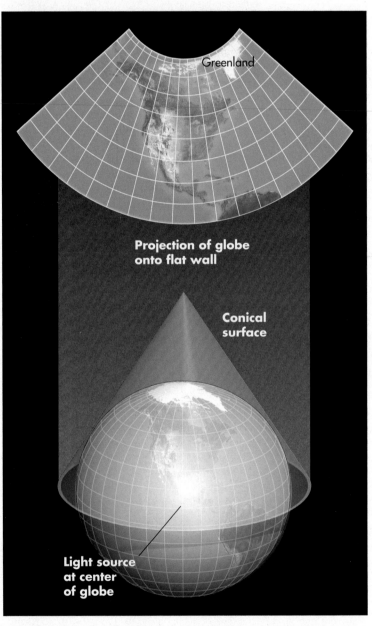

Figure 2.15 Constructing an equivalent projection. The resulting map in this diagram is a specific kind of equivalent projection called an Albers equal-area projection.

Equivalent projection *A map projection that accurately portrays size features throughout the map.*

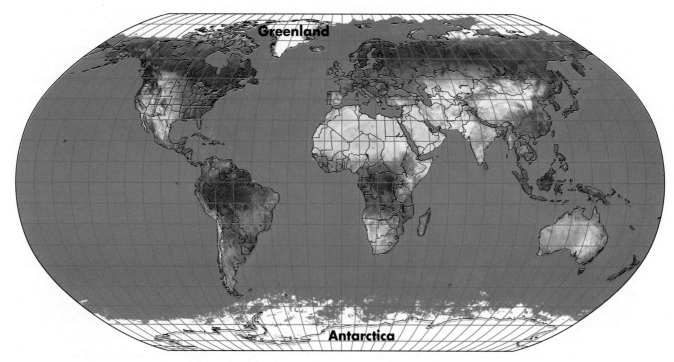

Figure 2.16 The Robinson projection. In an effort to balance the errors associated with projection, this map contains distortion in both shape and size.

that a map projection can preserve only one spatial variable at a time. In the case of the Albers projection, the shape of places at high latitudes is distorted so that the relative size of places can be more accurately observed.

You might ask, why does it matter whether or not the sizes of landmasses on Earth on a map are accurate? Remember that the Mercator projection, because it preserves shape, allows you to compare regional features such as the configuration of coastlines. However, would a Mercator projection provide an accurate comparison of the *length of* coastline between the United States and Greenland? The answer to this question is "no" because the size of continents at high latitudes is severely distorted by a conformal projection. Thus, the only way to compare accurately the length of coastline between Greenland and the United States would be to use an equivalent-area projection.

It is important to emphasize that many different kinds of projections exist, although it is beyond the scope of this text to describe each of them thoroughly. Depending on the kind of information you want to see, a projection exists for that purpose. Some projections intensively exaggerate the distortion of one property in order to represent another more accurately. Others attempt to show all spatial information as accurately as possible, given the limitations of the technique, and therefore moderate the distortion of both shape and size. A good example of such a projection is the Robinson projection (Figure 2.16), which

was developed in 1963 by the American cartographer Arthur H. Robinson. Notice how this particular projection seems to be evenly balanced relative to the Mercator (Figure 2.14) and Albers (Figure 2.15) projections. Greenland and Antarctica are still somewhat enlarged relative to their accurate size, but some semblance of the Earth's curvature exists.

Map Scale

Aside from understanding how the concept of map projection works, another important consideration involving maps is the scale. Map scale is a critical part of cartography because it relates the size of features on a map to their actual size in the real world. Similarly, the scale relates the distance between features on the map to the corresponding actual distances on Earth. In other words, **map scale** represents the ratio of the size/distance on the map to the size/distance on the ground.

Map scale can be depicted in several ways.

1. *Written (Verbal) Scale*. Often written as: One Inch = 2000 Feet, which means that 1 in. measured on the map represents 2000 ft on Earth.

2. *Representative Fraction*. An example of a representative fraction is the ratio 1:24,000 or 1/24,000. This scale means that one unit of measurement on the map represents 24,000 of the same unit on the ground. It does not matter what unit of measurement is used—it could be inches, centimeters, etc.—but the units must be the same within the ratio (inches to inches, centimeters to centimeters, etc.).

3. *Graphic Scale or Bar Scale*. These scales show the actual size of units on the map. The advantage of the graphic (or bar) scale is that the scale remains accurate if the map is enlarged or reduced when photocopying.

Maps are made at different scales depending on what geographic qualities the maps are designed to display, including the area they cover and spatial detail. In general terms, geographers think of maps in terms of being large-scale or small-scale. Although these designations appear easy to differentiate, the fact is that many nongeographers find them confusing because an inverse relationship exists between map scale and actual Earth features. You might think, for example, that a large-scale map is a map that shows a large part of Earth. Conversely, a small-scale map would be a map that focuses on an area of limited geographic extent, such as a city or park. The fact is that **small-scale maps** are maps that show large areas with relatively limited detail because the representative fraction is a small number; in other words, the size ratio of the map features to the real world is small. **Large-scale maps**, in contrast, represent much smaller geographic areas with a greater level of detail because the representative fraction is a larger number; that is, the size ratio on the map is closer to the actual world. In order to obtain a better feel for map scale, you should examine the information in Figure 2.17 and Table 2.2.

Isolines

Isolines are another important map component because they are lines that connect points of equal value. These lines are particularly useful, for example, when showing regional temperature and precipitation patterns, elevation, or atmospheric pressure. These data can be derived in a variety of ways and then used to illustrate fundamental geographic trends. For example, imagine that you want to create a map showing the pattern of high temperatures throughout the United States on any given day. The first thing you would do would be to access the temperature measurements taken at all of

Small-scale map *A map that shows a relatively large geographic area with a relatively low level of detail.*

Large-scale map *A map that shows a relatively small geographic area with a relatively high level of detail.*

Isolines *Lines on a map that connect data points of equal value.*

Map scale *The distance ratio that exists between features on a map and the real world.*

TABLE 2.2	Map Scale	
Large-Scale Maps	**Small-Scale Maps**
Used for maps of small geographic areas.	Used for maps of large geographic areas.
Used to illustrate great detail, such as road networks, locations of parks, etc.	Used to illustrate limited geographic detail, but show spatial relationships of large areas.
Representative fraction is a large number, such as 1:1000.	Representative fraction is a small number, such as 1:25,000.

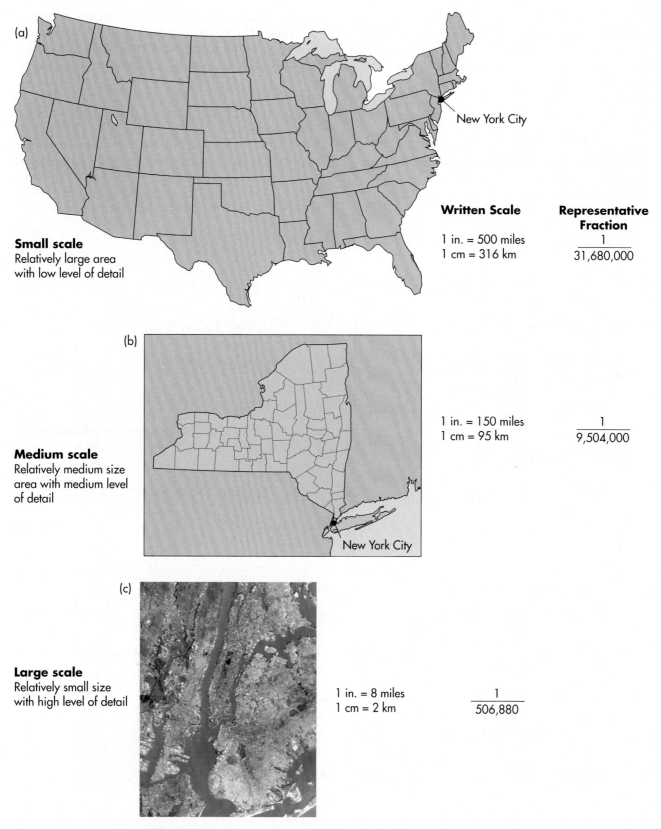

	Written Scale	Representative Fraction
Small scale Relatively large area with low level of detail	1 in. = 500 miles 1 cm = 316 km	$\dfrac{1}{31,680,000}$
Medium scale Relatively medium size area with medium level of detail	1 in. = 150 miles 1 cm = 95 km	$\dfrac{1}{9,504,000}$
Large scale Relatively small size with high level of detail	1 in. = 8 miles 1 cm = 2 km	$\dfrac{1}{506,880}$

Figure 2.17 Examples of map scale. Note that as scale increases, the amount of detail about a place increases. (a) Map of the United States that shows the location of New York City. (b) Map of the state of New York showing the location of New York City. Note that this map shows more detail (county boundaries) of the state of New York than (a). (c) Satellite image of New York City. Note how much more detail of New York City can be seen in this image than in the other maps.

the National Weather Service sites in the country and show them on a map. This map would contain an outline of the country, state boundaries, the various locations where temperature measurements were taken, and the temperature at each place (Figure 2.18a).

Although this map would contain the information you desire, it would be difficult to discern any geographic pattern because the map contains nothing but a series of numbers. By looking at the numbers closely, you could perhaps determine that it was generally warmer in the southern part of the United States than in the northern part of the country, but the overall geographic patterns would be vague. Geographers typically clarify

these patterns by using isolines, which are lines that connect points of equal value at predetermined intervals. In the context of our temperature map, you might decide that you wanted to show the temperature pattern in the country at 5°C (9°F) intervals, or you may choose another interval. Using a 5°C (9°F) interval as an example, you would draw an isoline connecting all points of equal value at, for example, 5°C (41°F), 10°C (50°F), 15°C (59°F), 20°C (68°F), etc. In many regions these distinct temperatures might not exist; that is, the temperature at one weather station might be 7°C (44.6°F), whereas at the adjacent station it could be 12°C (53.6°F). Where this kind of pattern exists, you would draw the line

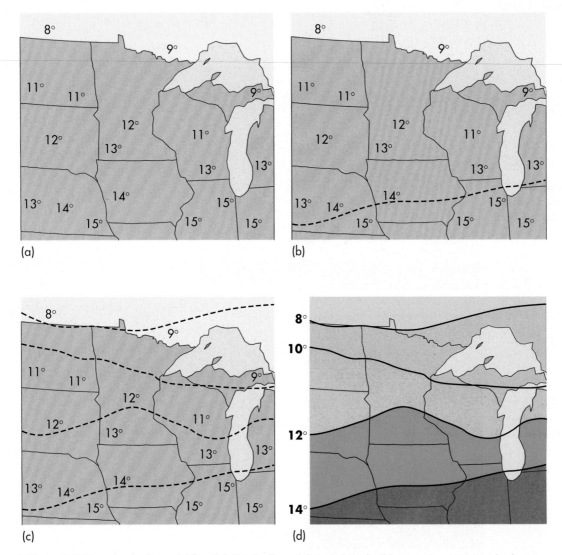

Figure 2.18 Drawing isolines. (a) Spatial distribution of data points, in this case temperature in degrees centigrade. (b) Interpolating the 14° isotherm. A line is drawn connecting points where 14° measurements were obtained. The position of the 14° isoline is estimated between locations where 13° and 15° temperatures were measured. (c) Isolines drawn at intervals of 2°C. (d) The resulting map shows the spatial distribution of temperature in the upper Midwest.

between the two weather stations in a process called *interpolation*, because you could assume that the 10°C line occurs somewhere midway between 7°C (44.6°F) and 12°C (53.6°F). Figure 2.18 shows this process more closely with some temperature data in the upper Midwest.

In the context of isolines, several different kinds are used when mapping specific types of geographic phenomena. The various kinds of isolines are:

1. **Isobars**—connect points of equal atmospheric pressure.
2. **Isotherms**—connect points of equal air temperature.
3. **Isohyets**—connect points of equal amounts of precipitation.
4. **Isopachs**—connect points of equal sedimentary thickness.
5. **Contours**—connect points of equal elevation.

Contour lines are particularly significant because they illustrate the configuration of the three-dimensional landscape, or the **topography**, on a **topographic map**. Topographic maps contain a great deal of information, especially at a 1:24,000 scale (Figure 2.19). Here are some simple rules to follow when reading them:

1. The closer the spacing of contour lines, the steeper the slope. Conversely, contours that have wide spacing represent terrain that is relatively flat.
2. Contours that form closed circles indicate hills. Closed circles with hatch marks indicate closed depressions.
3. Where contours cross a stream, they form V's pointing upstream (which is also uphill).

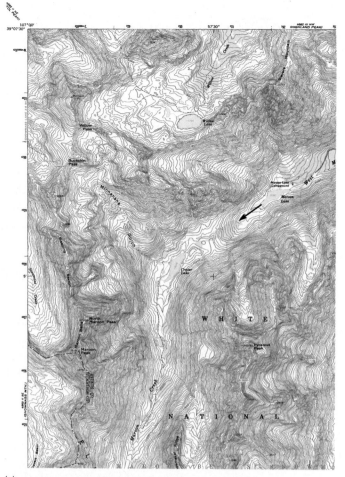

(a)

(b)

Figure 2.19 A typical topographic map. Topographic maps show elevation patterns of a given geographic area. (a) This topographic map focuses on the Maroon Bells area near Aspen, Colorado. (b) Photograph of the Maroon Bells from Maroon Lake. The arrow in the accompanying topographic map shows the approximate line of sight from Maroon Lake toward the mountain peaks.

Isobars *Isolines that connect points of equal atmospheric pressure.*

Isotherms *Isolines that connect points of equal temperature.*

Isohyets *Isolines that connect points of equal precipitation.*

Isopachs *Isolines that connect points of equal sediment or rock thickness.*

Contours *Isolines that connect points of equal elevation.*

Topography *The shape and configuration of the Earth's surface.*

Topographic map *A map that displays elevation data regarding the Earth's surface.*

Using Maps

Now that you have learned something about cartography, go to the *GeoDiscoveries* website and select the module *Using Maps*. The goal of this module is to increase your understanding of maps by interacting with them online. This particular module covers how maps are produced, issues relating to scale, and the various kinds of data that can be presented in a map. Once you finish this exercise, be sure to answer the questions at the end of the module to test your understanding about maps.

VISUAL CONCEPT CHECK 2.2

a) The process of map projection is the method through which three-dimensional features on Earth are presented as a two-dimensional image. Two basic projections, conformal and equivalent, are presented here. Which of these two maps is a conformal projection? Which one is the equivalent projection? How can you tell the difference between the two?

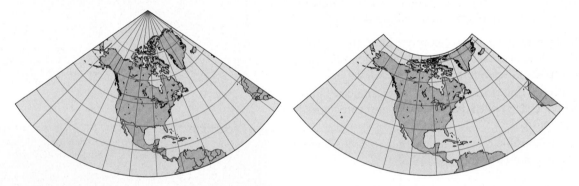

b) The scale of a map depends on the amount and kind of information that a cartographer wishes to present. In this case, two maps are presented that contain information about Seattle, Washington. Which one of the maps is a relatively small-scale map? Which one is the large-scale map? How can you tell the difference between the two?

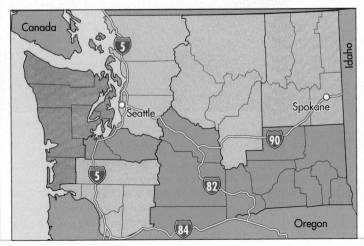

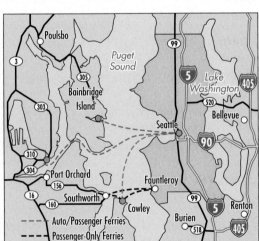

Digital Technology in Geography

As you are no doubt aware, digital technology is an essential part of our life in the industrialized world today, with personal computers, cell phones, the Internet, and satellite television common features in many households. These technological advances have also been very important in the recent evolution of geography as a scientific discipline, resulting in expanded work in the field that has increased the need for skilled people at all professional levels at excellent pay. Much current geographic research is focusing on the development and refinement of three digital techniques: Global Positioning Systems (GPS), remote sensing, and Geographic Information Systems (GIS). These techniques, used independently as well as together, enable geographers to use maps and other images in a digital format in order to increase the speed and efficiency of geographic research. In addition, the ability to display geographic information has improved remarkably. The remainder of this chapter focuses on these new and exciting tools that are revolutionizing how people are viewing the planet from a geographic perspective.

Remote Sensing

Until the early 20th century, the only way that a geographer could consistently learn anything about new places firsthand was to personally conduct fieldwork and collect samples or make observations. This limitation began to change somewhat in the early 1900s with the development of airplanes, which people could use to see places without actually setting foot on the ground. With the advent of this new technology, geographers began to use a new approach to spatial analysis called **remote sensing**, which means measuring properties of the environment without direct contact. In this manner, geographers could learn much information about the vegetation patterns of the landscape, the routes that rivers take across the countryside, or even the configuration and distribution of agricultural fields. Remote sensing began to be used in a widespread, systematic fashion in the 1930s when on-board cameras were developed that could take **aerial photographs** (Figure 2.20) at prescribed intervals of time. With these photographs, which could be taken in a succession of linear flight lines, entire counties or even states could be viewed in a systematic way.

Remote sensing *The method through which information is gathered about the Earth from a distance.*

Aerial photographs *Photographs taken of the Earth's surface from the air.*

Figure 2.20 A typical aerial photograph. With an aerial photograph you can see a variety of landscape patterns that cannot be viewed from the ground. This aerial photo, for example, shows the pattern of crop fields, trees, roads, and dwellings at Rajasthan, India.

Although aerial photography was a dramatic leap in the geographer's ability to study the landscape, this technology proved limiting because a complete set of photographs of large regions (like a state) was only taken once every decade, or at best, at irregular intervals of time. Although this temporal pattern allowed for some studies of landscape change, perhaps by taking photographs of the same place two or more times over, say, 25 years, geographers were still limited because they could not see the change as it was occurring. The fact is that major changes can occur on a landscape in a very short period of time and geographers usually want to see them quickly.

With the development of satellite technology in the 1950s and 1960s, our ability to remotely sense the landscape increased dramatically. As you know, satellites are platforms that continuously orbit Earth. These systems are designed for either Sun-synchronous or geostationary orbits (Figure 2.21). A satellite with a **Sun-synchronous orbit** has an orbit that keeps pace with the Sun's westward progress as Earth rotates. Such a position is maintained because the satellite's orbit is basically north–south between the poles with a slight (8°) angular inclination. As a result, a particular satellite may cross the Equator 12 times during the course of a day, with the local time perhaps being 3:00 P.M. at the time of each pass. A Sun-synchronous orbit is usually at an altitude of about 700 km to 800 km (430 mi to 500 mi).

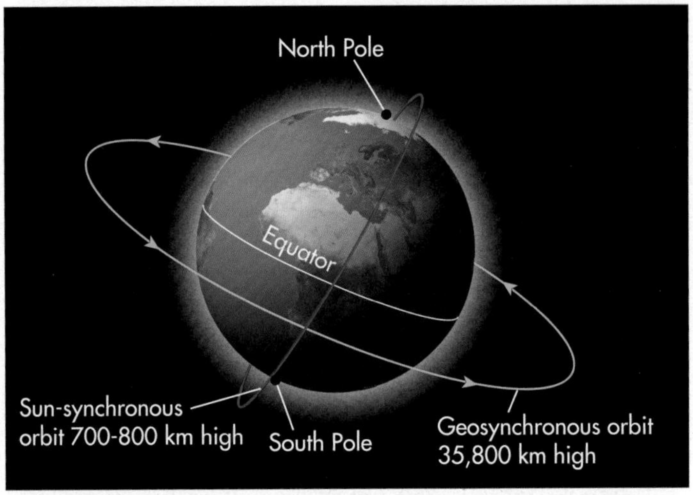

The orbit of Sun-synchronous satellites results in large gaps in image coverage between successive orbits on any given day. Given the slight inclination of the orbit, however, the satellite progresses slightly westward with each new day. In this way the satellite slightly overshoots the orbital path from the previous day, yielding imagery that overlaps a little so that no information is lost. Because satellites with Sun-synchronous orbits pass over the same part of Earth at about the same time each day, they are particularly good for monitoring landscape change over time.

Several well-known satellite systems have Sun-synchronous orbits. Perhaps the best known of these satellites is the Landsat system, which is jointly administered by the National Aeronautics and Space Administration (NASA), the National Oceanic and Atmospheric Administration (NOAA), and the U.S. Geological Survey (USGS). This satellite system is the longest continuous Earth-observing project in history, beginning with Landsat 1 in 1972 and continuing through the launch

Figure 2.21 Sun-synchronous and geosynchronous satellite orbits. Sun-synchronous orbits are designed to keep pace with the Sun's westward progress as Earth rotates. The orbit essentially extends between the poles with a small inclination so that the satellite moves slightly westward with each successive day. Geosynchronous orbits hover over the same point on Earth, but to do so, they must stay in a very high orbit over the Equator.

of Landsat 7 in 1999. Landsat 7 orbits Earth at an altitude of about 705 km (438 mi), has a swath width (it can "see") of 185 km (115 mi), and returns to the point of origin every 16 days (233 orbits). Landsat 7 carries two sensors: Thematic Mapper (TM) and Enhanced Multispectral Scanner Plus (EMSP+). Landsat TM has seven spectral bands, which means it can discriminate emitted energy in several parts of the electromagnetic spectrum, whereas the EMSP+ sensor has four spectral bands. The **spatial resolution** of the various Landsat sensors ranges from 15 m (49.2 ft) to 60 m (197 ft). Spatial resolution refers to the size of the ground area that the sensor can "see" in detail. A 15-m (49.2-ft) spatial resolution, for example, means that the satellite can discern objects that are at least 15 m × 15 m in size. Each of these 15 m × 15 m (49.2 ft × 49.2 ft) blocks is referred to as a **pixel**. In 2012 NASA is expected to launch the Landsat Data Continuity Mission, which will include revolutionary advances in technology and performance, including two new spectral bands.

In addition to the Landsat platform, another satellite with a Sun-synchronous orbit is the Advanced Very High Resolution Radiometer (AVHRR), which is man-

---

Sun-synchronous orbit *A slightly inclined polar orbit that keeps pace with the Sun's westward progress as Earth rotates, resulting in regular return intervals over every location on Earth.*

Spatial resolution *The area on the ground that can be viewed with detail from the air or space.*

Pixel *The smallest definable area of detail on an image; short for pixel element.*

aged by NOAA. This satellite orbits Earth 14 times per day from an altitude of 833 km (517 mi) and has a spatial resolution that ranges between 1.1 km (0.7 mi) and 4 km (2.5 mi), depending on whether the place viewed is directly below the satellite or on the fringe of the area seen.

A **geostationary orbit**, in contrast to a Sun-synchronous orbit, is designed to permanently remain in one place above Earth. This consistent position is accomplished by placing the satellite in an easterly orbit, directly over the Equator, and at a very high altitude. The satellite orbits in the same direction and at the same speed that Earth rotates. With such an orbit, the satellite can view nearly half of Earth at any given time. This kind of orbit is mostly used for observing weather and to facilitate communications. An excellent example of satellites with a geostationary orbit is NOAA's Geostationary Operational Environmental Satellites (GOES), which are used to monitor meteorological conditions (Figure 2.22). This system consists of an array of satellites that are positioned about 35,800 km (22,300 mi) above Earth, high enough to allow very large regions of Earth to be viewed at any given time. The GOES-12 (or GOES East) satellite, for example, is positioned above the Equator at 75° W longitude and views most of North America, all of South America, and most of the Atlantic Ocean. The GOES-10 (or GOES West) satellite, in contrast, is located above the Equator at 135° W longitude and views the western portion of North America and South America, as well as the vast majority of the Pacific Ocean. As with other forms of satellite remote sensing, the GOES-10 and GOES-12 images slightly overlap one another.

Remote sensing operates on the simple principle that objects on Earth emit electromagnetic energy that can be measured. Although the electromagnetic spectrum will be discussed in much greater detail in Chapter 4, it is sufficient for now to say that objects on Earth emit energy that is both visible and invisible to our unaided eyes. This energy can be measured by either satellites or airplane platforms and then stored digitally in a computer for detailed analysis.

Figure 2.23 shows the typical steps through which satellite imagery is processed, with this example focusing on a false-color image of Cape Cod in Massachusetts. The sequence begins with a measurement of

Figure 2.22 A typical GOES image. This example is from the GOES East satellite, which provides daily imagery of the eastern United States. Clouds are bright white, whereas darker areas are zones of clear sky. This image shows that clear skies dominated the eastern part of the country on this particular day.

emissivity by the satellite. These data are then sent to a receiving station on Earth, such as at NASA, where they are archived until distributed to the appropriate users in the scientific community as a black-and-white image. The image is then modified by the users through the application of colorizing filters to produce a false-color image that highlights particular geographic features of interest.

Although much can be learned by viewing Earth as humans normally see it, a great deal more information can be acquired by investigating the landscape in the invisible parts of the spectrum. This analysis often occurs by examining the thermal infrared energy emitted by Earth. All objects emit this kind of energy, but the relative amount varies depending on the temperature—warm objects emit more energy than cold ones—and the overall emissivity of the object. With a thorough understanding of these properties, geographers can see very subtle differences on the landscape and thus determine the spatial distribution

---

Geostationary orbit *An orbit where satellites remain over the same place on Earth every day. This orbit is achieved because the satellite is placed very high above Earth and travels at the same speed as the Earth's rotation.*

Emissivity *The amount of electromagnetic energy released by some aspect of the Earth's surface.*

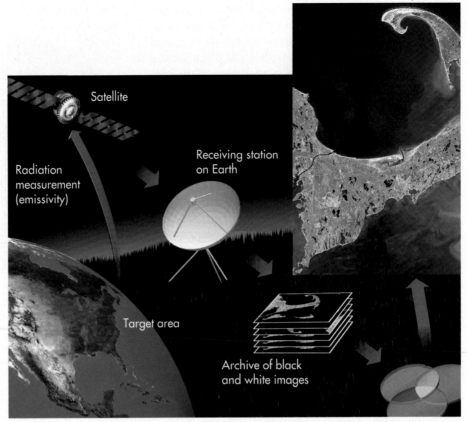

Figure 2.23 Processing of remotely sensed images. This sequence of events produces a false-color image of Cape Cod, Massachusetts, that can be geographically analyzed.

Satellite

Radiation measurement (emissivity)

Receiving station on Earth

Target area

Archive of black and white images

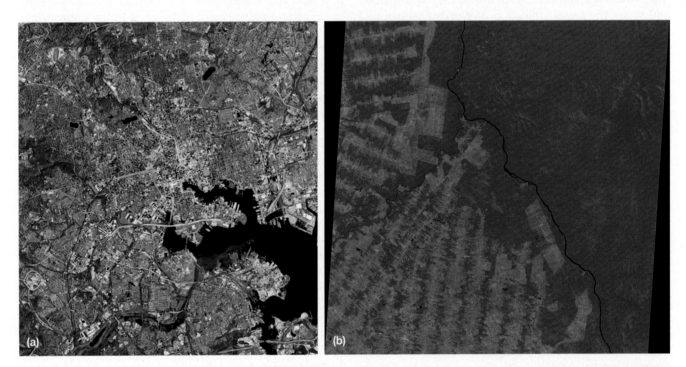

(a)

(b)

Figure 2.24 Infrared satellite images. (a) This image of the region around Baltimore, Maryland, shows areas of vegetation in bright red and urban areas as bluish gray. Areas of dark shading consist of water, specifically part of Chesapeake Bay. (b) Deforestation in the Amazon rainforest as seen from NASA's Terra Satellite. Dark red represents uncut forest, whereas the lighter shades of red indicate the pattern of forest clearing. The dark wavy line in the image is the Jiparaná River. Note how the image path appears to be tilted slightly. This shape reflects the slight inclination of the satellite's Sun-synchronous orbit.

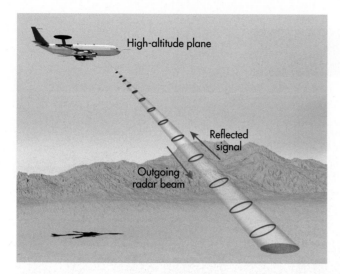

High-altitude plane

Reflected signal

Outgoing radar beam

Figure 2.25 Acquisition of Synthetic Aperture Radar (SAR) imagery from an airplane. In this system, a radar unit at the base of the airplane sends radar waves to the ground in a sweeping motion across the landscape. Subsequently, these waves bounce back to the airplane where their timing (from and to the airplane) and strength are calculated and converted into an image of the landscape.

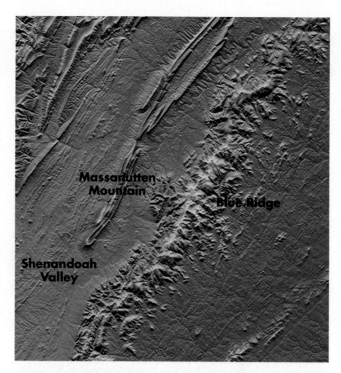

Massanutten Mountain

Blue Ridge

Shenandoah Valley

Figure 2.26 Radar image of Shenandoah National Park in the Appalachian Mountains. This image was acquired with C-Band Interferometric Radar during the Shuttle Radar Topography Mission (SRTM) and subsequently color-coded to show elevation. Greens represent topographically lower areas, rising through yellow, red, and magenta, to bluish white at the highest elevations.

of specific features such as vegetation, rock, and water. Areas such as these can be studied frequently in this fashion with infrared aerial photography, while larger regions can be investigated on an annual basis with Sun-synchronous satellites. Figure 2.24 shows a pair of infrared satellite images. This kind of analysis is especially useful, for example, as a means to show urban patterns (Figure 2.24a) and to monitor deforestation in tropical rainforests (Figure 2.24b).

The methods described so far are passive systems in that they depend on energy emitted by objects on Earth to obtain measurements. Another form of remote sensing relies on active systems that send a beam of wave energy toward Earth. This beam is then partially reflected back to the sensor, which can make measurements about the amount of energy emitted. A good example of this kind of system is *radar*, which uses energy in the microwave part of the electromagnetic spectrum. Radar systems emit short pulses of energy toward Earth and then measure the time it takes for the pulse to return, plus its strength, to create a picture of the landscape. Such imagery can be obtained either from an airplane (Figure 2.25) or from space. In the latter instance, the most extensive radar imagery was acquired in the Shuttle Radar Topography Mission (SRTM), during which virtually all the surfaces on Earth were mapped with radar. In contrast to passive systems, which can be limited when cloud cover is extensive, radar can "see" through the clouds and

can therefore be used all the time. As a result, radar has been extensively used to map the surface of Venus, which is masked by continuous cloud cover. On Earth, radar is especially effective at mapping geological formations in mountainous landscapes such as the Appalachians (Figure 2.26).

Global Positioning Systems

Another way in which digital technology has benefited geographers is with respect to determining location within the geographic grid. Until recently, the most common way that people determined their position was by looking at the grid coordinates in a large-scale map. Although this method provided a general location, usually within a degree or two, it was nevertheless a rough approximation of position.

With the proliferation of satellites in the late 20th century, the U.S. military developed a system that enables people to determine their geographic location quickly and with much greater precision than with the use of a map. This system is called the Global Positioning System, or GPS, and is based on a network of 24 satellites that orbit Earth every 12 hours (Figure 2.27). These satellites

INFRARED IMAGES

Infrared satellite images can be used to obtain many different kinds of information, depending on what range of radiation the satellite cameras are set to detect. For example, this image is an infrared satellite image of the Imperial Valley in southern California. The many red squares and rectangles are farm fields that contain maturing crops. The distinct horizontal line at the south end of the reddish zone is the border between the United States and Mexico. Although the agricultural area extends into Mexico, the reduced infrared output from those fields suggests that the Mexicans are using a different crop rotation than the Americans to the north. The areas of bluish gray indicate the presence of two cities, with El Centro, California, north of the border and Mexicali, Mexico, immediately to the south.

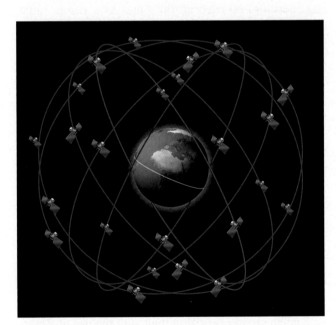

Figure 2.27 The GPS satellite system. Satellites are positioned in six orbital planes, with four satellites in each orbit (24 total) and are about 20,200 km above Earth's surface. This array provides complete coverage of Earth.

continuously transmit a radio signal known as *Pseudo-Random Code* (PRC) that is unique to the GPS. This code consists of a complicated sequence of on/off pulses that travel at the speed of light (299,300 km/sec; 186,000 mi/sec) in space and can be acquired by receivers on Earth. Figure 2.28 shows an example of a hand-held GPS receiver. The fundamental principle of GPS is that by knowing the positions of satellites in space, the exact time that a signal is sent from a given satellite and received by a receiver, and the speed of the incoming PRC, you can use this artificial constellation of satellites as locational references through a system of triangulation.

So how does this system work (Figure 2.29)? Imagine your car contains a GPS receiver and receives a signal from a single satellite indicating that it is 7000 km (4350 mi) away from your location. This lone signal indicates that you are anywhere on the surface of a sphere that is 7000 km (4350 mi) from the satellite. If the signal from a second satellite is added, for example, one that is 8000 km (4971 mi) away from the receiver, your location can be narrowed to the region contained within the space where both spheres

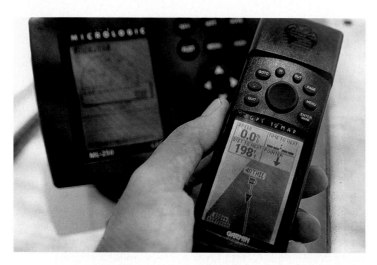

Figure 2.28 A hand-held GPS receiver. This unit receives the PRC signal from satellites in the GPS constellation.

third satellite, say, one that is 9000 km (5592 mi) away, the location is further narrowed to a very small area where all three spheres intersect. To further increase the confidence of the measured location, the distance to a fourth satellite can be measured as a check. In this way, the system is more immune to fluctuations in the speed of the PRC signal as it travels through the atmosphere and inconsistencies that may exist between the time indicated by the incredibly accurate atomic clocks on satellites and less reliable clocks in receivers.

Through this process of triangulation, it is possible to pinpoint geographic location within about 20 m (about 90 ft) with inexpensive (less than $100) systems and to less than 1 m (3.3 ft) with more expensive systems (more than $10,000). This degree of accuracy is especially useful in the military because it has facilitated the development of weapons that can be precisely delivered, as you may have seen in television footage from recent conflicts. Within the geographic community, high-resolution GPS has been used in a wide variety of circumstances, including coastal navigation, determining the precise location of study sites,

intersect. Unfortunately, the size of this overlapping region is still considerable and might tell you that your location, for example, is someplace within the eastern United States. However, by adding the distance to a

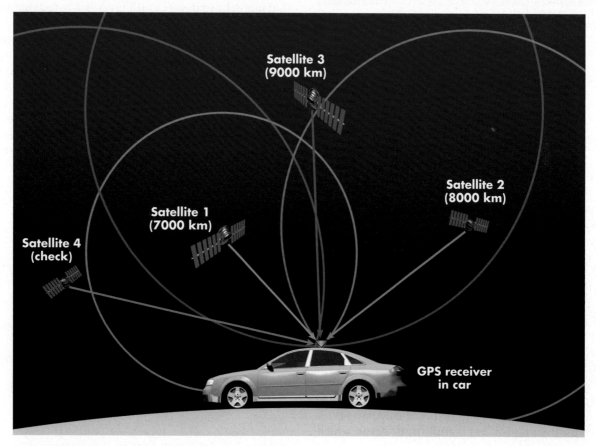

Figure 2.29 GPS triangulation. By determining the distance from three different satellites, the GPS receiver narrows its location to the region where three different spheres intersect. The distance to a fourth satellite can be determined as a check. Locations can be determined to within one meter with expensive receivers.

Figure 2.30 Using GPS to measure landscape change. Given the high accuracy of some GPS systems, it is possible to monitor changes in ground elevation, such as those that may occur near the Augustine Volcano in Alaska due to renewed volcanic activity.

and the monitoring of plate movement along faults in earthquake zones (Figure 2.30). It is now used at an everyday level by millions of ordinary people to find their way around as they travel. Perhaps you have a GPS yourself.

Although GPS is an exciting tool that has great potential for geographic analysis, it has several potential sources of error with which you should be familiar. One potential source of error is that the altitude of individual satellite orbits may vary from time to time. This variation is a problem because the receiver bases its distance calculations on a predetermined satellite altitude. Another source of error is the effect of the atmosphere on the satellite signal as it beams toward Earth. This effect is especially a problem when dust, water vapor, and ionized particles in the atmosphere interfere with the speed of the GPS signal, which leads to errors in distance calculations. A final transmission problem is **multipath error**, which

---

Multipath error *Disruption of the GPS signal from satellites due to obstructions such as trees and buildings.*

results when obstructions such as buildings and trees cause the incoming GPS signal to be deflected before it reaches the receiver. This deflection confuses the receiver because two slightly different signals arrive at the same time.

Geographic Information Systems

Another tool that has recently become very important in the way that geographers do their work is Geographic Information Systems, or GIS. A GIS is a system for storing, analyzing, and manipulating spatially referenced data, usually in digital form in a computer. Most GIS databases consist of a series of individual data layers that are considered to be relevant for the study being conducted. A data layer contains measurements obtained with respect to a specific geographic variable, such as vegetation, soils, road networks, municipal boundaries, and the distribution of surface water (hydrology) in lakes and rivers.

Let's choose vegetation as an example. A geographer may choose to map the spatial distribution of trees and grassland in a region, which can be determined through infrared remote sensing or by digitally tracing old maps. In so doing, a GIS layer is created that can be identified as "vegetation" in the overall database. A variety of attributes can be subsequently assigned to this layer, such as species composition, height, the slope of the land, and condition. Subsequently, similar maps can be created for other desired variables, resulting in layers for each of them. Now look at Figure 2.31, which shows a potential series of GIS layers from a theoretical study area. Notice how these layers exist as distinct units that can stand alone. In addition, the layers can be combined to illustrate the overall geographic character of the region. It is beyond the scope of this book

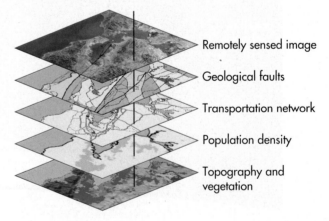

Remotely sensed image

Geological faults

Transportation network

Population density

Topography and vegetation

Figure 2.31 GIS layers. Each layer represents a distinctive part of the landscape, such as topography, vegetation, population density, or the local road network. When these layers are integrated, they form a detailed image of the landscape.

to describe GIS in detail, but you should understand that it is an extremely powerful tool that is used in a wide variety of ways, including:

- **Environmental management**—GIS can be used to manage spatial information associated with soils, wetlands, vegetation species, topography, and the location of data collection sites.

- **Municipal planning**—GIS is used in virtually all cities, large towns, and counties to manage spatial information such as road networks, location of sewer and utility lines, and emergency traffic routes.

- **Business needs**—A growing number of companies are incorporating GIS into their operations. GIS can be used, for example, to preselect the most efficient delivery routes that will save fuel costs. In addition, GIS can be used to identify the best location for new shopping centers based on surrounding demographic characteristics.

www.wiley.com/college/arbogast

Using a Geographic Information System

To get a feel for GIS, go to the *GeoDiscoveries* website and select the module *Using a Geographic Information System*. This module focuses on the recent oil spill in the Gulf of Mexico and illustrates the way that GIS could be used to assess the character of the spill and its geographical relationship to various aspects of the regional environment and economy. After you complete the module, answer the questions at the end of the module to test your understanding of Geographic Information Systems.

VISUAL CONCEPT CHECK 2.3

A variety of geographic information can be acquired, stored, and analyzed using digital technology. This image is an infrared satellite image of a portion of Washington, D.C. What do the red, black, and bluish-gray areas represent in this image? What are at least four data layers that could be constructed in a GIS?

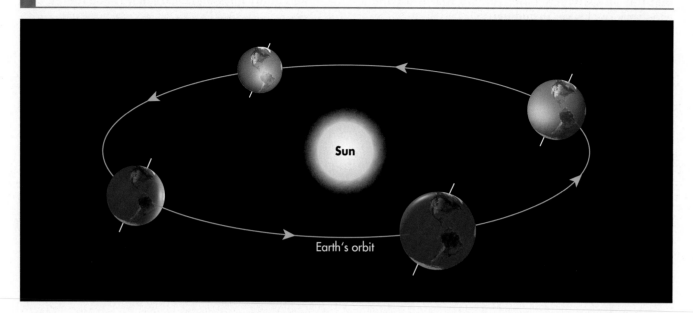

Now that we have introduced the fundamental tools that geographers use, and how they can be applied to spatial analysis, we can begin the investigation of natural processes and patterns associated with physical geography. The logical place to begin is by examining the relationship that Earth has with the Sun. You will discover that this relationship is not only critical to life on Earth, but also drives many of the processes on the planet, such as atmospheric circulation, climate, and even the character and distribution of soils. The next chapter begins this discussion by looking at the geometric relationship of the Earth and Sun and how that influences our seasons. This is a very important chapter because it provides the foundation for subsequent discussions throughout the book.

SUMMARY OF KEY CONCEPTS

1. Lines of latitude are called parallels and determine location north and south relative to the Equator. Latitude designations are calculated using simple geometric principles and extend to points 90° N and S, which are the North and South Poles, respectively. Lines of longitude, in contrast, are called meridians and determine location east and west of the Prime Meridian in Greenwich, England. These designations are also calculated using the geometry of a circle and extend to 180° E and W. Latitude and longitude overlay to form a geographic grid.

2. Maps are two-dimensional representations of three-dimensional aspects of Earth. They are created through the process of map projection, which creates distortion in some aspect of the map. A conformal projection, for example, maintains the correct shape of features at the expense of their relative size.

3. Map scale refers to the ratio of the size of features on a map relative to the real world. Small-scale maps show relatively large areas, but with less detail. A large-scale map, in contrast, shows a relatively small area but with greater detail.

4. "Remote sensing" refers to the ability to observe and monitor aspects of Earth from a distance. The most systematic way to accomplish this task is with satellite platforms such as the Landsat system. These systems measure emitted energy that can be stored as digital data that can be ultimately accessed and manipulated by scientific users.

5. A Geographic Information System (GIS) is a digital database that contains relevant geographic information such as the density of roads, location of human structures, and vegetation type, to name but a few. These various data sets are stored as individual layers that can be individually or collectively manipulated and analyzed. In contrast to GIS, GPS stands for Global Positioning System and is the method through which precise locations can be determined through triangulation with a constellation of satellites.

CHECK YOUR UNDERSTANDING

1. What are the two components of the geographic grid and how are they calculated?

2. Where are the low latitudes on Earth and how do they differ from the mid- and high latitudes?

3. Discuss the concept of map projection and why it is necessary when constructing maps.

4. What kind of map projection would provide the most accurate comparison of the shape of national boundaries between countries in the low and high latitudes?

5. Which would have the larger scale: a map showing the detail of New York City, or one that showed the position of New York City within the state of New York?

6. Define the concept of remote sensing.

7. What is the difference between active and passive remote sensing?

8. Compare and contrast a Sun-synchronous orbit with a geostationary orbit.

9. What is a GPS satellite array and how is it used to determine location on Earth?

10. What are the three potential sources of error associated with GPS?

11. Describe the character of a geographical information system.

12. Compare and contrast three GIS data layers.

ANSWERS TO VISUAL CONCEPT CHECKS

VISUAL CONCEPT CHECK 2.1

a) Mont Blanc is located in the Alps in France.
b) Lake Chad is located in central Africa.
c) Valle de Luna (Valley of the Moon) is located in eastern Chile.

VISUAL CONCEPT CHECK 2.2

a) The map on the left-hand side of the figure is a conformal projection, whereas the one on the right is an equivalent projection. You can tell the difference because the conformal projection maintains the shape of features on Earth, but distorts their relative size, especially at the high latitudes. The equivalent projection, in contrast, maintains relative shape, with size being distorted. To see the difference between the two, note the relative shape and size of Greenland compared to the rest of North America.

b) The map of the state of Washington has the relatively small scale, whereas the map of just the Seattle area has the larger scale. You can tell the difference because the map of Washington shows a relatively large geographic area that contains very little detail about Seattle. The map of Seattle, in contrast, shows much more detail about the city, including the road pattern and shape of water bodies.

VISUAL CONCEPT CHECK 2.3

This infrared satellite image contains a variety of geographic information. The red, black, and bluish-gray areas represent vegetation, water (mostly the Potomac River), and urban development, respectively. Potential GIS layers include (1) the configuration of water bodies, (2) distribution of vegetation, (3) the road network, and (4) the pattern of urban development.

CHAPTER THREE

EARTH–SUN GEOMETRY AND THE SEASONS

Of all the spatial relationships that geographers study, the Earth–Sun relationship has the greatest impact on the way Earth functions and the existence and character of life as we know it. In the context of physical geography, this relationship is especially important because most of the physical processes on Earth are powered by incoming radiation received from the Sun. Although Earth receives only about one-half of one-billionth the amount of energy that the Sun radiates, this amount is nonetheless sufficient to power physical processes and life on the planet.

At first glance it might appear that the Earth–Sun relationship is simple and easy to comprehend. In fact, this relationship is very complex because it depends on many different geographic variables such as latitude, the position of the Sun in the sky, time of year, and the nature of the Earth's rotation and axial tilt. It is essential to understand the information in this chapter because it provides the foundation for many geographic patterns you will see in the remainder of the book.

A multiple exposure such as this one shows the change in Sun position during a beautiful Arctic sunset in the summer months. This chapter focuses on the character and significance of the Earth/Sun geometric relationship.

CHAPTER PREVIEW

Our Place in Space

The Seasons
On the Web: *GeoDiscoveries Orbital View and Earth as Viewed from the Sun*

Human Interactions: How We See Earth–Sun Geometry on Earth
On the Web: *GeoDiscoveries Celestial Dome*
On the Web: *GeoDiscoveries Sun Angle and Length of Day*
On the Web: *GeoDiscoveries Earth–Sun Geometry and Ancient Humans*

The Big Picture

LEARNING OBJECTIVES

1. Understand the Earth's place in space.

2. Describe the nature of Earth's axial tilt and orbit around the Sun.

3. Discuss the concept of the subsolar point and its seasonal movement.

4. Explain why the Equator, Tropic of Cancer, and Tropic of Capricorn are important features associated with Earth–Sun geometry.

5. Explain the concept of Sun angle and how it varies both by latitude and seasonally.

6. Understand how we sense the Earth–Sun geometric relationship.

7. Demonstrate understanding of Earth–Sun relationships by accurately predicting outcomes based on variable tilt scenarios.

Our Place in Space

A good way to begin a discussion of Earth–Sun relationships is with a brief discussion of the Universe. The origin of the Universe seems to be best explained by the **Big Bang theory**, which is based on the concept that the Universe emerged from a singular, enormously dense and hot state about 14 billion years ago. The primary evidence for this theory comes in two forms, specifically that (1) the most distant star clusters are moving away from us at faster speeds than those that are relatively close, and (2) the amount of cosmic microwave background radiation is remarkably uniform throughout the Universe, which implies that it represents the leftover energy from an early period of rapid expansion.

According to present estimates, the Universe is approximately 20 billion light years across, with a light year being the distance that light travels over the course of a year at the speed of 1,079,252,848.8 km/h (670,616,629.4 mi/h). With respect to the ultimate fate of the Universe, it will either continue to expand for all of eternity or will one day collapse upon itself, depending on the average density of matter and energy within the entire Universe. Given that the speed of distant star clusters seems to be accelerating, it appears that the Universe may indeed expand forever.

The largest clearly definable unit within the Universe is a galaxy, which is a collection of billions of stars. According to astronomers, the Universe has approximately 50 billion galaxies, with each galaxy containing billions of stars. We happen to live in the Milky Way Galaxy (Figure 3.1), which includes approximately 400 billion stars. The Milky Way is a typical spiral galaxy, consisting of a bright central region with a high density of stars and a flat circular region containing most of the other stars. Younger, brighter stars form in long spiral arms that extend out from the galactic center. Our solar system lies in one of these spiral arms.

The center of our particular solar system is the star that we refer to as the Sun. Eight official planets orbit the Sun (Figure 3.2), with Pluto now considered a dwarf planet. To give you an idea of the size of the solar system, consider that Pluto is approximately 5.9 billion km (3.6 billion mi) from the Sun. A probe launched in January 2006 to study Pluto is currently zooming through space at about 14,500 km/h (9000 mi/h). Even at that remarkable speed, the probe will not reach Pluto until July 2015! Given this great distance, Pluto is much too far away from the Sun to receive sufficient energy to support life. Of the remaining planets, most are either too far away from the Sun (and thus are too cold) or too close (and thus are too hot) for life as we know it to exist. The lone exception is Earth, the third planet from the Sun, whose orbit happens to be a perfect distance from the Sun for supporting life as we know it.

The Shape of Earth

Let's now turn our attention to the shape of Earth. Although Earth appears to be a perfect sphere, it bulges slightly at the Equator and is flattened somewhat at the poles to form a shape known as an *oblate spheroid*. The circumference of Earth measured at the Equator is 40,075 km (24,902 mi), while the circumference measured through the poles is slightly less: 40,008 km (24,860 mi). The bulge at the Equator, which gives Earth a slightly thicker middle, is caused mainly by the centrifugal force of the Earth's rotation and by differences in density of the Earth's crust and gravitational field. This centrifugal force is similar to the sideways push you feel when rounding a curve quickly in your car.

The roughly spherical shape of Earth is an important factor to consider regarding the Earth–Sun relationship.

Big Bang theory   *The theory that the Universe originated about 14 billion years ago when all matter and energy erupted from a singular mass of extremely high density and temperature.*

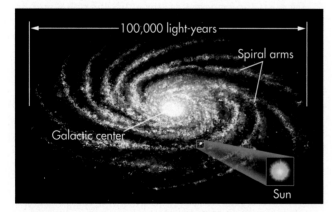

Figure 3.1 The Milky Way Galaxy. Our Sun is but one of approximately 400 billion stars in this galaxy. Note our Sun's location in one of the spiral arms of the galaxy.

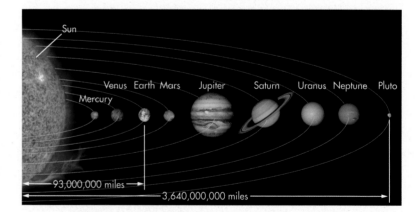

Figure 3.2 Our solar system. The solar system consists of eight official planets (Pluto is now considered a dwarf planet) that orbit the Sun. Earth is the third planet from the Sun. (Not drawn to scale.)

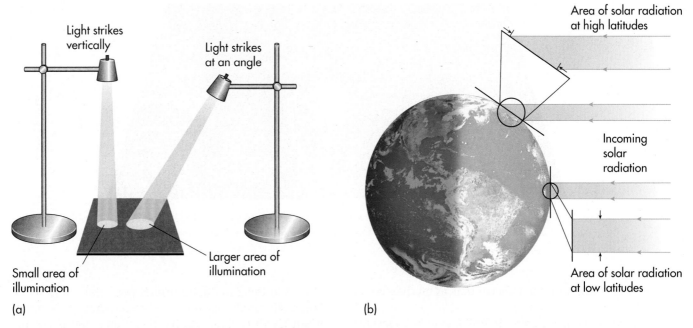

Figure 3.3 Interaction of solar radiation and the curved Earth. (a) You can use a desk lamp to show that light striking a surface at an angle illuminates a larger area (and is more diffuse) than light striking the surface vertically. (b) Solar radiation strikes Earth most directly at low latitudes because the Sun angle is high, resulting in concentrated solar radiation in these regions. In contrast, solar radiation is more diffuse at high latitudes because the surface curves away from the radiation and the Sun angle is relatively low.

Because Earth presents a curved surface to the arriving rays of the Sun, the intensity of solar radiation received varies by latitude. This difference is a reflection of the **Sun angle**, which is the angle at which the Sun's rays strike Earth's surface. In general, the Sun angle is relatively high at low latitudes and becomes progressively less toward the poles because the Earth's surface is curved. The net effect of this variation is that the solar energy received at higher latitudes is spread over a relatively large surface area compared to the same amount of energy received at lower latitudes. As a result, the amount of incoming energy per unit area is relatively low in areas where the Sun angle is low.

Figure 3.3 illustrates this difference. Lower latitudes generally receive more intense solar radiation because the Sun angle is high (between 75° and 90°). The point on Earth where the Sun's rays are perpendicular to the surface (Sun angle = 90°) at any given point in time, and therefore most intense, is known as the **subsolar point**. Higher latitudes, in contrast, receive less intense radiation because the Sun's rays hit the Earth's surface at angles much less than 90° (a more oblique angle) due to the Earth's curvature. You can explore this relation-

ship yourself by shining a desk lamp onto a surface, as in Figure 3.3a, both directly and at an angle. Note that the light is spread more diffusely when shone at a lower angle than when the angle is 90°. The Earth's curvature has an analogous effect on incoming solar radiation, which, in turn, contributes greatly to many geographic patterns on Earth, including atmospheric circulation and the global distribution of climate, vegetation, and even soils.

Earth's Orbit Around the Sun

Let's now turn to the character of Earth's orbit. As Earth revolves around the Sun, it moves in a counterclockwise direction relative to a view above the North Pole on a flat (imaginary) plane referred to as the **plane of the ecliptic** (Figure 3.4). Although it may be difficult to visualize this plane, imagine that the Sun is in the middle of a large, round cake pan. If the planets were marbles on the cake pan, they would roll around the Sun only on the surface of the pan, not above or below it. In a very similar way, Earth orbits the Sun on a flat plane, except that the plane of the ecliptic passes through the

---

Sun angle *The angle at which the Sun's rays strike the Earth's surface at any given point and time. This angle is high at low latitudes and is progressively less at higher latitudes.*

Subsolar point *The point on Earth where the Sun angle is 90° and solar radiation strikes the surface most directly at any given point in time.*

Plane of the ecliptic *The flat plane on which the Earth travels as it revolves around the Sun.*

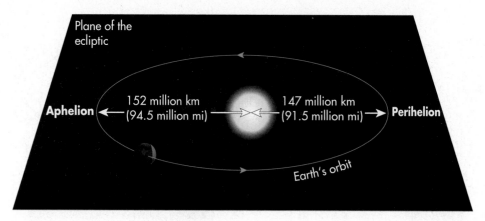

Figure 3.4 The elliptical Earth orbit. As Earth travels on the plane of the ecliptic, it is closest to the Sun at perihelion on January 3. Earth is farthest away from the Sun at aphelion on July 4.

center of Earth rather than beneath it, as marbles do on a cake pan.

It takes about 365 days (365.24 days to be exact) for Earth to make one full revolution around the Sun. In comparison, Pluto requires 248 of our years to complete one orbit. Early astronomers who calculated Earth's orbital time length configured our calendar year to reflect this period of time. You may wonder why February has a variable number of days, depending on the year. Normally, February contains 28 days. Every fourth year (known as *Leap Year*), however, February contains 29 days. This day is added in order to correct for the uneven length of time (365.24 vs. 365 days) it takes for Earth to complete its orbit.

As Earth orbits the Sun, it does not follow a perfectly circular path, but rather an elliptical one (see Figure 3.4). Note also that the Sun is not located in the exact center of our orbit and that the distance between Earth and Sun varies over the course of the year, with an average distance of approximately 150 million km (93 million mi). The point where Earth is closest to the Sun, approximately 147 million km (91.5 million mi) away, is called **perihelion**. In contrast, the term **aphelion** refers to the point where Earth is farthest from the Sun, about 152 million km (94.5 million mi) away. At the present time, perihelion and aphelion occur in early January and July, respectively.

Given the basic geometric relationship shown in Figure 3.4, what seems odd about the timing of our seasons in the Northern Hemisphere? The answer is that winter in the Northern Hemisphere occurs when we are closest to the Sun during orbital perihelion. Intuitively, you might think that we would experience our summer when Earth is closest to the Sun. Since that is not the case, we can conclude that the distance of Earth from the Sun does not cause our seasons. In other words, a different variable must be responsible for the seasonality we experience. That variable is the Earth's axial tilt.

The Earth's Rotation and Axial Tilt

As Earth revolves around the Sun, it also rotates (spins) on its **axis**, which is an imaginary line that extends through the center of Earth from pole to pole. When viewed from above the North Pole, Earth rotates in a counterclockwise direction on its axis (Figure 3.5). It takes 24 hours for Earth to make one complete rotation, which is the origin of our day length. During this interval, one-half of Earth is always illuminated, whereas the other is in shadow. The boundary between day and night is known as the **circle of illumination** and is constantly moving across the surface as Earth rotates. You can see a good example of the circle of illumination by looking at a half Moon. Notice that one-half of the Moon is brightly illuminated, whereas the other half is completely dark. The dividing line between those two portions of the Moon is the circle of illumination.

In association with Earth's rotational cycle, humans established 24 time zones that span specific portions of the planet (Figure 3.6). The contiguous United States contains four such areas, including (from east to west) the Eastern, Central, Mountain, and Pacific time zones

Perihelion   *The point of the Earth's orbit where the distance between the Earth and Sun is least (~147 million km or 91.5 million mi).*

Aphelion   *The point of the Earth's orbit where the distance between the Earth and Sun is greatest (~152 million km or 94.5 million mi).*

Axis   *The line around which the Earth rotates, extending through the poles.*

Circle of illumination   *The great circle on Earth that is the border between night and day.*

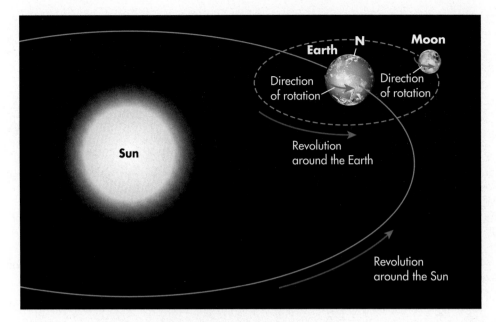

Figure 3.5 The Earth's revolution and rotation. As Earth revolves around the Sun, it rotates in a counterclockwise direction on its axis, when viewed from above the North Pole. Also note the geometrical relationship of Earth and Moon.

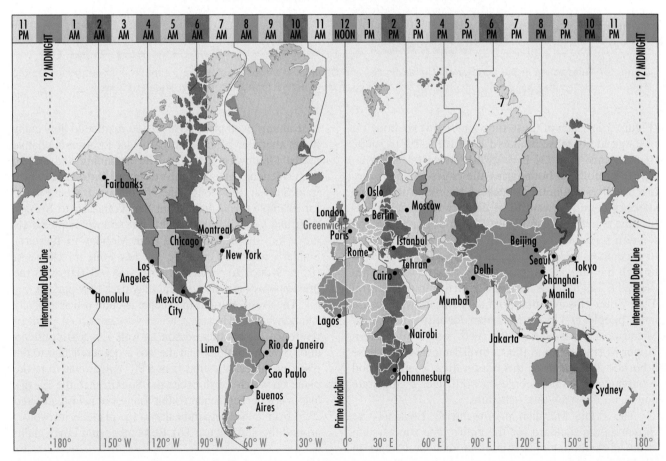

Figure 3.6 World time zones. Earth is subdivided into 24 time zones. The International Date Line, which lies at approximately 180° E and W longitude, is the arbitrary designation used to mark the beginning of each new day.

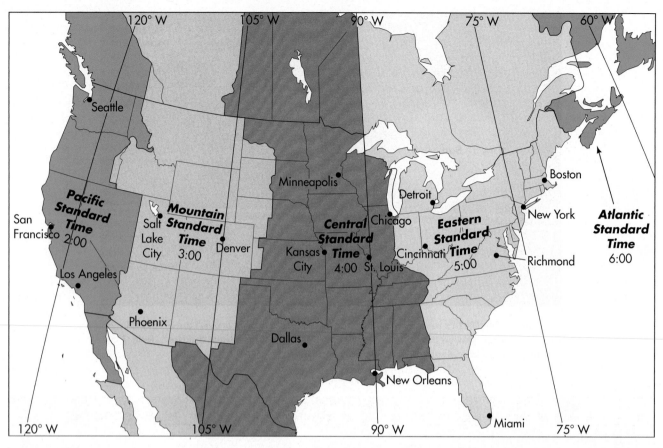

Figure 3.7 Time zones in the contiguous United States and Canada. The most eastern time zone in the contiguous United States is Eastern Standard Time, which is 3 h ahead of the most western time zone, Pacific Standard Time.

(Figure 3.7). Each of these time zones covers about 15° of longitude because it takes 1 hour for Earth to rotate that distance (360°/24 h = 15°/h). Given the irregular nature of political boundaries, the edges of time zones are usually asymmetrical because people in individual states or countries want to be entirely within a particular time zone. Notice in Figure 3.7, for example, that the western boundary of the Eastern Time Zone jogs west to encompass most of Michigan's Upper Peninsula even though the far western part of the state nearly reaches the same longitude as St. Louis, which is in the Central Time Zone. This irregular boundary was drawn because most people in Upper Michigan identify with Lower Michigan as far as time is concerned. A sliver of western Upper Michigan lies in the Central Time Zone because it borders Wisconsin. In this border area, it makes good economic sense that people on both sides of the state line be on the same time schedule.

The Prime Meridian in Greenwich, England, was chosen as the standard for the entire time zone system at the 1884 International Prime Meridian Conference in Washington, D.C. This system, referred to as Universal Time Coordinated (UTC), is used by the vast majority of countries on Earth, although some (such as Saudi Arabia) adhere to their own time systems. In the UTC system,

time at any given place is calculated relative to how many hours ahead or behind that particular location is relative to the Greenwich meridian. The beginning and end of each calendar day is the **International Date Line**, which is located at 180° longitude. If you are traveling west, for example, on a flight from San Francisco to Sydney, Australia, you will cross the date line and will immediately move into the next day (say, from Monday to Tuesday). Similarly, when you cross the date line while traveling east (from Sydney to San Francisco), you will return to the previous day (such as from Tuesday back to Monday).

The axis of Earth is the anchor around which the daily rotation occurs. However, it is a very important point of reference for another reason as well. On first consideration, you might think that the axis is perpendicular to the plane of the ecliptic, that is, at a 90° angle relative to the plane on which Earth orbits the Sun. In fact, the Earth's axis is not perpendicular to this plane, but is instead tilted 23.5° from a line perpendicular to the plane as it revolves around the Sun (Figure 3.8). Remember that Earth main-

International Date Line *This line generally occurs at 180° longitude, with some variations due to political boundaries, and marks the transition from one day to another on Earth.*

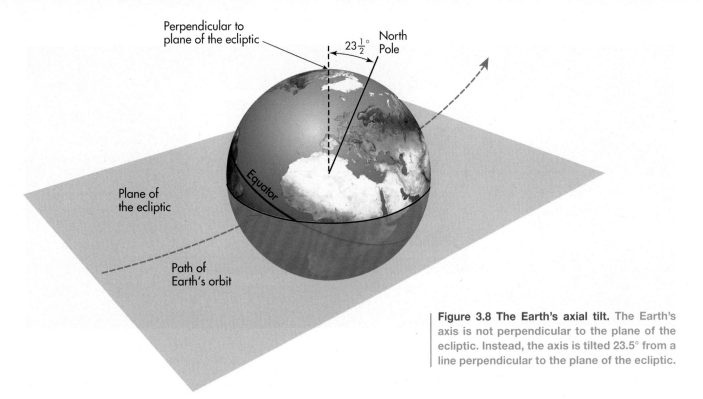

Figure 3.8 The Earth's axial tilt. The Earth's axis is not perpendicular to the plane of the ecliptic. Instead, the axis is tilted 23.5° from a line perpendicular to the plane of the ecliptic.

tains this degree of tilt *and* orientation with respect to the Sun throughout the year. A common misconception is that the tilt of Earth somehow changes over the course of the year; *it does not*! This geometric relationship is important to understand because it is the reason for the seasons that we experience, as explained in the next section.

The Seasons

To understand how the Earth's axial tilt causes the seasons, first imagine what would happen if the Earth's axis were *not* tilted (Figure 3.9). Under these circumstances, as Earth traveled around the Sun on the plane of the ecliptic, the Sun's rays would always strike Earth most directly at the Equator. As a result, the subsolar point

would perpetually be at the Equator and the Sun's rays would always strike Earth at progressively lower angles with increasingly higher latitudes (greater than 0°). In this scenario, as Earth revolves around the Sun, all places on Earth would receive a consistent intensity of solar radiation (although this amount would differ with latitude) over the course of the year and distinct seasons would not be experienced anywhere on the planet.

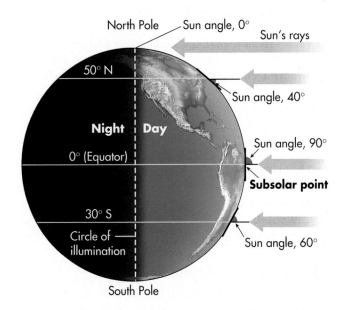

Figure 3.9 Geometric relationship if the Earth's axis were perpendicular to the plane of the ecliptic. If these conditions existed, the subsolar point would always be located at the Equator and the circle of illumination would extend from the North to South Poles. At 50° N the Sun angle would be 40°, whereas it would be 0° at 90° N (the North Pole).

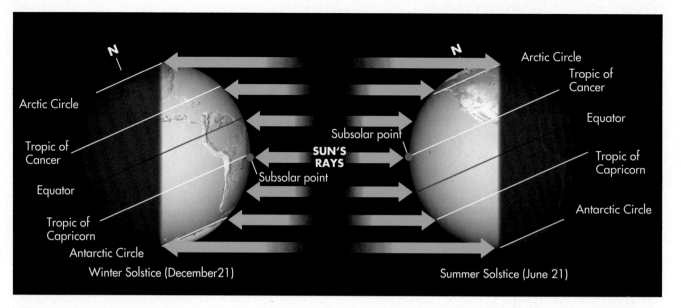

Figure 3.10 Axial geometry of Earth during June and December. Note that the axis is tilted at 23.5°, which causes the orientation of the hemispheres to change with respect to the location of the subsolar point. On December 21, the subsolar point is at the Tropic of Capricorn, whereas it is at the Tropic of Cancer on June 21.

Now picture the Earth's actual axial tilt, which is 23.5° from a line perpendicular to the plane of the ecliptic. Keep in mind that the axis maintains a constant tilt angle *and* orientation as Earth travels around the Sun (Figure 3.10). These geometric relationships—angle and orientation—are the fundamental cause of our seasons because one of the hemispheres is always in the gradual process of tilting toward the Sun, while the other one is slowly tilting away from the Sun. You might be thinking, "Wait a minute, you just said that the axial tilt and orientation never change; isn't that contradictory?" To see why not, examine Figure 3.10 and focus on the Northern Hemisphere. Notice that during June, the Northern Hemisphere is tilted *toward* the Sun, whereas in December, it is tilted *away*. The opposite is true for the Southern Hemisphere. In both cases the tilt did not change, nor did the orientation of the tilt. Instead, *the position of Earth relative to the Sun changed as its orbit progressed*, which has the effect of moving each hemisphere either toward or away from the Sun's rays. This apparent movement of the hemispheres toward or away from the Sun is the fundamental cause of the seasons because it results in the migration of the subsolar point 23.5° N or S of the Equator over the course of the year.

Solstice and Equinox

If you live in the Northern Hemisphere, you probably know that the recognized first day of summer is sometime toward the end of June. Conversely, in the Southern Hemisphere, the first day of summer is toward the end of December. You also probably know that certain dates mark the first days of autumn, spring, or winter on the calendar. Why do these dates exist and what is their significance?

These dates are approximately the same every year because they indicate key periods within the Earth–Sun geometrical relationship. Remember that the Earth's axial tilt, in combination with its orbit, causes the subsolar point to move between 23.5° N and 23.5° S over the course of a year (Figure 3.11). In other words, the Sun angle is 90° at some point on the Earth's surface between these latitudes every day. Although this migration is a seamless process that we barely notice from day to day, early astronomers found it useful to mark the passage of the seasons by noting the date when the Sun's rays are perpendicular to the Equator and to the Tropic of Cancer (23.5° N) and Tropic of Capricorn (23.5° S). These two latitudes represent the highest latitude that the subsolar point reaches in each hemisphere during its seasonal cycle. The term *equinox* refers to the time when the subsolar point is at the Equator and all localities on

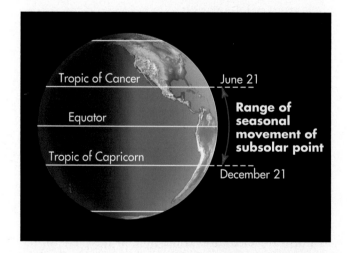

Figure 3.11 Range of seasonal subsolar point movement. The subsolar point migrates between the Tropic of Cancer and Tropic of Capricorn over the course of the year.

Earth experience equal hours of daylight and darkness (12 of each). In contrast, the term *solstice* is used to denote when the Sun angle is 90° at either of the tropical boundaries.

In an effort to better understand when these dates occur and how they are related to Earth–Sun geometry, let's examine the passage of the seasons by beginning in March and moving forward in time. It will help if you refer to Figures 3.12 and 3.13 to view the process and assume that your viewpoint of time is from the perspective of living in the Northern Hemisphere.

March 20–21 This date is the **Spring Equinox** (also called the *vernal equinox*) and represents the official beginning of the spring season in the Northern Hemisphere and the fall season in the Southern Hemisphere. This noteworthy date occurs because the subsolar point is located at the Equator and neither the Northern nor the Southern Hemisphere is tilted toward the Sun on this date. Notice that the circle of illumination extends through each pole in Figures 3.12 and 3.13.

June 20–21 This date is the **Summer Solstice** and represents the first official day of summer in the Northern Hemisphere and winter in the Southern Hemisphere. The Summer Solstice occurs because, following the Spring Equinox, Earth continues to orbit the Sun in a counterclockwise direction when viewed from above. As the orbit progresses, the Northern Hemisphere slowly reaches its position of greatest tilt toward the Sun. Meanwhile, the Southern Hemisphere is tilted away from the Sun. During this solstice the subsolar point is located at the **Tropic of Cancer** (23.5° N). In Figure 3.13, notice that all latitudes above 66.5° N (the Arctic Circle) receive 24 hours of daylight during this period of time; this is why these high latitudes are called the *land of the midnight Sun*. At the same time the northern arctic regions are in perpetual sunlight, all locations above 66.5° S (the Antarctic Circle) experience 24 hours of darkness. This continual darkness occurs because the Southern Hemisphere is tilted away from the Sun during this time and those latitudes above the Antarctic Circle never rotate into the illuminated part of Earth.

September 22–23 This date is the **Fall Equinox** (also called the *autumnal equinox*) and represents the first of-ficial day of fall in the Northern Hemisphere and spring in the Southern Hemisphere. At this time the Earth's orbit has progressed such that the subsolar point is now once again at the Equator (Figures 3.12 and 3.13). As at the Spring Equinox, all locations on Earth get equal hours of day and night (12 hours daylight, 12 hours darkness) because neither the Northern nor the Southern Hemisphere is tilted toward the Sun and the great circle of illumination extends from pole to pole.

December 21–22 This date is the **Winter Solstice** and represents the first official day of winter in the Northern Hemisphere and summer in the Southern Hemisphere. Given the progression of the orbit since the Fall Equinox, Earth is now positioned such that the Northern Hemisphere is at its greatest tilt away from the Sun, while the Southern Hemisphere is tilted toward it. At this time the subsolar point is located at 23.5° S (Figures 3.12 and 3.13), which is also known as the **Tropic of Capricorn**. As a result of this subsolar position, all latitudes above the Arctic Circle experience 24 hours of darkness, whereas those higher than the Antarctic Circle experience 24 hours of daylight.

KEY CONCEPTS TO REMEMBER ABOUT THE SEASONS

1. If the axis of Earth were not tilted, there would be no seasons.

2. The subsolar point is the place where the Sun's rays strike Earth most directly. This occurs because the Sun angle is 90° at that point and the rays are perpendicular to the surface.

3. Due to the axial tilt, the subsolar point migrates between the Tropic of Cancer and Tropic of Capricorn over the course of the year.

4. A solstice occurs whenever the subsolar point is at either 23.5° N or S latitude. An equinox, in contrast, occurs when the Sun is directly over the Equator.

5. The position of Earth relative to the Sun changes as the orbit progresses.

Spring Equinox *Assuming a Northern Hemisphere seasonal reference, the Spring (or vernal) Equinox occurs on March 20 or 21, when the subsolar point is located at the Equator (0°).*

Summer Solstice *Assuming a Northern Hemisphere seasonal reference, the Summer Solstice occurs on June 20 or 21, when the subsolar point is located at the Tropic of Cancer (23.5° N).*

Tropic of Cancer *The line of latitude at 23.5° N where the subsolar point is located (and Sun angle is thus 90°) on the Summer Solstice in the Northern Hemisphere.*

Fall Equinox *Assuming a Northern Hemisphere seasonal reference, the Fall (or autumnal) Equinox occurs on September 22 or 23, when the subsolar point is located at the Equator (0°).*

Winter Solstice *Assuming a Northern Hemisphere seasonal reference, the Winter Solstice occurs on December 21 or 22, when the subsolar point is at the Tropic of Capricorn (23.5° S).*

Tropic of Capricorn *The line of latitude at 23.5° S where the subsolar point is located on the Winter Solstice (and Sun angle is thus 90°) in the Northern Hemisphere.*

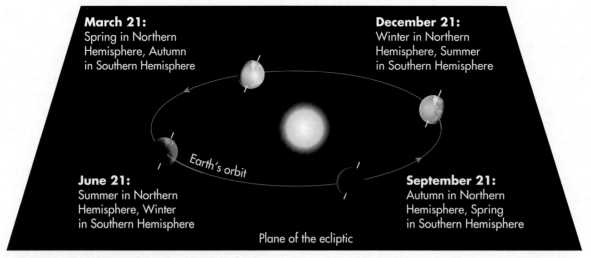

Figure 3.12 The orbital positions of Earth during the Solstices and Equinoxes. Note that the tilt of Earth does not change; instead, the position of Earth relative to the Sun changes. Due to this relationship, the intensity of the solar radiation that strikes any particular place on Earth varies over the course of the year.

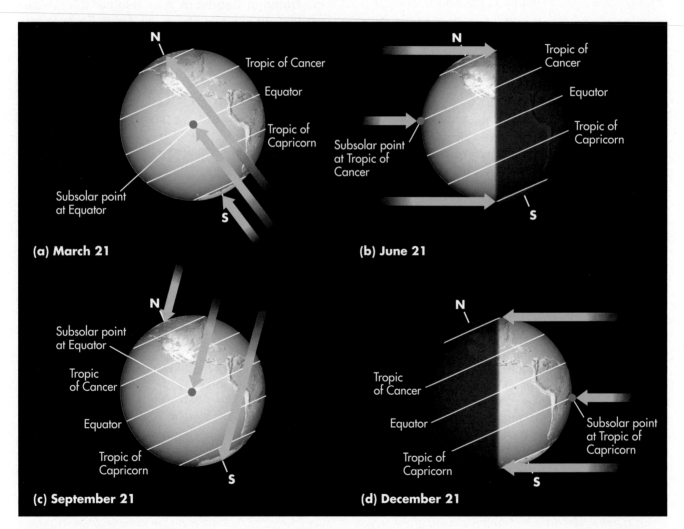

Figure 3.13 Seasonal migration of the subsolar point. Due to the tilt of the Earth's axis, the subsolar point migrates between the Tropic of Cancer and Tropic of Capricorn over the course of the year. For perspective, imagine that the Sun is in front of Earth in (a) and behind Earth in (c). Further, imagine that it is to the left of Earth in (b) and right of Earth in (d). Also notice the orientation of the circle of illumination at different times of the year.

Suppose the subsolar point is at 10° S latitude. In other words, the Sun angle at that latitude is 90° at solar noon. Which of the following months are logical times of the year when the subsolar point would be at this latitude? Explain.

a) October
b) December
c) February
d) June

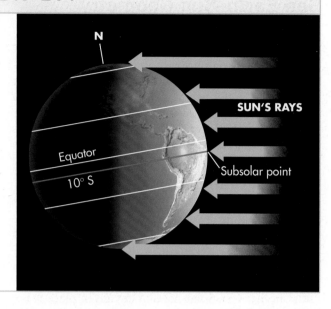

 www.wiley.com/college/arbogast

Orbital View and Earth as Viewed from the Sun

Let's reinforce the important concept of axial tilt by interacting with some simulations. First, go to the *GeoDiscoveries* website and select the module *Orbital View*. This simulation shows Earth orbiting the Sun, with the Earth's axis indicated. This simulation has close-up views of Earth at the solstices and equinoxes, showing Earth rotating (so lengths of daylight periods are indicated), the Earth's axis (indicating axial tilt), and the portion of Earth that is illuminated at each of the four times. Click on the "Orbit," "Fall Equinox," "Winter Solstice," "Spring Equinox," and "Summer Solstice" buttons to see different perspectives. Pay particularly close attention to the tilt and orientation of Earth, and the fact that these variables do not change during the orbit. In addition, be sure to notice where the subsolar point is at any given point in time and compare this migration to Figure 3.13 in the text. Last, change the axial tilt and explore how the migration of the subsolar point changes. Remember that the subsolar point represents the place where solar radiation is striking Earth most intensely. Notice how, with increased tilt,

the subsolar point migrates a great deal. With less tilt, the subsolar point migrates less.

After you interact with the Orbital View simulation, select the module *Earth as Viewed from the Sun*. This part of the module provides a different view of the Earth–Sun relationship by simulating Earth as it would appear if viewed from the Sun. As the months progress in time, notice how the *apparent* tilt of Earth changes, relative to your view (on the Sun). Remember that the *actual* tilt does not change; instead, the position of Earth in its orbit around the Sun progresses. In this simulation, you can also adjust the amount of axial tilt on Earth and see how the migration of the subsolar point changes as viewed from the Sun. Again, notice that the subsolar point migrates more with increased tilt and less with decreased tilt. Once you complete this simulation, examine Figure 3.13 to review where the subsolar point strikes Earth, with the tilt at 23.5° on each of the solstices and equinoxes. Be sure to answer the questions at the end of the module to make sure you understand this concept.

Human Interactions: How We See Earth–Sun Geometry on Earth

In the previous section you examined Earth–Sun geometry as it looks when viewed from space. Now it is time to explore how this geometry looks from the surface of Earth, in other words, how *we* see and experience it. Let's begin with the basic principle of day and night, which is related to the Earth's rotation on its axis.

Day and Night

Over the course of the day, you can "see" the Earth's rotation by noting the position of the Sun in the sky. Everyone knows that the Sun "rises" in the east and then arcs across the sky toward the west. As the arc progresses over the course of the day, the Sun reaches its highest position at **solar noon** and then gradually lowers as it "sets" in the west. Solar noon at any given place is not necessarily equivalent to standard clock time because the time of sunrise and sunset varies across a time zone due to the curvature of Earth. Although it appears that the Sun is actually rising in the morning and moving across the sky during the course of the day, the fact is that the apparent motion of the Sun is caused by the rotation of Earth on its axis. In short, the Sun rises, wherever you may live, because your part of Earth has rotated to a position where the Sun begins to illuminate your home. If you ever have the chance to watch the Sun rise, watch very closely and you can actually see Earth rotate. (Never look directly at the Sun, however, because it can cause serious eye damage!)

At the same time that your day is beginning on one side of the circle of illumination, night has begun on the opposite side of Earth because that place has rotated such that the Sun is no longer illuminating it. Remember, the Sun is always illuminating half of Earth at any given time, whereas the other half is in shadow.

Seasonal Changes in Sun Position (Angle) and Length of Day

In addition to the daily or **diurnal cycle** of day and night, which is related to the Earth's rotation, we can also see a seasonal cycle in the Sun's position in the sky, which depends on orbital progression. Whereas the day–night cycle is obvious and can be easily seen because of the apparent east–west arc of the Sun, the seasonal migration of the Sun's position is a slow progression in north–south directions relative to the Equator. This migration results in a seasonal variation in the angle of the Sun's rays with respect to any location on Earth. To follow this migration, the best point of reference is the daily position of the Sun at solar noon at any given place because that is when the Sun is at its highest point in the arc at that locality. As we examine the seasonal migration of the noontime Sun, imagine that you are living at 45° N (but remember that what you see is opposite for those who reside in the Southern Hemisphere). Also, keep in mind that the overall seasonal north–south migration of the Sun is occurring in combination with the Sun's daily arc.

If you live in the middle latitudes of the Northern Hemisphere, you will notice that the Sun arcs across the southern part of the sky over the course of the day. At solar noon, it is directly to the south rather than overhead because the subsolar point never reaches a point higher than 23.5° N at any time in the year. Given that the Sun is in the southern sky at solar noon, shadows from buildings and trees project toward the north. In the context of the seasons, the noon Sun is highest in the sky during summer. In other words, the Sun angle is greatest during summer. This high angle occurs because the Sun is in the northern hemisphere during this time of year.

As the season moves from summer to fall, the position of the noontime Sun (as you see it) slides farther south in the sky on a daily basis. Why? This southerly migration occurs because the subsolar point moves toward the Equator (which is to the south) as fall approaches and Sun angle decreases. With the gradual approach of winter, the noontime Sun moves still farther to the south because the subsolar point is migrating toward the Tropic of Capricorn. At this time, Earth is in a position within its orbit where the Northern Hemisphere is tilted away from the Sun. If you happen to live in the Northern Hemisphere, the noontime Sun is lowest in the southern sky (in other words, the Sun angle is the least) on December 21, or the Winter Solstice. From this point on, the noontime Sun slowly migrates back to the north, as Sun angle increases, until the Summer Solstice.

You can see the combined daily and seasonal migrations of the Sun, with respect to the Earth's surface, in a graphical way by examining Figure 3.14. This diagram is called a **celestial dome** and shows the Sun's daily arc and seasonal migration relative to the Earth's surface. In this particular diagram, the dome is oriented with respect to a person standing at 45° N latitude. The surface of Earth is presented as a two-dimensional feature that extends from pole to pole, with the arc and position of the Sun shown at three periods: Summer Solstice, Winter Solstice, and Equinox. In other words, this diagram

---

Solar noon   *The time of day when the Sun angle reaches its highest point as the Sun arcs across the sky.*

Diurnal cycle   *A 24-hour cycle.*

Celestial dome   *A sphere that shows the Sun's arc, relative to the Earth, in the sky.*

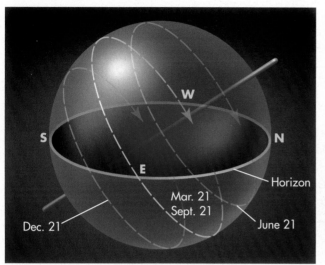

(a)

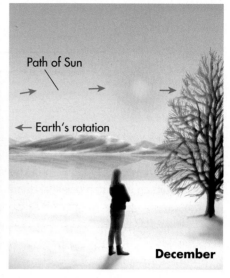

(b)

Figure 3.14 The celestial dome. (a) The Sun's arc at the Winter Solstice, Equinoxes, and Summer Solstice when viewed from 45° N latitude. (b) The position of the Sun at solar noon varies dramatically between the winter and summer seasons. It is much higher in the sky in summer and lower in the sky in winter. This difference occurs due to the Earth's axial tilt.

illustrates how you would see the Sun as the day progresses at three different times of year.

One of the most obvious ways in which we notice the seasonal change in Sun angle is with respect to the number of daylight hours, or the one length of day. Look again at the celestial dome in Figure 3.14 or the one in the *Celestial Dome* module and notice how the curvature of Earth influences when the Sun emerges from the horizon in the east or disappears beneath it in the west. You should be able to see that the Sun emerges from the horizon earlier and disappears later in the summer months than during winter. It should make sense, too, that this seasonal variation is more pronounced at higher latitudes than at lower ones. Table 3.1 illustrates this concept by showing the number of daylight hours at five different latitudes in the Northern Hemisphere at the equinoxes and solstices. Notice how the day length varies among the five sites.

KEY CONCEPTS TO REMEMBER ABOUT HOW WE SEE EARTH–SUN GEOMETRY

1. We experience day and night because Earth rotates on its axis.

2. Due to the Earth's rotation, the Sun arcs from east to west across the sky.

3. As the Sun arcs across the sky, it is at its highest point at solar noon.

4. Sun angle refers to the angle at which solar radiation strikes Earth. Sun angle is generally highest at the low latitudes and progressively lessens toward the poles.

5. Sun angle at solar noon changes over the course of the year as the Earth's orbit progresses.

TABLE 3.1	Variations in Day Length on Equinoxes and Solstices across Latitude				
		Day Length on Equinox or Solstice			
Location	Latitude (N)	March 20	June 21	September 22	December 21
Bogota, Colombia	4°32′	12 h, 06 min	12 h, 24 min	12 h, 07 min	11 h, 51 min
Miami, Florida	25°46′	12 h, 08 min	13 h, 45 min	12 h, 09 min	10 h, 32 min
Topeka, Kansas	39°07′	12 h, 07 min	14 h, 55 min	12 h, 13 min	9 h, 26 min
Calgary, Alberta	45°01′	12 h, 12 min	16 h, 33 min	12 h, 15 min	7 h, 56 min
Barrow, Alaska	71°03′	12 h, 23 min	24 h, 00 min	12 h, 22 min	00 h, 00 min

www.wiley.com/college/arbogast

Celestial Dome

To enhance your understanding of Sun angle and seasonality, go to the *GeoDiscoveries* website and select the module, *Celestial Dome*. This module provides an opportunity to interact with an animated celestial dome. Change the latitude from several options to observe seasonal change in the arc of the Sun, and its height above the horizon at equinox and solstice. Be sure to notice how the Sun arc and angle vary between low- and high-latitude locations, as well as between the hemispheres. Be sure to answer the questions at the end of the module to ensure you understand this concept.

www.wiley.com/college/arbogast

Sun Angle and Length of Day

Let's now take a comprehensive look at Earth–Sun geometric relationships and how they influence the length of day. Go once again to the *GeoDiscoveries* website and select the module *Sun Angle and Length of Day*. As you watch this animation, notice how and why day length at three different latitudes varies over the course of the year. Relate your observations with the *Celestial Dome* animation and Table 3.1. After you complete the animation, be sure to answer the questions at the end of the module to test your understanding of this concept.

www.wiley.com/college/arbogast

Earth–Sun Geometry and Ancient Humans

Earth–Sun geometric relationships have been important to people for a long time. Although ancient humans did not know that Earth revolves around the Sun, they nevertheless understood that Sun angle and day length changed over the course of the year, and was somehow associated with the seasons. In many prehistoric human cultures, these changes had important spiritual significance because they were related to agricultural cycles. To learn more about how ancient humans celebrated the Sun, go to the *GeoDiscoveries* website and select the module *Earth–Sun Geometry and Ancient Humans*. After you watch the video, be sure to answer the questions at the end of the module to test your understanding of this concept.

VISUAL CONCEPT CHECK 3.2

This cross section of the Eastern Time Zone in the United States ranges from Boston, Massachusetts, on the east to Holland, Michigan, on the west. Although both cities occur at the same line of latitude (42° N), the timing of sunrise and sunset varies between the two locations. The Sun rises about an hour earlier in Boston than it does in Holland. Conversely, the Sun sets about an hour later in Holland than it does in Boston. Considering the rotation of Earth and basic Earth–Sun geometry, why does this variation occur?

THE BIG PICTURE

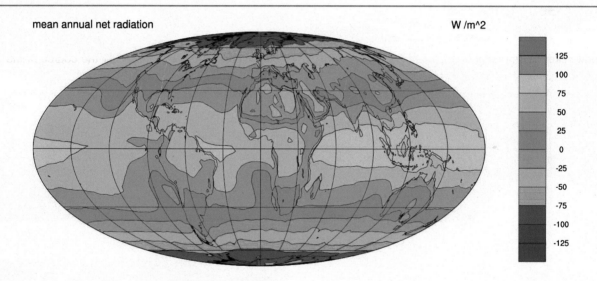

mean annual net radiation   W /m^2

The Earth–Sun relationship—especially the tilt of the Earth's rotation axis as it travels around the Sun—causes variations in the spatial distribution of solar radiation on Earth. This variability has a profound impact on a variety of geographic patterns on Earth, including atmospheric circulation, the distribution of global climates, and even vegetation. In the next chapter, we explore how this solar radiation interacts with the atmosphere around the planet and how this interaction affects surface temperatures. To see a preview of the material covered in the next chapter, examine the accompanying figure, which shows the geography of net radiation on Earth. Net radiation is the difference between incoming and outgoing solar radiation. Notice that the highest amounts of net radiation are generally at low latitudes, whereas the lowest amounts are at high latitudes. Given what was covered in this chapter, you should have a basic idea of why this pattern occurs. In the next chapter we will investigate these relationships, as well as others, more closely.

DISCOVER...

SUNRISE IN THE SOUTHERN HEMISPHERE

This beautiful sunrise is seen from the top of Te Mata Peak, which lies on the east coast of New Zealand, about one hour south of the city of Napier. Given New Zealand's location is in the Southern Hemisphere, south of the Tropic of Capricorn, the Sun arcs across the northern part of the sky, rather than the southern sky as it does in the Northern Hemisphere. Thus, in the Southern Hemisphere, as you face the sunrise,

you are looking northeast. In contrast, if you watched the sunrise from the top of a mountain in the northern United States you would be facing southeast.

Te Mata Peak has an interesting geographic distinction. Because it is just to the west of the International Date Line, it is one of the first places on Earth where the Sun can be seen on any given day.

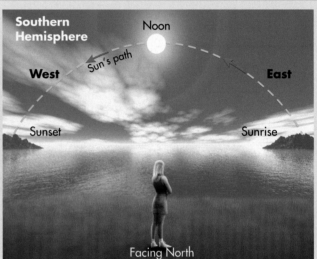

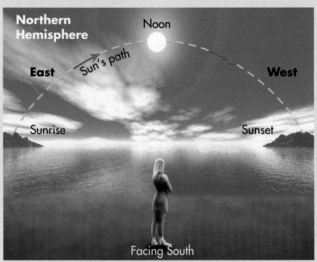

SUMMARY OF KEY CONCEPTS

1. Earth revolves around the Sun within an elliptical orbit in the plane of the ecliptic. As the Earth revolves, it rotates on its axis, which is tilted at 23.5° from a line perpendicular to the plane of the ecliptic. It takes approximately 365 days for one complete revolution to occur.

2. The subsolar point is the line of latitude on Earth where the Sun angle is 90° and thus, where solar radiation strikes Earth most directly. Due to the axial tilt, the subsolar point migrates between the Tropic of Cancer and Tropic of Capricorn between June 21–22 (Northern Hemisphere Summer Solstice) and December 21–22 (Northern Hemisphere Winter Solstice), respectively. The Sun is directly over the Equator at solar noon on both the Spring and Fall Equinoxes (March 21–22 and September 21–22, respectively).

3. Day and night on Earth occur because Earth rotates on its axis. It takes 24 hours for a complete rotation to occur.

CHECK YOUR UNDERSTANDING

1. Describe Earth's place in the solar system.

2. What is the shape of Earth's orbit around the Sun and how do the terms *aphelion* and *perihelion* apply?

3. What is the tilt of Earth's axis relative to the plane of the ecliptic?

4. Define the concept of the subsolar point and how it relates to Sun angle. Describe why it moves on a seasonal basis. What are the four important dates in this movement and how do they relate to latitude?

5. Chicago, Illinois, and Omaha, Nebraska, are both located in the Central Time Zone, with Chicago being on the east side of the zone and Omaha on the west. With respect to time, which of these two cities is the first to "see" the sunrise? Why does this variation occur?

6. If the axis of Earth were perpendicular to the plane of the ecliptic, where would the subsolar point be on December 21?

7. Imagine that the axis of Earth were tilted at 45°. In this case, where would the Tropic of Cancer be located?

8. If the axis of Earth were tilted at 40°, what would the Sun angle be at 40° S latitude on December 21–22?

9. Which direction, north or south, do shadows project during solar noon on December 21–22 at 40° S latitude?

10. Where does the greatest range in daylight hours occur, at high or low latitudes?

11. Assume that you are standing on the Equator and it is June 21–22. Given your perspective, is the noontime Sun in the northern or southern part of the sky?

ANSWERS TO VISUAL CONCEPT CHECKS

VISUAL CONCEPT CHECK 3.1

Both October and February are logical times of the year when the subsolar point would be at 10° S latitude. In October, the Sun is in the process of moving from the Equator (0°) to the Tropic of Capricorn (23.5° S). After the subsolar point reaches 23.5° S in late December, it begins to move back to the Equator, which it will reach on or about March 21. To do so, however, it must pass by 10° S, which it does in February.

VISUAL CONCEPT CHECK 3.2

The Sun rises and sets earlier in Boston, Massachusetts, than in Holland, Michigan, because the surface of Earth is curved. When the Sun rises in Boston, Holland lies on the part of Earth that is curved away from the Sun's light. As Earth rotates, the Sun ultimately "rises" in Holland, but about an hour later than in Boston. The Sun sets later in Holland than in Boston because Boston rotates out of sunlight about an hour before Holland.

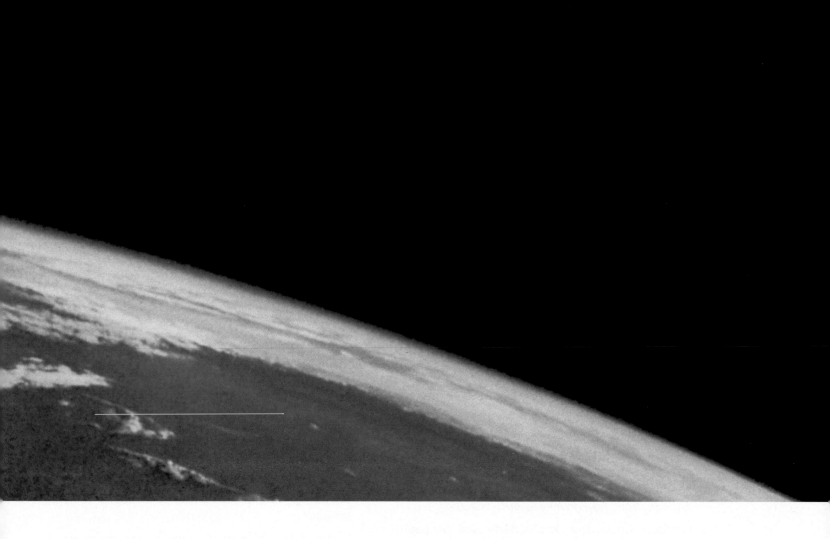

CHAPTER FOUR | THE GLOBAL ENERGY SYSTEM

You have now examined Earth's geometric relationship with the Sun and how it influences the distribution of solar radiation seasonally and by latitude. Now we will look more closely at the way this radiation is intercepted by Earth, how it flows through the atmosphere, how it interacts with the land and ocean surfaces, and how it maintains a balance with outgoing radiation on a global scale. Understanding these relationships is critical to comprehending many Earth processes discussed in later chapters, such as wind, weather, ocean currents, and rivers, because the Sun provides the energy required to power them, either directly or indirectly. We begin by investigating the concept of radiation energy and how it is measured.

62

Illumination of clouds over the Pacific Ocean by the Sun. The Sun sends a continuous stream of electromagnetic energy to Earth. This energy interacts with Earth in a wide variety of ways that are discussed in this chapter.

CHAPTER PREVIEW

The Electromagnetic Spectrum and Solar Energy
On the Web: *GeoDiscoveries The Electromagnetic Spectrum*

Composition of the Atmosphere

The Flow of Solar Radiation on Earth
On the Web: *GeoDiscoveries The Angle of Incidence*

The Global Radiation Budget
On the Web: *GeoDiscoveries The Global Energy Budget*

Human Interactions: Solar Energy Production

The Big Picture

LEARNING OBJECTIVES

1. Describe the fundamental components of the electromagnetic spectrum and how they differ from one another.

2. Discuss the various components of the atmosphere.

3. Compare and contrast the processes of conduction, convection, and radiation as methods of heat transfer.

4. Explain the flow of radiation from the Sun to Earth and how absorption, reflection, scattering, and albedo influence the system.

5. Discuss the various components of the global radiation budget and how the system achieves balance.

6. Understand why solar energy is a renewable energy source and the challenges associated with its production.

The Electromagnetic Spectrum and Solar Energy

The Electromagnetic Spectrum

Radiation is electromagnetic energy that is transmitted in the form of waves. These waves can be measured in terms of their length and amplitude, with **wavelength** being the distance between wave crests, and the term **wave amplitude** referring to half the height between the wave crest and wave trough (Figure 4.1). The entire wavelength range of electromagnetic energy is called the **electromagnetic spectrum**.

Within the electromagnetic spectrum, wavelengths range from *gamma rays* that are less than a nanometer long (one billionth or 10^{-9} of a meter) to *radio waves* that are tens of meters in length (Figure 4.2). Humans can see radiation directly only in the *visible light* part of the spectrum, which ranges from violet with the shortest wavelength of 375 nanometers (nm) to red with the longest wavelength (740 nm).

You should remember two very important principles about electromagnetic radiation. The first principle is that an *inverse* relationship exists between the temperature of an object and the wavelength of the electromagnetic radiation it emits. In other words, hotter objects emit radiation with shorter wavelengths than cooler objects. For example, the Sun has a surface temperature of about 6000°C (11,000°F) and emits energy as **shortwave radiation**, which includes gamma rays, X-rays, ultraviolet (UV) radiation, visible light, and near-infrared radiation. The Earth, in contrast, is a much cooler object (~16°C, 61°F) and thus emits **longwave radiation** in the thermal infrared part of the spectrum (Figure 4.3). Keep in mind, though, that the vast majority of energy released by Earth initially originated at the Sun.

Wavelength   *The distance between adjacent wave crests or wave troughs.*

Wave amplitude   *The overall height of any given wave as measured from the wave trough to the wave crest.*

Electromagnetic spectrum   *The radiant energy produced by the Sun that is measured in progressive wavelengths.*

Shortwave radiation   *The portion of the electromagnetic spectrum that includes gamma rays, X-rays, ultraviolet radiation, visible light, and near-infrared radiation.*

Longwave radiation   *The portion of the electromagnetic spectrum that includes thermal infrared radiation.*

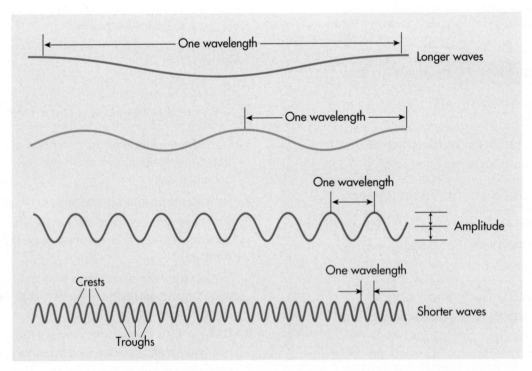

Figure 4.1 Relative sizes of electromagnetic waves. The distance from one crest to the other, or from one trough to another, is called the wavelength. Wave amplitude refers to half the distance from the wave crest to the wave trough.

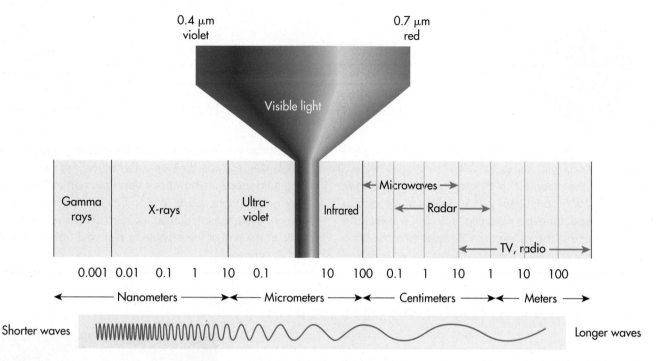

0.4 μm
violet

0.7 μm
red

Visible light

| Gamma rays | X-rays | Ultra-violet | Infrared | Microwaves → / Radar | TV, radio |

0.001   0.01   0.1   1   10   0.1   10   100   0.1   1   10   1   10   100

← Nanometers → ← Micrometers → ← Centimeters → ← Meters →

Shorter waves   〰〰〰   Longer waves

Figure 4.2 The electromagnetic spectrum. Wavelengths range from gamma rays, which are the shortest, to long radio waves.

The second principle of electromagnetic radiation is that a *direct* relationship exists between the absolute temperature of the object and the amount of radiation it emits. This relationship is described by the Stefan–Boltzmann law and means that hotter objects emit more radiation than cooler ones. As a result, Earth emits much less radiation than the Sun. This temperature/emitted radiation relationship is exponential, meaning that a small temperature increase in an object results in very large increases in emitted radiation. For example, you have probably experienced how a blacktop or concrete surface emits much more radiation, and is therefore

much hotter, than a plowed field or grassy park that it might surround. This higher release of radiation takes place because the highway or sidewalk is warmer than the field or park.

Solar Energy and the Solar Constant

As we discussed in Chapter 3, solar energy is created in vast quantities by nuclear fusion within the Sun. This energy works its way to the surface of the Sun, where it is emitted as electromagnetic radiation. From this point this energy travels along straight lines (or rays) through space at the speed of light.

Given that the Sun produces energy at a nearly constant rate, the output of solar radiation is also nearly constant. Although some variability in solar output occurs due to the waxing and waning of energy production within an 11-year solar cycle, the overall output of solar energy is considered to be consistent over time. As a result of this generally constant production and emission, the amount of solar radiation received in an area of fixed size in space (outside Earth's atmosphere), and at right angles to the Sun, is also constant. This amount of received energy, which is referred to as the **solar constant**, has a value of about 1370 watts per square meter (W/m^2) at the top of the atmosphere.

Shortwave radiation from Sun (visible, UV, near infrared)

Longwave radiation from Earth (thermal infrared)

Figure 4.3 Radiation to and from Earth. Earth receives shortwave radiation from the Sun. Some of this radiation reaches the surface of Earth where it is absorbed and then emitted as longwave energy.

Solar constant *The average amount of solar radiation (~1370 W/m^2) received at the top of the atmosphere.*

The Electromagnetic Spectrum

To better visualize the various components of the electromagnetic spectrum, go to the *GeoDiscoveries* website and select the module *The Electromagnetic Spectrum*. You will be able to interact with elements of the electromagnetic spectrum by selecting different wavelengths, such as those in the infrared and X-ray parts of the spectrum. When you select a specific wavelength, an image will appear that is representative of that part of the spectrum. In this way you will develop a better feel for the electromagnetic spectrum, how it is a pervasive part of Earth, and how geographers use it to study the environment. Once you complete the interaction, be sure to answer the questions at the end of the module to test your understanding of this concept.

VISUAL CONCEPT CHECK 4.1

Although we take the Sun for granted, it is by far the most important variable with respect to life on Earth. How is solar energy produced? What form does solar energy take as it travels through space? How long does this energy take to reach the Earth?

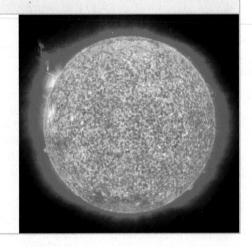

KEY CONCEPTS TO REMEMBER ABOUT THE ELECTROMAGNETIC SPECTRUM AND SOLAR ENERGY

1. The Sun produces energy that radiates to Earth as various sizes of electromagnetic waves.

2. The electromagnetic spectrum is the total wavelength range of electromagnetic energy.

3. An inverse relationship exists between temperature and radiation wavelength.

4. A direct relationship exists between the temperature of an object and the amount of electromagnetic radiation it produces.

5. The Sun produces electromagnetic energy at a generally constant rate over time. The amount of energy that reaches the top of Earth's atmosphere is known as the solar constant.

Composition of the Atmosphere

Before examining how solar radiation interacts with the atmosphere, you first need to know the fundamental composition of the atmosphere because it is the medium through which radiation flows on its way to the surface. The atmosphere is unique in our solar system because it supports life by providing oxygen and carbon dioxide for animals and photosynthesis, respectively. In addition, the atmosphere serves as a buffer that shields the planet from the potentially harmful effects of UV radiation from the Sun, allowing mostly visible and infrared wavelengths to reach Earth. The atmosphere behaves as a fluid in much the same manner as water, with flowing currents and eddies. Fluctuations in these currents and eddies shape the course of environmental conditions on the Earth's surface at every moment. Knowing how solar radiation flows in the atmosphere is key to understanding Earth's temperature, atmospheric circulation, and precipitation patterns, which are discussed in later chapters.

Figure 4.4 The Earth's atmosphere as viewed from space. This image nicely demonstrates the thinness of the atmosphere, which is the faint blue haze that appears on the horizon. Note the infinite blackness of deep space in the background.

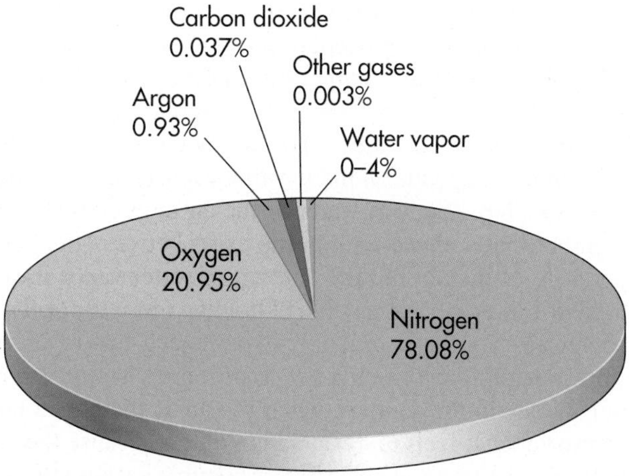

Carbon dioxide 0.037%
Other gases 0.003%
Argon 0.93%
Water vapor 0-4%
Oxygen 20.95%
Nitrogen 78.08%

Figure 4.5 Proportion of various gases in the atmosphere. Most of the atmosphere consists of nitrogen and oxygen, which are constant gases. Other gases, such as carbon dioxide and water vapor, are variable gases.

In general terms the atmosphere consists of air, an invisible medium that surrounds and protects Earth (Figure 4.4). The fundamental components of the atmosphere can be divided into three categories: (1) constant gases, (2) variable gases, and (3) particulates. Each of these constituents is critical to the way the atmosphere functions because it performs a unique role essential to life on Earth.

Constant Gases

Constant gases maintain more or less the same proportion in the atmosphere. In our atmosphere, nitrogen and oxygen are the primary constant gases and together make up 99% of the atmosphere (Figure 4.5). Argon is the other constant gas, composing approximately 1% of the atmosphere. It is inert (chemically inactive) and of little importance in natural processes.

Nitrogen occurs in molecular form as two nitrogen atoms bonded together (N_2). This gas makes up 78% of the atmosphere and is derived from the decay and burning of organic material, volcanic eruptions, and the chemical breakdown of specific kinds of rocks. Although nitrogen is largely inert in the atmosphere, it is critical to plant life because it can be transformed, or fixed, into chemical compounds (ammonia or nitrates) in the soil. These compounds are absorbed by plants and incorporated in their

tissues as proteins. Nitrogen maintains a constant proportion of the total atmosphere because what is added is balanced by what is removed through precipitation and various biological processes.

Oxygen makes up 21% of the atmosphere and is a by-product of photosynthesis. In contrast to the inert nature of nitrogen, oxygen gas (O_2) is very active and can combine with a variety of other elements through the process of oxidation. Oxygen is essential to animal respiration because it is required to convert foods into energy. Oxygen is a constant gas because the amount produced by plants balances the amount absorbed by various organisms through respiration.

Variable Gases

Variable gases differ in their proportion of the atmosphere over time and space, depending on environmental conditions. Although some of these gases are very important to life, they make up only a tiny portion (less than 1%) of the atmosphere. The most important variable gases are carbon dioxide, water vapor, and ozone.

Water Vapor An important variable gas in the atmosphere is water vapor. Water is found in three physical states: liquid, solid (ice), and gas (water vapor). The amount of water vapor in the atmosphere near Earth's surface is about 2% in most parts of the planet, but can range from just less than 1% over deserts and polar regions to about 4% in tropical areas. Atmospheric water vapor is especially vital because it absorbs and stores heat energy from the Sun and is thus an important component,

Constant gases *Atmospheric gases such as nitrogen, oxygen, and argon that maintain relatively consistent levels in space and time.*

Variable gases *Atmospheric gases such as carbon dioxide, water vapor, and ozone that vary in concentration in space and time.*

along with carbon dioxide, of the greenhouse effect. As airflow within the atmosphere moves vapor around, the effect is to moderate temperature and transport energy around Earth.

The amount of water vapor at any given place in the atmosphere depends on many variables, such as the proximity to a large body of water or the air temperature. Remember that a direct relationship exists between the temperature of the air and the amount of water vapor it can hold, with warmer air capable of holding more vapor than cooler air. You have probably experienced this relationship yourself. Try to remember what it feels like on a hot, muggy day in the summer when the air is sticky and uncomfortable. It feels that way in large part because the air holds a lot of water vapor, which is possible because the air is so warm. This vapor/temperature relationship is an important principle in physical geography because it directly influences the process of precipitation. As we will discuss more thoroughly in Chapter 7, precipitation happens when bodies of air cool and water changes from vapor to liquid.

Figure 4.6 shows atmospheric water vapor over the North American continent in February 2009. Notice that on this particular day, the highest concentration of water vapor was in the central and eastern United States. If you happened to be in those areas, you would have likely noticed numerous clouds in the sky. The water vapor in those clouds flowed into the region in association with a body of warm air originating over the tropical Pacific Ocean. Areas of relatively dry air appeared over northwestern Mexico and the upper Midwest.

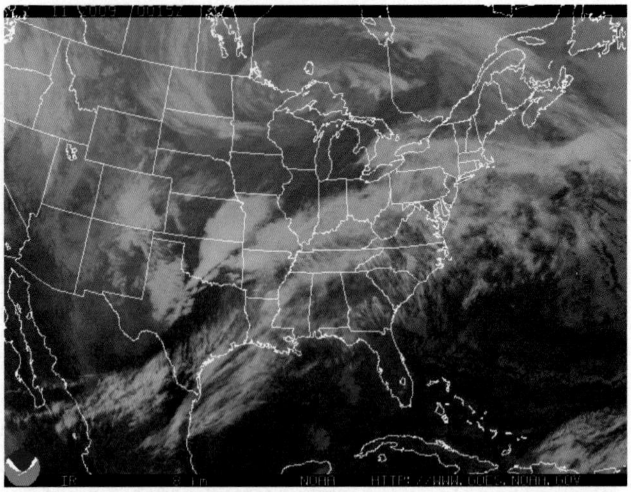

Carbon Dioxide Another important variable gas is carbon dioxide (CO_2), which currently comprises about 0.039% of the atmosphere (or 392 parts CO_2 for every 1 million parts of the atmosphere). Despite this very small percentage, CO_2 is a critical part of the atmosphere for two very important reasons. First, plants absorb CO_2 and release oxygen as a by-product. Second, atmospheric CO_2 contributes significantly to the **greenhouse effect**, which is the process through which the atmosphere traps longwave radiation. This process warms the atmosphere, which, in turn, warms Earth. Another significant greenhouse gas is methane, which is derived from natural gas and decaying organic matter. Methane currently makes up about 0.00017% of the atmosphere. You may associate the greenhouse effect most closely with the concept of global warming, an environmental issue that concerns environmental scientists today. In fact, a common misconception among many people is that the greenhouse effect is *entirely* a problem because it is associated with the impor-

Figure 4.6 Water vapor over the North American continent. This image was taken by the Geostationary Operational Environmental Satellite (GOES) on February 11, 2009. Dark zones represent regions in the atmosphere about 6 km to 10 km (~3.7 mi to 6.2 mi) above the Earth's surface containing very little water vapor, whereas bright areas represent relatively high concentrations of water vapor. In areas other than the tropics and poles, the average water vapor content is about 2% of the air.

tant issue of global warming. Although the greenhouse effect is indeed linked to this issue, it has historically been a positive part of the Earth's system because it has helped maintain a constant atmospheric temperature that is hospitable to life. The problem now is that CO_2 levels are rapidly rising beyond recent historical norms due to human industrial activity, specifically the consumption of fossil fuels. As a result, Earth appears to be in the midst of a rapid warming trend that may have major environmental consequences in the future.

You can examine the processes associated with the greenhouse effect more closely in Figure 4.7. Although we will discuss the flow of solar radiation more thoroughly later in the chapter, this diagram shows the significance of CO_2 in the atmosphere and is relevant here. As illustrated in Figure 4.7, Earth receives shortwave radiation from the Sun, which passes through the atmosphere and is absorbed by the surface. Subsequently, the Earth releases energy as longwave radiation. Some of this radiation flows directly back into space, but most is absorbed by the atmosphere, where it is held in large part by CO_2 (water vapor is another greenhouse gas). The atmosphere then redirects some longwave energy back to the surface as **counterradiation**. Some atmospheric longwave energy escapes to space.

Greenhouse effect *The process through which the lower part of the atmosphere is warmed because longwave radiation from Earth is trapped by carbon dioxide (CO_2) and other greenhouse gases.*

Counterradiation *Longwave radiation that is emitted toward the Earth's surface from the atmosphere.*

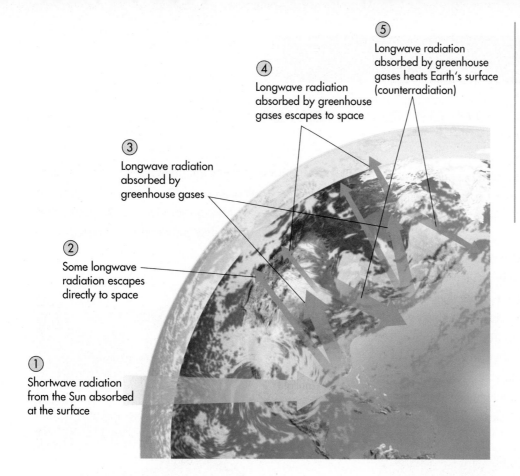

④ Longwave radiation absorbed by greenhouse gases escapes to space

⑤ Longwave radiation absorbed by greenhouse gases heats Earth's surface (counterradiation)

③ Longwave radiation absorbed by greenhouse gases

② Some longwave radiation escapes directly to space

① Shortwave radiation from the Sun absorbed at the surface

Figure 4.7 The greenhouse effect. Earth absorbs shortwave radiation from the Sun and emits it as longwave energy. Some of this emitted longwave radiation flows directly to space, but most is absorbed by carbon dioxide (CO_2) and water vapor in the atmosphere. Subsequently, this longwave radiation either escapes into space or reflects back to the Earth's surface through the process of counterradiation.

The greenhouse effect is the process through which counterradiation is returned to the surface. The really fascinating aspect of the greenhouse effect, as far as life on Earth is concerned, is that the balance is just right for the planet to be neither too hot nor too cold. To see the significance of this balance, consider the planets Mars and Venus. In both places, CO_2 is the dominant component of the atmosphere, comprising about 96% of the volume on each planet. On Mars, the overall density (measured by pressure) of the atmosphere at the surface is about 1000 times less than that of Earth; thus, there is comparatively much less CO_2 on Mars than on Earth. As a result, Mars is much colder than Earth, with an average temperature of $-63°C$ ($-81°F$). With respect to Venus, the atmosphere there is about 90 times denser at the surface than Earth's atmosphere; thus, the concentration of CO_2 on Venus is much higher than on Earth. Consequently, Venus is considered to have a runaway greenhouse effect, with an average temperature of $500°C$ ($932°F$)!

Although Venus provides an extreme example of the greenhouse effect, it is a context through which we can view the concerns that many scientists have regarding the ongoing human impact on the greenhouse process through the combustion of fossil fuels for energy. Since the middle of the 19th century, levels of atmospheric CO_2 have steadily increased largely through our consumption of fossil fuels. To see how these levels have changed, examine Figure 4.8, which

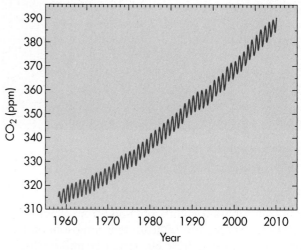

Sources: 1958–1974 Scripps Institute of Oceanography; NOAA
1974–2006 National Oceanic and Atmospheric Administration; NOAA

Figure 4.8 Changes in atmospheric carbon dioxide from 1958 to 2005 as measured at Mauna Loa, Hawaii. The vertical axis is the concentration of CO_2 in parts per million (ppm). Note the steady increase during this period of time, which has been attributed to human consumption of fossil fuels. Peaks and valleys in the curve represent seasonal fluctuations, with low points taking place during the Northern Hemisphere summer when plants take up CO_2. (*Sources: C. D. Keeling and T. P. Whorf. Atmospheric CO_2 records from sites in the SIO Air Sampling Network. In Trends: A Compendium of Data on Global Change. U.S. Department of Energy; NOAA; NASA.*)

THE FORMATION OF RAINBOWS

In addition to producing clouds, the presence of water vapor in the air can have a very beautiful side effect: it causes rainbows. A rainbow develops when white light from the Sun peeks through the clouds and strikes water droplets in the air from a rainstorm. The water droplets bend the light rays, but by a different amount for each color of light. The result is that the white light is spread out into bands of colors. These colors are reflected back down toward Earth by the water droplets, at an average angle of about 42°. If you are standing in the right place at the right time, you see these colors as a rainbow.

Rainbows require bright sunlight, so you cannot see one during the height of a rainstorm. On the other hand, a rainbow requires the presence of water droplets in the air, so it will disappear after the rain has gone. In other words, the sight of a rainbow usually means that the rain is clearing out and fair weather is ahead.

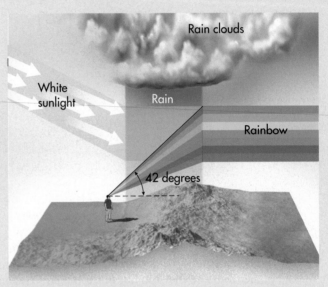

Water droplets in the air bend sunlight by an angle that depends on the wavelength (color) of the light. We see the result as a rainbow. Notice that red appears at the top of the rainbow and blue at the bottom because red light is bent more than blue light.

| A rainbow at Banff Park in Canada.

shows CO_2 measurements obtained from Mauna Loa in Hawaii. These data indicate that atmospheric CO_2 increased approximately 80 parts per million (ppm) between 1958 and 2010. This increase is thought to be a primary cause of global climate change and is the negative side of the greenhouse effect that many people associate with the relationship. This important issue will be covered in greater detail in Chapter 9.

Ozone Ozone is a form of oxygen that has three oxygen atoms (O_3), rather than the two found in normal oxygen gas (O_2). Atmospheric ozone forms when gaseous chemicals react in the upper atmosphere (Figure 4.9) with light energy. In the initial stage of this reaction, O_2 absorbs UV energy, which causes the oxygen molecule to split into two oxygen atoms (O + O). Subsequently, one of the O atoms combines with an O_2 molecule to form ozone. Ozone is naturally destroyed when it absorbs UV radiation and splits from O_3 into O_2 + O. The single oxygen atom can then recombine with another O_2 molecule to form ozone. Ultimately, O_3, O_2, and oxygen atoms (O) are repeatedly formed, destroyed, and reformed in the ozone layer in a way that absorbs UV radiation each time a transformation takes place.

Ozone primarily occurs in two layers in the atmosphere: at ground level and within the stratosphere. Although diffuse amounts of ozone exist as high as 50 km (32 mi) above Earth, it is most concentrated in the ozone layer between about 15 km and 30 km (9 mi to 19 mi; see Figure 4.10). Ground-level ozone is a form of pollution (Figure 4.11) created when nitrogen

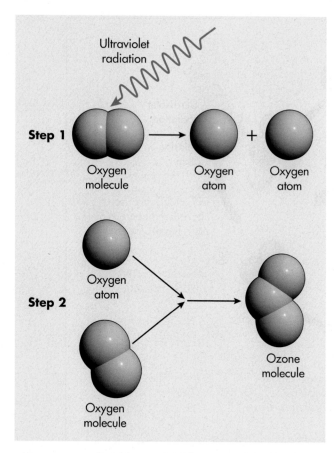

Figure 4.9 Formation of atmospheric ozone. In the first step, UV radiation is absorbed by oxygen molecules, creating free oxygen atoms. In the second step, these free oxygen atoms combine with oxygen molecules to form ozone.

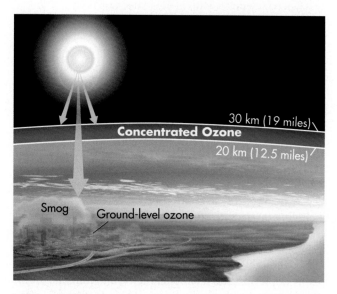

Figure 4.10 Ozone concentrations in the atmosphere. One layer is at high altitudes and absorbs UV radiation from the Sun. Most of this ozone is concentrated in the ozone layer. The second concentration occurs in the lower part of the atmosphere and is associated with pollution, particularly chemical smog in cities.

Figure 4.11 Smog in Singapore. A large component of this haze is ozone created by the interaction of nitrogen oxides and organic gases emitted by automobiles and factories.

oxides and organic gases emitted by automobiles and industrial sources react. This form of ozone has been linked to cardiorespiratory illness and may also cause crop damage.

Human Interactions: The Ozone Hole The greatest concentrations of ozone are found in the **ozone layer**. This layer is critical to life because it absorbs harmful UV radiation from the Sun that would otherwise reach the Earth's surface. A major environmental issue that developed in the last half of the 20th century is the depletion of the ozone layer. This depletion happened on a global basis due, in part, to the release of chlorofluorocarbons (CFCs) associated with air conditioners and other cooling systems. CFC molecules are very stable in the atmosphere and gradually diffuse upward from the surface to the ozone layer. Once they reach this altitude, CFCs absorb UV radiation and break down into chlorine oxide (ClO) molecules that, in turn, attack ozone molecules and convert them into oxygen atoms (Figure 4.12). The net effect of this process is that the ozone layer is depleted and more UV radiation reaches the ground because less is absorbed in the atmosphere.

In the 1980s it was discovered through satellite remote sensing of the atmosphere that the ozone layer was being depleted at an alarming rate. Although some depletion occurs naturally by volcanic aerosols periodically injected into the atmosphere, most of the reduction was attributed to industrial CFC production. The most striking example of ozone depletion is

Ozone layer *The layer of the atmosphere that contains high concentrations of ozone, which protect the Earth from ultraviolet (UV) radiation.*

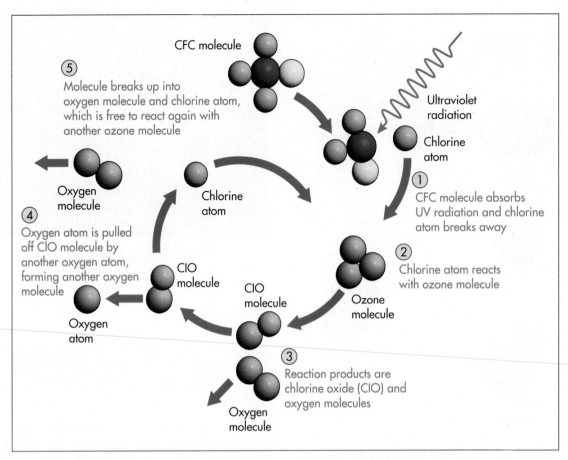

Figure 4.12 Destruction of the ozone layer. Industrial chlorofluorocarbons (CFCs) interact with UV radiation in the ozone layer and release chlorine atoms. Follow the numbered steps to see how this process works. It begins with the breaking away of a chlorine atom from a CFC molecule due to the absorption of UV radiation. The resultant free chlorine atom then begins to interact with an ozone molecule. The chlorine atoms transform ozone molecules into ordinary oxygen molecules without being used up themselves. Thus, one CFC molecule can destroy many ozone molecules.

the Antarctic **ozone hole,** which was discovered in the mid-1980s. This feature forms every spring in the Southern Hemisphere (Figure 4.13), because a polar vortex develops during the previous winter that traps the air within it. This vortex is somewhat analogous to a whirlpool in water that does not allow a floating object out as it spins.

Over the course of each winter, air becomes very cold and thin clouds of ice form within the atmospheric whirlpool. The floating ice crystals, in turn, facilitate the transformation of chlorine to chlorine oxide that destroys ozone, with the peak loss taking place in September and October. As a result of this decrease in ozone, the amount of surface exposure to UV radiation at 55° S has increased about 10% per decade since the late 1970s. In

Ozone hole *The decrease in stratospheric ozone observed on a seasonal basis over Antarctica, and to a lesser extent over the Arctic.*

addition to the Antarctic ozone hole, a similar but weaker hole develops over the Arctic region in the Northern Hemisphere winter. Significant depletions of ozone have also been measured in the midlatitudes of the Northern Hemisphere. As a result, the average amount of annual exposure to UV radiation at 55° N has increased about 7% per decade since the late 1970s. These increases in UV exposure are of great concern for several reasons, including damage to some forms of aquatic life, reduced crop yields, and increases in the incidence of skin cancer in humans.

Although scientists have legitimate concerns about ozone depletion and its impact on life and human health, the threat may diminish in our lifetime. Shortly after the ozone issue emerged, the world community began working diligently to solve the problem of CFC emission. These efforts culminated in the 1987 Montreal Protocol, an environmental treaty sponsored by the United Nations and signed by 23 nations. This treaty called for

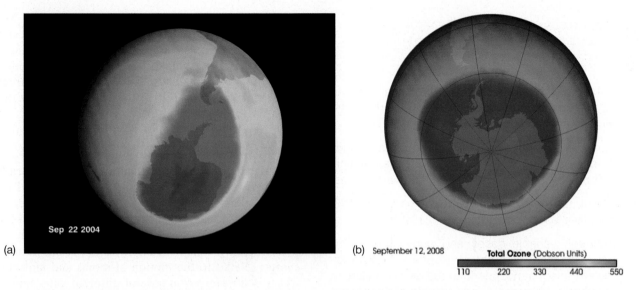

Figure 4.13 Antarctic ozone hole. (a) Ozone hole during September 2004. The area of the hole was 24,200,000 km² (9,400,000 mi²), or larger than the combined area of the United States, Canada, and Mexico. Dark blue represents levels of ozone that are about 20% less than normal. (b) Ozone hole in September 2008. During 2008 the hole was about 27,000,000 km² (~10,400,000 mi²) in size, which was slightly more than in 2007 but slightly less than in 2006.

developed countries such as the United States and the nations of the European Union to cut CFC emissions by 50% by 1999 and for developing countries such as various African nations and India to halt CFC use by 2010. The effect of this treaty has been generally positive, with stratospheric CFC concentrations peaking in 1997 and slowly falling since that time. If the current rate of decline continues, the ozone layer could be completely restored by the latter part of the 21st century.

Particulates

The last important variable component of the atmosphere is particulates, which are microscopic bodies carried in the air, existing in both liquid and solid form. The liquid variety comes in the form of clouds and rain, which develop when water vapor changes its physical state due to temperature changes in the atmosphere. Solid particulates come in an assortment of forms, including snow, hail, pollutants, wind-blown soil (dust), smoke from wildfires, volcanic ash, pollen grains from plants, and salt spray from breaking waves. Most of the time particulate concentrations are densest near their place of origin, such as during an intense dust storm (Figure 4.14) or near a volcano. Overall, however, they comprise less than 1% of the Earth's atmosphere.

Despite their small proportion of the atmosphere, particulates nevertheless play an important role in weather and climate. Precipitation would not occur if dust particles were not present in the atmosphere because they provide a nucleus around which water condenses in the first step of cloud formation. You can see

this relationship the next time it rains right after you wash your car, because your car will be coated with fine dust. Other particulates, such as smoke and volcanic ash, are important because they either absorb or reflect

Figure 4.14 Dust storm in New South Wales, Australia. The dust in this cloud was derived from poorly vegetated soils that were severely eroded by the wind.

1. Although the atmosphere is technically about 10,000 km (6000 mi) thick, the vast majority of the air is found in the lowermost 30 km (10 mi).

2. The atmosphere contains constant gases (primarily nitrogen and oxygen), variable gases (mainly carbon dioxide and water vapor), and particulates.

3. The atmosphere consists largely of nitrogen (78%) and, to a lesser extent, oxygen (21%).

4. Although variable gases and particulates compose less than 1% of the atmosphere, they nevertheless affect atmospheric processes and climate in important ways. A good example of the impact of these gases is the greenhouse effect and the way it moderates temperature on Earth.

solar energy. These combined processes influence local weather and regional climate by moderating the temperature of the atmosphere.

Although most particulates are a positive component of the atmosphere, some can cause negative environmental and health effects, especially the toxic air pollutants associated with human activities. According to the U.S. Environmental Protection Agency, over 100 pollutants are of particular concern. Benzene and perchlorethylene, for example, are pollutants derived from gasoline and dry cleaning facilities, respectively. Other toxic pollutants include dioxin, asbestos, toluene, and metals such as cadmium, mercury, chromium, and lead compounds. These pollutants can originate from stationary locations, such as factories and power plants, or mobile sources, such as automobiles. When ingested in large concentrations over a long period of time, these pollutants can accumulate in human tissue and ultimately cause cancer, immune disorders, and a variety of neurological, reproductive (e.g., reduced fertility), developmental, respiratory, and other health problems.

The Flow of Solar Radiation on Earth

Once solar radiation reaches Earth, it flows along several pathways within the atmosphere and to the surface. Some radiation flows straight from the Sun to the surface of Earth as **insolation**, which is most accurately defined as the amount of solar radiation measured in watts per square meter (W/m^2) that strikes a surface perpendicular to the Sun's incoming rays. Most solar radiation, however, follows a very indirect path. Some of it bounces around like a ping-pong ball, while some of it is directly absorbed by various components of the atmosphere. These various pathways directly influence climate on Earth and will be discussed thoroughly later in the chapter.

Heat Transfer

Before exploring the various pathways that solar radiation follows on Earth, you must first understand how heat is transferred to Earth's atmosphere, land, and oceans. This discussion is based on the first law of thermodynamics, which states that energy cannot be created or destroyed; it can only change forms. Heat is a form of energy, related to the motion of atoms and molecules, which is transferred in several different ways that are easy to imagine (Figure 4.15). This transfer follows the second law of thermodynamics, which can be stated as follows: heat can spontaneously flow from a region of higher temperature to an area of colder temperature, but not the other way around.

The next time that you are around a campfire, consider that heat is transferred by **radiation**. This process involves the creation of electromagnetic waves, either as visible light or invisible to the naked eye. Because this radiation exists in wave form, it carries energy and thus can move from one place to another without requiring an intervening medium. When this radiation reaches you, part of the energy of the wave is converted back into heat, which is why you feel warm sitting beside a campfire. Some of the radiation may exist in the form of visible light that we can see, but a great deal of the radiation emitted is infrared waves, whose longer wavelength is usually detectable only with special infrared detectors.

Another form of heat transfer is **conduction**, which involves the diffusion of energy through molecules that are in contact with one another. This diffusion occurs because, as the temperature of molecules increases, they begin to vibrate more rapidly, causing collisions that produce similar motions in adjacent molecules. In this fashion, sensible heat always moves from areas of relatively high temperature to zones of relatively low temperature. You can experience conduction yourself if you pick up a heated pot from a stove or other fire.

Insolation *Amount of solar radiation measured in watts per square meter (W/m^2) that strikes a surface perpendicular to the Sun's incoming rays.*

Radiation *Energy that is transmitted in the form of rays or waves.*

Conduction *The transfer of heat energy from one substance to another by direct physical contact.*

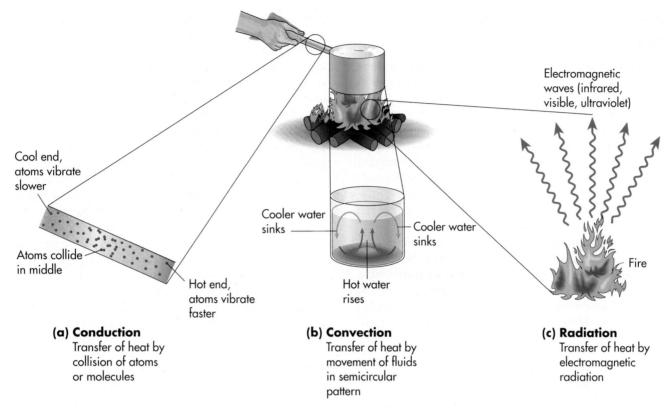

Cool end,
atoms vibrate
slower

Atoms collide
in middle

Hot end,
atoms vibrate
faster

Electromagnetic
waves (infrared,
visible, ultraviolet)

Cooler water
sinks

Cooler water
sinks

Hot water
rises

Fire

(a) Conduction
Transfer of heat by
collision of atoms
or molecules

(b) Convection
Transfer of heat by
movement of fluids
in semicircular
pattern

(c) Radiation
Transfer of heat by
electromagnetic
radiation

Figure 4.15 Mechanisms of heat transfer. (a) Conduction transfers heat by collisions between fast-vibrating atoms or molecules (hot) and slower-vibrating atoms or molecules (cool). (b) Convection is the transfer of heat by the large-scale movement of matter. (c) Radiation is the transfer of heat by electromagnetic radiation, usually in the form of infrared, visible, or UV waves.

The third form of heat transfer is **convection**, which involves the upward movement of heat. The next time you boil water on the stove, notice how a small circular current of water arises within the pot where hot water moves up and is replaced at the bottom by relatively cooler water. As you will see in later chapters, convection is a very impor-

tant mechanism of heat transfer on Earth and is associated with atmospheric circulation and precipitation.

Convection *A circular cell of moving matter that contains warm material moving up and cooler matter moving down.*

VISUAL CONCEPT CHECK 4.2

The atmosphere contains a wide variety of natural constant gases, variable gases, and particulates. People also contribute to the overall character of the atmosphere. What potential impacts do you think this factory complex has on the atmosphere?

Flow of Solar Radiation in the Atmosphere

With the mechanisms of heat transfer in mind, it is time to explore the various ways that solar radiation interacts with Earth. As you work your way through this discussion, refer to Figure 4.16, which illustrates the flow of solar radiation on the planet. Of all the incoming radiation that reaches Earth, about 25% flows uninterrupted to the surface as **direct radiation**. However, this amount can vary greatly depending on local geographic variables such as cloud cover or density of atmospheric dust. Figure 4.17 shows, for example, the impact that cloud cover can have on direct solar radiation.

The remaining 75% of incoming insolation is either absorbed or otherwise redirected in the atmosphere (see Figure 4.16). **Absorption** takes place when variable gases and particulates in the atmosphere interrupt the flow of solar radiation by absorbing specific wavelengths. For exam-ple, almost all UV wavelengths (those less than 0.3 μm or micrometer in length) are absorbed by oxygen and ozone. Similarly, radiation at the 1.3 μm and 1.9 μm wavelengths is absorbed very strongly by water vapor and CO_2. Overall, approximately 24% of incoming solar radiation is absorbed, with 18% absorbed by atmospheric water vapor and dust, 3% by ozone, and 3% by clouds. Remember, this absorption of solar radiation is important because it helps to moderate temperature in the atmosphere.

Some incoming solar radiation in the atmosphere is also redirected by **reflection**. The amount of radiation reflected depends on the **albedo** of a surface and takes place in the atmosphere when insolation bounces off bright cloud tops. (Albedo refers to the amount of reflection a given surface can cause; a high albedo is typical of bright surfaces such as a mirror. We will come back to this factor shortly.) Approximately 21% of radiation is reflected back to space in this manner and clouds are by far the most important reflector.

Direct radiation *Solar radiation that flows directly to the surface of Earth and is absorbed.*

Absorption *The assimilation and conversion of solar radiation into another form of energy by a medium such as water vapor. In this process, the temperature of the absorbing medium is raised.*

Reflection *The process through which solar radiation is returned directly to space without being absorbed by Earth.*

Albedo *The reflectivity of features on the Earth's surface or in the atmosphere.*

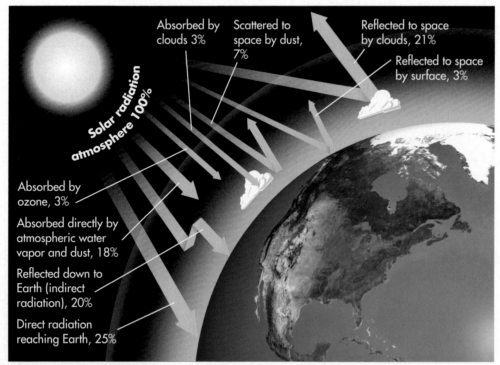

Figure 4.16 Interaction of solar radiation with various components of the atmosphere. Note that solar radiation is absorbed, stored, and reflected in many ways, and that about half the overall amount actually reaches the Earth's surface.

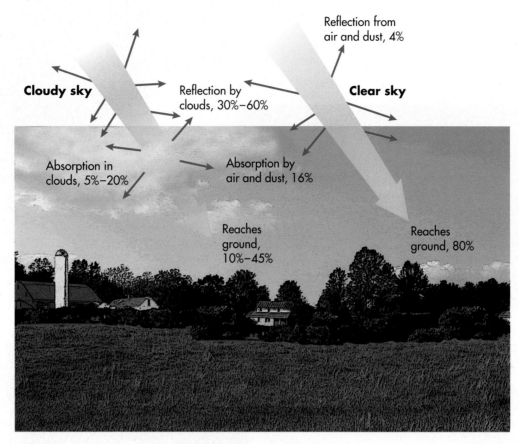

Reflection from
air and dust, 4%

Cloudy sky

Reflection by
clouds, 30%–60%

Clear sky

Absorption in
clouds, 5%–20%

Absorption by
air and dust, 16%

Reaches
ground,
10%–45%

Reaches
ground, 80%

Figure 4.17 Variation in the receipt of solar radiation based on cloud cover. Areas of dense cloud reflect more radiation than areas of clear sky.

In addition to the amount of radiation reflected, an additional component of incoming radiation is redirected within the atmosphere by a process called **scattering**. Although scattering is a process that cannot be directly observed, you can see the indirect effects very clearly in the apparent colors of the sky (Figure 4.18). Have you ever wondered why the sky is blue? The reason is that oxygen and nitrogen molecules cause solar radiation to bounce around in the air in a process called *Rayleigh scattering*. This effect is more pronounced during the middle part of the day when the Sun is high in the sky and solar radiation is streaming more or less directly toward you. The sky appears to be blue because the blue wavelength is preferentially scattered, whereas the longer wavelengths pass on through the atmosphere. Approximately 7% of this scattered radiation is redirected back into space.

You may also wonder why sunsets are so colorful, with the orange and red parts of the spectrum visible.

This change happens because solar radiation is streaming toward your eye from a much lower angle at dawn or dusk than when the Sun is high in the sky. As a result, the incoming radiation is passing through a thicker slice (when viewed horizontally) of the atmosphere to reach your line of sight. This thickened atmosphere causes the blue wavelengths to be scattered out entirely before they reach you, leaving the longer wavelengths (oranges and red) for you to observe.

As you can see, the effects of atmospheric reflection and scattering have a strong influence on the way you perceive Earth. The brightness of cloud tops and the colors of the sky are visible environmental indicators of these processes. A by-product of these interactions is that some reflected and scattered radiation is also redirected downward toward the surface of Earth as **indirect radiation**. Approximately 20% of solar radiation that reaches Earth does so in this indirect fashion.

Scattering *The redirection and deflection of solar radiation by atmospheric gases or particulates.*

Indirect radiation *Radiation that reaches Earth after it has been scattered or reflected.*

(a)

(b)

White light from Sun

Observer sees white Sun, blue sky at noon

North Pole

Direction of rotation

White light from Sun

Blue light scattered

Observer sees red sunset

Figure 4.18 Colors in the sky. (a) The sky is blue because oxygen and nitrogen molecules scatter light in the blue wavelength. (b) Geometric relationship of insolation, atmospheric thickness, and sky color. (c) Colors at sunset along the Atlantic coast at Rhode Island. Oranges and reds appear because the blue wavelength is completely scattered before it reaches your eye.

Interaction of Solar Radiation and the Earth's Surface

Let's now examine the approximately 45% of all solar energy (direct and indirect) that strikes the Earth's surface. What happens to this portion of incoming radiation is important because it influences variables such as temperature, atmospheric circulation, the density and kind of vegetation in a region, soils, and even where glaciers occur. Relationships such as these will be discussed in later chapters. In general, either of two things happens when solar radiation strikes the ground: (1) it is absorbed, or (2) it is reflected.

Absorbed Radiation Of the amount of solar energy that reaches the ground, 96% is absorbed by the various land and water bodies on the surface, thus heating Earth. One way this heat energy can be stored is in the form of **sensible heat**, which is heat that can be sensed by touching or feeling and can be measured by a thermometer. A second way it can be stored is as latent heat, when water from land and ocean/lake surfaces is transformed into water vapor within the atmosphere through evaporation. **Latent heat** is a form of heat that is hidden and cannot be measured with a thermometer

(c)

Sensible heat   *Heat that can be felt and measured with a thermometer.*

Latent heat   *Heat stored in molecular bonds that cannot be measured.*

because it goes into breaking bonds between molecules when a substance changes physical state, such as from a liquid to a gas.

Stored energy can be lost from Earth in several ways. Sensible heat can be transferred from the Earth's surface to the atmosphere through the process of convection. In this case, warm air rises upward from Earth and cooler air descends to replace it. Heat can also be removed through the process of **evaporation**, which is the change of liquid water to water vapor by absorption of heat. In addition, absorbed radiation can simply be re-radiated back into space as longwave radiation. When longwave radiation is released, it can either escape directly back into space, be absorbed by atmospheric CO_2 and water vapor, or be backscattered by dust particles.

Reflected Radiation Reflected radiation is energy that bounces off the Earth's surface in such a way that it does not provide heat. Overall, approximately 3% of incoming radiation is reflected in this way. The proportion of incoming radiation reflected by the surface is, just as for the atmosphere, dependent on the albedo of an object. Although you may not be directly aware

of surface albedo, you are probably indirectly aware of its effects. Almost everyone knows how uncomfortable it feels to wear a dark-colored shirt on a very hot day with bright sunshine. The reason for this is that dark objects absorb radiation and therefore increase in temperature. A dark asphalt highway, for example, reflects only about 5% to 10% of incoming solar energy. A snow-capped mountain peak, in contrast, reflects 80% to 95% of incoming solar radiation and thus absorbs little heat. Figures 4.19 and 4.20 show how albedo can influence the amount of energy reflection on different surfaces.

The amount of radiation reflected depends not only on albedo, but also on the Sun angle in the sky. Remember from Chapter 3 that the Sun is always at some angle in the sky relative to any location on the planet. This concept was illustrated in Figure 3.3, if you wish to review. With respect to incoming solar radiation, this angle is called the **angle of incidence**. Think of the angle of incidence as being similar to the angle at which you might throw a rock into the water. If you drop the rock straight down into the water, it is immediately absorbed by the water and sinks to the bottom. However, if you throw the rock at a shallower angle (relative to the water), it might skip across

Evaporation *The process by which atoms and molecules of liquid water gain sufficient energy to enter the gaseous phase.*

Angle of incidence *The angle at which the Sun strikes Earth at any given place and time.*

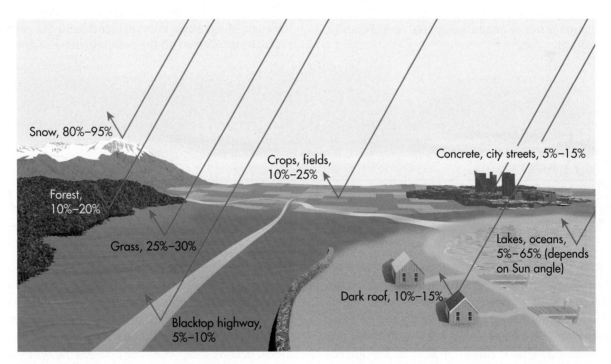

Figure 4.19 Variation of surface albedo. In general, albedo values are higher on brighter surfaces than on darker ones.

mean annual albedo

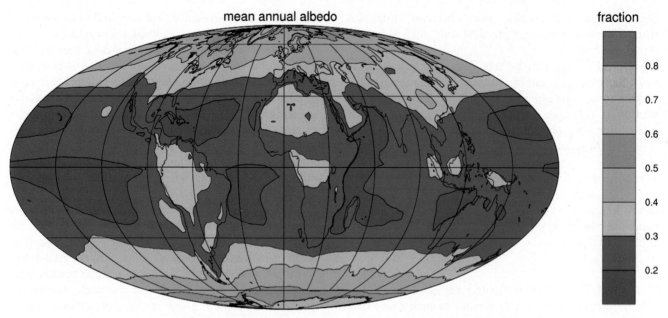

fraction

Figure 4.20 Mean annual albedo across the Earth's surface. The bar code at the right represents percentage albedo. Blue areas indicate zones of relatively low albedo and are concentrated in the oceans where radiation is absorbed. Polar ice caps, in contrast, have relatively high albedo, almost as much as 70%.

the water. The angle of incidence works in much the same fashion in that solar radiation from a high-angle Sun interacts directly with Earth; in addition, the Sun's rays are confined to a relatively small area and are thus intense at that point. When the Sun is lower in the sky, however, more radiation is reflected when it strikes Earth; also, the radiation is spread over a larger area. When this takes place, the radiation is not as intense compared to higher-angle Sun locations.

When the Sun is directly overhead, water has an albedo value of around 5%, but when the Sun is near the horizon, the albedo value is around 65% because radiation reflects off of the surface more readily. You can see this difference clearly in Figure 4.20, which shows that oceans at higher latitudes have a higher albedo than those at lower latitudes. This variability occurs because the angle of incidence is less at higher latitudes and more radiation is reflected off the oceans in those regions.

www.wiley.com/college/arbogast

The Angle of Incidence

With the concept of solar radiation and Earth interactions in mind, you can further explore this important concept by interacting with a simulation on the *GeoDiscoveries* website. Go to the website and select the module *The Angle of Incidence*. This module allows you to interact with the angle of incidence as it relates to latitude. The foundation of this simulation is Figure 3.3, which shows how the angle of incidence influences how much solar radiation is received at any given location. You will be able to choose

between several latitude locations in the Northern and Southern Hemispheres to see how the angle of incidence varies between locations. As you work through the various scenarios, notice not only how the flow of solar radiation changes, but also the direction from which it comes. You should observe a big difference between the Northern and Southern Hemispheres. Pay attention to these differences and relate them back to what you have learned about Earth–Sun geometry and seasons.

1. Once solar radiation reaches the atmosphere, it begins to flow along several different pathways. Some of it flows directly to the surface, whereas other parts are reflected off clouds or scattered by particulates.

2. Insolation that reaches the surface is either absorbed or reflected (a function of albedo).

3. Radiation that is absorbed is re-radiated as long-wave radiation.

4. The amount of energy reflected or absorbed depends in large part on the angle of incidence, which, in turn, varies by latitude and season.

The Global Radiation Budget

To fit our radiation discussion into a coherent model, think of Earth's energy flow as being analogous to money that flows in and out of your bank account. Sometimes you have a net surplus of cash because you have made a lot of money, whereas at other times you might even be a little in the hole because you spend more than you earn. Like your bank account, the Earth's **radiation budget** refers to the balance between incoming (shortwave) radiation and outgoing radiation, which is either re-radiated through reflecting and scattering processes described earlier or released from the Earth in longwave form. The difference between incoming and outgoing values is the **net radiation.** For Earth as a whole, the long-term radiation budget must be balanced; otherwise, Earth would become progressively warmer or cooler. This balance is achieved because incoming radiation flows are matched by outgoing flows.

Figure 4.21 illustrates the fundamental nature of the Earth's radiation budget by incorporating the data from Figure 4.16. To simplify the concept, remember that solar energy is in the form of shortwave radiation and that 69% of it is actually absorbed by some aspect of Earth, either by clouds, water, or the surface. The remaining 31% of shortwave energy is reflected directly back into space due to albedo or scattering. For the global radiation budget to remain in balance, therefore, Earth must emit all the absorbed energy back to space, which it does in longwave form. This release, plus the 31% of reflected shortwave radiation, keeps the system balanced. Earth's average temperature is moderated because greenhouse gases such as

---

Radiation budget *The overall balance between incoming and outgoing radiation on Earth.*

Net radiation *The difference between incoming and outgoing flows of radiation.*

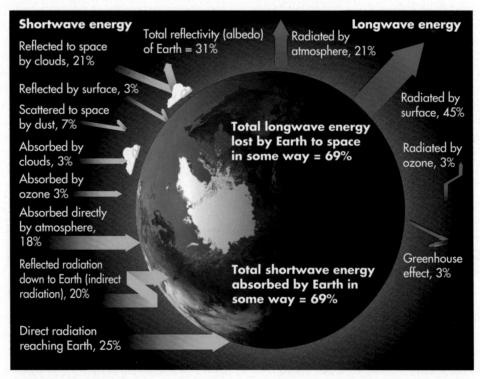

Figure 4.21 The global radiation budget. The difference between incoming (shortwave) radiation and outgoing (longwave) radiation must be balanced; otherwise, the temperature of Earth would either cool or warm dramatically.

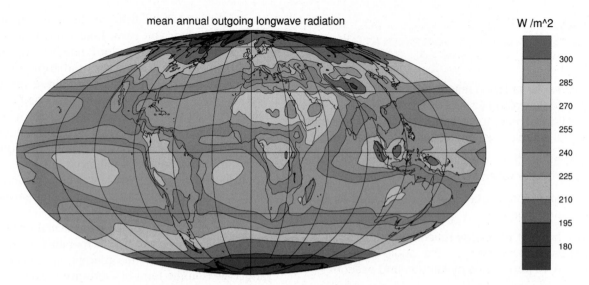

mean annual outgoing longwave radiation

W /m^2

	300
	285
	270
	255
	240
	225
	210
	195
	180

Figure 4.22 Mean annual outgoing longwave radiation on Earth in watts per cubic meter. The original source of this radiation was the Sun. The radiation was then absorbed and stored by the Earth until its release as longwave energy. Note the geography of this process, with most longwave radiation being released in the tropical latitudes.

CO_2 and water vapor trap a small amount of longwave radiation.

Aside from the amount of longwave energy that is trapped by the atmosphere, the bulk of it is lost in a way that balances the global radiation budget. This balance is accomplished in a variety of ways (see Figure 4.21). The majority (45%) of longwave energy is lost to space by direct radiation from the surface. This form of loss does not happen evenly across the Earth's surface, but rather has a distinct geographic pattern of its own, with lower latitudes emitting more energy than higher latitudes (Figure 4.22). This pattern makes sense, of course, because low latitudes receive more shortwave radiation than high latitudes. In addition to direct energy loss from the surface, 21% of longwave energy is lost directly by radiation from the atmosphere and another 3% is emitted by ozone.

Although the long-term energy budget for Earth balances on a global scale, it can vary markedly over the short term in different regions, due to the interaction of various environmental factors. Because of this variability, certain locations have a net surplus of radiation, whereas others have a net deficit. In addition, a particular place may have a net surplus of radiation during one time of year but a net deficit at another. These net differences are important because they are the driving mechanism behind climate and many physical processes, such as atmospheric circulation and evaporation. Four primary factors influence net radiation around the globe: (1) the Sun's angle of incidence, (2) latitude, (3) seasonality, and (4) length of day. Other secondary factors include the varying output of the Sun due to sunspots and solar flares, the elliptical nature of the Earth's orbit, changes in the thickness and properties of the atmosphere, and variability in the length of day. The best way to understand the changes in net radiation across the globe is by focusing on the four primary factors because they work

Geo Discoveries

www.wiley.com/college/arbogast

WILEY PLUS

The Global Energy Budget

After you have examined Figure 4.21, go to the *GeoDiscoveries* website and select the module *The Global Energy Budget*. This module allows you to better visualize the flow of solar radiation from the Sun to Earth. As you watch the animation, notice the various pathways that radiation takes as

it interacts with the atmosphere. In addition, try to integrate all the concepts so far in this chapter to solidify your understanding about the global energy budget. Once you complete the animation, be sure to answer the questions at the end of the module to test your understanding of this concept.

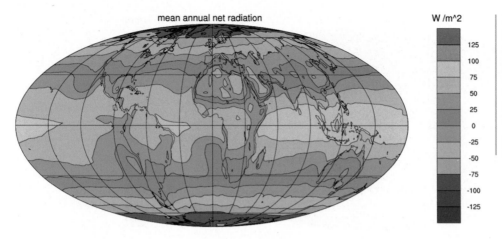

mean annual net radiation

W /m^2

	125
	100
	75
	50
	25
	0
	-25
	-50
	-75
	-100
	-125

Figure 4.23 Mean annual net radiation at the top of the atmosphere on Earth. Note the relationship between color and value (watts per square meter), which is indicated in the key to the right of the image. Low to middle latitudes have a net surplus of radiation, whereas high latitudes have a net deficit.

in a holistic way. This discussion will begin with the angle of incidence.

Remember that the angle at which solar radiation strikes Earth directly affects absorption and reflection. This angle is primarily a function of latitude, with low latitudes having high Sun angles and high latitudes having low Sun angles. Therefore, low latitudes receive high amounts of radiation and thus have a net annual surplus, whereas the high latitudes have a net deficit because they receive less direct radiation (Figure 4.23).

How does this variation in net radiation between latitudes influence natural processes on Earth? We will cover the significance of this spatial relationship in greater detail later in the text, but for now it is sufficient to say that the difference in net radiation is a primary driving force behind many atmospheric circulatory processes, especially large-scale wind patterns. Figure 4.24 shows how this transfer works.

What happens to net radiation patterns when we add the effect of the seasons to the mix of variables? Recall that this seasonal effect takes place because of the geometric relationship between Earth and the Sun, which results in changes in the angle of incidence over the course of the year. Remember that at any given point on Earth, the position of the Sun at solar noon migrates over the course of the year because of axial tilt and orbital position. This seasonal migration of the Sun, in turn, causes the subsolar point (on Earth) to change position with respect to latitude. This shift of the subsolar point north and south of the Equator has a major influence on the

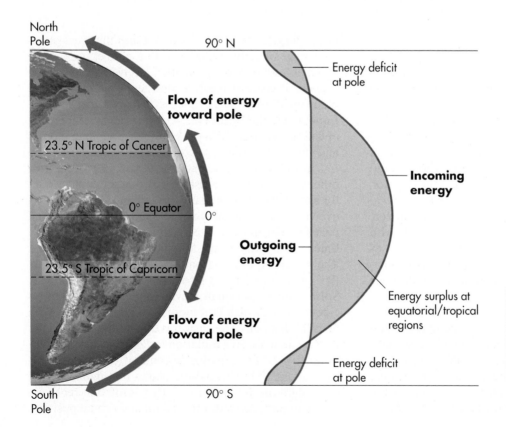

North Pole

90° N

Energy deficit at pole

Flow of energy toward pole

23.5° N Tropic of Cancer

Incoming energy

0° Equator

0°

Outgoing energy

23.5° S Tropic of Capricorn

Energy surplus at equatorial/tropical regions

Flow of energy toward pole

Energy deficit at pole

South Pole

90° S

Figure 4.24 Net radiation and the transfer of heat energy on Earth. Compare this diagram to Figure 4.23 and note where areas of net surpluses and deficits are found and how energy flows on the planet.

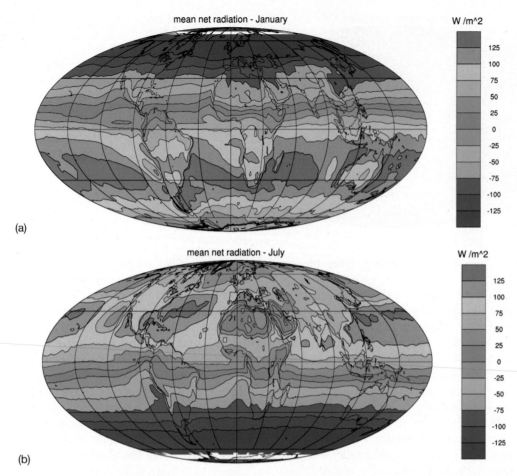

mean net radiation - January

W /m^2

125
100
75
50
25
0
-25
-50
-75
-100
-125

(a)

mean net radiation - July

W /m^2

125
100
75
50
25
0
-25
-50
-75
-100
-125

(b)

Figure 4.25 Average seasonal changes in net radiation on Earth. (a) Mean January net radiation. (b) Mean July net radiation. In both figures, reds indicate high values of net radiation, measured in watts per square meter, whereas blues equal low or deficit amounts.

seasonal distribution of net radiation, as you can see in Figure 4.25.

Figure 4.26 compares two curves of annual insolation by latitude—one with Earth's actual axial tilt (23.5°) and the other if the axis were not tilted. With a tilted Earth, you can see that annual radiation at low latitudes is about 40% greater than at high latitudes, but that high latitudes still receive a considerable amount. However, if the Earth's axis were perpendicular to the plane of the ecliptic, annual radiation would range from a high value at the Equator, where the Sun angle would consistently be 90°, to *zero* at the poles, where the Sun would always be at the horizon. You can see that axial tilt is an important variable with respect to daily insolation at various latitudes, and that as a result, the poles receive nearly half the amount of radiation as the Equator, even though they are in darkness for 6 months of the year.

A good way to conclude this discussion about the global radiation balance is to examine the effects when all the variables—angle of incidence, latitude, sesonality, and length of day—are considered together. Figure 4.27 graphically illustrates these combined effects on seasonal insolation at the top of the atmosphere at three places: the Equator (0°), 45° N, and the North Pole (90° N). Notice in this diagram that the North Pole has the greatest annual range of daily insolation, with a distinct peak at the Summer Solstice and no radiation received in winter. The Equator, in contrast, receives consistently high amounts of radiation all year. Given its location midway between the Equator and the North Pole, 45° N has a moderate range of daily insolation, with a peak during the summer months and lower values during winter.

In addition to these patterns, Figure 4.27 reveals two other interesting relationships. The first is that the Equator shows two peaks in daily insolation, rather than one, over the course of the year. The reason for these dual peaks is that the Sun is directly over the Equator two times during the year—specifically, at each Equinox. The second interesting pattern is that the North Pole (90° N) receives more insolation around the Summer Solstice than the Equator, even though the angle of incidence at very high latitudes is still relatively low. Why does this happen? The answer is that latitudes above the Arctic Circle receive radiation for 24 h because the Sun never sets. At the Equator, in contrast, day length is always about 12 h. This pattern shows the significance of day length on the amount of daily insolation received, which, in turn, depends on axial tilt and orbital position.

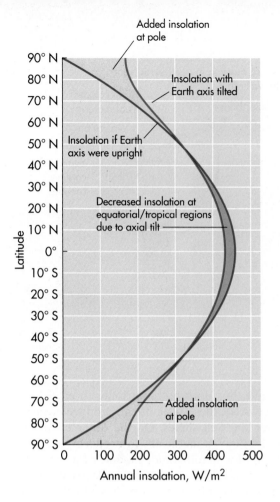

Figure 4.26 Variation in annual insolation (by latitude) on Earth. The red line shows the variability that exists as a function of the actual axial tilt. The blue line indicates what insolation would be if the axis were not tilted. Note the added insolation at high latitudes due to the tilt of Earth.

Figure 4.27 Daily insolation through the year at the top of the atmosphere at select latitudes in the Northern Hemisphere. Note how insolation changes at each of these locations on a seasonal basis and that the annual range increases at progressively higher latitudes.

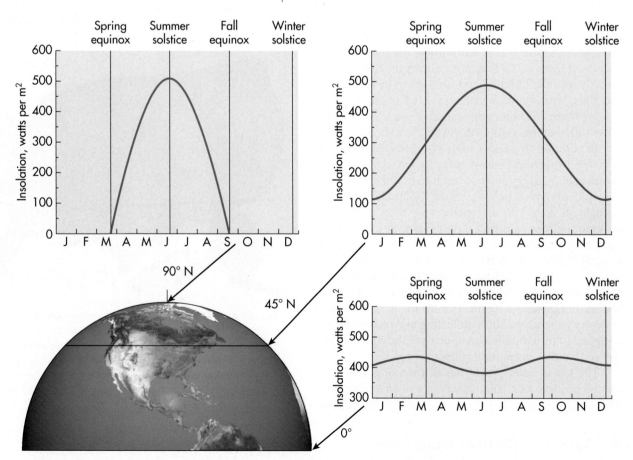

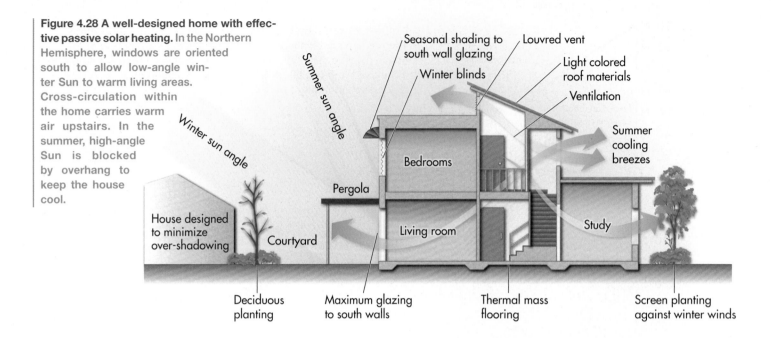

Figure 4.28 A well-designed home with effective passive solar heating. In the Northern Hemisphere, windows are oriented south to allow low-angle winter Sun to warm living areas. Cross-circulation within the home carries warm air upstairs. In the summer, high-angle Sun is blocked by overhang to keep the house cool.

Labels: Summer sun angle · Winter sun angle · Seasonal shading to south wall glazing · Louvred vent · Winter blinds · Light colored roof materials · Ventilation · Summer cooling breezes · Bedrooms · Pergola · House designed to minimize over-shadowing · Courtyard · Living room · Study · Deciduous planting · Maximum glazing to south walls · Thermal mass flooring · Screen planting against winter winds

Human Interactions: Solar Energy Production

Humans have attempted to harness solar energy for practical purposes for thousands of years. The Anasazi Indians of the U.S. Southwest, for example, built their cliff dwellings on the south-facing side of canyons so the Sun would shine into their homes and warm them during the winter. Since the oil price shocks of the 1970s, and to lessen our reliance on fossil fuels, more office buildings and homes today are designed with windows that face the Sun when it is at a low angle during the winter months (Figure 4.28). This angle allows the Sun to shine directly into the building and for heat to be trapped by the glass. You can experience this kind of *passive solar heating* when you enter your car after you leave the windows rolled up on a sunny day while you shop.

In addition to passive solar systems, efforts are under way to develop active solar systems that directly capture solar energy and convert it to useful outputs. For example, photovoltaic systems, or "solar cells," change sunlight directly into electricity; they tend to be used in remote places to power devices such as lighted road signs that are not connected to the electrical grid. On a much larger scale, solar power plants contain a number of mirrors, or "heliostats," that indirectly generate electricity because they redirect sunlight into a fluid that warms and produces steam. This steam, in turn, is used to power generators. These reflecting mirror systems are designed to turn to follow the Sun over the course of the day so that they constantly collect energy.

The use of solar power has distinct pros and cons. On the positive side, solar power is a clean and sustainable energy source that does not produce greenhouse emissions and thus does not contribute to global climate change. Also, these systems can produce electricity wherever the Sun shines. On the other hand, the current technology is inefficient in producing an adequate supply. Most photovoltaic modules, for example, convert only about 10% of sunlight to electricity. Solar power plants are inefficient too, which is why they typically contain dozens (or even hundreds) of collectors that funnel solar energy into a common focal point.

Perhaps the biggest problem associated with capturing solar energy is the simple fact that the

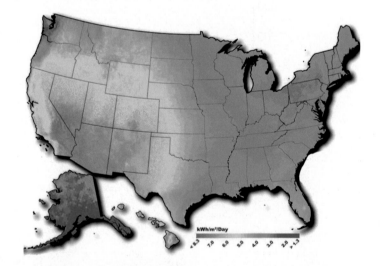

Figure 4.29 Map of solar energy potential in the United States. The Northeast is a poor place for solar energy production because it is often cloudy, especially in winter. In contrast, the Southwest has a very high potential for producing solar energy because it is a sunny desert environment. Map units are in number of kilowatt hours of production per square meter per day.

Sun does not shine everywhere on a consistent basis (Figure 4.29). The Great Lakes region of North America, for example, is cloudy much of the time and is thus a poor place to produce solar energy on a large scale. In contrast, the best place in the United States for solar energy production is the desert region of the Southwest where the Sun shines twice as much as in other parts of the country. A number of solar plants have been built here since the early 1980s, including nine Solar Energy Generating Systems (SEGS) that collectively have over 936,000 trough energy collectors (Figure 4.30) that cover more than 6.5 km². In contrast to less efficient flat mirrors, trough mirror sys-

Figure 4.30 The Solar Energy Generating System in California's Mojave Desert. This photograph shows a portion of the complex. Over 936,000 trough energy collectors are contained within the entire system, producing enough electricity for 500,000 people per year.

tems can focus the Sun at 30 to 60 times its normal intensity and contain synthetic oil that captures heat, reaching temperatures of 390°C (735°F). These plants can collectively produce 354 mW of electricity, enough to meet the power needs of approximately 500,000 people a year. Despite the many advances in solar energy production, it remains a supplementary supply when compared to energy produced by fossil fuels.

KEY CONCEPTS TO REMEMBER ABOUT THE GLOBAL RADIATION BUDGET

1. The global radiation budget refers to the balance between incoming and outgoing radiation on Earth.

2. Over the long term, the global radiation budget is balanced. However, even though the long-term radiation budget is balanced, a great deal of variability occurs across Earth as a whole.

3. In general, low latitudes have a net surplus of radiation, whereas high latitudes have a net deficit. This imbalance is important because it drives atmospheric circulatory processes.

4. Net radiation depends on a complex interaction of several variables, including angle of incidence, latitude, season, day length, and albedo.

5. In an effort to supplement current energy supplies, efforts are under way to increase the production of renewable solar energy. The major limitations are that production of solar energy with current technology is inefficient and the Sun does not consistently shine in many places.

VISUAL CONCEPT CHECK 4.3

Net radiation refers to the total amount of solar radiation received by Earth at any given place and time. This image shows the geography of mean net radiation during a particular time of year, with yellows and reddish hues representing high values in watts per square meter. Darker colors, in contrast, represent low values of net radiation. By observing this diagram, can you determine what time of year it must be? What must the Earth–Sun geometric relationship be like during this time of year?

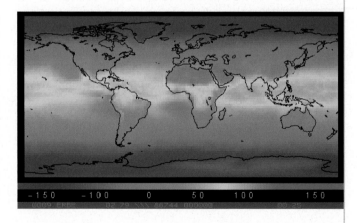

THE BIG PICTURE

(a)

(b)

Now that we have followed the pathways of solar radiation and how it interacts with Earth, subsequent chapters will examine how natural processes on the planet respond, such as global wind patterns and the geography of plants. In the next chapter, for example, we will explain how the heat generated by global radiation is measured in the form of temperature, and what causes temperature to fluctuate both at the surface and in the various layers of the atmosphere. For example, the coldest place on Earth is in Antarctica, whereas the hottest is the Sahara Desert. The next chapter will help explain why these geographic patterns are found.

SUMMARY OF KEY CONCEPTS

1. The atmosphere serves as a protective shield that filters the potentially deadly effects of ultraviolet radiation from the Sun, allowing mostly visible and infrared wavelengths to reach Earth.

2. The lower part of the atmosphere is warmed primarily due to the greenhouse effect. This warming occurs because variable gases in the atmosphere, such as carbon dioxide and water vapor (as well as others), trap longwave radiation that is emitted by Earth.

3. Insolation refers to the amount of solar radiation received at the top of the atmosphere. From this point, radiation follows several paths. Approximately 50% of all radiation reaches Earth, with some arriving directly and some indirectly. The remaining radiation is either absorbed in the atmosphere, scattered by dust, or reflected directly back into space.

4. Albedo refers to the reflectivity of surfaces on Earth. Snowy surfaces have the highest albedo, whereas darker surfaces such as roads and oceans have the lowest albedo. Overall, the Earth reflects about 31% of all incoming solar radiation through albedo.

5. The global radiation budget refers to the overall balance of incoming versus outgoing radiation. Earth receives shortwave radiation from the Sun. This radiation is either absorbed by the atmosphere and the various Earth surfaces, or reflected and scattered back to the surface or back into space. The radiation that reaches Earth's surface is absorbed and then reradiated as longwave radiation. The overall radiation budget must be in balance; otherwise, Earth would become progressively hotter or colder.

6. Efforts are currently under way to more systematically develop solar energy to supplement energy supplies. This supply of energy is renewable and environmentally friendly because its production does not result in CO_2 emissions. Unfortunately, current technology is inefficient and the production of solar energy is hampered in many places because the Sun shines inconsistently.

1. Discuss the various components of the electromagnetic spectrum.

2. Are wavelengths in the visible part of the electromagnetic spectrum short or long?

3. Why does the Sun emit shortwave radiation, whereas Earth emits longwave radiation?

4. What are the constant gases in the atmosphere and why are they important?

5. List the variable gases and describe why they change with respect to their proportion of the atmosphere.

6. Why is the sky blue during the day, but red and orange at dawn and dusk?

7. Compare and contrast the methods of heat transfer.

8. Describe some of the variables that influence the amount of radiation that is absorbed on Earth.

9. How does the angle of incidence vary by latitude?

10. If the axis of the Earth were perpendicular to the plane of ecliptic, how would net radiation change at high latitudes? Would these regions have a greater net surplus or net deficit of radiation?

11. Why does the zone of net radiation surplus migrate (on Earth) over the course of the year?

12. Describe the global radiation budget. What are some of the things that happen to solar radiation when it reaches Earth? Why is there a global balance in the radiation budget?

13. Describe the pros and cons of solar energy production. Why is it considered to be renewable and environmentally friendly? What are the major limitations to the production of solar energy?

ANSWERS TO VISUAL CONCEPT CHECKS

VISUAL CONCEPT CHECK 4.1

Solar energy is produced by the process of nuclear fusion within the Sun. In this fashion, hydrogen is converted to helium at very high temperatures and pressures. Solar radiation travels in the form of electromagnetic energy at 300,000 km (186,000 mi) per second. It reaches Earth in about 8 min.

VISUAL CONCEPT CHECK 4.2

This factory is producing a variety of particulates and variable gases that are streaming into the atmosphere. Some of these industrial by-products will cause solar radiation to be reflected, whereas others will cause absorption or scattering of electromagnetic radiation.

VISUAL CONCEPT CHECK 4.3

This map of net radiation reflects patterns observed at either Equinox. Note that the highest values of net radiation are received in the tropical regions, which reflect the location of the subsolar point and a maximum (90°) Sun angle. Such relationships can only occur along the Equator during the Spring and Fall Equinoxes.

CHAPTER FIVE

GLOBAL TEMPERATURE PATTERNS

In the previous chapter we examined the various ways that energy flows in the atmosphere. Now it is time to investigate its primary effects on atmospheric temperature patterns. The word *atmosphere* is derived from the Greek phrase meaning "sphere of air," which implies that the air is uniform in its consistency and character. In fact, the atmosphere consists of several layers, each having different characteristics and interactions with solar radiation. Each of these layers will be examined in this chapter, including how the temperature varies between and within them. Most of the discussion will focus on the interaction of the environmental variables that influence temperature in the lower part of the atmosphere and how these variables interact to form distinct geographic temperature patterns on Earth.

A profile of the atmosphere can be seen in this sunset image taken from the space shuttle. The deep red band near the horizon is probably caused by smoke from biomass burning at the surface. The purple layer marks the tropopause, which contains a layer of dust at an altitude of 14 km (~ 8.7 mi). Above the tropopause is the stratosphere, which appears pale in this image.

CHAPTER PREVIEW

Layered Structure of the Atmosphere

Surface and Air Temperatures
On the Web: *GeoDiscoveries Surface Temperature*
On the Web: *GeoDiscoveries Maritime vs. Continental Effect*
On the Web: *GeoDiscoveries Global Temperature Patterns*
On the Web: *GeoDiscoveries Temperature and Location*

The Big Picture

LEARNING OBJECTIVES

1. Compare and contrast the various layers of the atmosphere and the vertical change in temperature that occurs within them.

2. Describe the way that the various large-scale factors influence surface air temperature on Earth.

3. Explain the character of maritime and continental climates and why they differ from one another.

4. Discuss the urban heat island effect and how human interactions in these environments influence air temperature.

Layered Structure of the Atmosphere

The place to begin a discussion of global temperature patterns is in the atmosphere, which is the medium through which solar radiation flows. The atmosphere consists of several distinct layers, which differ in density and composition. Within the context of air temperature, the configuration of the atmosphere is important because of the way that solar energy moves through and interacts with the various layers. These interactions affect air temperature in both the horizontal and vertical dimensions of the atmosphere.

The atmosphere extends from a shallow depth within Earth (because soil contains air and water) to a height of about 480 km (300 mi). Most of the atmosphere's mass, however, lies below an altitude of 30 km

(18.6 mi). Although we usually think of the atmosphere as a uniform gaseous medium that envelops Earth, it actually contains four distinct layers: the troposphere, stratosphere, mesosphere, and thermosphere. These layers are distinguished by temperature and also by what elements they contain. Figure 5.1 illustrates these major layers and the temperature trends within them.

The Troposphere

The **troposphere** is the lowest layer of the atmosphere, extending from the surface to an average altitude of about 12 km (~7.5 mi). The name of this layer is derived from the Greek word *tropos* for "turning" or "mixing" because it is the most active zone of the atmosphere, with vigorously moving currents of air. Physical geographers and meteorologists are most interested in this atmospheric layer because it contains the vast majority

Troposphere   *The lowermost layer of the atmosphere, which lies between the Earth's surface and an altitude of about 12 km (~7.5 mi).*

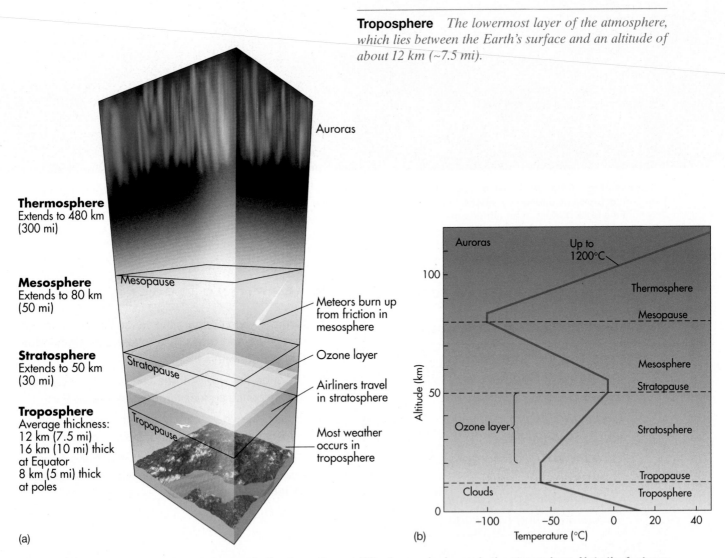

(a)

(b)

Figure 5.1 **Layers and temperature patterns in the atmosphere.** (a) The four major layers in the atmosphere. Note the features such as clouds and auroras that occur within specific layers. (b) Temperature changes with respect to altitude in each of the four major layers of the atmosphere.

of nonmarine living organisms and is the zone where most weather occurs, such as rain, wind, and snow. Much of this weather occurs because the troposphere contains most of the atmospheric water vapor and particulates.

The troposphere is warmed by longwave radiation emitted from Earth as part of the global energy balance discussed in Chapter 4. Because the source of heat in the troposphere is Earth, temperature decreases with increasing altitude (Figure 5.2). This change in temperature is known as the **environmental lapse rate**. As altitude (from the Earth) increases in the troposphere, the average temperature drops 6.4°C per 1000 m or 3.5°F per 1000 ft (a positive lapse rate). At the upper limit of the troposphere, which is called the **tropopause**, the temperature stops decreasing and begins to increase into the next layer of the atmosphere. The tropopause is found at the altitude where the air temperature is –57°C (–70°F), which again varies depending on latitude.

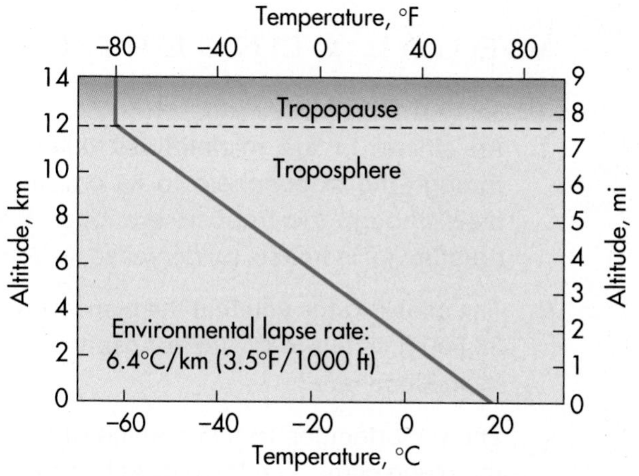

The thickness of the troposphere depends largely on temperatures at the Earth's surface; therefore, it varies with season and latitude, ranging in thickness from about 16 km (10 mi) at the Equator to approximately 8 km (5 mi) at the poles. Why is this so? Recall that the low latitudes have a net surplus of radiation because these regions receive more intense solar radiation. Thus, near the Equator, more radiation is transferred back to the atmosphere as longwave radiation, which heats the lower atmosphere. The higher temperature causes the atmospheric gases to expand, making the troposphere extend higher from the Earth's surface at the Equator than at the poles.

The Stratosphere

The layer of the atmosphere that lies immediately above the troposphere is the **stratosphere** (see Figure 5.1), which ranges in altitude between about 12 km and 50 km (between ~7.5 mi and 30 mi). This portion of the atmosphere derives its name from the Greek word *stratos*, which means "layer." The stratosphere is critically important to life on Earth because it contains the ozone layer, which occurs at an altitude between 20 km and 50 km (12 mi to 31 mi). The concentration of ozone in this portion of the atmosphere is about 10 parts per million by volume (ppmv), whereas it is only 0.04 ppmv in the troposphere.

As discussed in Chapter 4, the ozone layer filters UV radiation from the Sun and re-radiates it as infrared energy. Stratospheric temperature trends reflect this filtering and the overall thickness of the ozone layer. From

Figure 5.2 Environmental lapse rate in the troposphere. The average temperature decreases with increasing altitude in the troposphere until it reaches –57°C (–70°F).

the top of the tropopause to the base of the ozone layer, temperatures are consistently about –57°C (–70°F) (see Figure 5.1b). Above that altitude, however, temperatures increase with altitude because of the way ozone absorbs UV energy from the Sun. At the lower altitudes in the stratosphere, ozone absorbs UV radiation at wavelengths between 44 nm and 80 nm. It absorbs these wavelengths inefficiently, however. At higher altitudes in the stratosphere, ozone very efficiently absorbs UV at wavelengths between 200 nm and 350 nm. This difference in absorption efficiency is the reason why temperature increases with altitude in the stratosphere. The top of the stratosphere, or **stratopause**, is marked by the altitude where temperature stops increasing. At this altitude, the average temperature is about –5°C (–23°F).

In addition to its filtering effects, the stratosphere is also important in the context of human travel because it is the portion of the atmosphere in which commercial jets fly. The stratosphere is a perfect medium through which to fly jet aircraft because it contains very little water vapor and impurities; thus, pilots find relatively few clouds and good visibility. In addition, the air is relatively calm compared to the turbulent troposphere below because air in the stratosphere usually flows parallel to the surface of Earth. Understanding these kinds of facts makes flying more fun and can make your future trips more interesting.

Environmental lapse rate   *The decrease in temperature that generally occurs with respect to altitude in the troposphere. This rate is 6.4°C per kilometer or 3.5°F per 1000 ft (a negative lapse rate).*

Tropopause   *The top part of the troposphere, which is identified by where the air temperature is –57°C (–70°F).*

Stratosphere   *The layer of the atmosphere, between the troposphere and mesosphere, that ranges between about 12 km and 50 km (between ~7.5 mi and 31 mi) in altitude.*

Stratopause   *The upper boundary of the stratosphere where temperature reaches its highest point.*

The Mesosphere

The **mesosphere** is a layer of decreasing temperature that occurs from about 50 km to 80 km (30 mi to 50 mi) in altitude (see Figure 5.1); it is the coldest of the atmospheric layers. Its name comes from the Greek word *mesos*, which means "middle." Vertical temperature trends in the mesosphere have a positive lapse rate because temperature decreases with increasing distance from the ozone layer located in the stratosphere below. The altitude at which temperature stops decreasing is known as the **mesopause** and is the upper boundary of the mesosphere. At this altitude, air temperature is about –100°C (–148°F). In the mesosphere, solar radiation reduces gas molecules to individual electrically charged particles called ions. This deep layer of charged particles can disrupt communications between astronauts and ground control and interfere with various satellite communications, such as transmission of television signals.

The mesosphere can indirectly be seen, particularly at night, because most meteors burn up in this layer as they fall through the atmosphere. Most of these *shooting stars* are about the size of one sand grain and are destroyed because they collide with billions of ions and gas particles as they fall through the mesosphere. These collisions create sufficient heat to burn the tiny rock fragments, creating a short-lived streaking path, long before they strike the ground. Occasionally, the largest rock fragments reach the Earth's surface. If they do, they are called *meteorites*.

The Thermosphere

The **thermosphere** is the upper layer of the atmosphere and occurs between about 80 km and 480 km (50 mi to 300 mi) in altitude. In this portion of the atmosphere, atmospheric gases are sorted into a variety of sublayers based on their molecular mass. Oxygen molecules are few and literally many kilometers (miles) apart from one another. In fact, they are so widely spaced that the boundary with space is very diffuse and thus difficult to precisely determine.

The name of this atmospheric layer is derived from the Greek *thermo*, which implies "heat," and makes perfect sense if you look at the temperature patterns of the thermosphere in Figure 5.1b. Notice that they increase drastically about 10 km (6 mi) above the mesopause, ultimately reaching 1200°C (2200°F) and higher. These high temperatures occur because intense solar radiation interacts with the upper part of the atmosphere, causing the few oxygen molecules present to vibrate at tremendously high speeds, which creates **kinetic energy**. This form of energy is specifi-

---

Mesosphere *A layer of decreasing temperature in the atmosphere that occurs from about 50 km to 80 km (~30 mi to 50 mi) in altitude.*

Mesopause *The upper boundary of the mesosphere where temperature reaches its lowest point.*

Thermosphere *The upper layer of the atmosphere, which occurs between about 80 km and 480 km (~50 mi and 300 mi) in altitude.*

Kinetic energy *The energy of motion in a body, measured as temperature, that is derived from movement of molecules within the body.*

cally contained within the molecules by virtue of their spatial relationship to other oxygen molecules. The thermosphere is important for human communications because it enables radiowaves from one location at the surface to bounce off and be received at locations beyond the horizon.

Despite the very high temperatures of the thermosphere, this atmospheric layer would not feel "hot" to you in the same way you would feel heat on the Earth's surface. The reason for this apparent oddity is that individual oxygen molecules in the thermosphere are so far apart from one another. Because of these great distances, the molecules hardly ever come into contact with each other, which means that very little heat is transferred from one to another. At lower levels of the atmosphere, in contrast, temperatures are felt to be genuinely higher because trillions of molecules are constantly colliding and a great deal of heat is exchanged.

Surface and Air Temperatures

Most of the atmospheric behavior of interest to geographers occurs near the Earth's surface. Of particular significance are global surface and air temperatures, which measure the amount of sensible heat at the surface or in the atmosphere. Atmospheric temperature is a measure of the kinetic energy contained within a unit of geographical space within the air. Surface temperature, in contrast, is a measure of the kinetic energy contained in a region very close to the Earth's surface. Thus, changes in surface temperature measure the ebb and flow of energy at ground level. As noted in Chapter 4, this variation in energy on Earth depends largely on net radiation. When a net surplus of radiation occurs, temperature increases because the surface absorbs more radiant energy than it is emitting in longwave form. Conversely, temperature decreases with a net deficit of radiation because the surface emits more energy than it absorbs. Recall from Chapter 4 that surface temperature can also change through the processes of conduction, evaporation, and convection.

In contrast to surface temperature, the term *air temperature* refers to the degree of warmth of a portion of the atmosphere. You are no doubt aware of air temperature on a hot sunny day or cold winter morning. The standard altitude at which air temperature is measured is usually about 1.2 m (4 ft) above the ground surface. Although air temperature usually differs from surface temperature, the amount of heat energy in the ground influences the temperature of the air above it. As you know, air temperature is measured with a thermometer, which is traditionally a hollow glass tube containing a liquid, often mercury, that expands or contracts depending on the amount of energy present. More recently, digital thermometers have become common, relying on small electrical devices that sense temperature.

Three temperature scales are used around the world (Figure 5.3). You are probably most familiar with the Fahrenheit scale, which was named after the German physicist Gabriel Daniel Fahrenheit, who devised the scale in the 18th century. This scale is used by the United States National Weather Service and American news media to report temperature. The common reference temperatures used in the Fahrenheit scale are the freezing and boiling points of water, defined as 32°F and 212°F, respectively.

Another temperature scale that you may be somewhat familiar with is the Celsius scale, named after Anders Celsius, the 18th-century Swedish astronomer who devised it. The Celsius scale is part of the International System of Measurement (SI) because it is a decimal scale, with 0°C and 100°C being the freezing and boiling points of water, respectively. The vast majority of the world uses the Celsius scale, including scientists in the United States, which is why measurement units in this text are first presented in this way. Because most of the world uses the Celsius scale, the stated goal of the U.S. government is to one day fully convert to this form of measurement rather than the Fahrenheit scale.

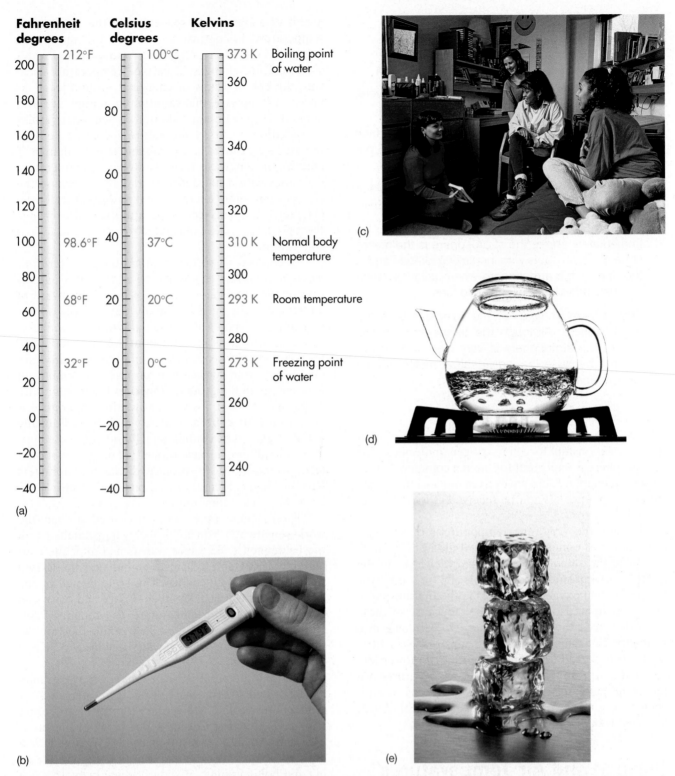

Fahrenheit degrees

- 212°F
- 200
- 180
- 160
- 140
- 120
- 100 — 98.6°F
- 80
- 68°F
- 60
- 40
- 32°F
- 20
- 0
- −20
- −40

(a)

Celsius degrees

- 100°C
- 80
- 60
- 40 — 37°C
- 20 — 20°C
- 0 — 0°C
- −20
- −40

Kelvins

- 373 K — Boiling point of water
- 360
- 340
- 320
- 310 K — Normal body temperature
- 300
- 293 K — Room temperature
- 280
- 273 K — Freezing point of water
- 260
- 240

(b)

(c)

(d)

(e)

Figure 5.3 Measures of temperature. (a) The Fahrenheit, Celsius, and Kelvin temperature scales. The size of one degree on the Celsius scale is the same as one kelvin; a degree on the Fahrenheit scale is smaller. (Note that units on the Kelvin scale are not called degrees, but rather kelvins.) (b) Temperature is measured with a thermometer. Until recently, most thermometers were filled with alcohol or mercury, which rose or fell within a glass tube if warmed or cooled, respectively. Today, most thermometers contain temperature-sensitive electrodes and display information digitally. (c) For most people a comfortable room temperature is about 21°C to 23°C (69°F to 73°F; 294 K to 296 K). (d) Water boils at a temperature of 100°C (212°F; 373 K). (e) The freezing/melting point of water is 0°C (32°F; 273 K).

AURORAS

Some of the most beautiful sights on Earth can usually be seen only at the high latitudes, around 60° and higher. These sights are called auroras, specifically the *Aurora Borealis* (Northern Lights) and *Aurora Australis* (Southern Lights). They appear as sheets of color shimmering high in the atmosphere. What are they?

The Sun emits more than just electromagnetic radiation; it also emits streams of high-energy particles. These particles are often referred to as the "solar wind" and extend to the farthest reaches of the solar system. When these particles reach Earth, they encounter the planet's magnetic field, which surrounds the planet high above the atmosphere. The charged particles are trapped within the magnetic field, especially within the two regions called the Van Allen belts, and do not reach Earth. However, the Van Allen belts bend down close to Earth at its poles, and the Sun's charged particles interact with atmospheric gases high in the thermosphere. The result of the interaction is the colorful glow of gases we see as auroras.

Auroras have been photographed from the space shuttle, where they can be seen following the Earth's magnetic field lines. In fact, space probes to Jupiter and Saturn have shown that auroras occur on those planets as well, again following the magnetic field lines.

(a)

| Colorful Aurora Borealis near Fairbanks, Alaska.

The Earth's magnetic field extends outward to several hundred thousand kilometers. However, it bends into the upper atmosphere toward the North and South Poles.

(b)

(c)

Aurora Australis, seen from the space shuttle in May 1991, follows a line of the Earth's magnetic field.

The conversions required to convert from Fahrenheit to Celsius and from Celsius to Fahrenheit are, respectively,

$$F = 9/5C + 32°$$
$$C = 5/9 \ (F - 32°)$$

Thus, 1°C equals 1.8°F and 1°F equals 0.56°C.

The third temperature scale in wide use is the Kelvin scale, named for the 19th-century British physicist William Thomson, Lord Kelvin. This scale is used in a great deal of scientific research because it measures absolute temperature, with absolute zero theoretically being the point where an object has no measurable temperature. On the Celsius scale, absolute zero is −273°C. The Kelvin scale has no negative values as the other scales do, but it is similar to the Celsius scale because it maintains a 100° temperature range between the boiling and freezing points of water. In other words, 1 K = 1°C. Given that the Kelvin scale is used by climatologists and meteorologists only for more advanced work, it will not be used further in this book. For reference purposes, however, the conversions between Kelvin and Celsius scales are

$$C° = K - 273$$
$$K = C° + 273$$

Human Interactions: Calculating the Heat Index and Wind Chill

Have you ever noticed that the air occasionally seems warmer or colder than the temperature that is given? This kind of variation occurs when additional environmental factors, such as wind speed and the amount of atmospheric water vapor, come into play. The combined impact of these factors with air temperature has implications for human comfort because they can make the air feel much warmer or colder than it really is.

The measures of human comfort that we use are the *wind chill index* in winter and the *heat index* in summer. The wind chill index is calculated using a variety of parameters, such as actual air temperature, wind speed, average face height, and components of modern heat loss theory. The wind chill index chart in Table 5.1 presents temperature data in the Fahrenheit scale and wind speed in miles per hour. In contrast to wind chill, the heat index measures the apparent temperature based on the combined variables of actual air temperature and relative humidity. (As we will discuss in Chapter 7, relative humidity is the ratio of the specific amount of vapor relative to the amount the air could hold at a given temperature.) The resulting heat index chart is shown in Table 5.2 and also uses the Fahrenheit temperature scale.

| TABLE 5.1 | The Wind Chill Index |

		Temperature (°F)												
W		40	35	30	25	20	15	10	5	0	−5	−10	−15	−20	−25
i	5	36	31	25	19	13	7	1	−5	−11	−16	−22	−28	−34	−40
n	10	34	27	21	15	9	3	−4	−10	−16	−22	−28	−35	−41	−47
d	15	32	25	19	13	6	0	−7	−13	−19	−26	−32	−39	−45	−53
	20	30	24	17	11	4	−2	−9	−15	−22	−29	−35	−42	−48	−55
s	25	29	23	16	9	3	−4	−11	−17	−24	−31	−37	−44	−51	−58
p	30	28	22	15	8	1	−5	−12	−19	−26	−33	−39	−46	−53	−60
e	35	28	21	14	7	0	−7	−14	−21	−27	−34	−41	−48	−55	−62
e	40	27	20	13	6	−1	−8	−15	−22	−29	−36	−43	−50	−57	−64
d	45	26	19	12	5	−2	−9	−16	−23	−30	−37	−44	−51	−59	−65
(mph)	50	26	19	12	4	−3	−10	−17	−24	−31	−38	−45	−52	−60	−67
	55	25	18	11	4	−3	−11	−18	−25	−32	−39	−46	−54	−61	−68

Frostbite occurs in 10 minutes or less.
Source: NOAA.

TABLE 5.2 The Heat Index

Relative Humidity (%)

Air Temperature (°F)	20	25	30	35	40	45	50	55	60	65	70	75	80	85	90	95	100
125	141																
120	130	139	148														
115	120	127	135	143	151												
110	112	117	123	130	137	143	150										
105	105	109	113	118	123	129	135	142	149								
100	99	101	104	107	110	115	120	126	132	138	144						
95	93	94	96	98	101	104	107	110	114	119	124	130	136				
90	87	88	90	91	93	95	96	98	100	102	106	109	113	117	122		
85	82	83	84	85	86	87	88	89	90	91	93	95	97	99	102	105	106
80	77	77	78	79	79	80	81	81	82	83	85	86	86	87	88	89	91
75	72	72	73	73	74	74	75	75	76	76	77	77	78	78	79	79	80
70	66	66	67	67	68	68	69	69	70	70	70	70	71	71	71	71	72

Celsius	Fahrenheit	Impact on Humans
27°C–32°C	80°F–90°F	Caution: Fatigue is possible with continued exposure.
32°C–41°C	90°F–105°F	Extreme Caution: Possible sunstroke, heat cramps, and exhaustion.
41°C–54°C	105°F–125°F	Danger: Likely occurrence of sunstroke and heat cramps.
54°C	above 130°F	Extreme Danger: Heat stroke or sunstroke likely.

Source: NOAA.

Large-Scale Geographic Factors That Influence Air Temperature

A common theme in this text is the holistic interaction of geographical variables. These interrelationships are especially important for understanding how temperature varies across Earth, which incorporates some of the concepts covered in Chapters 3 and 4. Some of these factors may be intuitive for you at this stage, but it is nevertheless a good time to review them and their relationships. Later in this section we will look at some factors that can influence temperature at a local scale. For now, however, let's focus on three large-scale factors that influence temperature across Earth, no matter the location.

Latitude Recall from Chapter 4 that the shape of Earth causes rays hitting the surface to differ in the area they cover, according to the latitude at which they strike. In other words, differences in the angle of incidence cause the same amount of energy to be directed at a smaller or larger area on the Earth's surface. When a larger area is covered at a lower incidence angle, as occurs at high latitudes, it results in less energy per unit area on that surface compared to a smaller area struck at a higher incidence angle, like what occurs at low latitudes. This difference is most pronounced when comparing high and low latitudes and results in distinct temperature differences between two such regions (see Figure 4.24).

Seasons and Length of Day As discussed in earlier chapters, Earth's axial tilt causes seasonal migration of the subsolar point because the Northern and Southern Hemispheres are tilted either toward or away from the Sun, depending on the time of year. This migration of the subsolar point occurs only in the tropics, but results in dramatic changes in the angle of incidence at all latitudes, which, in turn, influences net radiation. If you need to review this concept again, examine Figures 4.24 and 4.26. A secondary impact of axial tilt and seasons is that length of day can fluctuate a great deal, depending on latitude. Longer days mean that more radiation is received, which, in turn, also influences temperature. Think of your own experience regarding the temperatures where you live during winter and summer.

Time of Day As the day progresses from the morning sunrise, the Sun arcs westward across the sky. Recall from Chapter 3 that the Sun is at its highest point, resulting in the day's most intense radiation, at solar noon. Temperature usually follows the Sun in that it increases as the Sun rises and decreases when it sets. The relationship is not quite that simple, however, because a **temporal lag** occurs between the highest Sun angle and the warmest temperature of the day. In other words, the warmest part of the day occurs later in the day than the peak period of insolation and net radiation.

A good way to see how these variables affect one another is in graphical form. For example, Figure 5.4 shows how season and time of day are interrelated with respect to temperature. These diagrams show three sets of data from a typical observing station at 40° to 45° N latitude in the interior of North America: (a) insolation, (b) net radiation, and (c) air temperature. Parameter values, such as the amount of net radiation, are presented on the vertical (or "y") axis of the respective diagrams, whereas time of day ranges across the horizontal (or "x") axis.

Notice in Figure 5.4 that each of the measured variables at the observation station varies both by season and time of day. With respect to insolation (Figure 5.4a), the amount of insolation on the June solstice is greater than during the December solstice. This seasonal variation in insolation results in a greater surplus (above the bold horizontal line) of net radiation during the summer months than in the winter (Figure 5.4b). Notice also that the number of hours of daily insolation and net radiation also vary per day. During the summer, for example, insolation is received for 16 h of the day, whereas during the winter it is only 8 h. This decrease occurs because Earth's orbital position has changed. As a result of this seasonal insolation variation, the time of a net deficit of radiation (below the bold horizontal line) increases from about 8 h in summer to 16 h in winter.

To see how annual insolation/net radiation patterns influence temperature, study Figure 5.4c. As you can imagine, temperature is related to insolation and

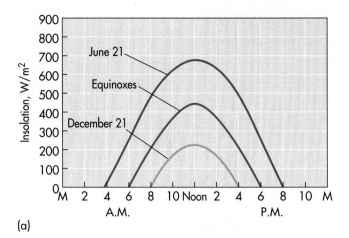

(a)

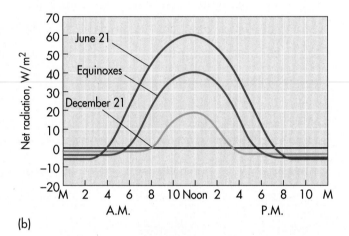

(b)

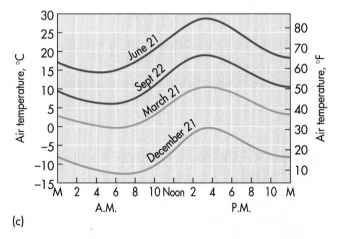

(c)

Figure 5.4 Interaction of (a) insolation, (b) net radiation, and (c) air temperature. These graphs were made from data taken at a midlatitude location in the Northern Hemisphere. Notice the close relationship among these three variables. Also note that a distinct lag occurs with respect to the temperature response.

Temporal lag *The difference in time between two events, such as when peak insolation and temperature occur.*

net radiation in that maxima and minima of each factor occur at about the same time. The pattern is predictable from a seasonal perspective in that temperatures are warmer during the summer and colder during the winter. Examining Figures 5.4a and 5.4b shows a clear correlation between temperature, insolation, and net radiation. Notice, however, that the maximum and minimum temperatures occur at different times than insolation/net radiation fluctuations on a daily basis. In other words, a temporal lag occurs. With respect to maximum temperature, this lag exists because the peak release of longwave energy by Earth occurs sometime after the insolation peak. Similarly, the minimum temperature occurs only after all the stored heat from the day is released.

Local Factors That Influence Air Temperature

In the previous discussion we examined the important factors that influence air temperature on a large scale across Earth. Besides these factors, additional, smaller-scale factors can influence air temperature in specific places. This section of the chapter focuses on some of these kinds of geographic patterns.

Maritime vs. Continental Locations Although the radiation budget is the largest influence on air and surface temperatures, other factors yet to be discussed can play an important role. A good example of a local or regional variable that influences temperature is the **maritime vs.**

Maritime vs. continental effect *The difference in annual and daily temperature that exists between coastal locations and those that are surrounded by large bodies of land.*

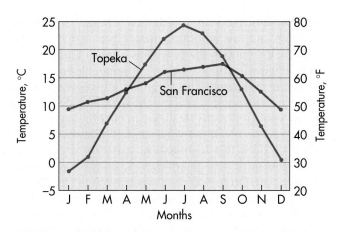

Figure 5.5 Annual temperature data from Topeka, Kansas, and San Francisco, California. A distinct peak in temperature occurs in Topeka during the summer, whereas San Francisco is comparatively moderate all year round.

continental effect. By definition, **maritime** places are located within or near a very large body of water, such as San Francisco, California, which is on the coast of the Pacific Ocean. In contrast, **continental** localities are surrounded by large landmasses, such as Topeka, Kansas, which is on the eastern edge of the Great Plains.

If you compare Topeka and San Francisco, observe that the annual temperature curves vary significantly (Figure 5.5). Notice that the San Francisco curve does not have a steep peak, but rather changes gradually over

---

Maritime *A place that is close to a large body of water that moderates temperature.*

Continental *A place that is surrounded by a large body of land and that experiences a large annual range of temperature.*

 www.wiley.com/college/arbogast

Surface Temperature

Here is a chance to examine the factors that affect temperature more closely. Go to the *GeoDiscoveries* website and select the module *Surface Temperature*. This module allows you to observe how the variables insolation, net radiation, and time of day influence temperature. The focus of this animation is a hypothetical location at 45° N latitude, one that lies deep within a continent. As you watch this animation, observe how

air temperature near the Earth's surface changes over the course of the day on June 21, which is the Summer Solstice in the Northern Hemisphere. In particular, notice the lag that exists with respect to increases and decreases in insolation and temperature over the course of the day. After you complete the animation, be sure to answer the questions at the end of the module to test your understanding of this concept.

the course of the year. In addition, the range between the high and low monthly temperatures is narrow. At Topeka, in contrast, the curve has a much higher peak with a more distinct change between seasons and a greater overall range. This variability exists even though both places are located at about the same latitude (38° N), which would (correctly) lead you to believe that each locality receives about the same amount of insolation.

Why does this temperature variation occur? The main reason is the maritime vs. continental effect (Figure 5.6). Large bodies of water, such as the Pacific Ocean, store tremendous amounts of thermal energy, primarily because water has a high specific heat; that is, water absorbs a relatively large amount of energy before its temperature rises. In addition, solar radiation can penetrate to great depths in the ocean, heating water below the surface. The ocean currents constantly mix this warm water with the cooler water not exposed to sunlight. As a result, large water bodies maintain a more or less constant temperature for most of the year (Figure 5.7). This effect can be additionally enhanced if the ocean current that flows along the coast is a cold or warm current. Such currents will be described further in Chapter 6.

As water evaporates from the ocean, energy is transferred to the atmosphere in the form of latent heat that, in turn, moderates the air temperature of coastal places such as San Francisco. This moderating effect is magnified on the west coast of North America because the prevailing winds are westerly; that is, they blow from west to east. This circulatory pattern will be discussed in more detail in Chapter 6, but for now it is sufficient for you to know that these westerly winds blow relatively mild, moist air off the ocean some distance into the interior. In this way, San Francisco maintains a moderate air temperature all year round.

At continental localities like Topeka, however, the surrounding landmass does not store as much thermal energy as oceans do, largely because of their low specific heat; that is, the land temperature starts to rise after absorbing relatively small amounts of energy. In addition, radiation does not penetrate to great depths in landmasses. As a result, continental localities exhibit dramatic annual temperature variations compared to maritime places. This pattern not only exists over the course of the year, but can also be seen in the daily cycle of temperature (Figure 5.8).

Human Interactions: The Urban Effect on Temperature In addition to the maritime vs. continental effect, another factor that influences temperature on a local scale is the presence of large cities. Cities are growing all around the world in response to increased human population and migrations to these locations. As cities grow, more of the natural landscape is being developed, and this results in new buildings, homes, and

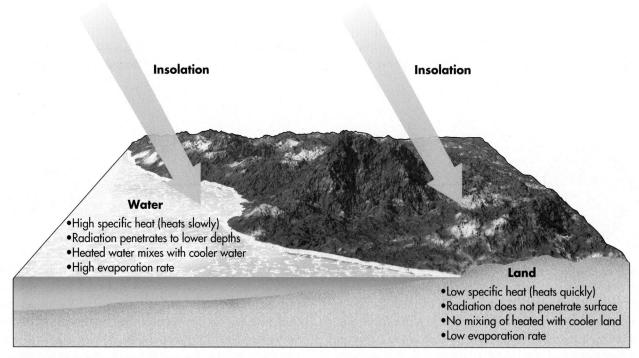

Figure 5.6 Maritime vs. continental contrasts. Several factors explain why temperature differences occur between coastal places and locations deep within continents. Note the different ways in which insolation interacts with land and water.

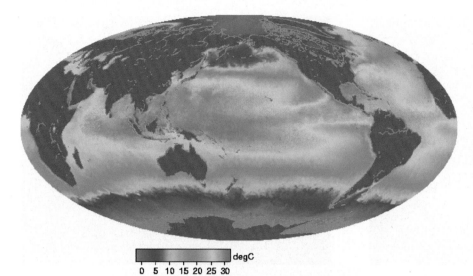

degC
0 5 10 15 20 25 30

Figure 5.7 Global sea surface temp–eratures during June and December 2000. Although the seasons differ dramatically between the images, notice the similarity in the geographic pattern. For example, the sea surface temperature in the eastern Pacific shows a small change between the months. Temperatures are in the Celsius scale and are coded by color, as indicated by the horizontal bar between the maps.

parking lots (Figure 5.9). In this process, vegetation is removed and soils are covered by pavement or other structures. This human-induced change in land cover can cause a measurable temperature increase in and around a city through a variety of ways, resulting in a distinct geographic feature known as an **urban heat island** where temperatures may be as much as 3°C to 4°C (6°F to 8°F) warmer (or more) than the surrounding countryside.

Urban environments tend to be warmer than rural areas for several reasons. One of the most important reasons for this effect is that urban surfaces consist of metal, glass, asphalt, and concrete, whereas rural areas are covered with soil in which forest and grass grow. As a result, urban surfaces tend to be darker than rural surfaces, which leads to greater net radiation in cities

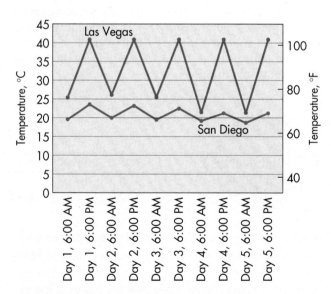

Figure 5.8 Daily cycle of temperature at San Diego, California, and Las Vegas, Nevada, from July 1 to July 5, 2003. San Diego is a maritime city, whereas Las Vegas is located deep within the desert. Notice how pronounced the daily cycle of temperature is in Las Vegas compared to San Diego.

Urban heat island   *The relatively warm temperatures associated with cities that occur because paved surfaces and urban structures absorb and release radiation differently than the surrounding countryside.*

Surface and Air Temperatures • 103

(a)   (b)

Figure 5.9 Urban vs. rural landscapes. The different characteristics of city and rural surfaces can significantly influence temperature. (a) Rural surfaces such as this Scottish landscape are cooler because less radiation strikes the ground due to tree cover, and more water is found in soil and on vegetation surfaces. (b) Urban surfaces, in contrast, receive more direct radiation and absorb more radiation. This image shows a portion of the cityscape in Tokyo, Japan.

because albedo is lower. In the context of this higher net radiation, up to 70% of net radiation in cities is converted to sensible heat, much more so than in the countryside. Urban surfaces also conduct more energy than rural soils and are thus warmer. This effect is magnified by the fact that forest canopies in rural areas shade the ground beneath them, keeping them relatively cool. In urban environments, on the other hand, solar radiation strikes the surface directly and is absorbed, later to be released as warming longwave radiation. This re-radiation tends to be trapped for longer periods of time in cities, further warming them, because the irregular geometry of city buildings limits wind speed by about 20% to 30% over the course of the year. This decreased flow of air reduces the loss of heat that would otherwise occur.

Another reason why cities are warmer than rural places is related to the way water is stored and moves in the two areas. In rural areas where soils cover the landscape, rainfall and snowmelt gradually soak into the ground. As described earlier, water has a modifying effect on temperature, which results in soils being cooler than urban surfaces. The presence of water in rural soils also reduces the conductivity of those surfaces, keeping them relatively cool. In the context of water absorption, it is useful to think of urban surfaces as being sealed because they are paved or covered with buildings. In the core of cities, for example, as much as 75% of surfaces are covered in this way. The effect of this change in land cover on the way water is stored and moves is significant because water does not absorb into the ground in cities like it does in the countryside. Instead, precipitation rapidly runs across these surfaces and into storm

 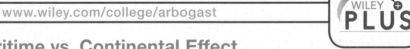

www.wiley.com/college/arbogast

WILEY PLUS

Maritime vs. Continental Effect

To see how the martime vs. continental effect occurs, go to the *GeoDiscoveries* website and select the module *Maritime vs. Continental Effect.* In this animation, you can observe how temperature varies over the course of the day at Yuma, Arizona, and San Diego, California. Although San Diego and Yuma are both located at about the same latitude (~33° N), and thus receive about the same amount of daily

insolation, the range in daily temperature differs dramatically between the two places. Watch how air temperature varies between these two places over the course of the day and year and relate what you see with the previous discussion in this text. After you complete the animation, be sure to answer the questions at the end of the module to test your understanding of this concept.

One of the most beautiful cities in North America is Vancouver, British Columbia. This Canadian city is particularly scenic because it lies between the Haro Strait, which is connected to the Pacific Ocean to the west, and the Cascade Mountains to the east. Although Vancouver lies at a fairly high latitude (49° N), it has a very moderate climate with average high temperatures that range from 22°C (71°F) in July to 6°C (43°F) in December. Which one of the following reasons accounts for this temperature pattern?

a) Vancouver is a continental location.
b) Vancouver lies next to a large body of water that has a consistent annual temperature.
c) The wind in Vancouver generally flows from east to west.
d) The Pacific Ocean has a wide range of annual temperature.

drains that carry it away. Thus, the modifying effect that water has on the temperature of soils in rural areas does not exist in cities, causing them to be warmer.

Yet another reason why cities are warmer than rural areas is the impact that human activities have on urban environments. In hot summer months, the production of electricity to cool homes and businesses releases a great deal of energy, perhaps as much as 25% to 50% of insolation. During the winter months, in contrast, a great deal of sensible heat is generated through artificial heating activities. Humans also alter the heating characteristics of cities through the production of pollutants such as ground-level ozone and other aerosols. In general, about 10 times more human-produced particulates are present in urban environments than in rural areas. Although these pollutants increase the reflectivity of the atmosphere above cities, thus reducing insolation, they cause an increase in the amount of infrared energy re-radiated downward to the surface.

An excellent example of an urban heat island is Atlanta, Georgia. This southeastern city has grown rapidly since the middle of the 20th century, becoming the leading commercial, industrial, and transportation area of the region. During this time, Atlanta has been one of the fastest-growing metropolitan areas in the United States, with population increasing approximately 30% between 1970 and 1990. In association with this explosive growth has been a dramatic expansion of the urban environment at the expense of agricultural land and forest. As a result, the air quality of the Atlanta region has decreased significantly, with increased amounts of ozone and volatile organic compounds polluting the air. Another impact has been the well-defined urban heat island that has evolved (Figure 5.10). As you examine this image, note that May

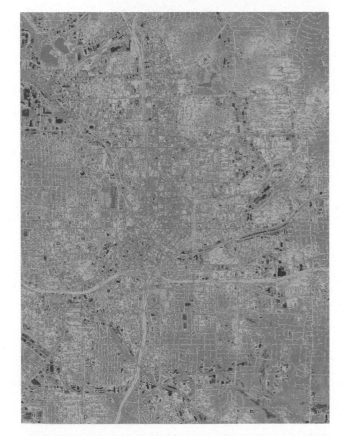

Figure 5.10 Urban heat island in Atlanta, Georgia, May 1997. This thermal infrared image was acquired during Project ATLANTA (ATlanta Land-use ANalysis: Temperature and Air-quality) with an airborne sensor. The inner-city temperatures (orange) are significantly warmer than those in the surrounding countryside, which are represented in greens and yellows. (*Source:* NASA.)

temperatures in the central business district reached 45°C (113°F) in some areas, whereas they were as low as 22°C (71°F) in the surrounding countryside. This wide disparity is the primary reason that the city is known as "Hotlanta" by many people who live there.

Other Local Factors That Influence Temperature In addition to the maritime vs. continental effect and urban impact on temperature, a variety of other variables can influence temperature. These variables include altitude, the position of topographic barriers, and windflow patterns. Altitude influences temperature because, as we saw in Figure 5.1b, temperature tends to decrease with increased elevation. As you will see in Chapter 7, topographic barriers such as mountain ranges can have a significant impact on temperature that goes beyond the simple variable of increased altitude. In areas where wind descends a mountain range, temperature can warm considerably due to increased molecular friction as air is compacted.

The Annual Range of Surface Temperature (A Holistic Assessment)

Let's now bring together some of the variables that influence temperature around the world in an effort to see some fundamental global geographic patterns. A good place to begin is by considering Figure 5.11, which shows a hypothetical continent bordered on the east and west by oceans. This theoretical continent straddles the Equator and contains, as a point of reference, the hypothetical position of the 15°C (59°F) isotherm in January and July. This diagram shows the combined effects of the maritime vs. continental relationship and seasonality. Due to seasonality, the isotherm shifts into the Southern Hemisphere during that hemisphere's summer season (January) and back into the Northern Hemisphere during July. This migration occurs because a net surplus of radiation occurs in each hemisphere during the respective summer months due to high Sun angles and increased insolation. The maritime vs. continental effect is evident in the range of latitude that the isotherm shifts in each hemisphere. Notice that in the oceans, the isotherm does not move much compared to how it migrates on the landmass. This difference is due to the fact that large water bodies heat and cool much more slowly than do continents.

Given the seasonal and geographic pattern observed in Figure 5.11, consider the broad-scale range cycle of temperature on Earth. Once again, a good place to begin is by examining some illustrations that show the fundamental patterns. With this in mind, take a look at Figure 5.12, which shows mean air temperatures on Earth during January and July.

Beginning with the January image, observe that the landmasses in the Northern Hemisphere are quite cold, with temperatures in northeastern Asia (Siberia) of –50°C (–58°F) and –35°C (–31°F) in northern Canada. An interesting pattern exists in North America, where the 0°C (32°F) isotherm extends across the center of the United States in the interior of the continent, but crosses into the Pacific Ocean much farther to the north. In other words, January temperatures are much warmer at higher latitudes along the west coast of North America than in the continental interior.

As expected for the Southern Hemisphere summer, mid- to high-latitude temperatures in that part of the world are also relatively warm, ranging from 10°C (50°F) on the southern tip of South America to 30°C (86°F) in the core of Australia. Predictably, temperatures in most of the tropical regions are warm, with an average of about 25°C (77°F) over a large part of the globe. The only significant deviation from this overall tropical pattern exists in western South America, where cool temperatures penetrate into very low latitudes. This pattern exists in part because the Andes Mountains follow much of the west coast of South America and are over 6000 m (19,680 ft) high. In addition, the cold Humboldt Current flows northward along the coast, which contributes to cooler temperatures in this area.

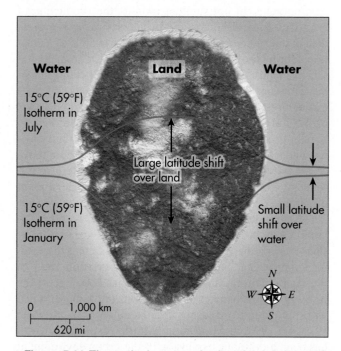

Figure 5.11 Theoretical seasonal migration of the 15°C (59°F) isotherm on a hypothetical continent. This hypothetical continent straddles the Equator, with 15°C (59°F) isotherms in each hemisphere. Notice how far the isotherm migrates over the continent compared to its movement.

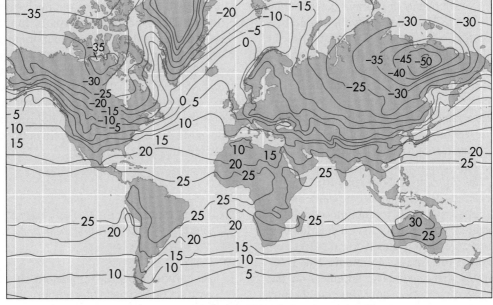

January

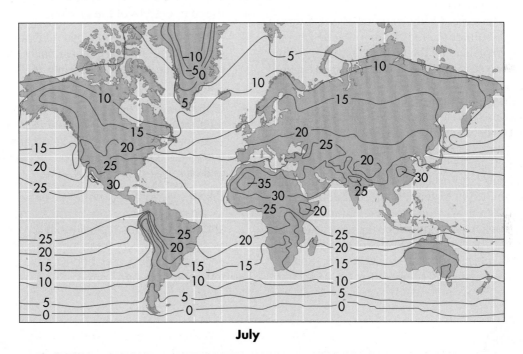

July

Figure 5.12 January and July distribution of surface air temperatures on Earth. Notice the geographic patterns and variations that occur between months. Blue isotherms indicate temperatures below 0°C (32°F), whereas red isotherms represent temperatures above 0°C (32°F).

Now turn your attention to the July image in Figure 5.12. What different patterns emerge in this diagram? For one, notice that the Northern Hemisphere landmasses are much warmer than they were in January, with temperatures reaching 10°C (50°F) in northeastern Asia and northern Canada. In other words, a 60°C (108°F)

annual range of temperature exists at high latitudes in the Northern Hemisphere. Similarly, the interior of the United States is much warmer as well. Interestingly, the high latitudes in the Southern Hemisphere do not experience such a range of temperature. Winter temperatures on the southern tip of South America, for example, cool to only about 0°C (32°F), which means the annual range there is approximately 10°C (18°F). Predictably, the tropical regions are warm, with a slight northerly migration of the 25°C (77°F) isotherm. Again, the only variation from this overall tropical pattern is the relatively cool temperatures that follow the Andes Mountains and Humboldt Current in South America.

Why do these annual patterns exist? The answer lies in the holistic interactions of the factors discussed so far, including seasonality, insolation, net radiation, latitude, the maritime vs. continental effect, and the environmental lapse rate in the troposphere. Some patterns are easy to explain. For example, the consistently warm temperatures in the tropics are naturally a function of high Sun angles and associated intense radiation all year long. As already noted, the exception to this tropical pattern exists along the axis of the Andes Mountains. This geography demonstrates that secondary variables—in this case, the troposphere's negative lapse rate and its cooling effect at progressively higher altitudes—also influence temperature patterns.

Another clear pattern is the extreme range of temperature at high latitudes in the Northern Hemisphere, which is easily explained by the highly variable amount of radiation received over the year due to changes in orbital position and the angle of incidence. It should also make sense that temperatures along the west coast of North America are warmer than in the continental interior in winter. This pattern is represented by the variation in temperature between San Francisco and Topeka, as illustrated in Figure 5.5.

Although the Northern Hemisphere pattern may seem logical, you might also ask yourself the following (very good) question: Why do the high latitudes in the Southern Hemisphere not experience the same extreme temperature range as those in the Northern Hemisphere? To answer this question, note the size of the Southern Hemisphere continents compared to those in the

Northern Hemisphere; they are much smaller than their Northern Hemisphere counterparts. In other words, the Southern Hemisphere is covered more by water than the Northern Hemisphere. Thus, more of a maritime effect occurs in the Southern Hemisphere than in the Northern Hemisphere, where the continental effect dominates. A good example of the continental effect in the Southern Hemisphere, however, is the core of Australia, where the annual range shows more variability due to the relatively large landmass of that country.

 www.wiley.com/college/arbogast

Global Temperature Patterns

In an effort to better visualize temperature geography on Earth, go to the *GeoDiscoveries* website and select the module *Global Temperature Patterns*. This module examines specific regions on Earth, such as the tropics and large landmasses, and how air flow influences seasonal temperature changes. After you interact with the module, answer the questions at the end to ensure you understand this concept.

Temperature and Location

To explore temperature patterns in another way, go to the *GeoDiscoveries* website and select the module *Temperature and Location*. In this module you can "visit" five cities that experience the kinds of seasonal temperature patterns discussed in this chapter. These cities are Cordoba, Argentina; Yakutsk, Russia; Manaus, Brazil; St. Louis, Missouri; and San Francisco, California. Visit each of these cities during each season and access insolation and temperature data from each location. As you work through this simulation, note how the environmental factors we have discussed influence the seasonal range of temperature at these places. Once you complete the simulation, be sure to answer the questions at the end of the module to test your understanding of this concept.

VISUAL CONCEPT CHECK 5.3

This image shows global temperature at a particular time of year, with the key showing the range of temperature in degrees Celsius. Given your understanding of the factors that influence temperature, which one of the following choices best explains the pattern you see?

a) It must be summer in the Northern Hemisphere.
b) The Northern Hemisphere must be tilted away from the Sun.
c) The interior of North America and Asia is influenced significantly by the maritime effect on temperature.
d) The Southern Hemisphere is receiving the most insolation.

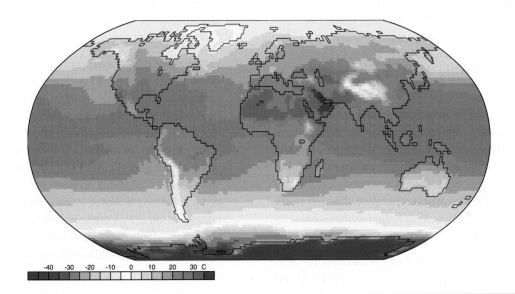

-40 -30 -20 -10 0 10 20 30 C

THE BIG PICTURE

Now that we have discussed how the factors of insolation and net radiation influence air temperature, we can examine how various atmospheric processes operate, such as wind patterns and the development of storm systems. A good place to begin this discussion is Chapter 6, which focuses on atmospheric air pressure and circulation and the way these concepts relate to Earth–Sun relationships and global temperature patterns. This image is a nice example of such a relationship. Notice the stream of clouds across the image? This line of clouds is associated with an atmospheric feature called a jet stream, which is a band of strong winds that exists in the lower stratosphere. These winds can be particularly strong in the midlatitudes when large differences exist with respect to temperature at higher and lower latitudes. This pattern, as well as a variety of others related

to the flow of air in the atmosphere, are discussed in Chapter 6. As in previous chapters, be sure to consider how all these concepts are interrelated and how they relate to previous discussions.

SUMMARY OF KEY CONCEPTS

1. The atmosphere contains four major layers: the troposphere, stratosphere, mesosphere, and thermosphere. Measurable temperature trends appear in each of these layers. One important change in temperature occurs in the troposphere, which cools with increasing altitude at a rate of 6.4°C per kilometer or 3.5°F per 1000 ft. This rate is called the environmental lapse rate.

2. On a large scale, a very close relationship exists between insolation, net radiation, and air temperature. In spite of this close relationship, a temporal lag exists on an annual and daily basis between when the receipt of the highest amount of radiation occurs and the timing of the warmest temperature. This lag exists because insolation is first absorbed by the surface of Earth before it is re-radiated as longwave radiation.

3. The maritime vs. continental effect explains the temperature difference between coastal and land-locked locations that are otherwise on or about the same latitude. Maritime locations such as San Francisco, California, have a relatively narrow annual temperature range due to the thermal characteristics of large

bodies of water like the Pacific Ocean. In contrast, continental locations such as Topeka, Kansas, have a much broader annual temperature range because landmasses do not absorb or release heat as consistently as water bodies.

4. Measurable differences occur with respect to temperature between urban and rural localities. Cities are typically warmer than the countryside due to the urban heat island effect, because paved surfaces and buildings have relatively low albedo and thus absorb more solar radiation. In addition, the moderating effect of water is reduced in cities because urban surfaces are sealed. Rural locations, in contrast, are covered with soil that absorbs water, which cools the landscape relative to an urban environment. Wind patterns also vary markedly between the two regions and are another cause for the relative difference in temperature.

5. Distinct seasonal temperature patterns occur on Earth. This geography is dependent on the interaction of several factors, including net radiation, season, latitude, the maritime vs. continental effect, and altitude.

1. Describe the four layers of the atmosphere and the temperature patterns that occur within them.

2. Why is latitude an important consideration when it comes to determining air temperature?

3. Where does the greatest range in annual temperature occur, at high or low latitudes? Why does this pattern exist?

4. Which location—Los Angeles, California, or Phoenix, Arizona—would experience the warmest peak temperature? Why does this pattern occur?

5. Central Canada and the southern tip of South America are both located at high latitudes. Despite this similarity, central Canada experiences a much greater range in temperature than the southern tip of South America. Why does this occur?

6. Which place would have a greater range in temperature, central Australia or Hawaii? Why?

7. Why is the tropopause at a higher altitude at the Equator than in northern Canada?

8. List three reasons why cities are typically warmer than rural landscapes.

9. Which place would be colder, the top or bottom of a high mountain? Why?

10. Why are temperatures in the lower stratosphere warmer than the upper part of the troposphere?

ANSWERS TO VISUAL CONCEPT CHECKS

VISUAL CONCEPT CHECK 5.1

1. The answer is *b*; temperature decreases as you rise through the troposphere. This decrease occurs because a negative environmental lapse rate exists from the surface of Earth to the tropopause.

2. The correct choice is *a*. This altitude occurs in the upper part of the troposphere.

3. The answer is *a*. At this altitude the temperature will begin to increase because the airplane is climbing closer to the ozone layer, which is found in the lower part of the stratosphere.

VISUAL CONCEPT CHECK 5.2

The answer is *b*. Vancouver lies next to a large body of water (the Pacific Ocean) that has a consistent temperature over the course of the year. As a result, the air temperature at Vancouver remains cool in the summer and warm in the winter relative to continental locations.

VISUAL CONCEPT CHECK 5.3

The answer is *a*. This image represents summer conditions in the Northern Hemisphere, because the high latitudes in Canada, Europe, and Asia are warm. Given that these are continental locations, which have a large annual temperature range, it must be summer and the Northern Hemisphere is tilted toward the Sun.

CHAPTER SIX

ATMOSPHERIC PRESSURE, WIND, AND GLOBAL CIRCULATION

Have you ever gone outside to discover the wind was blowing very hard and wondered why it was happening? In this chapter, you will investigate the way that air flows through the process of atmospheric circulation. The circulatory processes described in this chapter have dramatic effects that will be seen in later chapters, including daily weather (Chapter 8) and global climate patterns (Chapter 9). Flowing air also has significant impact on the circulation of ocean currents, coastal erosion by waves (Chapter 19), and the formation of sand dunes (Chapter 18). Many factors influence the direction and speed of airflow, including pressure and temperature differences, the Earth's rotation, and surface friction. After these variables are discussed,

the focus will turn to the geographic patterns of airflow around the globe. Toward the end of the chapter, we turn to how winds affect oceanic circulation.

CHAPTER PREVIEW

Atmospheric Pressure

Atmospheric Pressure Systems

The Direction of Airflow
  On the Web: *GeoDiscoveries Fluctuations in the Pressure Gradient*
  On the Web: *GeoDiscoveries The Coriolis Force*

Flowing air is usually invisible to the naked eye. However, in some cases, such as this spectacular sandstorm in Eritrea, Africa, strong winds blow dust in a way that allows us to visualize air movement. Flowing air is the focus of this chapter.

Global Pressure and Atmospheric Circulation

    On the Web: *GeoDiscoveries Global Atmospheric Circulation*

    On the Web: *GeoDiscoveries The Asian Monsoon*

    On the Web: *GeoDiscoveries Global Atmospheric Circulation and Water Vapor Movement*

Human Interactions: Harnessing Wind Energy

Local Wind Systems

Oceanic Circulation

    On the Web: *GeoDiscoveries El Niño*

The Big Picture

LEARNING OBJECTIVES

1. Understand the concept of air pressure and how it is measured.

2. Compare and contrast high- and low-pressure systems and how air flows within and between them.

3. Describe the factors that influence large-scale wind patterns on Earth.

4. Discuss the various components of the global circulation model.

5. Discuss the seasonal migration of atmospheric pressure systems on Earth.

6. Understand why people are developing wind as a supplementary energy system, where in the United States the potential is high for this development, and the challenges facing the effort.

7. Describe the various local circulatory systems and why they occur.

8. Discuss the nature of oceanic circulation and how it compares with atmospheric circulatory patterns.

Atmospheric Pressure

Atmospheric circulation is an incredibly important process on Earth for a variety of reasons. The primary impact of airflow is to move heat energy around the globe in a way that moderates temperature on Earth. Airflow also affects global and local air quality and pollution levels. For example, winds can carry ash and gases from volcanic eruptions many kilometers from the volcano (Figure 6.1), or they can clear smog from large cities. Similarly, dust from eroded agricultural fields can be transported great distances before it settles back to Earth. Before we can begin to describe the nature of airflow, however, the concept of air pressure must first be introduced because it directly influences the character of large- and small-scale wind patterns.

As discussed in Chapter 4, the atmosphere is made up of a variety of gases that we collectively call "air." Like anything else, air is held to Earth by gravity and thus has weight. The weight of the air exerts pressure on the Earth's surface, which is measured as **air pressure** (also called *atmospheric pressure* or *barometric pressure*).

Factors That Influence Air Pressure

Atmospheric pressure is most closely associated with the temperature and density of air. The concept of air density is easy to understand if you consider that air expands or contracts depending on the environmental setting. Imagine, for example, you have a specific number of air molecules that are contained within a shoebox. If you can somehow move all the molecules in the shoebox to a larger box, the gas will expand to fit comfortably within

---

Air pressure *The force that air molecules exert on a surface due to their weight.*

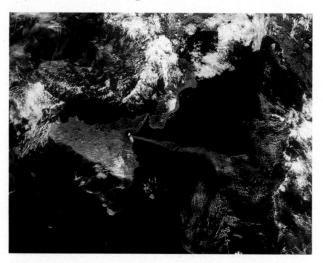

Figure 6.1 Ash plume from Mount Etna in Sicily. Northwesterly winds carry volcanic ash from the erupting volcano toward the southeast, in an image taken on July 22, 2001, from Space Station *Alpha*.

the new container. In this example, the air molecules are closer together in the shoebox, and thus the density is greater, than when they are in the larger container.

You can see the effect of density on air pressure by noting how pressure changes with altitude (Figure 6.2).

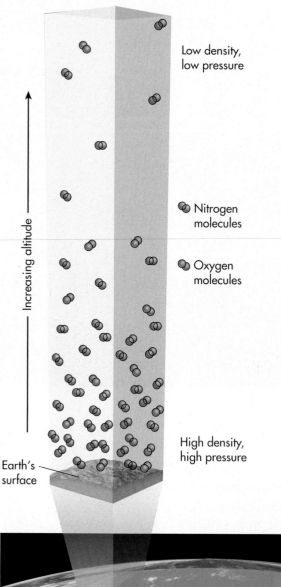

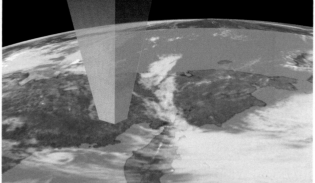

Figure 6.2 Air density, altitude, and atmospheric pressure. At low altitudes air molecules are held close to Earth by gravity and thus are more dense, resulting in high atmospheric pressure. In contrast, the density of air molecules is low at high altitudes and air pressure is thus relatively low.

In general, air pressure decreases with increasing altitude because most air molecules are held close to the surface by gravity. As a result, the density of air molecules is greater closer to the Earth's surface, which means that air pressure is relatively high in that part of the atmosphere. With increased altitude the density of air molecules becomes progressively less, resulting in progressively lower air pressure.

In addition to the impact of gravity on the molecular density of air, atmospheric pressure is also strongly influenced by air temperature. The most obvious way that air temperature influences atmospheric pressure occurs when air close to the Earth's surface is warmed a great deal. Such warming causes air molecules to scatter and density thus decreases, resulting in relatively low atmospheric pressure. Consider the analogy of a hot air balloon. When heat is added to the air within the balloon, this causes the air to expand and lift within the relatively cooler (more dense) air that surrounds it. Low atmospheric pressure also results when air is forced to rise vigorously. Very cold surface air is usually associated with high atmospheric pressure because cold air is dense and thus sinks. In some instances, air from the upper atmosphere descends vigorously toward the Earth's surface. This process also results in high pressure.

Measuring and Mapping Air Pressure

Air pressure is often measured in units called millibars (mb) with an instrument called a barometer (Figure 6.3). A common type of barometer consists of a long glass tube, closed at one end, that is filled with a liquid (usually mercury) and inverted into a dish containing the same liquid. The liquid in the tube drops down slightly, leaving a vacuum at the closed end of the tube. When the liquid in the tube comes to rest, the force due to atmospheric pressure, pressing down on the liquid in the dish, exactly balances the weight of the column of liquid. The tube can then be calibrated to measure atmospheric pressure in terms of inches or millimeters of mercury, which can be converted to millibars or any other pressure unit.

You can observe how air pressure changes with altitude by looking at three separate locations at different elevations (Figure 6.4). At sea level, for example, the average pressure of the air is 1013.25 mb. In contrast, at Denver, Colorado (5280 ft—"the Mile High City"), the air pressure is 840 mb. At a still higher elevation, such as at the top of Mount Everest, the Earth's tallest mountain at 8850 m (29,035 ft), the average air pressure is only 320 mb.

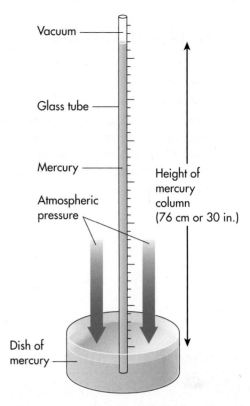

Figure 6.3 Measurement of atmospheric pressure. The pressure of the atmosphere is measured by the height of a column of mercury that can be supported by that pressure.

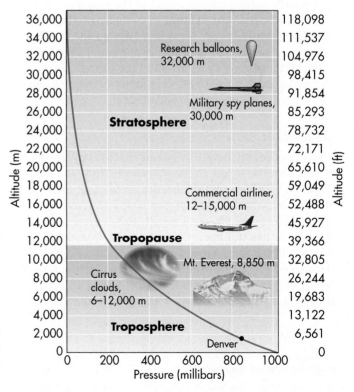

Figure 6.4 Atmospheric pressure and altitude. Average atmospheric pressure decreases with increasing elevation and altitude above the Earth's surface.

Atmospheric Pressure • 115

Mount McKinley is the tallest mountain in North America, rising 6194 m (20,320 ft) in Alaska's Denali National Park. How do you suppose air pressure changes from the base of the mountain to the top—does it increase or decrease? Why does this change occur?)

Atmospheric Pressure Systems

In addition to the pressure changes that occur with respect to altitude, air pressure also varies horizontally across the Earth's surface. A **high-pressure system** is a circulating body of air that exerts relatively high pressure on the surface of Earth because air descends (toward the surface) in the center of the system. In contrast, a **low-pressure system** is a circulatory body of air where relatively less pressure exists on the Earth's surface because the air is rising (away from the surface) in the system's core. This portion of the chapter focuses on the nature of these large-scale pressure systems and how air flows within and between them.

Low-Pressure Systems

Low-pressure systems are often referred to as **cyclones**. If you observe a low-pressure system from the side (Figure 6.5), notice that the vertical flow of air consists of rising air, with the most vigorous upward flow in the center of the system. As a result, the central part of the system has the lowest pressure and is designated with the letter "L" on a weather map. In the Northern Hemisphere, the horizontal flow of air around the center of a low is counterclockwise (looking at it from above) as the air flows toward the core of the system.

In the Southern Hemisphere, the horizontal flow of air is clockwise. This inward flow occurs because the rising air at the center of the system creates a void into which air must flow. Low-pressure centers are generally associated with cloudy or stormy weather because rising air cools with altitude and thus squeezes water out of the atmosphere. This relationship will be discussed in much greater detail in Chapter 7.

High-Pressure Systems

In contrast to low-pressure systems, large areas of relatively high air pressure are referred to as **anticyclones**. In these systems, the vertical flow of air consists of descending air, which diverges (spreads apart) at the surface (Figure 6.5). Because air sinks most vigorously in the center of a high, the highest pressure is in the central part of the system, which is designated with the letter "H." The horizontal airflow around the center of a high is clockwise in the Northern Hemisphere and counterclockwise in the Southern Hemisphere, just opposite to the direction of flow around a low-pressure system. High-pressure centers are generally associated with fair weather because descending air warms as it approaches the surface. As mentioned earlier, warm air can hold more moisture than cold air and this makes precipitation less likely. Again, more will be said about this relationship in Chapter 7.

High-pressure system *A rotating column of air that descends toward the surface of Earth where it diverges.*

Low-pressure system *A rotating column of air where air converges at the surface and subsequently lifts.*

Cyclones *Low-pressure systems.*

Anticyclones *High-pressure systems.*

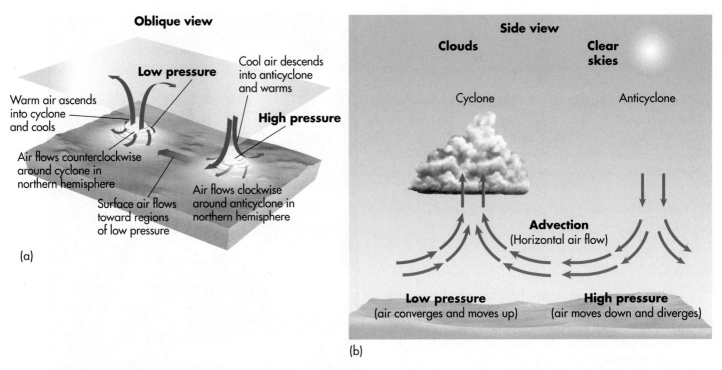

Oblique view

Warm air ascends into cyclone and cools

Low pressure

Cool air descends into anticyclone and warms

High pressure

Air flows counterclockwise around cyclone in northern hemisphere

Surface air flows toward regions of low pressure

Air flows clockwise around anticyclone in northern hemisphere

(a)

Side view

Clouds

Clear skies

Cyclone

Anticyclone

Advection (Horizontal air flow)

Low pressure (air converges and moves up)

High pressure (air moves down and diverges)

(b)

Figure 6.5 Atmospheric pressure systems. Oblique view (a) and side view (b) of typical low- and high-pressure systems. In a low-pressure system, air converging at the surface rises and forms clouds. In a high-pressure system, air descends and diverges at the surface; these systems are usually associated with clear skies.

In general, high- and low-pressure systems (anticyclones and cyclones) create large-scale circulatory systems that are interconnected by airflow. Look again at Figure 6.5 and see how the air flows horizontally, in a process called **advection**, from the high-pressure system to the low-pressure system at the surface. You can see clearly the geographic pattern of low- and high-pressure systems on a barometric pressure map. Figure 6.6 illustrates pressure variations in the upper part of the atmosphere across the North Atlantic. The (gray) lines on the map are the isobars. The red arrows illustrate the way that air is moving relative to the pressure systems.

Notice in Figure 6.6 that the isobars in the eastern Atlantic form a rough oval or egg-shaped area. This region is the center of a low-pressure system (or the area of lowest air pressure) at this particular point in time. From the wind direction arrows, observe that the air is moving counterclockwise around this low. Another low exists in the western part of the North Atlantic, over eastern Canada, where the winds are also circulating counterclockwise. In the central Atlantic, between the two lows, a broad area of high pressure exists that is rotating clockwise. On the western side of this system the air is flowing southwest to northeast, whereas on the eastern side of the system the air

is moving northwest to southeast. Another way to describe these patterns is that the winds on the western side of the high are southwesterly and that they are northwesterly on the system's eastern side. You will learn more about the direction of wind flow later in this chapter, but a simple rule to remember is that wind direction is noted by the direction from where the wind is originating.

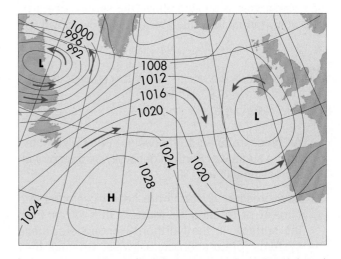

Figure 6.6 Atmospheric pressure map of the North Atlantic. The red arrows represent the direction of the winds. Notice the pressure variability across the ocean and the way air flows.

Advection *The horizontal transfer of air.*

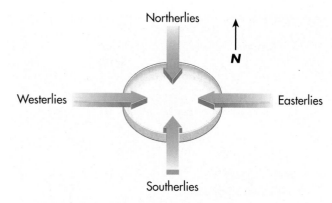

Northerlies

N

Westerlies

Easterlies

Southerlies

Figure 6.7 Compass headings and wind directions. Winds are named for the direction in which they originate. Westerly winds, for example, originate in the west and flow toward the east.

Many people mistakenly believe that the named direction of wind reflects the direction in which the air is moving. Instead, the name of the wind direction, such as "westerlies," reflects the direction from which the air is coming (Figure 6.7). Westerly winds, for example, originate in the west and flow toward the east. Similarly, a north wind in the Northern Hemisphere brings in air from the north toward the south.

Although atmospheric pressure maps such as Figure 6.6 are very useful to illustrate detailed information, you can also observe basic barometric patterns by viewing satellite images such as the view of western Europe in Figure 6.8. Notice that France, Spain, and Germany are cloud-free, whereas Great Britain and Ireland are shrouded in a swirling band of clouds. Zones of clear sky, such as the large region in eastern Europe, are places dominated by high pressure. Remember that air pressure is low in the center of a cyclone and high in the middle of an anticyclone. Because air rises in the center of a low, a void is created above the surface of Earth. This void is ultimately filled by air that flows down and

Figure 6.8 Atmospheric pressure systems in Europe. The low is indicated by the clouds that cover Great Britain and Ireland, whereas the high is the clear sky to the east over France, Spain, and Germany.

outward from the center of the high (remember that air descends in a high-pressure system) and into the low. We feel this exchange of air as wind. Look at Figure 6.5 again to visualize how this process works and be sure you understand it thoroughly.

KEY CONCEPTS TO REMEMBER ABOUT ATMOSPHERIC PRESSURE SYSTEMS

1. Air pressure refers to the weight of air distributed on the surface of Earth. It generally decreases with increasing altitude.

2. Low-pressure systems are called cyclones and consist of rotating air masses that lift air from the surface. In the Northern Hemisphere, these systems rotate counterclockwise, whereas they rotate clockwise in the Southern Hemisphere.

3. Low-pressure systems are usually associated with clouds and precipitation.

4. High-pressure systems are called anticyclones and consist of rotating air masses that descend toward the surface. In the Northern Hemisphere, these systems rotate clockwise, whereas they rotate counterclockwise in the Southern Hemisphere.

5. High-pressure systems are usually associated with clear skies.

6. On weather maps, isobars indicate areas of equal air pressure.

7. Winds flow from high to low pressure.

The Direction of Airflow

Now that we have discussed the fundamentals of air pressure, let's look more closely at the concept of airflow and the factors that govern its movement. We will examine several factors, including unequal heating of land surfaces, the pressure gradient force, the Coriolis force, and various frictional forces.

Unequal Heating of Land Surfaces

The ultimate cause for all wind patterns on Earth is the unequal heating of surfaces that results from variations in the amount of solar radiation received between latitudes. This spatial variation in surface air temperature means that air density, and thus pressure, differ from place to place. At a fundamental level, surface air flows from areas of high pressure to low pressure because the atmosphere works to balance the difference between the two areas.

The best example of unequal heating on Earth is the difference that exists between the tropics and the poles. Recall from Chapters 4 and 5 that the tropics are much warmer than the poles because the equatorial regions receive the most direct insolation throughout the year. If no mechanism existed to balance this difference, specifically through atmospheric circulation, then the tropical and polar regions would become excessively hot and cold, respectively. As a result, the differences in their respective temperatures would be even greater than they presently are.

Instead, this process of unequal heating causes motion of the atmosphere through the process of convection. Recall from Chapter 4 that convection is the vertical mixing of fluid material (in this case, air) due to differences in temperature. In contrast, remember that the term *advection* refers to the horizontal movement of air or water. Figure 6.9 shows how the processes of

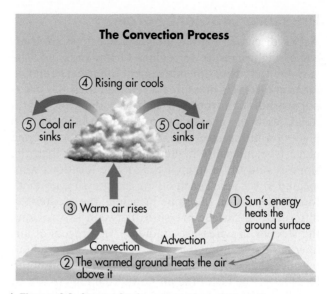

The Convection Process

④ Rising air cools
⑤ Cool air sinks
⑤ Cool air sinks
③ Warm air rises
Convection
Advection
① Sun's energy heats the ground surface
② The warmed ground heats the air above it

Figure 6.9 Atmospheric convection. Convection occurs when one portion of the Earth's surface is heated relative to another. When this happens, a large "bubble" of air lifts from the surface.

convection and advection relate to air movement. On a global scale, air heated near the Equator results in low air pressure as air rises within the upward-moving part of the convection process. Subsequently, it travels to higher latitudes in both hemispheres by advection, where the air cools and descends as a high-pressure system at some point in the downward part of the convection process. The combined processes of convection and advection represent the first stage of atmospheric circulation on Earth and are the overall method through which heat is distributed around the planet. Once the air is set in motion in this manner, the other forces—pressure gradient, Coriolis, frictional—then directly influence the movement of the air.

VISUAL CONCEPT CHECK 6.2

Atmospheric pressure systems are often easy to see on satellite images. This particular image focuses on Europe in October 2001, with Spain in the lower left part of the image and clouds showing up in bright white. Given your understanding of pressure systems, where is the high-pressure system in this image? Where is the approximate center of the high? In what direction is this system spinning and where are the southerly and northerly winds in the system?

Pressure Gradient Force

As we discussed earlier, air generally flows from areas of high pressure to regions of low pressure. The variable that drives the movement of air between two areas at different pressures is referred to as the **pressure gradient force**. It is useful to think of a "gradient" as being analogous to the slope between two places on the surface of Earth. If the elevation of one place is significantly higher than the other, then the slope is said to be steeper than if the locations were located at (or near) each other. The pressure gradient force operates on the same principle: the greater the difference in surface pressures between two regions, the steeper the pressure gradient. The process of airflow is significantly affected by the pressure gradient because the greater the pressure difference between two places, the faster the air flows from high to low to equalize the pressure.

Figure 6.10 is a barometric map that illustrates the pressure gradient force. The map focuses on a hypothetical difference in surface air pressure between Oklahoma City, Oklahoma, and Nashville, Tennessee. In this image, the center of the high pressure (indicated by the

"H") is located in Oklahoma City, whereas the center of the low (indicated by the "L") is located in Nashville. Air pressure in this figure ranges from 1028 mb at the center of the high to 993 mb at the center of the low. It should make sense to you that as a result of this variation, the air is generally flowing from west to east (left to right on the map)—that is, from the Oklahoma City area toward Nashville.

Close to the center of the high—say, from Oklahoma City to central Arkansas—the isobars are widely spaced. This means that very limited pressure change occurs across this portion of the Earth's surface at this particular point in time. Thus, the pressure gradient in this part of the circulatory system is shallow. If you were in this area on this particular day, you would find the winds to be light, because air flowing into that region would be moving slowly. As you approach the center of the low, however, the isobars become closer together. Given that isobars represent locations of equal atmospheric pressure, this close spacing can only mean a rapid change in surface pressure occurs over a relatively small geographical space—for example, from Memphis to Nashville. In other words, this part of the circulatory system has a steep pressure gradient and the air flows faster as it moves toward the center of the low, to fill the relative void created by the less dense air at the surface. If you happened to be in this area on this particular day, you would notice strong winds.

Pressure gradient force   *The difference in barometric pressure that exists between adjacent zones of low and high pressure that results in airflow.*

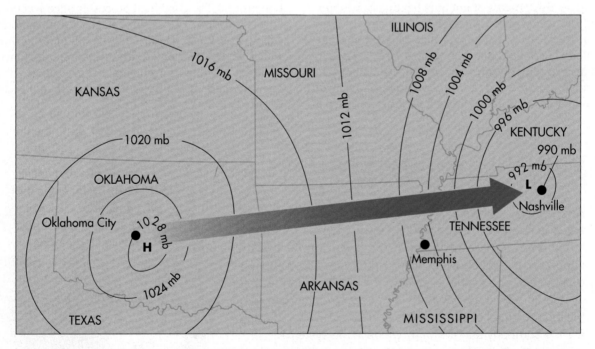

Figure 6.10 Hypothetical pressure gradient between Oklahoma City, Oklahoma, and Nashville, Tennessee. Notice the direction of airflow and the spacing of isobars across the region. A steep pressure gradient exists in regions where isobars are closely spaced together, whereas the gradient is shallow in places where the isobars are far apart.

DISCOVER...

MIGRATING PRESSURE SYSTEMS

Atmospheric pressure systems in the midlatitudes generally migrate from west to east. An atmospheric scientist looking at this GOES satellite image sees that clouds and perhaps rain, associated with a low-pressure system, occurred in the southeastern part of the United States in early June 2004. At the same time, a strong high-pressure system and sunny skies dominated the western part of the country. Given the migration of these pressure systems, a geographer knows that, in all probability, the western high will move slowly to the east, ultimately bringing clear skies over the eastern part of the country. This motion of air pressure systems is the basic idea behind satellite weather forecasting. In the midlatitudes, for example, you can forecast tomorrow's weather by looking at today's weather farther to the west.

A GOES visible satellite image of the contiguous United States in June 2004.

Coriolis Force

In addition to the pressure gradient force, another factor strongly influences the process of airflow in the atmosphere: the **Coriolis force**. The Coriolis force is a very simple effect that is often difficult to explain and comprehend. Whereas the pressure gradient force arises because of differences in atmospheric pressure between regions, the Coriolis force is related to the rotation of the Earth on its axis. Given this rotation, objects in the atmosphere, including the air, appear to be deflected or pulled sideways as Earth rotates under them. In the Northern Hemisphere, the direction of deflection for an object moving toward the Equator is to the right when viewed from above the North Pole. In the Southern Hemisphere, the deflection of an object moving toward the Equator is to the left when viewed from below the South Pole.

A good way to visualize how the Coriolis force works is to study the apparent path that a hypothetical rocket would take between the North Pole (90° N) and New York City (about 40° N). Look at Figure 6.11 as you work through this discussion. During the initial period of flight, the rocket follows the 74° W meridian as it flies south. The course appears to change, however, as the rocket continues southward. Why does the course change occur? It is not because the rocket's direction actually varies from its original destination; instead, the change occurs because Earth rotates *under* (or eastward of) the rocket while it is in the air. In this fashion, the rocket would land somewhere to the *west* of New York City unless the course is corrected in some way to account for the Earth's rotation. Figure 6.11 also shows the apparent path of a rocket fired from the South Pole

---

Coriolis force *The force created by the Earth's rotation that causes winds to be deflected to the right in the Northern Hemisphere and to the left in the Southern Hemisphere.*

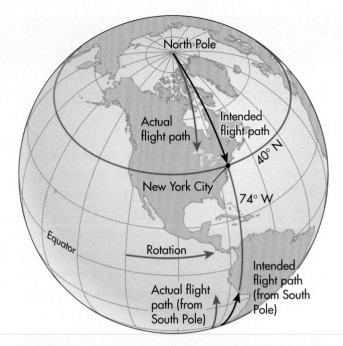

Figure 6.11 The Coriolis force. The Coriolis force influences the path of a rocket traveling from the North Pole to New York City, deflecting it to the west. Notice the direction of the Earth's rotation and how the rocket path is diverted more as it approaches the Equator. Also note that a rocket traveling from the South Pole to the Equator is deflected west as well.

(90° S) toward the Equator. Note that this rocket is deflected to the west as well, but this time from right to left relative to the path of initial motion.

We can observe how a rocket's path might be altered by the Earth's rotation, but how does the Coriolis force influence the wind? Remember that at ground level the direction of airflow is influenced by the pressure gradient force and results in airflow that is per-

---

www.wiley.com/college/arbogast

Fluctuations in the Pressure Gradient

In order to better understand how the pressure gradient force influences airflow, go to the *GeoDiscoveries* website and select the module *Fluctuations in the Pressure Gradient*. This simulation is based on Figure 6.10, which shows a hypothetical pressure gradient between Oklahoma City, Oklahoma, and Nashville, Tennessee. You can adjust the atmospheric

pressure between the two places to see how these changes impact the flow of air between them. As you make these adjustments, notice the impact that they have on wind speed and the overall direction of the winds. After you complete the simulation, answer the questions at the end of the module to test your understanding of this concept.

What you see on a weather map

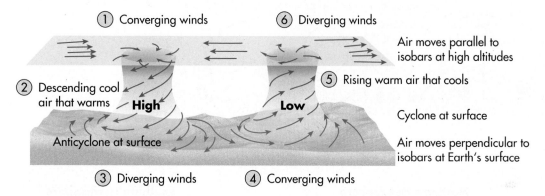

What is happening in the atmosphere

Figure 6.12 A dynamic convection loop. Cyclones and anticyclones are linked together in a convection loop consisting of air masses that spiral due to the Coriolis force. Note how the air masses move vertically within the high- and low-pressure systems and horizontally between them.

pendicular to the isobars (as in Figure 6.10). Once the air rises into the upper troposphere, however, its speed increases because it flows freely without obstruction. Once this altitude is reached, the air is influenced most directly by the Coriolis force, which causes winds to spiral to the right in the Northern Hemisphere and to the left in the Southern Hemisphere.

In this context, remember that convection loops consist of spiraling masses of descending or rising air that are linked horizontally by advection (Figure 6.12). This process is most pronounced at higher latitudes and generally results in westerly airflow from the subtropics to the poles. Once again, you might be inclined to think that the air is flowing toward the west; instead, westerly winds are those that flow from west to east. At this point the Coriolis force and the pressure gradient force effectively balance each other, resulting in upper airflow that is *parallel* to the isobars rather than perpendicular, as seen at the surface. The net result of this flow pattern is that air moves *around* pressure systems in the upper atmosphere. Such winds are called **geostrophic winds**.

Geostrophic winds *Airflow that moves parallel to isobars because of the combined effect of the pressure gradient force and Coriolis force.*

Frictional Forces

As briefly mentioned earlier, a third force influences the process of airflow in the atmosphere, one that occurs at ground level and that operates in direct opposition to the winds. This force is the force of friction and occurs because of the drag and impediments created by features on the surface of Earth, such as mountains, trees, and even buildings. As you can imagine, these features cause winds to slow down and move in irregular ways (Figure 6.13). Remember from the discussion about the urban heat island in Chapter 5, for example, that winds typically flow less strongly in cities than they do in the surrounding countryside. The force of friction results in airflow that is somewhere between the flow driven by the pressure gradient (i.e., perpendicular to isobars) and the Coriolis force (which is parallel to isobars). As a general rule, the effect of friction is strongest at the surface and diminishes progressively to an altitude of about 1500 m (approximately 5000 ft). In response to this variability, the wind flows at an angle relative to isobars at ground level. At higher altitudes, however, winds follow a geostrophic course that is parallel to isobars.

In an effort to integrate all the major factors that influence atmospheric circulatory processes, let's now turn to Figure 6.14 to review. Recall that the pressure

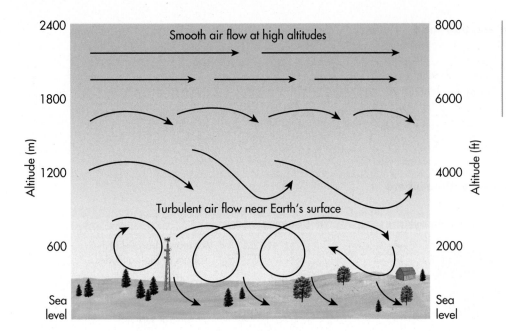

Figure 6.13 The effect of friction on wind flow near the Earth's surface. Compared to winds at higher altitudes, the flow of surface air is significantly modified by features on the Earth's surface.

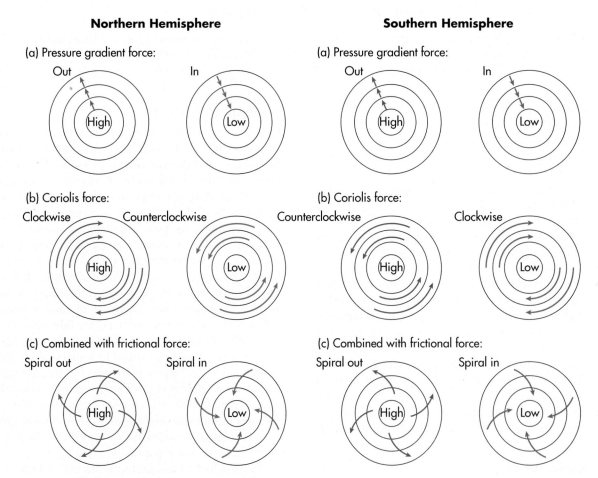

Figure 6.14 The factors that influence large-scale atmospheric circulation. (a) The pressure gradient force causes air to flow perpendicular to isobars. (b) The Coriolis force causes movement of air that is parallel to isobars. (c) In combination with the pressure gradient and Coriolis forces, frictional forces result in winds that flow somewhere intermediately between 0° and 90° of isobars.

gradient force causes winds to flow at right angles to isobars in the direction of the lower pressure. The effect of this force can be seen in Figure 6.14a. When the influence of the Coriolis force is also included (Figure 6.14b), winds then flow parallel to the isobars as geostrophic winds. This process occurs because of the balancing effect that the Coriolis force and pressure gradient force have on one another. In other words, the Coriolis force keeps wind from flowing across isobars, whereas the pressure gradient force stops winds from curving up the pressure slope. Finally, when the force of friction is taken into account (Figure 6.14c), the end result is winds that follow an intermediate course relative to the isobars, somewhere between perpendicular (due to the pressure gradient force) and parallel (due to the Coriolis force) to those lines of equal atmospheric pressure.

www.wiley.com/college/arbogast

The Coriolis Force

To see how the Coriolis force looks in animation, go to the *GeoDiscoveries* website and select the module *The Coriolis Force*. This module examines how the Coriolis force influences atmospheric circulation. Be sure to watch how the rotation of Earth on its axis impacts the way that air flows on the planet. After you finish watching the animation, be sure to answer the questions at the end of the module to test your understanding of this concept.

KEY CONCEPTS TO REMEMBER ABOUT THE VARIABLES THAT INFLUENCE LARGE-SCALE WINDS

1. Air generally flows from high pressure to low pressure.

2. Large-scale atmospheric circulation is caused by the unequal heating of the tropics and poles. The process of airflow begins at the Equator because of convection.

3. Isobars are isolines that connect points of equal atmospheric pressure.

4. The speed of airflow is determined by the pressure gradient force. The steeper the gradient, the stronger the winds.

5. The Coriolis force causes air moving toward the Equator to be deflected to the right in the Northern Hemisphere and to the left in the Southern Hemisphere.

6. Features on the Earth's surface, such as mountains, forests, and buildings, create a frictional force that acts opposite to the wind's direction.

7. The combined effect of the pressure gradient force, Coriolis force, and frictional force causes a spiral motion of air in both low- and high-pressure systems.

Global Pressure and Atmospheric Circulation

In the preceding sections, we looked at the fundamentals associated with air pressure systems and the variables that influence the process of airflow in the atmosphere. Within that context, let's now examine the general circulation of air around the globe. This discussion will refer to the flow of air both in the upper part of the atmosphere and at the surface. The difference between the flow in these parts of the atmosphere is often noticeable on the ground. If you want to see this difference yourself sometime, look for a partly cloudy day where two distinct layers of clouds occur, one low and another high. When these conditions exist, the clouds in the upper part of the atmosphere are often moving in a slightly different direction or speed than those closer to the surface.

As discussed previously, the primary driver of global circulation is the unequal heating of the tropics and the poles. Because of this energy imbalance, the atmosphere works to balance the system through the process of airflow. If the Earth's surface had a uniform character (i.e., no distinction between continents and oceans), did not rotate, and the axis was not tilted, the basic circulatory system would be very easy to understand. In this simplistic scenario, which is illustrated in Figure 6.15,

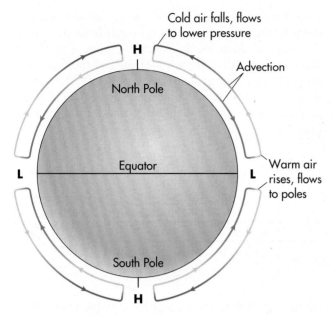

Figure 6.15 Global circulation on a nonrotating, untilted Earth with a consistent surface composition. In this model, a few simple convection loops would dominate the system, with rising air (low pressure) at the Equator, descending air (high pressure) at the poles, and horizontal flow of air by advection in between.

low pressure would occur at the Equator because the air is very warm and high pressure would occur at the poles because the air is very cold. Very simply, air would rise away from the surface at the Equator, within a low-pressure system, and would then flow toward the poles in the upper atmosphere by advection. Once it reached the poles, the air would descend toward the surface where it would then diverge and flow back toward the Equator by advection.

As you know, however, Earth does not have a uniform surface, it does rotate, and it is tilted with respect to the plane of the ecliptic. Given these factors, the global circulatory system is more complex than the simplified model in Figure 6.15. Figure 6.16 illustrates the general global circulation model as it truly functions, showing the major wind systems on Earth. Let's look at the model in more detail in order to explain how these wind patterns occur. We will begin at the Equator and proceed systematically toward higher latitudes, focusing on the directions of winds at both the surface and the upper part of the troposphere.

Tropical Circulation

The logical place to begin discussion of global atmospheric circulation is the tropics. As noted previously in this chapter, atmospheric circulatory processes are set in motion in the tropics because of the extremely warm temperatures that occur there. Although tropical air eventually finds its way into the midlatitudes, it is useful to think of the tropical circulatory system as being a convection loop, with air flowing between the Equator and about 30° N and S latitude. Each of these loops is referred to as a Hadley cell. In that context, the discussion of tropical circulation focuses on the Intertropical Convergence Zone and the Subtropical High Pressure System.

Intertropical Convergence Zone Tropical circulatory processes begin at the Equator, where air is warmed due to year-round receipt of direct sunlight. This warming helps create a zone of low pressure, called the **equatorial trough**, because warm air is less dense and more buoyant. As a result of the rising air mass, air from higher tropical latitudes in both hemispheres flows toward the equatorial trough along the surface by advection, converging in a narrow band known as the **Intertropical Convergence Zone** or **ITCZ** (Figure 6.16). The converging winds are known as **trade winds** because they were systematically used to power sailing vessels during the Age of Exploration between the 1400s and 1800s. Notice in Figure 6.16b that the trade winds north of the ITCZ are northeasterly, whereas trade winds south of the region are southeasterly. Together, they form the **tropical easterlies**. At the point where the northeasterly and southeasterly trade winds converge at the ITCZ, the winds can become relatively calm and highly variable because the pressure gradient is very weak.

An important characteristic of the ITCZ is that it is a region of cloudiness, frequent rains, and even volatile weather. This occurs for two primary reasons. One is that the trade winds flow over warm oceans across much of their length; therefore, a great deal of evaporation from these surfaces takes place and the air contains abundant moisture. In this context, high levels of atmospheric moisture can be reached because the air is warm and expanding. A second reason is that, due to the warmth of the air, the ITCZ is a place where the air lifts from the Earth's surface. As the air rises, it cools and promotes clouds and rainfall because water condenses at lower temperatures.

Equatorial trough   *Core of low-pressure zone associated with the Intertropical Convergence Zone.*

Intertropical Convergence Zone (ITCZ)   *Band of low pressure, calm winds, and clouds in tropical latitudes where air converges from the Southern and Northern Hemispheres.*

Trade winds   *The primary wind system in the tropics that flows toward the Intertropical Convergence Zone on the equatorial side of the Subtropical High-Pressure System. These winds flow to the southwest in the Northern Hemisphere and to the northwest in the Southern Hemisphere.*

Tropical easterlies   *Band of easterly winds that exists where northern and southern trade winds converge.*

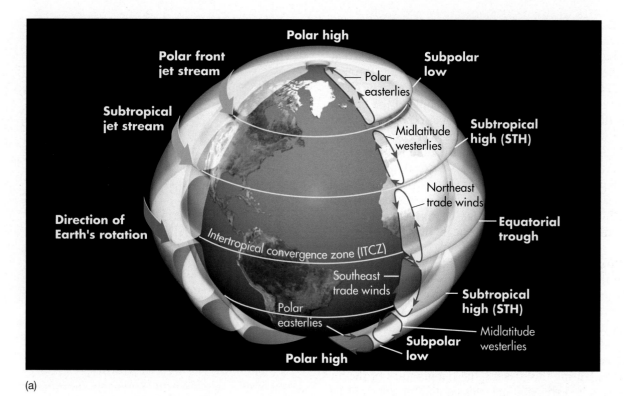

(a)

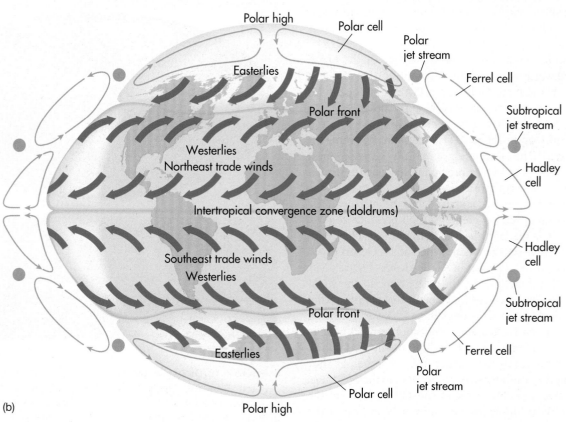

(b)

Figure 6.16 The global circulation model. (a) This three-dimensional diagram illustrates the basic flow of air in cross section as well as top views. Notice the various convection loops that exist around the surface of Earth. (b) A two-dimensional map of the world shows the directions of winds across various bodies of land and water.

Due to these combined processes, the ITCZ is usually easy to find on a satellite image because it consists of a band of clouds and thunderstorms (Figure 6.17) that can be extremely violent. Airline pilots flying through the ITCZ are thus wary of the weather because it can be quite turbulent. An unfortunate example of the impact that the ITCZ can have on air travel occurred in June 2009 when a French airliner flying at cruising altitude suddenly crashed as it flew from Brazil to France across the Atlantic Ocean. Although the exact cause of the crash remained unknown at this writing, pilots in nearby planes reported very strong storms in that part of the ITCZ (when they flew through it) and suggested that turbulence may have simply brought down the plane.

Subtropical High Pressure System You have just seen that the ITCZ is an area of low pressure present only at very low latitudes. As we move our way through the tropical circulatory loop, let's see what happens at slightly higher latitudes.

As air in the equatorial trough rises from the Earth's surface, it spirals upward to the upper part of the troposphere. During this process, the temperature of the air drops as it rises. Although this cooling process will be discussed more thoroughly in Chapter 7, for now it is sufficient to know the process occurs because the air is expanding as it rises.

High-altitude air on the northern side of the ITCZ flows northward by advection, while air on the southern side of the ITCZ flows to the south. Given that the air has cooled, it must sink, and does so at approximately 25° to 30° N or S latitude. These zones of sinking air are referred to as the **Subtropical High (STH) Pressure System** (see Figure 6.16a). The descending air is dry because much of its moisture was lost as precipitation over the Equator. The air is also compressed as it descends, making the air denser and warmer as it approaches the surface of the Earth. This compression creates a high-pressure zone of hot, dry air. On the Earth's surface, this zone is characterized by extensive deserts, such as the Sahara Desert in Africa and the Arabian Desert in Saudi Arabia (Figure 6.18).

The STH functions in the same way as the high-pressure systems discussed earlier in this chapter. That is, as the air descends, it rotates in a clockwise direction in the Northern Hemisphere and counterclockwise in the Southern Hemisphere. Once the downward spiraling airstreams reach the surface, they diverge. As they diverge on the southern side of the STH in the Northern Hemisphere, they flow back to the ITCZ, forming the northeasterly trade winds. On the northern side of the STH in the Southern Hemisphere, the airstreams form

---

Subtropical High (STH) Pressure System *Band of high air pressure, calm winds, and clear skies that exists at about 25° to 30° N and S latitude.*

Figure 6.17 The Intertropical Convergence Zone. The ITCZ is a zone of low pressure, as indicated by the band of dense clouds that extends from Central America west into the Pacific Ocean in this satellite image.

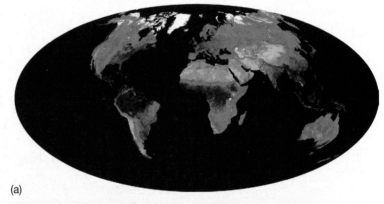

(a)

(b)

Figure 6.18 Global vegetation and the Subtropical High (STH) Pressure System. (a) Brown shades represent sparse vegetation in deserts, whereas progressively darker shades of green represent progressively denser vegetation such as the Amazon rainforest. The extensive deserts in northern Africa and the Arabian Peninsula exist because these regions are dominated by the STH. (b) The Rub al-Khali, or Empty Quarter, of the Arabian Peninsula is a large sea of sand, as large as all of France, with no roads and hardly any rainfall.

Which one of the following answers correctly describes the location of the ITCZ in this June 2004, GOES east infrared satellite image? White areas represent clouds.

a) The ITCZ is the zone of high pressure seen in the center of the Atlantic Ocean east of Florida.

b) The ITCZ is the band of clouds in the United States that extends from Texas into the Great Lakes region.

c) The ITCZ is the band of clouds that extends east and west through the northern part of South America.

d) The ITCZ is the large zone of clear skies in the southern part of South America.

the southeasterly trade winds. At low latitudes, this flow to and from the ITCZ and STH forms a convection loop known as a **Hadley cell** (see Figure 6.16 again).

Midlatitude Circulation

With a basic understanding of tropical circulation, it is time to examine the way the atmosphere circulates in the midlatitudes. In general, the primary purpose of midlatitude circulation is to mix the cool polar air that originates at high latitudes and the warm tropical air that develops at lower latitudes. With this mixing in mind, the midlatitudes are the regions where these contrasting air masses converge. The following discussion describes the circulatory processes in the midlatitudes and how they function in the context of balancing temperature differences.

The focal point of midlatitude circulation is the **polar front**, which on average occurs at about 60° N and S latitudes (Figure 6.19). This atmospheric feature is the contact between cold air that originates at very high latitudes and the relatively warm air that streams

---

Hadley cell   *Large-scale convection loop in the tropical latitudes that connects the Intertropical Convergence Zone (ITCZ) and the Subtropical High (STH).*

Polar front   *The contact in the midlatitudes between warm, tropical air and colder polar air.*

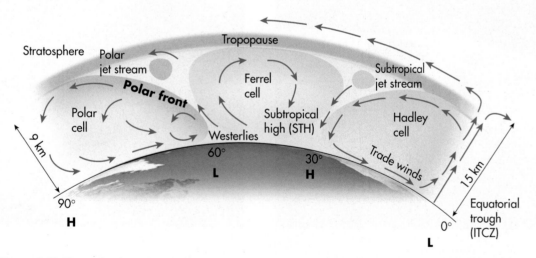

Figure 6.19 The polar front in the Northern Hemisphere. The polar front is the boundary between warm air (to the south) and cold polar air (to the north). Note how air from the STH flows toward the polar front.

northward from the tropical latitudes. You can see how tropical air is pumped into the midlatitudes by looking again at Figure 6.16. For the purposes of this discussion, simply focus your attention on the Northern Hemisphere and notice that the midlatitudes are north of the STH. Remember that this pressure system is centered at about 30° N and consists of warm, dry air that spirals downward in a clockwise fashion. As this descending air reaches the surface, it diverges, with air on the southern side of the system flowing toward the ITCZ in the form of northeasterly winds. Air on the north side of the STH, in contrast, flows northward in the form of southwesterly winds. It is this southwesterly air that flows toward the polar front, where it converges with air flowing southward from the highest latitudes. This southwesterly flow of air contributes to the westerly winds, or **westerlies**, that prevail in the midlatitudes.

The midlatitude westerlies are especially important because they often flow at very high speeds, reaching velocities of between 350 km/h and 450 km/h (about 200 mi/h to 250 mi/h) in the upper troposphere. The winds reach these extreme speeds due to a distinct temperature gradient along the polar front,

which results in a very steep pressure gradient as well. Given the seasonal effect created by the Earth–Sun geometric relationship, this temperature/pressure gradient is very weak during the summer months but strengthens with the progression of winter. The resulting river of rapidly moving air is known as the **polar front jet stream** and occurs at altitudes of 10 km to 12 km (about 30,000 ft to 40,000 ft; see Figures 6.19 and 6.20). Like the ITCZ, the polar jet stream is often easy to see on satellite imagery because a band of clouds forms along this line (such as in Figure 6.20b).

Although the term *westerly winds* implies that midlatitude winds are flowing straight from west to east, this is not always the case. In fact, the smooth westward flow in the upper-air circulatory system frequently develops distinct undulations, called **Rossby waves**. Rossby waves along the polar front are the mechanism through which significant temperature differences on either side of the front are moderated, especially during the winter months. Figure 6.21 shows how the process develops.

At any given location in the midlatitudes, the flow of the polar jet stream essentially follows a smooth west-to-east path for several weeks. This circulatory pattern is typically called **zonal flow** because cold Arctic air is

Westerlies   *Midlatitude winds that generally flow from west to east.*

Polar front jet stream   *River of high-speed air in the upper atmosphere that flows along the polar front.*

Rossby waves   *Undulations that develop in the polar front jet stream when significant temperature differences exist between tropical and polar air masses.*

Zonal flow   *Jet stream pattern that is tightly confined to the high latitudes and is thus circular to semicircular in polar view.*

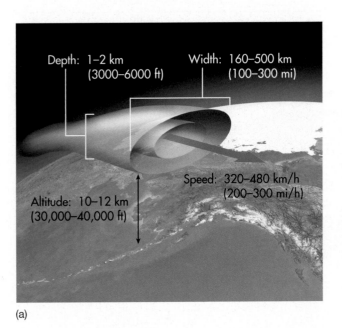

Depth: 1–2 km
(3000–6000 ft)

Width: 160–500 km
(100–300 mi)

Altitude: 10–12 km
(30,000–40,000 ft)

Speed: 320–480 km/h
(200–300 mi/h)

(a)

(b)

Figure 6.20 Jet streams. (a) Jet streams consist of rivers of air that flow at very high speeds. (b) Several jet streams flow in the upper atmosphere and can frequently be identified by a band of clouds, such as in this image of the Nile/Red Sea area in the Middle East.

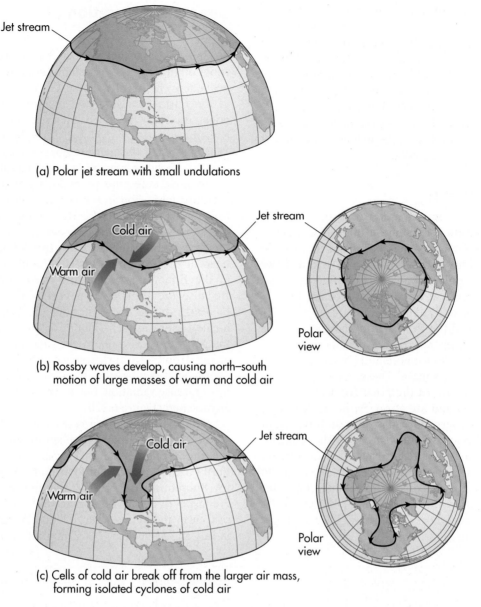

(a) Polar jet stream with small undulations

(b) Rossby waves develop, causing north–south
motion of large masses of warm and cold air

Polar
view

(c) Cells of cold air break off from the larger air mass,
forming isolated cyclones of cold air

Polar
view

Figure 6.21 Development of Rossby waves in the midlatitudes of the Northern Hemisphere. The progressive development of jet stream undulations along the polar front ultimately results in pinching off of cold air pools and the reestablishment of zonal flow.

confined to a small zone at very high latitudes. As the temperature contrast on either side of the polar front increases, the midlatitude atmosphere responds by forming an undulation in the jet stream, beginning the Rossby wave. The core of this undulation becomes a midlatitude cyclone rotating counterclockwise. With the continued development of this Rossby wave, the westerly winds no longer flow directly west to east, but have significant northerly and southerly components as well, due to the counterclockwise circulation. In contrast to zonal flow, the variable flow associated with a developing Rossby wave is called **meridional flow**, because the air frequently flows parallel to the meridians.

With the onset of meridional flow, warm air on the east side of the wave pushes poleward because the winds are southerly. At the same time that this northward push of warm air occurs, cold air from the north plunges southward on the west side of the system, where the winds are northerly. Such an influx of cold air into a region is called an *Arctic outbreak* and can result in extremely cold air reaching latitudes far south of its origin. If you have spent any winters in the northern part of the

Meridional flow *Jet stream pattern that develops when strong Rossby waves exist and the polar front jet stream flows parallel to the meridians in many places.*

United States, you probably know that an Arctic out-break can bring extremely cold temperatures into the region for a long time. An extreme Arctic outbreak can even bring freezing temperatures as far south as Florida. As time progresses, the cold pool of air can literally be pinched off from the main body of Arctic air, resulting in the reestablishment of zonal flow conditions. Such a pool of cold air can persist in a region for several weeks, gradually warming because of the higher Sun angles at these somewhat lower latitudes.

Polar Circulation

In contrast to the complex patterns associated with tropical and midlatitude circulation, atmospheric circulation in the polar regions is associated with a simple circulatory loop known as a *polar cell*. Air that flows northward at the polar front cools considerably and subsequently descends at very high latitudes, producing a weak high-pressure system (see Figure 6.16 again). This system is called the **Polar High** and consists of a mass of descending air that rotates in a clockwise fashion in the Northern Hemisphere and counterclockwise in the Southern Hemisphere. These rotating systems direct cold, dry air toward the polar front in the form of **polar easterlies** and are strongest in the Northern Hemisphere, where large landmasses exist. If you want to review why this geographic pattern exists, return to Figure 5.12 to view the annual range of temperature on Earth and note the variation that exists between the hemispheres. As a result of this variability, the Polar High located over northern continental regions of the Earth makes northernmost Canada and Siberia bitterly cold and dry during the winter.

Seasonal Migration of Pressure Systems

So far in this discussion, the focus has been on the basic distribution of air pressure systems and how they distribute heat energy in the atmosphere. Given your understanding of the Earth–Sun geometric relationship, however, you might also suspect that a distinct seasonal component occurs in association with global atmospheric circulation. In fact, a powerful seasonal component does come into play. Figure 6.22 shows how the seasonal migration occurs.

A good reference point for studying the seasonal pressure migration is the subsolar point and its relationship with the ITCZ. Remember that the ITCZ migrates with the subsolar point. For example, during the Spring and Fall Equinoxes, the Sun is directly overhead at the Equator. This geographic position results in the Equator receiving the most intense radiation. Because the Equator is generally the warmest zone on Earth during these months, the ITCZ is generally located there as well (Figure 6.22a). During winter in the Northern Hemisphere, the subsolar point is located at the Tropic of Capricorn because the Sun is directly overhead at that latitude. Because this zone receives the most intense radiation, this is where the ITCZ is generally located (Figure 6.22b).

Similarly, the pattern is reversed with the approach of the Northern Hemisphere spring and summer. With the coming of the Northern Hemisphere Spring Equinox, the ITCZ follows the subsolar point back to the Equator. Subsequently, the ITCZ migrates into the Northern Hemisphere during April and May, reaching the Tropic of Cancer in the latter part of June. As the days shorten in July and August, the ITCZ migrates back to the south.

Polar High   *Zone of high atmospheric pressure at high latitudes.*

Polar easterlies   *Band of easterly winds at high latitudes.*

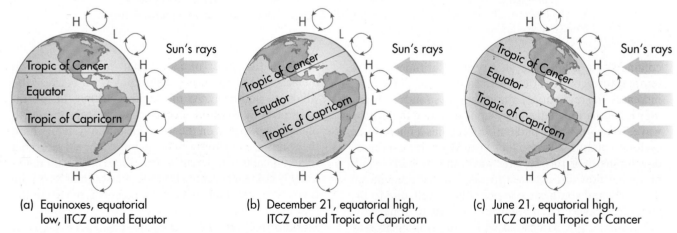

(a) Equinoxes, equatorial low, ITCZ around Equator

(b) December 21, equatorial high, ITCZ around Tropic of Capricorn

(c) June 21, equatorial high, ITCZ around Tropic of Cancer

Figure 6.22 Seasonal migration of atmospheric pressure. The positions of the major high- and low-pressure systems during (a) Equinoxes, (b) Northern Hemisphere Winter Solstice, and (c) Northern Hemisphere Summer Solstice.

Global Atmospheric Circulation

Now that you have investigated atmospheric circulation and the seasonal migration of pressure systems, a good review is to observe the patterns in animated form. Go to the *GeoDiscoveries* website and select the module *Global Atmospheric Circulation*. This module demonstrates the specific components of the global circulatory system and how they move. Once you view this animation, be sure to answer the questions at the end of the module to test your understanding of this process.

Remember, however, that the ITCZ is not the only large pressure system that moves seasonally. *All of them do*. This migration occurs because the pressure systems maintain a more or less consistent distance from one another across the surface of Earth. Thus, when the ITCZ migrates, all the systems move (see Figure 6.22). You will discover in Chapter 9 that this seasonal migration of pressure systems influences the distribution of global climates in a major way.

Monsoonal Winds One of the most important outcomes of the seasonal migration of air pressure systems is the **monsoon**. The monsoon is a cyclical shift of the prevailing wind direction that occurs at a subcontinental scale over the course of a year. This process is a good example of how closely geographic variables are integrated because it is related both to the seasonal migration of the ITCZ and the maritime vs. continental

effect described in Chapter 5. Although a form of monsoon occurs in many subtropical places around Earth, including the southwestern United States, it is most pronounced in Southeast Asia because this is where the greatest seasonal fluctuation of the ITCZ occurs.

In order to understand how the monsoon functions, let's compare the geographic variability that exists with respect to the seasonal migration of the ITCZ (Figure 6.23). Beginning first with the Americas, notice the ITCZ is located in South America throughout the year, and migrates from the central part of the continent in January to the northern part of the continent in

---

Monsoon *The seasonal change in wind direction that occurs in subtropical locations due to the migration of the Intertropical Convergence Zone (ITCZ) and Subtropical High (STH) Pressure System.*

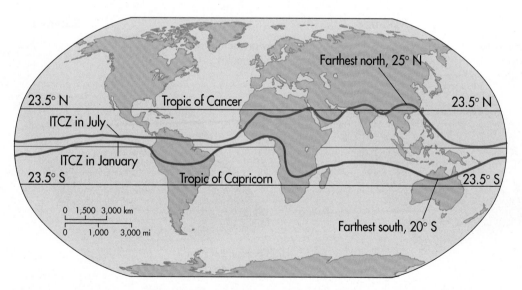

Figure 6.23 Position of ITCZ during the Solstices. The red line marks the position of the ITCZ in January (bottom line) and the blue line indicates its position in July (top line). Note the extent of seasonal migration in Asia as compared to South America.

The Asian Monsoon

As with many geographic processes, the Asian monsoon is easy to visualize in animation. Go to the *GeoDiscoveries* website and select the module *The Asian Monsoon*. This module clearly illustrates the movement of pressure systems over the course of the year and how they contribute to precipitation patterns. After you view this animation, be sure to answer the following questions to test your understanding of this concept.

July. In other words, the ITCZ migrates approximately 20° latitude in South America over the course of the year. How does this migration range compare with Asia? Figure 6.23 shows that the position of the ITCZ varies dramatically in Asia, ranging from northern Australia in January to northern India in July. This is a migration of about 45° latitude, or twice that which occurs in South America. The primary reason for this tremendous differential is the huge size or continentality of the Asian continent (as compared to the Americas), which results in the extreme range of temperature you saw in Figure 5.12.

How does this large temperature range and seasonal ITCZ migration create the Asian monsoon? A good place to examine the monsoon is India (Figure 6.24). In the winter a strong outflow of very dry, continental air moves from central Asia south across India. This *winter monsoon* occurs because the pressure gradient slopes steeply from the very strong Siberian high-pressure

system (in central Asia) toward the ITCZ, which is far to the south over the Indian Ocean (Figure 6.24a).

During the summer, the direction of surface winds reverses, causing the *summer monsoon*. This reversal occurs because temperatures over the Asian landmass increase significantly due to the high Sun angle in summer. As a result, the ITCZ shifts northward over Asia (see Figure 6.23 again), bringing low pressure into the region. Given the seasonal shift in the large pressure systems, a subtropical high is now located over the Indian Ocean and the pressure gradient slopes steeply toward the continent (Figure 6.24b). The warm winds from this high-pressure anticyclone pick up moisture as they pass over the Indian Ocean. As the air moves north along the pressure gradient, it is lifted initially by the warmth of the Indian landmass and then intensively by the high Himalaya Mountains. As the air rises, it begins the process of cooling and precipitation occurs. The high precipitation of the monsoonal rainy season lasts from June to September.

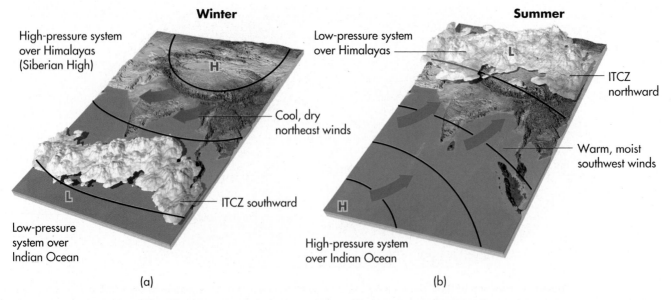

Figure 6.24 Monsoon flow in India during (a) winter and (b) summer. The flow of air is distinctly related to the seasonal position of the ITCZ and the relative temperature of the Asian landmass to the north.

Global Atmospheric Circulation and Water Vapor Movement

An excellent way to visualize global atmospheric circulation is to observe the movement of water vapor. In this context, go to the *GeoDiscoveries* website and select the module *Global Atmospheric Circulation and Water Vapor Movement*. This excellent simulation models the movement of water vapor on Earth over the course of the year as it relates to the wind patterns discussed so far. After you watch this simulation, be sure to answer the questions at the end of the module to test your understanding of this concept.

Human Interactions: Harnessing Wind Energy

People have used flowing air as an energy source for thousands of years. Ancient Egyptians, for example, used wind to sail ships on the Nile River. In the late 12th century, windmills were developed in northwestern Europe to grind flour and pump water out of low-lying areas. Many such windmills still dot the landscape in Holland (Figure 6.25). Wind energy was the power source behind the great sailing voyages during the Age of Exploration that set the stage for the modern world.

Given increasing demands for electricity and for renewable, clean sources of energy, wind power technology has advanced dramatically in the past 30 years. This shift began with the development of the modern wind turbine in the early 1980s following the oil shortages of the 1970s. Like the windmills of old, modern wind machines use large blades that slow wind speed and collect kinetic energy. As the wind flows over the top of the curved blade, it causes lift and a turning motion that spins a drive shaft connected to a turbine. This turbine generates electricity. To give you an idea of how much energy a turbine can produce, consider that a 5 megawatt turbine can produce more than 15 million kilowatt hours (kWh) of electricity in a year, which is enough to power more than 1400 households.

Although wind energy currently comprises only about 1% of global and U.S. energy production, a major push is currently under way to increase its development. What are the pros and cons associated with this source of electricity? The good things about wind energy are fairly obvious—specifically, that it is a clean and an efficient way to produce electricity. Unlike coal power plants, it produces no greenhouse gas emissions and thus does not contribute to climate change. On the other hand, some problems exist. The blades of wind turbines can strike migratory birds, for example, and turbines can be damaged by strong storms and by lightning. To some people, wind turbines are aesthetically unattractive and are too loud, although quieter turbines are currently being developed.

The biggest problem associated with wind turbines is that they are useless if the wind does not blow. In order to reach maximum efficiency, the wind needs to blow at over 10 mph for long periods of time. Given this requirement, many parts of the world are not suitable for developing a sustainable wind-energy program. In the United States, for example, the southeastern part of the country is poorly suited for wind energy, as are significant parts of the American West. Much of the Midwest is also less than

Figure 6.25 A Dutch windmill. Windmills like this were first built in the Netherlands in the 12th century to grind flour and pump water from swampy areas.

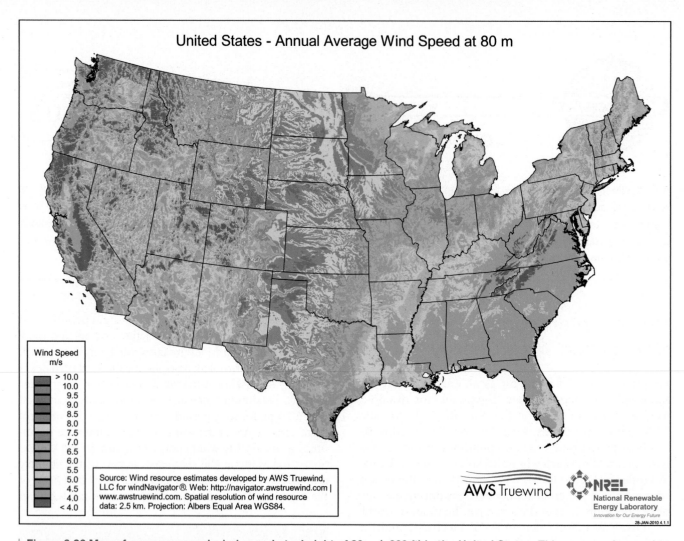

United States - Annual Average Wind Speed at 80 m

Wind Speed
m/s

> 10.0
10.0
9.5
9.0
8.5
8.0
7.5
7.0
6.5
6.0
5.5
5.0
4.5
4.0
< 4.0

Source: Wind resource estimates developed by AWS Truewind,
LLC for windNavigator®. Web: http://navigator.awstruewind.com |
www.awstruewind.com. Spatial resolution of wind resource
data: 2.5 km. Projection: Albers Equal Area WGS84.

AWS Truewind

NREL
National Renewable
Energy Laboratory
Innovation for Our Energy Future
28-JAN-2010 4.1.1

Figure 6.26 Map of average annual wind speed at a height of 80 m (~260 ft) in the United States. This map can be used to predict the suitability of regions for wind energy potential. Note that the central United States is particularly well suited for wind energy production, whereas much of the southeastern and western parts of the country have low potential. (*Source:* NREL.)

Figure 6.27 The Horse Hollow Wind Energy Center in Texas. This facility has over 400 turbines, which collectively produce enough energy for 220,000 homes per year.

ideal (Figure 6.26). On the other hand, many parts of the country, such as the Great Plains and along many mountain ranges, are excellent places to develop wind energy.

To maximize the production of wind energy in these places, extensive wind farms are being developed in many localities. The world's largest wind farm is the Horse Hollow Wind Energy Center in north central Texas (Figure 6.27). This facility contains over 400 turbines spread over 190 km² (73 mi²) and generates enough power for about 220,000 homes per year. Similar large wind farms occur elsewhere in the Great Plains, as well as California, which produces more than twice as much wind energy as any other state in the United States. An interesting development to watch is the Cape Wind project off the coast of Cape Cod, Massachusetts. In November 2008 voters along the south shore of the state approved a nonbinding ballot to allow the construction of 130 large turbines far offshore. Similar wind farms may one day be built in Lake Michigan where the potential for energy production is high.

1. The fundamental consequence of global air movement is to equalize the differences in temperature and pressure between the tropics and the poles.

2. The global circulatory system consists of a series of separate, but connected, pressure systems that distribute air around Earth.

3. Tropical circulation occurs in association with convection processes in Hadley cells, forming the Intertropical Convergence Zone (ITCZ) and the Subtropical High (STH) Pressure System. The ITCZ is a zone of low pressure where the winds are easterly. In contrast, the STH is a high-pressure system that rotates clockwise in the Northern Hemisphere and counterclockwise in the Southern Hemisphere.

4. Midlatitude circulation is driven by contrasting air temperatures on either side of the polar front. A major feature of midlatitude circulation is the jet stream, which consists of high-speed westerly winds that flow along the polar front. During the summer, the jet stream exhibits zonal flow that is confined to high latitude. As fall approaches, the jet stream begins to develop distinct undulations that are associated with Rossby waves.

5. The monsoon is a distinct circulatory feature that exists in tropical areas where large land bodies border oceans. Landmasses are warm relative to water in summer, resulting in low pressure over land and onshore airflow. During winter, water bodies are warm relative to land and the winds reverse their flow direction.

6. People are now systematically harnessing wind power to supplement current energy supplies. Although environmentally friendly, challenges remain because the production of wind energy is less efficient than conventional energy sources and is unpredictable in many places.

Local Wind Systems

So far, we have examined circulatory systems and associated wind patterns that occur over regional and global scales. On a much smaller scale, local wind systems result from temperature differences created by topography, proximity to large bodies of water, or other geographic features.

Land–Sea Breezes

Land–sea wind systems are localized along the shores of major water bodies, such as oceans or large lakes. These winds form due to the different heating and cooling characteristics of continents and water, as we discussed in Chapter 5 (the maritime vs. continental effect). During the day, air over land heats up more rapidly than air over water, resulting in a small zone of low pressure that forms over the land near the shore. Over water, in contrast, the air pressure is relatively high because the air is cooler and denser as it sinks. This difference sets in motion a **sea breeze**: the air flows from the sea toward the land; that is, from high to low pressure (Figure 6.28a).

Sea breeze *Daytime circulatory system along coasts where winds flow from a zone of high pressure over water to a zone of relatively low pressure over land.*

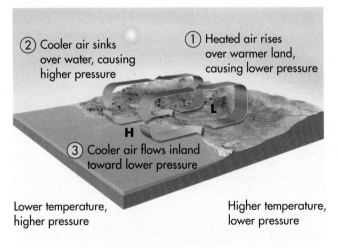

(a) Daytime sea breeze

② Cooler air sinks over water, causing higher pressure

① Heated air rises over warmer land, causing lower pressure

③ Cooler air flows inland toward lower pressure

Lower temperature, higher pressure

Higher temperature, lower pressure

(b) Nighttime land breeze

② Cooler air sinks over land, causing higher pressure

① Heated air rises over warmer water, causing lower pressure

③ Air flows offshore toward lower pressure

Higher temperature, lower pressure

Lower temperature, higher pressure

Figure 6.28 Coastal wind systems and the maritime vs. continental effect. (a) A sea breeze develops when air flows inland from the water during the daytime. (b) A land breeze occurs at night when air flows toward the water.

This local circulatory pattern reverses at night because air over the land cools more rapidly than air over the sea. As a result, the pressure systems switch locations, with relatively higher pressure over land and low pressure over the water. When this adjustment occurs, the air reverses direction and moves from land toward sea in a **land breeze** (Figure 6.28b).

Topographic Winds

In addition to the winds that develop along coastlines, several kinds of winds form due to variations in topography (Figure 6.29). These winds generally form because large temperature differences can occur with elevation in hilly or mountainous landscapes. For example, a **valley breeze** is a wind system that develops when mountain slopes heat up due to re-radiation and conduction over the course of the day. When this occurs, a zone of relatively low pressure develops on the mountain slopes, whereas high pressure is found in the lowlands below. As a result, air can flow upslope to fill the pressure void created by the upward movement of air over the heated slopes. Like the coastline breezes,

the situation in mountainous regions reverses at night, resulting in a downslope **mountain breeze**.

A more extreme example of mountain breezes is called **katabatic winds**, which form in particularly cold places such as Greenland and Antarctica. From the Greek *katabatik*, which means "descending," this process is most common during the winter months, when extremely cold air accumulates over higher-altitude regions covered by ice sheets. This extremely cold, dense air then flows downhill under the force of gravity. These winds sometimes flow at great speeds that can be quite destructive, especially where the air is funneled down a narrow valley. Although the air associated with katabatic winds typically warms when descending, it usually remains colder than the air that it replaced.

The last kind of topographically related wind process is the **Chinook winds** (also called *foehn winds*). These winds also form in mountainous regions, but occur only when a steep pressure gradient develops along the range (Figure 6.30). For this gradient to develop, a high-pressure system must be on the side of the range that faces the direction of oncoming winds; this side is called the **windward side**. On the other side of the

Land breeze *Nighttime circulatory system along coasts where winds from a zone of high pressure over land flow to a zone of relatively low pressure over water.*

Valley breeze *Upslope airflow that develops when mountain slopes heat up due to re-radiation and conduction over the course of the day.*

Mountain breeze *Downslope airflow that develops when mountain slopes cool off at night and relatively low pressure exists in valleys.*

Katabatic winds *Downslope airflow that evolves when pools of cold air develop over ice caps and subsequently descend into valleys.*

Chinook winds *Downslope airflow that results when a zone of high air pressure exists on one side of a mountain range and a zone of low pressure exists on the other.*

Windward side *The side of a mountain range that faces oncoming winds.*

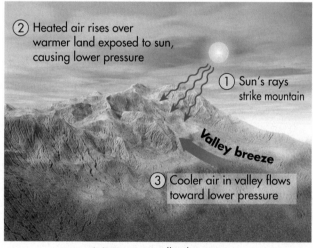

(a) Daytime valley breeze

(b) Nighttime mountain breeze

Figure 6.29 Local wind flow caused by differences in elevation. (a) During the daytime, air flows upslope in a valley breeze. (b) At night, the flow reverses in a mountain breeze.

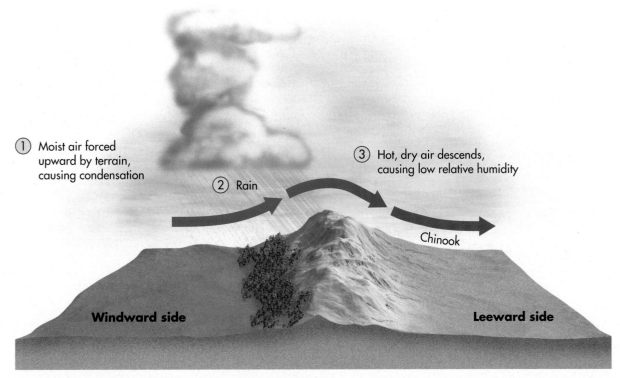

Figure 6.30 The development of Chinook (foehn) winds. For these winds to develop, the pressure gradient must slope from the windward to leeward side of a mountain range.

① Moist air forced upward by terrain, causing condensation

② Rain

③ Hot, dry air descends, causing low relative humidity

Chinook

Windward side

Leeward side

range, or **leeward side**, a low-pressure system must be in place. When this combination of systems occurs, air flows along the pressure gradient—that is, from the windward to leeward side of the mountain range. The resulting airflow moves downslope on the leeward side, resulting in dry air that warms as it descends.

An excellent example of a valley wind is the *mistral wind*, which occurs along the eastern Mediterranean coast of France during the winter months. This wind develops when cold air sweeps across France from the English Channel, hits the Alps, and spills down the Rhône Valley into the coastal region. On days when this process is especially active, winds can gust to over 120 km/h (75 mph). Another great example of downslope winds is the *Santa Ana winds* in Southern California. These winds occur in winter when high-pressure systems dominate over Nevada and Utah. On the southern side of the high, air flows west toward California, rising up and over the Sierra Nevada Mountains. When the air descends on the western side of the Sierra, it picks up speed as it rushes into the Los Angeles area. These winds are often beneficial because they clear smog from the city. On the other hand, they can fan wildfires and pose a danger in that regard.

Leeward side *The side of a mountain range that faces away from prevailing winds.*

Oceanic Circulation

Gyres and Thermohaline Circulation

So far in this chapter, we have seen how atmospheric circulatory systems operate and how they distribute heat energy around Earth. Another way that heat energy is moved is through oceanic circulation (Figure 6.31), which is strongly related to atmospheric circulation in many ways. In fact, surface currents in the ocean are driven by winds through the transfer of energy from the air to the water by friction. As a result, the direction of surface ocean currents is related to the same pressure factors that influence atmospheric circulation. For example, the tropical easterly winds produce easterly oceanic currents in the low latitudes.

The Coriolis force also plays a major role in the movement of oceanic currents. Notice in Figure 6.31, for example, that ocean currents in the Northern Hemisphere generally move in a clockwise direction, moving warm tropical water into higher latitudes. A great example of such a current is the Gulf Stream, which transports warm water north from the tropical latitudes along the east coast of the United States (Figure 6.32). Currents in the Southern Hemisphere perform the same function, but circulate counterclockwise. A great example of such a current is the Humboldt Current, which flows northward along the west coast of South America. This current delivers cold water from

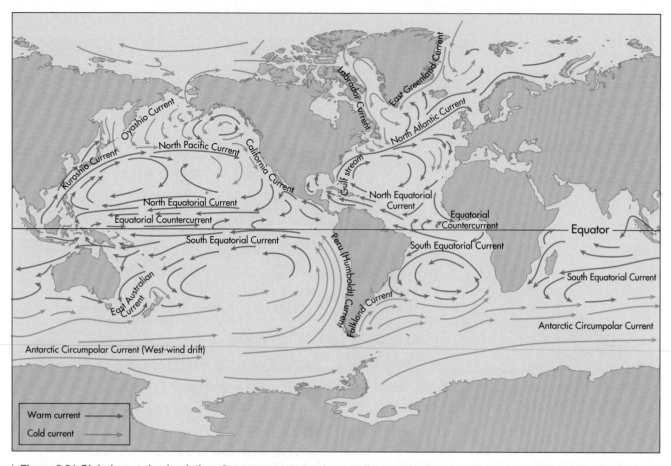

Figure 6.31 Global oceanic circulation. Ocean currents fundamentally move in the same direction as the winds. Note, for example, the easterly current along the Equator and how it relates to the easterly winds. The primary difference between ocean and air currents is that the continents interrupt the ocean currents, forming distinct gyres.

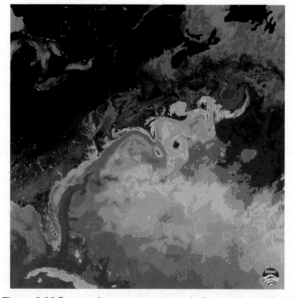

Figure 6.32 Sea-surface temperatures in the western Atlantic. NOAA-7 orbiting satellite image showing sea-surface temperatures for a week in April. Red represents the warmest water, which is flowing northeastward along the east coast of North America in the Gulf Stream. Greens, blues, and purples are progressively cooler sea-surface temperatures.

the Antarctic latitudes to the equatorial region where it warms and then flows west as the South Equatorial Current.

The primary difference between oceanic and atmospheric circulatory processes is that the continents block the movement of water. This blockage results in distinct circulatory systems, called **gyres**, such as the one in the Atlantic basin. In addition to the oceanic circulation across the surface, a slow mix of water also occurs vertically between layers of the ocean. These currents, which together make up the **thermohaline circulation**, also called the *oceanic conveyor belt*, that links all the world's oceans, are generated because of regional differences in water density that evolve through variations in temperature and salinity. In general, water at the surface of the ocean is typically warmer and less salty than deeper water. A good place to witness the beginning of

Gyres *Large oceanic circulatory systems that form because currents are deflected by landmasses.*

Thermohaline circulation *The global oceanic circulatory system that is driven by differences in salinity.*

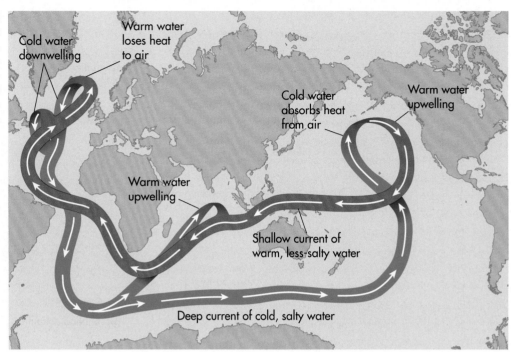

Figure 6.33 Surface and deep water circulation loop in the world oceans. Note the locations where water upwells and downwells. These combined processes drive the circulation of ocean currents around the world.

Cold water downwelling

Warm water loses heat to air

Cold water absorbs heat from air

Warm water upwelling

Warm water upwelling

Shallow current of warm, less-salty water

Deep current of cold, salty water

this system is in the western part of the tropical Atlantic basin, shown in Figure 6.33. Here, as part of the North Atlantic gyre, warm water flows northward in the Gulf Stream (as illustrated in Figure 6.32). As this warm water travels to higher latitudes, extensive evaporation occurs, which increases the relative salinity of the current. Along with this increased salinity, the surface waters gradually cool as they continue northward.

Due to decreasing temperature and high salinity, the water density increases and the water sinks in the northern Atlantic in Figure 6.33. In other words, the sinking water becomes a **downwelling current**. This downward motion is enhanced by the development of sea ice at high latitudes, which pushes most salt into the water below the ice and thus further increases the salinity of the water and its density compared to the water around it. The resulting downwelling current subsequently flows at great depths to the southern part of the Atlantic basin. From there, it flows to the east between Antarctica and Australia to the Pacific Ocean, where it then flows northward toward Alaska. As the water moves into the tropical regions of the Pacific, it warms and becomes an **upwelling current** that moves back to the surface. This pattern is

repeated in several places around the world's oceans. To give you an idea of the time involved for ocean currents like this to move, consider recent research demonstrating that it takes about 2000 years for water to flow from Great Britain in the northern Atlantic Ocean to the southern Pacific Ocean!

El Niño

As with many other Earth processes that can depart from the normal pattern, the oceanic circulatory system sometimes functions in unusual ways. One of the most important anomolies in oceanic circulation is the *El Niño* phenomenon, which is a reversal of the "normal" flow in the tropical Pacific Ocean. During most years, the dominant current in the tropical Pacific is easterly along the South Equatorial Current (see Figure 6.31). As you might suspect, this easterly flow is caused by the easterly trade winds that flow from strong subtropical high-pressure systems to the ITCZ (Figure 6.34).

When the easterly pattern is in place, the flow of water away from the South American coast allows cold water from the deep-flowing Humboldt Current (which flows northward from Antarctica along the coast of Chile) to upwell (rise to the surface) in the eastern Pacific near the Equator. Due to this upwelling, the overlying air is cool and descending, which, in turn, results in little evaporation of moisture from the ocean and, consequently, little precipitation. You can see the effect of this low precipitation in the lack of vegetation along the west coast of South America in Figure 6.18. As the water flows westward along the Equator, however, it warms because of the high Sun angle. By the time the

---

Downwelling current   *A current that sinks to great depths within the ocean because water temperature drops and salinity increases.*

Upwelling current   *A current that ascends to the surface of the ocean because water temperature warms and salinity decreases.*

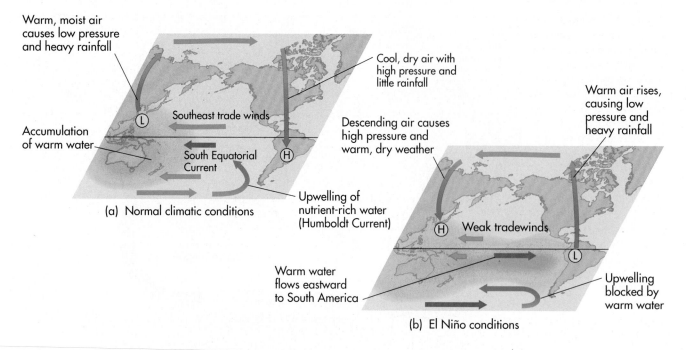

Warm, moist air causes low pressure and heavy rainfall

Cool, dry air with high pressure and little rainfall

Accumulation of warm water

Southeast trade winds

South Equatorial Current

Descending air causes high pressure and warm, dry weather

Upwelling of nutrient-rich water (Humboldt Current)

(a) Normal climatic conditions

Warm air rises, causing low pressure and heavy rainfall

Weak tradewinds

Warm water flows eastward to South America

Upwelling blocked by warm water

(b) El Niño conditions

Figure 6.34 Atmospheric circulation and surface water flow in the Equatorial Pacific. (a) Normal conditions occur when strong easterly flow pushes warm water into the western Pacific. (b) An El Niño occurs when easterly flow weakens, allowing warm water to collect along the South American coast. Note the relationship between precipitation and the location of pressure systems.

current reaches the western Pacific, the water is quite warm, which results in overlying (atmospheric) low pressure and heavy rains in that part of the basin.

For some as yet unexplained reason, the "normal" circulatory pattern in the Pacific Ocean reverses every 3 to 8 years. The apparent cause of this reversal is that the normally strong tropical easterlies weaken, allowing a distinct westerly flow to develop in surface ocean waters. This opposing pattern is called El Niño, derived from the expression *Corriente del Niño*, or the "Current of the Christ Child," used by Peruvian fishermen to describe the influx of warm waters into the eastern Pacific Ocean. This reversal of surface flow usually occurs around Christmas time (hence the name) and is particularly noticeable to fishermen because the warmer waters are relatively sterile when compared to the nutrient-rich upwelling current (Humboldt Current) that normally supports superb fishing. In addition to the depleted fish industry, an El Niño brings intense storms to the normally dry parts of South America (and even the southwestern United States) and drought to the western Pacific. This weather reversal occurs because the atmospheric pressure in the eastern Pacific is relatively low, whereas it is relatively high in the western Pacific. Remember that air flows from areas of high pressure to low pressure, which, in this case, pushes the warm surface waters to the east side of the Pacific.

KEY CONCEPTS TO REMEMBER ABOUT OCEANIC CIRCULATION

1. Oceanic circulation is driven by winds and generally flows in the same direction as atmospheric circulation. Due to the presence of the continents, however, distinct circulatory cells, or gyres, exist in the oceans.

2. Like the atmosphere, the primary consequence of oceanic circulation is to redistribute heat energy around Earth.

3. Warm, generally low-salinity currents flow along the top of the oceans. In contrast, cooler, saltier water flows at great depths.

4. Normal equatorial circulation in the Pacific is easterly, resulting in extremely warm waters in the western Pacific and upwelling cold waters in the eastern part of the basin. This normal pattern causes precipitation in the west and dry conditions in the east.

5. An El Niño occurs when the normal (equatorial) circulatory system in the Pacific is reversed. This reversal results in drought in the western Pacific and strong storms in the eastern Pacific.

El Niño

To investigate El Niño more thoroughly, go to the *GeoDiscoveries* website and select the module *El Niño*. This module contains four parts that demonstrate how the patterns evolve, recover, and impact the coast of Peru. Once you complete the animation, be sure to answer the questions at the end of the module to test your understanding of this concept.

THE BIG PICTURE

In this chapter, you learned about how the atmosphere flows and the various factors that cause distinct circulatory patterns and their geographic distributions. Not only is this understanding critical to why winds fundamentally occur, but it also helps explain other physical processes on Earth, such as the distribution of vegetation, the evolution of soils, and the location of major rivers.

One of the most important processes associated with atmospheric circulation is condensation and precipitation. You have undoubtedly witnessed a rainshower many times and probably have experienced heavy rainfall or snowfall at other times in your life. Were you not amazed by these events, which may have resulted in flash flooding, impassable roads, and school and business shutdowns? Even so, did you really understand why the rain or snow was falling? With these ideas in mind, Chapter 7 discusses the various ways that water flows in the atmosphere, including why clouds form and precipitation occurs.

1. Air pressure refers to the weight of air distributed on the surface of Earth. The two basic kinds of pressure systems are high (anticyclones) and low (cyclones). High-pressure systems in the Northern Hemisphere are columns of air that rotate clockwise and descend toward the surface of Earth within their core. In contrast, low-pressure systems in the Northern Hemisphere are columns of air that lift from the Earth's surface within their core as they rotate counterclockwise. The rotation of these respective systems is opposite in the Southern Hemisphere.

2. The speed and direction of airflow depend on many variables. The pressure gradient force refers to the difference in air pressure that exists between adjacent areas. The Coriolis force refers to the deflection that occurs in winds due to the rotation of Earth on its axis. Features on the Earth's surface, such as mountains, forests, and buildings, create a frictional force that acts opposite to the wind's direction. The combined effect of the pressure gradient force, Coriolis force, and frictional force causes a spiral motion of air in both low- and high-pressure systems.

3. The global circulatory system consists of a series of separate, but connected, pressure systems that distribute air around Earth. Humans are now trying to harness this energy source in a systematic way to supplement current energy supplies.

4. In addition to large-scale wind systems, several regional and local-scale circulation systems occur, including monsoon winds, land–sea breezes, katabatic winds, and Chinook/foehn winds.

5. Oceanic circulation is driven by winds and generally flows in the same direction as atmospheric circulation. Due to the presence of the continents, however, distinct circulatory cells, or gyres, exist in the oceans.

6. An El Niño occurs when the normal (equatorial) circulatory system in the Pacific is reversed. This reversal results in drought in the western Pacific and strong storms in the eastern Pacific.

CHECK YOUR UNDERSTANDING

1. Define the concept of air pressure and discuss how it is measured.

2. Why is air pressure greater at low altitudes rather than high altitudes?

3. Compare and contrast airflow in a high-pressure system with that of a low-pressure system in the Northern Hemisphere.

4. Discuss the three factors that influence large-scale atmospheric circulation.

5. Compare and contrast the various circulatory systems within the global circulation model.

6. Which place is more likely to be a zone of convection, the ITCZ or STH? Why does this pattern occur?

7. Why is the polar front an important feature in the midlatitudes?

8. What are Rossby waves and why are they more likely to form during the winter months than in the summer months?

9. How does Earth–Sun geometry influence the position of atmospheric pressure systems?

10. What are the challenges we face in our attempt to systematically develop wind energy?

11. How is a land–sea circulatory system like a tropical Hadley cell?

12. How are land–sea breezes and mountain breezes similar? What conditions are required for these circulatory systems to develop?

13. Within the Atlantic gyre, what is the primary purpose of the Gulf Stream?

14. Discuss the El Niño pattern, how it differs from the "normal" pattern, and how it impacts weather.

VISUAL CONCEPT CHECK 6.1

Air pressure decreases from the base of Mount McKinley to the top of the mountain. This occurs because air molecules tend to be held closely to Earth due to the force of gravity. Given that more air molecules occur at the base of the mountain than the top, the air pressure (or weight of the air) is greater at the base of the mountain. Conversely, air pressure decreases with altitude because progressively fewer air molecules exist.

VISUAL CONCEPT CHECK 6.2

The high-pressure system in this image is centered over central Europe. This can be determined because that is the approximate center of the zone of clear (or sunny) skies. Given that this anticyclone is in the Northern Hemisphere, it spins clockwise. Thus, northerly winds exist on the eastern side of the system, whereas southerly breezes are on the west. Spain is under the influence of a low-pressure system, as indicated by the dense cloud cover.

VISUAL CONCEPT CHECK 6.3

The answer to this question is c. The ITCZ consists of the band of clouds that extends east and west through the northern part of South America.

CHAPTER SEVEN

ATMOSPHERIC MOISTURE AND PRECIPITATION

Chapter 6 focused on air pressure and the processes associated with atmospheric circulation. As air flows, it carries many things, including dust, volcanic ash, and even small organisms. One of the most important of these atmospheric constituents is water, which moves within the air in vapor, liquid, and solid forms.

Water is a critical part of the global ecosystem because it is necessary for all living things. In addition, water serves an important regulatory function in climate patterns and processes because it moderates temperature. Water vapor in the atmosphere is especially vital because it absorbs heat from the Sun. As water vapor moves around the planet in atmospheric currents, the effect is to moderate surface temperature.

The first part of this chapter focuses on the chemical and physical composition of water and how it changes from one phase to another. From there, the discussion moves on to the way that clouds form and how precipitation occurs. You will find this information useful the next time it snows or an intense storm occurs where you live.

CHAPTER PREVIEW

Physical Properties of Water
   On the Web: *GeoDiscoveries*
   Latent Heat

The Hydrosphere and the Hydrologic Cycle

Have you ever wondered why intense thunderstorms develop? This beautiful image from Montana shows what can happen when convection leads to the development of a thunderstorm. This chapter focuses on the various ways that water changes physical state and how it can occur.

Humidity

    On the Web: *GeoDiscoveries*
    Atmospheric Humidity

Evaporation

Adiabatic Processes

    On the Web: *GeoDiscoveries*
    Adiabatic Processes

Cloud Formation and Classification

Precipitation

    On the Web: *GeoDiscoveries*
    Orographic Processes
    On the Web: *GeoDiscoveries*
    Convectional Precipitation

The Big Picture

LEARNING OBJECTIVES

1. Discuss the physical properties of water and its three phases.

2. Understand the concept of the hydrosphere and hydrologic cycle.

3. Compare and contrast the various kinds of humidity.

4. Explain the concept of dew-point temperature and why it is an important variable in the hydrologic cycle.

5. Describe the nature of adiabatic processes and why the dry and wet adiabatic rates are used in moving bodies of air.

6. Explain the various types of clouds and why they occur.

7. Compare and contrast orographic and convectional uplift.

Physical Properties of Water

Have you ever really thought about the character of water, the way it feels, flows, and freezes? It is a fascinating substance if you really stop to think about it. Water is found almost everywhere on Earth and is absolutely crucial to life as we know it. Water covers 71% of our planet's surface and even comprises about 70% of the human body by weight. It is a very important component of the atmosphere because it stores energy that contributes to global wind patterns. In addition, it plays a major role in the regulation of climate. In order to understand how water behaves in the physical environment, let's first review its basic properties.

Hydrogen Bonding

The logical place to begin a discussion of water is the way that water molecules are bound to each other because the nature of this bond gives water its unique properties.

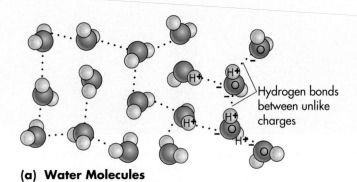

(a) Water Molecules

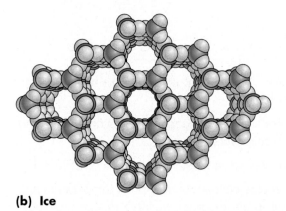

(b) Ice

Figure 7.1 Chemical composition of water. (a) Water molecules are composed of two hydrogen atoms and one oxygen atom, which form partially positive ends of molecules (hydrogen) and partially negative ends (oxygen). Given the attraction between positive and negative charges, relatively weak hydrogen bonds are produced between molecules. (b) Hydrogen bonds are the reason liquid water molecules form solid ice at a comparatively high temperature; it takes more energy to break these bonds than is the case for most other common liquids.

A drop of water is made up of billions of molecules, with each consisting of two hydrogen atoms combined with one oxygen atom (Figure 7.1a). This combination is the basis for water's well-known chemical formula: H_2O. Within a water molecule, oxygen attracts the bonding electrons more strongly than do the hydrogen atoms. Because of this unequal attraction, the oxygen end of the molecule has a partial negative charge, whereas the hydrogen side has a partial positive charge.

Water molecules are attracted to each other because of these contrasting positive and negative charges. The negative end of one molecule is attracted toward the positive end of an adjacent molecule. This is why water molecules stick together in a drop of water and why water occurs in liquid form at normal surface temperatures. (Most chemical compounds are solids or gases at normal surface temperatures.) The attraction between molecules of water is known as *hydrogen bonding* because hydrogen atoms form particularly good bonds between molecules. These bonds are strongest, of course, when water is frozen as ice. In these circumstances, the molecules firmly bond to each other in distinct hexagonal forms (Figure 7.1b).

Another interesting attribute of water is that it has high surface tension. Surface tension results when molecules at the surface of a liquid have a strong attachment to each other but not to the molecules of air above them. This attraction between molecules is particularly strong in water because of hydrogen bonding. Thus, the water molecules pull harder to the sides to create the smallest possible amount of surface area, forming spherical bubbles and droplets. The high surface tension also creates an elastic "skin" on the surface of water, strong enough to allow some kinds of insects (such as water striders) and even small lizards to walk on the surface.

Yet another important feature of water is its ability to move upward in thin openings (or capillaries) against the force of gravity within the soil and plants in a process called **capillary action**. This motion occurs because, through hydrogen bonding, water molecules pull other water molecules along. This process is very important because it enables plants to transport nutrients from their roots up into their stems and leaves. Without this process, trees and other tall vegetation could not exist.

Thermal Properties of Water and Its Physical States

One of the most important characteristics of water is that it absorbs and releases latent heat, which, if you recall from Chapter 4, is hidden energy stored in molecular bonds. This form of hidden energy, which cannot be detected, contrasts with sensible heat that is felt (or "sensed"). The

---

Capillary action *The process through which water is able to move upward against the force of gravity.*

capability of water to store and release latent heat is important because it contributes significantly to atmospheric circulation and helps regulate climate. Understanding this property is also important because it explains how water can exist in the three physical states, or phases, that you are familiar with: solid (as *ice*), liquid (as *water*), and gas (as *water vapor*). The movement of water from one phase to another occurs when hydrogen bonds are formed, loosened, broken entirely, or tightened.

The transformation of these bonds is directly related to the ways that heat energy interacts with water molecules. A simple rule of thumb is that heat energy must be applied to molecules if hydrogen bonds are to be loosened or broken, and extracted from molecules if they are to be formed or tightened. When bonds are loosened, energy is transformed from sensible heat to latent heat, which has a net cooling effect because less sensible heat is "sensed." In contrast, when bonds are strengthened, energy is converted from latent heat to sensible heat, which has a net warming effect.

The thermal properties of water are important because they strongly influence weather patterns on Earth, providing over 30% of the energy required to drive atmospheric processes. For example, much of the energy contained within a thunderstorm is produced when latent heat is released, because water changes rapidly from gas to liquid form (this will be discussed in more detail later). This section of the chapter focuses on the way in which water changes physical states and the amount of latent heat energy released or absorbed during these transformations. As you follow this discussion, refer frequently to Figure 7.2, which shows the various

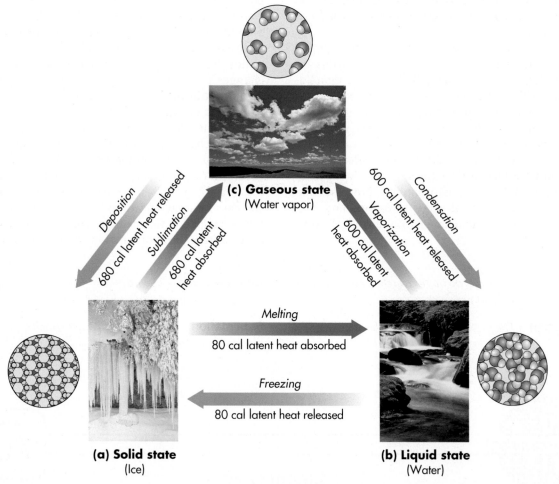

Figure 7.2 Physical states of water. Arrows indicate the pathways that water follows in changing between various states, when latent heat is either absorbed or released. Arrow colors reflect the net cooling or warming effect due to the transfer of sensible heat to latent heat or latent heat to sensible heat, respectively. (a) Water exists as ice at temperatures below 0°C (32°F) under normal air pressure because the molecules slow down enough to lock into hexagonal crystals. (b) At temperatures between 0°C (32°F) and 100°C (212°F), water molecules move freely, but slowly enough to remain attached to one another in the liquid phase. (c) When temperatures exceed 100°C (212°F), all water molecules move fast enough to become completely detached from one another in the vapor phase. (Note that clouds actually contain liquid water as well as water vapor; we see them because the liquid water reflects light.) The number of calories absorbed or released in any given phase change assumes 1 g of water.

pathways that water follows and the amount of latent heat that is absorbed or released.

Let's begin the examination of water phases and latent heat transfers by considering what happens when liquid water turns to ice in the process of **freezing**. Think of this process as one in which progressively more hydrogen bonds develop and tighten. It actually starts when water cools to a temperature below 4°C (39°F) because the motion of water molecules begins to slow considerably, which allows hydrogen bonds to strengthen. When liquid water cools down to 0°C (32°F), the motion of the water molecules slows even further, still more hydrogen bonds develop and tighten, and ice forms. The exact temperature at which water crystallizes to form ice depends on the air pressure. Water has to cool a bit more to freeze at higher air pressure, for example, because as it is squeezed by the descending air, more and more nitrogen and oxygen molecules go into solution. During the process of freezing, 80 calories (cal) of heat energy are released for every gram of water as the *latent heat of freezing.*

Have you ever wondered why ice floats on liquid water, even though it is solid? The reason for this apparent oddity is that ice is actually less dense than liquid water. This decreased density occurs because as more hydrogen bonds form through freezing, the medium expands and volume increases. In fact, water can increase by 9% in volume, which is why water pipes sometimes break in homes during the winter. As you will see in Chapters 14 (Weathering and Mass Movement) and 17 (Glacial Geomorphology: Processes and Landforms), the ability of water to expand and contract has had a profound impact on shaping Earth's surface.

When heat energy is applied to and absorbed by ice at 0°C (32°F), the motion of the water molecules increases and some of the hydrogen bonds begin to break. As a result, the ice melts and becomes liquid water, which is a noncompressible fluid that conforms to the shape of its container (such as a glass). During this transformation 80 cal of latent heat are absorbed as the *latent heat of melting* to change 1 gram (g) of ice to 1 g of water. At this point the water molecules are still bound together, but the bonds are sufficiently weak that water flows.

Now let's examine what happens when liquid water changes to water vapor. This process begins at room temperature when added energy causes hydrogen bonds to further loosen, resulting in the liberation of some water molecules to the vapor phase in the process of **evaporation**. The added heat is referred to as the *latent heat of vaporization* and amounts to 600 cal for every 1 g of water. If the temperature of water increases to 100°C (212°F), the tremendous amount of added energy causes all molecular bonds to break and all water molecules thus move into the atmosphere as vapor. Conversely, when water vapor changes back to liquid form through the process of **condensation**, 600 cal of energy are released as the *latent heat of condensation.*

In somewhat rare circumstances, it is possible for water to change directly from the ice phase to vapor in a process called **sublimation**. This transformation occurs if the temperature of ice rapidly changes from 0°C (32°F) to 100°C (212°F). Under these conditions, 680 cal of energy known as the *latent heat of sublimation* are added to the medium. This amount of energy is derived from the addition of the 600 cal associated with

---

Freezing *The process through which water changes from the liquid to solid phase.*

Evaporation *The process through which water changes from the liquid to vapor phase.*

Condensation *The process through which water changes from the vapor to liquid phase.*

Sublimation *The process through which water changes directly from ice to the vapor phase.*

www.wiley.com/college/arbogast

Latent Heat

Let's review the concept of latent heat and how it relates to the phases of water. Go to the *GeoDiscoveries* website and select the module *Latent Heat*. The animation nicely illustrates the various phases of water and the way

that latent heat is either absorbed or released. After you view the animation, be sure to answer the questions at the end of the module to test your understanding of this concept.

---

vaporization *plus* the 80 cal absorbed when ice changes to liquid water. In other words, the energy associated with both the melting and vaporization processes must be accounted for. Conversely, water can change directly from the vapor phase to ice through the process of **deposition**. When this transformation occurs, 680 cal of latent heat are released, which reflects the addition of the energy released *both* by condensation and freezing.

The Hydrosphere and the Hydrologic Cycle

An implication of the previous discussion is that water is stored in various places on Earth such as the atmosphere, lakes, and oceans. Overall, the total water realm of Earth is known as the **hydrosphere**. Within the hydrosphere, approximately 97.2% of water is stored in the oceans as salt water (Figure 7.3). The remaining 2.8% is largely freshwater, most of which is stored in massive polar ice sheets and smaller mountain glaciers. This frozen water accounts for about 2.15% of the total global water volume. Approximately 0.63% is stored within Earth as groundwater.

What about the remaining 0.02% of hydrosphere water? Although this amount is analogous to *only one drop* of liquid in a liter of water, it is the most important to us because it includes the water available for plants, animals, and human use. The right-hand pie chart in Figure 7.3 shows the geographic distribution of this part of Earth's water realm. Of this small amount, most (0.009% of the total volume) is stored in freshwater lakes. Another 0.008% is contained within salty

lakes and inland seas. Soil water, which is held at shallow depths within reach of plant roots, comprises 0.005%. Of the remaining 0.002% of the total water volume, the atmosphere and streams contain 0.001% and 0.0001%, respectively.

As you probably know on some level, water moves from one place to another within the hydrosphere. You have no doubt seen water move from the atmosphere to the ground as rain to form street puddles. And then, over the course of several warm days, the water evaporates from the puddles as it moves back to the atmosphere. Water not only moves back and forth from the air to the ground, but also to and from places on and within the ground. Taken together, this movement of water between the various storage locations (or reservoirs) constitutes the **hydrologic cycle**.

Overall, the hydrologic cycle is balanced in the sense that Earth has a finite amount of water, and the amount evaporated equals the amount precipitated on a global scale. However, a variety of local and regional imbalances occur. To see how the hydrologic cycle fits into the context of the global water balance, examine Figure 7.4, which shows the flow of water on Earth in thousands of cubic kilometers (km^3) per year. With the notion of a global water budget in mind, consider that water flows to the land and ocean surfaces are positive inputs, whereas those leaving are negative. As you examine these geographic patterns, note the positive balance (+36) on landmasses. This surplus reflects the fact that land gains 36,000 km^3 more water through precipitation than it loses through evaporation. In contrast, oceans have a negative balance (−36). This negative balance exists because oceans lose 36,000 km^3 more water by evaporation than they gain through

Deposition   *The process by which water vapor changes directly to ice.*

Hydrosphere   *The water realm on Earth.*

Hydrologic cycle   *A model that illustrates the way that water is stored and moves on Earth from one reservoir to another.*

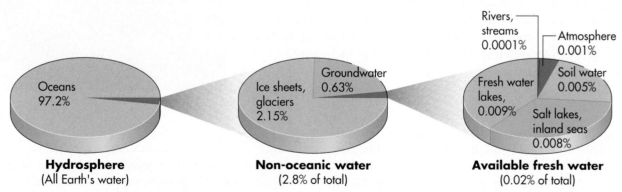

Figure 7.3 Geographic distribution of water within the hydrosphere. Most water is contained within the oceans, with proportionately smaller amounts stored in polar ice sheets and as fresh surface and soil water.

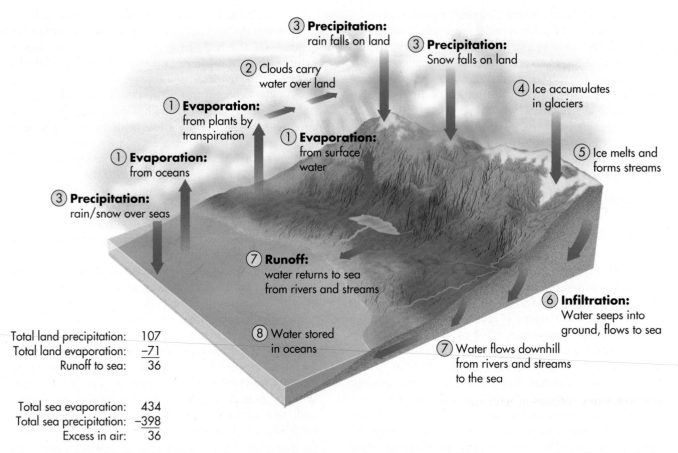

① **Evaporation:** from plants by transpiration

② Clouds carry water over land

③ **Precipitation:** rain falls on land

③ **Precipitation:** Snow falls on land

① **Evaporation:** from oceans

① **Evaporation:** from surface water

④ Ice accumulates in glaciers

③ **Precipitation:** rain/snow over seas

⑤ Ice melts and forms streams

⑦ **Runoff:** water returns to sea from rivers and streams

⑥ **Infiltration:** Water seeps into ground, flows to sea

⑧ Water stored in oceans

⑦ Water flows downhill from rivers and streams to the sea

Total land precipitation:	107
Total land evaporation:	–71
Runoff to sea:	36

Total sea evaporation:	434
Total sea precipitation:	–398
Excess in air:	36

Figure 7.4 The global water balance. The global water balance refers to the way that water moves within the hydrologic cycle and the relationship between evaporation, precipitation, and runoff. Oceans are a net source of evaporation, whereas more precipitation occurs on landmasses than evaporation. The system stays balanced because water runs off the land and returns to the ocean. Values are in cubic kilometers (km³) and indicate average annual water flows to and from the Earth's land and ocean areas.

VISUAL CONCEPT CHECK 7.1

The hydrologic cycle has many different components. This image shows a typical farm pond in the Midwest. Explain how this pond is part of the hydrologic cycle. What are two ways in which water could fill the pond? What happens to the water once it becomes stored within the pond? How does water return to the atmosphere from the pond?

precipitation. Obviously, if there were no direct connection between the land and the oceans, the landmasses would gradually become submerged by water and the oceans would empty. This reversal does not occur because 36,000 km³ of water run off the land per year and return to the oceans.

Humidity

Let's now look more closely at how water actually changes physical state from vapor to liquid and the factors that govern this process. In the atmosphere, this change of state ultimately results in precipitation.

Maximum, Specific, and Relative Humidity

We begin this discussion with the concept of **humidity**, which is the concentration of water vapor within the air. This concentration is expressed in three ways: *maximum humidity*, *specific humidity*, and *relative humidity*. Although at first it may seem confusing that three different humidity types are recognized, their definitions and interrelationships are really quite simple and easy to follow. Here is a simple analogy to keep in mind as you work your way through this section. Imagine that you are about to fill a drinking glass to some level with water. Given that the glass can hold only so much water, you probably fill it to some level below the maximum amount of water that the glass can hold. If you try to put more water in the glass than it can hold, the water will spill over the top. This analogy is useful because the atmosphere works in somewhat the same way as water moves from vapor to liquid. Sometimes the air can hold a lot of vapor, similar to a large glass, and other times the amount of vapor the air can contain is less.

Maximum humidity refers to the maximum amount of water vapor that a definable body of air can hold. Although the amount of liquid that a drinking glass can hold depends on the size of the glass, the maximum

---

Humidity *A measure of how much water vapor is in the air. The ability of air to hold water vapor is dependent on temperature.*

Maximum humidity *The maximum amount of water vapor that a definable body of air can hold at a given temperature.*

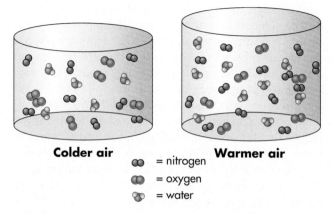

Colder air   **Warmer air**

∞ = nitrogen

∞ = oxygen

∞ = water

Figure 7.5 Maximum humidity and air temperature. A given parcel of warmer air can hold more water vapor than a similar parcel of colder air. The reason is that molecules at higher temperature are farther apart, so more water molecules can fit in a given amount of air.

humidity of an air parcel is completely governed by the air temperature. *Warmer air is capable of holding much more water vapor than colder air*, because air expands in the process of warming, which leaves room for more water vapor. This is a very simple, but important rule in physical geography that explains the processes of evaporation and precipitation. Figure 7.5 presents a simplified example of this concept.

The relationship between air temperature and maximum humidity is described by a *saturation curve*, such as the one presented in Figure 7.6. In the case of the atmosphere, the term *saturation* refers to the point

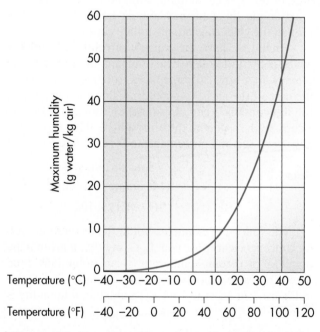

Figure 7.6 The saturation curve. Maximum humidity increases as temperature increases. In other words, warmer air can hold more water vapor than cooler air.

where the air cannot hold any more water vapor based on the temperature at that time. To help make sense of this concept, simply think of what happens when a sponge absorbs water. The sponge absorbs water until it is saturated and unable to hold any more. The atmosphere works on much the same principle. The line on the saturation curve merely illustrates the respective intersections of maximum humidity and air temperature. Notice that maximum humidity is measured as grams of water per kilogram of air, or g/kg in abbreviated form.

You can see, for example, that at an air temperature of −10°C (14°F) the maximum humidity is approximately 2 g/kg. Notice that the maximum humidity increases exponentially with increasing temperature. At 20°C (68°F) the maximum humidity is about 15 g/kg, and at 30°C (86°F) the maximum humidity nearly doubles again to about 26 g/kg. If you are having difficulty relating this concept to the water glass, try to imagine what would happen if the glass could expand and contract with warmer and colder temperatures, respectively. In this scenario, the glass can hold a larger amount of *total* water when it is warm, and less as it contracts with progressively decreasing temperature. The atmosphere operates on the same principle in that it expands when warm and contracts when cold.

Whereas the term *maximum humidity* refers to the maximum amount of water vapor that a mass of air can hold, the term **specific humidity** refers to how much water vapor is *actually* in the air. Think back to the glass analogy. Recall that the glass can hold a maximum amount of water, which, in fact, rarely occurs in the course of everyday use. Instead, you usually fill the glass to some point that is short of the container lip. The volume of this *specific* amount of liquid can be measured; for example, the glass may contain 1 cup of water, even though the glass can actually hold 2 cups.

Finally, we have the concept of **relative humidity**, which is the ratio of specific to maximum humidity—in other words, how close the air is to being saturated. In the case of the glass, you could say that the glass is half full if it contains 1 cup of water even though it can hold 2 cups. With respect to the atmosphere, this ratio is usually expressed as a percentage and is calculated as

$$\text{Relative humidity (\%)} = \left(\frac{\text{Specific humidity}}{\text{Maximum humidity}} \right) \times 100$$

$$\text{RH (\%)} = (\text{SH / MH}) \times 100$$

The following example should make these humidity concepts clearer for you. Let's say that a given mass of air at a certain temperature can potentially hold 2 units of water vapor, as seen in the simplified diagram in Figure 7.7a. In other words, its maximum humidity is 2 units.

(a) Maximum humidity = 2 units

(b) Specific humidity = $\frac{1}{2}$ unit

(c) Relative humidity = 25%

Figure 7.7 Hypothetical humidity values. (a) In this example, the maximum humidity is 2 units of water vapor. (b) In this example, the specific humidity is $\frac{1}{2}$ unit. (c) The mass of air can hold 2 units of water vapor, but only $\frac{1}{2}$ of a unit is present. Thus, the relative humidity is 25%.

But how much water vapor is actually there—in other words, what is the specific amount of vapor present? Well, imagine that the mass of air contains a unit of water vapor, as in Figure 7.7b. Under these circumstances, you could say that the specific humidity is a unit. In this simplified example, you can conclude that the volume of air actually has a unit but has a potential of 2 units, so the air contains the water vapor that it can *potentially* hold (see Figure 7.7c). In other words, *the relative humidity is 25%*.

Humidity is usually measured with an instrument known as a *sling psychrometer* (Figure 7.8). This measurement system uses two thermometers—a wet bulb and a dry bulb—mounted side by side on a small platform. The wet bulb is covered with a cotton sleeve and moistened with water of room temperature, whereas the dry bulb remains dry. At this point the temperature of the two bulbs is the same. The two thermometers are then whirled in the air. If the surrounding air is not saturated, water from the moist cotton sleeve on the wet bulb will evaporate. As a result, heat is absorbed from the bulb and thus the temperature of the bulb drops. In these conditions, the wet-bulb and dry-bulb temperatures are set on a scale and the relative humidity is then calculated. The amount of evaporation, and subsequent

Specific humidity *The measurable amount of water vapor that is in a definable body of air.*

Relative humidity *The ratio between the specific and maximum humidity of a definable body of air.*

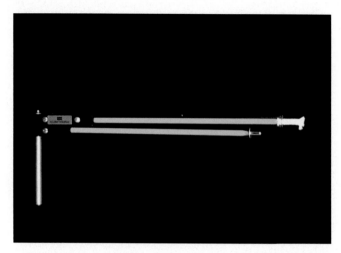

Figure 7.8 Sling psychrometer. This system uses a pair of thermometers that detect the amount of cooling due to evaporation. The dry-bulb thermometer shows the actual temperature, whereas the wet-bulb thermometer indicates the amount of cooling.

cooling of the wet bulb, increase when the air is progressively drier. On the other hand, if the air is saturated, then no cooling of the wet bulb takes place because no water is evaporated from the cotton sleeve, and the thermometers show the same temperature. Other devices can measure relative humidity directly. One such system absorbs an amount of water vapor that depends on the relative humidity.

No matter how relative humidity is measured, when the relative humidity is low, the air is relatively dry. When relative humidity is high, the air is relatively moist and close to being saturated. Remember that warm air can hold more water vapor than cold air because warm air is rising and expanding and has room for more water molecules. Cold air, on the other hand, is descending and contracting, so it has less room for molecules of water.

You can see how specific humidity changes with air temperature by examining Figure 7.9a, which shows the geographic relationship of specific humidity to latitude in the middle atmosphere during January 2004. Specific humidity is greater in the lower latitudes where the temperatures are generally warm. In contrast, the specific humidity is relatively low at the higher latitudes because air temperature is relatively cold.

A common mistake many students make is to assume that the relationship illustrated in Figure 7.9a means that relative humidity varies in the same geographic pattern, with relative humidity being higher at lower latitudes than at high latitudes. Although this pattern may sometimes follow, it is not necessarily so. In fact, the relative humidity at higher latitudes is higher, on average, than in warmer ones. Figure 7.9b shows the nature of this geographic distribution. Why do you suppose this inverse relationship exists?

(a)
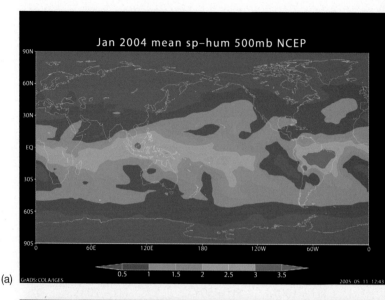

(b)
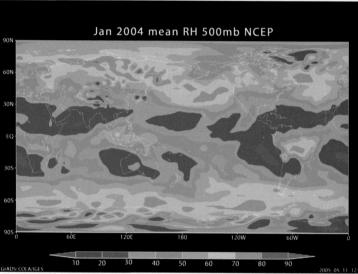

Figure 7.9 Humidity maps. (a) A direct correlation appears between specific humidity and latitude in the lower atmosphere (as reflected at the 500-mb pressure surface). Reds and yellows indicate high values of average specific humidity, whereas greens, blues, and purples are progressively lower levels of specific humidity. (b) Higher latitudes usually have higher relative humidity, as indicated by the orange and yellow shades. Greens, blues, and purples represent areas with lower amounts of relative humidity.

The reason is that cold air can hold less water vapor than warm air. So even though the high latitudes typically contain less water vapor, the inherently colder air is proportionately closer to saturation than the warmer air at lower latitudes. Thus, the relative humidity is usually greater at high latitudes compared to low latitudes.

The same kinds of changes in relative humidity associated with latitude can also be seen over a typical day at any location that experiences strong diurnal changes in temperature; in other words, changes that

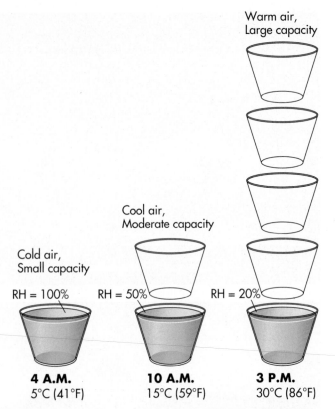

Figure 7.10. Relationship between air temperature, time of day, and relative humidity. Although specific humidity does not change in this example, maximum humidity increases while relative humidity decreases.

Warm air,
Large capacity

Cool air,
Moderate capacity

Cold air,
Small capacity

RH = 100%   RH = 50%   RH = 20%

4 A.M.   **10 A.M.**   **3 P.M.**
5°C (41°F)   15°C (59°F)   30°C (86°F)

Figure 7.11 Formation of dew. Water condenses on grass when nighttime temperatures cool to the dew-point temperature.

occur over the course of a day and night. Figure 7.10 shows air temperature warming over the course of the day from 4:00 A.M. to 3:00 P.M. As the warming process continues, the maximum humidity increases because the capacity to hold vapor becomes greater. In contrast, the specific humidity, represented by the blue shading, does not change between the morning and afternoon hours. What happens to relative humidity as the day progresses? It gradually decreases, as the ratio of specific humidity to maximum humidity becomes less. If you lived in a place where such a daily temperature/humidity change occurred, you would notice that the air might become slightly less sticky as it warmed over the course of the day.

Dew-Point Temperature

As shown in Figure 7.10, the relative humidity of a hypothetical air mass decreased over the course of an imaginary day because the temperature warmed from 5°C (41°F) to 30°C (86°F), creating a larger capacity to hold water vapor in the afternoon. This decrease in relative humidity occurred despite that fact that the specific humidity *did not change*. Assuming that the air mass would cool over the course of the evening, and by early morning back to

5°C (41°F), without any change in specific humidity, what would the relative humidity be? To calculate this value, simply work backward in the process and watch how the capacity of the air mass to hold moisture decreases until it is saturated. In this particular example, if the air cooled to 15°C (59°F), the relative humidity would increase to 50%. If the air continued to cool to 5°C (41°F), then the relative humidity would rise to 100%.

Another way to look at this example is to say that the **dew-point temperature** of this air mass is 5°C (41°F). The dew-point temperature, therefore, is an important designation and is defined as the temperature at which a mass of air becomes saturated. The term *dew point* originates from the formation of dew on grass (Figure 7.11). This occurs when air at ground level cools to the dew-point temperature at night and water condenses on blades of grass. As the year progresses, dew typically forms during the early fall and later part of the spring, when nighttime temperatures cool sufficiently for condensation to occur at ground level, but not so much that water crystallizes as it does in winter. Dew is less likely in the summer months because nighttime temperatures remain relatively high and the dew-point temperature is not reached at ground level.

Higher in the atmosphere, the dew-point temperature is important because it marks the temperature at which water begins to condense into liquid form as clouds, fog, and ultimately various forms of precipitation. This condensation occurs because the air mass can no longer hold any more water vapor, either because the air cools to the dew-point temperature or because more water vapor is added to the air mass through

---

Dew-point temperature *The temperature at which condensation occurs in a definable body of air.*

evaporation of surface water. Whereas the maximum humidity of an air mass is a function of its temperature, the dew-point temperature depends on the specific humidity, which is another way of referring to how much water vapor is actually in the air.

You can observe the trend in dew-point temperature by reexamining the saturation curve in Figure 7.6. This time, consider first the humidity values on the vertical (y) axis, then temperature data on the horizontal (x) axis. In this fashion, you are simply reversing the association you made earlier when you determined the maximum humidity at certain temperatures. For example, if the specific humidity of the air is 2 g/kg, then the dew-point temperature is –10°C (14°F). If the specific humidity increases to 15 g/kg or 26 g/kg, then the dew-point temperature also rises, in this case, to 20°C (68°F) and 30°C (86°F), respectively.

In an effort to integrate all the humidity concepts into a coherent mathematical model that explains how precipitation occurs, let's look at some examples in Figure 7.12. First consider an air mass with a temperature of 20°C (68°F), which you can see in Figure 7.12a. At this temperature, the maximum humidity is 17.3 g/kg. That is, the air mass has the *potential* to contain 17.3 g/kg of water vapor. If the specific humidity is also 17.3 g/kg, that means the parcel is saturated, the relative humidity is 100%, and the dew-point temperature is 20°C (68°F). On the other hand, if the specific humidity is only 9.4 g/kg (Figure 7.12b), then the relative humidity is about 54% and the dew-point temperature is approximately 10°C (50°F). If the air temperature were to increase without a change in specific humidity, the relative humidity would further decrease.

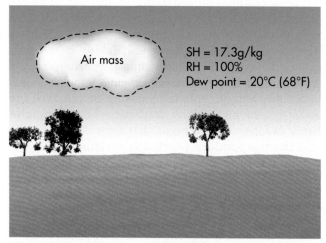

(a) Air mass temperature = 20°C (68°F); MH = 17.3 g/kg

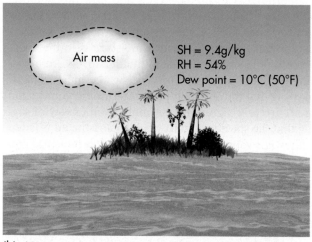

(b) Air mass temperature = 20°C (68°F); MH = 17.3 g/kg

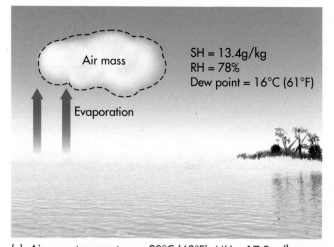

(c) Air mass temperature = 20°C (68°F); MH = 17.3 g/kg

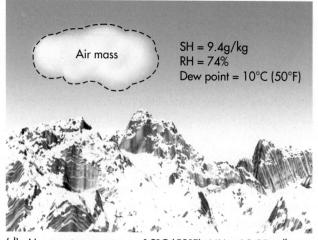

(d) Air mass temperature = 15°C (59°F); MH = 12.83 g/kg

Figure 7.12 Humidity examples. (a) Hypothetical air mass with air temperature of 20°C (68°F) and relative humidity of 100%. (b) Same air mass, now with a relative humidity of 54%. (c) Same air mass, but now with increased relative humidity due to higher specific humidity. (d) Same air mass with a cooled air temperature of 15°C (59°F) and relative humidity of 74%.

How could we get the relative humidity of the drier air mass to increase? This kind of increase can happen in one of two ways. One way would be to somehow increase the amount of water vapor in the air mass, perhaps through the process of evaporation at a nearby ocean source. For example, if another 4 g/kg of vapor were added to the air mass through this process (Figure 7.12c), then the specific humidity would be 13.4 g/kg and the relative humidity would increase to 78%. At the same time, the dew-point temperature would increase to about 16°C (61°F). The second way would be to lower the temperature (Figure 7.12d), let's say to 15°C (59°F). If this cooling process occurred at the same time that specific humidity remained 9.4 g/kg, then the relative humidity would increase to about 74% because the air could hold less moisture; in other words, the maximum humidity would be less.

An important reason why understanding atmospheric humidity is relevant to your life is that it allows meteorologists to predict when and how precipitation will occur. You may have encountered such a discussion on your local weather station, when the weatherperson talks about how much moisture is in the air at a given time, especially when thunderstorms are a possibility. When these scenarios occur, the meteorologist may say that the dew-point temperatures in your region are "currently high." This is simply another way of saying that the specific humidity is high and the air is laden with water vapor that may be released as moisture in a storm should the right atmospheric conditions develop.

At another level, understanding atmospheric humidity is critical for agriculture because it is indirectly associated with crop production. Farmers are keenly aware of atmospheric moisture conditions because their crops depend on the appropriate amount of water, which, in turn, is relevant to you because the price of food is directly related to supply. If an extended period of drought occurs in a major agricultural region, production yields decrease dramatically and the cost of food increases.

Atmospheric humidity conditions also have ramifications for human comfort and health, especially during the warm, humid days of summer when dew-point temperatures are elevated. When the heat index is high, most people are very uncomfortable working outside for any length of time. The reason for this discomfort is that our bodies do not regulate internal temperature effectively when the air is warm and muggy. The primary way that we regulate internal temperature is through the production of sweat. When the air is dry, sweat evaporates and body heat is removed in a manner consistent with the earlier discussion in this chapter about latent heat transfers and the wet bulb on a sling psychrometer. On warm/humid days, however, sweat evaporates less efficiently and our core temperature increases. If the sweat-producing activity continues, body temperatures can rise to the point that heat stroke occurs and the temperature regulation system shuts down completely.

www.wiley.com/college/arbogast

Atmospheric Humidity

Although the concept of humidity is fundamentally a simple one, students often have difficulty with it because the terms can seem to be interchangeable. Another problem is that it is sometimes hard to understand these processes when simply reading about them. Like many concepts in physical geography, humidity is particularly well suited for simulation to enhance understanding. In this context, go to the *GeoDiscoveries* website and select the module *Atmospheric Humidity*. This module is designed so you can interact with the concept of humidity and assess how

it varies depending on environmental circumstances. You will be presented with a variety of scenarios in which you can manipulate the amount of maximum humidity and specific humidity in a hypothetical air mass such that changes in the relative humidity and dew-point temperature occur. As you work your way through these scenarios, remember that they are representative of an almost infinite number of combinations. When you complete the simulation, be sure to answer the questions at the end of the module to test your understanding of this important concept.

1. Three kinds of humidity exist: maximum, specific, and relative. Maximum humidity refers to how much vapor a parcel of air can hold. This variable depends on temperature, with warm air having a higher maximum humidity than colder air. Specific humidity measures how much water vapor is in the air. Relative humidity is the ratio of specific humidity to maximum humidity.

2. When the relative humidity is 100%, the air is saturated and can hold no more water vapor.

3. Although the specific humidity may not change on any given day, the relative humidity can change dramatically due to rising and falling temperature. In this case, the warmer the temperature, the lower the relative humidity.

4. The dew-point temperature is the temperature at which a mass of air becomes saturated. This is the temperature at which condensation occurs in that air mass. As specific humidity increases, so does dew-point temperature.

5. Once the dew-point temperature is reached, the specific humidity begins to decrease as water vapor condenses to liquid form. This decrease in specific humidity continues as long as the process of cooling does.

Evaporation

Where does water vapor in the atmosphere come from? Liquid water is transformed into water vapor through two processes, evaporation and transpiration. As described earlier in the chapter, the term *evaporation* is used when water is lost from a surface such as soil, water, or pavement. Liquid water is also transformed into vapor when it flows through leaf pores in plants to the atmosphere. This process of water loss is called **transpiration** and can result in the transformation of tremendous amounts of water. In tropical areas, for example, some trees can transpire over 300 liters (L) (~80 gal) of water a day! You can indirectly see the process of transpiration by noting the rigid upright posture of plants. This posture is maintained by water flowing upward through the plant by capillary action toward the leaf surface. A plant wilts when the amount of water in the soil reduces and therefore less flows upward through the plant.

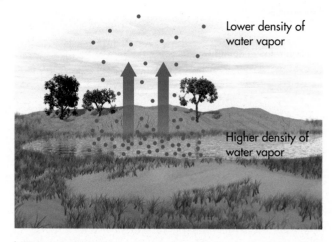

Lower density of water vapor

Higher density of water vapor

Figure 7.13 Water vapor gradient. Water vapor flows from areas of high-density toward areas of low-density vapor content. This flow creates room for additional water to be evaporated from the source area.

When geographers consider the humidity characteristics of a particular region, they typically combine the processes of evaporation and transpiration into the singular concept of **evapotranspiration**. Evapotranspiration rates are dependent on many factors, including:

- **Net Radiation:** Heating of plants or ground surface. Evaporation rates are higher when a lot of energy (sunlight) exists.

- **Air Temperature:** Warm air can hold more moisture than cold air.

- **Relative Humidity:** How much moisture the air is already holding relative to its capacity to contain water vapor. Evaporation rates are higher when the air is dry because plenty of room exists for more water vapor molecules in the air.

You have previously seen that air flows along a pressure gradient from high pressure to low pressure. Similarly, water vapor flows along a vapor pressure gradient from areas of high vapor pressure to areas of low vapor pressure (Figure 7.13). Close to the water surface, the air contains more molecules of water vapor. As you move higher above the water, however, the air contains fewer water vapor molecules. Water vapor flows upward from the areas with more water vapor to the areas with less water vapor.

A final factor that influences evapotranspiration rates is wind speed. Wind moves moist air away from the source of the water vapor. Evaporation rates are higher

---

Transpiration *The passage of water from leaf pores to the atmosphere.*

Evapotranspiration *The combined processes of evaporation and transpiration.*

No wind

High relative humidity over water

Lower relative humidity over land

(a)

With wind

Water molecules blown over land, away from water

(b)

Figure 7.14 Wind and evaporation. (a) When the wind is calm, little evaporation occurs because the relative humidity of the air mass above the water is high. (b) More evaporation takes place during windy days because water vapor continually flows away from the moisture source.

in windy conditions when the air containing water vapor molecules can be moved away from the source, making room for more water molecules to move into the air. Figure 7.14 shows an example of this concept. This is also why fans are useful cooling devices; they create a wind to increase evaporation.

Adiabatic Processes

When water evaporates into vapor, it moves into the atmosphere where it exists in the gas phase. It can, however, recondense into liquid water to form clouds where it is temporarily stored until it either vaporizes again or falls to Earth as precipitation. This section deals with the processes through which condensation occurs, with the ultimate goal of providing the foundation for understanding the precipitation process. Although many of the terms in this section may seem like jargon to you at first, they are everyday terms to many geographers because they are used when we monitor atmospheric conditions and predict the weather on any given day.

From the discussion about relative humidity, you might surmise that condensation must be somehow related to the cooling of air to the dew-point temperature. This is a correct assumption, but, as you might also imagine, the process is more complex than it first appears. After all, nighttime cooling can cause condensation of near-surface water vapor if the saturation point of the air is reached immediately above the ground, resulting in dew or even frost if vapor changes directly to ice crystals. Neither of these processes, however, is associated with precipitation.

Precipitation occurs only when a substantial parcel of air experiences a steady drop in temperature below the dew-point temperature, which can only occur if the air parcel is somehow lifted enough for it to cool adiabatically. For purposes of discussion, think of a parcel of air as having a diameter of about 300 m (1000 ft) with uniform humidity and temperature characteristics.

Recall from Chapter 6 that the dominant way in which a parcel of air rises or descends is through changes in atmospheric pressure. When this kind of movement occurs, the air either expands as it rises or compresses as it descends (Figure 7.15). Of particular relevance here is that air cools as it rises and expands and warms when it descends and compresses. It is important to note that these temperature changes are *internal* to the rising or descending air parcel and occur without any exchange of heat with the surrounding atmosphere. These kinds of temperature changes are called **adiabatic processes** because they result solely from expansion and compression of air due to pressure fluctuations.

The Dry Adiabatic Lapse Rate

In most cases that involve a parcel of surface air, the air is not saturated and the relative humidity is thus less than 100%. Imagine that this parcel begins to rise and expand because it is warmer than the surrounding air. When these conditions occur, the air temperature decreases at what is called the **dry adiabatic lapse rate (DAR)** as the air starts to rise. This rate is not to be confused with the *environmental lapse rate* that was discussed in Chapter 5. The environmental lapse rate refers to the average change in temperature of *still* air with altitude, which can

---

Adiabatic processes *Changes in temperature that occur due to variations in air pressure.*

Dry adiabatic lapse rate (DAR) *The rate at which an unsaturated body of air cools while lifting or warms while descending. This rate is 10°C/1000 m (5.5°F/1000 ft).*

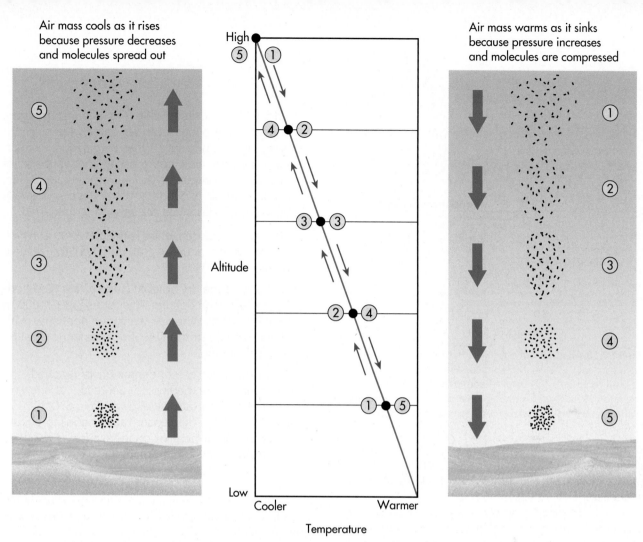

Air mass cools as it rises because pressure decreases and molecules spread out

Air mass warms as it sinks because pressure increases and molecules are compressed

Altitude

High

Low

Cooler   Warmer

Temperature

Figure 7.15 Adiabatic heating and cooling. Air cools as it rises because the pressure drops and air molecules disperse. When air descends, it warms adiabatically because the air molecules are compressed by higher pressure.

change with time or place. In contrast, the DAR is always constant and is applied to a parcel of rising air according to physical laws. The DAR has a value of about 10°C/1000 m (5.5°F/1000 ft) of vertical lift and is applied *only* when the air mass is not saturated. *This is a very important rule to remember.* You can see this rate of change if you examine Figure 7.16 and focus on the area at the bottom of the graph, which indicates altitudes of less than 1000 m.

The DAR is relevant to not only lifting air masses that are not saturated but also descending ones. In the case of a descending air mass, the air *warms* at the DAR. That is, for every 1000 m that the air descends, it warms 10°C (5.5°F/1000 ft) due to compression of the air under pressure.

The Wet Adiabatic Lapse Rate

The DAR applies when air is not saturated and the relative humidity is less than 100%. How does temperature change once the dew-point temperature is reached, saturation occurs, and the relative humidity is 100%? The altitude at which the saturation point is reached is known as the **level of condensation**, which varies

depending on the particular temperature and humidity characteristics in any given air parcel. Air parcels that have relatively high humidity when they begin to lift reach the level of condensation at a lower elevation than air parcels that are comparatively dry.

Once the level of condensation is reached, water droplets begin to form if the air continues to cool. In this fashion, the air is analogous to a saturated sponge that releases water as you squeeze it at the same time that you lift it. Give it a try. All you have to do is take a sponge and let it absorb a small amount of water in the sink. Starting at the surface of the sink, slowly raise the sponge and squeeze it slowly as you lift it. At some point, the sponge will begin to release water. After you reach this point, start over but let the sponge absorb more water before you begin. Raise the sponge as before and try to exert the same amount of force as you did previously. You should notice that the sponge begins to release water at a lower

---

Level of condensation *The altitude at which water changes from the vapor to liquid phase.*

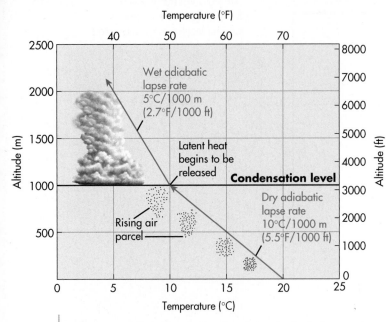

Temperature (°F)

Figure 7.16 Adiabatic cooling in a hypothetical air parcel.
The air cools at the dry adiabatic lapse rate (DAR) as it rises until it reaches the level of condensation. At that point the rate of temperature change switches to the wet adiabatic lapse rate (WAR) due to the release of latent heat.

1. Adiabatic processes refer to the temperature changes that occur in an air parcel due solely to changes in air pressure. When air pressure increases, a parcel of air is compressed and warms adiabatically. In contrast, when air pressure decreases, a parcel of air expands and thus cools internally because air molecules are spaced farther apart.

2. The level of condensation is the elevation at which condensation occurs, that is, when the relative humidity is 100%.

3. Air cools at the dry adiabatic lapse rate (DAR) until condensation occurs. This rate is 10°C/1000 m (5.5°F/1000 ft). When air warms adiabatically, it always does so at the DAR because condensation is not occurring.

4. Once the level of condensation is reached, the air temperature begins to decrease at the wet adiabatic lapse rate (WAR), which is 5°C/1000 m (2.7°F/1000 ft). This lesser rate of cooling occurs because latent heat of condensation is released when water changes from vapor to liquid.

elevation than before. This occurs because the sponge was closer to being saturated the second time you ran this simple experiment. The atmosphere operates on much the same principle as this sponge test.

If the air parcel continues to rise and cool, increasing amounts of water vapor condense into liquid water. During this process, another principle comes into effect—the release of latent heat of condensation that was described earlier in this chapter. This heat energy has the effect of warming the air at the same time that it is cooling due to the reduction in atmospheric pressure and expansion. The cooling effect is stronger overall, however, so the overall temperature continues to decrease if the air continues to rise. Nevertheless, the rate of cooling in this saturated state is less than in the unsaturated state because of the release of latent heat.

Look at Figure 7.16 again and focus this time on the area above the condensation level to see how this change appears graphically. The cooling rate of saturated air is called the **wet adiabatic lapse rate (WAR)**. It is also known as the moist or saturated lapse rate. Although the WAR varies with moisture content and temperature, the average rate is about 5°C/1000 m (2.7°F/1000 ft) and this value will be used hereafter in this text.

Wet adiabatic lapse rate (WAR) *The rate at which a saturated body of air cools as it lifts. The average rate is about 5°C/1000 m (2.7°F/1000 ft).*

www.wiley.com/college/arbogast

Adiabatic Processes

In an effort to integrate these concepts into an animated format, go to the *GeoDiscoveries* website and select the module *Adiabatic Processes.* This module reviews all the concepts that have been discussed in this section of the text. As you watch the animation, pay particularly close attention to when the DAR and WAR are used. Also, watch what happens to the air parcel as it descends and warms at the DAR. When this process occurs, the effect is to produce a warmer air mass at the surface than the original parcel of air. After you complete the animation, be sure to answer the questions at the end of the module to test your understanding of this concept.

Cloud Formation and Classification

As noted in the preceding discussion on adiabatic processes, once air reaches the dew-point temperature, through cooling at the DAR, water vapor begins to condense into liquid droplets. We see the results of this process through the development of clouds, with the base of any particular cloud representing the level of condensation in the atmosphere.

Clouds are visible masses of suspended, minute water droplets or ice crystals. Recall from Chapter 4 that clouds have a very important role in regulating the way solar radiation interacts with Earth. Some insolation is reflected immediately back into space from bright cloud tops. At the same time, clouds cause insolation to be scattered and absorbed. They also absorb longwave radiation from the Earth's surface. In addition, the character of clouds is a clue to atmospheric dynamics at any given point in time.

Figure 7.17 shows the close relationship between atmospheric water vapor and the geographic distribution of clouds in the United States on April 24, 2005. Notice in part (a) that high concentrations of vapor (shown in white) are present in the western part of the country, as well as over the Great Lakes region. These concentrations of vapor are closely associated with the band of clouds in part (b) that cover the same geographic areas.

Two conditions are necessary in order for clouds to form:

1. The air must be saturated. Air can become saturated two different ways: when air cools below the dew-point temperature or when more water vapor is added to the air.

2. There must be a substantial quantity of small particles, such as dust or pollutants, about which water vapor can collect or condense. These particles are called **condensation nuclei**. Have you ever no-

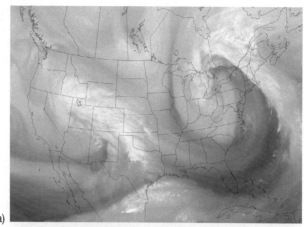
(a)

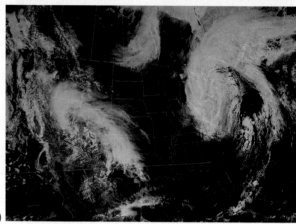
(b)

Figure 7.17 Satellite images showing atmospheric moisture on April 24, 2005. (a) GOES water vapor image. Notice the particularly high concentrations of water vapor (in white) in the west and northeast. (b) GOES visible image. Note how the concentrations of clouds are closely associated with the vapor patterns in (a).

Condensation nuclei   *Microscopic dust particles around which atmospheric water coalesces to form raindrops.*

VISUAL CONCEPT CHECK 7.2

Water vapor is produced in the exhaust of jet engines, such as those on large airliners. As the vapor cools in the air behind the plane, it often forms condensation trails, or contrails. What conditions must be present for the contrail to form?

a) Air must be close to saturation.
b) Condensation nuclei (dust) must be present.
c) The plane must be above the condensation level of the air.
d) All of the above.

ticed that, if it rains after you washed your car, little circles of dust exist on your vehicle that resulted from the rain? The tiny particles of dust were suspended in the atmosphere prior to the rainfall and served as the nuclei for condensation to occur.

Cloud Classification

Like many other things in nature, it is possible to classify clouds based on particular characteristics, specifically with respect to altitude and form (Figure 7.18). The international cloud classification scheme recognizes three categories based on their form. **Cirrus clouds** are thin and wispy clouds composed of ice crystals rather than water droplets. **Cumulus clouds** consist of individual, puffy clouds with a flat, horizontal base. **Stratus clouds** consist of layer-like, grayish sheets that cover most or all of the sky. As you can see in Figure 7.18, these clouds occur in various combinations. For example, stratocumulus clouds have characteristics of both stratus and cumulus clouds because they are individual clouds that spread out more than regular cumulus clouds.

Clouds can be further classified based on their altitude, with three categories present that you can see in Figure 7.18. *High clouds* are usually found at altitudes greater than 6 km (20,000 ft) and generally consist of cirrus clouds because very little water vapor is present and the temperatures are cold. *Middle clouds* range between approximately 2 km and 6 km (6500 ft and 20,000 ft) and include cumulus and stratus clouds that form during intervals of stable and changing weather, respectively. *Low clouds* typically occur below 2 km (6500 ft), usually in association with stratus and cumulus clouds. These low clouds are the source of most precipitation, with *nimbostratus clouds* affiliated with long-term rain or snow events and *cumulonimbus clouds* growing to great heights during short-term, severe storms.

With a fundamental understanding of clouds and their formation, it is possible to determine atmospheric conditions and to even predict weather a few days in advance. For example, if the sky is filled with individual puffy, cumulous clouds, it means that the atmosphere is fairly stable and fair weather will probably last for a little while. If you happen to notice that high-level cirrus clouds are becoming more numerous and dense, it probably means that a storm system is approaching. These clouds often give way to altocumulus clouds, which indicate that the storm system is closing in and that moisture is flowing into the region from hundreds of kilometers away. If the clouds begin to thicken and expand into nimbostratus clouds, it means that some form of precipitation will most likely occur soon.

Fog

Although condensation of water vapor usually occurs at fairly high levels of the atmosphere, sometimes conditions are right for it to happen very close to the ground. When this relationship occurs, the hazy, moist air is known as *fog* and is one of the more distinctive atmospheric phenomena people encounter. As you probably know, foggy conditions can be rather eerie because visibility is poor and the Sun and street lights have an unusual glow. For these same reasons, fog is a travel hazard because drivers have difficulty seeing oncoming cars and sharp bends in the road.

So how does fog develop? Actually, it can form in several different ways. Perhaps the most common type of fog on autumn or spring days is **radiation fog**, which develops at night when a **temperature inversion** exists in the lower troposphere. Recall from Chapter 5 that temperature generally cools at a consistent rate (the environmental lapse rate) from the surface up into the higher levels of the troposphere. Occasionally, a temperature inversion develops whereby a body of cooler air lies beneath warmer air. If this cooler air reaches the dew-point temperature, radiation fog develops. This kind of scenario sometimes develops in deep valleys when cool air collects at the bottom of the valley (Figure 7.19a). You can see this type of fog when you are driving and popping in and out of fog while passing through valleys and over hills, respectively.

In addition to radiation fog, another type of fog is **advection fog**. This type of fog develops when warm air flows over a cooler surface, such as snow or a body of water. When this process occurs, the warmer air cools to the dew-point temperature. Yet a third type of fog is **sea fog**, which develops when cool marine air comes into direct contact with the colder ocean water. In the United States, a place where sea fog frequently develops is the coast of California, along which

---

Cirrus clouds   *Thin, wispy clouds that develop high in the troposphere.*

Cumulus clouds   *Individual puffy clouds that develop due to convection.*

Stratus clouds   *Layered sheets of clouds that have a thick and dark appearance.*

Radiation fog   *Fog that develops at night when a temperature inversion exists.*

Temperature inversion   *A layer of the atmosphere in which the air temperature increases, rather than cools, with altitude.*

Advection fog   *Fog that develops when warm air flows over cooler air.*

Sea fog   *Fog that develops when cool, marine air comes into direct contact with colder ocean water.*

UNUSUAL CLOUDS

Some cloud forms are typical of certain conditions, but are very rarely seen. An example is a lenticular cloud, so-called because of its lens-like appearance. Lenticular clouds are actually a variation of altocumulus clouds that form when an air parcel rises very rapidly to pass over a mountain peak. Although they are very common over the Rocky Mountains when weather systems pass, you will rarely see this kind of cloud in the eastern part of the country.

Another unusual type of cloud is called a noctilucent, or nacreous, cloud. These clouds form at high altitudes—so high, in fact, that they are in the mesosphere, not the troposphere. They become visible when they reflect the light of the Sun, which has usually set below the horizon at ground level. The result is a bright cloud seen at night. Noctilucent clouds are most often seen at high latitudes during the summer, when the Sun is frequently not far below the horizon.

the cool California current flows. If you happen to live near the Bay area, or have visited in the summer, you know that San Francisco can be socked in by sea fog (Figure 7.19b).

Regardless of the type of fog that develops, you have probably noticed that fog typically dissipates or *burns off* during the day. This process occurs because the cool air in which the fog forms warms above the dew-point temperature as the day progresses. Next time you experience morning fog, notice that by the late morning or middle of the day it will probably be gone.

Precipitation

As we just discussed, condensation of water vapor leads to the formation of clouds. Cloud formation does not always lead to precipitation, as clouds usually appear in the sky in one form or another and it certainly does not rain or snow every time they do. This logically leads to the question, *why does precipitation occur?*

You probably already know that precipitation must involve some kind of a phase change from vapor to liquid or from liquid or vapor to solid. On the other hand,

the nature of the process might be a mystery to you, as it is with most people. The bottom line is that precipitation does not occur until droplets of water are heavy enough for them to fall under the influence of gravity. Until then they remain suspended in the atmosphere.

Types of Precipitation

Droplets grow in size and become heavier by two processes: ice-crystal formation and coalescence of water droplets. Ice crystallization is the dominant process in most places outside the tropical regions and occurs because most clouds or portions of clouds extend to altitudes where air temperatures are below the freezing point of liquid water. In these places, ice crystals occur in a matrix of water vapor and super-cooled water droplets that form around condensation nuclei (Figure 7.20a). The ice crystals "feed" off of the water droplets in that they cause droplets to rapidly evaporate when they are in close proximity to one another. Subsequently, the ice crystals absorb the water vapor that was released through the evaporation of the droplets, causing them to grow. Assuming the crystals grow sufficiently large in order to fall, they precipitate either

Cirrocumulus: appears as white patches; usually formed from cirrus or cirrostratus torn by winds; indicates approaching surface winds

Cirrus: thin and whispy high-level clouds; known as mares' tails

Altocumulus: puffy with dark undersides; sometimes known as a mackerel sky (resembling fish scales); usually a sign of fair weather

Altostratus: thick sheets of pale gray; enough to obscure Sun or Moon; usually a sign of rain or snow

Cumulus: puffy, fleecy, fair-weather clouds, most common in summer, with blue sky visible between clouds; often evaporate at night

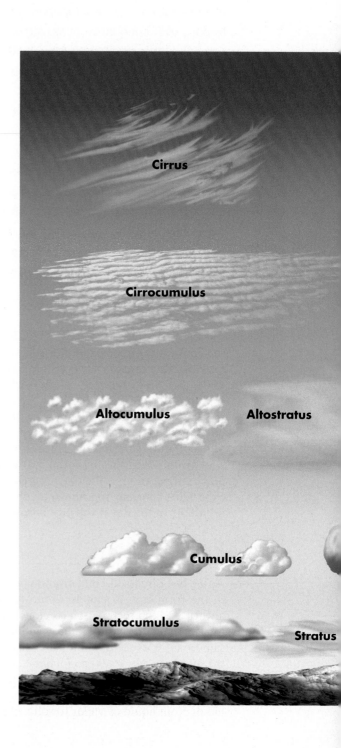

Cirrus

Cirrocumulus

Altocumulus

Altostratus

Cumulus

Stratocumulus

Stratus

Cirrostratus: thin, gauze-like sheets; usually gray and featureless but produce spectacular colors at sunsets; usually a sign of approaching rain

Cumulonimbus: thunderheads, extending to the tropopause or beyond, with typical anvil-shaped top; most storm clouds are cumulonimbus

Tropopause

Cirrostratus

High level
above 6 km
(above 20,000 feet)

Cumulonimbus

Middle level
2–6 km
(6,500 to 20,000 ft)

Low level
below 2 km
(0 to 6,500 ft)

Nimbostratus

Stratus: thick, dull gray, low-lying layers; occurs as fog at ground level, often produces mist or drizzle

Nimbostratus: dark rain clouds; form near the ground but usually extend upward

Stratocumulus: forms in patches, sheets, or layers of white or gray; can produce overcast winter skies but clears quickly in summer

Figure 7.18 Cloud classification. Clouds are classified according to their height and form. Most of these categories are variations on the three basic forms of clouds: cirrus, stratus, and cumulus.

Figure 7.19 Types of fog. (a) An excellent example of radiation fog in the Blue Mountains of Australia. This fog developed when nighttime cold air collected in the valley bottom. It was entirely gone by late morning because the air in the valley had warmed above the dew-point temperature. (b) Sea fog at San Francisco, California. This type of fog develops when moisture-laden air flows over cold water, causing the air temperature to drop.

as snowflakes or melt on their way to the surface and become raindrops that can grow further through the coalescence process.

In contrast to ice crystallization, which is a process of vapor attraction, raindrop coalescence is a process that merges small water droplets into large ones (Figure 7.20b).

This process occurs mostly in tropical regions where the cloud tops do not reach sufficiently high for liquid water to freeze. Raindrop coalescence is the simple growth of water droplets through the persistent collision of small droplets in the upper atmosphere. One way in which this occurs is that large raindrops fall faster than smaller ones, allowing

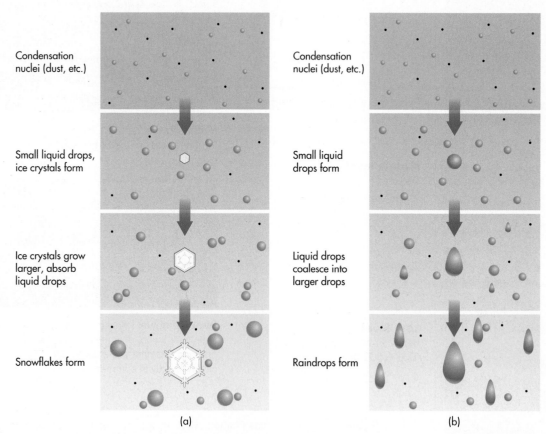

Condensation nuclei (dust, etc.)

Small liquid drops, ice crystals form

Ice crystals grow larger, absorb liquid drops

Snowflakes form

(a)

Condensation nuclei (dust, etc.)

Small liquid drops form

Liquid drops coalesce into larger drops

Raindrops form

(b)

Figure 7.20 Types of precipitation. (a) Snowflakes form by the crystallization of ice particles. (b) Raindrops form through coalescence of small water droplets.

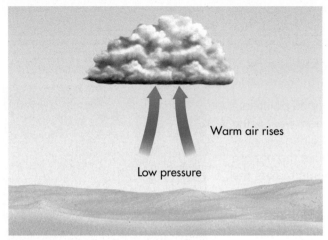
(a) Convectional

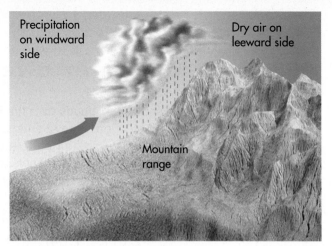
(b) Orographic

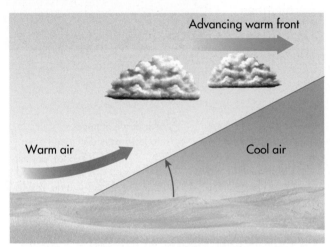
(c) Frontal

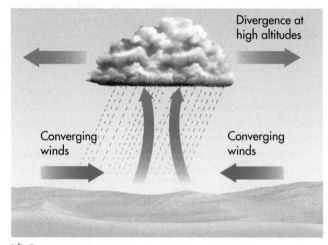
(d) Convergent

Figure 7.21 Uplift processes that can cause precipitation. (a) Convectional, (b) Orographic, (c) Frontal, and (d) Convergent.

the larger raindrops to absorb the smaller droplets as they overtake them on their fall to Earth.

Precipitation can occur in several forms:

- Rain—large, unfrozen water droplets
- Snow—ice crystals that do not melt before reaching the surface
- Sleet—rain that freezes before hitting the ground
- Freezing rain—rain that freezes on impact with ground that is below the freezing point of water
- Hail—ice crystals that are continually pulled back up and grow in size during violent thunderstorms

This process is discussed in more detail in Chapter 8.

Precipitation Processes

As just described, precipitation occurs when water drops or ice crystals become sufficiently large to fall under the force of gravity. The only way that this can happen is for an air mass to rise sufficiently high to condense large quantities of water. After all, the simple formation of clouds does not necessarily mean that precipitation will occur. How, then, does an air parcel rise to the point where precipitation does occur?

This kind of air lifting takes place in one of four ways (Figure 7.21). One way that air rises is through the process of **convectional uplift**, which results in distinct bubbles of air rising through denser surrounding air. Another way is called **orographic uplift**, which occurs when air is forced to flow up and over mountains. A third way is through the collision of large air masses along frontal boundaries. This process is called **frontal uplift** and typically occurs when contrasting bodies of air collide. The fourth way in which large parcels of air can uplift is through the process of **convergent uplift**. Convergent uplift occurs whenever bodies of air meet at a central location, forcing air upward at that point. Although this process is sometimes

---

Convectional uplift *Uplift of air that occurs when bubbles of warm air rise within an unstable body of air.*

Orographic uplift *Uplift that occurs when a flowing body of air encounters a mountain range.*

Frontal uplift *Uplift of air that occurs along the boundary of contrasting bodies of air.*

Convergent uplift *Uplift of air that occurs when large bodies of air meet in a central location.*

This satellite image shows the clouds over the Florida peninsula in January 2002. East of Florida is the Atlantic Ocean; west is the Gulf of Mexico. Notice that a distinct pattern of clouds is specifically associated with the Florida peninsula. Given the pattern that you see, which one of the following statements makes the most sense?

a) No convection is occurring over the Florida landmass.

b) Lots of condensation is occurring over the Gulf of Mexico.

c) Strong convection is occurring over the Atlantic Ocean.

d) The temperature of the Florida landmass is warmer than that of the surrounding oceans.

associated with low-pressure systems, it is most common in the low latitudes along the Intertropical Convergence Zone (ITCZ), where air from the Northern and Southern Hemispheres converges.

The remainder of this chapter focuses on convectional and orographic processes because they are relatively limited in a geographic extent and beautifully exemplify adiabatic processes in a way that is easy to understand. Frontal and cyclonic precipitation processes are more complex because they are associated with major weather systems that influence entire regions of the country in many different ways. These larger systems are discussed in Chapter 8.

Orographic Uplift A good place to begin a holistic discussion of rising air, adiabatic processes, and precipitation is with the concept of orographic uplift because it makes so much visual sense. Simply put, when airflow is interrupted by a mountain range, the air must flow up and over the barrier. For point of reference, the side of the mountain range that faces the direction of the oncoming wind is called the *windward side*, whereas the opposite, or downwind side of the range, is referred to as the *leeward side*. If the range is sufficiently high for the air parcel to reach its dew-point temperature through adiabatic cooling on the windward side of the range, clouds form and precipitation can potentially occur.

Figure 7.22 shows a hypothetical flow of air over a mountain and a graph of the associated temperature changes, which occur with altitude. The numbers in both images represent specific points for comparisons in the process. In this example, wind is flowing off the sea, let's say the Pacific Ocean, on the left side of the diagram (point 1). As the air encounters the mountain range, it begins to rise up the windward slope, cooling at the DAR until it reaches the level of condensation at point 2.

From point 2, the air begins to cool at the WAR because condensation, cloud formation, and precipitation

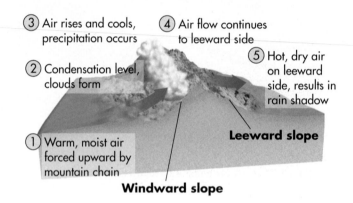

(a)

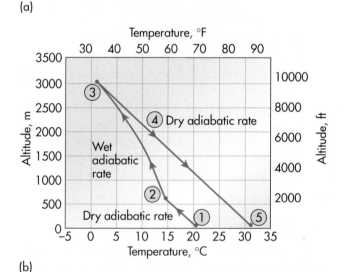

(b)

Figure 7.22 Orographic uplift. (a) When flowing air encounters a mountain range, it must flow over the top. As it does, the air adiabatically cools on the windward side and warms on the leeward side of the range, respectively. (b) Graph of the associated changes in temperature with altitude.

(a) (b)

Figure 7.23 Orographic clouds. (a) These clouds formed on the windward side of a small mountain range along the northeastern coast of Australia. (b) Orographic precipitation along the Inyo Mountains in eastern California.

are occurring, resulting in a landscape that may look very much like those pictured in Figure 7.23. The air continues to cool at this lesser rate, due to the release of latent heat of condensation, until the mountain crest is reached at point 3 (Figure 7.22). Here, the specific humidity of the air mass is less than it was at the level of condensation because moisture was lost through precipitation. This moisture loss is somewhat analogous to the water that is squeezed out of a sponge. Subsequently, the parcel of air continues over the mountain and descends on the leeward side (point 4).

As the air descends down the leeward side of the mountain range, a very interesting thing happens—the air dries rapidly. Recall that when air descends, it warms adiabatically at the DAR, which is 10°C/1000 m (5.5°F/1000 ft). This rate of warming takes place because as soon as the air begins its descent from the mountain crest, its maximum humidity increases while the specific humidity remains constant. In other words, the air is no longer saturated and thus warms at the DAR. The air warms at this rate from the crest of the mountain range at point 3, to an elevation of 0 m (0 ft) in the leeward valley at point 5. Given the high temperature of the air mass at this point, coupled with the very low relative humidity, the leeward side of the mountain range is dry and therefore called the **rain shadow**.

The orographic pattern just described can result in a fascinating range of landscape variability in a relatively small geographical area, from windward slopes that receive abundant precipitation to leeward slopes that are very dry. A good place to see the results of these processes over a

short distance is in California. Figure 7.24a shows the assemblage of landform regions of the state, including (from west to east) the low Coast Ranges, broad Great Valley, and high Sierra Nevada (mountains). The prevailing winds are westerly in the region, bringing in moisture-laden air with high dew points off the Pacific Ocean to the west. Notice that air flows over the Coast Ranges, down into the Great Valley, and then up and over the Sierra Nevada.

Figure 7.24b shows mean annual precipitation in the region. What patterns do you see? For one, observe that annual precipitation is high along the Coast Ranges and Sierra Nevada, with values reaching 180 cm (about 70 in.) in the central Sierras. Much of this precipitation falls as snow, which can reach depths of over 20 m (68 ft) at the highest elevations during a snowy winter! In contrast, the Great Valley is a distinct rain shadow, with only 40 cm (16 in.) compared to the 200 cm (80 in.) that falls in some parts of the mountains to the west. This valley will be discussed in more detail in Chapter 20 when agriculture in California is examined. Another rain shadow exists east of the Sierras, with only 25 cm (10 in.) of precipitation in northeastern California.

Probably the best known example of a snowy winter in the Sierras occurred in 1846–47 when the ill-fated Donner Party tried to cross the mountains in October on their way to California in a wagon train. Just before they reached the summit at what is now Donner Pass, an unseasonably intense snow fell and they became stranded short of the pass. A series of powerful winter storms subsequently swept across the range that winter, forcing them to stay put and resulting in a snow pack about 4.5 m to 6 m (15 ft to 20 ft) thick. Of the 82 pioneers who were stranded, 34 died either at their camps or trying to escape the mountains in the harsh conditions. Many of the survivors, who were rescued in April, had lived only because they resorted to cannibalism due to lack of food.

Rain shadow *The body of land on the leeward side of a mountain range that is relatively dry and hot (compared to the windward side) due to adiabatic warming and drying.*

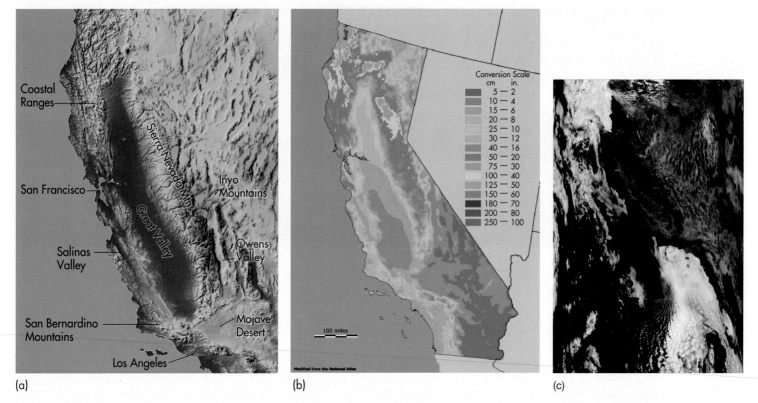

(a)   (b)   (c)

Figure 7.24 Airflow and precipitation patterns in California. (a) Map showing the major landform regions in California. (b) Geography of mean annual precipitation. The pattern of precipitation follows the geography of the land, with greater rainfall on the windward sides of the mountain ranges and less rainfall on the leeward sides. (c) You can see these patterns on the satellite image, with green areas representing relatively dense vegetation where high rainfall occurs, and brown zones reflecting less vegetation in areas of relatively low precipitation.

www.wiley.com/college/arbogast

Orographic Processes

Now that you have a fundamental understanding of orographic processes, we can integrate all the concepts that have been discussed to this point in the text in a simulated format. At this time, go to the *GeoDiscoveries* website and select the module *Orographic Processes*. This module is an opportunity to interact with the flow of air across a mountain range and the fluctuating humidity levels that occur on its ascent on the windward side and descent on the leeward side. The first part of the module is an animated review of the orographic process that shows the basic patterns. The second part of the module is a simulation where you can interact with the orographic process along a fictional mountain range by adjusting variables such as temperature and specific humidity before an air mass begins its ascent on the windward slope. You can also change the elevation of the mountains to see how this factor influences adiabatic processes. In this context, you will be presented with several scenarios that represent a minute fraction of the nearly infinite number of possibilities in nature. As you interact with this simulation, notice how the level of condensation changes as you adjust the variables. Also, pay attention to the changes that occur with respect to the various forms of humidity on both the windward and leeward mountain slopes. After you complete the animation and simulation, be sure to answer the questions at the end of the module to test your understanding of this process.

Convectional Uplift As just described, one way in which the uplift of air can result in precipitation is through the process of convection. Recall from the discussion of atmospheric circulation in Chapter 6 that convection is the process through which heated air rises. Convection is the initial driver of global atmospheric circulation due to the unequal heating that exists between low and high latitudes. In addition to being a primary cause of atmospheric airflow, convection is a process that contributes greatly to cloud formation and precipitation.

Have you ever noticed that, on a hot summer day, the surface of an asphalt parking lot is much hotter than the surrounding grass-covered ground? This unequal heating creates a bubble of warm air over the parking lot. Warm air rises within the larger body of air, which is relatively still (Figure 7.25a). If the air cools sufficiently, the dew-point temperature will be reached, clouds will form, and precipitation may occur. A good indicator that convection is occurring on any given day is the presence of cumulus clouds, which are the individual puffy clouds pictured in Figure 7.25b. When you see these kinds of clouds, think of each as representing a bubble of air that lifted from the surface into the atmosphere, with the base of the clouds being the level of condensation. In most circumstances, a cumulus cloud will float along for some distance downwind and then evaporate.

Although most cumulus clouds eventually dissipate before rain falls, some grow into cumulonimbus clouds that produce rain because convection continues until the water drops become sufficiently large to fall. If convection is very strong, a thunderstorm may develop. Whether or not this evolution occurs depends on the overall stability of the air in which the convection bubble rises. At a fundamental level, air stability refers to the potential for convection within a body of air, with **stable air** being air in which strong convection cannot occur. In

contrast, **unstable air** is air in which strong convection of air bubbles can occur. As we work through this part of the discussion, refer to Figure 7.26, which illustrates air stability in graphical form.

Let's first consider an unstable air mass in which convection can occur to the point that clouds develop (Figure 7.26a). In other words, a bubble of warm air rises within a main body of relatively cool air that is stationary. This larger body of air cools with increasing altitude at the environmental lapse rate, which, on average, is 6.4°C/1000 m (3.5°F/1000 ft). Imagine, however, that the environmental lapse rate is 12°C/1000 m (6.6°F/1000 ft) in this particular body of air. If we further imagine that the temperature of this body of air is 32°C (89.6°F) at ground level, this lapse rate means that the temperature is 20°C (68°F) at 1000 m and 8°C (46.4°F) at 2000 m. In Figure 7.26a, you can see this temperature change with elevation in the green graph line on the left as well as in the vertical list of blue temperature readings on the right.

Now, imagine that a bubble of relatively warm air develops over a large dark surface. This bubble has a ground temperature of 33°C (91.4°F), as you can see in the black temperature readings. Because it is warmer at ground level than the surrounding air, the bubble begins to convect, which means that it cools at the dry adiabatic lapse rate of 10°C/1000 m (5.5°F/1000 ft) due

---

Stable air *A body of air that has a relatively low environmental lapse rate compared to potential uplifting air; thus, strong convection cannot occur.*

Unstable air *A body of air that has a relatively high environmental lapse rate compared to uplifting air within it; thus, strong convection can occur.*

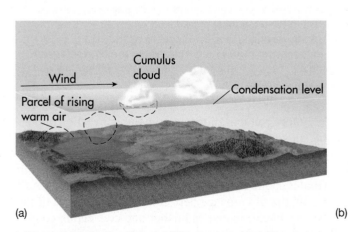

Figure 7.25 Convection. (a) When a warm bubble of air forms due to unequal heating at the surface, it rises within a body of relatively still air, causing the formation of cumulus clouds. (b) Cumulus clouds formed by convection on the Great Plains.

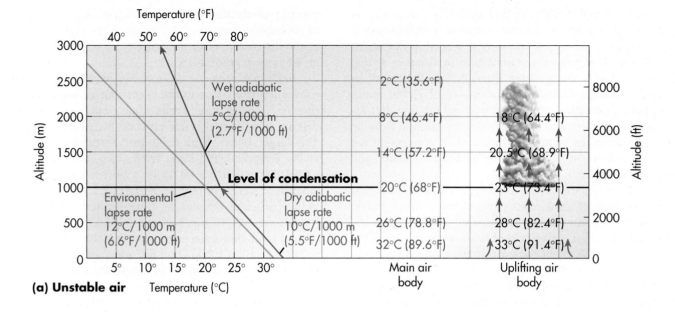

(a) Unstable air

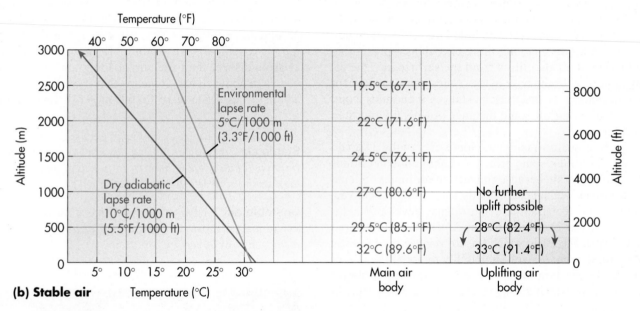

(b) Stable air

Figure 7.26 Graphical representations of unstable and stable bodies of air. (a) In this hypothetical body of unstable air, strong convection occurs when the environmental lapse rate is high and a bubble of relatively warm air forms at the surface. The graphical representation of temperature on the left shows changes with respect to altitude in the bubble and surrounding air mass; these changes can also be seen in numerical form to the right within the convecting air bubble and main air body. (b) In this hypothetical body of stable air, the environmental lapse rate is relatively low and the convecting air cannot lift because it cools more quickly than the surrounding air.

to its decreasing air pressure as it lifts. Given that this rate of cooling is less than the rate of cooling in the primary body of air that surrounds the bubble, the bubble continues to lift. Note, for example, that the convecting bubble represented by the arrows in Figure 7.26a has a temperature of 23°C (73.4°F) at 1000 m. If we assume that this bubble of air had a specific humidity of 20.6 g/kg at ground level, it would reach the level of condensation at 1000 m and a cloud would begin to form at that elevation.

Once condensation begins in our convection bubble, the air would then start to cool at the wet adiabatic lapse rate of 5°C/1000 m (2.7°F/1000 ft), which is still less of a rate than that observed in the surrounding body of air (that is cooling at a high environmental lapse rate). As a result, the discrepancy between the temperature of the two air bodies increases further and convection continues. This is why the air temperature of the convective bubble is 18°C (64.4°F) at 2000 m, whereas it is only 8°C (46.4°F) in the surrounding air.

Now, let's look at what happens in a stable body of air by examining Figure 7.26b. Imagine in this scenario that the environmental lapse rate of the main body of air is 5°C/1000 m (3.3°F/1000 ft) and that the temperature at ground level is 32°C (89.6°F), which is the same as in the previous example. In this case, the temperature of the main body of air would be 27°C (80.6°F) at 1000 m and 22°C (71.6°F) at 2000 m.

Consider what happens over the warm parking lot described in the example of unstable air. Let's imagine that the air immediately above this parking lot warms to the same temperature it did previously; that is, 33°C (91.4°F). Given that this bubble of air is warmer than the surrounding air at the surface, it begins to convect. As it does, it cools at the dry adiabatic lapse rate 10°C/1000 m (5.5°F/1000 ft) because the air pressure decreases as the air lifts. In this example, however, this rate of cooling is *greater* than that observed in the surrounding body of air. Thus, at 500 m the temperature of the convecting air bubble is 28°C (82.4°F), whereas it is 29.5°C (89.1°F) in the surrounding body of air. Under these conditions, convection cannot continue because the air bubble is cooler than the main body of air. As a result, the air sinks toward the surface and cloud formation cannot occur.

Geo Discoveries www.wiley.com/college/arbogast **WILEY PLUS**

Convectional Precipitation

Now that you have completed the discussion of convectional uplift, let's see what the effects of this process are in the real world. Go to the *GeoDiscoveries* website and select the module *Convectional Precipitation*. This module consists of a video that shows how convection results in the formation of clouds and precipitation. After you complete the video, be sure to answer the questions at the end of the module to test your comprehension of this concept.

KEY CONCEPTS TO REMEMBER ABOUT PRECIPITATION AND PRECIPITATION PROCESSES

1. In order for precipitation to occur, condensation nuclei must be present in the form of dust particles or other small solids.

2. Water droplets remain in suspension as clouds until they grow sufficiently large to fall under the force of gravity.

3. For precipitation to occur, some mechanism must be present to cause sufficient uplift of an air parcel such that high amounts of condensation take place.

4. Convectional uplift occurs when a bubble of air rises due to unequal heating of the Earth's surface. The bubble of air continues to rise as long as the air is unstable, that is, as long as the temperature of the bubble is warmer than the surrounding air. When these temperatures become equal, the air is stable.

5. Orographic uplift refers to the rising air that occurs due to a blocking mountain range. The windward side of the range faces the oncoming winds, whereas the leeward side is the downwind side. Condensation occurs on the windward side; the leeward side is typically in the rain shadow and is dry.

In this chapter, you learned about the nature of the hydrologic cycle and humidity. Through the course of this discussion, a model of adiabatic processes described how precipitation occurs due to orographic and convectional uplift, which are easy concepts to visualize. Adiabatic processes are also associated with more complex weather systems, specifically midlatitude cyclones and hurricanes, which can influence large regions in many different ways. These systems integrate the atmospheric circulatory processes discussed in Chapter 6 with the adiabatic processes presented in Chapter 7. The interaction of these various processes can result in powerful weather-makers that release tremendous amounts of energy, as this great image of lightning demonstrates.

SUMMARY OF KEY CONCEPTS

1. Water molecules are attracted to one another through hydrogen bonding. Given the unique nature of this bond, water can exist in three physical states: liquid (water), solid (ice), and gas (water vapor).

2. Latent heat is stored in water molecules. When water changes phase, latent heat is either absorbed or released, depending on the transformation.

3. The hydrologic cycle refers to the ways in which water moves on Earth through the combined processes of evaporation, precipitation, and runoff. Most water is stored in the oceans, which are a net source of evaporation in the context of the global water balance. On landmasses, more precipitation occurs than evaporation.

4. Three kinds of humidity occur: maximum, specific, and relative. Maximum humidity refers to how much water vapor a parcel of air can hold. This variable depends on temperature, with warm air having a higher maximum humidity than colder air. Specific humidity measures how much water vapor is in the air. Relative humidity is the ratio of specific humidity to maximum humidity. The dew-point temperature is the temperature at which a mass of air becomes saturated. This is the temperature at which condensation occurs in that air mass. As specific humidity increases, so does dew-point temperature.

5. Adiabatic processes refer to the temperature changes that occur in an air parcel due solely to changes in air pressure. When air pressure increases, a parcel of air is compressed and warms adiabatically. In contrast, when air pressure decreases, a parcel of air expands and thus cools internally because air molecules are spaced farther apart. Air cools at the dry adiabatic lapse rate (DAR) until condensation occurs. This rate is 10°C/1000 m (5.5°F/1000 ft). When air warms adiabatically, it always does so at the DAR because condensation is not occurring. Once the level of condensation is reached, the air temperature begins to decrease at the wet adiabatic lapse rate (WAR), which

is 5°C/1000 m (2.7°F/1000 ft). This lesser rate of cooling occurs because latent heat of condensation is released when water changes from vapor to liquid.

6. Air is uplifted such that cloud formation and precipitation occur in any of four different ways. The extent of uplift and condensation depends greatly on the stability of the main body of air. Orographic lifting occurs when air flows over a mountain range. Convection occurs when bubbles of warm air develop in a body of relatively cool air that has a high environmental lapse rate. Frontal uplift occurs when two contrasting bodies of air collide. Convergence occurs when air flows into the center of a low-pressure system.

CHECK YOUR UNDERSTANDING

1. Describe the physical properties of water and its three phases.

2. Compare and contrast the various ways that water changes phase.

3. Discuss the hydrologic cycle and the concept of a *reservoir*.

4. What are the various kinds of atmospheric humidity?

5. Which body of air would have the higher maximum humidity: an air parcel that has a temperature of 25°C (77°F), or one that has an air temperature of 15°C (59°F)?

6. Of the two air masses described in the previous question, which of the two will likely have the higher specific humidity? Which one will probably have the higher relative humidity?

7. Imagine that the dew-point temperature of an air mass is 30°C (86°F). What would the specific humidity be? (*Hint:* Look at the saturation curve, Figure 7.6.)

Would this be a humid air mass, or dry? Would the actual temperature of the air mass likely be more or less than the dew point?

8. Imagine that the temperature of an air parcel is 20°C (68°F) and the specific humidity is 10 g/kg. What would the relative humidity be? To what temperature would the air have to cool to reach a point of saturation?

9. Why are condensation nuclei important for precipitation to occur?

10. Compare and contrast the wet adiabatic lapse rate (WAR) with the dry adiabatic lapse rate (DAR). When do you use the DAR as opposed to the WAR?

11. Discuss the three fundamental cloud types—cirrus, cumulus, and stratus—and the kinds of atmospheric conditions in which they form.

12. Why is the specific humidity of an air parcel less on the leeward side of a mountain range than on the windward side?

ANSWERS TO VISUAL CONCEPT CHECKS

VISUAL CONCEPT CHECK 7.1

The pond is part of the hydrologic cycle because it is a place where water is temporarily stored. Water can enter the pond either through precipitation or from within the ground itself. Once the water is stored in the pond, it can either flow back into the ground or evaporate into the atmosphere.

VISUAL CONCEPT CHECK 7.2

The answer is *d*, "all of the above." For contrails to form, the air must be close to saturation. Condensation nuclei must be present for water droplets to form, and the plane must be above the level of condensation.

VISUAL CONCEPT CHECK 7.3

The answer is *d*. Given that the cloud pattern is distinctly associated with the peninsula, strong convection must be occurring over the land. For this to happen, the Florida landmass must be warmer than the surrounding water.

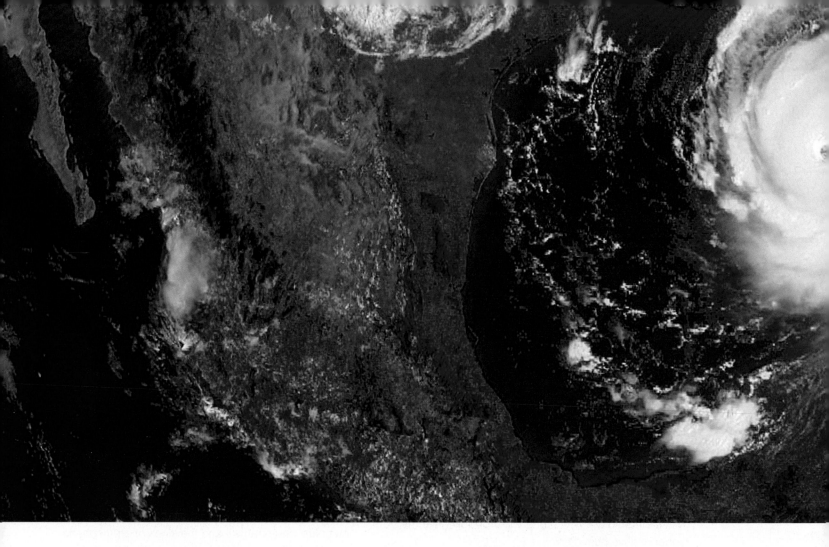

CHAPTER EIGHT

AIR MASSES AND CYCLONIC WEATHER SYSTEMS

Chapter 7 focused on atmospheric humidity and two uplift mechanisms that result in precipitation—orographic and convectional. These processes involve air parcels contained within a definable unit of geographic space, such as above a mountain range or over a relatively warm surface, respectively. In this chapter, the discussion broadens to large air masses and complex atmospheric systems that influence weather over entire regions. The chapter focuses on cyclonic weather systems that occur in the tropics and midlatitudes, including thunderstorms, tornadoes, and hurricanes. These are the kinds of storms you hear about all the time, or even have experienced yourself. As a result, most students find this to be an especially fun chapter to read.

CHAPTER PREVIEW

Air Masses and Fronts

Evolution and Character of Midlatitude Cyclones
    On the Web: *GeoDiscoveries*
    Formation of a Midlatitude Cyclone

Hurricanes are the largest storms on Earth. This satellite image shows Hurricane Katrina shortly before it struck the Gulf Coast in August 2005. Katrina caused billions of dollars in damage and killed thousands of people, many of whom died when New Orleans flooded. This chapter focuses on midlatitude and tropical circulatory processes, including the evolution of hurricanes and tornadoes.

On the Web: *GeoDiscoveries*
Migration of a Midlatitude Cyclone

Thunderstorms
On the Web: *GeoDiscoveries*
Formation of Thunderstorms
On the Web: *GeoDiscoveries*
Tornadoes

Tropical Cyclones
On the Web: *GeoDiscoveries*
Migration of Hurricane Katrina

The Big Picture

1. Define the concept of an air mass and describe the various air masses that influence North American climate.

2. Compare and contrast warm and cold fronts, including their character and movement.

3. Explain the evolution of midlatitude cyclones, their component parts, and how they typically migrate.

4. Describe the evolution of thunderstorms and how they can become severe.

5. Discuss how tornadoes form, how they are classified, and how people monitor them.

6. Define a tropical cyclone and explain how it forms and migrates.

7. Explain the evolution of a hurricane in the Atlantic Ocean, including its various stages, migration, and how it is classified and monitored.

Air Masses and Fronts

Have you ever experienced hot and humid weather for 2 or 3 days and then, *boom*, a strong storm occurred and the temperature dropped sharply? You might have even talked with your friends about how the weather changed so fast. Most likely, what you experienced was the passage of a midlatitude cyclone through your region and the associated air masses that came with it. The first air mass was warm with high dew points that made it uncomfortable to be outside, whereas the second air mass was relatively cool (maybe even cold) and dry. Although you may not have known the reasons why these weather fluctuations occurred, you certainly noticed the change. In this section, we examine the characteristics of the various air masses.

Air Masses

Before we investigate midlatitude circulatory processes and how they influence weather, the concept of an air mass must first be defined. This definition is essential because the formation of midlatitude cyclones depends on the interaction of very large bodies of air, thousands of square kilometers in size, that have distinctly contrasting physical properties.

An **air mass** is any large body of the lower atmosphere that has fairly uniform conditions of temperature and moisture. Although the terminology and definition sound like scientific jargon, they actually represent conditions that you have experienced and probably noted in conversations. The source region of an air mass is any large body of land or water from which the air derives these characteristics. Imagine that a maritime air mass moves from the northern Pacific Ocean to northern Canada, where it stagnates for a couple of weeks in January. What will happen to that air mass? How will it change character? From previous discussions, you should realize that the temperature of the air mass will drop while it is over Canada because the landmass is cold during winter. In addition, the air mass would lose a lot of its moisture because cold air holds less water vapor than warm air. If the air mass then moved south into the United States, perhaps due to changes in the pressure gradient or the overall circulatory system, it would be cold and dry by the time it reached your town.

Air masses are categorized by a combination of letter designations. The first letter is lowercase and designates the moisture source of the air mass:

c = continental (dry)
m = maritime (moist)

Air mass *A large body of air in the lower atmosphere that has distinct temperature and humidity characteristics.*

The second letter is uppercase and designates the latitude position of the source region:

A = Arctic
P = Polar—from between 50° and 60° N or S latitude
T = Tropical—from between 20° and 35° N or S latitude

Using these designations, five principal types of air masses can be identified that most often affect North America (Figure 8.1). The most common cold air mass to affect North America is the *continental Polar* (cP), which forms over northern Canada and consists of stable, dry air. You have probably experienced this type of an air mass when a very cold snap occurred in the middle of January or February where you live. This weather pattern likely occurred because a cP air mass moved into your region of the country. Such an air mass can be further cooled when it is infused with *continental Arctic* (cA) air, which is extremely cold. In northern Minnesota, for example, cA air has produced a record low temperature of –43°C (–46°F). This is the kind of air that makes you run from one building to another. Although cP air masses are most intense in winter, they sometimes affect the weather in the summer. Such an influence is particularly noticeable when it has been hot and muggy for a few days and then the temperature cools to perhaps 24°C (75°F) with a very comfortable northwest breeze.

Another dry, stable air mass that influences the continent is the *continental Tropical* (cT). This air mass forms over places like Mexico and Arizona and thus is hot and extremely arid. Temperatures in these air masses can reach 45°C (~115°F), with low relative humidity of ~40%. If you have ever been to the southwestern deserts of the United States in the summer, at places such as Phoenix, Arizona, or Las Vegas, Nevada, you probably experienced such an air mass that is just . . . *brutally hot*. The air feels like it will take your breath away.

In contrast to the continental air masses, two maritime air masses periodically influence parts of the continent. One of these air masses is the *maritime Polar* (mP), which consists of cool and moist bodies of air that form over the northern Pacific Ocean and northwest Atlantic Ocean in the Northern Hemisphere. Given the influence of westerly winds in the midlatitudes, this kind of air has a particularly strong influence in the Pacific Northwest. If you happen to live there, or have visited places such as Seattle, Washington, or Portland, Oregon, in the winter, you certainly experienced the cool, damp air associated with an mP air mass. This air often has temperatures of about 12°C (~53°F) and relative humidity of 70% to 80%. The other maritime air mass is the *maritime Tropical* (mT), which forms over places like the Gulf of Mexico and is distinctive because it is warm and moist.

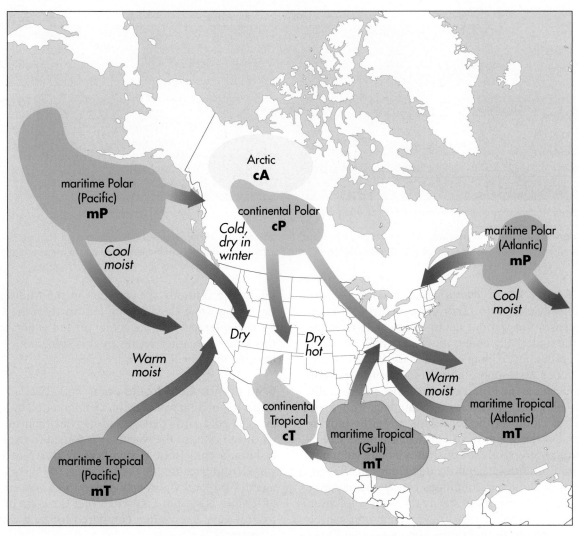

Figure 8.1 Principal air masses in North America. Five major types of air masses periodically flow into the continent. Note the character of these air masses and the geographical regions in which they originate.

To associate this air mass with your experience, try to remember when the summer air was hot and sticky, with temperatures of about 33°C (~95°F) and ~60% relative humidity. This was an mT air mass and was the kind of air that kept you inside an air-conditioned room much of the day.

Fronts

We have just discussed the character of air masses, where they originate, and how they move into North America. In all likelihood, you have experience with each of these air masses. Now it is time to examine how they interact with each other. The logical place to begin is by investigating the boundaries of these air masses on the Earth's surface. These boundaries are called *fronts*. The dominant front in the Northern Hemisphere mid-latitudes is the polar front, which, if you recall, is the boundary between cold, dry (cP) air to the north and warmer, moist (mT) air to the south. These contrasting air masses flow parallel to one another along a **stationary front** most of the time. Sometimes, however, atmospheric conditions arise that cause one particular air mass to advance into another along distinct frontal boundaries. It is somewhat helpful to think of these advancing air masses as being analogous to an army that is advancing into a particular area, with the lead columns of the force being the front. This part of the chapter focuses on these atmospheric boundaries and the types of precipitation that occur along them as air is uplifted. Recall from Chapter 7 that this type of uplift is called *frontal uplift*.

Stationary front *A boundary where contrasting air masses are flowing parallel to one another.*

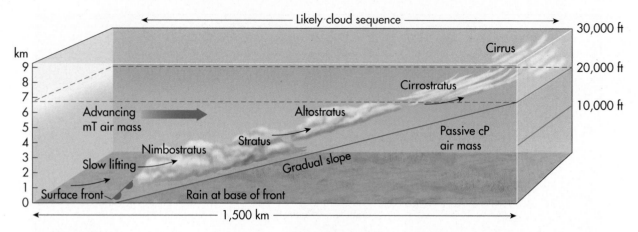

Figure 8.2 Uplift of air at a warm front. At a warm front, the advancing warm air gradually slides over the top of underlying, passive cooler air, causing the formation of stratiform clouds as the warm air slowly cools adiabatically.

Warm Fronts Warm fronts occur in places where warm air advances into relatively cool air. This interaction causes the warmer air to slowly slide over the top of the underlying cooler air (Figure 8.2). This process of gradual overriding of cooler air actually occurs in the upper atmosphere ahead of the surface warm front and causes the lifting mT air to cool adiabatically. As the air cools, clouds form, beginning with high-level cirrus clouds at the top of the upwardly moving air mass. As the warm front approaches, these clouds change to progressively lower stratus clouds, culminating in rain-producing nimbostratus clouds at the surface front. When these conditions evolve, the sky may be overcast with a slow but steady rate of precipitation that may last a day or two. This precipitation can be in the form of drizzle, light rain, and even snow. These

kinds of conditions are common in the Midwest during spring when slow-moving mT air from the Gulf of Mexico interacts with the cooler air to the north. When this kind of weather settles in, it is sometimes hard to imagine that it will ever end.

Cold Fronts Cold fronts occur when cool air moves into a region that was previously dominated by warmer air. The cold air is denser and heavier than the warm air ahead of it, so the warm air is forced to rise. In this fashion, a cold front is significantly different from a warm front because the cold air hugs the surface along a cold front and vigorously drives the warm air ahead of it aloft. Notice in Figure 8.3 that the edge of a cold front is very steep when compared to the warm front in Figure 8.2.

Warm front   *A frontal boundary where warm air is advancing into relatively cool air. This front is typically associated with slow, steady precipitation.*

Cold front   *A frontal boundary where cold air is advancing into relatively warm air. This front is typically associated with intense rain of short duration.*

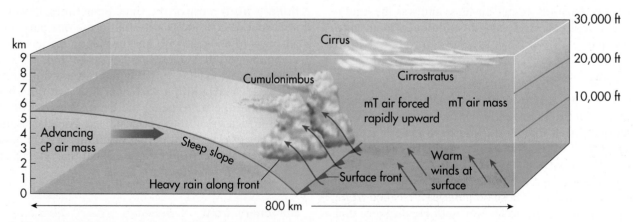

Figure 8.3 Uplift of air along a cold front. In contrast to a warm front, warm air ahead of a cold front is abruptly lifted and quickly cooled, causing cumulonimbus clouds to form.

Once the warm air begins to lift along the cold front, adiabatic cooling starts and vapor condenses, forming clouds. Because the air rises rapidly, it quickly cools adiabatically, which means that large amounts of latent heat energy are rapidly released as the water condenses. As a result, rainfall is intense and of short duration. If sufficient moisture is present and enough latent heat is quickly released, an intense **thunderstorm** can form, with the latent heat being the fuel for its development. Although thunderstorms can develop through the simple process of convection, they are prone to be more severe and widespread along a cold front because of the rapid cooling and latent heat release that occur. For example, Figure 8.4 shows how widespread frontal precipitation can be. This radar image shows a portion of a cold front that stretched from Canada into Oklahoma. A line of thunderstorms extends along this section of the front from southwestern Lower Michigan into southeastern Kansas. We will describe the evolution of thunderstorms in more detail later in the chapter.

Thunderstorm   *A brief, but intense storm that contains strong winds, lightning, thunder, and perhaps hail.*

1. An air mass is a large body of air that has distinctive characteristics and forms in specific geographic regions.

2. Five principal types of air masses affect North America. Continental air masses include continental Polar (cP), continental Arctic (cA), and continental Tropical (cT). Maritime air masses are maritime Tropical (mT) and maritime Polar (mP).

3. Air masses have distinct boundaries called fronts. A stationary front is a place where contrasting air masses are flowing parallel to one another.

4. A warm front is a place where warm air is advancing into relatively cool air. Warm air slowly slides over the top of the cooler air along a warm front, causing slow and steady rainfall.

5. A cold front is a place where cold air is advancing into relatively warm air. Given the higher density of colder air, the warm air is rapidly forced aloft, where it cools quickly. As a result, rainfall is intense and of short duration along the front.

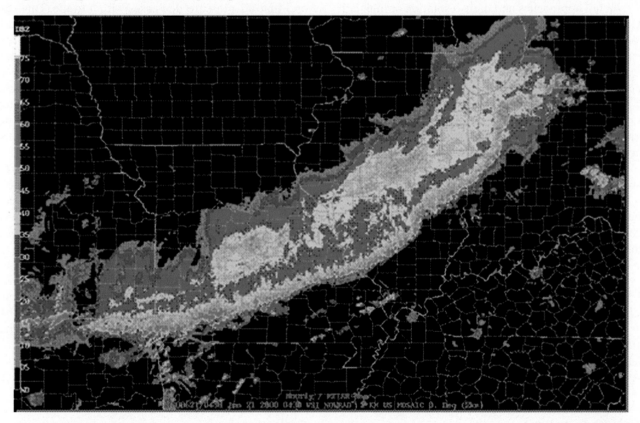

Figure 8.4 Frontal precipitation. This radar image shows a line of thunderstorms extending from southwestern Lower Michigan to southeastern Kansas. This line of storms developed along a strong cold front that passed through the Midwest. Orange and yellow colors represent areas of heavy rain, whereas the areas of lightest precipitation are in blue.

Evolution and Character of Midlatitude Cyclones

Now let's closely examine the atmospheric circulatory systems in which warm and cold fronts develop. Such a system is called a **midlatitude cyclone** and is a well-organized low-pressure system that migrates across a region while it spins. In order to explain how midlatitude cyclones develop along the polar front and how they influence weather, we must first briefly review how the polar front and associated jet stream move on a seasonal basis. This geographical pattern was examined in Chapter 6, so you might want to review Figure 6.21.

Let's begin this review in the summer when high latitudes receive high amounts of solar radiation. During this time of year, the polar front and jet stream retreat to a location very close to the poles and exhibit a zonal-flow pattern much like that illustrated in Figure 6.21a. This confinement of the polar front to these high latitudes occurs, of course, because temperatures are warm well into the higher middle latitudes. Although midlatitude cyclones may develop along the polar front during this time of year, they are generally weak systems because no strong temperature contrast occurs north and south of the polar front.

As fall approaches and daily insolation at high latitudes decreases, large masses of cP air begin to develop north of the polar front. This increase of cold, dry air north of the polar front causes the front and associated jet stream to move south where it begins to encroach upon the warmer air. Essentially, the buildup of cold, dense air at high, northerly latitudes forces the polar front south. As this migration proceeds, the path of the jet stream gradually changes from a zonal pattern to an undulating meridional pattern with well-developed Rossby waves in the upper atmosphere (see Figure 6.21 again). This development occurs because the atmosphere is reacting to the growing temperature difference on either side of the polar front. These waves allow tongues of warm air (south of the front) to penetrate northward at the same time that wedges of cold air (north of the front) expand to the south.

On a global scale, the development of Rossby waves is how the atmosphere begins the process of mixing cold and warm air. The most vigorous mixing of these air masses occurs on a more regional scale, within midlatitude cyclones that form along distinct sections of the polar front in any given Rossby wave. The term *cyclone* is sometimes confusing because tropical storms near southern Asia are referred to by the same name. In addition, tornadoes are sometimes called *cyclones*. Although these specific kinds of storms are investigated later in this chapter, for now the term *cyclone* is used only in the context of midlatitude circulatory processes.

Formation of midlatitude cyclones is called **cyclogenesis**. It encompasses a complex set of processes that are most often associated with an undulating polar jet stream pattern in the upper atmosphere and the interaction of these winds with airflow at ground level. To better understand this interaction, think of the atmosphere as consisting vertically of many different pressure surfaces, with the highest air pressure exerted at ground level and progressively lower air pressure at progressively higher altitudes. Just like a particular isobar (a line of equal air pressure) can be followed on the ground, it can also be traced horizontally within different levels of the atmosphere.

Upper Airflow and the 500-mb Pressure Surface

A good example of such an upper air-pressure surface is the 500-mb surface, which occurs at a specific but varying altitude at any given place on Earth. Recall that the average surface air pressure at sea level is 1013.2 mb and that air pressure decreases with altitude. Thus, the 500-mb pressure surface is at a relatively high altitude compared to the surface air pressure. From a meteorologic standpoint, the 500-mb pressure surface is important because it vertically divides the atmosphere in two, if we assume that the average surface air pressure is 1013 mb and 0 mb of air pressure occurs at the top of the atmosphere. In addition, the wind patterns at the 500-mb surface exert a strong steering influence on the circulatory patterns at the surface.

To understand the significance of the 500-mb pressure surface and how its height variation is important to airflow, consider Figure 8.5. Figure 8.5a shows two columns of air (Air columns 1 and 2) that contain the same mass of atmosphere. Given that each column contains the same number of molecules, the surface pressure is the same. In addition, the altitude of the 500-mb surface in both columns is equal, in this case 5460 m (17,900 ft).

What happens if we warm one of the columns of air, say Air column 2, relative to the other (Figure 8.5b)? When this warming occurs, the column of air stretches vertically—because warm air rises—and the height of the 500-mb surface in that column moves to a higher altitude; let's say, for example, 5760 m (18,900 ft). At that

---

Midlatitude cyclone *A well-organized low-pressure system in the midlatitudes that contains warm and cold fronts.*

Cyclogenesis *The sequence of atmospheric events along the polar jet stream that produces midlatitude cyclones.*

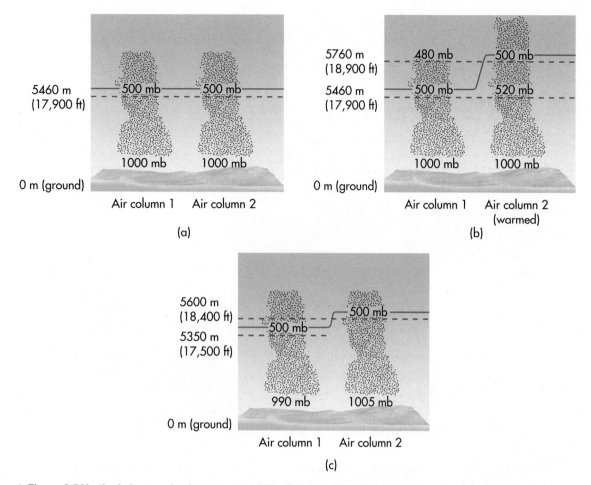

Figure 8.5 Vertical changes in air pressure. (a) The height of the 500-mb pressure surface in two columns of air with the same mass and temperature. (b) If one column of air (column 2) is warmed relative to the other, then the atmosphere stretches vertically and the height of the 500-mb surface rises. (c) The height of the 500-mb surface also fluctuates when the surface air pressure changes.

same altitude in Air column 1, the air pressure is 480 mb. Air column 1 has this relatively lower pressure at 5760 m (18,900 ft) because the elevation of the 500-mb surface in that column did not change from the previous example (Figure 8.5a). This pressure/altitude difference between the two air columns reflects the fact that the warmer column of air now has a lower overall density than the first (cooler) one. No change occurs in the mass of Air column 2, which is why the surface air pressure is still 1000 mb. Another way of comparing the two air masses in the second example is to say that Air column 2 has a smaller vertical pressure gradient than Air column 1.

The same kind of altitude change in the 500-mb surface also occurs when air pressure at the surface varies between two regions. Using Figure 8.5a again as a hypothetical baseline scenario, let's say that the surface air pressure in Air column 1 decreases to 990 mb at the same time it increases to 1005 mb in Air column 2 (Figure 8.5c). What happens? Notice that the altitude of the 500-mb surface drops to, let's say, 5350 m (17,500 ft)

in Air column 1, whereas it rises to 5600 m (18,400 ft) in Air column 2. This change occurs because the mass of air beneath the 500-mb surface decreases in Air column 1 at the same time it increases in Air column 2.

This example shows how pressure can vary within two distinct columns of air. Now examine how the altitude of the 500-mb surface might vary horizontally in the continuous medium of the actual atmosphere. Figure 8.6a presents a typical isoline map that shows the change in altitude of the 500-mb surface across the United States on a particular day. This reflects the fact that surface air pressure varies across the nation in a manner similar to Figure 8.5c. The map shows a strong high-pressure system (anticyclone) in the western United States and two areas of low pressure—one over the northeastern part of the country and another off the Pacific Northwest coast. You can see how this information translates into an actual topographic map of the 500-mb surface in Figure 8.6b. Notice that the height of the pressure surface is higher and forms a

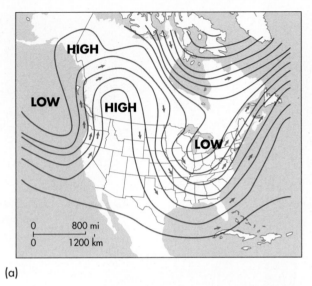

(a)

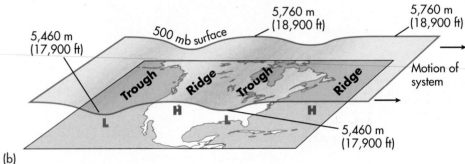

(b)

Figure 8.6 Mapping the 500-mb pressure surface. (a) Isolines of constant pressure across the United States on a particular day. (b) A topographic map showing the variation in the 500-mb surface.

ridge-like feature where the anticyclone occurs. Such an atmospheric feature is called a **high-pressure ridge** by meteorologists. In contrast, the height of the pressure surface is lower in the areas associated with the low-pressure systems along each coast. Meteorologists call these valley-like features in the 500-mb surface **low-pressure troughs**. Regardless of whether the upper atmospheric pressure system is a ridge or a trough, each shares the common reference of an axis, which is the imaginary line that extends along their length, that is, from north to south.

High-pressure ridge *An elongated area of elevated air pressure in the upper atmosphere that is typically associated with sunny skies and calm winds.*

Low-pressure trough *An elongated area of depressed air pressure in the upper atmosphere that is typically associated with cloudy skies and rain.*

Interaction of Upper Airflow and Surface Airflow

In the context of understanding how midlatitude cyclones form, it is useful to examine the interaction of airflow on the ground with airflow at the 500-mb pressure surface. As we begin this part of the discussion, follow along by studying Figure 8.7, which shows airflow both on the ground and at the 500-mb altitude. This figure is very similar to Figure 6.12, because it shows a dynamic convection loop in the atmosphere, but differs because it is viewed within the context of the 500-mb surface.

Let's imagine a scenario in which a prominent Rossby wave exists in the upper atmosphere and a low-pressure trough has developed poleward of the jet stream at the 500-mb level. When this condition develops, the upper airflow becomes disturbed, with a region of converging airflow developing behind (to the left of) the trough axis in Figure 8.7. At the same time, a zone of diverging air forms east (to the right) of the trough axis. As these airflow patterns continue to develop, the

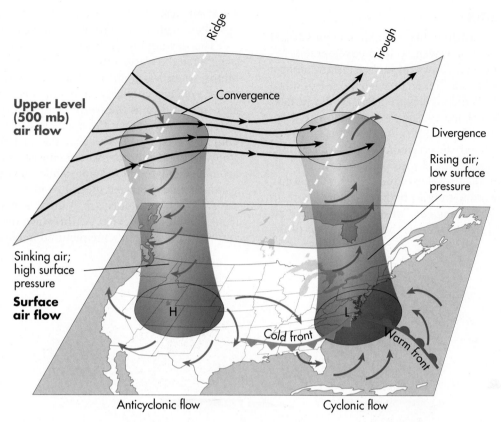

Figure 8.7 Linkage of upper air flow and surface air flow in a midlatitude cyclone. These systems form through a complex interaction of upper air and surface air flow. Under the high-pressure system, the weather is stable with clear sky. Where the cyclone forms, however, the atmosphere is unstable.

air pressure beneath them (at ground level) changes. Where the air converges aloft, more air is flowing into a given unit of space at the 500-mb level, which forces air toward the surface as a zone of high pressure. In contrast, where the air diverges aloft on the downwind side of the trough, less air is occupying a given unit of space, which has the effect of pulling air up from beneath to fill the void. When this situation occurs, the surface and upper airflows become closely linked as columns of spinning air—in other words, a dynamic convection loop—that are set in motion by the Coriolis force, as described in Chapter 6.

To understand better how midlatitude cyclones form and how they influence the weather on the ground, let's investigate the system evolution at the surface by studying the section of the polar front outlined in Figure 8.8. Notice that prior to the development of a midlatitude cyclone, the polar front is linear, with parallel-flowing but opposite-direction winds— cold, easterly winds to the north of the front and warm westerly winds to the south.

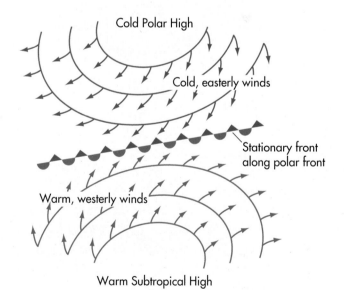

Figure 8.8 Section of the polar front. Prior to the development of a midlatitude cyclone, a stationary front exists along the polar front. Winds in contrasting air masses flow parallel and opposite to each other.

Cyclogenesis

As previously described in the discussion about vertical airflow within Rossby waves, cyclone development begins when the upper airflow and surface airflow along a stationary front become linked through the rise of warm air and descent of cold air (Figure 8.9a). At the surface, this developing linkage causes an undulation to form in the polar front (Figure 8.9b). As this undulation becomes better defined, the convergence of surface winds intensifies in a counterclockwise pattern, with cold air flowing south on the backside of the system, while warm air flows northward to the east of the circulatory core. In this situation, the leading edge of the inflowing cold air mass is the cold front, whereas the leading edge of the

warm air mass is the warm front. Notice in Figure 8.7 how these fronts relate to the dynamic convection loop.

The developing cyclone is said to be in the *open stage* when the warm and cold fronts are fully separated; that is, when warm air exists at the surface all the way to the system's core (Figure 8.9c). This region of warm air is referred to as the *warm sector*. Imagine that such a system develops over the northern Great Plains in the United States. In this scenario, the air behind the cold front (i.e., to the north) would consist of cold, dry (cP) air that originated in Canada, while the air in the warm sector that lies between the cold and warm fronts is relatively warm and moist (mT) air that may have come from the Gulf of Mexico. At this stage the cyclone may very well look like the one pictured in Figure 8.10.

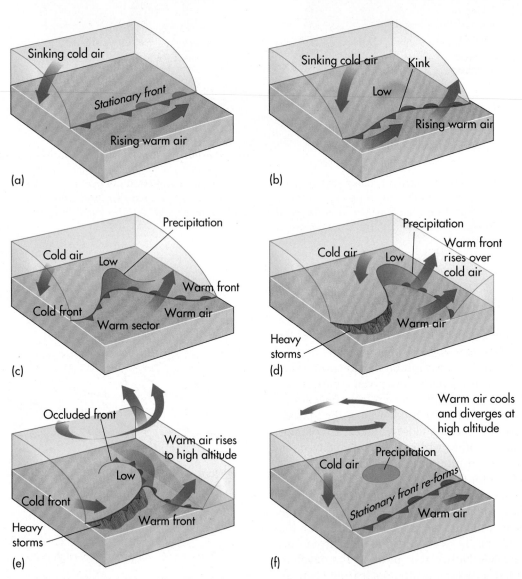

Figure 8.9 Evolution of a midlatitude cyclone. (a) Stationary front stage; (b) early stage when undulation develops along stationary front; (c) open-wave stage with prominent cold and warm fronts; (d) closing of open wave due to rapid speed of advancing cold front; (e) occluded stage when cold front begins to overtake the warm front; (f) dissolving stage when the warm sector is completely pinched off by overtaking cold front.

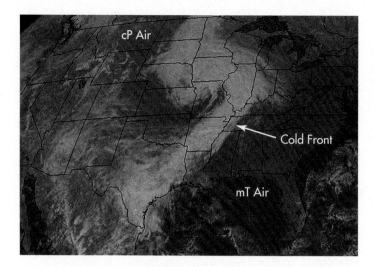

Figure 8.10 A classic midlatitude cyclone. A cyclone centered over the midwestern United States. Notice the prominent line of clouds, which developed along a strong cold front, extending from Illinois to eastern Texas. To the east of the cold front the air was warm and muggy, whereas west of the cold front it was relatively cold and dry, as indicated by the clear skies in western Kansas and eastern Colorado.

When the cyclone is in the open stage, it is fully mature. This stage does not last very long, however, because the cold front moves faster than the warm front ahead of it. These contrasting frontal speeds occur because the air behind the cold front is denser than the warm air ahead of it and thus flows more quickly. In this fashion, the cold front drives into the warm, moist air ahead of it, whereas the warm front gradually slides over the top of the stationary cold air ahead of it

(Figure 8.9d). As a result of this difference in speeds, the cold front begins to overtake the warm front in the *occluded stage* of development (Figure 8.9e). Although a distinct warm sector still exists in the southerly parts of the system, the overtaking cold air lifts warm air aloft near the center of the low. An **occluded front** develops when the cold front overtakes the warm front and begins to drive the warm air at the surface to a higher altitude. This process of occlusion continues as the system enters the *dissolving stage* (Figure 8.9f), at which time all the warm air is pinched aloft and a stationary front is reestablished.

Remember that the evolution of midlatitude cyclones occurs within the context of the overall westerly wind pattern in the jet stream. You can think of the jet stream as the medium that steers cyclones across the continent. Figure 8.11 shows an example of this migration, beginning with an open-stage cyclone centered in the central United States (Figure 8.11a). This view of the system shows the position of the warm and cold fronts and their relationship to the air mass source regions. In the open stage of development, mT air in the warm sector is clearly flowing north from the Gulf of Mexico into Louisiana and Arkansas. If you were there on that day, the weather would be warm and muggy, with southerly winds. To the north and west of the cold front, in contrast, the air is flowing in a southerly direction from Canada into places like the panhandle of Texas and Oklahoma. In these places, the temperature would probably be cold, with clear sky and northwest winds.

Occluded front *The area where a cold front begins to overtake a warm front and thus lift warm surface air aloft.*

VISUAL CONCEPT CHECK 8.1

This image shows temperature in the United States. Where is the approximate location of the cold front?

a) Western Texas.
b) Florida.
c) It cuts across the states of Washington and Oregon.
d) It extends from western Michigan to Central Texas.

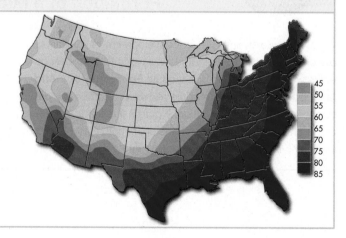

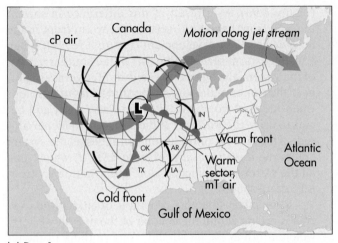

(a) Day 1, open stage

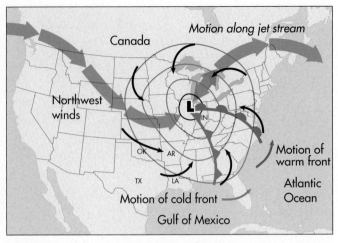

(b) Day 2, cold front overtaking warm front

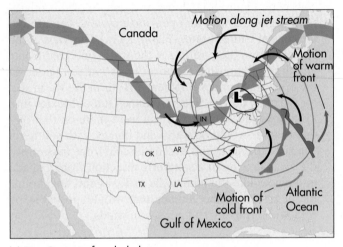

(c) Day 3, start of occluded stage

Figure 8.11 Evolution and migration of a cyclone. Midlatitude cyclones migrate from west to east as they spin counterclockwise and evolve from initial formation, through the mature stage, and into the occluded stage.

www.wiley.com/college/arbogast

Formation of a Midlatitude Cyclone

We can now examine how the development of midlatitude cyclones occurs in an animated way. Go to the *GeoDiscoveries* website and select the module *Formation of a Midlatitude Cyclone*. This module allows you to see the process of cyclogenesis in motion. Figure 8.9 is the foundation of this animation. As you watch the animation, follow how an initial kink in the atmosphere evolves into a mature cyclone that spins counterclockwise in the Northern Hemisphere. This animation will help you better understand how these systems and associated processes cause highly variable weather when they migrate through a region. Once you have completed the animation, be sure to answer the questions at the end of the module to test your understanding of this concept.

It is now time to visually integrate the various moving components of a midlatitude cyclone. Go to the **GeoDiscoveries** website and select the module **Migration of a Midlatitude Cyclone**. This module contains a video that shows the migration of a typical midlatitude storm system in North America over the course of several days. As you watch this system evolve, pay attention to *both* its counterclockwise circulation and movement from west to east. See if you can identify the center of the system and associated cold and warm fronts. After you complete the video, be sure to answer the questions at the end of the module to test your understanding of this concept.

Now let's move on in our hypothetical sequence of cyclonic evolution to Day 2 (Figure 8.11b). Two important things are seen in this diagram. First, notice that the cold front is beginning to overtake the warm front, as you might have predicted. As a result, the temperature in Louisiana and Arkansas, which had been warm the previous day due to southerly airflow, is now colder because the winds became northwesterly. Second, the center of the system has migrated slightly to the northeast along with the wave in the jet stream. Eastward migration continues into Day 3 (Figure 8.11c), with the onset of the occluded stage as the system moves into the northeast part of the country.

KEY CONCEPTS TO REMEMBER ABOUT THE FORMATION OF MIDLATITUDE CYCLONES

1. Midlatitude cyclones spin in a counterclockwise fashion in the Northern Hemisphere as seen from above.

2. Midlatitude cyclones occur when undulations form in the polar front jet stream. These undulations are related to the altitude of the 500-mb pressure surface.

3. As a cyclone spins in the Northern Hemisphere, it pulls in mT air from the south on its eastern side. This warm, moist air encounters cold air as it moves to the north.

4. As a cyclone spins in the Northern Hemisphere, it pulls cP air down from the north on its western side. This cold, dry air encounters warm air as it moves to the south.

5. As the system spins, it also migrates from west to east.

Thunderstorms

You have probably experienced a severe thunderstorm at some point in your life. Maybe you had a sense that a storm was brewing and then suddenly it hit with heavy rain, strong winds, booming thunder, lightning, perhaps even some hail. The storm lasted for a while and then gradually tapered off as it moved away. If you have ever been impressed by such a storm, talked about it with your friends, or wondered why it happened, then this next section will be interesting because it focuses on the evolution and characteristics of thunderstorms. It immediately follows the discussion about midlatitude cyclones because the most severe storms in the midlatitudes are most often associated with low-pressure systems. As you work your way through, just keep in mind that storm development is related to the way air rises and cools, and its interactions with latent heat.

Evolution of Thunderstorms

Like midlatitude cyclones, thunderstorms have a distinct life cycle, illustrated in Figure 8.12. The first stage in the development of a thunderstorm is the *cumulus stage* (Figure 8.12a), which begins with convection or the advancement of a cold front into mT air. This kind of disturbance causes air to rise, which, in turn, results in the formation of ice crystals and raindrops in growing cumulus clouds. Given their small size in this early stage of thunderstorm development, the ice crystals and raindrops continue to be uplifted. Condensation adds additional latent heat energy to the air during the *developing stage* and the atmosphere becomes very unstable with strong updrafts.

As the thunderstorm continues to develop, it enters the *mature stage* (Figure 8.12b). In this stage, the atmosphere is very unstable and continued convection

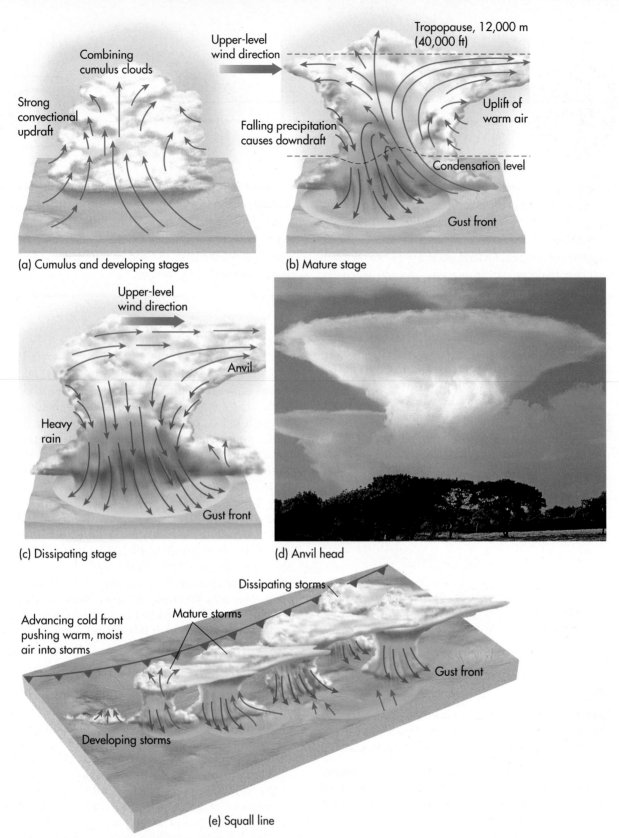

(a) Cumulus and developing stages

Strong convectional updraft

Combining cumulus clouds

(b) Mature stage

Upper-level wind direction

Tropopause, 12,000 m (40,000 ft)

Falling precipitation causes downdraft

Uplift of warm air

Condensation level

Gust front

(c) Dissipating stage

Upper-level wind direction

Anvil

Heavy rain

Gust front

(d) Anvil head

(e) Squall line

Advancing cold front pushing warm, moist air into storms

Mature storms

Dissipating storms

Developing storms

Gust front

Figure 8.12 Stages of thunderstorm development. (a) Convection or advancement of a cold front forces air to rise during the cumulus stage. As the storm evolves, convection strengthens and the system moves into the developing stage. (b) The storm is most intense during the mature stage when strong updrafts occur next to strong downdrafts of cold air. (c) Development of an anvil head. This feature develops in strong storms at the altitude where convection can no longer continue, causing the upper clouds to be sheared downwind. (d) Anvil head of a thunderstorm. Note how the cloud appears to be spreading outward at the top. (e) Squall lines consist of several individual thunderstorms that form along an advancing cold front.

or frontal advancement results in the massive uplift of air. Winds are blustery during this stage, with perhaps some lightning and thunder. As cold raindrops begin to fall to the ground, however, they pull cold air toward the surface with them at high speeds in a **downdraft** at the front of the storm. If you happen to be in this place as a storm forms, you can often feel a distinct increase in wind speed at this *gust front*. Where convection is most intense, an anvil head cloud forms at very high altitude (Figure 8.12c), which may be 6 km to 12 km (20,000 ft to 40,000 ft). This cloud completely develops when the top of the convection bubble is pulled (or sheared) downwind by upper airflow. You can see an example of a forming anvil cloud in Figure 8.12d. If thunderstorm development occurs along an advancing cold front, a group of mature storms and associated anvils will form a feature called a *squall line* (Figure 8.12e). These features show up clearly on radar images and are one of the many atmospheric phenomena about which meteorologists become excited.

Following the mature stage of storm development, thunderstorms begin to weaken. This weakening occurs during the *dissipation stage* and really begins as strong downdrafts develop when the storm has its maximum intensity. Look for these downdrafts in Figure 8.12c. These downdrafts cause the lower part of the atmosphere to cool, which serves to stabilize the air such that further convection cannot occur on a large scale. When these circumstances develop, rain can continue to fall for a short period of time, but will consist of a shower that gradually tapers off.

Severe Thunderstorms

In many cases, thunderstorms are short-lived events that are not particularly intense, consisting of brief downpours of rain, short gusts of moderately strong winds, some isolated lightning, and a few low rumbles of thunder. At other times, however, thunderstorms develop into very severe storms that include extremely heavy rain along with strong winds, intense lightning, numerous loud cracks of thunder, and hail. The most intense thunderstorms contain tornadoes that have incredibly strong winds that can cause great damage. Although these kinds of events are relatively rare, it is nonetheless important to understand their development because you may encounter such a storm in the future. At the very least, you will appreciate them more the next time one occurs where you live.

Severe midlatitude storms form during the mature stage of thunderstorm development. Along with very strong winds, one of the characteristics of a severe thunderstorm is abundant lightning. The production of lightning begins when collisions among ice crystals and rain droplets cause a separation to develop in the electrical charge within clouds (Figure 8.13a). The result is that the tops and bottoms of clouds have positive and negative charges, respectively, while the ground is positively charged. In stage 2 the negative charge on the bottom of the cloud increases to the point where it overcomes the air's resistance to electrical flow. At this point, negatively charged electrons then begin flowing toward Earth along a zigzag, forked path, called a *leader,* at about 97 km/sec (60 mi/sec). As these electrons approach the ground, a positive charge collects in the ground. These positive charges are attracted to the negative charges of the downward-flowing leader and thus move upward through a channel (called a *streamer*) that begins at any conducting object, such as a tree, a house, and even people. Both leaders and streamers are invisible to the naked eye.

The electrical circuit becomes complete in stage 3 (Figure 8.13c), when the upward-flowing charges meet those moving downward. This meeting typically occurs at an altitude of about 30 m (~100 ft) and is the place where the actual lightning bolt you see begins. Less than a millisecond later, millions of volts of electricity reach the ground. The visible bolt, however, is the return stroke from this initial ground impact (Figures 8.13d and 8.13e). This stroke moves upward at a velocity of about 57,910 m/sec (190,000 ft/sec) and can reach altitudes of about 5 km to 6.4 km (~3 mi to 4 mi). Once the return stroke reaches the cloud, it drains the cloud of its excess negative charge and the cycle begins again.

Thunder occurs in association with lightning because lightning heats the air in the path of the stroke to above 45,000°F. This incredible heat creates a shock wave, which disintegrates within a few meters (feet) of the stroke but generates sound waves heard as thunder. Have you ever noticed that thunder seems to rumble? This occurs because sound waves travel from different distances within individual lightning bolts, from lightning strikes, and from sound bouncing off of clouds. If you want to approximate your distance to a lightning strike, count the number of seconds between when you see the lightning flash and hear the thunder. A good rule of thumb is that a lightning strike is about 1.6 km (1 mi) away for every 5 sec it takes between when you see the flash and subsequently hear the thunder.

Another indicator that a thunderstorm is severe is hail, which consists of fragments of ice that plummet from the sky to Earth. Hail indicates that strong **updrafts** are present at the rear of the storm, which repeatedly pull ice crystals back into the upper part of the storm, where they add layers of ice through condensation and grow larger. Once these hailstones grow sufficiently large that they cannot be pulled up anymore, they fall to Earth

---

Downdraft *A rapidly moving current of cool air that flows downward in a thunderstorm.*

Updrafts *An area of rapidly flowing air that is moving upward within a thunderstorm.*

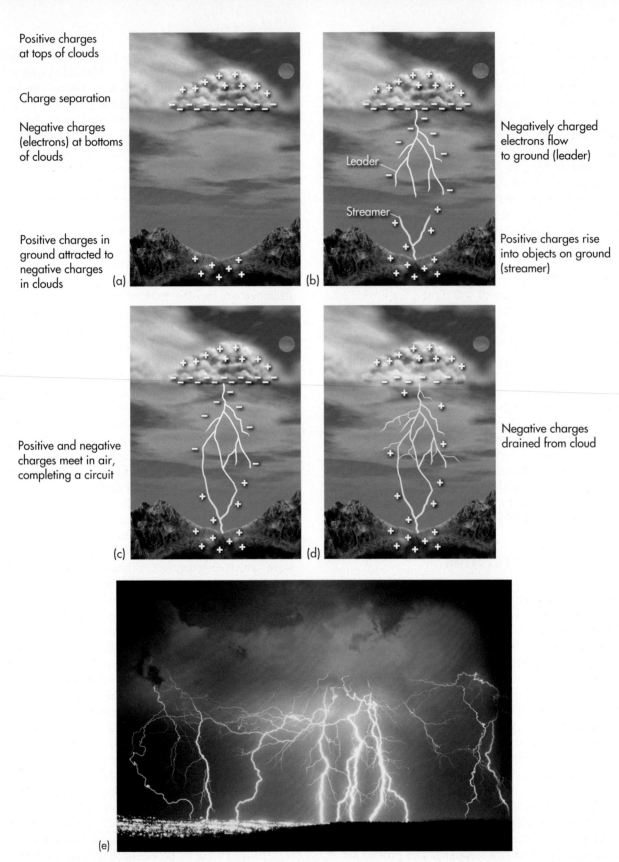

Positive charges
at tops of clouds

Charge separation

Negative charges
(electrons) at bottoms
of clouds

Positive charges in
ground attracted to
negative charges
in clouds (a)

Negatively charged
electrons flow
to ground (leader)

Leader

Streamer

Positive charges rise
into objects on ground
(streamer) (b)

Positive and negative
charges meet in air,
completing a circuit (c)

Negative charges
drained from cloud (d)

(e)

Figure 8.13 Evolution of a lightning strike. (a) Lightning begins when a separation develops in the electric charge between the bottom and top of a cloud. (b) Negatively charged electrons begin to move down toward the ground from the base of the cloud. At the same time, positively charged electrons begin to move upward through conducting material on the ground. (c) The circuit completes when the upward- and downward-moving electrons meet in the air. The process begins the visible lightning strike. (d) The lightning strike drains the cloud of excess negative electrons. (e) A massive lightning storm.

Figure 8.14 Hail damage. Hail can cause extensive damage to homes, crops, and automobiles, as this photo clearly demonstrates.

by the force of gravity, with large hailstones reaching speeds of 160 km (100 mi) an hour. Usually, hail ranges in size from *pea-sized* to *golf-ball-sized*. Occasionally, updrafts will be sufficiently strong to produce *softball-sized* hail. The largest hailstone ever recovered in the United States fell in south-central Nebraska during a severe thunderstorm in June 2003. This hailstone was 17.8 cm (7 in.) wide, which is almost as large as a soccer ball! A severe hailstorm can be very destructive, causing millions of dollars in damage to agricultural crops, buildings, and automobiles (Figure 8.14).

Tornadoes

When certain atmospheric conditions are present, very strong **supercell thunderstorms** can develop. These extremely violent storms are unique because they contain large rotating updrafts known as **mesocyclones** (Figure 8.15) that range from 3.2 km to 9.7 km (2 mi to 6 mi) in diameter. When these atmospheric conditions occur, they can easily evolve to include tornadoes, which are small, but deep, low-pressure cells surrounded by a violently spinning mass of air. Although generally less than 400 m (0.25 mi) in diameter, tornadoes are the most destructive of all atmospheric phenomena, with wind speeds ranging

Supercell thunderstorms *Large thunderstorms that contain winds moving in opposing directions and are associated with strong winds, lightning, thunder, and sometimes hail and tornadoes.*

Mesocyclones *Strong updrafts that rotate within a supercell thunderstorm.*

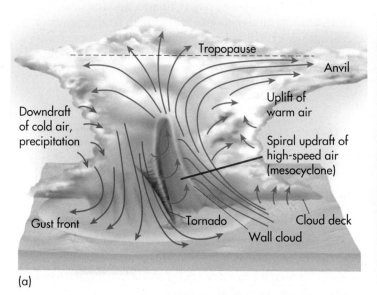

(a)

(b)

Figure 8.15 Supercell thunderstorms. (a) Typical development of a supercell thunderstorm. (b) Seen from the outside, a supercell thunderstorm is an ominous sight. The lower (circular) part of the cloud complex is rotating cyclonically.

Thunderstorms • 195

TABLE 8.1	The Enhanced Fujita Scale for Tornado Classification	
EF Number	3-sec Gust Speed (mph)	Potential Damage
0	65–85 mph	*Minor damage:* Some shingles peeled from roofs and some damage to gutters and siding; branches broken off trees; shallow-rooted trees blown over.
1	86–110 mph	*Moderate damage:* Roofs stripped significantly; mobile homes turned over; broken windows.
2	111–135 mph	*Considerable damage:* Roofs ripped off homes; frame homes moved on foundation; mobile homes destroyed; large trees uprooted.
3	136–165 mph	*Severe damage:* Entire portions of well-constructed homes destroyed; severe damage to shopping centers; trains overturned and cars lifted from ground.
4	166–200 mph	*Devastating damage:* Well-constructed homes entirely destroyed; cars thrown and small missiles created.
5	Over 200 mph	*Extreme damage*: Strong frame houses leveled and blown away; cars thrown farther than 100 m (300 ft); steel-reinforced concrete damaged.

from 320 to 800 km/h (200 to 500 mph). These high speeds occur because tornadoes have extremely tight pressure gradients that may differ by 100 mb between the inside and outside of the funnel. Between 1971 and early 2007, tornadoes were classified according to the Fujita scale based on the amount of damage caused. Acknowledging that the quality of building construction can vary and thus give misleading conclusions about wind speed, the *Enhanced Fujita Scale* was implemented in February 2007. This classification system is based on the speed of winds at 3-sec gusts at the point of damage (Table 8.1), ranging from the EF0 tornado to the rare EF5 tornado.

Figure 8.16 shows the evolution of a tornado. Tornadoes usually form at the rear of a storm when strong updrafts occur in conjunction with wind shear at higher altitudes. Wind shear occurs when winds are moving in different directions at variable altitudes. This wind shear causes a horizontal vortex of air to form (Figure 8.16a) that is then pulled vertically in the updraft (Figure 8.16b).

Assuming this process is not interrupted, a funnel develops that becomes a full-fledged tornado if it reaches the ground (Figure 8.16c). Figure 8.16d shows a supercell tornado (arrow) that formed near Spearman, Texas, in May 1990. Notice how similar this tornado and spawning supercell system are to the diagram in Figure 8.16c. The airflow in a fully developed tornado can generate fierce winds of over 480 km/h (300 mph). Fortunately, tornadoes seldom last very long, only 15 min on average. The air sucked into the funnel slows down the winds—the tornado becomes clogged with the air it takes in. Eventually, the funnel stretches out into a thin shape and disappears.

Occasionally, however, a tornado can be on the ground for an hour at a time. The best-known example of such a storm was the famous Tri-State Tornado in 1925 that devastated Missouri, Illinois, and Indiana. This tornado was on the ground for over 3 h straight and cut a swath over 370 km (~230 mi) long! The funnel was about 1.6 km (1 mi) wide and killed nearly 700 people.

www.wiley.com/college/arbogast

Formation of Thunderstorms

Thunderstorms are fascinating atmospheric phenomena that involve complex patterns of air flow. These patterns, including updrafts and downdrafts, are particularly well suited for animation to visualize how they really move. To do so, go to the *GeoDiscoveries* website and select the module

Formation of Thunderstorms to observe these processes in motion and to watch the lifecycle of a thunderstorm. Be sure to answer the questions at the end of the module to test your understanding of this concept.

Figure 8.16 Evolution of a tornado. (a) Air begins to rotate due to wind shear; (b) updrafts pull a rotating cylinder of air upward; (c) a tornado develops when the funnel reaches the ground. (d) Supercell tornado near Spearman, Texas, in 1990.

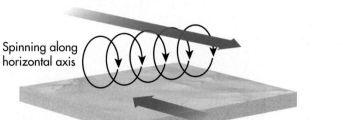

Stronger westerly winds

Spinning along horizontal axis

Weaker southeasterly winds

(a)

Developing thunderstorm

Rotating air is shifted from horizontal to vertical

Updraft of warm air

(b)

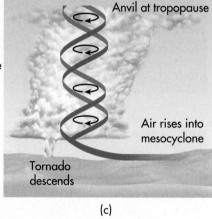

Anvil at tropopause

Mesocyclone, 3–10 km wide (2–6 mi wide)

Air rises into mesocyclone

Tornado descends

(c)

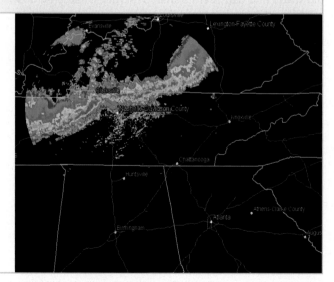

(d)

VISUAL CONCEPT CHECK 8.2

This Doppler radar image shows a line of thunderstorms along a cold front extending across Kentucky and Tennessee. Which one of the following choices best explains the geography of air along the front?

a) The warmest air temperatures were northwest of the front.

b) The cold front caused mT air to rapidly uplift along the front.

c) The air southeast of the front consisted of cP air.

d) The most humid air was located northwest of the front.

Thunderstorms • 197

FASCINATING CLOUDS ASSOCIATED WITH THUNDERSTORMS

In addition to strong wind, lightning, and thunder, powerful thunderstorms sometimes produce fascinating cloud patterns diagnostic of distinct meteorological processes. One such cloud formation is a *shelf cloud* like the one pictured here. A shelf cloud develops when a cold downdraft within a storm surges outward along the ground after it reaches the surface. As it does so, it lifts the relatively warm and moist air ahead of the storm up to the level of condensation, forming this shelf-like feature. Such a cloud is often accompanied by a cold gust front.

Another interesting and diagnostic cloud formation associated with thunderstorms is *mammatus clouds*. Also known as *mammatocumulus*, which means "mammary cloud," mammatus clouds look like pouches or lobes hanging eerily upside down in the air like those shown here. These clouds are usually found on the base of a spreading anvil downwind of a large thunderstorm. Their formation is poorly understood but may be somehow related to overturning pockets of colder air that descend into warmer air at high altitudes. Although they are usually composed of ice, they also contain liquid water and can seem translucent. Mammatus clouds are more common during warm months and more often seen in the Midwest and eastern parts of the country.

Although tornadoes occur every month of the year in the continental United States, they most frequently develop between April and July. This time interval is peak tornado season because the strongest contrast in air masses exists at this time in the midlatitudes. Although tornadoes statistically occur at least once per year in every state of the United States (except Alaska), they are by far most common in the central part of the country (Figure 8.17), which is why this area is known as *Tornado Alley*. This region is prime tornado country because the relatively flat terrain provides an ideal place for cool, dry Canadian air to mix violently with warm, moist air from the Gulf of Mexico. These conditions are ideal for the development of the massive thunderstorms that contain tornadoes. Some of the most intense tornadoes in history have struck this region, such as the F-5 storm that struck Topeka, Kansas, in June 1966 (Figure 8.18) that killed 16 people and caused $100 million (1966 dollars) in losses by damaging 3000 homes and buildings. This storm led to my lifelong interest in geography because I grew up there and remember it vividly. Sometimes atmospheric conditions are so ripe that a tornado outbreak occurs, with numerous funnels in a small region over several hours. Such an outbreak occurred in early May 2007 when over 100 tornadoes were reported in the central Great Plains over the course of 3 days. The most severe of these storms hit Greensburg, Kansas, and caused over $153 million worth of damage. This storm was the first EF5

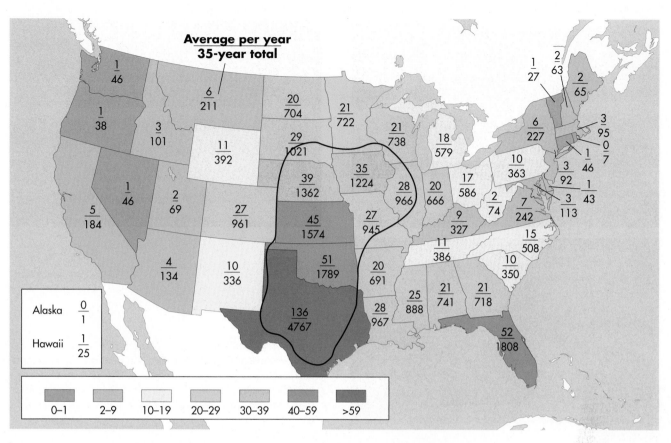

Average per year
35-year total

Alaska	$\frac{0}{1}$
Hawaii	$\frac{1}{25}$

0–1 2–9 10–19 20–29 30–39 40–59 >59

Figure 8.17 Distribution of tornadoes in the mainland United States. Note the yearly average and total number observed for over 35 years per state. The approximate boundary of Tornado Alley is outlined in black.

Figure 8.18 The Topeka Tornado. This storm was an F-5 tornado that traveled through much of the city of Topeka, Kansas in June 1966. This tornado remains one of the five costliest tornadoes in U.S. history.

Figure 8.19 The Greensburg Tornado. (a) In May 2007, an EF5 tornado struck Greensburg, Kansas, killing 11 people and destroying 95% of the town. (b) Air photo of Greensburg before the tornado. (c) Air photo of Greensburg after the tornado. Note that most of the town was essentially destroyed.

tornado ever reported, with estimated winds of 330 km/h (205 mph) in the core of the system. Approximately 95% of the town was destroyed in this storm (Figure 8.19). Since the Greensburg tornado, another EF5 storm struck the town of Parkersburg, Iowa, in May 2008. This storm killed 6 people and leveled half of the town.

Human Interactions: Tornado Monitoring Although studying tornadoes is interesting purely from a meteorologic perspective, the primary goal of tornado research is to improve public safety. In this context, the National Weather Service constantly monitors atmospheric conditions and makes public announcements as weather events unfold on any given day. A *tornado watch* is issued for an area, such as the eastern part of Oklahoma or central Texas, when atmospheric conditions are favorable for the development of tornadoes. When an actual funnel or tornado has been identified, a *tornado warning* is issued for a specific place or, perhaps, county. At this time sirens are sounded and people are advised to seek shelter, either in a low-lying ditch or, preferably, in a basement.

With respect to tornado monitoring, the most significant technological development in the past few decades has been Doppler radar, such as the Next Generation

www.wiley.com/college/arbogast

WILEY PLUS

Tornadoes

Tornadoes are fascinating natural phenomena that are often caught on film by storm chasers and regular citizens alike. To learn more about tornadoes, go to the *GeoDiscoveries* website and select the module *Tornadoes*. The first part of this module demonstrates how tornadoes form and shows real storm video. The last part of the module is an interactive exercise, involving the Enhanced Fujita Scale, that allows you to determine the strength of a tornado based on the damage it caused. After you interact with this module, be sure to answer the questions at the end to test your understanding of this concept.

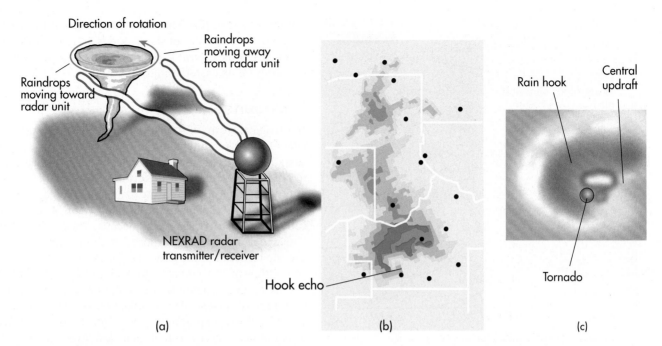

Direction of rotation

Raindrops moving away from radar unit

Raindrops moving toward radar unit

NEXRAD radar transmitter/receiver

(a)

Hook echo

(b)

Rain hook

Central updraft

Tornado

(c)

Figure 8.20 Doppler radar. (a) Doppler radar "sees" the rotation of a tornado because of the Doppler Effect, which, in this case, reflects the variation in returning wave frequency from either side of the system. (b) The rotation of a tornado is displayed as a hook echo on the computer screen. (c) A hook echo is indicative of aggressive cloud rotation, which usually means that a tornado is present.

Weather Radar (NEXRAD) used by the National Weather Service (Figure 8.20). Prior to this development, tornado warnings were based largely on direct observations, which were often too late to save lives. In contrast to conventional radar, which shows the intensity of precipitation in any given storm, Doppler radar can detect *both* rainfall patterns and the actual rotation of a tornado. It does so by making use of the Doppler Effect, described first by the 19th-century physicist Christian Doppler, which states that the frequency of energy waves generated by a moving source changes relative to an observer.

A classic example of this effect is the sound a train makes as it approaches you. What happens? The pitch of the train's whistle rises with the approach of the train and then lowers after it passes. Using this effect, Doppler radar can determine that raindrops on one side of a tornado are moving toward the radar detector, while those on the other side are rotating away because of the shift in energy frequency that occurs relative to each side of the twister (Figure 8.20a). This rotation is seen by the meteorologist in the weather laboratory as a distinct **hook echo** on the computer screen (Figure 8.20b). If such a feature is identified, a tornado warning is quickly issued for the area in the path of the storm and people usually have sufficient time to seek shelter before the storm strikes.

Tropical Cyclones

Now it is time to turn our attention to cyclonic storms that form in the tropical regions. Although tropical systems are similar to midlatitude systems because they consist of rotating air masses of low pressure, they differ significantly because they form in regions where there are no fronts or contrasting air masses. Instead, a tropical cyclone develops entirely within a homogeneous air mass at very low latitudes. As a result, abundant water vapor and latent heat are present to fuel these systems. About 80 tropical cyclones develop around the world every year, with the strongest becoming extremely powerful storms that, depending on the region, are called hurricanes, typhoons, or cyclones.

Most Atlantic hurricanes form in association with an atmospheric feature called an **easterly wave**, which is a slow-moving trough of low pressure that migrates along the belt of tropical easterlies. Recall that this belt is associated with the easterly trade winds and occurs between 5° and 25° N and S latitude. This low-pressure system causes air to converge on the windward (east) side of the trough axis, while air diverges in the lee (west) side of the system. The convergence of upper air on the eastern side of the wave

Hook echo *The diagnostic feature in Doppler radar indicating strong rotation is occurring within a thunderstorm and tornado development is thus possible.*

Easterly wave *A slow-moving trough of low pressure that develops within the tropical latitudes.*

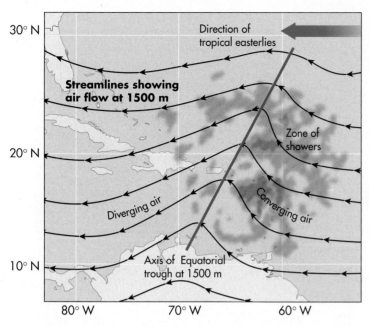

Figure 8.21 Characteristics of a tropical wave. Tropical waves consist of a center of surface low pressure that results in the convergence and divergence of air aloft. Showers develop on the eastern side of the wave axis because converging air causes convectional uplift.

causes further convectional uplift within the system and a zone of showers develops parallel to the trough axis. Figure 8.21 shows the basic structure of an easterly wave, including the axis, wind flow patterns, and precipitation zone.

If the storm strengthens into a **tropical depression**, it begins to rotate cyclonically around a definable center of low pressure at the same time that a complex of strong thunderstorms develops. At this time, sustained winds in the center of the system are 20 knots to 34 knots (23 mph to 39 mph). If the storm continues to strengthen, it becomes a **tropical storm** and the maximum sustained winds reach 35 knots to 63 knots

(39 mph to 73 mph). The storm at this point begins to become more circular in shape and is assigned a formal name (such as *Charlie*) for monitoring purposes as it proceeds on its easterly migratory path.

Hurricanes

If a tropical storm in the Atlantic Ocean or eastern Pacific Ocean strengthens to the point that the maximum sustained winds are greater than 63 knots (73 mph), then it officially becomes a **hurricane**. A storm of similar strength in the western Pacific Ocean is called a *typhoon* or a *cyclone* if it forms in the Indian Ocean. Australians call such a storm a *willy nilly* because they twist about and wreak havoc. For purposes of discussion here, we use the term *hurricane* and focus on Atlantic storms.

Although it is common to hear the word *hurricane*, storms of this magnitude are really quite rare because they require a particular combination of environmental variables. Most importantly, they require an abundance of latent heat energy to fuel their development. Such an abundance of energy can occur only when a great deal of evaporation from the underlying ocean takes place, which means that the surface waters must be warmer than about 27°C (80°F). This required temperature is the primary reason why hurricane season in the Atlantic Ocean largely occurs from August to October (Figure 8.22), in other words, in the period following the highest Sun and insolation.

Tropical depression *A tropical low-pressure system with central sustained winds ranging between 20 knots and 34 knots (23 mph and 39 mph).*

Tropical storm *A tropical low-pressure system with maximum sustained winds between 35 knots and 63 knots (39 mph and 73 mph).*

Hurricane *A tropical circulatory system with maximum sustained winds greater than 63 knots (73 mph).*

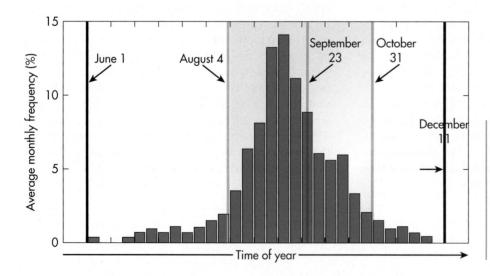

Figure 8.22 Average frequency of monthly hurricanes in the Atlantic Ocean from 1896 through 1996. Hurricanes are most likely to develop from the beginning of August to the end of October because sea-surface temperatures are warmest this time of year in the Northern Hemisphere. (*Source:* NOAA.)

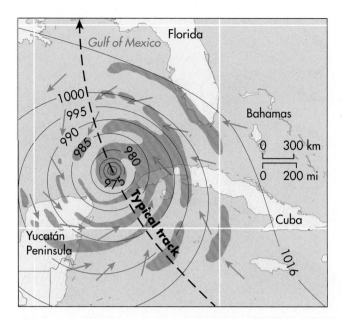

Figure 8.23 Hurricane weather map. This simplified weather map of a typical Atlantic hurricane illustrates the cyclonic rotation and pressure gradient associated with the system. Shaded areas are regions of heavy rainfall.

In addition to the warm ocean waters, another requirement for hurricane development is that the upper air wind pattern must be favorable, with high air pressure aloft. This relationship may seem strange because hurricanes are zones of low air pressure, which might lead you to wonder how both low pressure at the surface and high pressure in the upper atmosphere could exist at the same time. Remember that the atmosphere has many different layers and that air can flow differently in each of them. In the case of a hurricane, high pressure aloft is required because it serves to "cap" the top of the storm so it can continue to strengthen at the surface. Otherwise, if the surface air can freely convect into the upper atmosphere, or if shearing winds aloft blow the top off the storm, then the system does not fully organize.

If all the necessary ingredients are in place for hurricane development, the system can grow to immense size and strength. Depending on the specific atmospheric conditions, hurricanes can range in diameter from 150 km to 500 km (100 mi to 300 mi). Hurricanes acquire their strength because of the steep pressure gradient that develops from the outside of the storm to a center of extremely low pressure, which can depress to 900 mb or even lower. As a result of this steep pressure gradient, winds spiral into the core of the system at very high speed. Figure 8.23 shows an example of the pressure gradient and storm size.

Hurricanes are classified based on their wind-speed intensity on the Saffir–Simpson scale. Table 8.2 shows this scale and how it relates to the central pressure of a storm and its wind speed. Notice the inverse relationship between these two variables. The fourth column of the table shows storm surge, which is the rise in sea level that occurs because strong winds cause water to pile up ahead of the storm. Storm surge is typically highest on the right front quadrant of a hurricane, where onshore winds flow as the storm approaches. These winds tend to push the water ahead of the storm and onto the coast. In addition to wind speed, the height of a storm surge depends on the timing of the tides and configuration of the coastline, with shallow coastlines typically more susceptible because water has a tendency to pile up at those locations.

Figure 8.24 shows the typical anatomy of a hurricane. Note the distinct bands of cumulonimbus clouds, which can drop prodigious amounts of rainfall, surrounding the core of the system. Also observe that air flows inward at the surface to a point where it rapidly flows upward at the eye wall. Remember that the entire time the air is flowing inward and upward, it is spinning cyclonically at intense speed. A unique characteristic of a well-developed hurricane is the eye, which occurs in the center of the storm and is associated with high-pressure air that rapidly flows toward the surface and warms adiabatically as it descends. As a result, this region of the storm is cloud-free with no winds and can be seen in satellite images of well-developed hurricanes.

Because hurricanes evolve in the easterly trade belt, they migrate in a fairly predictable way. Figure 8.25 shows typical tracks that hurricanes take in August. Notice that many of the storms originate in the eastern Atlantic Ocean. These storms actually begin to organize when they exit western Africa as easterly waves in the trade-wind belt. When they reach the open warm waters of the

TABLE 8.2	Saffir–Simpson Scale of Hurricane Intensity		
Category	**Central Pressure, mb**	**Mean Wind, m/sec (mph)**	**Storm Surge, m (ft)**
1 Weak	> 980	33–42 (74–95)	1.2–1.7 (4–5)
2 Moderate	965–979	43–49 (96–110)	1.8–2.6 (6–8)
3 Strong	945–964	50–58 (111–130)	2.7–3.8 (9–12)
4 Very strong	920–944	59–69 (131–155)	3.9–5.6 (13–18)
5 Devastating	< 920	> 69 (>155)	> 5.6 (>18)

Figure 8.24 Components of a typical hurricane. Hurricanes contain numerous spiral rain bands of cumulonimbus clouds. Air spirals inward until it reaches the eye wall, where it circles upward. The eye of the storm is clear and calm, which is the result of air descending and adiabatically warming in the core of the system.

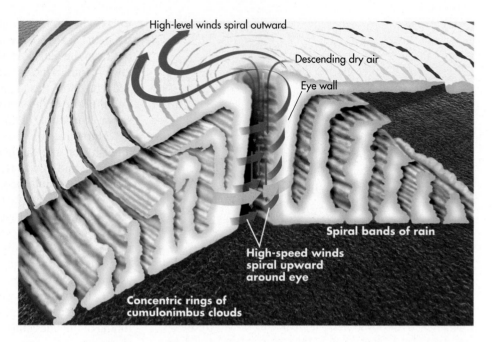

High-level winds spiral outward

Descending dry air

Eye wall

Spiral bands of rain

High-speed winds spiral upward around eye

Concentric rings of cumulonimbus clouds

eastern Atlantic, they begin to intensify, first as tropical depressions, then as tropical storms, and finally, if conditions are favorable, into hurricanes. The systems continue to migrate westerly, sometimes strengthening, other times weakening. If the storm remains organized for several days, it ultimately becomes influenced by the midlatitude westerlies and is deflected to the northeast. As you can see, some storms remain over open water throughout their entire history. Other storms, however, strike land at some place in the Caribbean Sea, Gulf of Mexico, or the Atlantic seaboard in the United States. When such a landfall occurs, it can result in significant damage and loss of life.

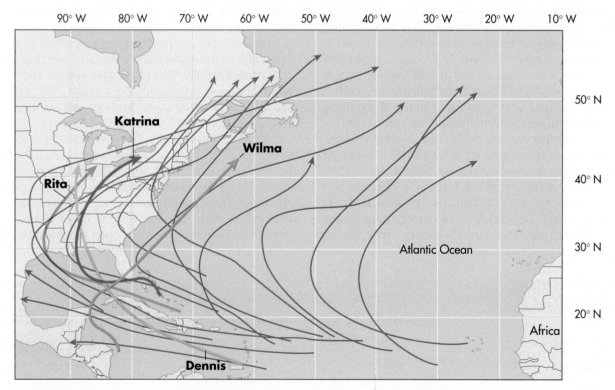

Figure 8.25 Typical track of August hurricanes in the Atlantic Ocean. During this time of year, hurricanes begin to develop over the warm waters of the eastern Atlantic. Subsequently, they migrate with the easterly trade winds until they become influenced by the midlatitude westerlies. This map also includes the tracks of the four major hurricanes that struck the U.S. in 2005. (*Source:* NOAA.)

VISUAL CONCEPT CHECK 8.3

This figure is a satellite image of Hurricane Floyd in September 1999, which at one point developed into a Category 5 storm. Note the position of Floyd in the central Atlantic Ocean in this image. Imagine that you are a hurricane forecaster who has become aware of the hurricane's presence for the first time. Describe how you would expect Floyd to migrate in the future, beginning at this position and ending when the storm would most likely dissipate.

Human Interactions: Recent Hurricane Activity Although some hurricanes occur every year, a particularly high number of intense hurricanes has occurred in the past decade. This increase may be linked in part to global climate change, which is thought by some to cause warmer ocean temperatures that fuel stronger storms. It may also be that we are simply in a period of greater natural hurricane occurrence, one similar to the period of higher activity in the 1930s and 1940s. Perhaps the increased recent activity is a combination of both global warming and a natural cycle. No one knows for sure.

Regardless of the cause, the 2004 and 2005 hurricane seasons were very intense. Florida was particularly hard hit in 2004, when four strong hurricanes struck the state between August and late September. The combined cost of these storms was about $30 billion in Florida alone and caused almost 100 deaths. In 2005 a record 27 tropical storms formed, with 15 reaching hurricane status. Seven of these hurricanes strengthened into major hurricanes (≥Category 3). Five of these major hurricanes reached Category 4 status and three (another record!) strengthened into Category 5 storms. Of these, Hurricane Wilma was the most intense storm ever recorded in the Atlantic Ocean, with a central pressure of 882 mb. Four major hurricanes struck the United States, as you can see in Figure 8.25. After a pair of relatively calm years in 2006 and 2007, 2008 was another very active year, with a total of 16 named storms. It was the fourth

Geo Discoveries   www.wiley.com/college/arbogast

Migration of Hurricane Katrina

An excellent example of the need for monitoring can be seen in videos showing the path of Hurricane Katrina in 2005. To see this migration, go to the *GeoDiscoveries* website and access the module *Migration of Hurricane Katrina.* The first part of this module is an animated cartoon illustrating the storm's path. Subsequently, a pair of videos follow the path of this catastrophic hurricane as it moved across the Gulf of Mexico. Watch for several things. First, look for the overall rotation of the system and how its path changed over time. Also note the rapid intensification of the system as it migrated across the Gulf. This intensification is obvious because the eye of the

storm suddenly becomes very prominent. At this point, the storm was at its most powerful, specifically a Category 5 hurricane. Next, observe how the size of the eye fluctuated as it approached the coast; this occurred because the storm's strength varied somewhat before it made landfall. Finally, notice how the eye suddenly disappeared after the storm reached land. As you watch the video, think about why satellite images such as these are essential for good monitoring and how they save lives. Once you finish the video, be sure to answer the questions at the end of the module to test your understanding of this concept.

(a)

(c)

(b)

Figure 8.26 Hurricane Katrina. (a) Satellite image of Hurricane Katrina shortly before it struck Louisiana. Note the incredibly well-defined eye, which developed when the storm reached maximum strength. (b) Extensive damage along the Mississippi coast was caused by a combination of strong winds and storm surge, which in some places was over 9 m (30 ft) high. (c) Failure of protective levees in New Orleans caused widespread flooding.

most active year since 1944 and the first in which major storms (≥Category 3) occurred every month of the season. The most significant of these hurricanes in the United States was Ike, which made landfall at Galveston, Texas, as a Category 2 storm. At one point in its evolution, the diameter of Ike's tropical storm was over 965 km (600 mi), making it the largest storm ever recorded.

Of all the most recent hurricanes, easily the most destructive was Hurricane Katrina, which struck the Gulf Coast near New Orleans on August 29, 2005. Katrina formed over the Bahamas on August 23 and first made landfall on the east coast of Florida as a Category 1 storm. It subsequently moved west over the Gulf of Mexico, where surface water temperatures were about 32°C (90°F). These incredibly warm waters caused Katrina to rapidly intensify to Category 5 status with sustained winds of 280 km/h (175 mph). At this time, the storm had an extremely well-defined eye, as you can see in Figure 8.26a. After such a period of intensification, the storm weakened slightly to a strong Category 3 shortly before it struck the Louisiana coast.

The tragic images of the storm's aftermath may be indelibly imprinted in your mind. Along the coast of Mississippi and Alabama, for example, thousands of homes and businesses were literally flattened (Figure 8.26b) from the combined power of the strong winds and storm surge. The storm surge was highest in this region, up to 9 m (30 ft) in some places, because it was in the right front quadrant of the storm as it approached. Additional extensive damage occurred in New Orleans, which actually lies below sea level and is protected from the Mississippi River (to the south) and Lake Pontchartrain (to the north) by a system of levees designed to keep high water out of the city. The storm surge in Lake Pontchartrain caused several of these protective walls to fail, resulting in severe flooding (Figure 8.26c) and the displacement of tens of thousands of people. The confirmed death toll from Hurricane Katrina was over 1,800 people and hundreds of thousands across the region were made homeless. The total cost of the storm was about $100 billion, making it by far the costliest hurricane ever to hit the United States.

Human Interactions: Monitoring Hurricanes The intensity of recent hurricane seasons underscores the need for accurate hurricane forecasting and monitoring. Had forecasting not been in place, it is very possible that thousands more people would have died during Hurricane Katrina alone. Such a high death toll occurred during the infamous Galveston Hurricane that struck Galveston, Texas, in 1900. Because accurate forecasting and monitoring methods did not exist at that time, the storm basically caught people unaware, resulting in somewhere between 6000 and 12,000 deaths.

Fortunately, in this day of real-time satellite imagery, it is possible to monitor the development of any individual storm as it migrates across the Atlantic or Caribbean. As a result, forecasters can predict the general path a storm will take several days in advance and provide warning of the storm's approach to residents in its path. If you happen to live in the southeastern part of the United States, you are probably very aware of hurricane monitoring. Should you happen to live elsewhere, try paying attention to storm events in the Atlantic Ocean during the hurricane season late next summer and early fall. These systems are easy to follow, can be quite dramatic, and serve as an excellent window into the way in which the atmosphere functions.

KEY CONCEPTS TO REMEMBER ABOUT SEVERE MIDLATITUDE AND TROPICAL STORMS

1. In general, the most severe midlatitude storms form along strong cold fronts when warm, moist (mT) air ahead of the front is rapidly forced aloft.

2. Thunderstorms evolve in predictable stages, including the cumulus stage, mature stage, and dissipating stage. These stages are related to the upward and downward flow of air.

3. Strong winds, lightning, and thunder typically accompany severe storms. Lightning occurs when opposing electronic charges develop between the base of clouds and the ground. Thunder is created by the shockwave produced when lightning superheats the atmosphere locally.

4. The strongest storms associated with midlatitude weather are tornadoes, which are localized centers of intense low pressure that develop in association with supercell thunderstorms.

5. The strongest tropical storms are hurricanes, which develop when easterly waves strengthen beyond the depression and tropical storm phases to produce sustained winds greater than 63 knots (73 mph).

THE BIG PICTURE

So far in this book a very wide range of topics have been covered, including Earth–Sun relations, insolation, temperature, global wind patterns, adiabatic processes, and weather systems. These concepts and associated processes can be holistically integrated into a geographic framework that helps explain a variety of phenomena on Earth, including vegetation patterns, soils, the weathering of rocks, and even the location of large sand dunes. A good place to begin this integration is Chapter 9, which focuses on global climate processes and patterns.

A satellite image of Africa nicely reflects how this integration occurs because it contains several noticeable geographic patterns. For example, a zone of dark green at the Equator represents the tropical rainforest, which correlates to the region that receives abundant precipitation over the entire year. In contrast, the broad expanse of tan in the northern part of Africa is the Sahara Desert, which corresponds to the part of the continent that receives very little annual rainfall. These patterns, as well as numerous others in Africa and the rest of the world, are related to the wide variety of processes presented so far in this text. In the case of Africa, the geographic patterns are closely related to the seasonal migration of Hadley cells presented in Chapter 6. Can you imagine how they are related? As you work through the next chapter, think about how all the factors you have learned about so far fit into the overall climate pattern.

SUMMARY OF KEY CONCEPTS

1. An air mass is a large body of air that forms in specific geographic regions and thus has distinctive characteristics. Five principal air masses affect North America. Continental air masses include continental polar (cP), continental arctic (cA), and continental tropical (cT). The maritime air masses are maritime tropical (mT) and maritime polar (mP).

2. Air masses have distinct boundaries called fronts. At a stationary front, contrasting air masses are flowing parallel to one another. A warm front is a place where warm air is advancing into relatively cool air. Given that warm air slowly slides over the top of the cooler air along a warm front, rainfall is slow and steady. A cold front is a place where cold air is advancing into relatively warm air. Given the higher density of colder air, rainfall is intense and of short duration along the front because warm air cools quickly when it is rapidly forced aloft.

3. A midlatitude cyclone in the Northern Hemisphere spins in a counterclockwise fashion as seen from above. These atmospheric features develop when undulations form in the polar front jet stream at the 500-mb pressure level. As a cyclone spins, it pulls warm (mT) air up from the south on its eastern side. This warm, moist air encounters cold air as it moves to the north. The cyclone also pulls cold air (cP) down from the north on its western side. This cold, dry air encounters warm air as it moves to the south.

4. In general, the most severe midlatitude storms form along strong cold fronts when warm, moist (mT) air ahead of the front is rapidly forced aloft. Thunderstorms evolve in predictable stages, including the cumulus stage, mature stage, and dissipating stage, that are related to the upward and downward flow of air. The strongest storms associated with midlatitude weather are tornadoes, which are localized bodies of intense low pressure that develop in association with supercell thunderstorms.

5. The strongest tropical storms are hurricanes, which develop when easterly waves strengthen beyond the depression and tropical storm phases to produce sustained winds greater than 64 knots (75 mph).

CHECK YOUR UNDERSTANDING

1. Define the concept of an air mass.

2. What are the specific characteristics of an mT air mass and how do they differ from the characteristics of a cP air mass?

3. Which air mass is most likely to be associated with precipitation—an mT air mass or a cP air mass? Why?

4. Discuss the evolution and migration of a midlatitude cyclone.

5. Why are midlatitude cyclones a mechanism through which contrasting air masses are mixed?

6. How does the formation of an upper air trough at the 500-mb level result in the development of a midlatitude cyclone?

7. What is the basic difference between a warm front and a cold front? Why is the term *front* used in association with these concepts?

8. Precipitation along a warm front is gradual and long-lasting, whereas it is short-lived and often violent along a cold front. Why does this difference exist?

9. Describe the evolution of a thunderstorm and the various stages it goes through during its life cycle.

10. What is a downdraft and why is it the first step in the dissipation of a thunderstorm?

11. Discuss the evolution of a hurricane in the Northern Hemisphere, including its various stages, movement, and relationship with ocean temperature.

ANSWERS TO VISUAL CONCEPT CHECKS

VISUAL CONCEPT CHECK 8.1

The answer is *d.* The cold front extends from western Michigan to Central Texas. You can tell the location of the cold front because of a sharp temperature difference between the east and west sides of the frontal boundary.

VISUAL CONCEPT CHECK 8.2

The answer is *b.* The cold front caused mT air to rapidly uplift. Remember that a cold front plows aggressively into relatively warm, moist air. This interaction causes the mT air to lift rapidly.

VISUAL CONCEPT CHECK 8.3

One would expect Hurricane Floyd to move to the west because the prevailing winds are easterly at low latitudes. At some point, the hurricane would begin to interact with the midlatitude westerlies, causing the hurricane to veer northeastward. In fact, Hurricane Floyd migrated in just this manner, moving westward toward Florida. Shortly before it reached Florida, however, Floyd veered to the northeast and just grazed the mainland of the United States.

CHAPTER NINE

GLOBAL CLIMATES AND GLOBAL CLIMATE CHANGE

The previous chapters examined features associated with Earth's atmosphere, including the flow of radiation within it, air circulation, and weather systems. Although these concepts can be viewed alone, they can and should also be considered holistically. The development of midlatitude cyclones, for example, is related to atmospheric circulation, precipitation, and the seasonal effects associated with Earth–Sun geometry. This chapter integrates all of the processes discussed so far in this text by focusing on the global distribution of climates.

CHAPTER PREVIEW

Climate and the Factors That Affect It

Köppen Climate Classification

Geography of Köppen Climates
On the Web: *GeoDiscoveries Tropical Savanna Climate (Aw)*
On the Web: *GeoDiscoveries Humid Subtropical Hot-Summer Climate (Cfa, Cwa)*
On the Web: *GeoDiscoveries Marine West-Coast Climates (Cfb, Cfc)*
On the Web: *GeoDiscoveries Global Climates*

Climate refers to the general temperature and precipitation characteristics of an area. This image of the Maroon Bells near Aspen, Colorado, shows two climate regions, one in the high mountains where cold, snowy conditions prevail, and another at a lower altitude that is sufficiently warm for trees to grow.

On the Web: *GeoDiscoveries* Remote Sensing and Climate

Reconstructing Past Climates
On the Web: *GeoDiscoveries* Reconstructing Past Climates Using Oxygen Isotopes

Causes of Past Climate Change
On the Web: *GeoDiscoveries* The Milankovitch Theory

Human Interactions and Future Climate Change
On the Web: *GeoDiscoveries* The Greenhouse Effect and Global Climate Change

The Big Picture

LEARNING OBJECTIVES

1. Define the concept of climate and describe the large-scale factors that influence it around Earth.

2. Explain the Köppen climate system and how it is used to differentiate climate regions on Earth.

3. Discuss the characteristics of the various Köppen climate regions.

4. Describe the various methods used to reconstruct past climate change, the patterns of climate change in the past 2 million years, and why it is important to understand those fluctuations.

5. Understand the factors that have caused climate change in the past.

6. Discuss the issue of global warming and why the majority of climate scientists feel it is caused by human industrial activity.

7. Describe the projected rates of warming in the 21st century and the impact it is expected to have.

211

Climate and the Factors That Affect It

When most people think of what "climate" means, they frequently think it is another word for "weather." This is a misconception. The term **weather** refers to the state of the atmosphere at a specific place and time on the Earth's surface. For example, the weather is described when you say something like "it's cold and rainy today." Think about it for a moment—what *is* the weather as you are reading this chapter? The concept of **climate** is different from that of daily weather because it refers to the long-term average values of weather elements, such as temperature and precipitation. These averages are based on a 30-year period of time.

Climate is a very geographic concept because Earth exhibits distinct climate patterns. The study of this geographic distribution and the character of Earth's climates is a subfield of geography called *climatology*. Do not confuse climatology with *meteorology*, which is the study of short-term atmospheric phenomena such as thunderstorms that constitute day-to-day weather in any particular place. In contrast to this short-term view, climatologists study long-term temperature and precipitation patterns. Their research is important to many disciplines outside the field of geography, including agriculture, architecture, ecology, forestry, and economics because climate is a factor that influences human behavior and natural processes in a variety of ways.

Before studying world climates, we should review all the factors that, in some way, play a role in the long-term weather patterns for a region (Figure 9.1). Given what has been covered so far in this book, these variables should come as no surprise to you.

Weather *Day-to-day changes that occur with respect to temperature and precipitation.*

Climate *Average precipitation and temperature characteristics for a region that are based on long-term records.*

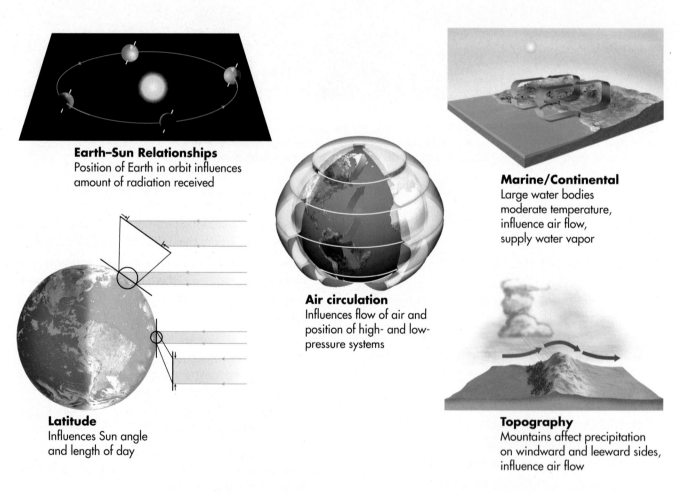

Earth–Sun Relationships
Position of Earth in orbit influences amount of radiation received

Latitude
Influences Sun angle and length of day

Air circulation
Influences flow of air and position of high- and low-pressure systems

Marine/Continental
Large water bodies moderate temperature, influence air flow, supply water vapor

Topography
Mountains affect precipitation on windward and leeward sides, influence air flow

Figure 9.1 Variables that influence climate. The climate of any given place is determined by the interaction of a variety of factors, such as Earth–Sun relationships, latitude, marine/continental relationships, air circulation, and topography.

1. *Latitude*—The two latitudinal variables critical to climate are the intensity of radiation and the length of the day, which are intimately related to the Earth–Sun geometric relationship. On average, lower latitudes receive more insolation over the course of the year than high latitudes because the Sun shines more directly on these locations. This insolation variability has a direct impact on ambient (overall) air temperature.

2. *Seasonality*—Seasonality is a critical variable in the character and distribution of global climates for two reasons related to Sun angle: (a) the number of hours of sunlight in a day changes in many places over the course of the year, and (b) insolation and temperature can be highly variable between the winter and summer seasons. In places like the tropical regions, however, very little annual difference in these variables is present, which is reflected in the relatively consistent temperature and precipitation characteristics of tropical climates.

3. *Air mass circulation*—As discussed in Chapter 6, the atmosphere flows in predictable ways, with distinct pressure systems associated with specific zones on Earth. The unique distribution of these pressure systems results in regions with heavy or persistent precipitation, such as those associated with the Intertropical Convergence Zone, whereas others (such as those near the Subtropical High [STH] Pressure System) are relatively dry. Still other regions have distinct weather-producing systems, such as midlatitude cyclones that result in highly variable weather from day to day.

4. *Maritime vs. continental relationships*—Proximity to a large body of water can have a big impact on temperature and the direction of airflow, as we saw in Chapter 7. In the interior of continents the annual temperature range can fluctuate a great deal, whereas it is moderated along ocean coasts. Large bodies of water can also be great sources of atmospheric water vapor through the process of evaporation, and airflow can transport water vapor inland where it falls as precipitation. Evaporation of large amounts of water vapor is usually associated with warm oceans, whereas cold oceans can have a drying effect on adjacent continental locations.

5. *Topographic effects*—Topography can dramatically influence atmospheric processes in a region, as shown in Chapters 6 and 7. For example, the windward side of a mountain range is often a place of heavy precipitation because air cools adiabatically there. In contrast, the leeward side of a mountain range is often a rain shadow due to the descent of air and associated adiabatic warming. Similarly, the topography of a region can also influence the flow of air, resulting in Chinook winds and cool-air drainages that otherwise would not occur.

All these variables interact with one another in a holistic way to influence climate. As you study the next section on the geography of global climates, remember to look for the ways in which these factors interact to form a regional climate pattern. To assist you with this process, you will frequently be prompted with questions that are intended to make you think about the interrelationship of variables. If you have trouble seeing how any particular variable fits into the picture, review parts of the preceding chapters.

Köppen Climate Classification

The purpose of climate classification is to identify certain characteristics, such as temperature and precipitation, that have definable regional patterns. Given the discussions in the previous chapters, you may already have some regional climate characteristics in mind. For example, tropical regions tend to be wet due to the presence of the ITCZ, whereas latitudes around 30° N and S are drier due to the influence of the STH Pressure System. Nevertheless, it is difficult to classify all global climates because so many factors contribute to these patterns. Also, the geographic distribution of climates is a spatial continuum; in other words, sharp breaks from one climate region to the next rarely occur.

Despite these difficulties, a variety of climate classification systems have been devised. One such system is the Thornthwaite system, which was developed by the American climatologist C. Warren Thornthwaite. This system focuses on the local scale and is used frequently in the United States to assess soil moisture characteristics as they relate to vegetation and agriculture. It is based on the concept of **potential evapotranspiration (potential ET)**, which is an estimate of the amount of water used by plants with an unlimited water supply. Potential ET increases with increasing temperature, daylight length, and wind strength and decreases when humidity increases. Can you determine why these changes take place? Think about factors such as insolation and the vapor gradient. In contrast, **actual evapotranspiration (actual ET)** reflects the actual amount of water used by plants.

The Thornthwaite system identifies three basic climate zones on Earth. Low-latitude climates have potential ET >130 cm (51 in.), whereas middle-latitude

---

Potential evapotranspiration (potential ET) *A measure of the maximum possible water loss from a given land area assuming sufficient water is available.*

Actual evapotranspiration (actual ET) *The quantity of water actually removed from a given land area by evaporation and transpiration.*

climates are those with potential ET that ranges between 130 cm and 52.5 cm (20.5 in.). High-latitude climates are those with potential ET that is <52.5 cm. The Thornthwaite system can also be subdivided on the basis of the length of time, and by what amount, actual ET falls below potential ET. In this context, moist climates are those with a surplus or small deficit (15 cm; 6 in.) of water, whereas dry climates have a yearly deficit >15 cm.

Although the Thornthwaite climate system is used frequently in the United States, it is not commonly used elsewhere in the world. Instead, the most widely used classification system is the Köppen (pronounced *Kepun*) Climate Classification System, which was developed by the German botanist and climatologist Wladimir Köppen, who recognized the relationship between major vegetation regions on Earth and regional climate characteristics. The Köppen system describes world climates based on average monthly temperature, average monthly precipitation, and total annual precipitation. It is such a widely used system in part because these variables are measured more frequently in many more places around Earth than any other climate variables. Although the system is great for characterizing regional climate conditions, it ignores more local factors such as wind speed and cloud cover, which often influence conditions dramatically in specific areas.

The best way to view the Köppen system is as a hierarchical classification scheme of categories and subcategories. At its most general level, the system recognizes six major climate groups (*A, B, C, D, E, H*). Climate groups *A, B, C, D*, and *E* range from tropical latitudes near the Equator (*A*) to high latitudes at the poles (*E*) and cover huge geographical areas. Climate group *H* is the designation for high-altitude regions, regardless of latitude, and may be found in isolated places. Each of these major climate groups is distinguished on the basis of temperature, except for category *B*, which is based on moisture characteristics. The Köppen *A* climate zones, for example, have a mean temperature each month >18°C (64°F), whereas the *D* climate regions have average temperature >10°C (50°F) in 4 months to 8 months. *B* climate regions are recognized as those with mean annual precipitation <76 cm (30 in.).

Although it is often sufficient to note the fundamental climate characteristics over broad areas, geographers are usually interested in the more specific attributes of smaller regions. To accomplish this task, the major Köppen groups are further subdivided on the basis of temperature and moisture by use of a second letter, and sometimes even a third. Climate regions with a second letter *s*, for example, have a distinct summer dry season. In contrast, areas with a second letter *w* have a distinct winter dry season. Those areas further classified with a third letter *a* have a warmest month that has an average temperature >22°C (71.6°F). Table 9.1 outlines the various Köppen letter designations and their meanings as far as specific environmental characteristics are concerned. The full range of Köppen climate types and subtypes is presented in Table 9.2, with the geographic distribution of these classes shown in Figure 9.2.

As you might imagine, the Köppen classification system can be very cumbersome to use. At this introductory level, it is important to understand the fundamental characteristics of the 23 major climate regions and avoid becoming overwhelmed by all the specific reasons why the many subcategories occur. With this strategy in mind, the following discussion is a simplified presentation of the Köppen system that focuses on the defining characteristics of each climate region, such as the annual range of temperature and precipitation patterns, along with the environmental variables that cause them. The description of each Köppen region will include a photograph of the landscape to provide a feel for how the climate affects the natural vegetation of the area.

Geography of Köppen Climates

This section of the chapter focuses on the global geography of Köppen climates. The discussion begins with the climates found in tropical regions and progresses to those occurring in very cold regions of the world. As you work your way through this discussion, refer frequently to the global map of Köppen regions in Figure 9.2 to improve your understanding of climate geography. You should also go to the *GeoDiscoveries* website at the end of this discussion to view the climograph for each region. This site also includes a selection of photographs of each area to give you a better idea of the landscape and, in some cases, how people interact with it.

Tropical (A) Climates

Tropical *A* climates are the climates found at low latitudes that straddle the Equator, extending to approximately 25° latitude in both the Northern and Southern Hemispheres (Figure 9.2). *A* climates are warm in that the average monthly temperature exceeds 18°C (64°F). This climate region is further divided into three subcategories on the basis of precipitation.

The tropical rainforest climate (*Af*) occurs at very low latitudes and is most closely associated with consistently high amounts of monthly solar radiation and the strong influence of the ITCZ. As a result, abundant

TABLE 9.1	Köppen Climate Designations	
First Letters	**Derivation**	**Distinguishing Characteristic**
A	Alphabetical	Mean temperature each month >18°C (64°F).
B	Alphabetical	Mean annual precipitation <76 cm (30 in.).
C	Alphabetical	Mean monthly warmest temperature >10°C (50°F) in warmest month; mean monthly coldest temperature between 18°C and −3°C (64°F–27°F) in coldest month.
D	Alphabetical	Mean temperature >10°C (50°F) in 4–8 months.
E	Alphabetical	Mean temperature <10°C (50°F) in all months.
H	Alphabetical	Significant climate changes due to altitude variations.
Second Letters		
F	German *feucht*, "moist"	6 cm (2.5 in.) of mean rainfall in each month.
M	Monsoon	Only 1–3 months with mean rainfall <6 cm (2.5 in.).
S	Summer dry	Summer dry season, with driest month having $< \frac{1}{3}$ the mean precipitation of wettest winter month.
S	Steppe (semi-arid)	Mean annual precipitation in low latitudes is 38 cm–76 cm (15 in.–30 in.), whereas it is between 25 cm–64 cm in midlatitudes. No distinct seasonal trend in either latitude range.
T	Tundra	At least 1 month with mean temperature between 0°C–10°C (32°F–50°F).
W	Winter dry	Winter dry season, with 3–6 months of <6 cm (2.5 in.) of mean rainfall in *A* climates. In *C* and *D* climates, the driest month has $< \frac{1}{10}$ the mean precipitation of the wettest summer month.
W	German *wüste*, "desert"	Mean annual precipitation <38 cm (15 in.) in low latitudes and <25 cm (10 in.) in midlatitudes.
Third Letters		
A	Alphabetical	Warmest month has mean temperature >22°C (71.6°F).
B	Alphabetical	Warmest month has mean temperature <22°C (71.6°F), but has 4 months with mean temperature >10°C (50.0°F).
C	Alphabetical	Warmest month has mean temperature <22°C (71.6°F); fewer than 4 months with mean temperature >10°C (50.0°F).
D	Alphabetical	Same as c, but coldest mean monthly temperature <−38°C (−36.4°F).
H	German *heiss*, "hot"	Mean annual temperature >18°C (64.4°F).
K	German *kalt,* "cold"	Mean annual temperature <18°C (64.4°F).

TABLE 9.2 Climate Types and Subtypes in the Köppen Climate System

First Letter	Subcategory	Köppen Designation	Köppen Climate Name	Defining Characteristic
A	Tropical humid	Af	Tropical Rainforest	No dry season
		Am	Tropical Monsoon	Short dry season; heavy monsoonal rains in other months
		Aw	Tropical Savanna	Winter dry season
B	Arid/semi-arid	BWh	Hot Low-Latitude Desert	Low-latitude desert
		BSh	Hot Low-Latitude Steppe	Low-latitude dry
		BSk	Cold Midlatitude Steppe	Midlatitude dry
		BWk	Cold Midlatitude Desert	Midlatitude desert
C	Mesothermal	Cfa	Humid Subtropical Hot-Summer	Mild with no dry season, hot summer
		Csa	Mediterranean Dry-Summer	Mild with dry, hot summer
		Csb	Mediterranean Dry-Summer	Mild with dry, warm summer
		Cwa	Humid Subtropical Hot-Summer	Mild with dry winter, hot summer
		Cfb	Marine West-Coast	Mild with no dry season, warm summer
		Cfc	Marine West-Coast	Mild with no dry season, cool summer
D	Microthermal	Dfa	Humid Continental Hot-Summer	Humid with severe winter, no dry season, hot summer
		Dfb	Humid Continental Mild-Summer	Humid with severe winter, no dry season, warm summer
		Dwa	Humid Continental Hot-Summer	Humid with severe, dry winter, hot summer
		Dwb	Humid Continental Mild-Summer	Humid with severe, dry winter, warm summer
		Dfc	Subarctic	Severe winter, no dry season, cool summer
		Dwc	Subarctic	Severe, dry winter, cool summer
		Dwd	Subarctic	Severe, very cold and dry winter, cool summer
E	Polar	ET	Tundra	Polar tundra, no true summer
		EF	Ice Cap	Perennial ice
H	Highland	Created after Köppen system was devised		High elevation with cool to cold temperature

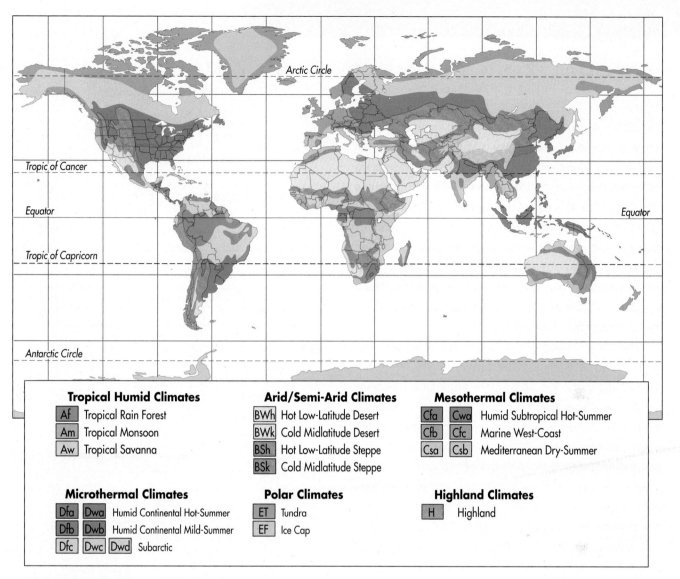

Tropical Humid Climates
Af	Tropical Rain Forest
Am	Tropical Monsoon
Aw	Tropical Savanna

Arid/Semi-Arid Climates
BWh	Hot Low-Latitude Desert
BWk	Cold Midlatitude Desert
BSh	Hot Low-Latitude Steppe
BSk	Cold Midlatitude Steppe

Mesothermal Climates
Cfa	Cwa	Humid Subtropical Hot-Summer
Cfb	Cfc	Marine West-Coast
Csa	Csb	Mediterranean Dry-Summer

Microthermal Climates
Dfa	Dwa		Humid Continental Hot-Summer
Dfb	Dwb		Humid Continental Mild-Summer
Dfc	Dwc	Dwd	Subarctic

Polar Climates
ET	Tundra
EF	Ice Cap

Highland Climates
H	Highland

Figure 9.2 Map of Köppen climates on Earth. The Köppen system contains six major climate groups that are usually broken into as many as 23 subcategories.

rainfall occurs every month due to afternoon convection, the relative humidity is always high, and vegetation consists of dense rainforest (Figure 9.3).

The tropical monsoon climate (*Am*) is very closely related to the tropical rainforest climate (*Af*) because it also receives abundant precipitation. This climate region differs, however, because it has a more prominent seasonal pattern that is associated with migrating Sun position and the flow of air. Thus, average monthly precipitation drops below 6 cm (2.5 in.) for a month or two. Nevertheless, this slightly reduced season precipitation is not sufficient to change the nature of the vegetation, which is also rainforest in these regions.

The tropical monsoon climate (*Am*) is directly related to the onshore flow of mT air that occurs in places like southeast Asia and southwest India (Figure 9.2), which we discussed in detail in Chapter 6. This onshore flow

Figure 9.3 The tropical rainforest climate (*Af*). Tropical rainforest in the African Congo. Due to the very warm temperatures, high humidity, and persistent rainfall in this region, the vegetation is very dense.

VISUALIZING CLIMATES WITH CLIMOGRAPHS

Vegetation is one way to visualize the climate characteristics of a region. Another way is to summarize regional climate conditions with a **climograph**. A climograph is really two graphs in one that shows the following climate characteristics of a region: (1) monthly average precipitation (at the bottom of the graph); (2) mean monthly high temperature (at the top of the graph); and (3) the seasonal position of the Sun (in the middle). When geographers see a climograph, they can visualize the natural vegetation of the area and the variables that influence the environment in that region.

Here is an example of a climograph illustrating the climate characteristics at Moscow, Russia. Note that the red line in the top part of the graph indicates temperature in degrees Celsius (left side) and Fahrenheit (right side), whereas the blue bars at the base of the graph represent precipitation in centimeters (left side) and inches (right side). Months of the year are listed across the base of the diagram. As you can see, annual temperature at Moscow has a distinct seasonal component, with an average high of 24°C (77°F) in July and an average high of −6°C (21°F) in January. Average precipitation ranges from 7.5 cm (~3 in.) in July to about 2.5 cm (~1 in.) in February. Winter (low) precipitation is associated with the dry Polar High pressure system, whereas the relatively wet summer season correlates with the migration of the cyclonic storm track into the region from the south. This climate zone results in a deciduous forest that looks very much like the forests of the eastern United States, with oak, maple, and elm trees dominating the landscape.

Climograph *A graphical representation of climate that shows average annual precipitation and temperature characteristics by month.*

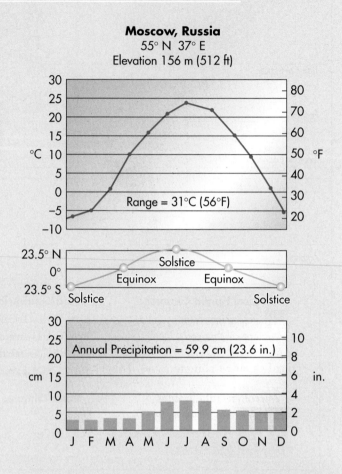

Moscow, Russia
55° N 37° E
Elevation 156 m (512 ft)

Range = 31°C (56°F)

Annual Precipitation = 59.9 cm (23.6 in.)

Figure 9.4 The tropical savanna climate (Aw). Tropical savanna in northern Australia consists of a mix of trees and grass due to the seasonal wet and dry cycle associated with the tropical savanna climate.

occurs in summer when the ITCZ is overhead and reverses in winter when the landmass cools. Because the onshore monsoon winds are southwesterly, the climate is largely associated with the west coasts of landmasses. The tropical monsoon climate (*Am*) also occurs on the eastern side of landmasses in some places, such as the northeast coast of South America (Figure 9.2), where northeast trade winds bring humid air into the region on a seasonal basis.

The third subdivision of the tropical *A* climates is the tropical savanna climate (*Aw*). This climate region is poleward of the tropical rainforest climate (*Af*) in both hemispheres and is found over large regions of the world (Figure 9.2). The primary difference between this climate zone and the tropical rainforest (*Af*) and tropical monsoon (*Am*) climates is that the tropical savanna climate (*Aw*) has very distinct wet and dry seasons. As you might imagine, this cycle is distinctly related to the impact that Earth–Sun geometry has on the migration of large-scale atmospheric pressure systems. The wet season occurs during the summer months because the ITCZ is overhead due to the high Sun angle that time of

year. In contrast, the dry season occurs during the winter months because the ITCZ is in the opposite hemisphere during that time of year and is replaced by the STH. As a result of this distinct wet/dry cycle, the vegetation is a mix of grass and trees known as *savanna* (Figure 9.4).

Arid and Semi-Arid (*B*) Climates

Arid and semi-arid *B* climates are regions with both hot and cold temperatures that are relatively dry, with annual precipitation <76 cm (30 in.). These climate regions are poleward of the *A* climates and are the most widespread on Earth, with *B* climates occurring virtually on every continent (Figure 9.2). From a geographical perspective, the arid climates can be first subdivided on the basis of

www.wiley.com/college/arbogast

Tropical Savanna Climate (*Aw*)

To better visualize how the tropical humid climate functions, go to the *GeoDiscoveries* website and select the module *Tropical Savanna Climate (Aw)*. This module presents an oblique view (an elevated view from the side) of Timbo, New Guinea, from the Atlantic Ocean. In this fashion, you can see how the migration of pressure systems influences the climate of the re-

gion. You will also be able to manipulate the migration of the ITCZ, which will allow you to better see the importance of this pressure system on precipitation in any specific geographic region. Once you complete this simulation, be sure to answer the questions at the end of the module to test your understanding of this concept.

latitude into low-latitude arid climates and middle-latitude arid climates. The low-latitude arid climates tend to be the hottest and driest and are closely associated with the subsiding air of the subtropical high-pressure belt between 20° and 30° N and S latitude. To see this distribution, examine Figure 9.2 and note the extensive deserts in Australia, southern and northern Africa, and Asia. In contrast, middle-latitude arid climates are less dry, semi-arid climates that are related both to the STH and rain shadow effects in continental interiors.

Low-Latitude Dry Climates The Köppen system recognizes two arid climate regions in the low latitudes of Earth. The most extreme of this pair of climate zones is the hot low-latitude desert climate (*BWh*). This climate zone is found in the center and eastern sides of the STH Pressure Systems, spanning 15° to 30° latitude in both hemispheres, and is characterized by mean annual precipitation <38 cm (15 in.) and average annual temperature >18°C (64.4°F). These conditions fundamentally occur due to subsiding air in the center of the STH that promotes extensive adiabatic warming and very little rainfall. The ground is further heated because the skies are usually clear and solar radiation strikes the surface directly because the Sun angle is high. On the ground, the hot low-latitude desert climate (*BWh*) is the classic tropical desert that you have seen pictured in many places (Figure 9.5a). The vegetation in these regions is extremely sparse and drought-tolerant, with waxy and spiny leaves designed to reduce water loss and trunks able to store water.

The second arid climate zone in the low latitudes is the hot low-latitude steppe climate (*BSh*) region. This climate zone is very closely related to the hot low-latitude

desert climate (*BWh*) and results from similar air mass patterns. Because these regions are located slightly poleward of the hot low-latitude desert climate (see Figure 9.2 again), they have a greater annual temperature range and an annual precipitation that ranges between 38 cm to 76 cm (15 in.–30 in.). The greater temperature range results from the occasional incursion of cP air masses, related to strong midlatitude cyclones, in the winter. As a result, the periodic influence of these storm systems causes some precipitation to fall during the winter months. Vegetation in this climate region looks very similar to what occurs in the Mojave Desert (Figure 9.5b) in the southwestern part of the United States.

Midlatitude Dry Climates In addition to the pair of arid climates that occur in the low latitudes, two dry climates are also recognized in Earth's middle latitudes. The most extensive of these middle-latitude dry zones is the cold midlatitude steppe climate (*BSk*). This climate region is most often found deep within the continental interiors of North America and Eurasia (Figure 9.2), and is most often associated with the rain shadow effect that occurs in the lee of major mountain ranges, such as the Rocky and Himalayan Mountains. As discussed in Chapter 7, this effect occurs because maritime air masses are effectively blocked by the mountains, allowing relatively dry continental air masses to dominate the region. Most precipitation falls in summer and is related to convectional processes. When precipitation does fall in the winter, it is usually associated with the midlatitude storm track. Given the relatively high amount of precipitation in this dry climate compared to the desert-like environments examined earlier,

(a)

(b)

Figure 9.5 Low-latitude desert climates. (a) Vegetation in the hot low-latitude desert climate (*BWh*) is very sparse due to the drying effects of subtropical high-pressure cells. (b) Vegetation in the hot low-latitude steppe climate (*BSh*) is very similar to that seen in the Mojave Desert in the southwestern United States. Although the vegetation here is sparse, some trees like this Joshua tree can grow.

(a)

(b)

Figure 9.6 Midlatitude dry climates. (a) Short-grass prairie in eastern Colorado. This low-growing vegetation is a response to the low rainfall in this climate region (*BSk*). (b) The Hongoryn Els sand dunes in Mongolia's Gobi Desert. This sparsely vegetated landscape occurs because of cold/dry conditions in the interior of Asia (*BWk*). This dune field is about 200 km (125 mi) long.

vegetation in this region consists of short-grass prairie (Figure 9.6a). Given your understanding of continental air masses, it should be no surprise that the annual temperature range in the dry midlatitude climate is large, ranging from hot to quite cold. (*Do you know why this pattern occurs?*)

The second midlatitude dry climate on Earth is the cold midlatitude desert climate (*BWk*). This climate zone encompasses one of the smallest overall areas of any climate group, occurring only in central Asia, north-central China, Patagonia in Argentina, and at high elevations in the southwestern United States. The primary distinguishing characteristic of this climate is that it receives 15 cm (5.9 in.) or less of rainfall per year. The annual monthly maximum temperature range in this zone can be high, ranging from ~0°C in the winter months to ~30°C (85°F) in July. The dry character of this climate results in typical desert vegetation (Figure 9.6b).

VISUAL CONCEPT CHECK 9.1

This satellite image focuses on the Middle East, including northern Saudi Arabia, Iraq, Jordan, and Israel. The land bodies are tan-colored because they are deserts with very little vegetation. Which of the following statements best explains the presence of deserts in northern Saudi Arabia and southern Iraq? (*Hint:* The latitude of Baghdad is 33° N.)

a) The region is predominantly influenced by the ITCZ.
b) The region is influenced largely by low-pressure systems.
c) The region is dominated by the STH Pressure System.
d) The region lies on the windward side of a major mountain range.

1. Four major dry arid and semi-arid climates are recognized: (a) hot low-latitude desert climate (*BWh*); (b) hot low-latitude steppe climate (*BSh*); (c) cold midlatitude steppe climate (*BSk*); and (d) cold midlatitude desert climate (*BWk*).

2. The hot low-latitude desert climate (*BWh*) occurs in places like northern Africa and is dominated largely by the STH Pressure System. Vegetation is very sparse.

3. The hot low-latitude steppe climate (*BSh*) lies slightly poleward of the hot low-latitude desert climate (*BWh*) in places like Saudi Arabia and Iraq. It thus has a wider temperature range than the *BWh* climate and slightly more precipitation. This greater precipitation occurs because periodic midlatitude circulatory systems move through the region in winter. Vegetation is sparse, consisting of some grasses, cactus, and isolated trees.

4. The cold midlatitude steppe climate (*BSk*) occurs in places like Denver, Colorado, which lie in the rain shadow of major mountain ranges. Large temperature ranges and seasonal precipitation variations occur in these areas. Vegetation consists largely of short to medium grasses.

5. The cold midlatitude desert climate (*BWk*) occurs in places like the Gobi Desert in north-central China. These regions are dry because they either lie adjacent to cold ocean currents (like northern Chile) or are deep within continents. Vegetation is limited.

Mesothermal (C) Climates

Now that the tropical and desert climates of Earth have been discussed, let's turn our attention to the climate regions characterized by a somewhat more seasonal temperature pattern. These temperature zones are known as mesothermal (*C*) climates because they have both a warm and cool season, with fairly abundant precipitation. Mesothermal (*C*) climates are situated in the midlatitudes, typically between about 20° and 60° N and S latitude (Figure 9.2). Most people live in this climate zone.

If you look again at the global climate map in Figure 9.2, notice that the distribution of *C* climates varies noticeably between the Northern and Southern Hemispheres. (*How can this variability be related to the hemispheric land-water differential that exists?*) In the Southern Hemisphere, *C* climates extend across much of the South American landmass south of 40° S latitude; this occurs because the continent is narrow. North of the Equator, in contrast, the east-west belt of *C* climate is interrupted by a zone of arid climate in the lee of the Rocky Mountains in North America and by the Himalayas in Asia. These breaks occur in the Northern Hemisphere because the continents in that part of the world are very large, which limits the amount of oceanic-derived moisture that can reach the interior.

Three broad mesothermal climate zones occur on Earth. The most humid of these climate regions are the humid subtropical hot-summer climates (*Cfa*, *Cwa*), which are found on the eastern side of continents between about 20° to 35° N and S latitude (Figure 9.2). The distinction between *Cfa* and *Cwa* is based on the fact that *Cwa* climate regions have a distinct winter dry period—hence, the *w* designation. Otherwise, the two regions are climatically similar because they have a hot and humid summer season.

Cfa regions are humid, with average annual precipitation ranging from 100 cm to 200 cm (40 in.–80 in.). This humid climate arises because these regions are affected by mT air masses on the western side of the adjoining ocean's subtropical high-pressure zone. A good example of this geographical pattern occurs in the southeastern United States, ranging from the Carolinas to east Texas (Figure 9.2). During a typical summer, a strong subtropical high-pressure system known as the "Bermuda High" develops off the southeast coast of the United States. Because of the anticyclonic circulation of the system, it pumps warm, moist air from the Atlantic into the *Cfa* region, where it falls as convectional rain due to land heating. Winter precipitation is caused by frontal systems associated with strong midlatitude cyclones. (*Why is it logical that in the winter midlatitude cyclones bring precipitation to the American Southeast, whereas they do not in summer?*) Total annual rainfall is about 162 cm (63 in.), with ample precipitation every month. Although winters are mild, a distinct annual high-temperature cycle is present, ranging from 16°C (62°F) in January to 33°C (91°F) in July.

Vegetation in *Cfa* regions is largely characterized by dense forests of broadleaf deciduous trees (like oak) that drop their leaves during extended cold or dry periods. In particularly warm places, such as southern China and the Gulf Coast of the United States, the native forest consists of broadleaf evergreen trees that remain green throughout the year (see Figure 9.7 as an example).

The second climate within the *C* classification is the Mediterranean dry-summer climate (*Csa*, *Csb*). In contrast to the humid subtropical hot-summer climates (*Cfa*, *Cwa*), *Cs* climates are located along the west coast of continents, at an average latitude of about 35° N and S (Figure 9.2). The difference between the *Csa* and *Csb*

Figure 9.7 The humid subtropical hot-summer climate (*Cfa, Cwa*). Along the Gulf Coast of the United States, the vegetation in many places is the kind of broadleaf evergreen forest shown here.

Figure 9.8 The Mediterranean dry-summer climates (*Csa, Csb*). The California chaparral is associated with the *Cs* climate and consists of wiry shrubs that retain moisture during summer dry periods.

climate zone is that the former has a hotter summer than the latter, with the warmest month having an average temperature >22°C (71.6°F). Otherwise, they are climatically very similar because the wet season occurs in the winter, as opposed to the summer when the air is dry. This pattern occurs because in the winter strong midlatitude cyclones bring moisture into the region from the nearby ocean to the west. During the summer months, however, the midlatitude storm track migrates north and is replaced by the STH Pressure System. The amount of rainfall any particular *Cs* region receives is essentially related to latitude, with lower-latitude locations usually being arid to semi-arid and those at higher latitudes more humid. (*Why do you think this pattern occurs?*) Regardless of the amount of rainfall received, the temperature in these regions is usually moderate with a very mild seasonal variation that may range from about 13°C (55°F) in January to 28°C (83°F) in August. The vegetation in this region is characterized by shrubs and trees that have thick, hard leaves (Figure 9.8) designed to retain moisture during long summer droughts.

The third type of mesothermal climates are the marine west-coast climates (*Cfb, Cfc*), which are found dominantly on the west coast of continents between 35° to 60° N and S latitude (see Figure 9.2). In North America this climate occurs from Oregon to northern British Columbia and also occurs on the east side of the continent in the Appalachian Mountains. Elsewhere, this climate is found in the British Isles, the west coast of Europe, New Zealand, the southern tip of Australia, and southern Chile. The distinction between the *Cfb* and *Cfc* subcategories is that the *Cfb* zones have 4 months with temperatures >10°C (50.0°F), whereas *Cfc* climate zones have fewer than 4 months that warm.

Given the higher latitude of these locations, they are more directly, and more frequently, in the path of the westerly storm track that brings cool, moist mP air masses from the large oceans to the west. Therefore, marine west-coast locations receive more precipitation than Mediterranean regions. This precipitation is magnified by orographic processes when mP air masses encounter large mountain ranges, such as the Cascades in the northwest United States, the Appalachians in the eastern United States, or the Andes in Chile. In regions where this geographical interplay occurs, the amount of annual rainfall is prodigious, exceeding 254 cm/year (100 in./year). As you might expect, marine west-coast regions have a greater annual temperature range than their Mediterranean counterparts at lower latitudes, but the winters are still very mild compared to continental locations. Due to this heavy precipitation, the native vegetation in these areas consists of dense needle-leaf forests (Figure 9.9).

 **Geo Discoveries** www.wiley.com/college/arbogast

Humid Subtropical Hot-Summer Climate (*Cfa*, *Cwa*)

In order to get a better feel for the variables that influence the humid subtropical hot-summer climate, go to the *Geo-Discoveries* website and select the module *Humid Subtropical Hot-Summer Climate (Cfa, Cwa)*. This module shows how the STH known as the Bermuda High pumps moisture into the southeastern United States during the summer. The animation will begin with a map that shows the rotating pressure system that is migrating on a seasonal basis due to the effects of Earth–Sun geometry. As spring progresses into summer, the Bermuda High forms off of the southeastern coast. This system is rotating clockwise and, in so doing, pumps moisture into places like Louisiana, Georgia, and North Carolina. This can be shown by clouds forming over land with precipitation. Once you complete this animation, be sure to answer the questions at the end of the module to test your understanding of this concept.

KEY CONCEPTS TO REMEMBER ABOUT MESOTHERMAL (C) CLIMATES

1. Three major mesothermal *(C)* climate subcategories are recognized: (a) humid subtropical hot-summer climates (*Cfa*, *Cwa*); (b) Mediterranean dry-summer climates (*Csa*, *Csb*); and (c) marine west-coast climates (*Cfb*, *Cfc*).

2. The humid subtropical hot-summer climates (*Cfa*, *Cwa*) occur on the southeastern side of continents in places like South Carolina and China. These regions have ample year-round precipitation, but have a distinct summer peak due to the influx of mT air by off-shore STH Pressure Systems. Vegetation consists predominantly of pines and deciduous trees.

3. The Mediterranean dry-summer climates (*Csa*, *Csb*) occur on the southwest coast of North America and in the Mediterranean region. These climates have a warm but narrow temperature range due to their relatively low-latitude coastal location. The summer dry season is associated with dominance by the STH Pressure System. Vegetation consists of a mix of trees and grass.

4. The marine west-coast climates (*Cfb*, *Cfc*) occur in coastal zones poleward of their Mediterranean counterparts. Thus, they have a wider temperature range that is, on average, cooler. These regions receive ample annual precipitation, but have a distinct winter wet season due to passage of midlatitude cyclones. Vegetation consists largely of needle-leaf trees.

This map shows the average annual precipitation in the contiguous United States. Explain the patterns for these three regions: (1) the Southeast, including Mississippi and Tennessee; (2) the Intermountain West, including Nevada and Utah; and (3) the Pacific Northwest coast, including western Washington and Oregon.

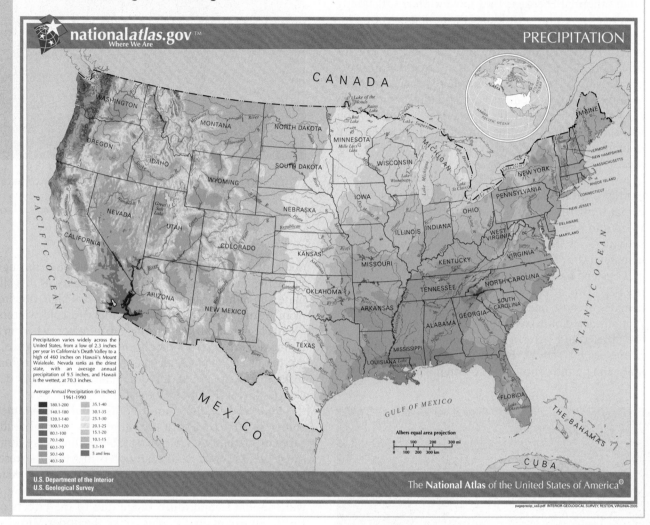

PRECIPITATION

nationalatlas.gov™
Where We Are

Precipitation varies widely across the United States, from a low of 2.3 inches per year in California's Death Valley to a high of 460 inches on Hawaii's Mount Waialeale. Nevada ranks as the driest state, with an average annual precipitation of 9.5 inches, and Hawaii is the wettest, at 70.3 inches.

Average Annual Precipitation (in inches) 1961-1990

180.1-200	35.1-40
140.1-180	30.1-35
120.1-140	25.1-30
100.1-120	20.1-25
80.1-100	15.1-20
70.1-80	10.1-15
60.1-70	5.1-10
50.1-60	5 and less
40.1-50	

U.S. Department of the Interior
U.S. Geological Survey

The **National Atlas** of the United States of America®

Albers equal area projection

 Geo Discoveries

www.wiley.com/college/arbogast

 WILEY PLUS

Marine West-Coast Climates (*Cfb, Cfc*)

In order to visualize how the marine west-coast climate functions, go to the *GeoDiscoveries* website and select the module *Marine West-Coast Climates (Cfb, Cfc)*. In this module, you will be able to see how the migrating midlatitude storm track brings moisture off of the eastern Pacific Ocean into the region. This animation is similar to the humid subtropical hot-summer climates (*Cwa, Cwb*) animation you previously viewed, in that it will show airflow and associated cloud development. As you watch the animation, notice the relationship between cyclones and moisture flow, as well as the effect that the Cascade Mountains have on precipitation intensity. Once you complete the animation, be sure to answer the questions at the end of the module to test your understanding of this concept.

Microthermal (*D*) Climates

At this point, let's turn to the climate zones that are more continental in their character by focusing on *D* climates. These climate zones are typically located poleward of *C* climates, ranging from about 35° to 60° N and S latitude (Figure 9.2). *D* climates are known as "microthermal" because they have winters sufficiently cold to ensure that snow can remain on the ground for extended periods of time. In contrast, summer temperatures are usually warm and can even be hot for brief periods of time. *D* climates are found almost exclusively in the Northern Hemisphere because no large landmasses are present at corresponding latitudes in the Southern Hemisphere. In fact, the only places you will see *D* climates in the Southern Hemisphere are in highland areas. (W*hy does this distribution of D climates make sense?*)

Humid Continental Climates Like the mesothermal *C* climates, microthermal *D* climates can be initially subdivided into three categories. The most southerly of these subcategories are the humid continental hot-summer climates (*Dfa*, *Dwa*), which are located in the central and eastern parts of North America and Eurasia (Figure 9.2). Two subclassifications are recognized in these climates because one is moist (≥6 cm [2.5 in.] of average rainfall each month) over the entire year (*Dfa*), whereas the other is dry (<6 cm [2.5 in.] of rainfall per month) in the winter (*Dwa*). The primary weather-producing system in humid continental hot-summer climates is the

Figure 9.11 The humid continental mild-summer climates (*Dfb*, *Dwb*). Pine forest in northern Michigan. This type of vegetation is closely associated with the wet-winter variant (*Dfb*) of this climate.

midlatitude jet stream, which steers cyclones through these regions. Given the continental locality of this climate, the annual range of temperature is high, ranging from warm, periodically hot summers to cold winters. Most precipitation falls during the summer, when mT air masses invade from the south. Winters are relatively dry because cP and cA air masses dominate the region. In these climate zones, the dominant vegetation is deciduous forest (Figure 9.10).

Poleward of the humid continental hot-summer climates (*Dfa*, *Dwa*) are the humid continental mild-summer climates (*Dfb*, *Dwb*). These climate regions range across central and eastern North America, western Europe, central Asia, and the Far East of Asia (Figure 9.2). These regions are known for having a wide annual temperature range due to the effects of Earth–Sun geometry. In places like the upper Midwest in the United States, average high temperatures in July are >27°C (80°F), with about 10 cm (4 in.) of rain. Average high temperatures in January, in contrast, are about −5°C (23°F) and most precipitation is snow. In northern Michigan this climate is associated with northern hardwood and pine forests (Figure 9.11).

Subarctic Climates The third major subdivision of the *D* climates is the subarctic climates (*Dfc*, *Dwc*, *Dwd*). These climate zones occur poleward of the humid continental climates (*Dfa*, *Dwa*, *Dfb*, *Dwb*), stretching across all of northern North America and from northern Scandinavia east across all of Asia (Figure 9.2). It is a continental climate characterized by long, bitterly cold winters and short, cool summers. In

Figure 9.10 The humid continental hot-summer climates (*Dfa, Dwa*). Deciduous forest in the humid continental hot-summer climate, including various oak and maple species, covered much of this climate region in the United States prior to European settlement.

Figure 9.12 The subarctic climate (*Dfc*). Typical vegetation in the subarctic climate (*Dfc*) consists of needle-leaf trees and scattered openings covered by moss and lichens.

1. Three major subcategories of microthermal (*D*) climates are recognized: (a) humid continental hot-summer climates (*Dfa*, *Dwa*); (b) humid continental mild-summer climates (*Dfb*, *Dwb*); and (c) subarctic climates (*Dfc*, *Dwc*, *Dwd*).

2. The humid continental hot-summer climates (*Dfa*, *Dwa*) occur in places like the midwestern United States and eastern Asia. Vegetation consists largely of deciduous trees. These regions have wide annual temperature ranges and moderate yearly precipitation. Most precipitation is associated with midlatitude cyclones.

3. The humid continental mild-summer climates (*Dfb*, *Dwb*) occur in places like the north-central and northeastern United States and eastern Asia. Vegetation consists dominantly of needle-leaf trees. These regions also have a wide annual temperature range, but with colder winters and milder summers than the humid continental hot-summer climates (*Dfa*, *Dwa*). Average annual precipitation is moderate and most closely associated with the midlatitude storm track.

4. The subarctic climates (*Dfc*, *Dwc*, *Dwd*) occur in high-latitude, continental locations that span North America and Asia. Vegetation is boreal forest. These regions have a very wide annual temperature range and have a summer wet season associated with the midlatitude storm track.

northwestern Canada, where the *Dfc* subdivision occurs, the average high January temperature is about –25°C (–13°F), whereas it approaches 20°C (69°F) in July. The impact of this extreme temperature range can be seen in the vegetation, which is dominated by boreal forests consisting of needle-leaf trees such as pine, spruce, and fir (Figure 9.12). The *Dwc* variant is found only in far eastern Russia and is characterized by a distinct winter drought. The *Dwd* variant occurs only in extreme northeastern Siberia and is distinctive because, in addition to a winter drought, its coldest month is below –38°C (–36.4°F).

The subarctic climates (*Dfc*, *Dwc*, *Dwd*) are the source regions of cold, dry cP air masses, and are often invaded by more frigid cA air masses that originate at still higher latitudes. Given that this region is often north of the midlatitude storm track, the annual amount of precipitation is relatively low. Most precipitation occurs during the summer months, when the belt of cyclonic storms migrates into the region from the south. Otherwise, the polar high-pressure system dominates the region, resulting in little rainfall. The annual range of temperature is greater within this climate than any other. (*Do you remember why this extreme range occurs? Think back to previous discussions about Earth–Sun geometry, seasonality, and insolation.*)

Polar (*E*) Climates

The most extreme climates in nonmountainous areas are the polar *E* climates, which occur at latitudes higher than 70° N and S latitude. The tundra climate (*ET*) is one of the two polar climate subcategories

and is found along the arctic coastal fringes, including such places as the island region of northern Canada, the north slope of Alaska, the Hudson Bay region, coastal Greenland, and all of northern Eurasia (Figure 9.2). Given this geographical location, a moderating climate effect occurs from the nearby oceans, resulting in winter temperatures less severe than in the more continental regions to the south. Nevertheless, winters are very long and cold and summers are short and cool. In northern Alaska, for example, the average high in January is –11°C (13°F) and an average high in July is 15°C (59°F). Very little precipitation falls in this region, largely because it is dominated by the polar high-pressure system the vast majority of the year. The vegetation in this climate region is characterized by short mosses and lichens that dominate the landscape (Figure 9.13a). These areas also contain well-developed bogs and zones of permafrost, which will be investigated in more detail in Chapter 17.

(a)

(b)

Figure 9.13 The polar climates (*E*). (a) Vegetation in the tundra landscape consists of low-lying lichens and mosses, which are the main diet of caribou. (b) A scene from the Antarctic ice cap where temperatures are frigid year round.

In contrast to the tundra climate, the ice-cap climate (*EF*) occurs in the interior of Greenland and Antarctica where enormous glaciers cover the landscape. Sun angle is consistently very low in these areas, resulting in little annual insolation. In addition, the ice-covered surfaces have high albedo, which means that most radiation is reflected and thus provides no heat. As a result, temperatures are consistently brutally cold to frigid (Figure 9.13b). At McMurdo Station in Antarctica, for example, average annual high temperatures range from −5°C (25°F) in December to −27°C (−17°F) in July and August. (*Why do the coldest temperatures occur in July and August?*) Annual precipitation is very low, <8 mm (0.3 in.). The reason for this low precipitation is that the air over the Antarctic ice cap is continental antarctic (cAA) air, which is very cold and dry. In short, this place is a polar desert.

Highland (*H*) Climates

Thus far the discussion about Köppen climates has largely focused on climate zones influenced by changes in latitude. The system also recognizes highland (*H*) climates, which are those associated with the world's large mountain ranges such as the Andes in South America, the Himalayas in Asia, the Swiss Alps in Europe, the Southern Alps in New Zealand, and the Rockies in North America (Figure 9.2). *H* climates are largely determined on the basis of the temperature changes associated with the environmental lapse rate as it relates to altitude. Recall from Chapter 5 that this value is 6.4°C/km (3.5°F/1000 ft). Perhaps you have observed when hiking in mountains that temperatures usually cool with increased elevation. Or perhaps you have noticed that mountain peaks may be covered with snow while it is hot in the surrounding lowlands.

www.wiley.com/college/arbogast

Global Climates

Now that the geography of global climates has been discussed, it is a good time to review them in an interactive way. Go to the *GeoDiscoveries* website and select the module *Global Climates*. This module is a tour of Köppen climate regions. It provides an example of each climate zone along with a representative city from that region. The discussion includes a climograph from that locality, as well as photographs of the city and associated vegetation. As a result, you should be able to better visualize the characteristics of each climate zone. Once you complete the module, be sure to answer the questions at the end to test your understanding of Köppen climate regions.

Global climates and the factors associated with them can be analyzed in many different ways. One way that the character of global climates can be viewed is through satellite remote sensing. Through remote sensing, we can see the pattern of global climates around Earth and some of the variables associated with them, such as atmospheric pressure systems and vegetation. To see some of these patterns, go to your *GeoDiscoveries* website and select the module *Remote Sensing and Climate*. This module allows you to see how remote sensing can be used to observe climate patterns on Earth. You will be asked, for example, to identify a variety of atmospheric features, such as the ITCZ, the STH, and mid-latitude cyclonic systems. You can also test your ability to locate hurricanes and determine the effect that cold ocean currents have on climate. As you work your way through this module, try to put together all the concepts discussed so far in this book and visualize how they fit together to form distinctive geographic patterns. Once you complete the module, be sure to answer the questions at the end to make sure you understand these concepts.

This climate zone also reflects the fact that the environmental characteristics in mountainous areas are extremely complex, with abrupt changes taking place over short horizontal distances and vertical slopes. These complexities are related to such factors as slope orientation (does the slope face the Sun?), time of day (does the Sun face the slope in the morning or afternoon?), and direction of airflow (windward or leeward?). If you recall from Chapter 7, orographic processes may be a very important part of a highland area, with abundant precipitation on the windward side of a mountain range and relatively dry conditions in the lee. Perhaps the best way to view highland (*H*) climates is that they can vary dramatically over short distances, and that local conditions can change very quickly in time. If you plan to spend time in the mountains, you should be aware of these characteristics.

Reconstructing Past Climates

The global climates discussed in the previous section on the Köppen system are essentially stable entities, with fairly predictable conditions occurring from one year to the next. For example, if you live in a continental mid-latitude location, chances are that every year the summers will be very warm and the winters cold. However, climate conditions can and do fluctuate, for a variety of reasons. Short-term fluctuations can be caused by volcanic eruptions, in which resulting ash in the atmosphere blocks insolation, causing lower temperatures.

Long-term fluctuations can also occur. For example, the global climate during the Jurassic period (a part of the dinosaur age), approximately 175 million years ago, was generally warmer and more humid than it is today.

> **KEY CONCEPTS TO REMEMBER ABOUT POLAR (ET, EF) AND HIGHLAND (H) CLIMATES**
>
> 1. Three major subcategories of very cold climates are recognized: (a) the tundra climate (*ET*); (b) the ice cap and sheets climate (*EF*); and (c) highland climates (*H*).
>
> 2. The tundra climate (*ET*) is a high-latitude maritime climate associated with tundra vegetation. It has a narrower average annual temperature range than lower-latitude subarctic climates (*Dfd, Dwd*). Average annual precipitation is low.
>
> 3. The ice cap and sheets climate (*EF*) is associated with large glaciers at high latitudes. Thus, average annual temperatures are very cold and very little or no vegetation is present.
>
> 4. The highland climate (*H*) occurs at high elevations in mountain ranges such as the Rockies, Himalayas, and Andes. Although these areas have an annual temperature range, it is much narrower and colder than nearby localities at lower altitudes.

As you probably know, a major environmental issue currently facing the world today is *global warming*, which is the belief among the vast majority of climate scientists around the world that human industrial activity is gradually causing the planet to warm. Global warming is the subject of numerous public discussions and news reports. Because so many interest groups are now involved, the issue has become much politicized. If you are to participate in future decisions regarding global warming and what to do about it, then it is essential to have some fundamental knowledge about the topic.

A good place to begin the discussion of future climate change is by investigating the magnitude and frequency of climate changes that have occurred during the *Quaternary Period,* the most recent 1.6 million years of Earth's history. You might be saying, Why do we look at Quaternary climate change to understand the future? The reason is that by understanding past climate changes, we gain insight into the range of possibilities that might occur in the future.

Humans have been keeping systematic climate records, including temperature and precipitation data, for only about 150 years. In some parts of the United States, these records span an even shorter amount of time. Although 150 years sounds like a long time in the context of the average human life span, it is really quite short from a geological perspective. In order to reconstruct the climate conditions of, say, 20,000 years ago, it is necessary to investigate clues about the character of the environment at any given place and time. Such clues are called **proxy data** because they are indirect evidence for past climate change. Scientists who are interested in prehistoric climate change are called *paleoclimatologists.*

Pollen Records

One kind of clue that provides indirect evidence of past climate change is ancient plant pollen, which reflects the kind of vegetation that occurred in a region at any given time. Given the demonstrable link between climate and vegetation, as we have seen earlier in this chapter, it is logical that significant changes in the prehistoric plant assemblage of an area would indirectly reflect climate change. Such change in a plant community can be reconstructed in regions (usually cool and moist) where pollen is well preserved in the deposits that slowly accumulate in lakes and marshes. Scientists who study such records are called *palynologists.*

Proxy data   *Indirect evidence of an event. For example, fossil pollen is a proxy indicator of climate change because vegetation reflects climate.*

The premise of palynology is that plants produce microscopic pollen that are unique in shape and size to specific species. You can see this kind of variation in Figure 9.14. Each year, plants within a region produce pollen as part of their reproductive cycle; this pollen is then spread by wind. If you happen to be allergic to some kind of pollen, you are acutely aware of this seasonal process because of the discomfort it may cause. Some of this pollen accumulates at the bottom of lakes and marshes in sediments that progressively thicken with time. Given that the deposits at the base of the deposit are the oldest, and progressively decrease in age toward the top of the sequence, the change in pollen composition through time can be reconstructed. To do so, a palynologist acquires a core (vertical sample) of the lake-bottom sediments and returns it to the laboratory where the pollen can be extracted from each layer of sediment and then analyzed.

Ultimately, a pollen diagram such as the one presented from Nelson Lake, Illinois, is created (Figure 9.15) that shows the change in pollen at the lake or bog over time. This pattern of change is then used to reconstruct the climate history. The diagram should be interpreted in this way. Begin by looking at the vertical axis, which shows the age in years before the present, beginning with 17,000 years ago and ending today (0). From left to right across the top of the graph are the various plant species identified in the core samples, ranging from sedge on the left to grass on the right. The relative percentages of each of these plant species are shown at the bottom of each

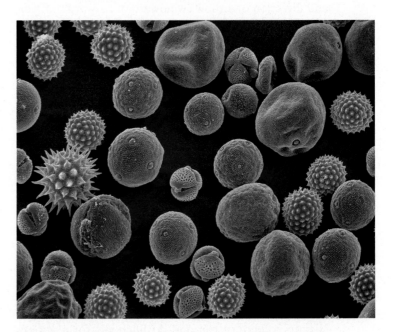

Figure 9.14 Pollen grains. Different plant species produce distinctive forms of pollen that can be seen under a microscope. Note the different forms of pollen in this image, which is magnified 400 times.

Nelson Lake, Kane County, Illinois

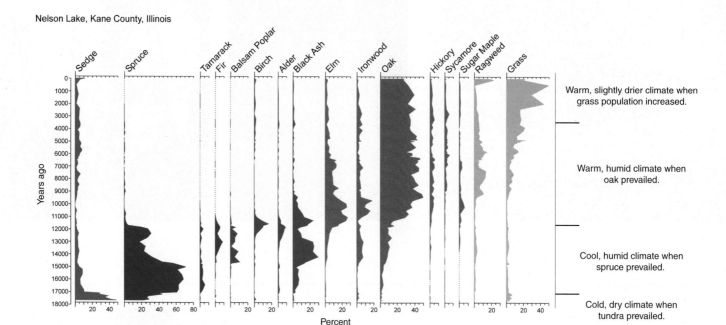

Figure 9.15 Pollen diagram from Nelson Lake, Illinois. Note the vegetation change through time and how it relates to climate fluctuations presented on the right-hand side of the graph. (*Source:* Eric Grimm, Illinois State Museum.)

plant column. For example, around 15,000 years before the present, spruce trees comprised about 60% of the overall sample, whereas the amount of oak pollen was very small. Approximately 10,000 years before the present, however, spruce pollen makes up 0% of the sample, whereas oak is over 40%.

Using these percentage changes through time, a palynologist can reconstruct the climate for the Nelson Lake area for the past 17,000 years. The high amount of sedges between 17,000 years and 14,000 years before the present reflects a cold, dry climate when tundra prevailed in the area. As you will learn in Chapter 17, this climate/vegetation combination existed because an enormous glacier was nearby. Some 14,000 years ago, the climate became warmer and more humid, which caused tundra to be replaced by spruce, which is a needle-leaf tree. Spruce dominated the local vegetation until about 11,000 years before the present when the climate apparently warmed sufficiently to allow the influx of oak (a deciduous tree) at the expense of spruce. Oak has dominated for the past 10,000 years at Nelson Lake, although grass has increased in the past 2000 years. This expansion of grass is probably related to the climate becoming slightly warmer and drier.

Tree Ring Patterns

Another way that past climate change is reconstructed is through the analysis of tree rings. The interior of most trees contains annual rings, such as those shown in Figure 9.16a, that reflect each year of growth in the tree's life. In trees that have annual rings, each ring consists of a dark and light wood couplet, with the dark wood representing the winter period when growth is slow or nonexistent and the light wood reflecting rapid growth in summer. When local climate conditions are favorable, the rings are wide, whereas they are narrower (reflecting less growth) when it is colder or drier, depending on the type of tree.

The study of tree rings is called **dendrochronology** and, in part, attempts to reconstruct the history of regional climate change by analyzing tree ring patterns. These ring patterns can be accessed by coring the tree with a device called an increment borer that does not harm the tree. To obtain a core, the implement is first screwed horizontally into the center of the tree (Figure 9.16b) and then an approximately 6-mm (0.2-in.) wide core is extracted that cuts across the tree rings, thus showing all of them. Some trees live a very long time and thus produce long, continuous tree ring records. An excellent example of such a tree is the famous bristlecone pines in the White Mountains of California (Figure 9.16c), which are the oldest living things on Earth. Some of these trees are over 4000 years old and still growing! Thus, by looking at the rings of one tree alone, it is possible to reconstruct the growth cycles, and therefore the history of climate change, for the past 4000 years in the area where the pine trees have lived.

Dendrochronology *The dating of past events and variations in the environment and climate by studying the annual growth rates of trees.*

(a)   (b)   (c)

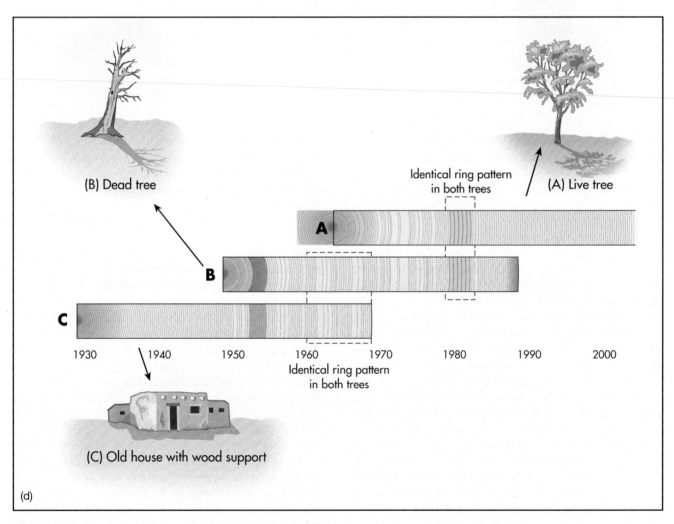

(d)

Figure 9.16 Using trees to reconstruct climate change. (a) The theoretical foundation of dendrochronology is that many trees produce annual growth rings, with dark and light wood representing winter and summer seasons, respectively. (b) Using an increment borer to core a tree. (c) A 4000-year-old bristlecone pine in the White Mountains of California. (d) The premise of cross dating is that the tree ring record can be extended back in time by overlapping ring patterns from a sequence of trees. In this simplified diagram, ring set *A* comes from a living tree, ring set *B* comes from a tree that recently died, and ring set *C* comes from wood that was used in an old house. This particular record permits climate reconstruction between about 1930 and the early 2000s.

Although the living bristlecone pines have yielded an excellent record of climate change, paleoclimatologists are constantly looking for ways to push the evidence of prehistoric climate change farther and farther back in time. In the context of dendrochronology, this goal is accomplished through a method called *cross dating*. The premise of cross dating is that ring patterns from similar trees that overlapped in their life spans can be matched to extend the record of ring variability back in time.

To see how this method works, look at Figure 9.16d. Imagine that a core is extracted from a living tree (*A*) that produces a distinctive ring record. Near this living tree is another tree (*B*) that is dead but still standing. Upon coring (or, in the case of a dead tree, cutting) the second tree, it is discovered that the ring pattern in the latter half of its life was the same as that during the early years of the first tree's (*A*) life. It is thus logical to conclude that the trees overlapped in their life spans, and that they responded to the same climate conditions as far as their growth was concerned. Imagine then that another tree (*C*) is discovered that was used as a support structure in an archaeological ruin. Analysis of its ring pattern indicates that in the latter part of its life it overlapped with the first part of the second (*B*) tree's life. We now have a record of climate change that extends back to a period before the life of the intermediate tree. Using this method, paleoclimatologists have extended the history of climate change in the White Mountain area to beyond 9000 years ago!

Ice Core Analysis

Yet another way that prehistoric climate records have been reconstructed is by investigating oxygen isotopes contained in ice cores recovered from glaciers in Greenland and Antarctica. These glaciers contain layers of ice that have accumulated annually for thousands of years. Assuming again that the more deeply buried ice layers are the oldest, it is possible to literally count layers to determine when they formed. The premise of using oxygen isotopes is that oxygen contains two primary isotopes, O-16 and O-18, that differ in their atomic weight. Although both isotopes are found in liquid water, O-16 water is almost 500 times more common and more easily evaporated than its heavier counterpart, O-18 water. This variation depends somewhat on the temperature of the water at the time it evaporates. When water is warmer, for example, proportionately more O-18 water is evaporated than when the water temperatures are colder. Thus, it is possible to examine the ratio of O-16 to O-18 in annual layers of ice in ice caps and yearly accumulations of marine sediments to indirectly reconstruct past climate cycles.

To see an example of how this method of climate reconstruction works, examine Figure 9.17, which illustrates the way in which O-18 and O-16 move relative to each other within the hydrologic cycle during a glacial period. The diagram shows a hypothetical air mass that flows away from its tropical source area toward a growing glacier at high latitudes. Note that relatively more O-16 is evaporated from the ocean than O-18 during

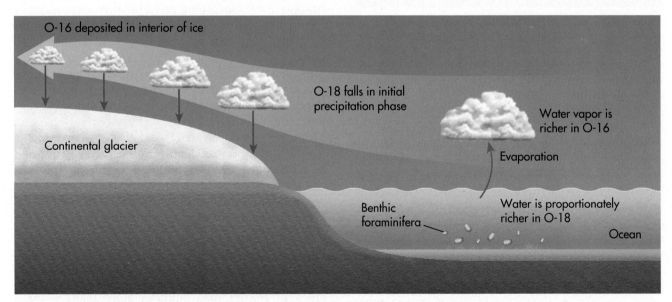

Figure 9.17 Using oxygen isotope ratios to reconstruct glacial and interglacial cycles. Seawater contains both O-16 and O-18 isotopes, which are preferentially evaporated (O-16) and precipitated (O-18) relative to one another during glacial cycles. Thus, microscopic marine organisms (benthic foraminifera) absorb relatively more of the heavier O-18 isotope during glacial cycles. At the same time, moisture-laden air flows toward growing ice sheets, where O-18 is precipitated first because it is heavier than O-16. By the time the air mass reaches the center of the ice sheet, it is enriched in O-16 relative to O-18.

(a)

(b)

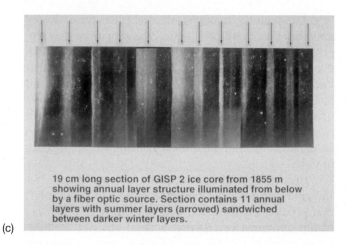

19 cm long section of GISP 2 ice core from 1855 m showing annual layer structure illuminated from below by a fiber optic source. Section contains 11 annual layers with summer layers (arrowed) sandwiched between darker winter layers.

(c)

Figure 9.18 Ice cores. (a) Coring the Greenland ice cap. The scientists pictured here are drilling into the ice to obtain a sample for detailed analysis. (b) An example of an ice core extracted from the Greenland ice cap. When this portion of the core was extracted, scientists marked its depth so its position in the ice would remain known. For preservation, it is kept in a store room at a consistent temperature of –25°C (–13°F). (c) An example of annual ice layers, showing distinct bands within the ice. Each of these bands represents an annual layer of snow that transformed to glacial ice. Each annual layer contains a sample of the atmosphere at the time the snow fell, including the relative amount of O-16 and O-18.

this glacial cycle. This leaves proportionately more O-18 in the seawater, which is absorbed by microscopic marine organisms called *benthic foraminifera* that live on the ocean bottom. As the air mass flows toward the center of the growing glacier, precipitation occurs and O-18 is the first oxygen isotope to be lost because it is the heavier isotope. By the time the air mass reaches the center of the ice mass, the oxygen isotopes are mostly O-16. In this way, the marine sediments (which contain the benthic foraminifera) become relatively enriched in O-18 (because O-16 is preferentially evaporated) at the same time that thickening glacial ice contains proportionately more of the O-16 isotope (because most O-18 was left behind or precipitated earlier).

So how do scientists reconstruct this record of climate change? It is actually very simple; they take cores of the marine sediments and of large ice masses. Both the marine sediments and glacial ice masses gradually thicken with time and the annual records of this growth are preserved. Although the coring process is essentially the same in both ocean and ice settings, the method is much easier to visualize on a glacier (Figure 9.18).

Members of a coring expedition pick a place where the ice is expected to contain the best annual sequence of isotopic data. Once they arrive at that location, they set up a station (Figure 9.18a) where a core tube can be drilled vertically into the glacier, through a large number of annual ice layers, to a great depth. With this tube, they can extract a cylinder of ice (Figure 9.18b) that contains a sample of all the annual snowfalls (and thus the annual ratio of O-16/O-18) they can penetrate. Once the core is extracted, scientists merely have to count the annual layers (Figure 9.18c) to record the length of time represented in the cylinder of ice and then determine changes that occur in the ratio of O-16 and O-18 over time.

A composite record of cores demonstrates a sequence of distinct **oxygen isotope stages** (Figure 9.19) that reflect major climate changes that have occurred through the Quaternary Period. This record is particularly relevant to reconstructing glacial cycles, indicating that massive ice sheets advanced and retreated across much of Earth's high latitudes more than 20 times during the Quaternary Period. The most recent of these

Oxygen isotope stages *Periods of time that have distinct O-18/O-16 ratios, which are used to reconstruct prehistoric climate change.*

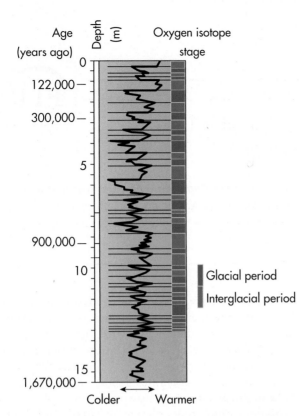

Figure 9.19 Oxygen isotope stages during the past approximately 1.9 million years. Each oxygen isotope stage reflects a glacial or interglacial period, as indicated through oxygen isotope ratios in marine sediments. This record indicates that glacial periods have been quite common during the past 1.9 million years. (*Source:* Adapted from N. J. Shackelton and N. D. Opdyke, 1976. Oxygen-Isotope and Paleomagnetic Stratigraphy of Pacific Core v28–239, Late Pliocene to Latest Pleistocene. Geological Society of America Memoir 145.)

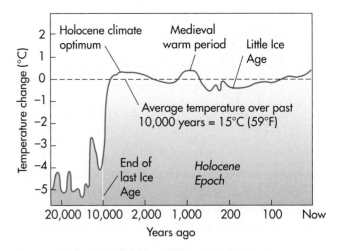

Figure 9.20 Holocene climate changes. This graph shows the changes in average surface temperature for the past 20,000 years relative to mean global temperature for the past 10,000 years. Note the dramatic temperature change at the end of the most recent ice age. Several notable climate fluctuations have occurred since that time, including the Holocene Climate Optimum, Medieval Warm Period, and Little Ice Age.

KEY CONCEPTS TO REMEMBER ABOUT RECONSTRUCTING CLIMATE CHANGES

1. Understanding past climate change is important because it provides a window into the potential nature of future climate changes.

2. The extent and timing of past climate changes can be reconstructed using indirect clues derived from ancient pollen, tree rings, and ice cores. Fossil pollen reflects the kind of vegetation that once lived in an area. Tree rings provide climate clues because they vary in width depending on environmental conditions. Ice cores can be used because they contain a record of oxygen isotopes (O-16 and O-18).

3. The oxygen isotope record indicates that many different glacial and interglacial periods have occurred during the past 1.9 million years.

4. The past 10,000 years are referred to as the Holocene Epoch and have consisted of "modern climate." Nevertheless, several distinct climate episodes occurred, including the Holocene Climate Optimum, Little Ice Age, and Medieval Warm Period.

glaciations reached its maximum extent about 18,000 years ago and ended about 10,000 years ago. We will discuss this history more thoroughly in Chapter 17.

The most recent 10,000 years of Earth history are generally considered to be the modern climate epoch known as the *Holocene*. Although Holocene climates have been remarkably stable compared to earlier glacial cycles, changes in average Earth temperature have nevertheless occurred (Figure 9.20). Although several fluctuations occurred during the Holocene, three are particularly notable. The *Holocene Climate Optimum* (a.k.a. Hypsithermal) occurred between 9000 and 6000 years ago and was a time when average summer temperatures in the Northern Hemisphere were 2°C to 4°C (~ 4°F–7°F) warmer than the present. Another Holocene warm interval was the *Medieval Warm Period*, which occurred between the 9th and 13th centuries. During this period, the average global temperature was about −0.5°C (~1°F) warmer than the present and the Vikings were able to establish colonies in southern Greenland. Following this warm spell, Earth cooled again during the *Little Ice Age*,

which occurred from the 13th to 19th centuries. Average global temperatures during this period were about 1°C to 1.5°C (2°F–3°F) cooler than now.

Reconstructing Past Climates Using Oxygen Isotopes

The sorting of oxygen isotopes across space by their size is a process that needs to be viewed in animated form to fully appreciate and understand. To do so, go to the *GeoDiscoveries* website and select the module *Reconstructing Past Climates Using Oxygen Isotopes*. Once you complete the animation, be sure to answer the questions at the end of the module to ensure you understand this concept.

Causes of Past Climate Change

As you have seen, a variety of proxy climate data indicate that Earth's climate has fluctuated greatly during the past 2 million years. Why did these fluctuations occur? One factor directly related to global cooling is extensive volcanic activity because increased levels of volcanic ash in the atmosphere block solar radiation. This effect was observed most recently in 1991 when Mount Pinatubo in the Philippines erupted catastrophically, resulting in a noticeably cool year globally in a sequence of otherwise warm years. Another factor that can cause climate change is asteroid impacts that dramatically increase levels of atmospheric dust, which also block sunlight. The best known example of such an impact occurred 65 million years ago when an asteroid about 10 km (6 mi) wide struck Earth near the modern Yucatan Peninsula in Mexico. Most geologists believe this impact, and subsequent extensive global cooling, caused the extinction of the dinosaurs. Still another factor that can cause global climate change is the movement of landmasses. A good example of how this variable influenced climate is the way North and South America became connected when the Isthmus of Panama formed. This joining of landmasses disrupted ocean circulation in the tropics, which, in turn, changed the circulation of the atmosphere.

Although the factors just discussed are linked with known periods of climate change on Earth, they are not considered to be responsible for the numerous glacial cycles that have occurred during the Quaternary Period. These cycles have been linked to rhythmic changes in the Earth–Sun geometric relationship. Milutin Milankovitch, a civil engineer and geophysicist, first proposed this theory in the 1920s and argued that fluctuations in the Earth's tilt, rotation, and orbit best explained the glacial cycles. This theory was largely ignored until the mid-1970s when, as discussed earlier, isotopic analyses of ice cores revealed that many more glacial periods occurred than previously believed. Most important, scientists discovered a close correlation between the orbital fluctuations proposed by Milankovitch and the timing of glacial and interglacial periods. Thus, the **Milankovitch theory** is now widely accepted as best explaining Quaternary climate fluctuations. Let's take a closer look at this important theory by discussing its three components:

1. *Orbital Eccentricity*. The *eccentricity* of Earth's orbit refers to the variation in the shape of the orbit over time (Figure 9.21). Recall from Chapter 3 that Earth's orbit is not perfectly circular in its current configuration, but is instead slightly elliptical. The elliptical nature of the orbit results in Earth being about 5 million km (about 3.1 million mi) closer to the Sun at perihelion on or about January 3 than it is during aphelion on or about July 4. This difference amounts to an approximate 6% increase in incoming solar radiation from July to January. Over the course of cycles that last approximately 90,000 to 100,000 years, however, the orbit becomes even more elliptical. When the orbit is more elliptical, the amount of insolation received at perihelion is about 20% to 30% greater than at aphelion. This kind of variability would result in a substantially different climate from what we experience today, with cooler summers in the high latitudes of the Northern Hemisphere than currently occur.

Milankovitch theory *The theory that best explains Pleistocene glacial/interglacial cycles through long-term variations in Earth's orbital eccentricity, tilt, and axial precession.*

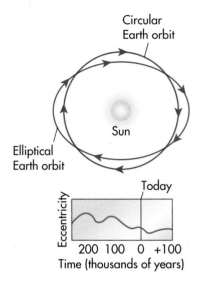

Figure 9.21 Orbital eccentricity. When Earth's orbit is highly elliptical, we receive more radiation at perihelion than at aphelion. This orbital variation occurs on cycles that last about 100,000 years and causes Earth to be approximately 18.27 million km (11.35 million mi) closer to the Sun during orbital perihelion than at aphelion.

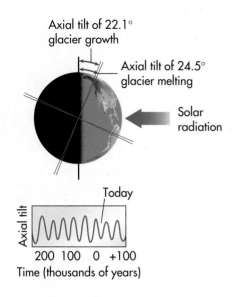

Figure 9.22 Tilt obliquity. The amount of Earth's axial tilt varies over 40,000-year cycles, ranging from 22.1° to 24.5°. When tilt is less, glacial source areas experience cooler summers, allowing glaciers to grow. Seasonality is greatest when axial tilt is high.

2. *Tilt Obliquity.* The term *obliquity* refers to the variations that occur with respect to the axial tilt of Earth. As you know, the axis of Earth is tilted 23.5° from the plane of its orbit around the Sun (the ecliptic). This tilt fundamentally explains the seasonal differences that we experience. During a cycle that averages about 40,000 years, the tilt of Earth's axis varies between 22.1° and 24.5° (Figure 9.22). As tilt increases, the seasonal contrast becomes greater, so that winters are colder and summers are warmer. A tilt of 24°, for example, results in about 8% more solar radiation received at high latitudes in summer. This greater variation occurs because the subsolar point reaches higher latitude when Earth is tilted more. Conversely, when tilt decreases, the difference in seasons is less extreme because the subsolar point is confined to a more narrow latitude range. In the context of continental glaciation, more tilt means warmer summers, which makes it less likely that glaciers can become established. This variable has been linked to the warm Climatic Optimum in the early Holocene (Figure 9.20). On the other hand, conditions are more favorable for glacier growth when tilt is less because summer temperatures are cooler at high latitudes.

3. *Orbital Precession.* The *precession* of Earth's orbit refers to the way Earth slowly wobbles on its axis as it orbits the Sun. Over the course of an approximately 23,000-year cycle, the axis slowly migrates, making

it appear as if the planet is wobbling, as when a spinning top slows down (Figure 9.23). This change alters the orientation of Earth with respect to perihelion and aphelion. If a particular hemisphere is pointed toward the Sun at perihelion, that hemisphere will be pointing away at aphelion, and the difference in seasons will be more extreme. This seasonal effect is reversed for the opposite hemisphere. Currently, northern summer occurs near aphelion.

As discussed previously, isotopic data from ice cores indicate a close correlation between orbital variations and periods of glaciation. Expansion of glaciers seems to occur on 23,000-, 46,000-, and 100,000-year cycles, with one, two, or all of the orbital variables playing a major role within any given glaciation. The 23,000- and 46,000-year cycles are minor fluctuations and are apparently most related to eccentricity of orbit and precession of the equinoxes; in other words, the time of year when a particular hemisphere is tilted toward or away from the Sun. Major glaciation seems to occur every 100,000 years, and although this 100,000-year cycle mirrors the length of the orbital eccentricity cycle, there appears to be no distinct relationship between the two. Instead, the 100,000-year rhythm seems to be most closely related to fluctuations in axial tilt and to the difference in the long time it takes a continental glacier to grow and the relatively short time it takes for the glacier to melt. When ice masses began to expand about 110,000 years ago, for example, axial tilt

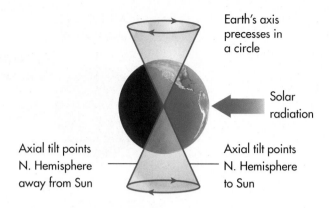

Earth's axis
precesses in
a circle

Solar
radiation

Axial tilt points
N. Hemisphere
away from Sun

Axial tilt points
N. Hemisphere
to Sun

Figure 9.23 Orbital precession. Over the course of a 23,000-year cycle, the Earth's axis slowly wobbles over time. The impact of this wobble is that a gradual change occurs with respect to the hemisphere that is tilted toward or away from the Sun at aphelion and perihelion.

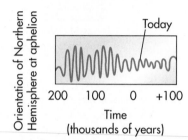

Orientation of Northern
Hemisphere at aphelion

Today

200   100   0   +100
Time
(thousands of years)

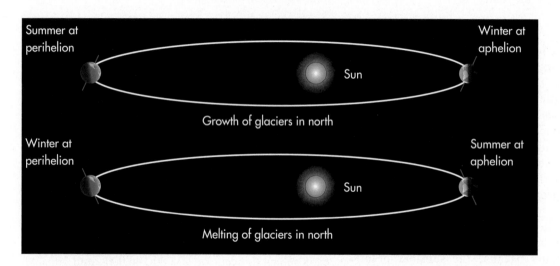

Summer at
perihelion

Winter at
aphelion

Sun

Growth of glaciers in north

Winter at
perihelion

Summer at
aphelion

Sun

Melting of glaciers in north

www.wiley.com/college/arbogast

WILEY PLUS

The Milankovitch Theory

To see how the Milankovitch theory looks in animation, go to the *GeoDiscoveries* website and open the module *The Milankovitch Theory*. This module shows the way that the Earth–Sun geometric relationship changes through time and how it contributes to glaciation. As you watch the animation, notice how variations in orbital eccentricity, axial tilt, and orbital precession result in cooler summers at high latitudes where continental glaciers begin to grow. After you complete the animation, be sure to answer the questions at the end of the module to test your understanding of this concept.

was 22°. Subsequently, it took 80,000 or 90,000 years for the ice sheet to reach its maximum size during the Wisconsin glaciation. About 15,000 years ago, tilt had increased to 24° and the high latitudes received much more summer-time radiation. In the ensuing 10,000 years, the great mass of ice melted almost entirely.

Human Interactions and Future Climate Change

As we saw earlier, numerous fluctuations in Earth's average temperature have occurred in the past. You have also seen how scientists can confidently know that these changes occurred by investigating ancient pollen records, tree rings, and oxygen isotope data derived from ice cores. Understanding these relationships is especially important in the context of ongoing and predicted future climate change on Earth. Although you have probably heard many points of view about this issue, the fact is that the *vast majority* of climate scientists around the world agree that some kind of global climate change is occurring and will continue for the foreseeable future. They further believe that the present warming trend is mainly due to **anthropogenic** (caused by humans) influences associated with greenhouse gas production. The remainder of this chapter is devoted to this topic.

The Carbon Cycle

In order to understand the essence of the global warming issue, it is first necessary to briefly review the carbon cycle and the "greenhouse effect." Recall from Chapter 4 that the various components of the atmosphere, including the constant and variable gases, were investigated. One of the variable gases is carbon dioxide, CO_2. Although this variable gas comprises less than 1% of the atmosphere, it is an important natural regulator of temperature because it traps longwave radiation emitted by Earth, similar to how a greenhouse functions.

Carbon is one of the essential elements of life and is a part of all living things. Simply put, carbon is stored in many different places on Earth and, over time, moves from one *reservoir* to another. Figure 9.24 is an idealized diagram of how the carbon cycle works, with storage and transfer amounts in billions of metric tons. A simple example of a transfer is the flow of carbon that occurs when a blade of grass dies and decomposes. During its life, carbon was stored within the grass fiber. After the blade of grass dies, however, it decomposes and the carbon within it (among other things, such as minerals) is gradually incorporated into the soil, where it is stored for a period of time.

Anthropogenic *Environmental changes caused by humans.*

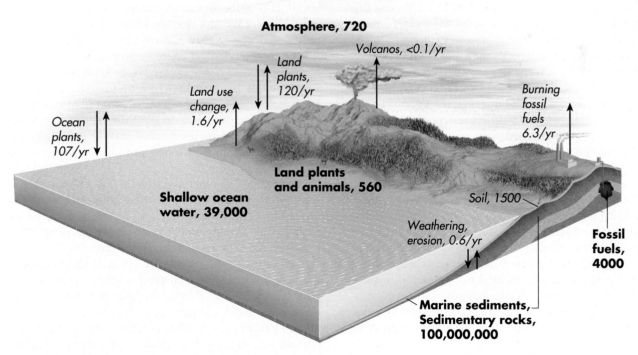

Figure 9.24 The carbon cycle. The carbon cycle represents the flow of carbon that occurs on Earth as it moves from one reservoir to another. Note the various places where carbon is stored and the amount of carbon contained within these reservoirs. (Reservoirs are noted in boldface type, transfers in italic type.)

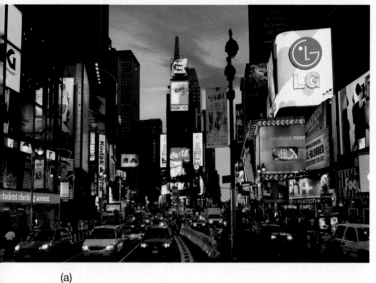

(a)   (b)

Figure 9.25 Powering the global economy. (a) Times Square, New York City, at night. A tremendous amount of electricity is required to power the urban landscape. (b) Traffic in Saõ Paulo, Brazil. This type of scene, with thousands of cars burning gasoline, is played out in countless places around the world at any given point in time.

The carbon cycle progresses fairly rapidly, by geologic standards of time, and can be measured in years to centuries. Nevertheless, it is thought that less than 1% of the total amount of Earth's carbon is actively within the cycle at any moment in time. If you examine Figure 9.24 again, you can see, for example, that about 120 billion metric tons of carbon transfer back and forth between plants and the atmosphere every year. (A metric ton is 1000 kg; multiply by 1.1 to get the equivalent number of English tons, each of which is 2000 lb.) Although this sounds like a large number, it is small when you consider that 100 million billion metric tons of carbon are stored in marine sediments and rocks, mostly as calcium carbonate ($CaCO_3$). These carbonates are largely the remains of dead organisms that accumulated on ocean bottoms millions of years ago. Other large quantities of carbon are contained within coal and petroleum deposits that essentially consist of decomposed plant and animal matter.

Fossil fuels are carbon-containing energy sources such as coal, natural gas, and petroleum that are extracted from the ground in mines and wells. The term *fossil fuels* comes from the fact that these compounds were once living organisms that died and slowly fossilized due to intense pressure and heat, turning into other carbon compounds over thousands and millions of years. Simply put, a tremendous amount of energy is required to power the complex and interconnected web of factories, automobiles, shopping malls, home appliances, and every other conceivable modern convenience in the world today (Figure 9.25). The vast majority of the energy required in

this industrial and technological age is derived from the consumption of fossil fuels in one form or another. To put this thought in perspective, consider that 98,000 kg (98 tons/216,000 lb) of prehistoric plant material is required to produce 3.8 L (1 gal) of gasoline. Looked at another way, you need to load the prehistoric equivalent of 16.2 hectares (40 acres) of wheat—stalks, roots, and all—into the tank of your car just to drive 20 mi! In this context, the United States currently leads the world by consuming about 25% of the annual global energy supply. This lead exists despite the fact that the United States is home to only 4% of the world's approximately 6 billion people. Of the total amount of energy consumed in the United States in 2008, 85% was derived from fossil fuels.

Through this global reliance on fossil fuels, humans have dramatically increased the amount of atmospheric carbon dioxide over time. In 1950, for example, approximately 1.6 billion metric tons of carbon were released into the atmosphere through industrial pathways. By 1997 this transformation of fossil fuel to atmospheric carbon dioxide had increased 400% to 6.3 billion metric tons. Approximately 50% of these total emissions has moved from the atmosphere and is stored in reservoirs such as the oceans and forests. The remaining 50%, however, has remained in the atmosphere.

You can see the cumulative effect of this pattern in the proportional change in the amount of preindustrial atmospheric carbon dioxide to current values. In 1850 there were about 288 parts of carbon dioxide per every million part of the overall atmosphere. By 2010 this proportion had increased to almost 390 parts per million (ppm) (Figure 9.26). This may not seem like much of an increase, but when you consider the importance of atmospheric carbon dioxide in the natural regulation of global

---

Fossil fuels *Carbon-based energy sources, such as gasoline and coal, that are derived from ancient organisms.*

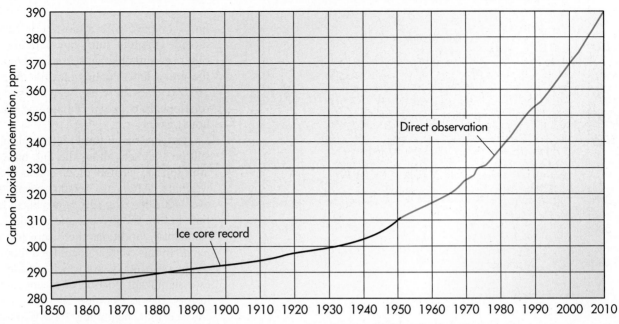

Figure 9.26 Changes in atmospheric carbon dioxide since 1850. Carbon dioxide levels have increased over 30% since 1870 and continue to rise. The black line represents atmospheric concentrations derived from ice cores. The red line is derived from actual atmospheric measurements taken by the Scripps Institute of Oceanography. (*Sources:* Pre-1958: A. Neftle et al., Historical CO_2 record from the Siple Station ice core, in *Trends: A Compendium of Data on Global Change*, 1994, U.S. Department of Energy. Post-1958: Earth Systems Research Laboratory (ESRL) National Oceanic and Atmospheric Administration (NOAA).)

temperature, it is significant. This increase is likely to continue in the foreseeable future when densely populated developing countries such as India and China, which between them contain about 2.3 billion people, fully industrialize and use the same fossil fuel power sources that have worked so well for us.

Is Anthropogenic Climate Change Really Occurring?

The "million dollar question," of course, is whether or not human-induced climate change is really occurring. In this context, you might first ask: "Is the climate really warming?" With respect to this initial question, the evidence strongly suggests that Earth is indeed in a warming phase. You can see the evidence of this warming by examining Figure 9.27. In this graph, the data clearly show that the observed global surface air temperature has increased about 1°C (1.6°F) since 1900. In fact, the 2000s was the warmest decade ever recorded. This increase in temperature correlates very closely with the overall increase in atmospheric carbon dioxide that has occurred at the same time. Much of this

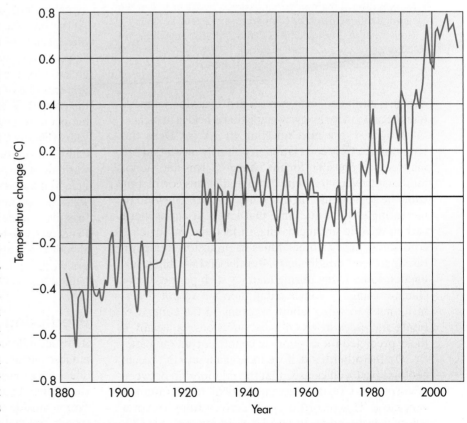

Figure 9.27 Global temperature change since 1880. This graph shows the change in average global temperature from 1880 to 2006. Note the rapid warming in the past 40 years. (*Source:* J. Hansen, M. Sato, R. Ruedy, K. Lo, D.W. Lea, and M. Medina-Elizade. "Global Temperature Change." *Proceedings of the National Academy of Sciences* 2006; 103:14288–14293.)

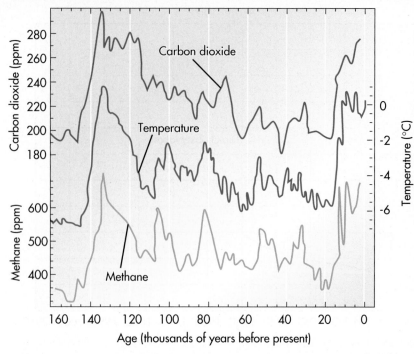

Figure 9.28 Greenhouse gases and prehistoric temperature from the Vostok ice core in Antarctica. This graph shows the close relationship between atmospheric temperature (red line), carbon dioxide (blue line), and another (lesser) greenhouse gas (methane), which is the green line. Temperature is the number of degrees above or below the average modern surface temperature of about 14°C (55°F). (*Source:* J. M. Barnola, D. Raynaud, Y. S. Korotkevich, and C. Lorius. Vostok ice core provides 160,000-year record of atmospheric CO$_2$. *Nature* 1987; 329: 408–414.)

change is attributed to an increase in average low temperatures because more longwave radiation is held in at night.

However, the next question to ask is, "Does this temperature/CO$_2$ correlation necessarily mean that humans are responsible for the change?" This question is more controversial. Skeptics argue that this correlation *could* be coincidental, in that temperature *happens* to be increasing at the same time that levels of atmospheric carbon dioxide are increasing. In fact, the geologic record shows that at other times in prehistory temperature increased rapidly *and* naturally, and we might just happen to be living during another such period. If this climate change is coincidental, then we do not necessarily need to worry about warming in the context of future increases in greenhouse emissions, especially if those predicted for developing countries come to pass.

On the other hand, if we look again into the recent geologic past, you can see that the relationship between atmospheric CO$_2$ and average global temperature is very close. This record is again derived from ice cores where measurements of past levels of atmospheric CO$_2$ can be measured from air bubbles trapped in layers of glacial ice. The graph in Figure 9.28 shows this comparison vividly. Look at this diagram carefully and notice how well peaks and valleys in temperature (the red line) compare with atmospheric carbon dioxide (the blue line) and methane (another greenhouse gas, represented by the green line). Notice that for the past 160,000 years there has been a very close relationship between climate and greenhouse gases.

After viewing these data, you might still be skeptical about the link between atmospheric carbon dioxide and global warming. After all, 160,000 years ago, even 100,000 years ago, seems like a long time before the present and maybe environmental conditions then were just *different*. In this context, focus for a moment on the past 1000 years, which encompasses a significant portion of recent human history. If you examine Figure 9.29, you can see that, beginning 1000 years ago, the temperature was about 0.2°C (0.4°F) cooler than the average between 1961 and 1990. Subsequently, between 1000 and 1900 the average temperature *further cooled* about another 0.2°C (0.4°F). This cooling was most intensive during the *Little Ice Age*, which occurred from the 14th to 19th centuries.

Now, look at the change that occurred around 1900. At this time the temperature began to warm quickly to its present state of being 0.7°C (1.3°F) warmer than the past 30-year average. In other words, the natural climate progression for the 850 years prior to the Industrial Revolution was one of gradual cooling, with a rapid warming in the past 100 years. Given the demonstrable link between temperature and greenhouse gases for the past 160,000 years (Figure 9.28), coupled with the fact that the climate had been *cooling* between 1000 and 1850 A.D., the rapid warming of the past 100 years seems to be best explained by human-induced greenhouse gas emissions. At least, that is what the vast majority of climatologists believe. The effort to explain this relationship is led by the Intergovernmental Panel on Climate Change (IPCC), which is an international body of scientific experts that was established in 1988 by the World Meteorological Organization (WMO) and the United Nations Environmental Programme (UNEP).

Predicting Future Climate Change

If the one million dollar question is whether or not human-induced climate change is occurring, then the two million dollar question is "How much future warming will occur?" Remember that the current proportion of atmospheric carbon dioxide is about 390 per million parts of the atmosphere, and that this proportion exists with the developed world producing 73% of the excessive carbon dioxide. The big unknown is how climate will respond when the developing countries increase their carbon dioxide emissions by the predicted 25% in the next few decades.

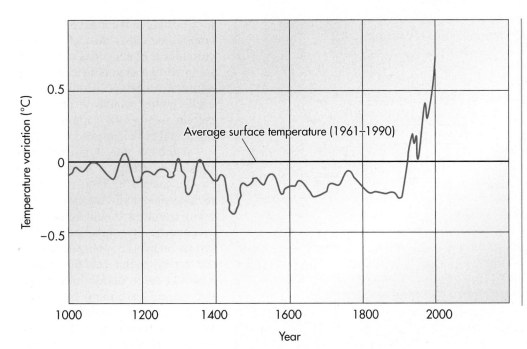

Figure 9.29 Average global temperature change in the past 1000 years. Temperature cooled slightly between 1000 years ago and approximately 1850. Since that time, temperature has been warming dramatically. The baseline (0) for these comparisons is the average temperature from 1961 to 1990. (*Source:* M. E. Mann, R. S. Bradley, and M. K. Hughes. Northern Hemisphere temperatures during the past millennium: Inferences, uncertainties, and limitations. *Geophysical Research Letters* 1999; 26:759–762.)

In the context of understanding future change in an enhanced greenhouse world, the scientific challenge is to somehow predict how the climate system will respond. This complex task is best accomplished with a general circulation model (GCM), which is a mathematical model that incorporates known climate variables, such as cloud cover, insolation, latitude, and proximity to oceans, to forecast future climate events. Using these variables, large computers can run simulations involving scenarios of low or high carbon dioxide levels. Models that run moderate-case scenarios use a doubling of atmospheric carbon dioxide as a benchmark.

With this benchmark in mind, the IPCC has predicted potential amounts of warming at 2100 based on a variety of global economic and social scenarios. One such scenario is based, in part, on a rapidly growing global population (to 15 billion) and limited international cooperation with respect to carbon emissions. Figure 9.30 illustrates the geography of warming in this scenario. Note that above average levels of warming are predicted over much of Earth, with much greater than average warming expected at high latitudes of the Northern Hemisphere, especially during the winter months.

What would happen if Earth really does warm to the levels indicated in Figure 9.30? Much of what occurs depends on the nature of climate feedbacks, which are indirect or secondary changes that take place within the overall climate system in response to the initial forcing mechanism—in this case, increased atmospheric carbon dioxide. For example, it is possible that increased warming will cause more water to evaporate from oceans, resulting in more atmospheric vapor. This increase, in turn, can have a larger effect on the global energy balance, both as a *positive climate feedback* that further increases warming or potentially as a *negative climate feedback* that causes

cooling. Increased atmospheric water vapor could cause a positive feedback because it traps solar energy, which would further warm Earth. On the other hand, increased atmospheric vapor will increase the number of clouds, which could increase the amount of sunlight that reflects back into space, thus resulting in a net cooling effect.

The ultimate positive feedback is a runaway greenhouse effect where significant warming causes the polar ice caps to melt and thus less radiation is reflected to space, which, in turn, results in even more warming. In contrast, the ultimate negative feedback is that melting ice caps cause warm-water currents such as the Gulf Stream to stop flowing as far north because of the rapid influx of cold water at high latitudes. Some models suggest that Earth might suddenly shift into another ice age if this feedback occurs. Fortunately, neither the ultimate positive nor negative feedback is considered likely.

Although the response of the Earth's future climate to feedbacks is uncertain, the fact is that the IPCC and vast majority of climatologists believe that warming will continue at least for the next century. Some scientists argue that no major environmental consequence will happen—that, in fact, a warmer world might be a good thing. In the context of this positive outlook, those who live in the northern tier of states in the United States might find some warming a welcome respite from the usually harsh winters that occur there. A potential benefit may have already occurred in the Great Lakes region, where the annual length of the growing season increased by about 1 week in the 1990s.

Although some "environmental benefits" may occur in a warmer world, these data suggest that several negative results will probably occur. One of these negative effects is a rising sea level, which would result from increased melting of polar ice caps and thermal expansion of water. According to current predictions, sea level will

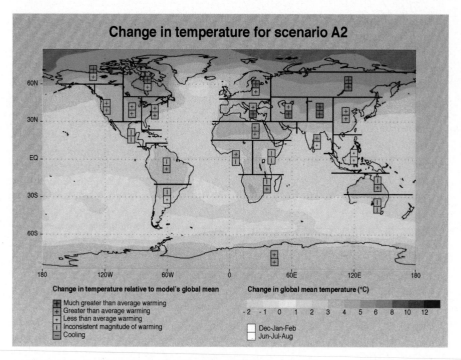

Change in temperature for scenario A2

Change in temperature relative to model's global mean

Change in global mean temperature (°C)

-2 -1 0 1 2 3 4 5 6 8 10 12

- ⊞ Much greater than average warming
- ⊞ Greater than average warming
- ⊞ Less than average warming
- ⊡ Inconsistent magnitude of warming
- ⊟ Cooling

☐ Dec-Jan-Feb
☐ Jun-Jul-Aug

Figure 9.30 Predicted climate change by 2100. This pre-dicted outcome is based on a specific social and economic scenario and shows the level of warming (relative to current average temperature) in December–February (upper box of box pair) and June–August (lower box of box pair). (*Source:* Intergovernmental Panel on Climate Change.)

rise 15 cm to 90 cm (5.11 in. to 35.4 in.) by 2100, which will place 92 million people around the globe within the risk of coastal flooding. Another potential negative fac-tor is that climate change may promote the spread of in-sect-borne diseases such as malaria. Climate boundaries may also shift, making some regions drier while others become wetter. This boundary shift may cause signifi-cant displacement of human populations due to changes in agricultural patterns. Still another potential change is that climate variability may be enhanced, with a higher frequency of extreme events such as snowstorms, rain-storms, and intense spells of hot and cold weather. In fact, data suggest that these kinds of extreme events have been occurring more frequently since 1980. Some scientists even see a link between the current warming phase and a higher incidence of hurricanes.

Because it appears that global warming really is oc-curring, and many negative consequences may result, world leaders have begun efforts to control greenhouse emissions. The first truly global effort occurred at the 1992 Earth Summit in Brazil, where 150 nations signed a treaty that limited the emissions of greenhouse gases. A subsequent, more comprehensive, global treaty was pro-posed in 1997 at Kyoto, Japan. The resulting Kyoto Proto-col calls for the 38 leading industrial nations, including the United States, nations of the European Union, and Japan, to reduce greenhouse emissions to 5% to 8% below 1990 levels. These reductions could be ac-complished either through outright curtailment of emissions or through using carbon credits as a currency between countries. In other words, each country would be allowed a certain number of carbon credits that could be bought, sold, or traded.

Although the United States signed the treaty, it was rejected by the Senate because it did not include guidelines for how devel-oping countries should adjust their greenhouse gas emissions. Accord-ing to many U.S. political leaders, the treaty is thus unfair because it would cause undue hardship to the national economy by forcing businesses and corporations to de-velop new technologies that would reduce carbon emissions. Develop-ing these technologies would add cost to products produced in the United States, which would theoretically put us at a competitive disadvantage compared to developing countries that have no carbon standards. In response to this argument, leaders of the developing countries argue that the increased levels of atmospheric carbon dioxide are the product of just a few countries such as the United States and Europe, and thus we should take the lead regarding carbon standards.

Several new factors emerged in 2009 to play a role in the climate issue. The U.S. House of Representatives passed the American Clean Energy and Security Act of 2009, which establishes a so-called cap and trade system that limits the amount of greenhouse emissions compa-nies can produce. As of this writing, the law had yet to be passed by the Senate and signed by the President. This bill assumes that some businesses can reach these limits more easily than others, which then allows them to trade emission permits to the less efficient companies. In other words, less efficient companies are allowed to emit more greenhouse gases because more efficient companies pro-duce fewer such emissions. Ironically, the legislation was passed during one of the coolest summers on record in the northeastern United States, leading climate skeptics to vigorously argue that the new law is unnecessary. Pro-ponents of the legislation noted, however, that the sum-mer of 2009 was the *second warmest* on record for the Earth as a whole and that Arctic sea ice continues to melt at alarming rates. In this context, another international climate conference occurred in Copenhagen, Denmark, in November 2009 to further address these issues. No firm commitments emanated from that conference. In the meantime, greenhouse emissions continue to rise and it is safe to say that the issue of climate change will be a prominent environmental debate for the rest of your life.

1. A major issue facing the world today is global warming, which most climatologists agree is related to human enhancement of the natural greenhouse effect.

2. The carbon cycle refers to the way that carbon is stored in various reservoirs (such as the atmosphere, ocean, and rocks) and how it moves between them.

3. In order to understand the extent of potential future climate change, it is necessary to understand prehistoric climate variability. This can be reconstructed by indirect methods, such as pollen analysis and dendrochronology.

4. Since the onset of the Industrial Revolution in the mid-1800s, atmospheric carbon dioxide has increased about 30%. This increase is almost entirely due to human activities.

5. Between the years 1000 and 1850 A.D., global climate cooled slightly. Since 1850 it has been warming.

6. General circulation models predict that average global temperature will increase 2°C to 4°C (4°F–7°F) this century.

www.wiley.com/college/arbogast

The Greenhouse Effect and Global Climate Change

The issue of ongoing and future climate change is complex and can be seen more clearly in animation. To do so, go to the *GeoDiscoveries* website and open the module *The Greenhouse Effect and Global Climate Change*. This module demonstrates how greenhouse gases such as carbon dioxide trap longwave radiation emitted by Earth. It subsequently walks you through the logic behind climate change science. The last part of the module is an excellent visualization of average global temperatures since the late 19th century and how they are projected to be in one scenario by 2100. It is impressive. After you complete the module, be sure to answer the questions at the end to ensure you understand this concept.

THE BIG PICTURE

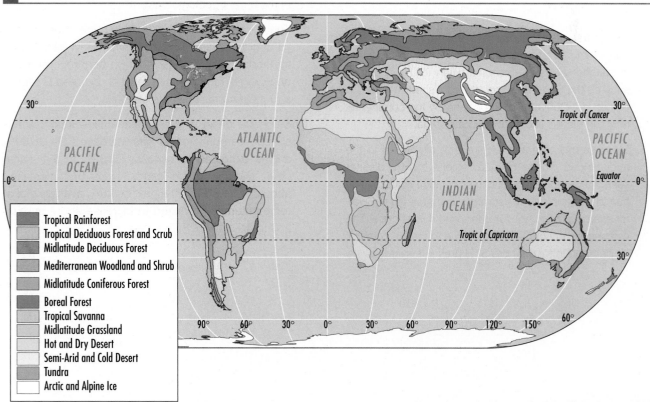

Tropical Rainforest
Tropical Deciduous Forest and Scrub
Midlatitude Deciduous Forest
Mediterranean Woodland and Shrub
Midlatitude Coniferous Forest
Boreal Forest
Tropical Savanna
Midlatitude Grassland
Hot and Dry Desert
Semi-Arid and Cold Desert
Tundra
Arctic and Alpine Ice

With this chapter, you have completed the focus on atmospheric processes. As noted earlier, this chapter marked the first time in this book that we began to see how geographic patterns can exist on the ground, here in the form of vegetation as it responds to climate. The next chapter will build on this transition by focusing on plant geography and the many variables that influence the distribution of plants, both at the global and local scales. You will find that a close correlation exists between global climate and vegetation patterns. Look at the map of global vegetation on previous page, for example, and see how the patterns resemble those in Figure 9.2. The next chapter will investigate these relationships, as well as more local influences.

SUMMARY OF KEY CONCEPTS

1. Climate refers to the average precipitation and temperature characteristics of a region. This depends on many interrelated variables, including latitude, insolation, seasonality, atmospheric circulation, and topographic effects.

2. The Köppen climate system categorizes global climates based on their average annual temperature and precipitation characteristics. Six major categories are recognized: tropical humid climates (A), arid/semi-arid climates (B), mesothermal climates (C), microthermal climates (D), polar climates (E), and highland climates (H).

3. The humid tropical climates (A), such as the tropical rainforest climate (Af), have a low annual temperature range and abundant precipitation caused by the Intertropical Convergence Zone (ITCZ). Precipitation becomes distinctly seasonal with increased latitude in these regions.

4. The arid/semi-arid climates (B) are regions of very low rainfall and sparse vegetation. This low precipitation is caused by the Subtropical High (STH) in places such as the Sahara Desert, the rain shadow effect in places like Denver, Colorado, and by the continental effect in places like north-central China.

5. Mesothermal climates (C) are typically found in the midlatitudes. These climate regions have a distinct seasonal temperature and precipitation pattern. The humid subtropical hot-summer climates (Cfa, Cwa), for example, have hot summers and mild winters. The wet season occurs in summer due to the influx of mT air pumped in by offshore STH pressure systems.

6. Microthermal (D) climates are largely continental climates that occur only in the Northern Hemisphere. These climate regions have hot to mild summers and cold winters. Average annual precipitation is moderate and is usually associated with midlatitude cyclones.

7. Polar (E) climates occur at very high latitudes and are associated with tundra and ice caps. Although these regions have a distinct annual temperature range, they are generally cold with low to moderate precipitation. Highland climates occur in mountainous areas and have characteristics of subarctic and polar climates.

8. The issue of climate change is currently highly visible on the national and international stage. In order to understand the nature of future climate changes, it is important to understand changes that have occurred in the recent geologic past.

9. The Quaternary Period includes the most recent ~1.6 million years of Earth's history. Several scientific methods, such as palynology, dendrochronology, and oxygen isotope analyses, can be used to reconstruct Quaternary climate changes.

10. Since the onset of the Industrial Revolution, atmospheric carbon dioxide has increased about 30% and is expected to increase further. In that same time, average global temperature has increased about 1°C (1.8°F). In the next century, average global temperature is predicted to increase about another 2°C to 4°C (4°F–7°F).

CHECK YOUR UNDERSTANDING

1. Compare and contrast the concepts of weather and climate.

2. How does latitude influence the distribution of global climates?

3. Discuss the variables that are included within the Köppen climate system.

4. Why does the wet-equatorial climate receive consistent rainfall throughout the year?

5. In contrast to the wet-equatorial climate, the tropical wet-dry climate has a distinct dry season. What time of year does the dry season occur and why does it happen?

6. The Mediterranean and marine west-coast climates are similar in that both have moderate annual temperatures. However, the marine west-coast climate receives more precipitation. Why?

7. Why is the tropical desert climate dry over the course of the entire year?

8. The Subtropical High (STH) Pressure System is usually associated with dry climates, but, in the case of the humid subtropical climate, it contributes to summer rainfall. Why does this pattern occur?

9. Both the moist-continental and boreal-forest climates have a large annual range of temperature. Why does this variability occur?

10. How can ancient plant pollen be used to reconstruct past climate changes?

11. How can tree rings be used to reconstruct past climate changes?

12. Describe the logic behind the use of oxygen isotopes to reconstruct Quaternary climate changes. What does this record indicate for the past ~1.9 million years?

13. Describe the three components of the Milankovitch theory. How does a change in each of these variables increase the possibility of glaciation?

14. Why do increased levels of atmospheric carbon dioxide contribute to global warming?

15. What is the evidence that there has been a close relationship between atmospheric carbon dioxide and climate change in the prehistoric past?

16. What role has the United States played in the enrichment of atmospheric carbon dioxide?

17. What are the projected levels of future global warming and some of the potential impacts?

ANSWERS TO VISUAL CONCEPT CHECKS

VISUAL CONCEPT CHECK 9.1

The answer is *c*; the area is dominated by the Subtropical High (STH) Pressure System. This pressure system consists of descending air that warms adiabatically. Thus, it is dry in the Middle East.

VISUAL CONCEPT CHECK 9.2

The lower 48 states can be subdivided into three general zones of precipitation. The southeastern United States is a zone of high precipitation that originates from moisture from the Gulf of Mexico. This moisture is pumped into the southeastern states by the Subtropical High (STH) Pressure System. The western part of the United States is fundamentally dry, mostly because this area lies in the rain shadow of the Rocky, Cascade, and Sierra Nevada Mountains. The West Coast of the United States is an area of high precipitation because orographic processes release moisture that flows inland from the Pacific Ocean.

| # PLANT GEOGRAPHY

The preceding chapter focused on the geographical distribution of global climates. One way this distribution was illustrated was by noting the type of vegetation in a particular climate zone. For example, dense forests occur in humid regions because it is sufficiently wet to support extensive stands of trees. In contrast, vegetation is quite sparse in a place like the Sahara Desert because it is very dry.

With the foundation of climate in mind, this chapter focuses more specifically on the geographical distribution of plants on Earth and the variables that influence these patterns. As a result, this point marks an important transition within this text because the discussion turns entirely to the patterns of environmental variables *on the ground*, rather

than within the atmosphere or oceans. The remainder of this book will focus on these kinds of geographic distributions through the study of variables such as soils, rivers, and the landscapes shaped by tectonic forces, wind, and ice. For now, this chapter will illustrate why particular kinds of plant communities occur where they do on Earth.

CHAPTER PREVIEW

Ecosystems and Biogeography

The Process of Photosynthesis
  On the Web: *GeoDiscoveries*
  Photosynthesis and Respiration

Some physical geographers study the spatial distribution of plants, including why some plants grow in some places and not in others. This aerial photograph shows the dense rainforest of French Guyana in South America. What geographical variables result in this kind of forest at this location? Certainly, the role of climate is critical, as you saw in Chapter 9. Other factors are important as well, which you will investigate in this chapter.

The Relationship of Climate and Vegetation: The Character and Distribution of Global Biomes

Local and Regional Factors That Influence the Geographic Distribution of Vegetation
    On the Web: *GeoDiscoveries Plant Succession*

Human Interactions: Human Influence on Vegetation Patterns
    On the Web: *GeoDiscoveries Deforestation*
    On the Web: *GeoDiscoveries Remote Sensing and the Biosphere*

The Big Picture

LEARNING OBJECTIVES

1. Define the concept of an ecosystem and its various components.

2. Understand the process of photosynthesis and how it relates to biomass construction.

3. Explain the concept of biomes.

4. Describe the geography and fundamental characteristics of the 12 global biomes.

5. Discuss how local and regional factors, including slope, aspect, vertical zonation, and riparian environments, influence vegetation.

6. Explain the concept of plant succession.

7. Describe the patterns of deforestation on Earth and some of its potential consequences.

Ecosystems and Biogeography

Recall from Chapter 1 that one of the subdisciplines of physical geography is *biogeography*. Biogeography examines the geographical distributions of organisms, their habitats, and the environmental or historical factors that produce them. As a result, biogeography is very closely related to ecology and thus focuses a great deal on **ecosystems**, which are communities of organisms that interact in interdependent ways (Figure 10.1). On a broad scale, the organisms within a particular land ecosystem can range from the smallest microbe living within the soil to the complex animals that roam the landscape. Ecosystems are not just confined to terrestrial settings, however, as they also occur within marine environments.

When considering the nature of any particular ecosystem, biogeographers often try to understand its various components. One important factor in an ecosystem is the *abiotic* or nonliving part of the system. This part of the system sets the environmental context for all the living organisms in the area. An excellent example of an abiotic factor is the climate of a particular region, or the types of rocks and minerals found in a specific place, to name but two.

As far as the living things are concerned, biogeographers may first consider the role of **autotrophs**, which are organisms that make their own food from light energy or chemical energy without eating. Although phytoplankton and some bacteria are autotrophs, the most important self-feeding species are plants. Yet a third component of an ecosystem is **heterotrophs**, which are animals that consume plants or other animals for survival. Heterotrophs, of course, are subdivided into three categories, including *herbivores* that feed on plants, *carnivores* that eat other animals, and *omnivores*, which consume both plants and animals.

The final component of any ecosystem is **decomposers**, which feed on dead plant and animal material. Decomposers are critical because they promote the return of minerals and organic material to soils and water that other plants and animals can subsequently consume. When ecosystems are viewed holistically, ecologists and biogeographers often consider the feeding

---

Ecosystems *A community of plants, animals, and microorganisms linked by energy and nutrient flows.*

Autotrophs *Organisms that synthesize their own food using heat or light as the source of energy.*

Heterotrophs *Organisms that consume complex organic substances for food.*

Decomposers *Organisms that consume dead or decaying organic substances for nutrition.*

Figure 10.1 Components of a forest ecosystem in northern North America. Grizzly bears hunt for salmon along a stream in Alaska. What are the various components of this ecosystem and how do they fit together within a holistic perspective?

Food Chain

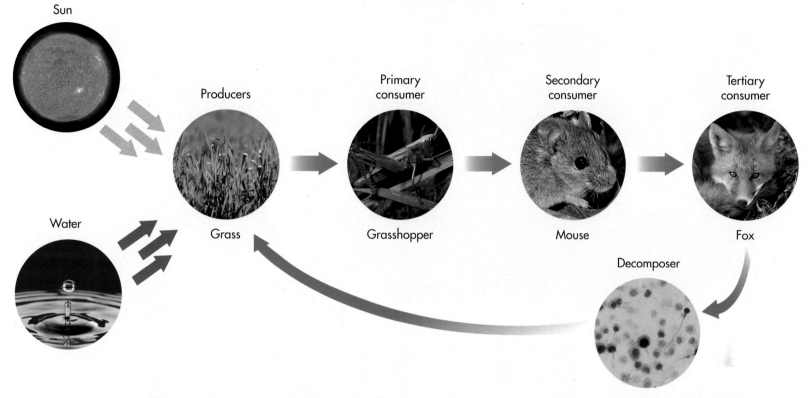

Figure 10.2 An example of a food chain. Note the various levels of consumption in this chain and how it ultimately cycles back to basic producers.

pattern (*or trophic structure*) within the community. The central concept within this pattern is the *food chain*, which can include many levels. An example of a food chain with five levels would start with grass followed by grasshopper, mouse, fox, and fungi that consume the fox when it dies (Figure 10.2).

Although all components of ecosystems are important, plants are usually the most visible living part of any landscape. They are certainly the most important autotrophs and are critical to life on Earth because they convert light energy to chemical energy through the process of **photosynthesis**. This production is the foundation of the food chain. Given the critical role that plants play on Earth, and their highly visible nature, the remainder of the chapter focuses on the global distribution of plant communities. The discussion begins with a more in-depth examination of the way that solar energy is converted to organic molecules through the process of photosynthesis.

Photosynthesis *The conversion of solar radiation into chemical energy. Sugars and starches are produced from carbon dioxide and water, through the interaction of light and chlorophyll in green plants. The process releases oxygen into the atmosphere.*

The Process of Photosynthesis

Photosynthesis requires the presence of water, sunlight, and carbon dioxide. Carbon dioxide is essential because it is converted to carbohydrate compounds known as sugars. Light energy is absolutely necessary because it is the power that drives the conversion of carbon dioxide. This process requires a water medium, which is contained within the plant fibers, because water provides the electrons required for the conversion to take place. In this fashion, light energy from the Sun is locked up in chemical forms that are subsequently used elsewhere in the food chain. This conversion is represented in the following equation:

$$6H_2O + 6CO_2 + \text{light energy} \rightarrow C_6H_{12}O_6 + 6O_2$$
$$\text{water} + \text{carbon dioxide} + \text{glucose (sugar)} \rightarrow \text{oxygen}$$

In other words, six molecules of water, plus six molecules of carbon dioxide, plus light energy produce one molecule of glucose (sugar), plus six molecules of oxygen. At such a scale, the result of this chemical reaction may not seem impressive, but approximately 91 billion metric tons (100 billion tons) of carbon dioxide are removed from Earth's atmosphere each year and each plant produces millions of these sugar molecules each second.

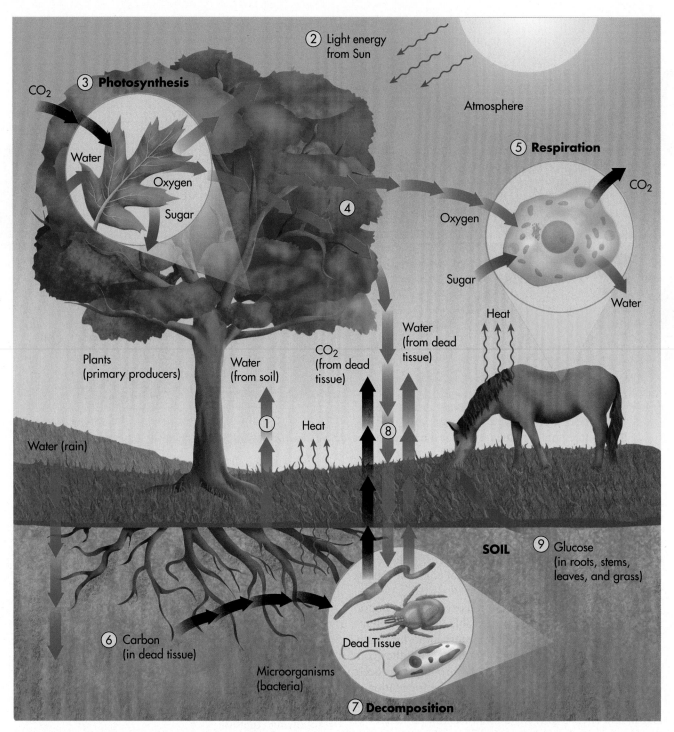

Figure 10.3 Photosynthesis, respiration, and decomposition cycles. Water, oxygen, and carbon dioxide take various pathways within the biosphere. Plants are part of these pathways through the processes of photosynthesis and decomposition, while animals (including people) are part of respiration and decomposition processes.

Figure 10.3 shows the photosynthetic and respiratory cycles, including the various pathways that oxygen, water, and carbon dioxide take within the system. As you work through this discussion, follow the numbered stages in the diagram. These numbers are not meant to imply that stages occur in a specific order, but rather, they help organize the flow of the presentation. For simplification, think of the cycle as broadly consisting of primary producers, such as plants, and decomposers, such as bacteria. Plant consumers are also present, but they are not really relevant to the photosynthetic cycle.

A logical place to begin to trace the various pathways is in the soil, which provides water to the living plant (1). At the same time the roots of the plant are absorbing water, the green leaves of the plant absorb solar radiation (2). Compounds in the plant's leaves called chlorophylls and carotenoids are important ingredients in the process because they absorb sunlight of specific wavelengths. Chlorophylls absorb wavelengths in the blue and red part of the electromagnetic spectrum, whereas carotenoids absorb blue-green light (carotenoids appear orange or yellow).

Figure 10.4 shows how these patterns of electromagnetic absorption take place. Notice that plants do not absorb energy in the green and orange parts of the spectrum. Instead, these wavelengths are either reflected or pass through the plant's leaves. The green wavelength is most actively reflected, which is why plants appear to be green to the human eye. This energy absorbed in the blue and red wavelengths is critical to photosynthesis because it excites chlorophyll and carotenoid electrons within the plant.

Turn back to Figure 10.3. Stage 3 (upper left) shows the photosynthetic process where carbon dioxide is absorbed by the plant and converted to sugars by light energy in the water medium. A by-product of this process is oxygen, which is released into the atmosphere (4). Excess CO_2 is also released to the atmosphere or used in the photosynthetic process. Oxygen that is released is then absorbed by animals, which produce CO_2 as a by-product of their own respiration (5). When plants and animals die, their remains fall to the ground and are consumed by decomposers (6). As the decompos-

ers absorb oxygen through respiration from the atmosphere or soil (7), they combine the oxygen with decomposing carbohydrates, releasing energy contained within CO_2 and water vapor that are emitted into the atmosphere (8).

Although most of the solar energy absorbed by plants is used for respiration, a great deal of it is used in photosynthesis to produce glucose, which is retained in the plant to build and sustain leaves, fruits, and seeds. In addition, plants convert glucose to cellulose, which is the structural material used to build cell walls. Although much of the produced glucose is used to sustain the plant, a great deal of it is stored as starches and other carbohydrates in roots, stems, leaves, and grass (9), where it can be used later. Some of these carbohydrates subsequently provide energy for animals after they are consumed.

Scientists who study plant production are called *plant ecologists*. Plant ecologists are often interested in the net amount of vegetation produced by photosynthesis. An important indicator of plant productivity is **biomass**, which is the dry weight of all living organisms in a given area. This measure is important because it reflects the amount of chemical energy that has been stored and is thus available for consumption. Biomass is typically measured in kilograms per square meter of ground or as metric tons per hectare. Given the distribution of global climates, it should be no surprise that biomass var-

---

Biomass *The amount of living matter in an area, including plants, animals, and insects.*

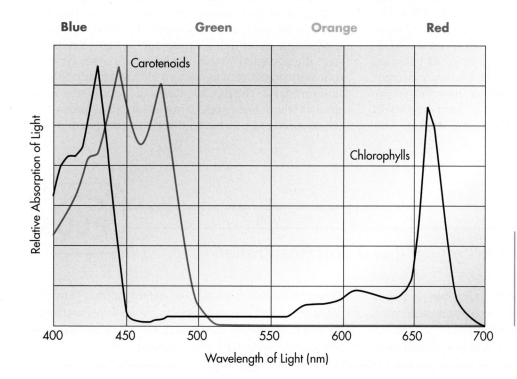

Figure 10.4 Preferential absorption of solar wavelengths in plants. Note that blue and red wavelengths are actively absorbed, whereas the green and orange parts of the spectrum are not.

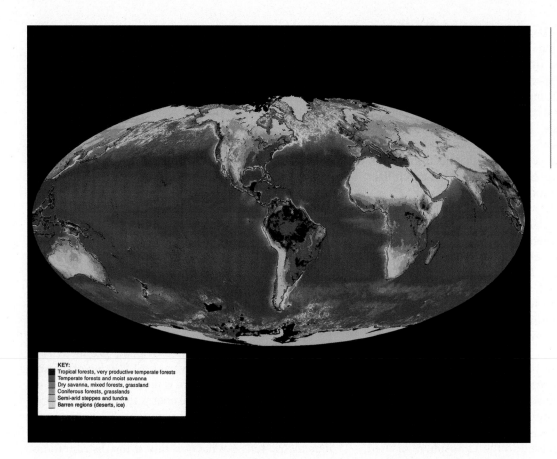

Figure 10.5 The global biosphere. This composite image was acquired by the NIMBUS-7 and AVHRR satellites over a period of 8 years. It shows the geographic distribution of potential global biomass.

KEY:
Tropical forests, very productive temperate forests
Temperate forests and moist savanna
Dry savanna, mixed forests, grassland
Coniferous forests, grasslands
Semi-arid steppes and tundra
Barren regions (deserts, ice)

ies greatly across geographic regions. Forests have the highest biomass, whereas grasslands have less.

In the context of global biomass, examine Figure 10.5, which is a composite satellite image that shows the geography of global biomass as constructed from data collected from 1978 to 1986. The ocean portion of the figure is a composite of more than 60,000 images collected by a sensor on the NIMBUS-7 satellite. This sensor measured the distribution and amount of phytoplankton, which are microscopic plants that grow in the upper part of the ocean where the Sun penetrates. Red and orange colors represent high concentrations of phytoplankton, whereas yel-

low to violet colors reflect progressively lower concentrations. The land vegetation image is a composite collected during 15,000 orbits by the Advanced Very High Resolution Radiometer (AVHRR). This sensor measured radiation from the land surface, which is then used to estimate vegetation production. The dark green areas are rainforests that have the highest growth potential, whereas the lighter greens reflect the potential vegetation growth in other tropical and subtropical forests, as well as midlatitude forests and farmland. The lightest shades of yellow represent deserts where little growth potential exists. Snow- and ice-covered regions have no vegetation potential.

www.wiley.com/college/arbogast

Photosynthesis and Respiration

Now that the process of photosynthesis has been examined in a static way, the concept can be reinforced by watching it in animated form. To do so, go to the *GeoDiscoveries* website and select the module *Photosynthesis and Respiration*.

This module nicely illustrates the various pathways involved in the photosynthetic process. After you view this animation, be sure to answer the questions at the end of the module to test your understanding of this concept.

1. Biogeography is a subdiscipline of physical geography that focuses on the geographical distribution of living organisms and the environmental factors that influence these patterns.

2. Ecosystems are communities of organisms that interact in interdependent ways.

3. The major components of ecosystems are abiotic factors, autotrophs, heterotrophs, and decomposers.

4. Photosynthesis is the process through which solar radiation is converted to chemical energy in plants.

The Relationship of Climate and Vegetation: The Character and Distribution of Global Biomes

Many different factors influence the geographic distribution of plants and biomass. Remember that environmental variables affect one another in holistic ways to form distinct geographic patterns. The study of plant geography is no exception. In order to understand these interactions, however, it is first essential to fully appreciate the impact certain individual variables have on these distributions.

The most important factor influencing the geographic distribution of vegetation is climate. Secondary factors such as geology, soils, landscape position, and human behavior also influence the pattern of vegetation on Earth and are discussed later in the chapter. For now, however, the focus of this part of the chapter is the relationship between climate and vegetation. You saw in Chapter 9 how vegetation reflects the geographic distribution of climates. These patterns, in turn, are related to many other primary factors, including latitude, proximity to oceans, and the global atmospheric circulatory system. In this context, to say that the geographic distribution of vegetation is related to climate is, in effect, the same as acknowledging the many different variables that influence the climate system. Remember these relationships in the following discussion.

A good way to study the relationship between climate and vegetation is through the concept of global biomes. A **biome** is a definable geographic region that is classified and mapped according to the predominant vegetation and organisms of that particular environment. Each biome is a distinct ecosystem and thus has a particular community of plants, animals, and microorganisms that are linked by energy and nutrient flows. Although ecosystems can be viewed at many different scales, this part of the chapter focuses on the plant communities in large areas that typically cover thousands of square kilometers. Although the biome concept also includes information about the character of soils, the major biomes are described here because vegetation is the most visible component of a biome. The relationship of soils to vegetation and climate will be covered in Chapter 11.

This section of the chapter focuses on geographic units as they relate to **potential natural vegetation**—that is, the vegetation you would find in a region if humans did not modify the landscape. Figure 10.6 is a map showing the distribution of the 12 biomes, and Table 10.1 outlines their fundamental characteristics. You should refer to this map and table frequently as you read this section. These biomes are organized into four categories: (1) forest biomes, (2) grassland biomes, (3) desert biomes, and (4) tundra biome. The alpine and arctic ice biome is not discussed in detail because it simply consists of landscapes covered by bare rock and glaciers and therefore has no plants.

As you study the remaining 11 biomes, remember that the boundary between vegetation communities is rarely distinct. Instead, a transition usually occurs from one vegetation type to another that may take place over several hundred kilometers. Such a vegetation transition is called an **ecotone**, because it contains elements of two distinct plant communities. These transitions occur because the boundaries between climate regions are usually difficult to precisely define.

Forest Biomes

Forest biomes are geographic regions where the vegetation is dominated by trees. These vegetation assemblages occur in areas where a net surplus of available

---

Biome *The complex of living communities maintained by the climate of a region and characterized by a distinctive type of vegetation.*

Potential natural vegetation *The vegetation that would occur naturally within a specific area if no human influence occurred.*

Ecotone *The transition area where two or more ecosystems merge.*

TABLE 10.1 Characteristics of the Major Global Biomes

Biome	Dominant Vegetation	Köppen Climate	Average Annual Precipitation and Moisture Patterns	Average Annual Temperature Patterns
Tropical Rainforest	Broadleaf evergreen trees. Dense canopy with open forest floor. Lianas (vines) and epiphytes.	Af, Am (wet side)	180 cm–400 cm (71 in.–158 in.)	20°C–30°C (68°F–86°F).
Tropical Deciduous Forest and Scrub	Ecotone between rainforest and grasslands; broadleaf trees with seasonal leaf fall.	Am, Aw	130 cm–200 cm (51 in.–79 in.), with distinct winter dry season.	Seasonal, but always warmer than 18°C (64°F).
Midlatitude Deciduous Forest	Broadleaf trees with seasonal leaf fall.	Cfa, Cwa, Dfa, Dwa, Dwb	75 cm–150 cm (30 in. –60 in.), with summer maximum; no seasonal water deficits.	Hot summers and cool to cold winters.
Mediterranean Woodland and Shrub	Individual short shrubs separated by grassy patches.	Csa, Csb	25 cm–65 cm (10 in.– 26 in.); winter surplus and summer deficit.	Warm summers and cool winters.
Midlatitude Coniferous Forest	Needle-leaf evergreen trees, including spruce, pine, fir.	Cfa, Cfb, Cfc, Cwa, Dfb,	75 cm–150 cm (30 in.–60 in.) in southeastern pines; 30 cm–100 cm (12 in.–39 in.) in northern conifers and highlands; 150 cm– 500 cm (60 in.–197 in.) in west coast rainforests.	Hot summers with cool winters in southeast; cold winters and cool summers in northern conifers and highlands. Cool winters and warm summers in west-coast rainforest.
Boreal Forest	Needle-leaf conifers, including spruce, pine, fir.	Dfc, Dwc, Dfd	30 cm–100 cm (12 in. –39 in.); no seasonal water deficit; ground frozen most of year.	Long, cold winters and short, cool summers.
Tropical Savanna	Ecotone between tropical forests and deserts; grasslands with isolated trees.	Aw	9 cm–150 cm (4 in.– 60 in.); short summer wet season and long dry season.	Seasonal, but always warmer than 18°C (64°F).
Midlatitude Grassland	Tall-grass prairie in humid areas to short-grass steppes in more arid regions.	BSk, Dfa, Dfb, Dwa, Dwb	25 cm–75 cm (10 in.– 30 in.); summer water stress.	Seasonal, with hot summers and cool to cold winters.
Hot and Dry Desert	Xerophytes such as succulents and cacti trending to large patches of bare ground.	BWh, BWk	Less than 2 cm (0.7 in.); perpetual moisture deficits.	Perpetually hot.
Semi-Arid and Cold Desert	Short-grass and dry shrubs; patches of bare ground.	BSh, BSk	2 cm–25 cm (0.7 in.– 10 in.).	Average annual temperature = 18°C (64°F).
Tundra	Stunted shrubs and sedges; mosses, lichens.	ET, Dwd	15 cm–180 cm (6 in.– 71 in.); no seasonal moisture deficit; frozen ground virtually all year.	Above freezing only 2 or 3 months of year, with warmest month <10°C (50°F).
Alpine and Arctic Ice	Some algae perhaps.	EF, H	Less than 10 cm (4 in.).	Frigid.

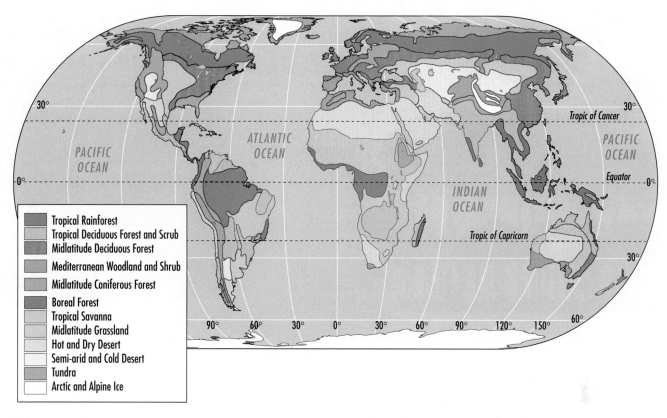

	Tropical Rainforest
	Tropical Deciduous Forest and Scrub
	Midlatitude Deciduous Forest
	Mediterranean Woodland and Shrub
	Midlatitude Coniferous Forest
	Boreal Forest
	Tropical Savanna
	Midlatitude Grassland
	Hot and Dry Desert
	Semi-arid and Cold Desert
	Tundra
	Arctic and Alpine Ice

Figure 10.6 Map of global biomes. Compare this figure with Figure 9.2, which shows the distribution of climate regions, and note some of the similarities and differences.

moisture occurs, either because of high average annual precipitation or because average temperatures are sufficiently cool that little evaporation takes place. Forest biomes occur from the tropical regions to the high midlatitudes. The forest biomes can broadly be categorized as the tropical forest biomes, midlatitude forest biomes, and the boreal forest biome. The tropical and midlatitude forest biomes are further subdivided into distinct geographical zones.

Tropical Rainforest Biome The tropical rainforest biome straddles the equatorial region between 23.5° N and 23.5° S (Figure 10.6) and is very closely related to the tropical rainforest climate (*Af*) zones. It is also associated with small areas of tropical monsoon climate (*Am*) that have a very limited dry season, such as in central Mexico. Average annual precipitation in the tropical rainforest biome ranges from 180 cm to 400 cm (70.1 in. to 158 in.), with monthly rainfall consistently greater than 6 cm (2.4 in.). These areas are always warm, with an average temperature ranging from 20°C to 30°C (68°F to 86°F).

The tropical rainforest biome contains a staggering assortment of trees and plants that form a very dense biomass. The trees are broadleaf and experience no seasonal leaf fall because of the consistently high temperature and abundant moisture; thus, these trees, like the needle-leaf

trees that you are probably familiar with, are *evergreen*. The rainforest also contains a variety of plants called *lianas* and *epiphytes*. Lianas are woody vines that use trees for support but are rooted in the ground. Epiphytes, in contrast, are plants that are rooted in a nook of a host plant and thus use the plant for support.

A mature tropical rainforest consists of four definable layers that can be seen vertically (Figure 10.7). The uppermost layer consists of randomly spaced emergent leaf crowns that exist at the tops of the tallest trees, some of which may be up to 60 m (200 ft) tall. Many of these trees are so large that they have *buttress roots* (Figure 10.8), which help support them. Directly below the emergent layer is the canopy, which is the densest layer of the tropical rainforest and consists of a near-continuous web of branches and leaves. This part of the tropical rainforest is from 15 m to 40 m (50 ft to 130 ft) above the forest floor.

Beneath the rainforest canopy is a third layer, called the *understory,* consisting of another dense network of interwoven branches and leaves that is 5 m to 15 m (15 ft to 50 ft) high (Figure 10.7). In conjunction with the upper canopy, this dense layer blocks sunlight from reaching the forest floor where the ground layer occurs. Given the shady character of the lower forest, the vegetation in this fourth layer of the rainforest is unevenly distributed and ranges from 0 m to 5 m (0 ft to 15 ft) above ground

(a)

Emergents
40–60 m (130–200 ft)

Canopy (densest layer)
15–40 m (50–130 ft)

Understory
5–15 m (15–50 ft)

Ground layer
0–5 m (0–15 ft)

(b)

Figure 10.7 The tropical rainforest. (a) The layered structure of the rainforest creates a continuous canopy that causes dense shading of the forest floor. (b) Do you see the emergent trees in this tropical rainforest?

level. In fact, contrary to the general assumption most have about the character of the ground-level rainforest, many places are open—that is, you could walk through them more easily than you might think. Only where distinct openings in the rainforest canopy are present, such as along streams, does sufficient light reach the forest floor so that dense *jungle* vegetation can grow.

Tropical Deciduous Forest and Scrub Biome The second major subdivision within the tropical forest biome is the tropical deciduous forest and scrub plant community. Extensive areas of tropical deciduous forest and scrub vegetation occur in Africa, South America, India, and Southeast Asia (see Figure 10.6). This plant subdivision is associated with the relatively dry tropical envi-

Figure 10.8 Buttress roots at the base of an emergent tree in the rainforest. These roots give support to the tall trees that extend to heights of 40 m to 60 m (130 ft to 200 ft).

ronments where annual precipitation is from 130 cm to 200 cm (50 in. to 80 in.), which occur due to a distinct seasonal dry cycle. These areas are found within both the tropical monsoon climates (*Am*) and the more humid parts of the tropical savanna climates (*Aw*). Although the temperature varies more in these areas than in the tropical rainforest regions, it is always warmer than 18°C (64°F).

The tropical deciduous forest and scrub plant assemblage reflects the progressive transition to a drier climate at higher tropical latitudes. In the wetter areas, where average annual precipitation is closer to 200 cm (80 in.), the vegetation is dominated by tropical deciduous forest (see Figure 10.6 again). Although these areas contain many of the same species seen in the tropical rainforest, they are not as tall and are not evergreen; that is, they drop their leaves during the dry season (hence, the name tropical *deciduous* forest). In addition, this environment has less overall species diversity and total biomass than the rainforest. Nevertheless, this biome has a greater array of shrub-like species and smaller plants at ground level because more sunlight reaches the forest floor (Figure 10.9). This enhanced ground-level sunlight is due to the upper canopy of the forest being less dense than in the tropical rainforest.

The vegetation gradually shifts from tropical deciduous forest to tropical scrub in areas where average annual rainfall is closer to 130 cm (50 in.). Tropical scrub consists mostly of low-growing, deciduous trees that reach heights ranging from 3 m to 9 m (10 ft to 30 ft; Figure 10.10). These trees tend to be ragged-looking due to the relative lack of moisture and they vary in density from being closely bunched to widely spaced. With increased distance between the trees, a lower level of tall bushes and

Figure 10.9 Example of the tropical deciduous forest. Although this is a low-latitude region, note that the trees are relatively short and the canopy is more open than in the tropical rainforest biome.

grasses appears. In contrast to the tropical rainforest and tropical deciduous forest, the species diversity within the tropical scrub regions is relatively low, with a few species dominating many of the taller growth areas.

Midlatitude Deciduous Forest Biome The midlatitude deciduous forest biome is found across the midlatitudes of North America, northeastern Asia, and western and central Europe (Figure 10.6) where the climate has a distinct seasonal temperature cycle with cold winters. This plant community consists of a dense network of broadleaf trees that forms a nearly continuous canopy in the summer. In the eastern United States, the dominant species are oak, elm, and maple, to name a few trees you may be familiar with (see Figure 9.10). These trees have a distinct seasonal cycle in which the leaves change color and fall during the autumn months.

Figure 10.10 Tropical scrub vegetation in the Australian Outback. Note the ragged-looking trees and the scrubby understory.

The midlatitude deciduous forest is associated with a variety of climate regions where average annual precipitation ranges between about 75 cm to 150 cm (30 in. to 60 in.), including the humid subtropical hot-summer climates (*Cfa, Cwa*), the humid continental hot-summer climates (*Dfa*), and the warmer parts of the humid continental mild-summer regions (*Dwa, Dwb*). Although these climates vary in some significant ways, each has sufficient moisture to support trees, either because average annual rainfall is abundant or evaporation is low because average annual temperatures are cool. In eastern North America, the midlatitude deciduous forest biome extends across a broad area from the Gulf of Mexico into southeastern Canada. Elsewhere, it occurs in an arc that extends from Europe to eastern Asia. This plant assemblage is also found in southeastern Australia and northern New Zealand (see Figure 10.6 again).

Mediterranean Woodland and Shrub Forest Biome A second subdivision in the midlatitude forest biome is the Mediterranean woodland and shrub forest. This plant assemblage is closely associated with the Mediterranean dry-summer climate (*Csa, Csb*) and is found on the west coast of continents in the lower middle latitudes (see Figure 10.6 again). Average annual precipitation in these areas ranges from 25 cm to 65 cm (10 in. to 26 in.). Winters are cool and wet, with a moisture surplus, whereas summers are hot and dry, with a water deficit.

The vegetation in the Mediterranean woodland and shrub biome is dominated by a dense cover of woody shrubs, known as the *chapparal* in North America (Figure 10.11). In many other places, the trees are somewhat larger and more widely scattered,

Figure 10.11 The Mediterranean woodland and shrub biome in southern California. This biome is adapted to dry summers and wet winters and consists of shrubby trees with intervening grass.

The Relationship of Climate and Vegetation: The Character and Distribution of Global Biomes • 259

with some patches of grass present. Although the trees vary greatly between continents—oak is most common in California—the overall appearance is quite similar regardless of the specific region. Given the distinct dry season that occurs in summer, many of the species in this biome are adapted to fire, which, as in the case of tropical savanna, recycles nutrients in the system. In fact, some plant species in the Mediterranean woodland and shrub biome have seeds that germinate only after they are stimulated by the heat of a fire.

Midlatitude Coniferous Forest Biome The last major subdivision in the midlatitude forest biome is the midlatitude coniferous forest. This plant community consists largely of needle-leaf evergreen trees and is found in places like the west coast and mountains of North America, the southeastern United States, and eastern Europe (see Figure 10.6). Coniferous forests in North America's mountainous areas are associated with humid continental mild-summer climate (*Dfb*) and have short growing seasons. Some of these coniferous forests occur at relatively low latitudes because temperatures are cool at high elevations. In North America, mountain forests include species such as Douglas fir and ponderosa pine (Figure 10.12a). On the west coast of North America, these forests are associated with the marine west coast climates (*Cfb*, *Cfc*), where annual precipitation is between 150 cm and 500 cm (60 in. and 197 in.) and average annual temperatures are mild for a midlatitude location. Given the cool, moist environment in this region, the forest is often referred to as *temperate rainforest* (see Figure 9.9). This forest also contains significant stands of deciduous trees, such as the giant redwoods in northern California and southern Oregon (Figure 10.12b).

Another kind of midlatitude coniferous forest exists in the southeastern part of the United States within the humid subtropical hot-summer climate (*Cfa*, *Cwa*) zone. This forest is called the *southeastern coniferous forest* and stretches across southeastern Louisiana, southern Mississippi, southern Alabama, central and southern Georgia, and the Florida panhandle on sandy sites that do not hold much water. The biological diversity in this region is very high, with the dominant tree species being longleaf pine that has an understory of wiregrass (see Figure 9.7 again). Needles in these trees come in bunches of three and are remarkably long, ranging from 20 cm to 45 cm (8 in. to 18 in.) in length.

(a)

(b)

Figure 10.12 Examples of the midlatitude coniferous forest biome. (a) This stand of ponderosa pine in California's Sierra Nevada is a great example of an alpine midlatitude coniferous forest. (b) Giant redwoods in northern California. These trees are over 100 m (350 ft) tall and can be over 6 m (20 ft) in diameter!

This is an image of the tropical deciduous forest biome. Which one of the following statements is best associated with this biome?

a) The trees in this forest are evergreen.
b) This forest is located in the midlatitudes.
c) This biome receives less average annual rainfall than the tropical rainforest biome.
d) The canopy in this forest is closed.
e) The wet season occurs in this biome when it is influenced by the Subtropical High (STH) Pressure System.

Boreal Forest Biome The last major forest biome is the boreal forest biome. The global distribution of this biome is closely associated with the subarctic climates (*Dfc*, *Dwc*, *Dwd*), and is the broad expanse of coniferous forest that occurs across Alaska, Canada, and Russia (see Figure 10.6). This vegetation assemblage is located entirely within the higher midlatitudes in the Northern Hemisphere because the large continents have very cold winter temperatures and relatively low rainfall. If you recall, the Southern Hemisphere contains no large landmasses, other than frigid Antarctica, at high latitudes. As a result, this hemisphere has no large expanse of subarctic climate and thus, no boreal forest biome.

The boreal forest biome, or *taiga* in the Russian language, has the least diverse assemblage of plants of any biome. Most of the trees in the boreal forest biome are coniferous needle-leaf evergreens such as spruce, pine, and fir, with large regions covered almost entirely by one or two of these species (Figure 10.13). These expanses of conifers are occasionally broken by small stands of broadleaf deciduous trees, specifically birch, poplar, and aspen. These trees are adapted to fire in that they are the first trees to move into a freshly burned area. Subsequently, they are succeeded by the coniferous trees, which also include the needle-dropping species tamarack and larch. The lower level of the boreal forest biome is sparsely vegetated, consisting of low shrubs, mosses, and lichens.

Although this region appears to be biologically productive, in fact it is not, due to the short growing season and the persistently wet soils in which the plants grow. The stress of this environment on plant growth is best seen in the northern parts of the boreal forest biome,

Figure 10.13 The boreal forest. This scene is from Alaska. Coniferous trees dot this landscape because it is near the northern margin of this biome.

where the growing season is very short and cold. Here, the trees are quite scraggly and stunted in comparison to their counterparts that live in the southern, warmer parts of the biome.

Grassland Biomes

Now let's turn our focus to the grassland biomes. These plant communities are distinct because they consist largely of grass with few trees. Two major grassland biomes are recognized: (1) tropical savanna biome and (2) midlatitude grassland biome.

WILDFIRES

When most people see a wildfire, they assume it is a negative thing that must be suppressed at all costs. Plant geographers and ecologists know, however, that fire is a natural part of many ecosystems and produces important ecological benefits. Regular fires reduce the amount of dead wood and thus lower the likelihood of potentially larger wildfires by removing their fuel. Fires often remove alien plants that compete with native species for nutrients and space. Fires also remove undergrowth in forests, which allows sunlight to reach the forest floor, thereby supporting the growth of native species. The organic debris produced by a fire adds nutrients to the soil to be used later by plants and animals. Fires can also provide a way of controlling insect pests by killing off the older or diseased trees and leaving the younger, healthier trees.

Overall, fire is a catalyst for promoting biological diversity and healthy ecosystems. Jack pine forests of the upper Midwest, for example, are highly dependent on fire for renewal. They have serotinous cones that require intense heat to open and release their seeds. If these forests did not burn periodically, the trees would gradually be replaced with less fire-adapted species and other plants and animals within the ecosystem would be threatened. Grasslands are also highly dependent on fire, as it prevents brush and trees from invading the prairie. Prairie grasses and flowers can survive a fire because their roots extend deeply into the ground.

Under natural circumstances, wildfires are usually started by lightning and are particularly likely in drought conditions. In grasslands, ranchers often set controlled fires to reduce brush and rejuvenate the prairie. Most of the wildfires you hear about, unfortunately, are the catastrophic events such as those in California nearly every fall that result in the loss of homes and even human life. Many of these begin with lightning, but some have been linked to arson. These fires are particularly intense due to a combination of factors, including previous fire suppression and buildup of dead wood that fuels the fire. Another factor is that more people have moved into drier areas that are naturally fire-prone, especially in hilly country where down-sloping winds fan a fire. Although these fires can certainly be tragic, remember that the overall ecological benefits of fire far outweigh the negative aspects often portrayed.

Wildfires are one of the most impressive natural phenomena on Earth, as this photograph from a fire in Montana attests. Although wildfires are most commonly associated with forests, they are a critical part of the prairie ecosystem because they recycle nutrients and remove woody vegetation. This image is of a prairie fire in the Great Plains.

The Tropical Savanna Biome The tropical savanna consists of grasslands at low latitudes that contain isolated trees. This biome is most closely associated with the tropical savanna climate (*Aw*) and is present in Africa, South America, India, and Australia (see Figure 10.6). In these regions, average annual precipitation is from 9 cm to 150 cm (4 in. to 60 in.), and average monthly temperature is greater than 18°C (64°F).

A good place to examine this biome is in Africa, where tropical savanna covers about 13,000,000 km² (5,000,000 mi²) of the continent. Recall from Chapter 9 that this region is most directly influenced by the seasonal migration of the tropical Hadley cell and associated ITCZ and STH pressure systems, with a summer wet season and winter dry season. In response to this distinct wet/dry cycle, the vegetation is a mix of forest and grass (Figure 10.14). Forest patches are denser where the conditions are more humid and thin considerably where it is drier. During the wet season, the savanna springs to life as grasses turn green and grow tall at the same time that leaves develop on trees. When the dry season comes, however, the grass withers and browns while the trees drop their leaves. A major factor that affects the savanna biome is the occurrence of wildfires, which are common in the dry season. These wildfires are critical to the ecosystem because they reduce the dead vegetation and recycle nutrients back into the system (see the *Discover...* feature in this chapter).

Midlatitude Grassland Biome Midlatitude grasslands are broad areas where the dominant vegetation is grass. The climate of these areas is seasonal, with warm to hot summers and cool to cold winters. Average annual precipitation ranges from 25 cm to 75 cm (10 in. to 30 in.). Several climate regions fall within this general description, including the dry portions of the humid continental

hot-summer (*Dfa, Dwa*) climates, the drier side of the humid continental mild-summer climates (*Dfb, Dwb*), and most of the cold midlatitude steppe climate (*BSk*). In many cases, the relative dryness of these climate zones is due to the continental effect in that the regions are located far from moisture sources. In other places, such as the interior of North America, distinct rain shadows exist in the lee of large mountain ranges such as the Rockies.

Extensive midlatitude grasslands occur in South Africa, eastern Europe, central Asia, and central North America (see Figure 10.6). In South America, grasslands are known as the *Pampa* (Figure 10.15a), whereas in North America they are most closely associated with the *Great Plains* (Figure 10.15b). In Russia, grasslands are referred to as the *steppe*. Although it may not look like it, the biomass in these areas is quite high, with a nearly continuous cover (except in the drier places) of grass and flowering plants

(a)

(b)

Figure 10.15 The midlatitude grassland biome. (a) The Pampas of Argentina is one of the great grasslands on Earth. (b) Mixed grass prairie in the central Great Plains. Note the combination of short and mid-level grasses on this landscape. Such a mix of grasses occurs because this region lies on the eastern part of the Rocky Mountain rain shadow, with annual precipitation of about 63 cm (25 in.). Grasses are taller to the east and shorter to the west of this area.

Figure 10.14 The tropical savanna. This photograph shows savanna in Brazil. Here, the vegetation consists of patches of grass that are dotted by isolated trees and woody shrubs.

that have an extensive root network. Most grasses are perennials that lie dormant through the winter and sprout in spring. This landscape is also adapted to fire, which usually occurs during the summer dry season. In fact, the common recurrence of fire historically contributed to the lack of trees and woody shrubs in this region, as did grazing by large animals such as bison (also known as "buffalo") in the Great Plains. Given that fire is suppressed by humans in many grasslands today, forest vegetation is actually expanding into the fringe of many grassland areas.

Desert Biomes

Desert biomes occur in areas where average annual precipitation is less than 25 cm (10 in.) and cover about one-fifth of Earth's land area. Although deserts most commonly occur in low latitudes, they also appear in colder regions such as in Nevada and Utah in the western United States and in portions of western Asia. Two major subdivisions within the desert biome are recognized: (1) the hot and dry desert biome, and (2) the semi-arid and cold desert biome.

Hot and Dry Desert Biome The hot and dry desert biome occurs in the subtropical regions and is centered on 30° latitude (see Figure 10.6). Major deserts of this kind include the Sahara in northern Africa, the Kalahari Desert in southern Africa, and the Mojave and Sonoran Deserts in the southwestern United States. Another large area of hot and dry desert occurs in central Australia. These areas are associated with the hot low-latitude desert climate (BWh), which is dominated by the STH Pressure System. Average annual precipitation can be less than 2 cm (0.8 in.), and average monthly temperature is greater than 18°C (64°F). As a result of the combined low rainfall and high temperatures, these areas often experience chronic moisture deficits.

The vegetation in the hot and dry desert consists of bare ground that grades into **xerophytic plants**, including cacti, low-growing shrubs, and scattered clusters of grass (Figure 10.16). These plants have evolved in ways that ensure their survival in very dry regions. One survival strategy used by many plants is to be *drought-resistant*, which means that they can conserve moisture either by developing very small waxy leaves that store moisture (these plants are called *succulents*), or by having roots that penetrate deeply into the ground. Other plants are *drought-evading* in that they rapidly flower and reproduce during the rare times of relatively high precipitation.

Xerophytic plants *Plants in very dry places that have a number of survival mechanisms in response to prolonged periods of drought.*

Figure 10.16 Example of the hot and dry desert biome. The Sonoran Desert in the southwestern United States features tall, columnar saguaro cactus, as well as a variety of smaller cacti and hard-leaved shrubs.

Semi-Arid and Cold Desert Biome The semiarid and cold desert biome occurs extensively in western North America and central Asia (see Figure 10.6). This biome is most closely associated with the cold midlatitude desert climate (BWk) but is also found in areas of the cold midlatitude steppe climate (BSk). In North America, this climate occurs in places like Utah and Nevada where extensive rain shadows form in the lee of the Sierra Nevada. This climate also occurs in places like the Gobi Desert in central Asia, which is far from a moisture source. Both places feature a distinct seasonal temperature pattern, with hot, dry summers and cold winters.

The vegetation of the semi-arid and cold desert consists of bare patches that grade to low-growing spiny and woolly plants that are randomly spaced. These uneven and sharp plant surfaces protect the plants from grazing animals and significantly reduce moisture loss. Some plants have a silver or glossy leaf, which give them a higher albedo and thus keep the plants cooler by reflecting more solar radiation. In the western United States, plants in this vegetation community include creosote bush, sagebrush, and mesquite (Figure 10.17).

Tundra Biome

The last of the major biomes on Earth is the tundra biome, which borders the Arctic Ocean at very high latitudes across North America and Russia (see Figure 10.6). The *arctic tundra* is found largely in association with the tundra climate (ET), which means that winters are long with an average temperature of −34°C (−29°F). It is also found in the colder and drier parts of the subarctic climate (Dwd). Although the summer growing season is only 60 to 80 days long, average temperature can reach 12°C (54°F) due to high Sun angle. Average annual precipitation is 15 cm to 25 cm (6 in. to 10 in.). Tundra also occurs to a limited extent at very high elevations

(a)

(b)

Figure 10.17 The semi-arid and cold desert biome. This sagebrush landscape in Monument Valley, Utah, is a nice example of vegetation in the semi-arid and cold desert biome.

Figure 10.18 The tundra biome. (a) The vegetation here consists of low-growing plants, primarily very short grasses, mosses, and lichens, which have adapted to the short, cold growing season. (b) Lichens are a mix of algae, bacteria, and fungi that favor rocky surfaces such as those pictured here. The central rock, which is about 15 cm (6 in.) wide at its thickest point, is surrounded by low-growing moss.

within the highland climate (*H*) as *alpine tundra*. In this context, alpine tundra is found as far south as the equatorial Andes Mountains.

Regardless of whether the tundra is alpine or arctic, the plant assemblage and thus the appearance are essentially the same. The tundra biome consists of a diverse array of grasses, shrubs, sedges, mosses, and lichens (Figure 10.18) that tend to be very short because the growing season is limited and the landscapes are wind-swept. Tundra

landscapes are very fragile and are easily disturbed by human impact. Although people have historically avoided tundra because of the harsh conditions, it is now a growing source of interest because of untapped reservoirs of fossil fuels and other minerals that may exist there.

VISUAL CONCEPT CHECK 10.2

This image shows a typical scene in one of the major biomes on Earth. Which biome is pictured here and how do migrating atmospheric pressure systems influence the seasonal climate and thus the vegetation in this place?

1. A biome is a definable assemblage of plants that occupy a large region of geographic space with generally recognizable boundaries.

2. The global distribution of biomes is closely related to the geography of climates.

3. Trees are typically found in humid regions, whereas dry to semi-arid regions are populated by shrubs and grasses.

4. The vegetation of deserts and that of the tundra biome are very similar in that both consist of low-growing plants that have adjusted to extreme environmental conditions.

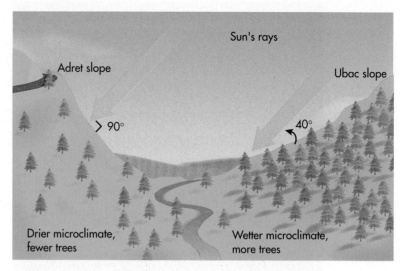

Figure 10.19 Slope aspect and Sun angle. Slopes that directly face the Sun (adret slopes) have higher Sun angles and thus a warmer and drier microclimate than opposite slopes (ubac slopes).

Local and Regional Factors That Influence the Geographic Distribution of Vegetation

In addition to the geographic distribution of biomes, other factors can influence the types of plants that live in a region or area. It is useful to think of these variables as being local or regional in scope because they may cause a particular pattern within the overall biome. The following discussion of local factors is not meant to be a complete review of all the variables that can influence the distribution of vegetation; instead, it is meant to provide a good sense of the most important of these factors.

Slope and Aspect

The **slope** of an area is its degree of steepness, with high slopes being places that are steep and low slopes being relatively flat. As the slope of a landscape increases, the amount of water that runs off the surface, rather than soaking into the ground, usually increases. As a result, steep slopes typically have less available soil moisture and the associated vegetation can be different than on shallow slopes. Soils also tend to be thinner on steep slopes than on shallow slopes, which influences the amount of nutrients available for plants.

Another factor related to the slopes within a particular landscape is the concept of aspect. *Slope aspect* refers to the orientation of the slope—that is, whether the slope faces north, south, east, or west. Aspect is an important variable in plant geography because it is related to how intensely the Sun strikes a given slope (Figure 10.19). In

Slope *The degree of steepness of a portion of the landscape.*

this context, the **adret slope** is the slope that faces the Sun most directly; that is, the slope at which solar radiation arrives at a high angle. Where this geographic situation occurs, the **microclimate** (climate over a small region) is usually warmer, more evaporation subsequently occurs, and the slope can be relatively dry. The opposing, or low-Sun slope, is called the **ubac slope**. This aspect is characterized by a relatively cooler micro-environment with less evaporation because less direct solar radiation is received. If you live in a region that receives abundant snowfall, notice the effect of aspect on hills that slope down on the north and south sides of highways. One slope will be free of snow due to melting, whereas the other slope will have snow on it for a longer period of time.

Now think about how aspect affects plant growth. In the Northern Hemisphere, where would you expect warmer temperatures and higher rates of evaporation: north- or south-facing slopes? It is warmer, with more evaporation, on the south-facing slopes because they face the Sun more directly. As a result, trees can grow (or are more frequent) on the north-facing ubac slope while they do not occur (or are less frequent) on the south-facing slope. Figure 10.20 shows a great example of this pattern by focusing on the growth of bristlecone pine in California.

Adret slope *The slope that faces the Sun most directly.*

Microclimate *Average temperature and precipitation characteristics within a small area within a larger climate region.*

Ubac slope *The slope that faces away from the Sun.*

(a)

(b)

Figure 10.20 Slope aspect and growth of bristlecone pine in the White Mountains, California. (a) Growth on the south-facing (adret) slope of a mountain. Note the wide spaces between trees. (b) Growth on the north-facing (ubac) slope. Bristlecone pine favors this slope, rather than the adret slope, because less evaporation on this surface leads to more available water.

Vertical Zonation

Another way that vegetation can vary locally from what you might initially expect is through the concept of vertical zonation. **Vertical zonation** refers to the change in vegetation that occurs with respect to elevation, rather than latitude. Recall that temperature usually decreases at a consistent rate (the environmental lapse rate) as you gain altitude in the troposphere. This overall trend shows up most clearly on the world climate map (see Figure 9.2) as highland climate (*H*) regions. Recall that these climate zones are known for tremendous variability over short horizontal and vertical distances. They are associated with the Earth's large mountain chains, including the tropical Andes where near-arctic conditions are found at high elevations.

In the context of vertical zonation, the key relationship is that environmental changes that occur with respect to elevation correlate to the same ones you see with latitude. Figure 10.21 illustrates how vertical zonation works in the European Alps. In the lower elevations of the mountains, the vegetation often consists of deciduous forest. As you move to about 1000 m (3300 ft), the vegetation gradually shifts to coniferous trees that are adapted to cooler conditions.

At about 2000 m (6600 ft), the vegetation shifts from trees to alpine grassland and scrub vegetation. The elevation where trees can no longer grow because conditions are too harsh is called the **tree line**. At still higher elevations, the vegetation becomes tundra (Figure 10.22), and, ultimately, the landscape is covered with snow and ice.

Vertical zonation *The change in environmental characteristics that occurs with respect to altitude.*

Tree line *The line that represents the upper limit in mountains and high latitudes where environmental conditions support the growth of trees.*

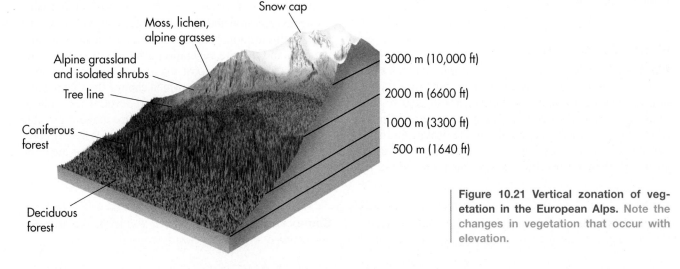

Snow cap

Moss, lichen, alpine grasses

Alpine grassland and isolated shrubs

Tree line

Coniferous forest

Deciduous forest

3000 m (10,000 ft)

2000 m (6600 ft)

1000 m (3300 ft)

500 m (1640 ft)

Figure 10.21 Vertical zonation of vegetation in the European Alps. Note the changes in vegetation that occur with elevation.

Figure 10.22 Alpine vegetation in the Snowy Range in Wyoming. The scattered trees, which are similar to high-latitude boreal forest, grade upward into tundra. This is an excellent example of vertical zonation.

Figure 10.23 A tree line landscape. Trees at circumpolar or alpine tree line are often dwarfed and scraggly because of the harsh environmental conditions. Can you tell the direction (left to right or right to left) of the prevailing winds in this photo?

When do environmental conditions become too harsh to support trees? The critical variable appears to be that seasonal mean temperature registers at least 6.5°C (43.7°F) at a depth of 10 cm below the ground, where roots grow. You are probably most familiar with *alpine tree line* that occurs at high elevations in the mountains, above which are only scattered tundra, bare rocks, and even perpetual snow. Although alpine tree line exists in all major mountain ranges on Earth, a distinct pattern exists with respect to latitude and altitude, with the elevation decreasing progressively with increased latitude. Another kind of tree line is called *circumpolar tree line* and is the highest latitude at which trees can grow. Regardless of whether the tree line is alpine or circumpolar, the appearance is the same, consisting of scattered trees that are stunted and deformed. Such an area is referred to as the *Krumholz zone*, which literally means "crippled wood." In these areas, trees are often bent in a particular direction that reflects the prevailing winds (Figure 10.23).

Plant Succession

Plant succession refers to the natural changes that occur in a biome of a particular place over time. Generally speaking, these natural changes progress to the point where the most complex array of vegetation possible arises, given the regional climate, water, and soil variables. If the succession begins on freshly deposited sediment, then we use the term *primary succession*. If,

on the other hand, the succession occurs in an area that was disturbed through fire or some other catastrophic event, then the term *secondary succession* applies.

Regardless of whether the succession is primary or secondary, the sequence of plant stages is usually predictable for a given area. The first plants that occupy the area are called *pioneer species*. These plants set the stage for future succession because they initiate the gradual change of the host soil through root penetration and the cycling of organic material and minerals as they spread across the area. In this fashion, bacteria and animals begin to inhabit the area and the site vegetation gradually becomes more complex. In time, the local micro-environment becomes hospitable for larger plants, which change the shading and associated microclimate at ground level. These adjustments in humidity and soil temperature make it possible for additional plants to move into the area. When this progression reaches the ultimate complexity of an area, the vegetation is said to be **climax vegetation**.

Plant succession *The predictable vegetation transition that occurs within a landscape over a period of time.*

Climax vegetation *The vegetation within an area that has reached its ultimate complexity.*

Plant Succession

Let's take a closer look at the concept of plant succession by watching the sequence of plants that populate a more complex landscape than the dunes just described. Go to the *GeoDiscoveries* website and select the module *Plant Succession*. This module illustrates the process of secondary plant succession in a cool, humid landscape in the higher midlatitudes. The environmental stage is set with a small stream that flows through a forested area about

7000 years ago. Subsequently, a colony of beavers builds a dam across the stream, causing a pond to form. From this point, watch how the succession of plants proceeds in this environment, culminating in the complete disappearance of the pond and return of climax vegetation. Once you complete the animation, be sure to answer the questions at the end of the module to test your understanding of plant succession.

A good example of primary succession can be seen on a sequence of sand dunes, which are mounds of sand that grow through wind deposition in humid environments. They are highly mobile landforms that move due to a complex interaction of wind and sand supply. These landforms are relatively infertile micro-environments because they are very well drained and nutrient-poor, due to high amounts of quartz and feldspar and low amounts of nitrogen, calcium, and phosphorus.

In the context of plant succession, the first plants to inhabit a fresh deposit of wind-blown sand are beach grasses. These grasses are adapted to the relatively infertile conditions of a new sand dune and continue to grow even if they are slightly buried by fresh deposits of sand. As they spread across the dune, they cause it to stabilize, which means that the sand stops moving. At this time, new plants move in that are adapted to the relatively infertile conditions of the dune. These plants, however, cannot tolerate the stress of being buried as the initial plants could. When this more established successional stage occurs, the soil chemistry and ground micro-environment continue to evolve, culminating in the establishment of trees on the older dunes.

Figure 10.24 is a scene from a coastal dune field along Lake Michigan and is an excellent example of this succession. Compare the vegetation on the low (young) dune in the foreground with that on the higher (older) dunes in the background. The younger dune is inhabited by beach grass, while those in back are covered with forest. In the Great Lakes region, this succession sequence is thought to take about 150 years.

Riparian Zones

Another important factor that can influence the geographic distribution of vegetation is proximity to water. This factor is especially important in semi-arid regions where water is often scarce. An example of this kind of relationship is the **riparian zone**, which is

Riparian zone *The strip of land, which borders a body of water, that supports plants and animals adapted to water systems.*

Figure 10.24 Primary succession on Great Lakes coastal dunes. Beach grass grows on the low dune in the foreground, which is the most recent deposit of wind-blown sand. The higher dunes in the background are covered by forest because these dunes are older and have been stable for a longer period of time.

Figure 10.25 An example of riparian vegetation. This aerial photograph of the Arikaree River in the western Great Plains shows a ribbon of trees along the stream. Although the overall environment of this region only supports grass, trees can grow along the river because more moisture is available there.

defined as the area immediately adjacent to a stream. More water is available for plant growth near the river than away from it, especially because elevation increases with distance away from the valley bottom. A good place to see how riparian vegetation may differ from the overall pattern can be seen in parts of the Great Plains of North America, shown in Figure 10.25. Notice in this image that trees are growing only along the river, while the surrounding landscape consists of grass and sagebrush.

Human Interactions: Human Influence on Vegetation Patterns

It would be a major omission to discuss the geographic distribution of vegetation without talking about the impact of humans. In fact, given the enormous impact that humans have had on the global distribution of vegetation, it is somewhat misleading to discuss the influence of people following a section of text that focuses on local and regional variables. Nevertheless, people are usually functioning at a local scale when it comes to decisions regarding vegetation. The combined impact of these local decisions by countless numbers of people over time often shows up as a distinctive regional pattern that occurs within the overall biome. This discussion is not meant to be comprehensive—far from it. Instead, it will provide a sense of how humans have modified the natural vegetation of regions all around the globe.

Deforestation and Its Consequences

A major issue that presently faces the global society is **deforestation**, which is the process through which large tracts of land are cleared of trees by humans, either for commercial use or to make way for agriculture. A good place to begin the discussion of deforestation is to examine the current extent of forest cover around the world. According to the United Nations *Global Forest Resources Assessment*, approximately 3.9 billion hectares (9.6 billion acres) of Earth were covered by forest in 2010. Of this total, 47% was in the tropics, 33% was in the boreal forest, 11% was in the middle latitudes and temperate areas like the Pacific Northwest, and 9% was in the subtropics (Figure 10.26). Overall, this extent of coverage amounts to about one-third of Earth's total landmass, excluding Antarctica and Greenland.

Ongoing monitoring by the United Nations indicates that the extent of global forest cover has decreased extensively in the past 20 years due to human activities. High amounts of loss have occurred even though forest is actually expanding in some areas due to natural processes or through *afforestation*, which is the planting of trees by people in previously nonforested areas. As you can see in Table 10.2, Europe and Asia had net increases in forest cover between 1990 and 2010. The pattern in Asia is particularly interesting, as the net gain this century reverses a trend of overall forest loss in the 1990s. These gains are largely the result of aggressive planting of trees in China, which have offset high amounts of deforestation in places such as Borneo and Indonesia. In North America, the rate of forest loss appears to have stabilized in the past 20 years.

Although some modest increases in forest cover have occurred recently in some areas, the average annual rate of global deforestation between 2000 and 2010 was about 13 million hectares (32 million acres). The good news is that this rate is actually slightly less than the 16 million hectares (14.6 million acres) lost each year globally in the 1990s, but, unfortunately, is still considered "alarmingly high." At a continental scale, this high rate of deforestation is clearly associated with Africa and South America. Over the course of the past 20 years Africa has lost about 12% of its forest, whereas about 5% of the forest has been removed in South America (Table 10.2). These high rates of forest loss are primarily linked with two land-use practices: (1) to clear land for cattle grazing and agriculture, and (2) logging of commercially valuable wood, such as mahogany.

In the context of agriculture, subsistence farmers in tropical regions practice a technique called *slash and burn agriculture*, in which the forest is burned both to

Deforestation *The removal of trees for economic or agricultural purposes.*

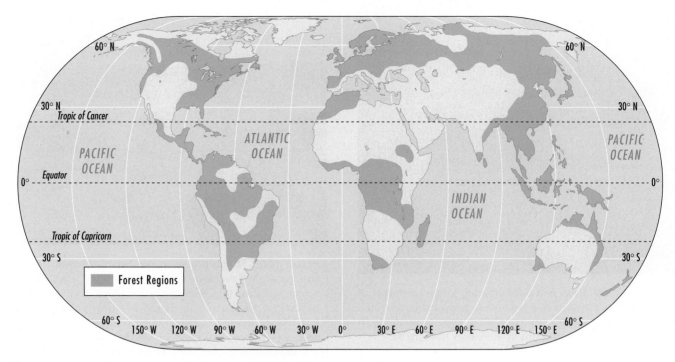

Figure 10.26 The extent of forest regions on Earth. Forests have adapted to every zone of latitude on Earth except for the polar regions.

clear the land of debris and to recycle nutrients to the soil (Figure 10.27). This method of farming is very widespread, as indicated by the satellite image (Figure 10.28) acquired in September 2001 of a portion of Bolivia, Brazil, and Paraguay in South America, showing the location of numerous fires. Note the extent of forest that had already been cleared prior to the fires.

Although the amount of deforestation appears to have slowed somewhat, tropical deforestation continues and the consequences may be far-reaching. Of major concern is the impact that extensive deforestation has on biodiversity within the tropical forest biomes. Given the complexity of these regions, it is unknown precisely how the effect of cumulative clearing ripples through the system. In fact, many scientists believe that it is possible that deforestation will cause extinction of plant and animal species that have yet to be discovered. Deforestation is also significant with respect to global warming. The tropical forest regions are sometimes called the "lungs of Earth" because they absorb carbon dioxide and release oxygen through photosynthesis. Increased atmospheric carbon dioxide is apparently related to global climate change, but its overall increase may be enhanced because the smaller tropical forests absorb less.

TABLE 10.2	Change in Forest Cover (in millions of hectares), 1990–2010			
Continent	Total Forest 1990	Total Forest 2000	Total Forest 2010	Percent Change 1990–2010
Africa	702	650	620	–11.7
Asia	551	548	580	+0.5
Oceania	201	198	195	–3.0
Europe	990	1000	1005	+1.0
North and Central America	555	549	550	–1.0
South America	923	886	875	–5.2
Total World	**3963**	**3869**	**3825**	**–3.5**

Source: U.N. Food and Agriculture Organization, Global Forest Resources Assessment 2010 (France, 2010).

Figure 10.27 Slash and burn clearing in the rainforest of Brazil. The forest in this region has been cut and burned to prepare the ground for cultivation. Organic matter from the burned forest temporarily increases soil fertility.

Figure 10.29 Stump prairie in northern Michigan. These stumps are the remnants of old-growth pine trees on the Kingston Plains that were cut in the 1890s. Following this period of deforestation, an abundant amount of dried limb and branch fragments lay on the ground. This *slash* subsequently burned in a very hot fire that torched seedlings and essentially cooked the soil. As a result, forest has not regenerated over much of the area.

Figure 10.28 Location of wildfires in central South America in September 2001. The red dots mark the location of fires, the largest of which produce distinctive smoke plumes. The extensive areas that are already cleared appear as tan zones in the sea of green forest.

Although deforestation is often considered a tropical issue, the history of forest management has also been mixed in the higher latitudes. In the Great Lakes region of North America, for example, virtually all the native pine forest was cut in the late 19th and early 20th centuries (Figure 10.29) to provide timber for buildings in growing cities such as Chicago, Milwaukee, and St. Louis. The forest in this region is still heavily managed, with large jack and red pine plantations in many parts of the region that are grown for pulpwood. A similar history has also transpired in the needle-leaf forests of the Pacific Northwest. Most of the old growth forests are gone in this area, and the second and third growth forests continue to be extensively logged (Figure 10.30). The ongoing demand for old growth timber in this region has controversial overtones regarding habitat loss and species extinction, most notably with respect to the spotted owl. Environmentalists fervently believe that the remaining forests should be preserved to protect important owl habitat. Logging firms, in contrast, argue that further cutting is necessary to meet timber demand and to protect jobs.

Agriculture in the Midlatitude Grassland Biome

In addition to the tremendous impact that people have had on global forests, another biome that has been severely modified by humans is the midlatitude grassland biome. Prior to European settlement of North America, as much as 162,000 hectares (400,000 acres) of the continental interior were grassland. You can see the potential

Figure 10.30 Clear cutting in the Pacific Northwest. Deforested areas on mountain hillslopes south of Olympic National Park in Washington.

Figure 10.31 Impact of grazing on vegetation in the American West. Overgrazing reduces the cover of vegetation on hillslopes, which can result in severe erosion when rain falls.

natural distribution of grassland in Figure 10.6. Notice how much of central North America, in particular, was covered by grass. As you will learn in the next chapter, the soils in the midlatitude grassland biome are some of the most fertile in the world. This fertility is ideal for the development of large-scale agriculture, which dominates the region today in the form of extensive corn, wheat, and soybean operations. Although the benefit to people of this form of intense agriculture is obvious, the impact on the midlatitude grassland biome has been immense. In that context, current estimates are that only 1% to 4% of virgin (unplowed) prairie remains throughout the entire biome.

Overgrazing

Another way that humans have impacted the network of natural vegetation is through overgrazing of the land-scape by livestock, such as cattle and sheep. This impact is most extensive in marginal landscapes that are extremely sensitive to disturbance. An excellent example of this kind of marginal landscape is the semi-arid Great Basin of the American West (Figure 10.31). Prior to European settlement, this region contained very few large herbivores that would have grazed on the vegetation there. In the 20th century, however, this region became available for ranchers because the U.S. Forest Service, Bureau of Land Management, and the National Park Service classified much of it as public lands. According to federal legislation, it is permissible to graze sheep and cattle in these areas. This practice continues to the present day and has dramatically affected this biome. Some of the potential impacts are weed invasion into disturbed areas, reduction in forest due to grazing of young saplings, and the increased likelihood of more intense and widespread fires.

www.wiley.com/college/arbogast

WILEY PLUS

Deforestation

The topic of deforestation is of particular concern to many scientists. One way to monitor the rates and patterns of deforestation is through satellite remote sensing. To see some examples of how this kind of monitoring occurs, go to the *GeoDiscoveries* website and select the module *Deforestation*. This interactive module illustrates the kind of patterns that geographers observe when they study deforestation around the world. As you work your way through the module, you will be asked to identify deforested and uncut areas using both true-color and infrared satellite imagery. A short animation also illustrates the rate of rainforest deforestation in the Amazon and Borneo. After you complete this module, be sure to answer the questions at the end of it to test your understanding of this concept.

VISUAL CONCEPT CHECK 10.3

This pair of infrared satellite images is of the same region in the Amazon rainforest at different times. Remember that red on this kind of image represents vegetation. The (a) image was acquired in 1975, whereas the (b) image dates to 2000. The difference that you see reflects extensive deforestation along road networks in the region. Which one of the following statements is accurate with respect to the impact that this kind of deforestation has on global climate change?

a) Fewer trees will result in the production of more oxygen.
b) The amount of atmospheric carbon dioxide will increase because fewer trees are present to absorb it.
c) Global temperature should decrease given that fewer trees are present in the rainforest.
d) There will be less atmospheric carbon dioxide because trees produce carbon dioxide as a by-product of respiration.

(a)

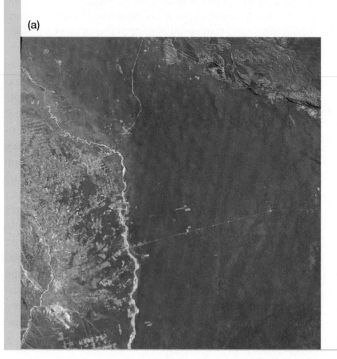

(b)

KEY CONCEPTS TO REMEMBER ABOUT THE FACTORS THAT INFLUENCE THE LOCAL GEOGRAPHIC DISTRIBUTION OF VEGETATION

1. Throughout major global biomes, significant variations can occur at local and regional scales.
2. The topography and soil characteristics of a region or area can significantly influence the vegetation patterns.

3. The same kind of vegetation changes that occur with latitude also occur with altitude.
4. Succession refers to the natural change in vegetation that occurs following disturbance or fresh deposition of sediment.
5. Humans have dramatically altered the natural vegetation in many regions of the world.

Remote Sensing and the Biosphere

Plant ecologists and geographers are very interested in the patterns and rates of vegetation change on Earth. In this context, one of the ways that plant geography is studied and monitored is with remote sensing. To provide a sense of how this form of monitoring can be done, go to the *GeoDiscoveries* website and select the module *Remote Sensing and the Biosphere*. Within this interactive module, you will be able to analyze (1) *Land Productivity*, (2) *Ocean Productivity*, and (3) *Fire Patterns*. In each of these submodules, you will be presented with a series of satellite images and prompts to correctly identify specific features on imagery. Once you complete the interactivity, be sure to answer the questions at the end of the module to test your understanding of the concepts contained within it.

THE BIG PICTURE

The next chapter focuses on the processes and geographic distribution of soils on Earth. You will notice a close relationship between soil and vegetation, with a dominant influence once again being the role of climate. Chapter 11 is also the first chapter where the kinds of processes and outcomes that occur beneath the surface of Earth will be examined. For example, this photograph shows a soil profile, which is the view of a soil from the surface to some depth within the ground. A distinct layering pattern appears in the profile, with a dark zone near the surface that overlies a lighter area that contains numerous white specks. The next chapter discusses how such profiles develop and how they vary geographically.

1. Biogeography examines the geographical distributions of organisms, ecosystems, and the environmental contexts in which they occur.

2. The photosynthetic, respiratory, and decomposition cycles explain how biomass is produced and consumed through the interaction of solar energy, carbon dioxide, water, and oxygen in plants, animals, and soil.

3. A biome is a definable geographic region that is classified and mapped according to the predominant vegetation and the adaptations of organisms to that particular environment. Twelve major biomes are recognized on Earth.

4. Six major forest biomes in the world occur in places where average annual precipitation is high. They also occur in places that have moderate amounts of precipitation and average temperatures that are sufficiently cool to maintain a good water balance.

5. The two major grassland biomes in the world are (1) tropical savanna and (2) midlatitude grassland.

The tropical savanna occurs in the tropical ecotone where the ITCZ dominates in the summer (bringing the wet season), and the STH dominates in winter (bringing the dry season). Midlatitude grasslands occur deep within continents or within rain shadows.

6. Two major desert biomes are recognized in the world: (1) hot and dry desert, and (2) semi-arid and cold desert biome. The hot and dry desert occurs in the higher tropical latitudes, dominated by the STH Pressure System. The semi-arid and cold desert biome occurs deep within continents and rain shadows within the midlatitudes.

7. The tundra biome occurs at very high latitudes and elevations where environmental conditions are too harsh for trees to grow. The plants here consist of low-growing woody shrubs, lichens, and mosses.

8. The human impact on the distribution and character of natural vegetation has been dramatic, including deforestation, extensive agriculture, and overgrazing.

CHECK YOUR UNDERSTANDING

1. Define the concept of ecosystems and the various components within them.

2. Explain the process of photosynthesis and why it is important to life on Earth.

3. Why is biomass higher in tropical regions than in desert regions?

4. Describe the concept of global biomes and the factors that influence them.

5. Why is the floor of the tropical rainforest biome considered to be "open"?

6. Many of the trees in the tropical deciduous forest biome can be found in the tropical rainforest biome. Nevertheless, the trees in the rainforest are evergreen, whereas they drop their leaves in the tropical deciduous forest biome. Why does this difference exist?

7. The ITCZ and STH Pressure System are very closely associated with the tropical savanna biome. What is this relationship?

8. Describe the change in the character of grasses in the Great Plains from the eastern part of the grassland to the west. Why does this difference exist?

9. Where would the warmest/driest slopes be in the Southern Hemisphere: on the north- or south-facing slopes?

10. Where is overgrazing likely to have the biggest impact: in the Great Plains or in the eastern United States? Why does this difference exist?

11. Why is there likely to be a denser forest on the windward side of a mountain range than on the leeward side?

12. Explain the concept of plant succession.

13. Tundra is found in the very low latitudes of South America, even though this region is commonly thought to be "tropical." Why?

14. Discuss the rates and geography of deforestation on Earth.

ANSWERS TO VISUAL CONCEPT CHECKS

VISUAL CONCEPT CHECK 10.1

The answer to this visual concept check is *c*, which states that this biome receives less annual precipitation than the tropical rainforest biome. The reason this answer makes sense is that the relative lack of rainfall causes a more open forest to develop than in the rainforest.

VISUAL CONCEPT CHECK 10.2

This biome is the African Serengeti, which exists in the part of Africa that has a distinct wet and dry season. The wet season occurs in summer when the insolation is high and the ITCZ dominates the weather. In contrast, the dry season occurs in winter when the STH Pressure System moves into the region.

VISUAL CONCEPT CHECK 10.3

The answer to this question is *b*; there will be more atmospheric carbon dioxide as a result of deforestation. This increase will occur because trees absorb carbon dioxide in the context of the global carbon cycle.

CHAPTER ELEVEN

THE GLOBAL DISTRIBUTION AND CHARACTER OF SOILS

If asked to name the most important natural resources on Earth, you might be inclined to say they are water, forests, oil/coal, and the air we breathe. Certainly, these resources are very important, but you might be surprised to learn that many people believe our most important natural resource is soil. Why? Because our food directly or indirectly comes from it, we build structures within it, and complex ecosystems are supported by it. This chapter will help you appreciate the character of soil and the important role it plays on Earth.

This chapter examines how soils form, what properties affect soil fertility and development, and where the basic types of soils are found on Earth. This chapter naturally follows the preceding chapters on climate and vegetation because a close relationship is usually found between these variables and soil characteristics. In this sense, this chapter integrates many components of the atmosphere, hydrosphere, biosphere, and lithosphere in a holistic way. This chapter also marks the point where we begin to examine the geographic patterns associated with parts of Earth that are within Earth, or the *lithosphere*, rather than being on top of it, as in the case of the *biosphere*, or above it in the *atmosphere*.

Soil is one of the most important natural resources on Earth and comes in many different forms. This image nicely shows the relationship between Earth's surface and the underlying soil. Notice the color of the sediment and the plant roots, and consider how water would flow downward. This chapter focuses on the evolution of soils and how they can differ around Earth.

CHAPTER PREVIEW

What Is Soil?

Measurable Soil Characteristics

Soil Chemistry
On the Web: *GeoDiscoveries Soil Colloids and pH*

Soil Profiles (Reading the Soil)
On the Web: *GeoDiscoveries Soil Horizon Development*

Soil Science and Classification
On the Web: *GeoDiscoveries African Climate, Vegetation, and Soils*
On the Web: *GeoDiscoveries Regional Pedogenic Processes*
On the Web: *GeoDiscoveries North American Climate, Vegetation, and Soils*

The Big Picture

LEARNING OBJECTIVES

1. Discuss the nature of soil and the factors that influence its development.

2. Describe the measurable soil characteristics and why they are important.

3. Explain the various aspects of soil chemistry and why they are important for soil fertility.

4. Define the concept of a soil horizon and explain how the classic sequence of horizons forms.

5. Describe the concept of soil classification and differentiate between the various soil orders in a holistic way.

6. Explain why soils are important natural resources and how they are managed in the United States.

What Is Soil?

Remember when you were a child and played in dirt? Did you ever stop to think about where the dirt had come from, and how long it had been there? Although you generally referred to the material as "dirt," in all probability you were playing in soil. You may wonder what difference it makes whether it is called "dirt" or "soil": Are they not the same thing? In everyday language the two terms can be used interchangeably, but in the context of physical geography and soil science, the concepts are indeed very different. The fact is that dirt is not necessarily soil, but soil *forms* in dirt. Another way to look at it is that soil is what happens to dirt.

Soil is the uppermost layer of the Earth's surface and contains mineral and organic matter capable of supporting plants (Figure 11.1). Another way to think

---

Soil *The uppermost layer of the Earth's surface that forms by the influence of parent material, climate, relief, and chemical and biological agents.*

Figure 11.1 Analyzing soil in the field. Soil scientists investigating a soil profile in a backhoe trench. Understanding soils is an important part of geography. Many geography students spend at least some time doing "hands-on" work like this.

about soil is that it is the outermost rind of Earth, similar to how an orange peel is the outermost part of the orange. Still another way to consider soil is as the transition between the atmosphere and the rocky Earth. Although all soils can be viewed in these various ways, they nevertheless vary dramatically in terms of their character and thickness, ranging from a few centimeters to several meters in depth.

Soils are indeed one of our most important natural resources. They are fundamental to life on Earth because of their interaction with climate and vegetation. Soils provide plants with physical support as well as nutrients and water for growth. Plants, in turn, support soils by anchoring them to Earth. Without a vegetative cover, soil is prone to erosion, which leads to reduced fertility for agriculture and can contribute to famines in the world's poorest nations.

Basic Soil Properties

The different variables that, taken together, comprise soil can be generally classified into the following four groups (Figure 11.2).

1. Inorganic materials Inorganic materials are naturally occurring chemical elements or compounds that possess a crystalline structure. Each specific mineral (element or compound) owes its structure to the particular rock from which it came. Some of the most common minerals found in soil contain the elements silicon, aluminum, iron, calcium, potassium, and magnesium. The element calcium, for example, is part of the rock limestone. As limestone disintegrates through a process called *weathering* (discussed in Chapter 14), the minerals become reduced in size enough for plants to use them as nutrients. The key point for right now is that, until the plants absorb the minerals, they are stored in the soil.

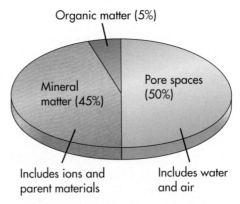

Figure 11.2 General composition of soil. Approximately 50% of soil is mineral and organic matter. The remaining 50% consists of pore spaces between grains. These spaces hold water and air.

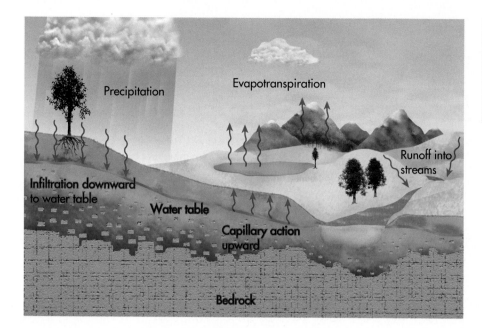

2. Organic matter Organic matter forms from living and decayed organisms and accumulates in the upper part of soil. As plants and animals die and decay, bacteria and fungi decompose their remains and produce **humus**, which is partially decomposed organic matter. Humus has a sponge-like quality, which allows soil to hold water. In addition, humus increases soil fertility because (a) it helps stabilize the soil by helping to hold it together and (b) microorganisms within humus give off compounds of nitrogen, phosphorus, sulfur, and other mineral substances that can be subsequently taken up by plants.

3. Water Water is critical to soils for a variety of reasons. The most obvious reason is that plants require water to grow. In turn, high plant production contributes to increased biomass, which generally results in more humus production and stable, more fertile soils. Water is delivered to soils in the form of rain and melting snow (Figure 11.3). After precipitation occurs, water can take several pathways with respect to the soil. If the precipitation is heavy, or if it falls on soils that are already very wet, some or most of the water will flow across the surface into local rivers and streams as runoff and never be absorbed into the soil.

When snow melts slowly, rain falls relatively slowly, or the soil is dry, water has a tendency to more readily infiltrate into soils. Once infiltration begins, some water will either evaporate into the atmosphere directly from the ground or when plants absorb it through their roots and carry it to their leaves through the process of transpiration. In the context of soil water loss, the combined processes of evaporation and transpiration are often lumped together in the single term *evapotranspiration*, which we described in Chapter 9.

Let's examine what happens to the soil water that is not lost immediately by evapotranspiration. A soil is *saturated* when water mostly fills all available pore spaces, except for a few small pockets of air here and there (Figure 11.4). Subsequently, within a few days, all excess water drains downward under the force of gravity toward the underlying groundwater zone, leaving larger pockets of air behind. Although most of this downward-moving water is lost as far as the soil is concerned, some of it will be pulled back up into the soil at a later time through the process of **capillary action** (see Figure 11.3). This occurs when the molecular attraction at the boundary between

---

Capillary action *The force that causes water to rise in the small tubular conduits within the soil.*

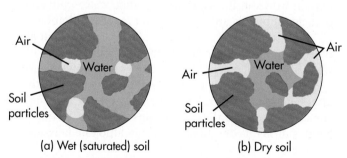

(a) Wet (saturated) soil   (b) Dry soil

Figure 11.4 Wet vs. dry soils. (a) A soil is considered to be saturated when pore spaces are nearly full of water, which occurs after extended periods of precipitation or snow melt. (b) Subsequently, water drains under the force of gravity, leaving water held to soil particles by surface tension. When most of the downward-moving water is lost, the soil is dry.

Humus *Decomposed organic matter, typically dark, that is contained within the soil.*

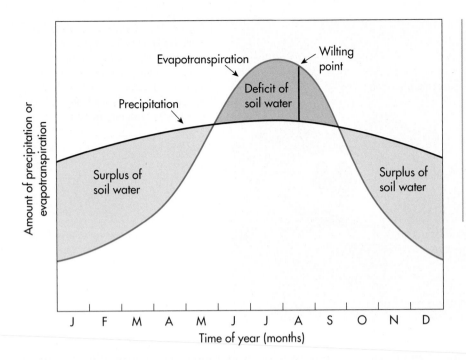

the underlying water and soil particles is stronger than the attraction between water molecules. To see how capillary action works, fill a transparent glass about half way with water. If you look at the side of the glass, where the water meets the inside surface, you can see that the edge of the water actually climbs a short way up the side of the glass. Water molecules have a slightly stronger attraction to the glass than they do to other water molecules. Capillary action in soils works the same way.

Although most water is flushed from the soil pores during the draining process, some remains in the soil–water belt because it is held to soil particles by the force of **surface tension**. Observe for yourself how surface tension works. Splash a few drops of water onto the surface of an upright bottle. Notice that each drop is enclosed within a film that pulls water inward into a rounded shape that withstands the force of gravity. Within soils, surface tension works in the same way, holding water onto soil particles until they are lost through evapotranspiration.

After the excess water has drained, the soil is left at **field capacity**, which refers to the maximum amount of water that a soil can hold after free gravitational drainage ceases. If additional precipitation occurs shortly after field capacity is reached, then the soil will again become saturated and the cycle begins anew. On the other hand, if drought conditions begin and little rain falls, soil

water will not be replenished. Even so, soil water will continue to be lost through evapotranspiration, resulting in progressively less water held by surface tension. If dry conditions persist, the soil reaches the **wilting point**, which represents the time when no soil water is available for plant use and plants literally begin to wilt.

The term **soil–water budget** reflects the balance between soil–water gains and losses. To understand the soil–water budget, Figure 11.5 presents a simple example that reflects changes over the course of the year in a typical midlatitude region. Imagine that the study location is a continental region in the midlatitudes. During most of the year, the region has a surplus of soil water. This surplus occurs from October through April and, interestingly, happens to correlate with the time of year when the least precipitation falls. Two reasons explain why the surplus occurs during this time of year. First, very little evaporation occurs because the Sun angle is low, temperatures are cool to cold, and it is frequently cloudy. Second, plants are not active during this time of year, so little soil water is lost due to transpiration. Taken together, more precipitation than evapotranspiration occurs during the cooler months of the year; thus, soils contain an excess of water. During the summer months, however, deficits of soil water frequently develop, even though precipitation is greater at that

Surface tension *The contracting force that occurs when the water surface meets the air and acts like an elastic skin.*

Field capacity *The amount of water remaining in the soil after the soil is completely drained of gravitational water.*

Wilting point *The threshold amount of soil water below which plants can no longer transpire water.*

Soil–water budget *The balance of soil water that involves the amount of precipitation, evapotranspiration, and water storage and loss.*

time of year, because the amount of evapotranspiration exceeds rainfall.

4. Air The air in soil is mostly carbon dioxide, which plants give off during respiration and then take in during photosynthesis. In fact, soils contain so much carbon dioxide that the soil layer is sometimes considered to be part of the atmosphere and thus a transition to the solid Earth below. The amount of air in the soil fluctuates depending on how wet the soils are. As you can see in Figure 11.4, when soils are saturated, the pore spaces are full of water and thus contain little room for air. In contrast, when soils dry out, the pore spaces contain less water and more air.

Soil-Forming (Pedogenic) Processes

A key concept to understand at this point is that soils form and evolve through a complex sequence of interrelated **pedogenic processes**. It is useful to think of the basic processes as consisting of *additions*, *depletions*, *translocations*, and *transformations* that occur within the soil (Figure 11.6). The following discussion illustrates how these processes operate individually.

Pedogenic processes *The natural processes of soil formation that involve additions, translocations, transformations, and losses.*

1) Soil Additions
- Organic matter
- Water (as precipitation, condensation, runoff)
- Oxygen, carbon dioxide (from atmosphere)
- Minerals (from precipitation)
- Sediments (from wind, water)
- Energy (from Sun)

2) Soil Translocations
- Clay, organic matter (carried by water)
- Nutrients (circulated by plants)
- Dissolved minerals (carried by water)
- Sediments (carried by animals)

3a) Upper Soil Depletions
- Water (by evapotranspiration)
- Carbon dioxide (from oxidation of organic matter)
- Nitrogen (by biological and chemical means)
- Sediments (by erosion)
- Energy (by longwave radiation)

4) Soil Transformations
- Organic matter (decomposed into humus)
- Particles made smaller (by weathering)
- Particles made into structures (by accretion)
- Minerals transformed (by weathering)

3b) Lower Soil Depletions
- Water and dissolved minerals

Figure 11.6 Basic soil-forming processes. Soils form through the complex interaction of several processes. Although each of these processes usually operates to some extent in all soils, the rate at which any given one operates varies from place to place.

What Is Soil? • 283

1. Soil additions Soil additions consist of material and energy that is added to the soil, such as organic material, water from precipitation, various minerals, solar radiation, oxygen and carbon dioxide from the atmosphere, or sediments deposited by wind or water. A good example of a soil addition is the leaf litter that accumulates during the fall under trees. In forest settings where the leaves are not raked, organic decomposers gradually break them down and their remains are incorporated into the soil by simple decay or with the aid of burrowing animals and worms. A similar process occurs when grasses die at the end of the growing season. Animal remains and animal excreta that accumulate on the ground also rapidly become part of the soil.

2. Soil depletions Soil depletions occur when some aspect of the soil is lost for one reason or another (see Figure 11.6). These losses can occur both in the upper and lower part of the soil. An example of a loss in the upper part of the soil is when erosion by wind, water, or glaciers removes sediment. Depletion in the upper part of the soil also occurs when plants take up nutrient minerals or water through evapotranspiration. Another form of depletion takes place when carbon is lost to the atmosphere as carbon dioxide when organic matter is oxidized. In a similar vein, nitrogen is lost to the atmosphere through biological and chemical means. The loss of longwave radiation to the atmosphere is another form of loss in the upper part of the soil. In the lower part of the soil, losses occur when water drains out of the soil and carries soluble minerals with it.

3. Translocations Translocations refer to the vertical movement of materials through the soil. Some movement occurs as nutrients are cycled through plants or when animals move soil around. A significant amount of additional movement occurs in the presence of water, which percolates downward through the soil during wet conditions. When water moves in this fashion, it dissolves minerals and carries them deeper within the soil, where they recrystallize and collect. When water moves materials, usually in the form of minerals, clays, and organic matter, *out* of the uppermost part of the soil, the process is called **eluviation**, or leaching. When leached materials from the upper part of the soil recrystallize *within* the lower part of the soil, the process is called **illuviation**. Figure 11.7 shows the results of the combined processes of eluviation and illuviation.

4. Transformations Transformations refer to the decomposition of minerals and organic matter into other forms within the soil. An excellent example of a transformation is when organic matter, such as leaf and animal litter, decomposes into humus. Another example is when minerals are transformed into other minerals or are made smaller by weathering. In addition, soil particles begin to clump together in a recognizable structure.

Eluviation   *The dissolution and downward mobilization of minerals by water in the soil.*

Illuviation   *The recrystallization of minerals that occurs directly below the zone of eluviation.*

Additions of dead plant matter from previous year

Living plant shoots/grass

Transformation of organic matter into humus by decay

Translocation of dissolved minerals by eluviation

Accumulation of eluviated minerals by illuviation

(a)

(b)

Figure 11.7 How additions and translocations affect soil development. (a) One way that additions occur in soil is through the accumulation of dead plant matter, which slowly transforms into humus by decay. The processes of eluviation and illuviation move materials vertically through soil. Eluviation moves material, such as dissolved minerals, out of the upper soil. These minerals recrystallize and accumulate in the lower part of the soil by illuviation. **(b)** In this photo of actual soil, the dense-looking whitish layer toward the base of the soil is calcium carbonate that was eluviated by water from the upper part of the soil.

Soil-Forming Factors

Let's now examine the factors that generally influence soil development. Five of these variables, traditionally called the **soil-forming factors**, consistently affect how soils evolve. These factors are climate, organisms, relief, parent material, and time; these variables can be remembered with the simple acronym *CLORPT*. Although you need to consider these factors in every soil, their character and relative significance vary considerably from a geographic perspective. In order to understand the discussion of specific soil types presented later in this chapter, you must first comprehend how soil-forming factors influence soil development individually and collectively. As you examine these variables, keep in mind that all of them contribute to soil development in some way in each and every soil.

1. Parent material An important concept to consider in soil science is that soils always form *within* sediment of some kind. This sediment is referred to as **parent material** and is directly related to the geology of the area in which a particular soil forms. The character of the parent material at any given locality is critical in the context of soils because it sets the stage for future development and affects the way in which additions, translocations, depletions, and transformations occur.

In general, soils form in two very broad types of parent material, residual and transported. **Residual parent material** is sediment that develops when underlying rock weathers *in place*; in other words, it has not been moved from someplace else. In this context it is important to remember that these rocks may be millions of years old, or even hundreds of millions of years old. The concept of weathering will be discussed more thoroughly in Chapter 14, but for now, think of weathering as the process by which rock slowly breaks up into smaller fragments and particles. This breaking up takes place through chemical and physical changes that are largely associated with water.

As this weathering occurs—in other words, as the rock fragments become smaller and change their chemical composition—a material called **regolith** begins to form. Figure 11.8 shows the relationship between bedrock, regolith, and soil. Notice how the rocks in the upper part of the bedrock break up into smaller pieces. Somewhere in this part of the vertical sequence is a transition in which the bedrock is sufficiently altered, through the weathering process, to be fundamentally different from the underlying (solid) rock from which it came. At the same time, it has not been modified by additions and translocations from above that occur in soil. In other words, regolith is not rock and is not soil. Rather, it lies somewhere between the two and is the material in which the soil forms. Often, regolith production at the bedrock interface occurs at the same time that soil development is occurring in the upper part of the weathered zone.

Soil-forming factors *The variables of climate, organisms, relief, parent material, and time that collectively influence the development of soil.*

Parent material *The mineral or organic material in which soil forms.*

Residual parent material *Parent material that forms by the weathering of bedrock directly beneath it.*

Regolith *The fragmented and weathered rock material that overlies solid bedrock.*

Figure 11.8 Relationship of regolith to soil. (a) Regolith consists of weathered bedrock that forms parent material in which soil forms. (b) Limestone bedrock and overlying regolith in the Great Plains. Note how the size of rock fragments becomes progressively smaller above the horizontal slab of solid rock. The soil is the dark zone at ground level.

The second kind of parent material is **transported parent material**. As the name implies, transported parent material differs from residual parent material because it has been moved from one place to another. Remember that residual parent material is regolith that forms *in place* as bedrock weathers. Although the sediments that comprise this bedrock were deposited after being moved from someplace else, the deposition happened such a long time ago (millions of years ago) that, as far as regolith production is concerned, it is considered to have occurred in one place. On the other hand, transported parent material is sediment that was moved relatively recently by water or glaciers. Or perhaps wind blew sand to form dunes like those pictured in Figure 11.9. If this sediment is left undisturbed for many years, plants and animals will begin to live in the uppermost part of the deposit and a soil will form through the combined processes of additions, depletions, translocations, and transformations.

2. Climate Climate is a critical variable in the context of soil development. Temperature and precipitation are the most important aspects of climate that affect soil formation because they influence the kind and rate of biological and chemical reactions that occur in the soil. In general, more reactions occur in warmer and wetter places than in colder and drier regions. As a result, soils

---

Transported parent material *Parent material such as glacial or stream sediments that has recently been deposited and in which soil forms.*

tend to be better developed and thicker in warm, humid regions than in cold, dry locations.

Climate also dramatically influences the process of translocation in a soil. Remember that translocation refers to the movement of organic and mineral matter from the upper part of the soil into a deeper zone. Most of this movement occurs when water percolates through the soil after it falls at the surface. To visualize this process,

Figure 11.9 Example of transported parent material. These sand dunes along the shore of Lake Michigan consist of sand deposits that were blown there by the wind. Note the grass growing on parts of the dunes. The establishment of grass begins the process of soil formation by providing organic material to be added to the sand parent material.

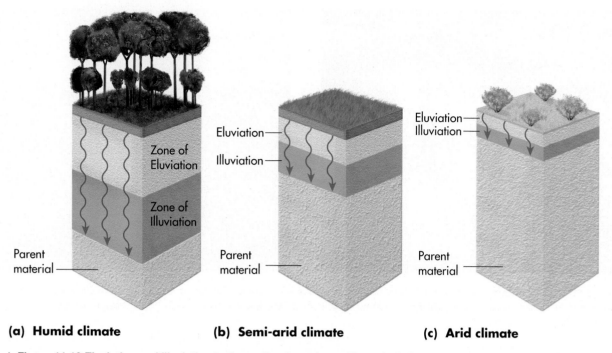

(a) Humid climate **(b) Semi-arid climate** **(c) Arid climate**

Figure 11.10 Eluviation and illuviation in three climate regimes. More eluviation takes place in (a) humid climates than in (b) semi-arid and (c) arid climates. In addition, the zone of illuviation (where minerals precipitate from solution) is deeper and thicker in more humid climates. This zone typically is progressively thinner and less deep as the climate becomes increasingly dry.

imagine, for example, the gradual melting of snow that occurs after a big winter storm passes through the area. As the snow melts, some of it evaporates into the atmosphere, but much of it is absorbed into the ground. When spring arrives, plants rapidly take up some of this soil water, but a certain percentage of the water continues to move slowly downward within the parent material under the force of gravity. As this water comes into contact with minerals in the soil or sediment, it dissolves them—much like how water dissolves sugar—and carries the dissolved particulates deeper into the soil through the process of eluviation. At some point, the water stops flowing downward and the dissolved minerals recrystallize in the zone of illuviation. Simply put, more eluviation occurs in humid climates than in arid regions because more water percolates through the soil. Figure 11.10 shows in general terms the depth to which eluviation penetrates and the resulting depth of the illuvial zone in humid, semi-arid, and arid climate regimes.

3. Organisms Organisms are the living things, both plant and animal, that reside within soil. The importance of organisms in soil formation simply cannot be overestimated. At a fundamental level, organisms give life to soil and contribute greatly to the notion of soil being significantly more than *just dirt*. At a more

complex level, organisms have a symbiotic relationship with soil in that they not only acquire their food from it, but also help regulate the soil environment in which they live.

Organisms contribute to soil health and formation in many different ways. One way is when plant roots penetrate deeply into the ground. This penetration creates passageways in the soil in which water and oxygen can flow. Many small animals, such as ants, ground squirrels, prairie dogs, and other rodents, spend the majority of their lives burrowing through the soil (Figure 11.11). Through this ongoing process, the soil is continually mixed in a process called **bioturbation**. The mixed soil is softened and aerated, which makes it easier for plants to grow. Earthworms are very important within the system because they recycle soil elements through consumption and waste production. Another important biological component is the huge number of micro-organisms, such as algae, fungi, and bacteria, that live within the soil. The primary function of these microorganisms is to assist with the decomposition of organic matter into humus.

Bioturbation *The mixing of soil by plants or animals.*

Figure 11.11 Soil bioturbation (a) Prairie dogs in the Great Plains constantly burrow in the soil, which recycles nutrients and keeps soil loose and well aerated. (b) Termite mounds in central Brazil are composed of earth brought to the surface by the termites. This process mixes the soil.

4. Relief Recall that topography refers to the configuration of the Earth's surface—in other words, the position, shape, and orientation of hills, valleys, and mountains. In a related vein, the term **relief** refers to the differences between the highs and lows of a landscape (Figure 11.12a). The Rocky Mountains, for example, is a region of very high relief because tall peaks are separated by lower valleys. In contrast, much of the Midwest has low relief because the landscape is relatively flat.

The most important variable associated with relief and soil development is slope. The slope of the landscape is loosely defined as the ground surface that connects the higher areas with lower areas. Slope is significant in the context of erosion because, in general, soils are thinner and less well developed where slopes are steeper. Figure 11.12b shows a hypothetical landscape with a steep slope next to a relatively flat surface. Note that the soil on the steep slope is thinner; that is, the bedrock is closer to the surface. This pattern occurs because more erosion takes place on the steeper slope, predominantly due to water flowing downhill under the force of gravity. As a result, the rate of erosion on the steeper hillslope may be equal to, or even greater than, the rate of soil formation (Figure 11.12c). On the more level surface, in contrast, the soil can develop more fully because it lies essentially undisturbed from erosion. In addition, the soil is thicker because sediments eroded from the steep slope are added where the slope is more level.

Relief *The difference between the high and low elevation of an area.*

5. Time Time is a critical variable because soils need it to develop. If it is assumed that all the other soil-forming factors (parent material, climate, organisms, relief) are equal, then the best-developed soils occur on surfaces that have been stable for the longest period of time. For example, where would you expect to find the better-developed soil: in a 3000-year-old sand dune or in a 100,000-year-old stream deposit? The answer is that the older stream deposit most likely has a better-developed soil because it had more time to develop. The significance of this greater amount of time is that collectively more additions, translocations, depletions, and transformations have taken place in the older soil than in the younger soil.

KEY CONCEPTS TO REMEMBER ABOUT PEDOGENIC PROCESSES AND SOIL-FORMING FACTORS

1. Four pedogenic processes—additions, translocations, deletions, and transformations—collectively promote soil development.
2. Some pedogenic processes cause physical changes to occur within the soil, whereas others cause soil components to move within the soil.
3. Pedogenic processes occur within the context of five soil-forming factors: parent material, climate, organisms, relief, and time.
4. Each soil-forming factor contributes to all soil development in some way.

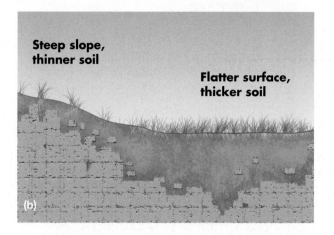

Figure 11.12 Relief and soil formation. (a) The term *relief* refers to the differences between the highs and lows of a landscape. In this scene, the area in the foreground is a zone of low relief, whereas the mountain in the background is an area of high relief. (b) Soils are generally thinner on steeper slopes than on flatter surfaces because steep slopes are more prone to erode. (c) Relief affects soil thickness on slopes, such as the exposed bedrock (arrow) on the steep slope shown here. The soil is very thin on this slope because of erosion; soil is thicker on top of the ridge and in the valley.

Measurable Soil Characteristics

To understand soil formation thoroughly, remember that the soil-forming factors just discussed are individual parts of a holistic system, in the sense that each contributes something to the development of any particular soil. As a soil forms under a particular set of environmental conditions, it develops properties that are unique to that particular setting. Using people as an analogy, each person whom you know has different characteristics that you can readily identify, such as brown hair, blue eyes, height, etc. With respect to soils, the properties are physical characteristics that can be measured to distinguish different soils from one another. By examining the physical properties of a soil, it is possible to learn something about how the soil formed. These properties are color, texture, structure, soil chemistry, and soil pH. Although these properties will now be discussed individually, remember that all work together to give any one soil a particular overall character.

1. Color Soils vary considerably with respect to color, both within and between soils. Color can be an indicator of the composition of the soil. For example, soils with dark colors usually have high humus or organic matter content (Figure 11.13a), which might indicate that the biomass of the region is high or that the environment is cool and wet or that organic matter decomposes slowly. Soils with a high content of iron-bearing minerals are often a reddish color, whereas soils (or parts of soils) that contain abundant calcium are whitish in color. These colors might reflect the color of the local parent material or they may indicate the amount and type of eluviation that has occurred over time.

Soil color is typically measured with a Munsell Color Chart (Figure 11.13b). This color-system uses three elements of color—hue, value, and chroma—to designate a specific color. *Hue* is a measure of the chromatic composition of light that reaches the eye. The Munsell system is based on five principal hues: red (R), yellow (Y), green (G), blue (B), and purple (P). In addition, five intermediate hues represent midpoints between each pair of principal hues, resulting in ten major hue names. The intermediate hues are yellow-red (YR), green-yellow (GY), blue-green (BG), purple-blue (PB), and red-purple (RP). *Value* indicates the degree of lightness or darkness of a color in relation to a neutral gray scale under natural lighting, ranging from pure black (0/) to pure white (10/). *Chroma* is the relative strength of the

Figure 11.13 Differences in soil color. (a) The lower part of this soil in Oklahoma is red because it formed in residual parent material (regolith) derived from iron-rich bedrock. In contrast, the upper part of the soil is relatively dark because of the addition of humus to the parent material. In other words, all the parent material was reddish at one time, but the upper part was transformed to a darker color through the addition process. (b) Soil color is measured with a Munsell Color Chart, which specifies hue, value, and chroma.

DISCOVER...

SOIL CONSERVATION ON STEEP SLOPES

Soils on steep slopes are prone to erosion because flowing water has a lot of energy and thus moves sediment. You can tell that hillslope erosion has been a problem in places where conservation terraces are found. Farmers build these features to reduce the effects of soil erosion by creating a series of stair-steps on steep hillslopes. These steps produce a series of essentially horizontal surfaces on which farmers plant their crops. Each of the terraces is separated by a built-up ridge of dirt, called a berm, which reduces the velocity of water and traps sediment. Although some soil erosion still occurs where terraces exist, the amount is significantly reduced. Perhaps the most impressive conservation terraces on Earth are the rice paddies shown here in the mountainous area of Yunnan Province, China. Note that each of these surfaces can hold a great deal of water, which falls as precipitation during the summer monsoon.

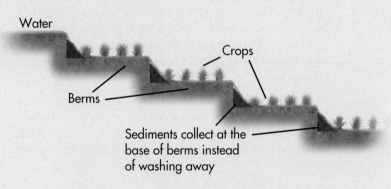

Water

Crops

Berms

Sediments collect at the base of berms instead of washing away

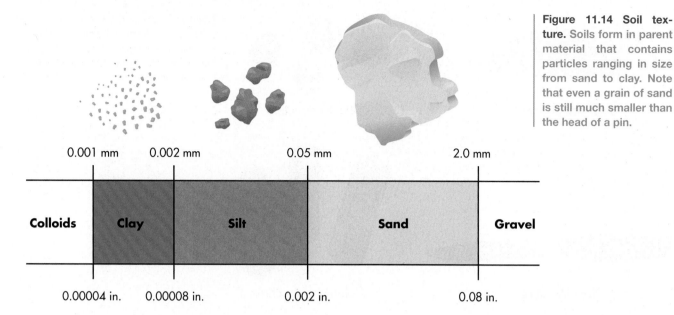

Figure 11.14 Soil texture. Soils form in parent material that contains particles ranging in size from sand to clay. Note that even a grain of sand is still much smaller than the head of a pin.

0.001 mm	0.002 mm	0.05 mm	2.0 mm

Colloids	Clay	Silt	Sand	Gravel

0.00004 in.	0.00008 in.	0.002 in.	0.08 in.

spectral color. The scales of chroma for soils extend from /0 for neutral colors to a chroma of /8 for the strongest soil color. To put it all together, a soil that is pale brown, for example, is designated 10YR 6/3. In contrast, a soil that is very dark brown has a Munsell designation of 10YR 2/2. Still another example is a soil that has a yellowish red (7.5YR) hue and is strong brown in color. This soil would have a 7.5YR 5/6 color designation.

2. Texture As discussed earlier, soil contains a multitude of inorganic particles that are typically related to the parent material. Most soils contain a continuum of three distinct size categories (Figure 11.14), which are (in order from largest to smallest) sand, silt, and clay. Some larger (gravel) and smaller (colloid) particles may also be present. The combined percentages of these soil particulates are referred to as the textural class of the soil, as shown in the soil textural triangle in Figure 11.15.

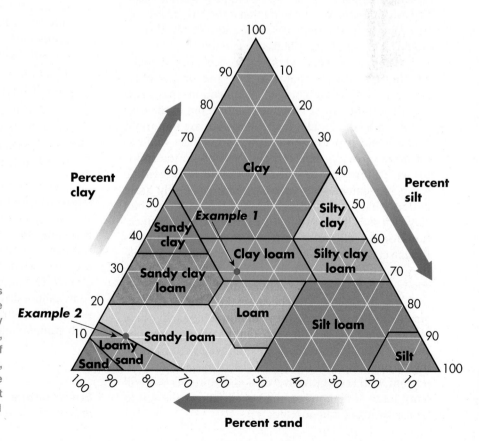

Figure 11.15 Soil textural triangle. Soils can be texturally classified based on the relative proportion of sand, silt, and clay in the parent material. For percent sand, use the lines slanting from the bottom of the triangle up to the left; for percent silt, use the lines slanting from the right side of the triangle down to the left; for percent clay, use the horizontal lines. Examples 1 and 2 described in the text are shown.

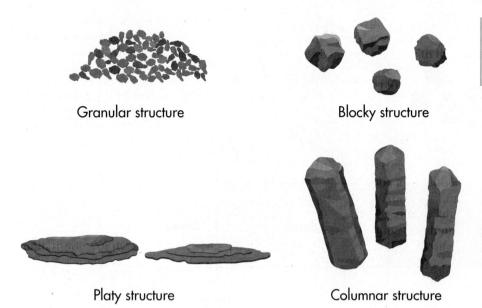

Granular structure

Blocky structure

Platy structure

Columnar structure

Figure 11.16 Basic soil structures. Particles within the soil clump together in various ways, resulting in a distinct classification system for soil structure.

So how does the textural triangle work? Imagine, for example, that a soil contains 40% sand, 30% silt, and 30% clay. To determine the textural class, simply follow the relative percentages of each variable to where the lines intersect within the triangle. For percent sand, use the lines slanting from the bottom of the triangle up to the left; for percent silt, use the lines slanting from the right side down to the left; for percent clay, use the horizontal lines. In this particular case, the soil would be texturally classified as *clay loam*. Now imagine that a soil has the relative percentages of 80% sand, 10% clay, and 10% silt. This soil would be texturally classified as *loamy sand*.

You may wonder what difference does it make whether a soil is clay loam, loamy sand, or any other textural class? These kinds of classifications may seem mind-numbing rather than informative, but in the case of soils individual textural classifications are important because they provide a rough measure of how well water flows through a soil or how appropriate the soil is for agriculture, just to name two examples. Regarding water movement, soils that are sandy generally drain better—that is, water flows through them faster—than clay-rich soils. Thus, soil texture has significant implications for the soil–water budget in any given locality. Although clay-rich soils hold more water than sandy soils, they are frequently difficult to work with because clays stick together so well. In general, the best agricultural soils are the loamy soils because they are nicely balanced. On the one hand, they contain some sand, which allows water to drain through. On the other hand, they contain some clay, which keeps the water from draining through the soil too fast. These clays are also important with respect to overall soil fertility, as you will learn later. A high percentage of silt allows the soil to be easily mixed and aerated.

3. Structure **Soil structure** refers to the way in which soil particles naturally clump together in *peds*, which are natural soil aggregates. Another way of looking at structure is that it is related to how clumps of soil break apart and their resulting shape, size, and arrangement. Soil structure is a difficult concept to illustrate; nevertheless, four primary types of structure are recognized (Figure 11.16).

- *Granular* structure occurs when the peds break up into very small, rounded clumps. You can recognize granular structure when the soil is slightly cohesive but still easily spills between your fingers.
- *Blocky* structure occurs when the clumps break up into distinct blocks that can be easily held.
- *Platy* structure occurs when the peds are stacked in distinct plate-like forms.
- *Columnar* structure occurs when the peds are organized vertically.

Again, you might be saying to yourself, "What difference does it make how soil particles group together?" The fact is that it is extremely important to how water percolates through the soil and how easily plant roots can penetrate it. Which soil structure presented in Figure 11.16 would be the most difficult for plants to penetrate? The answer is the platy structure because the roots cannot drive straight down through the soil. As far as water's infiltration is concerned, it flows best in soil with granular and blocky structure.

Soil structure *The way soil aggregates clump to form distinct physical characteristics.*

What is the texture of a soil that is 50% sand, 40% clay, and 10% silt?

a) Loam
b) Silt
c) Sandy clay loam
d) Sandy clay

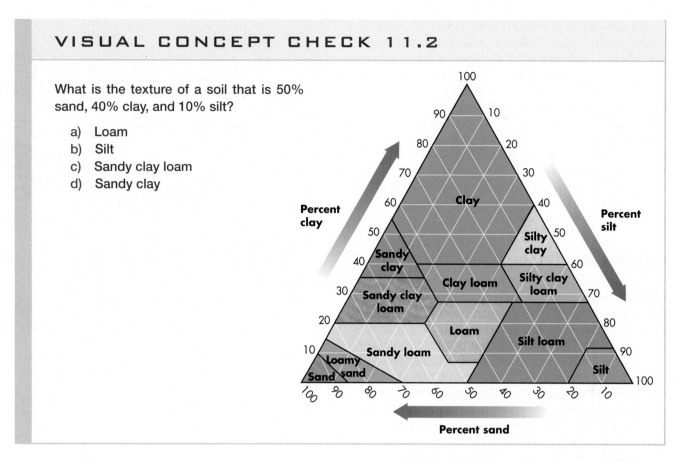

Soil Chemistry

Given that soils contain a variety of mineral, organic, water, and atmospheric components, each soil has a distinct chemical composition that is important regarding its formation and fertility. The chemical composition of any given soil is determined by a complex set of chemical reactions that occur within the soil solution. As noted before, when water comes into contact with minerals under the right conditions, it causes many minerals to dissolve. Similarly, when water mixes with carbon dioxide and various forms of organic matter, it creates carbonic and organic acids, respectively, that promote further changes within the soil. A primary product of these kinds of transformations is chemical ions that are free to be absorbed by plants.

Soil pH

One of the ways to characterize the chemical composition of soil is by measuring its **pH**, which is the relative concentration of hydrogen ions (H^+) present in solution. The pH scale ranges from 0 to 14, with lower values indicating acidic conditions and higher values indicating basic (alkaline) conditions. Neutral pH is about 7, which is the pH of pure water. To get a feel for pH and how it relates to various common products, see Figure 11.17.

pH *The measure of acidity or alkalinity of a solution, ranging from 0 to 14, based on the activity of hydrogen ions (H^+).*

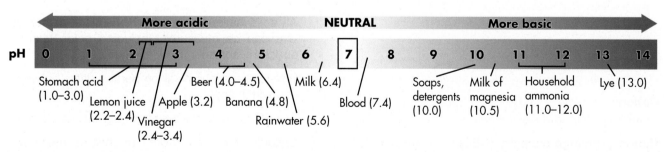

Figure 11.17 The pH scale. pH refers to the concentration of hydrogen ions (H^+) in solution. Note that most natural foods are slightly acidic and most common cleaning agents are basic.

In the context of soils, pH is an important indicator of soil fertility. Highly alkaline soils, for example, are not efficient with respect to the dissolution of minerals and making them available as nutrients for plants. Acidic soils, on the other hand, result in extensive leaching of minerals—so much so, in fact, that many nutrients are completely lost before they can ever be consumed by plants. In general, most plants, including agricultural crops, are adapted to soils that have a neutral or near-neutral pH. It is possible, however, to treat slightly acidic soils with alkaline fertilizers, such as slaked lime (calcium carbonate, $CaCO_3$), to raise the pH to the appropriate level.

Colloids and Cation Exchange

An important part of soil chemistry is the concept of **colloids**, which are extremely small (<0.001 millimeter) particles that exist in inorganic and organic form. Inorganic colloids consist of crystalline clays that are thin and platy. These microscopic bodies are created when larger particles are chemically altered through weathering. Organic colloids, in contrast, consist of humus derived from organic matter. Regardless of whether colloids are inorganic or organic, they are critical to soil chemistry because they are chemically active and have a high water-holding capacity.

Colloids are also important to soil chemistry because they hold and exchange cations. **Cations** are positively charged (atomic) ions that exist in soil solution due to the dissolution of soil minerals, such as calcium, magnesium, iron, and potassium. Because soil colloids are negatively charged, they attract the positively charged cations that are suspended in the soil solution. Although colloid particles are individually small, they collectively have a very large surface area on which cations can become attached. Some ions, such as metal and hydrogen ions, are bound tightly to colloids, whereas basic ions such as calcium move relatively freely. In many instances, the loosely bound (basic) cations are replaced by the metallic and hydrogen ions through the process of cation exchange. Soils that have a high **cation exchange capacity (CEC)** are typically the most fertile because they contain an

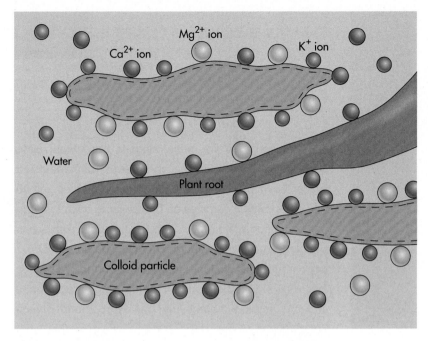

Figure 11.18 Microscopic view of colloids and soil fertility. Colloids are microscopic clays and organic particles that are negatively charged. Cations, which are positively charged atoms that exist in soil solution, are thus attracted to the colloid particles, which hold them until a plant root absorbs them.

abundance of colloids on which cations can be held and exchanged.

In the context of soil fertility, colloids are important because they keep minerals from being completely leached from the soil. Often, cations are held by the colloid until a plant root comes into contact with the colloid and absorbs the cation as a nutrient. Figure 11.18 illustrates how individual, negatively charged colloid particles (in this case, clays) attract the positively charged cations surrounding them until a plant root "consumes" the stored nutrients.

KEY CONCEPTS TO REMEMBER ABOUT MEASURABLE SOIL CHARACTERISTICS AND SOIL CHEMISTRY

1. All soils have physical and chemical properties that can be measured, including color, texture, structure, soil chemistry, and soil pH.

2. Physical and chemical properties differ between and within soils.

3. Soil pH refers to the acidity or basicity of the soil.

4. Colloids are collections of microscopic soil particles that hold and exchange cations.

Colloids *Very small (10 nanometers to 1 micrometer), evenly divided solids that do not settle in solution.*

Cations *Positively charged ions, such as sodium, potassium, calcium, and magnesium.*

Cation exchange capacity (CEC) *The total amount of exchangeable cations that a soil can absorb.*

Soil Colloids and pH

The chemical composition of any given soil is determined by a complex set of reactions taking place within the soil solution. The ultimate outcome is often dependent on the nature of soil colloids, which are microscopic particles that occur in organic and inorganic form. These particles have an important role as far as cation exchange and soil pH is concerned. To better understand these relationships, it is useful to view them in animated form. Go to the *GeoDiscoveries* website and select the module *Soil Colloids and pH*. Once you complete the module, be sure to answer the questions at the end to test your understanding of this concept.

Soil Profiles (Reading the Soil)

With the various soil components, processes, and chemistry in mind, let's now investigate the progressive development of soil that results from additions, transformations, translocations, and depletions. The really nice thing about these processes is that they cause the soil to become progressively organized through time in a way that can be easily seen by looking at a **soil profile**. Just as the profile of your face refers to a side view that ranges from the top of your head to your chin, a soil profile displays all the soil characteristics in a vertical "image." A soil profile can be obtained by cleaning a natural exposure such as a creek bank, or, more commonly, by excavating a soil pit with a shovel or backhoe. Although it may not seem like it, this kind of work is really a great deal of fun. Figure 11.19 shows examples of exposing a soil profile.

As soils form, they organize internally into distinct **soil horizons**. If the term "horizon" seems confusing, simply think of the term as being equivalent to "layer" or "band" for the time being. Look again at the soil profile in Figure 11.19a. Do you see the distinct layers that occur in the profile? These layers are referred to as soil horizons because they typically blend into one another; in other words, it is very rare to see abrupt boundaries between different parts of the soil.

Soil profile *A vertical exposure in which all soil components can be seen.*

Soil horizons *The distinct layers within a soil that result from pedogenesis.*

Figure 11.19 Soil profiles. (a) After a soil pit is excavated you can then see the soil profile and its specific characteristics, such as the dark zone near the surface of this particular soil and the reddish zone beneath it. (b) Sometimes soils are exposed naturally, such as along this stream cutbank. In these situations, they can easily be studied.

The specific character of soil horizons varies dramatically between different parts of the world, but, in general, any well-developed soil contains, from top to bottom, O, A, E, B, C, and R horizons (Figure 11.20). These horizons form by different processes and have the following six characteristics.

1. **O horizon**—The **O horizon** is the uppermost horizon of the soil, existing at ground level, and consisting mainly of undecomposed organic matter. A good example of an O horizon is the leaf litter that accumulates on the forest floor during the fall season. Thus, an O horizon forms through the addition of organic matter.

2. **A horizon**—The **A horizon** is the uppermost mineral horizon of the soil. This horizon is distinctive because it contains abundant humus, which consists of decomposed organic matter added to the soil. Given their relatively high humus content, A horizons are typically a darkish gray color. In general, A horizons have granular structure due to the high concentration of humus. In Figure 11.13, for example, the dark horizon at the top of the soil profiles is an A horizon.

3. **E horizon**—The **E horizon** lies immediately below the A horizon (see Figure 11.20) and is the zone of eluviation; thus, this horizon forms largely through the process of translocation. In other words, this horizon evolves due to the downward movement of minerals leaching from it. As a result of this eluvial process, the E horizon is usually relatively light in color compared to the horizons above and below it.

4. **B horizon**—The **B horizon** lies below the E horizon and is the zone of illuviation and thus forms largely through the process of translocation. In other words, this part of the soil evolves because minerals precipitate at this level. Recall that the zone of illuviation is deeper in more humid climates than in relatively dry ones, which is to say that the B horizon is deeper and thicker where precipitation is greater (see Figure 11.10). Given the accumulation of eluviated minerals in the B horizon, this horizon usually has stronger structure (such as "blocky") than the overlying horizons.

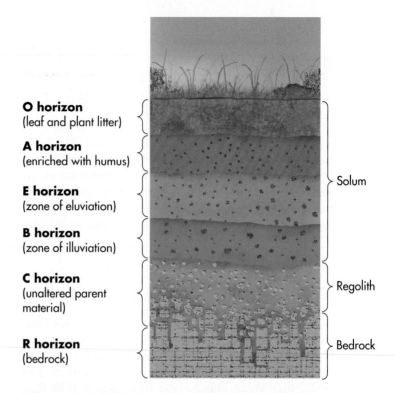

O horizon
(leaf and plant litter)

A horizon
(enriched with humus)

E horizon
(zone of eluviation)

B horizon
(zone of illuviation)

C horizon
(unaltered parent material)

R horizon
(bedrock)

Solum

Regolith

Bedrock

Figure 11.20 Soil horizons in a fully developed soil on a bedrock-dominated landscape. With time, soils evolve to produce a predictable sequence of horizons. Keep in mind that the soil in this illustration formed within regolith.

The base of this horizon corresponds to the base of the soil. The full thickness of the soil, from the top of the A horizon to the bottom of the B horizon, is referred to as the **solum**.

5. **C horizon**—The **C horizon** consists of unaltered parent material that has yet to be affected by soil-forming processes. In other words, this horizon is regolith and is not part of the solum.

6. **R horizon**—Where soils form in landscapes dominated by bedrock, such as those you find in the Appalachian Mountains or parts of the Great Plains, an R horizon occurs. Very simply, the **R horizon** is the bedrock that is the source of the regolith/parent material, which is the sediment in which the soil forms.

O horizon   *The uppermost soil horizon that consists of undecomposed plant litter.*

A horizon   *The soil horizon that is enriched with humus.*

E horizon   *The soil horizon that is progressively depleted of minerals through dissolution and translocation.*

B horizon   *The soil horizon that forms below the E horizon because translocated minerals recrystallize.*

Solum   *The A, E, and B horizons of a soil, which form through pedogenic processes.*

C horizon   *Unaltered soil parent material.*

R horizon   *Unweathered bedrock that underlies soil.*

Time and Soil Evolution

Although the profile in Figure 11.20 is typical, it is far from being the profile observed in every soil. In fact, not all horizons are present in every soil. Horizons also vary in color and configuration between different soils. These differences are due not only to the five soil-forming factors discussed previously, but also to geographic variability in the way that soil processes function.

To see how soil horizons generally evolve, consider Figure 11.21. This figure presents the formation of a soil in a hypothetical profile observed at four separate times: today (0), 500, 1000, and 5000 years into the future. Imagine that "time 0" begins with the deposition of sediment deposited along a stream during a big flood. This sediment is thus transported parent material within which a soil will form given the additions, translocations, transformations, and depletions that occur "in the future." To understand how this occurs, it is essential to think of the system in the holistic fashion presented in the next three paragraphs.

If you could visit this site at three future times, you would see that the soil evolves in a fairly predictable way. This is not to say that each soil would change in the same way with respect to horizon thicknesses or colors, or that it would evolve at the same rate given climate and textural differences. However, the overall pattern very well could be consistent. Following deposition of the fresh parent material, the first horizon that would become obvious in the first 500 years would likely be the A horizon. This horizon would develop due to the additions of organic matter that resulted from the plant and animal life at or near the surface. (Given your understanding of soil horizons, what would be the diagnostic characteristic of this horizon?)

At the same time that the A horizon is forming, translocations and transformations would also be occurring. The cumulative result of these combined processes might begin to be seen about 1000 years in the future by the development of a Bw horizon immediately beneath the A horizon. The "w" means that the horizon is weakly developed, and will probably be identified because the structure has become slightly blocky due to the illuviation of clays and other minerals at this level of the soil. Another thing to notice is that the A horizon has become thicker because of the continual additions that have occurred over the millennia. It is probably darker as well.

If you returned to examine the soil pit 5000 years into the future, you might very well discover that the soil had become fully developed. Remember, this development has occurred because plants, microbes, and animals have been interacting within this soil for a lengthy period of time. In addition, water has been percolating through the soil for the same length of time. As a result of these

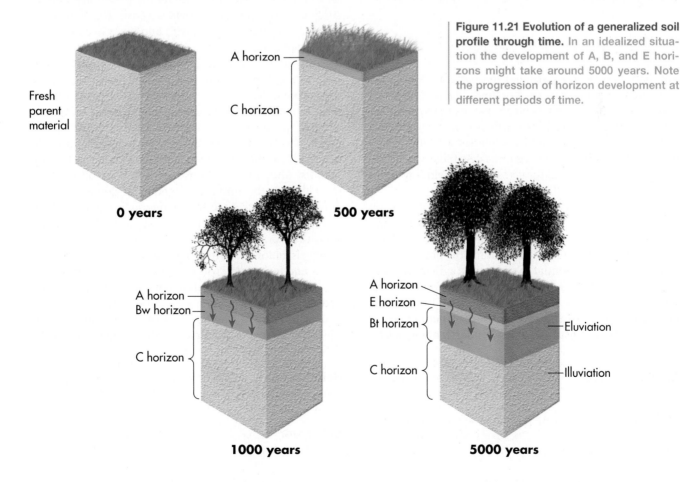

Figure 11.21 Evolution of a generalized soil profile through time. In an idealized situation the development of A, B, and E horizons might take around 5000 years. Note the progression of horizon development at different periods of time.

The formation of soil horizons is complex because it involves the internal organization of sediment through a combination of additions, depletions, transformations, and translocations over a period of time. This organization is best understood when it is viewed in an animated way that demonstrates the progressive development of soil. To do so, go to the *GeoDiscoveries* website and select the module *Soil Horizon Development*. Once you complete the module, be sure to answer the questions at the end to test your understanding of this concept.

---

processes operating day by day, month by month, and year by year for 5000 years, the soil has a distinctive E horizon and a thick Bt horizon. The E horizon is the zone of the soil that has been extensively leached of its mineral and clay components. These minerals and clays moved downward and recrystallized in the evolving B horizon. In this case, the inclusion of the lowercase "t" means that the horizon is texturally distinctive, through the illuviation of clay at that level, from the rest of the soil. Another way to make this distinction is to say that the B horizon is likely to be an **argillic horizon** with a strong blocky structure. This structure developed because abundant clay was added to this part of the soil through eluviation.

Soil Science and Classification

The previous section discussed the basic model of soil formation and the various horizons that evolve over time. Given that model, you can now examine some of the soil variability on Earth, which is known because of the research conducted by scientists working in the field of **soil science**. This scientific discipline specifically deals with soils as natural resources, including their physical, chemical, and biological properties.

Soil science is a discipline particularly relevant to humans because we use soils in countless ways. Of particular importance is that soils are the medium in which we grow our food. Thus, understanding the properties of soils is absolutely essential to systematic agriculture and directly affects the quality and price of the food we buy. Soils are also the fabric in which we build our homes, buildings, highways, and myriad other structures. As a result, engineers must have a good understanding of soil

characteristics in order to build structures that are properly supported. Soils are also important filters for surface and groundwaters and are thus highly relevant to hydrology, which focuses on the way that water moves on Earth. People who study soils are called *soil scientists* (Figure 11.1). The goal of soil scientists is to improve our understanding of the Earth's soils in order to preserve and efficiently utilize them.

One of the ways in which soil scientists study and manage soils is in association with the concept of soil classification. In general, classification is a common practice in many scientific disciplines because it makes it easier to remember attributes about the phenomena under study. Classification also provides a basis from which one category can be compared with another. Given the predictable pedogenic processes and factors that operate on Earth, soils are particularly well suited for classification.

Although many different classification systems are used around the world, the one used in the United States is called **soil taxonomy**. This system is based on the existing properties of a soil, such as color, texture, structure, and mineral content, which can be measured. Many other soil classification systems are genetic; that is, the soil scientist attempts to reconstruct how the soil evolved, even though the initial environmental conditions may not be known. Because soil taxonomy is a generic, logically based system, the classification scheme is hierarchical with several levels of general classifications, each with sublevels below it. With each successively lower level, fewer and fewer similarities appear between soil types.

The highest level within soil taxonomy is the **soil order**. Twelve soil orders occur around the world, each one distinguished on the basis of diagnostic horizons (Table 11.1) that meet certain criteria. These diagnostic

---

Argillic horizon *A B horizon that is enriched in eluviated clay.*

Soil science *The study of soil as a natural resource through understanding of its physical, chemical, and biological properties.*

Soil taxonomy *The method of soil classification that is based on the physical, chemical, and biological properties of the soil.*

Soil order *A group of 12 distinctive soils differentiated at the most general level.*

Well-Developed Mineral Soils

Order Name	Derivation	Characteristics
Alfisols	Nonsense syllable	Soils with high-base-content B horizon rich in eluviated clay. Found from low to subarctic latitudes.
Aridisols	Latin: *aridus* = dry	Soils in dry climates that are low in organic matter and frequently have subsurface horizons rich in calcium carbonate or soluble salts.
Mollisols	Latin: *mollis* = soft	Soils in semi-arid/subhumid grasslands in midlatitudes that have humus-rich A horizon and a B horizon that has a high base status.
Oxisols	French: *oxide* = oxide	Highly weathered soils in tropical environments that have low base status and subsurface horizon that has a high oxide concentration.
Ultisols	Latin: *ultimus* = last	Highly weathered soils in tropical and subtropical environments that have low base content and a subsurface horizon rich in eluviated clay.
Spodosols	Greek: *spodos* = wood ashes	Soils in cool, moist environments that have a B horizon rich in eluviated sesquioxides.
Vertisols	Latin: *verto* = turn	Tropical and subtropical soils with high base status that contain an abundant amount of expandable clay that swells when wet and shrinks when dry.

Organic Soils

Gelisols	Latin: *gelare* = freeze	Largely organic soils that form in extremely cold environments where permafrost is thick. Due to repeated freezing and thawing, soil horizons and surface expression are chaotic.
Histosols	Greek: *histos* = tissue	Very dark soils consisting mostly of organic matter. Typically found in cool/moist environments where organic decomposition is slow.

Weakly Developed Soils

Andisols	Spanish: Andes	Weakly developed soils formed within glassy volcanic sediments ejected by active volcanoes.
Entisols	Nonsense syllable	Horizonless soils usually formed within recently deposited sediments.
Inceptisols	Latin: *inceptum* = beginning	Soils with poorly developed horizons, but which may evolve further.

horizons are the **epipedon** (from the Greek word *epi*, meaning "over" or "upon"), which is the horizon formed at the soil surface, and the subsurface horizons formed by removal or accumulation of material.

Soil orders are subdivided into several categories, which are, from highest to lowest, *suborders, great groups, families,* and *series*. To give you an idea of how extensive this hierarchy is, 50 suborders and 180,000 soil series are recognized in the United States. It is well beyond the scope of this text to describe each of the soil series that has been identified, but we will outline the basic characteristics of the 12 soil orders. This will provide a good overview of how soils differ and how pedogenic processes and factors affect soil development.

Epipedon   *The uppermost horizon or horizons of a soil that is used for diagnostic purposes.*

The Twelve Soil Orders

This portion of the soils discussion focuses on the 12 soil orders that occur within the soil taxonomy classification system. These soils reflect the maximum development that occurs in these orders in association with various combinations of the soil-forming factors: (1) climate, (2) organisms, (3) relief, (4) parent material, and (5) time. Note that two of these factors, climate and organisms, are distinctly environmental in nature. In other words, they reflect the specific climate and vegetation relationships in any given place or region. Given this close correlation, it should not surprise you that the global distribution of soil orders reflects these environmental factors and thus builds directly on the previous two chapters. Figure 11.22 shows the geographic distribution of soil orders on Earth. Notice the similarity with respect to the patterns on this map and the global climate and vegetation

Global Soil Regions

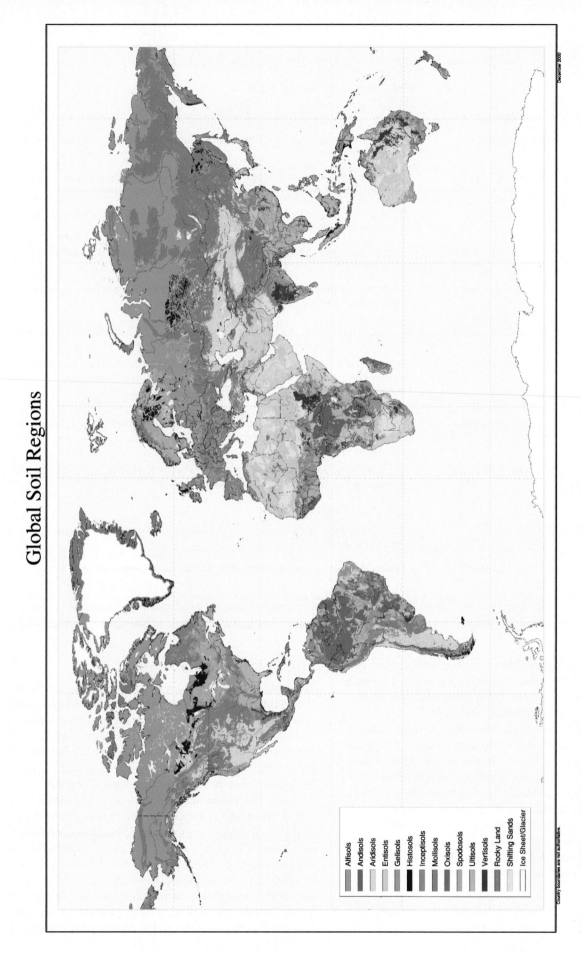

Alfisols
Andisols
Aridisols
Entisols
Gelisols
Histosols
Inceptisols
Mollisols
Oxisols
Spodosols
Ultisols
Vertisols
Rocky Land
Shifting Sands
Ice Sheet/Glacier

Country boundaries are not authoritative.

December 2000

Figure 11.22 The global distribution of major soil orders. This map is color-coded to reflect the geography of soil orders. Areas mapped as "Rocky Land," such as in central Asia, are places where large expanses of rock are exposed at the surface and soils have yet to form. Areas designated as "Shifting Sands," such as in the Sahara Desert, are places where the wind deposits sand in a way that precludes the development of soil. Areas mapped as "Ice Sheet/Glacier," such as in Antarctica and Greenland, are places where ice covers the ground and soils cannot form. (*Source:* U.S. Department of Agriculture.)

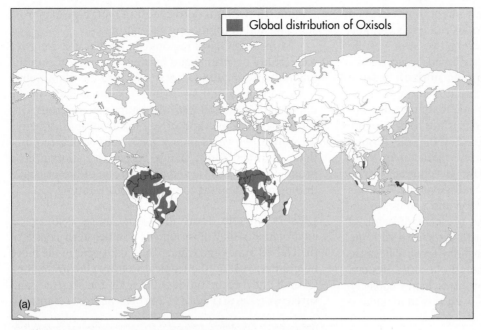

Global distribution of Oxisols

(a)

(c)

(b)

Figure 11.23 Oxisols. (a) Generalized map of Oxisols on Earth. These soils occur on about 8% of Earth's ice-free area. (*Source:* U.S. Department of Agriculture.) (b) Oxisol landscape in Hawaii. Note the reddish color of the soil, which occurs due to extensive weathering in this tropical environment. These soils are used in the growth of sugarcane. (c) A typical Oxisol. Notice the distinctive orange color of this soil, which occurs because soluble minerals such as calcium and magnesium are completely leached, leaving iron behind.

maps seen earlier (see Figures 9.2 and 10.6, respectively). As you work your way through this discussion, try to keep these geographic patterns in mind.

As with the descriptions of climate and vegetation, the discussion of soil orders begins by starting at the tropics and proceeding to higher latitudes. A few soil orders can occur anywhere under the proper conditions and will be dealt with at the end of the discussion. While you are studying the various soil orders, refer back to Figure 11.22, which shows the geographic distribution of soil orders and is similar to the global climate and vegetation maps in Figures 9.2 and 10.6, respectively. You might want to familiarize yourself with these illustrations again before you begin your tour of soil orders. Remember that many different variations of these soil orders arise, given the unique interactions of the five soil-forming factors; this discussion provides only a basic overview of their character and distribution. The discussion of each order is supported by a figure that includes a map of its global distribution, an example of a typical landscape associated with the order, and an example of a soil profile of that order.

Oxisols **Oxisols** are soils that form in tropical environments and are found in about 8% of the ice-free land area on Earth (Figure 11.23a). These soils form through a specific soil-forming process associated with tropical environments called **laterization**, which means "to become brick-like." The premise of laterization is that the hot and wet environment causes a great deal of mineral weathering to occur within soils. Bases such as calcium, magnesium, and potassium ions are rapidly leached because they are easily mobilized and water consistently percolates through the soil. In fact, percolating water flow is so steady that even otherwise-resistant minerals

---

Oxisols *Mineral soils in tropical and subtropical environments that form through laterization and thus have an oxic horizon within 2 m of the surface.*

Laterization *A regional soil-forming process in tropical and subtropical environments that results in extensive eluviation of minerals except for iron and aluminum.*

such as silica are leached. The only minerals not leached are iron and aluminum sesquioxides (minerals containing three atoms of oxygen for every two atoms of the metal), which results in highly weathered A and B horizons that together are called an **oxic horizon** because the sesquioxides are oxidized (i.e., atoms lose an electron). In addition, very little organic matter is added to the soil because decomposition of surface litter is rapid.

Oxisols are the most weathered of the soils formed by laterization. Given the oxidized character of these soils, they are usually red, yellow, or yellowish brown in color (Figures 11.23b and c). Well-developed Oxisols typically have a very thin A horizon and a thick Bo horizon, with the "o" reflecting the presence of oxidized sesquioxides. Although Oxisols are associated with the rainforest, they actually have very low fertility because most nutrients have been weathered out of the system. In addition, Oxisols have low cation exchange capacity, which further reduces fertility. Given this fundamental soil infertility, plants derive most of their nutrients from standing and decomposing vegetation.

Farmers in Oxisol regions can temporarily improve soil fertility by cutting and burning the rainforest in the manner described in Chapter 10. Once the plant debris is burned, the resulting organic residue collects in the upper part of the soil, providing nutrients for crops. Unfortunately, this method of soil fertilization is effective for only a few years in any given field, which requires that additional forest stands be cut to maintain crop production. This cause and effect is one of the principal driving forces leading to tropical deforestation.

Ultisols Ultisols form in warm, moist, subtropical environments like those on the southeastern coasts of the United States and China. They also occur in the more humid parts within the tropical wet/dry climate regions. Overall they occur on about 9% of the ice-free land surfaces on Earth (Figure 11.24a). Recall from Chapter 10

Oxic horizon *A diagnostic soil horizon in tropical and subtropical environments rich in iron oxides.*

Ultisols *Mineral soils in subtropical environments that formed through laterization and thus are depleted of calcium and have an argillic horizon.*

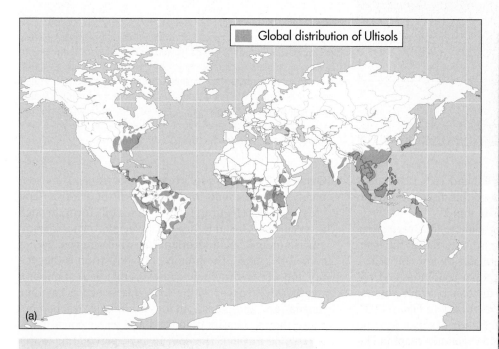

Figure 11.24 Ultisols. (a) Generalized map of Ultisols on Earth. These soils are found on 9% of Earth's ice-free land area. (*Source:* U.S. Department of Agriculture.) (b) An Ultisol landscape in central Brazil. Again, note the reddish color of this soil, which is due to extensive weathering in this subtropical environment. This region was once covered by forest and is now farmed. (c) A typical Ultisol. Note that most of the profile has an orange hue, somewhat similar to an Oxisol, because it also forms by laterization. Note the A horizon (dark zone) at the top of the soil and the increased structure in the underlying Bt horizon (reddish zone) due to eluviated clays.

that the vegetation in these tropical and subtropical areas consists of forest. As a result of this environmental association, these soils form through a variant of the laterization system and are similar to Oxisols in their reddish color (Figure 11.24b) and low base concentration.

The primary difference between Oxisols and Ultisols is that Ultisols are a bit more fertile because they are not as heavily leached. This relative lack of weathering happens because Ultisols occur in regions that experience a dry season. Ultisols also contain a distinct argillic (Bt) horizon, resulting from the translocation of clays. In fact, some Ultisols have Bt horizons that contain as much as 70% clay. They support productive forests, but require the frequent addition of lime fertilizers if they are used for continuous agriculture. In this manner, Ultisols in the southeastern United States have been farmed for over 200 years.

Vertisols A third soil order that frequently occurs in the tropical and subtropical regions is **Vertisols.** These soils are found on about 2% of the ice-free land surface on Earth, including some areas outside of the tropics (Figures 11.25a, b). The primary soil-forming factor associated with Vertisols is parent material because they contain an abundance of expandable clays, which are clays

---

Vertisols *Soils that contain an abundance of expandable clay and thus swell and shrink during wet and dry cycles, respectively.*

Figure 11.25 Vertisols. (a) Generalized map of Vertisols on Earth. These soils occur on about 2% of Earth's ice-free land area. (*Source:* U.S. Department of Agriculture.) (b) A Vertisol landscape in South Dakota. The small humps and ridges are known as "gilgai" and form due to frequent expansion and contraction of clays. (c) Ground cracks that resulted from contraction of expandable clays in a Vertisol. (d) A Vertisol in New Mexico. Note the strong structure (due to high clay content) in the upper part of the profile and the cracks in the ground. The horizontal grooves in the lower part of the profile were made by the backhoe that excavated the trench.

that swell like a sponge when they are wet and shrink when they dry. Because these soils contain so much clay, fields that lie within Vertisol soils are typically dark with very strong structure, which you can see as the prominent irregular forms in Figure 11.25c.

One of the distinctive characteristics of Vertisols is that prominent cracks can form in the ground (Figure 11.25d) during long dry periods because the clays within the soil shrink so much. Because of the constant shrinking and swelling that occurs in these soils, they are slowly, but constantly mixed, with much of the mixing occurring when fragments of soil fall into the cracks. The results of this shrinking and swelling can be seen in a profile through the presence of shiny clay skins on peds. These glossy surfaces, or *slickensides*, are created when the long-term expansion and contraction of the soil cause clays to be polished as they rub against one another. As a result, a diagnostic horizon of a Vertisol is a Bss horizon, with the "ss" reflecting the presence of slickensides.

In fact, if you have ever noticed that cracks form in the ground during long droughts around your home, then the soils in your neighborhood are likely Vertisols,

or closely related. In most places, Vertisols occur in isolated pockets that are too small to show up on the global soils map. Significant regions of Vertisols form in the tropical and subtropical regions of the world, however, including the Deccan Plateau of western India, along the Nile River in eastern Africa, and much of eastern Australia. Although these soils are often high in base content, they are difficult to farm because they are hard to cultivate.

Alfisols **Alfisols** form in cooler parts of the moist continental climate zones that lie poleward of the subtropical regions. These soils occur on about 10% of Earth's ice-free land area over an extremely wide range of latitude, extending from as high as 60° N in North America to the equatorial zone in South America and Africa (Figure 11.26a). Large areas of these soils occur

---

Alfisols   *Soils generally found in seasonal midlatitude regions that formed through podzolization and have an alkaline argillic horizon.*

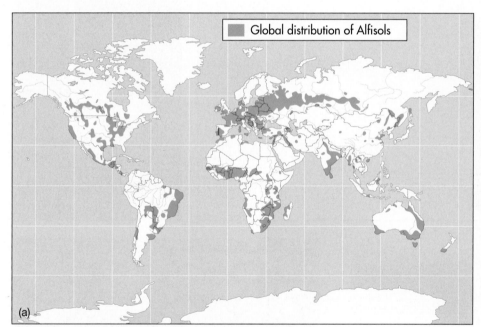

Figure 11.26 Alfisols. (a) Generalized map of Alfisols on Earth. These soils occur on about 10% of Earth's ice-free land area. (*Source:* U.S. Department of Agriculture.) (b) A typical Alfisol landscape in central Missouri. Prior to cultivation, the vegetation of this region was a mix of forest and tallgrass prairie. (c) An Alfisol from the northern Midwest. Note the light E horizon (white arrow), which is present in well-developed Alfisols. The reddish zone below the E is a Bt horizon rich in eluviated clays.

African Climate, Vegetation, and Soils

Now that you have examined the three tropical soils, you should review the relationship of climate, vegetation, and soil orders by going to the *GeoDiscoveries* website and selecting the module *African Climate, Vegetation, and Soils*. This module is a brief animation that illustrates how these variables are interrelated to one another to form a geographically distinct landscape. As you watch it, be sure to look at the spatial patterns that exist on the African continent. Once you complete the animation, be sure to answer the questions at the end of the module to test your understanding of this concept.

---

in places like the American Midwest (Figure 11.26b), northern Canada, Europe, Asia, and the east coast of China. Smaller pockets are found in Central America, Australia, and Africa. Given the relatively high moisture of these regions, the natural landscape is largely covered with deciduous forest, but extensive areas of boreal forest occur as well.

Alfisols form by a specific soil-forming process called **podzolization**. Podzolization is similar to laterization because both processes are associated with abundant leaching in their respective humid environments. The primary difference between the two processes, which is reflected in the character of the soils, is that organic matter decomposes slowly where podzolization occurs because the environment is cooler. Thus, a distinct O horizon forms in the regions where podzolization is the dominant pedogenic process. This O horizon consists mainly of deciduous leaves and coniferous needles. This litter layer is slightly acidic, which causes infiltrating soil water to become even more acidic. As a result, organic acids are incorporated into the upper part of the soil to form a distinct A horizon. From there, they are leached to the B horizon.

The leaching of organic acids causes translocation of the iron and aluminum sesquioxides, which are not moved where laterization occurs. In this environment, podzolization results in a very well-defined E horizon (Figure 11.26c), which is ashy gray in color and is thus called an **albic horizon** because it is so light. Below this horizon, a reddish B horizon forms where eluviated iron and aluminum collect. This horizon is called a **spodic horizon** because of the high sesquioxide content.

Alfisols are the soils that form in the warmer regions where podzolization occurs, specifically in the areas where the litter layer typically consists of deciduous leaves. Alfisols are similar to Ultisols in that they contain an argillic Bt horizon that is rich in clay. These soils are not as heavily weathered as Ultisols, however, which results in Alfisols generally being more fertile because they have higher concentrations of base ions. Nevertheless, Alfisols contain a distinct E horizon that is lighter, through extensive eluviation of minerals and clays, than the overlying A and underlying B horizons. Because the rate of decomposition is much less in this cooler environment, the A horizon of Alfisols is generally better developed and darker than in Ultisols.

Spodosols Spodosols are acidic soils that are also associated with the podzolization process. These soils occur in about 3% of the ice-free land surfaces on Earth (Figure 11.27a). In contrast to Alfisols, Spodosols form in somewhat cooler, more humid regions. As a result, the vegetation in these regions is typically coniferous and hardwood forest (Figure 11.27b) and the litter layer is thus a mix of needles and leaves. As with Alfisols this litter layer is acidic, which lowers the pH of percolating water

---

Podzolization *A regional soil-forming process in cool, humid environments that results in the eluviation of iron, aluminum, and organic acids to form well-developed E and Bs horizons.*

Albic horizon *A diagnostic horizon of podzolization from which clay and free iron oxides have been removed, resulting in a light-colored E horizon.*

Spodic horizon *A mineral soil horizon characterized by the illuvial accumulation of aluminum, iron, and organic carbon.*

Spodosols *Soils in cool, humid regions that form through podzolization and contain a spodic horizon enriched in eluviated iron, aluminum, and organic carbon.*

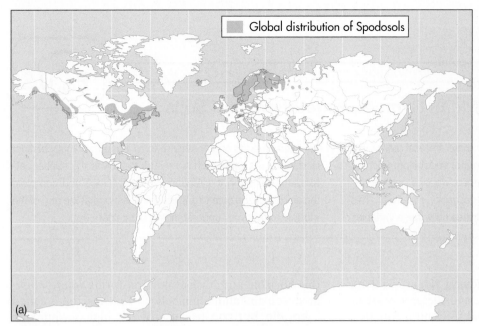

Figure 11.27 Spodosols. (a) Generalized map of Spodosols on Earth. These soils occur on about 3% of Earth's ice-free land area. (*Source:* U.S. Department of Agriculture.) (b) A typical Spodosol landscape in the Great Lakes region. (c) A well-developed Spodosol in the Great Lakes region. Note the thick E horizon, which has distinct tongues (black arrows) that extend deeper into the soil. These features developed because water preferentially followed tree roots as it moved downward. The dark horizon (white arrows) directly below the E horizon is a Bhs horizon rich in eluviated organic acids and iron.

sufficiently to mobilize organic acids and sesquioxides. In places where the parent material is sandy, soil formation occurs more rapidly because water percolates quickly through the solum.

Spodosols have a very distinct sequence of soil horizons that are readily visible (Figure 11.27c). The uppermost part of the soil usually contains a thin O horizon (needle-leaf litter) over a thin A horizon, which is very dark due to the slow decomposition of organic matter. In well-developed Spodosols, the E horizon is prominent because iron, aluminum, and base ions have all been eluviated, resulting in a distinct white or light-gray albic horizon. Directly beneath the E horizon is a particular type of spodic horizon, which is designated as a Bs horizon. This horizon is dark brown or reddish in color because it contains high amounts of illuviated organic matter, aluminum, and iron. If sufficient time or eluviation has transpired, the Bs horizon will actu-

ally become cemented with eluviated organic material called *ortstein* (Figure 11.27c). In these instances, the B horizon will have a Bhs designation to reflect the presence of both humic acids and sesquioxides.

Histosols Although the vast majority of soils have a distinct mineral component to them, some soils are almost entirely organic. The organic soils usually fall into the soil order **Histosols**. Histosols are found on about 1% of Earth's ice-free land surfaces (Figure 11.28a), most of which are in cool, moist environments. These soils form in isolated shallow lakes, ponds, wetlands, and bogs (Figure 11.28b) where the decomposition of organic matter is very slow as a result of the cool environmental conditions.

Histosols *Organic soils that form in cool, wet environments where organic carbon decomposes very slowly.*

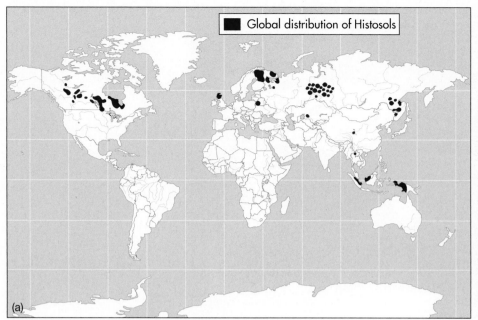

Figure 11.28 Histosols. (a) Generalized map of Histosols on Earth. These soils occur on about 1% of Earth's ice-free land area. (*Source:* U.S. Department of Agriculture.) (b) A typical Histosol landscape in Michigan forms in bogs like this where organic matter decomposes slowly. (c) A Histosol from Upper Michigan. Note the very dark color of this soil, which reflects the high amount of organic matter. The lighter sediment visible at a depth of about 40 cm is a mineral deposit on which the soil formed.

Given the organic nature of Hisotols, the major horizons of Histosols are O horizons that are very dark in color (Figure 11.28c). Because these soils are extremely fertile, they are actively cultivated for crops, such as mint or cranberries, in places where they are easily accessible. They are also excellent soils for sod farms and certified organic farms. These soils are also used as mulch and a low-grade fuel source. In the American Midwest, Histosols are commonly referred to as *peat* or *muck*.

Gelisols In 1998 the soil taxonomy system was revised to include a new soil order: **Gelisols**. Gelisols develop in very cold climates at high latitudes or high mountain elevations and are estimated to occupy about 9% of Earth's ice-free land area, including most of the state of Alaska (Figure 11.29a). Thus, these soils are frozen most of the year, resulting in permafrost within 2 m of the surface. Because these soils are very cold, the decomposition of organic matter proceeds even more slowly than where Histosols occur. As a result, Gelisols contain large amounts of organic carbon, some of which is in the form of muck or peat at the surface, and are thus dark in color (Figure 11.29b). Given that Gelisols regularly freeze and thaw, they are actively mixed and are thus frequently associated with deformed topography (Figure 11.29c).

Gelisols *Soils in subarctic and arctic environments that contain permafrost within 2 m of the surface.*

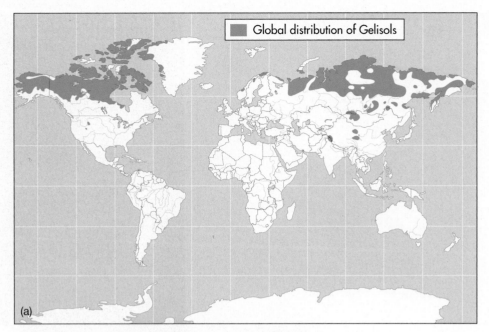

Figure 11.29 Gelisols. (a) Generalized map of Gelisols on Earth. These soils are found in about 9% of Earth's ice-free land area. (*Source:* U.S. Department of Agriculture.) (b) A Gelisol in Alaska. The lighter horizon at the base of the soil is permafrost. (c) A typical Gelisol landscape in Alaska. The distinct polygons on the ground, called *patterned ground*, form as a result of repeated freezing and thawing.

Mollisols Moving away from the humid zones, we next look at soils that form in more arid regions, beginning with **Mollisols**. These soils occur over about 7% of Earth's soil network, including most of the American Midwest and Great Plains, a broad belt across Eurasia (including the Russian steppe), and the Pampas in South America (Figure 11.30a). Mollisols are closely associated with the more humid parts of the dry midlatitude climate and associated grassland vegetation (Figure 11.30b). They are the most extensive soil order in the United States and are found on 21.5% of the land area (Figure 11.30a). In the Great Plains, the presence of Mollisols is directly linked to the fact that the region lies within the Rocky Mountain rain shadow.

Mollisols form in association with a specific pedogenic process called **calcification**. As with podzolization, vegetation plays an important role in the calcification process. Where podzolization occurs, deciduous and coniferous trees produce a needle-leaf litter layer that depresses the pH sufficiently to mobilize iron and aluminum sesquioxides. In grassland environments, where calcification takes place, the grasses play an important role in soil formation because they are base cyclers. The first part of the cycle occurs when grasses absorb bases as nutrients during the growing season. When the grass dies, however, the bases contained within the grass are cycled back into the soil as it decomposes.

Mollisols *Soils that form through calcification and have a mollic epipedon that overlies mineral matter which is more than 50% saturated with base ions.*

Calcification *A regional soil-forming process in which calcium carbonate is cycled within the soil.*

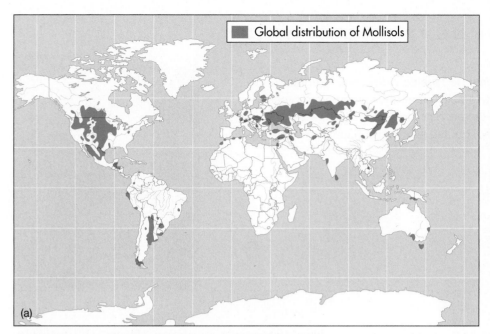

Figure 11.30 Mollisols. (a) Generalized map of Mollisols on Earth. These soils are found on about 7% of Earth's ice-free land area. (*Source:* U.S. Department of Agriculture.) (b) A typical Mollisol landscape in the central Great Plains. Note the tall-grass prairie. (c) A Mollisol in North Dakota. Note the thick A horizon near the surface and the prominent Bk horizon (arrow) that is rich in eluviated calcium.

Here is where the role of climate becomes important. If this base cycling somehow occurred in a humid environment, the bases would be leached completely out of the soil because they are so easily mobilized. Given the general moisture deficit that occurs in places like the Great Plains or the Russian steppe, however, the bases are not completely leached. Instead, they are merely translocated to the zone of illuviation, or B horizon, just like iron and aluminum where podzolization occurs. Given that this part of the soil is rich in calcium, it is called a **calcic horizon**. This process can ultimately result in the complete cementation (solidification) of the B horizon by calcium carbonate, resulting in material called *caliche*.

Calcic horizon *A diagnostic soil horizon of calcification enriched in illuviated calcium carbonate.*

Mollisols are also distinctive because they have thick, dark A horizons with high humus content (Figure 11.30c). This horizon forms in part because grasses decompose slowly in the relatively dry environment, therefore slowly and consistently adding humus to the soil. Grass also has a dense underground root system that contributes to the high content of organic matter and calcium minerals in these soils. Soil organisms, like earthworms, ants, and moles, mix the soils, carrying organic matter deeper below the surface. Given the association with the calcification process, the B horizons of Mollisols are rich in calcium; in fact, the B horizon of a Mollisol is saturated at least to a level of 50% by bases such as calcium ions. It also often contains an abundance of translocated clays. As a result, the B horizon of a well-developed Mollisol is designated as a Btk ("k" meaning abundant calcium carbonate) horizon.

Which one of the regional pedogenic processes is contributing to the development of this soil? What soil horizons are present? How can you tell? How did each of them form? Here's a hint: this soil is not fully developed, but the evolving horizons can be seen.

The combination of high amounts of organic matter and minerals makes Mollisols among the most fertile and productive soils on Earth because they contain abundant nutrients that plants need. In the United States, Mollisols support a wide variety of agricultural crops, including corn, wheat, and soybeans. The next loaf of bread you buy, in fact, will most likely contain wheat grown in Mollisols.

Aridisols As the name implies, **Aridisols** are soils that form in arid environments. These soils are found over about 12% of Earth's ice-free land area, including the Sahara Desert and significant parts of Australia, South America, the Middle East, Asia, and the southwestern United States (Figure 11.31a). These dry environments have very sparse vegetation (Figure 11.31b), which means that Aridisols have poorly developed A horizons, if they have them at all (Figure 11.31c).

Aridisols can form through either the process of calcification or another process called **salinization**, which entails the movement of sodium (Na) in the soil. In the other pedogenic processes, such as laterization and podzolization, sodium is easily leached from soils because it is one of the more soluble minerals, and water is sufficiently abundant to move it out of the solum. In semi-arid to arid environments, however, sodium is not completely leached from soils and instead is moved only part way down the soil profile when infrequent rains do fall. When the soil dries out during the extended dry periods that occur in this environment, the partially

Aridisols *Mineral soils that form in arid environments and thus are poorly developed.*

Salinization *A regional soil-forming process in which soluble salts are cycled within the soil.*

www.wiley.com/college/arbogast

WILEY PLUS

Regional Pedogenic Processes

At this point we have completed our discussion of soils that generally form in association with regional environmental variables and pedogenic processes. These processes are particularly well suited for animation to more completely illustrate how they work. Go to the *GeoDiscoveries* website and select the module *Regional Pedogenic Processes*.

This module illustrates how the soil-forming processes laterization, podzolization, calcification, and salinization influence the development of soils at a regional level. When you complete the animation, be sure to answer the questions at the end of the module to test your understanding of this concept.

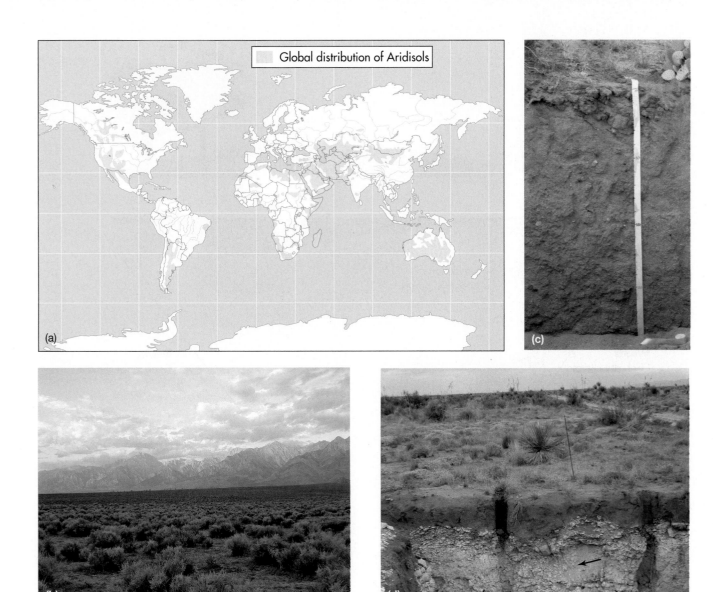

Figure 11.31 Aridisols. (a) Generalized map of Aridisols on Earth. These soils occur on about 12% of Earth's ice-free land area. (*Source:* U.S. Department of Agriculture.) (b) A typical Aridisol landscape in the southwestern United States. (c) An Aridisol in New Mexico. Note the lack of a well-developed A horizon, which is a reflection of the desert environment. (d) An Aridisol with a calcic horizon (arrow) in New Mexico. This horizon developed through extensive eluviation of calcium over a long time.

Global distribution of Aridisols

leached minerals rise back up through the profile in a process called *wicking*.

This wicking process is most active in enclosed basins where pools of water collect when it does rain. Sodium in these low spots can be leached many centimeters when the water percolates through the profile following a hard rain. When the site dries out, however, the salt rises back up and forms a salty crust at the surface called an Az horizon. If this crust reaches a thickness of 15 cm (6 in.) and contains at least 20 g/kg of salt, it is called a **salic horizon**. These salts are highly toxic to most plants. In regions where calcification is the dominant process associated with

Aridisols, the soils have thick Bk, or calcic, horizons that can become completely cemented with time. If they do, the horizon is designated as a Bkm horizon (Figure 11.31d).

At this time, each of the three major soil orders that occurs within the midcontinental United States—Alfisols, Mollisols, and Aridisols—have now been discussed. This juncture thus marks a good point to holistically

---

Salic horizon *The diagnostic horizon of salinization that forms due to the recrystallization of secondary salts.*

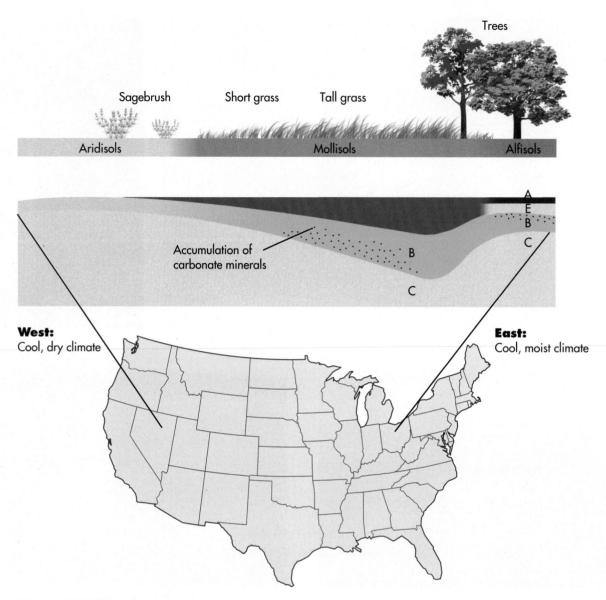

Figure 11.32 Midcontinent soil transect in the United States. The change in soil characteristics is reflected by soil order, from central Nevada in the west to Ohio in the east. (*Source:* Modified from C. E. Millar, L. M. Turk, and H. D. Foth, *Fundamentals of Soil Science.* New York: John Wiley & Sons, 1990.)

examine how soils vary along a study transect that extends from Ohio to central Colorado (Figure 11.32). Which side of this transect has the most moisture? Which side is drier? Do you know why? You should know that the eastern side of this transect is more humid due to the influx of mT air from the Gulf of Mexico. As you proceed westward, however, the climate generally becomes drier due to the influence of the Rocky Mountain rain shadow. The response of vegetation is that trees populate the eastern side of the transect. As you move westward, the vegetation shifts to tall grass and then to short grass.

How do soils differ along this transect? The primary change is that more leaching takes place on the eastern side of this transect than on the west. This geographic

pattern should make sense because the climate in Ohio is generally more humid, with more precipitation, than in eastern Colorado. As you can see in Figure 11.32, this distribution of soil water causes a distinct E horizon in Ohio, which fades away as you move farther west. The thickest A horizons are in the tall-grass region of the Mollisol belt because the biomass is so high, and thus decomposition and leaching are less than to the east. In this zone, the depth of illuviated carbonate (lime, $CaCO_3$) is the greatest and the B horizon is the thickest. As you move deeper into the rain shadow, in the western part of the Mollisol belt, the A horizon thins significantly and the depth of the B horizon becomes shallower. This pattern culminates within the zone of Aridisols with a very thin or nonexistent A horizon and a very shallow, carbonate-rich

B horizon. This discussion is an excellent example of how many geographic variables come together to produce a distinctive pattern—in this case, regarding soils.

Entisols Finally, it is time to examine the soils that are not related to any specific environmental variable or pedogenic process. In other words, these soils can be found in many different places on Earth and are a product of specific soil-forming factors. The most widespread of these more randomly distributed soils are **Entisols**,

---

Entisols *Soils that are very weakly developed and thus have no distinct horizonation.*

---

which means *young soils*. These soils occur on about 16% of Earth's ice-free land area (Figure 11.33a).

Although Entisols are soils in the sense that they can support plants, they are the least developed soils because they have no diagnostic horizons such as an E or a B horizon (Figure 11.33b). They may contain a thin A horizon. This overall lack of development can occur for several reasons: (1) the climate is extremely dry, making the processes of soil additions and translocations very weak; (2) the parent material is quartz sand, which is a resistant deposit that does not easily yield nutrients through weathering; or (3) the parent material has recently been transported, resulting in little time for a soil to develop (Figure 11.33c). Given that this range of soil-forming factors can occur just about anywhere, Entisols appear at all latitudes.

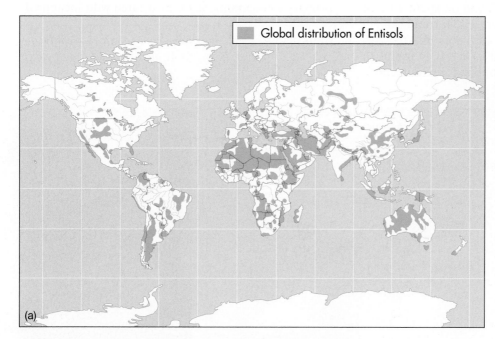

Global distribution of Entisols

(a)

(c)

(b)

Figure 11.33 Entisols. (a) Generalized map of Entisols on Earth. These soils occur on about 16% of Earth's ice-free land area. (*Source:* U.S. Department of Agriculture.) (b) An Entisol in southern Idaho. Note the very thin, if nonexistent, A horizon in this soil. (c) An Entisol landscape in Michigan. The vegetation is growing on a recent deposit of windblown sand along Lake Michigan.

At this point, you should have a basic understanding of mid-latitude soil orders and where they occur. In this context, it is a good time to review the close relationship between climate, vegetation, and soils in these regions. To do so, go to the *Geo-Discoveries* website and select the module *North American Climate, Vegetation, and Soils*. This module contains a brief animation that illustrates how these variables are inter-related with one another to form a geographically distinct landscape. Once you complete the animation, be sure to answer the questions at the end of the module to test your understanding of this concept.

Inceptisols **Inceptisols** are soils with weakly developed horizons. These soils are found on about 10% of Earth's ice-free land area (Figure 11.34a), and can occur just about anywhere, even in places like the Appalachian Mountains (Figure 11.34b). It is useful to think of Inceptisols as being a step in formation above

Entisols, which have limited horizon development. The primary soil-forming factor associated with Inceptisols

Inceptisols *Soils that have one or more weakly developed horizons due to some alteration and removal of soluble minerals.*

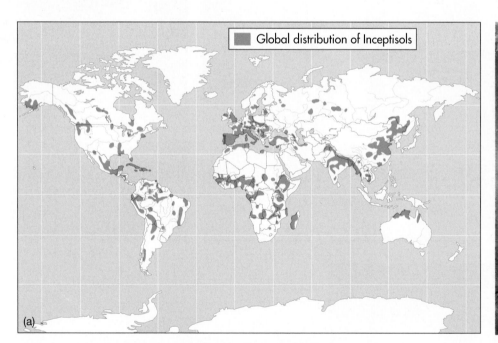

Global distribution of Inceptisols

Figure 11.34 Inceptisols. (a) Generalized map of Inceptisols on Earth. These soils occur on about 10% of Earth's ice-free land area. (*Source:* U.S. Department of Agriculture.) (b) An Inceptisol landscape. Soils on these steep slopes within the Appalachian Mountains have weakly developed horizons because they are easily eroded. (c) An Inceptisol in Wisconsin. Note the weak E horizon (arrow) that indicates some eluviation of iron has occurred.

is time. On the one hand, these soils are better developed than Entisols because they have had more time to develop. Thus, Inceptisols may have a weak E horizon (Figure 11.34c). On the other hand, insufficient time has elapsed for the soil to develop horizons that meet the specific criteria of a better-developed soil order such as Mollisols, Spodosols, or Ultisols. Another important soil-forming factor associated with Inceptisols is parent material because most of these soils form in sediments that have been recently transported, but not so recently that the soil shows no sign of development. As in the case of Entisols, the soil-forming factors that contribute to the formation of Inceptisols can occur just about anywhere.

Andisols **Andisols** are soils that form in parent material that is more than half volcanic ash. Like Entisols and Inceptisols, Andisols occur at random places on Earth, occupying about 1% of Earth's ice-free landmass (Figure 11.35a). Most of Hawaii, for example, lies within areas classified as Andisols (Figure 11.35b). Given their volcanic origin, it is accurate to say that the most important soil-forming factor associated with Andisols is parent material because fresh deposits of volcanic ash are deposited on the landscape following a volcanic eruption. These deposits consist mostly of glass-like shards that frequently contain a high proportion of carbon. As a result, these soils are usually dark in color (Figure 11.35c) and quite fertile. Given the high frequency of volcanic eruptions in some areas, Andisols may be poorly developed because the soils have had insufficient time to develop.

Andisols *Soils formed in parent material that is at least 50% volcanic ash.*

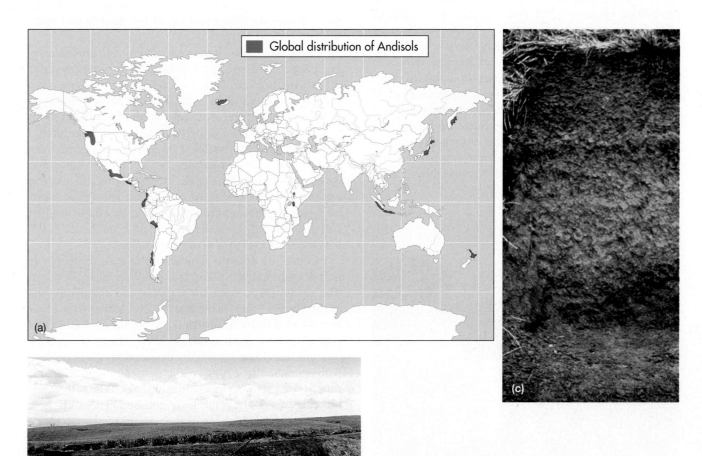

Figure 11.35 Andisols. (a) Generalized map of Andisols on Earth. These soils are associated with volcanic regions and occur on about 1% of Earth's ice-free land area. (*Source:* U.S. Department of Agriculture.) (b) Typical Andisol landscape in Hawaii. The Hawaiian Islands are covered in most places with volcanic sediments. (c) Andisol in Hawaii. Note the banding in this soil, which reflects periodic deposition of new volcanic parent material.

Human Interactions with Soils

Throughout much of recorded history, soil has been a key environmental factor around which human societies revolved. For most people living in the United States today, however, soil is a vague thing that lies beneath the ground on which we walk. This lack of awareness is due in large part to major demographic changes in the country over the past 200 years. In 1800 approximately 90% of the labor force were farmers intimately familiar with the soil because they worked with it every day. By 1900 the percentage of the labor force devoted to agriculture had dropped to about 38%. Today, farmers comprise less than 2% of the American workforce. In contrast, over 80% of Americans live in urban environments where soils are rarely considered by the average person.

In spite of the indifference most people have toward the ground beneath them, the fact is that soils are very important to everyone. In terms of agriculture, understanding the character of soil is essential for a variety of reasons directly related to crop yields and food production (Figure 11.36). For example, the texture of soil is important because it influences how rapidly water drains through the soil. In general, farmers desire loamy-textured soils like a silt-rich Mollisol because they hold draining water long enough (but not too long) for crops to be consistently well watered. Likewise, soils with good structure are usually preferred because they have higher porosity, better aeration, and crop roots can spread easily through the ground. The amount of organic matter in soil, as well as minerals such as calcium, magnesium, and potassium, is critical for crop yields because they directly

Figure 11.37 Cracked outside wall in a home built on clay-rich soils. This house was severely damaged because the surrounding soil contains expandable clays, which can either expand like a sponge when wet or lose cohesion and collapse when they dry.

influence soil fertility. Factors such as these are very important considerations that affect the quality and price of the food you eat.

Soils are also important from an engineering perspective because the foundations of buildings and other structures are often built within them. As we have seen, a critical variable is the amount and character of the clays that may be contained within the soil. Vertisols, in particular, contain a high percentage of expandable clays that swell and shrink during wet and dry periods, respectively. Structures built on Vertisols often have weak foundations and walls because the soil can be expanding or collapsing around them. This process can be so strong that it actually causes a basement wall to collapse or a wall to crack like that shown in Figure 11.37, which costs thousands of dollars to repair.

Given the obvious significance of soils to people, it is important to understand them well and care for them. The agency responsible for mapping and managing soils in the United States is the Natural Resources Conservation Service (NRCS), which is part of the United States Department of Agriculture (USDA). The NRCS has an office in each county within the country and employs field agents who consult with farmers, engineers, and other residents who have questions about how they are utilizing soil. A key reference published by the NRCS is the county *Soil Survey*, which discusses the character of soils in the county and presents its geographic distribution in a series of accessible maps (Figure 11.38). Soil surveys are often updated using refined mapping techniques that allow farmers to practice precision agriculture, which maximizes yields by targeting specific crops to particular soil variations within fields. These kinds of applications improve agricultural efficiency, which, in turn, enables farmers to produce increasingly greater amounts of food in a way that keeps costs low for you.

Figure 11.36 Soils and agriculture. A farmer tends a field in Wyoming. In order to produce food efficiently, farmers must have a keen understanding of the character and variability of soils.

United States
Department of
Agriculture

Natural
Resources
Conservation
Service

In cooperation with
the University of Georgia,
College of Agricultural and
Environmental Sciences,
Agricultural Experiment
Stations

Soil Survey of Pulaski and Wilcox Counties, Georgia

(a)

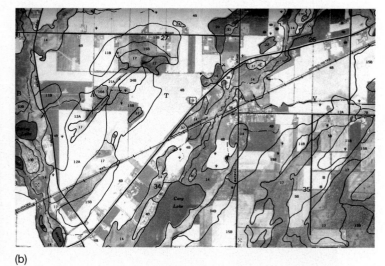

(b)

Figure 11.38 Country soil surveys. (a) The cover of the soil survey for Pulaski and Wilcox Counties, Georgia. Survey guidebooks such as this one are available for most counties in the United States. (b) A typical soil map in a soil survey. Note the labels in distinct map areas, or polygons. Each label represents a specific soil type (or series) in the area. Each series has distinctive characteristics, such as the degree of development, color, texture, and structure, to name a few. (*Source:* U.S. Department of Agriculture.)

KEY CONCEPTS TO REMEMBER ABOUT SOIL SCIENCE AND CLASSIFICATION

1. Soil science is a distinct subdiscipline of geography that focuses on the chemical, physical, and biological properties of soils.

2. One of the primary ways that soil scientists study and compare soils is through soil classification. Soils are classified (in the United States) through a hierarchical system called *soil taxonomy*. The most general classification in soil taxonomy is the soil order, of which 12 are recognized. These orders are distinguished on the basis of measured physical properties and horizons.

3. Oxisols and Ultisols form as a result of laterization, which involves the intense weathering of soils in tropical and subtropical environments. These soils are typically reddish in color because minerals besides oxidized iron and aluminum are leached.

4. Alfisols and Spodosols form due to podzolization, which involves extensive translocation of organic acids and sesquioxides in cool, humid environments. Alfisols and Spodosols have distinctive E horizons that are light gray in color. Spodosols also have a distinctive Bs horizon that forms due to the illuviation of iron and aluminum.

5. Histosols are soils that are entirely organic because they form in cool, humid places where the decomposition of organic matter is very slow.

6. Mollisols and Aridisols form in drier environments. Both can form because of calcification, which occurs where base ions are partially translocated in the soil. Aridisols can also form due to salinization, which involves the movement and surface recrystallization of sodium.

7. Andisols, Entisols, Gelisols, Inceptisols, and Vertisols are directly related to local environmental factors. Andisols form in volcanic environments, whereas Gelisols occur in subarctic regions where the soil is frozen most of the year. Entisols and Inceptisols can form anywhere and represent soils that are very weakly to slightly better developed, respectively. Vertisols are soils that contain abundant expandable clays that swell when wet and shrink when dry.

Up to this point, this text has focused on geographic processes and patterns that are distinctly interrelated with one another. It began with a discussion of Earth–Sun geometry, which naturally led to a section on radiation and climate. These chapters provided the context for the global distribution and character of vegetation and soil. Now that we have concluded the chapter on soils, we turn our attention to processes associated with the solid Earth or lithosphere. In other words, we now begin to investigate the processes and variables that influence the shape of the landscape.

A good example of this landscape building is the Teton Mountains, which are located in northwestern Wyoming.

This mountain range has evolved through internal forces within the solid Earth that have deformed rock on the surface. Although we have examined some relationships between the solid Earth and the atmosphere and oceans in the earlier chapters of the book, these relationships are incidental and even random in some instances. We begin exploring the lithosphere in more detail in the next chapter, which focuses on the nature of rocks on Earth, the configuration and character of the Earth's interior structure, and the topography (or geomorphology) of Earth.

SUMMARY OF KEY CONCEPTS

1. Soil consists of the outermost layer of Earth and forms through the complex interaction of additions, transformations, translocations, and losses. These processes occur in various combinations that depend on the five soil-forming factors: (a) climate, (b) organisms, (c) relief, (d) parent material, and (e) time.

2. Soils have a variety of distinctive characteristics, including color, texture, structure, pH, and cation exchange capacity that can be measured and compared from one soil to another.

3. Soils are organized into horizons that form by distinctive processes. The O horizon is the uppermost horizon and consists of freshly added organic matter,

such as leaves. The A horizon contains decomposed organic matter called humus. The E horizon forms through the eluviation of minerals, which recrystallize and collect through the process of illuviation in the underlying B horizon. The C horizon is unaltered parent material, which, in bedrock landscapes, is known as regolith.

4. Soils in the United States are classified on the basis of their genetic properties in a scheme called *soil taxonomy*. The highest level of this classification system is the soil order. Twelve soil orders are recognized: Oxisols, Ultisols, Alfisols, Spodosols, Histosols, Vertisols, Gelisols, Mollisols, Aridisols, Entisols, Inceptisols, and Andisols.

1. How is regolith related to bedrock and soil?

2. Describe the soil-forming factors and provide an example of how they interact to produce soil.

3. Compare and contrast the concepts of soil texture and soil structure.

4. Which texture will allow water to drain more rapidly: sand or clay? Why?

5. What are the various aspects of soil chemistry that are used to characterize soil?

6. Which parent material will have a higher cation exchange capacity—one that is sandy, or one that is rich in clay? Explain your answer.

7. Explain how the various soil horizons form.

8. Which pair of horizons is created through the process of translocation? In what way does this evolution occur?

9. It is conceivable that a soil could be classified as an Entisol even though it is very "old." In what environment—tropical, subtropical, or arid—would this kind of weak development most likely occur? Why?

10. What is a soil profile and why is it useful for determining the characteristics of a soil?

11. How are Spodosols and Oxisols similar? How are they different?

12. Why is grass an important part of the calcification pedogenic model?

13. In which parent material would you most likely find an Inceptisol: a sand dune that is 500 years old, or one that is 1000 years old? Why?

14. Provide an example of how soils are important to agriculture and how they are managed.

ANSWERS TO VISUAL CONCEPT CHECKS

VISUAL CONCEPT CHECK 11.1

The answer is *b*. The photograph shows an example of soil additions. When leaves fall to the ground, they slowly decompose and become incorporated into the soil.

VISUAL CONCEPT CHECK 11.2

The answer is *d*; this soil is classified as a sandy clay.

VISUAL CONCEPT CHECK 11.3

This soil is forming by the podzolization pedogenic process, which results in the eluviation of iron, aluminum, and organic acids. The soil contains developing A, E, and Bs horizons. You can identify the A horizon because it is dark, which occurs through the addition of humus. The E horizon is whitish and forms through the eluviation of iron and aluminum. The Bs horizon is visible beneath the E horizon as the light orange zone. This color reflects the illuviation of iron.

CHAPTER TWELVE

EARTH'S INTERNAL STRUCTURE, ROCK CYCLE, AND GEOLOGIC TIME

In the preceding chapter, we discussed the formation and character of soils, which can be viewed as the Earth's skin. Now it is time to look more intently at the solid Earth by focusing on its internal structure and composition. The innermost parts of Earth are important because they drive the movements of the outermost layer of Earth, the surface crust. Geographers are especially interested in the patterns and processes associated with the crust because this surface is where modern landforms occur and people live. All these topics will be presented in the concept of Earth history as it relates to deep geologic time. This chapter will help you understand the composition of Earth and give you an appreciation for geologic time and why it is relevant, especially when you consider how quickly Earth is now changing due to human impacts.

The Grand Canyon is one of the best known landscapes on Earth. In addition to being a beautiful place, the canyon is noteworthy because it contains rocks that span much of the Earth's history. This chapter focuses on the geologic history of Earth, the evolution and composition of rocks, and the character of continents and ocean basins.

CHAPTER PREVIEW

Earth's Inner Structure

Rocks and Minerals in the Earth's Crust
    On the Web: *GeoDiscoveries Identification of Igneous Environments*
    On the Web: *GeoDiscoveries Clastic Rocks*
    On the Web: *GeoDiscoveries The Rock Cycle*

Geologic Time
    On the Web: *GeoDiscoveries Geologic Time*

Human Interactions with the Rock Cycle: The Case of Petroleum

The Big Picture

LEARNING OBJECTIVES

1. Compare and contrast the various structural layers of Earth.

2. Explain the difference between igneous, sedimentary, and metamorphic rocks and discuss the various kinds of rocks that occur in each of these categories.

3. Discuss the concept of radiometric dating and how it is used to estimate the age of rocks.

4. Describe the geological timescale and how it is subdivided.

5. Holistically explain how the evolution of landforms is tied to the rock cycle and geologic time.

6. Describe the general history of petroleum exploration in the industrial period and how it may change in the future.

Earth's Inner Structure

Have you ever wondered what the inside of Earth is like? Science fiction writers have pondered this question for a long time, with the most famous example being Jules Verne's book *Journey to the Center of the Earth*. In this story, a professor leads a party of explorers on a wild adventure to the center of Earth through a volcano in Iceland. Although such an adventure is pure fiction, most geologists would certainly like to make the trip. Unfortunately, the only way to actually "see" a portion of the Earth's interior is to core into it with a large drill rig and collect samples. Although such studies are informative, the deepest geologists have ever been able to drill is only about 12 km (7.5 mi) into Earth. The distance to the center of Earth is thousands of kilometers, however, which means that geologists must rely on indirect evidence to construct models about the Earth's interior.

Much of what is known about the Earth's structure is based on how **seismic waves** travel through

Earth after earthquakes occur. Cooler areas within Earth transmit these waves at higher speeds than hotter zones because cooler regions are more rigid. Areas that are denser absorb seismic waves, whereas other variations in density cause seismic waves to be reflected or bent. These differences have allowed scientists to determine the nature of Earth's interior and map its basic structure, somewhat like the explorers in the Jules Verne novel. The following discussion outlines that character of the layers associated with the Earth's structure.

The Major Layers

With a cross-sectional view of Earth, you can see that the inner Earth contains several major layers (Figure 12.1). Because it is difficult to imagine the depths and distances associated with these layers, it is useful to compare them to the distances between places on a map of North America (Figure 12.2). For example, the distance to the center of the Earth's core—in other words, to the center of Earth—is about 6370 km (3963 mi), which happens to be the approximate distance from Anchorage, Alaska, to Miami, Florida.

Seismic waves *Vibrations that travel through the Earth when stress is released in an earthquake.*

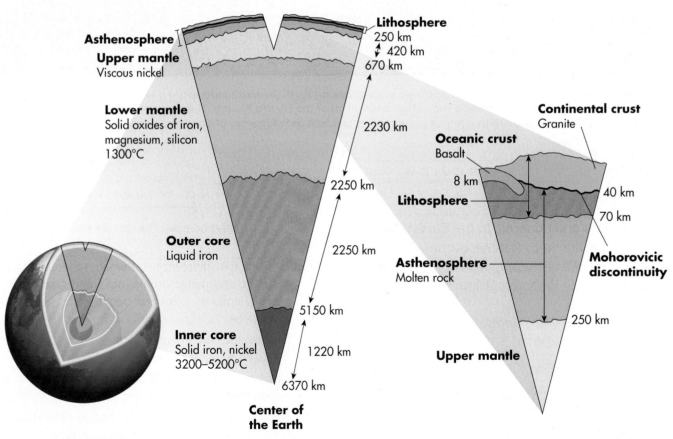

Figure 12.1 Earth's interior structure. The Earth's interior structure consists of several major layers, each with a distinct mineral composition and density. Note that the middle image is a blown-up diagram of the pie-shaped cross section in the left-hand diagram. The right-hand image, in turn, is a blown-up diagram of the pie-shaped cross section in the upper part of the middle image.

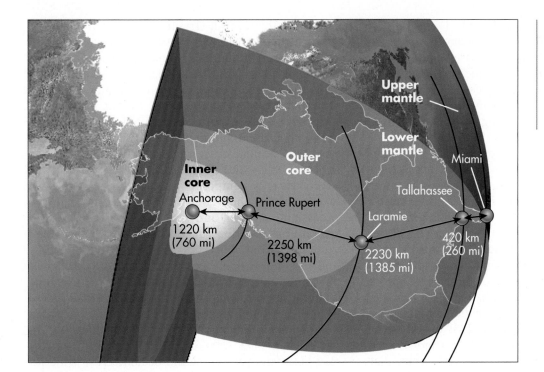

Figure 12.2 Distance to the center of Earth. The distance to the center of Earth is about the width of the North American continent between Anchorage, Alaska, and Miami, Florida.

The Earth's innermost layer is the **inner core**. This part of Earth has a radius of about 1220 km (760 mi), which is the approximate distance from Anchorage to Prince Rupert, British Columbia (Figure 12.2). The inner core is composed mostly of solid iron, along with a little bit of nickel. The pressure in this part of Earth is enormous, about 3 million times greater than the air pressure you feel at the surface, due to the weight of the overlying material comprising the rest of Earth. This high pressure explains why the inner core is solid even though the temperature there is well above the melting temperature of iron. Rocks simply cannot melt under such high pressure. In fact, geologists estimate that the temperature of the inner core ranges from about 3200°C to 5200°C (5800°F to 9400°F).

Surrounding the solid inner core is the **liquid outer core** (Figure 12.1). The top of this layer lies about 2900 km (1800 mi) below sea level. The liquid outer core is approximately 2250 km (1398 mi) thick, which is about the same distance as from Prince Rupert, British Columbia, to Laramie, Wyoming (Figure 12.2). In other words, it would take 2 or 3 days to drive that distance at a reasonable pace. The outer core is composed of the same material as the inner core, except

the iron is molten; that is, it is mostly liquid. This variation occurs because the temperatures of the outer core approach those of the inner core, hot enough to melt solid rock. The difference between the two layers is that the outer core has less internal pressure, which allows rock to melt. A key function of the outer core is that it generates at least 90% of the Earth's magnetic field and the resulting magnetosphere that protects Earth from the solar wind. The magnetic field may exist because the Earth's inner core rotates 19 km/year (12 mi/year), which is faster than the rest of the planet. This rotational difference is thought to cause circulation patterns in the outer core that generate electrical currents. These currents, in turn, may generate the magnetic field. Scientists are not certain of the exact processes that lead to the Earth's magnetic field, but circulation of the outer core is most likely involved.

Working out from the inner core, the next layer encountered is the **mantle**, which surrounds the core (Figure 12.1). This part of Earth is composed largely of solid iron, magnesium, and silicon oxides and is divided into two parts: the *lower mantle* and *upper mantle*. The lower mantle is about 2230 km (1385 mi) thick—in other words, about the same distance as

---

Inner core   *The inner part of the Earth's core. This area is about 1220 km (760 mi) thick and consists of solid iron and nickel.*

Liquid outer core   *The outer part of the Earth's core. This area is about 2250 km (1400 mi) thick and consists of molten iron and nickel.*

Mantle   *The layer of the Earth's interior that lies between the liquid outer core and the crust. This area is about 2900 km (1800 mi) thick and consists largely of silicate rock.*

from Laramie to Tallahassee, Florida (Figure 12.2). Again, this distance would take 2 or 3 days to drive. Although temperatures in the mantle are still quite hot, over 1300°C (2370°F), they are much cooler than in the core. These cooler temperatures, coupled with intense pressure, cause the lower mantle to be solid. This pressure lessens gradually, along with temperature, into the upper mantle.

The upper mantle is approximately 420 km (260 mi) thick, which is about the same distance as from Tallahassee to Miami, Florida. In contrast to the lower mantle, which is solid, the upper mantle consists mostly of viscous nickel; that is, it is a material like very thick syrup or a slowly flowing plastic. The upper part of the upper mantle is called the **asthenosphere**, which occurs between about 40 km and 250 km (25 mi and 105 mi) below Earth's surface. This part of the inner Earth is the least rigid portion of the mantle because it contains scattered zones of high temperature due to radioactive decay. These zones comprise about 10% of the asthenosphere and consist of molten rock that moves very slowly by convection. This process is highly influential on the Earth's surface, causing earthquakes, volcanoes, and the deformation of rocks to form mountain chains.

The uppermost layer of Earth is the **lithosphere** (Figure 12.1). This portion of Earth extends from the surface into the uppermost part of the asthenosphere at a depth of 70 km (44 mi). This portion of Earth contains the *crust*, which is the cool, stiff, and brittle exterior of Earth. The boundary between the crust and the asthenosphere is very well defined and is called the **Mohorovicic discontinuity** (pronounced Mo-ho-ro-vi-chich; "Moho" for short), named for the Yugoslavian scientist who first suggested its presence in 1909. This boundary is known because earthquake waves change speeds dramatically at the Moho level as a result of the dramatic density difference between the mantle and crust.

The crust is the thinnest of the major Earth layers, ranging in thickness from 8 km to 40 km (about 5 mi to 25 mi). This thickness represents about 1%

of the Earth's overall structure and is equivalent to driving from downtown Miami to a distant suburb in the metropolitan area. It floats on top of the mantle because its overall density is less. The crust contains two parts: oceanic crust and continental crust. **Oceanic crust** is about 8 km (about 5 mi) thick and consists mostly of a rock called *basalt*, which is very fine-grained and high in silica, magnesium, and iron; thus, oceanic crust is often called *sima*, which is short for silica and magnesium. **Continental crust**, in contrast, is composed largely of granite, which is a coarse-grained rock high in silica, aluminum, potassium, calcium, and sodium. As a result, continental crust is often called *sial*, short for silica and aluminum. Continental crust is fundamentally thicker and less dense than oceanic crust, averaging about 40 km (25 mi). It is thicker beneath mountain areas, where it can reach depths of 50 km to 60 km (31 mi to 37 mi). In nonmountainous areas, in contrast, continental crust is only about 30 km (19 mi) thick. As you will see in the next chapter, the difference in composition between continental and oceanic crust is a critical part of the overall history of Earth.

With the layers of Earth in mind, take a closer look at the relationship among the asthenosphere, lithosphere, and crust. Remember that these layers blend into one another, with the lithosphere being the transition between the asthenosphere and crust. Note in Figure 12.1 that the depth of the Moho level is greater beneath continents than under ocean basins. The reason for this pattern is fairly straightforward, as shown in the sequence of diagrams in Figure 12.3. Figure 12.3a represents a period of time during which intrusions of magma rise up from the mantle into the overlying continental crust. In this way, the overall elevation and mass of the continental crust increase, perhaps due to mountain building. This increased mass increases the pressure on the underlying asthenosphere, which is plastic, relative to the ocean basin. As a result, the depth of the Moho level increases beneath the continent.

Asthenosphere   *The layer of very soft rock that occurs in the upper part of the upper mantle. This region is about 40 km to 250 km (25 mi to 105 mi) below the surface of Earth. The soft character of this rock allows isostatic adjustments to occur.*

Lithosphere   *The outer, solid part of Earth that is about 70 km (44 mi) thick and includes the uppermost part of the asthenosphere and the crust.*

Mohorovicic discontinuity   *The boundary between the Earth's crust and the upper part of the asthenosphere; seismic waves change speed at this boundary.*

Oceanic crust   *Basaltic part of the Earth's crust that makes up the ocean basins. Oceanic crust is about 8 km (5 mi) thick and is also called* sima *because it consists largely of silica and magnesium.*

Continental crust   *Granitic part of the Earth's crust that makes up the continents. Continental crust averages about 40 km (25 mi) in thickness and is also called* sial *because it consists largely of silica and aluminum.*

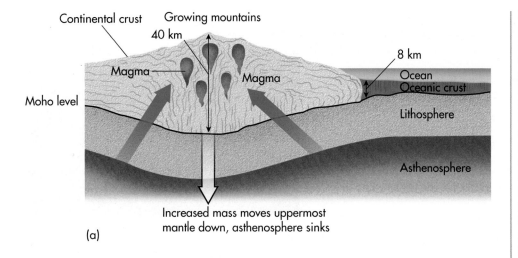

Continental crust Growing mountains
40 km

Magma

Moho level

Magma

8 km

Ocean
Oceanic crust

Lithosphere

Asthenosphere

Increased mass moves uppermost
mantle down, asthenosphere sinks

(a)

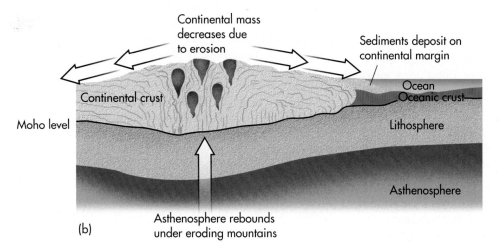

Continental mass
decreases due
to erosion

Sediments deposit on
continental margin

Continental crust

Moho level

Ocean
Oceanic crust

Lithosphere

Asthenosphere

Asthenosphere rebounds
under eroding mountains

(b)

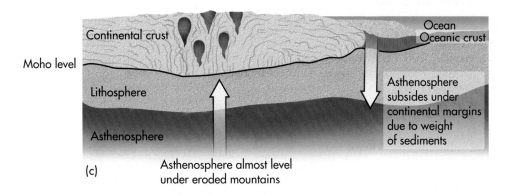

Continental crust

Moho level

Lithosphere

Asthenosphere

(c)

Asthenosphere almost level
under eroded mountains

Ocean
Oceanic crust

Asthenosphere
subsides under
continental margins
due to weight
of sediments

Figure 12.3 Isostatic adjustment. (a) Continental crust is thickest where mountain ranges occur. Given the extreme weight of these locations, the underlying asthenosphere subsides relative to the area beneath the oceanic crust. (b) As mountains weather, the weight of the landmass decreases, allowing isostatic rebound of the asthenosphere. At the same time, the asthenosphere beneath the continental margin begins to depress due to the increased weight of the freshly deposited sediments eroded from the continental interior. (c) As the continent weathers further, the isostatic rebound of the underlying asthenosphere continues. At the same time, the asthenosphere is further depressed due to the increased weight of margin sediments.

Over a very long period of time, the elevation of the continent gradually decreases due to erosion (Figure 12.3b). This process transports sediments to the continental margin, where they are deposited. As a result, the mass of the continent slowly decreases and the underlying asthenosphere begins to bounce back (or rebound). At the same time, the asthenosphere beneath the continental margin begins to subside because of the weight of the new sediments. This process continues until the continent is nearly leveled (Figure 12.3c), which results in the near complete rebound of the underlying asthenosphere. At the continental margin, however, the asthenosphere subsides even more due to the increased weight of sediments transported from the landmass. This overall process of subsidence and rebound of the asthenosphere is called *isostatic adjustment* or *isostacy* for short.

VISUAL CONCEPT CHECK 12.1

What two types of crust occur in landscapes that look like this? Which of the two is denser and why?

KEY CONCEPTS TO REMEMBER ABOUT THE EARTH'S INTERNAL STRUCTURE

1. Earth's interior is divided into several major layers. These layers are the inner core, outer core, lower mantle, upper mantle, asthenosphere, lithosphere, and crust.

2. The inner core and lower mantle are solid rock bodies composed mostly of iron. The outer core is liquid iron and the upper mantle viscous nickel.

3. The Earth's upper three layers are of most interest to geographers because these layers contribute directly to the development of surface landforms and thus affect peoples' lives. The asthenosphere is the lowermost of these layers and extends to a depth of 250 km (155 mi). The upper part of the asthenosphere forms the lithosphere, which contains the Earth's crust.

4. The Earth's crust is divided into oceanic and continental crusts. The oceanic crust is composed mostly of dense, fine-grained basalt, whereas continental crust consists mostly of coarse-grained granite.

Rocks and Minerals in the Earth's Crust

Have you ever picked up an interesting-looking rock when walking along the beach or hiking in the mountains? Have you ever noticed a fantastic exposure of rock along a highway? If so, perhaps you wondered how the rocks formed, where they came from, or how old they were. In most cases the rocks were very old and contained elements that had been recycled many times throughout Earth's history. Some of these minerals may have even been in the center of Earth at one time. This part of the chapter examines the basic kinds of rocks on Earth and how they form.

Rocks are composed of a variety of Earth elements. Figure 12.4 shows the approximate percentage by weight of the various elements, including oxygen and silicon, with lesser amounts of aluminum, iron, calcium, sodium, potassium, and magnesium. These elements combine in various ways to form **minerals**, which are naturally occurring substances with distinctive chemical and atomic configurations that usually take some kind of crystalline form. A great example of a crystalline mineral that you may be familiar with is quartz, which typically occurs as a clear, six-sided prism (Figure 12.5).

When two or more minerals become bonded in a solid state, they form a **rock**. Most rocks are very old by human standards, with the oldest being over 3.5 billion years old! Nevertheless, new rock is probably being

---

Minerals *Naturally occurring substances with distinctive chemical configurations that usually manifest themselves in some kind of crystalline form.*

Rock *An amorphous mass of consolidated mineral matter.*

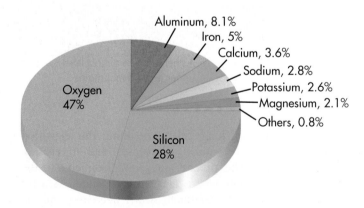

Figure 12.4 Chemical elements in the Earth's crust. Percentages refer to the percentage by weight of each element in the Earth's crust.

Figure 12.5 Quartz crystals. Quartz is an excellent example of a crystalline mineral. In small grains, it forms sediments in deserts, river deposits, and coastal areas. It is also a common part of most rocks formed in the continental crust.

created someplace at the very instant that you are reading this. Is a major volcanic eruption occurring on Earth right now that is spreading lava across the surface, or did a large flood recently happen that deposited mud in river valleys? If such an event is taking place, then the foundation is being set for the development of solid rock at those locations in the future.

Because of the many different chemical elements that occur on Earth, an extremely wide variety of rock types exist. It goes way beyond the scope of this book to describe each one of the many different kinds of rocks. Instead, we will focus on the three major rock classes—igneous, sedimentary, and metamorphic—and some fundamental kinds of rock that occur within these groups. After completing this section of the chapter, you should be able to identify these basic rocks, whether they lie randomly on the ground or are exposed at the surface in a **rock outcrop**. The presence of a particular rock type yields important clues about the geologic history of the area.

Igneous Rocks

One of the distinctive categories of rocks on Earth is that associated with liquid rock, known as **magma**, that originates in the mantle and moves upward into the Earth's crust or onto the surface (Figure 12.6a).

Rock outcrop   *A place where rocks are exposed at the surface of Earth.*

Magma   *Molten rock beneath the surface of Earth.*

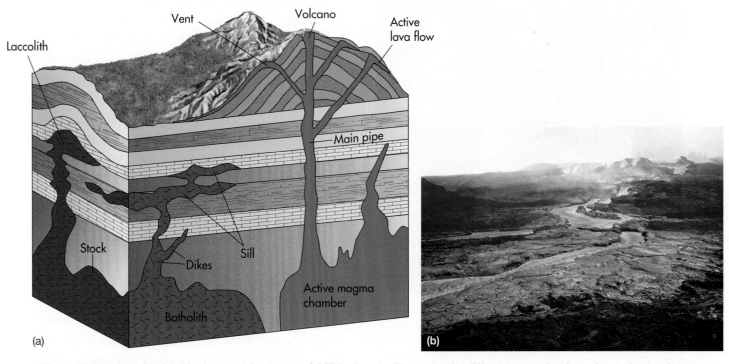

Figure 12.6 Various kinds of igneous environments. (a) This diagram illustrates the different ways that intrusive and extrusive igneous rocks form. (b) Lava flow at Mauna Loa Volcano in Hawaii. This material will ultimately cool and darken, consistent with the dark material surrounding it, forming extrusive igneous rock.

EXPOSED IGNEOUS INTRUSIONS

(a) Shiprock, New Mexico, as it appears today. Note the prominent dikes (arrows) that radiate away from the central core.

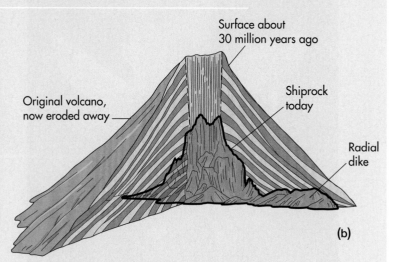

(b)

Surface about 30 million years ago

Original volcano, now eroded away

Shiprock today

Radial dike

Diagram showing how the original volcano may have appeared before it eroded away. Try to imagine how much time must have occurred for this landscape to evolve.

(c) The tall columns of Devil's Tower have made it a popular, if somewhat dangerous, route for serious rock climbers. It is also a sacred place for some Native American tribes and appeared in the popular movie *Close Encounters of the Third Kind* (1977) as a landing site for extraterrestrials. In 1906 it was designated the first national landmark in the United States.

Two of the most dramatic natural features in North America consist of tall rock monoliths arising out of otherwise flat landscapes. They are not isolated mountains; rather, they are exposed igneous rock intrusions.

Shiprock, in the New Mexico desert, is sacred to the Navajo Indians, who call it "Winged Rock." Standing 500 m (1700 ft) tall, the feature was once magma that cooled within the main pipe of a volcano. Geologists believe that this volcano existed about 30 million years ago. Since that time, the volcano has eroded extensively, leaving the central solid rock from the main pipe standing alone in the desert. Some of the dikes surrounding the volcanic pipe have also survived, forming radial rock fans about 3 m (10 ft) thick and 20 m (65 ft) high, extending as far as 3 km (2 mi) from the Shiprock formation.

Another rock intrusion is Devil's Tower, located in the Great Plains of eastern Wyoming. Devil's Tower is 264 m (867 ft) tall and is thought to have formed when magma intruded into older rocks about 40 million years ago. Since that time the overlying rocks eroded, leaving the central core standing above the surrounding plains. The rocks of Devil's Tower have a distinctive columnar orientation as a result of contraction that occurred during the cooling of the magma.

Identification of Igneous Environments

This section on igneous rocks contained a number of terms that relate to specific processes and places within the Earth's crust. Now is a good time to review these terms in an interactive way. Go to the *GeoDiscoveries* website and select the module *Identification of Igneous Environments*. This module is a drag-and-drop exercise that will test your ability to associate terms with respective igneous features. Once you complete the module, answer the questions at the end to test your understanding.

When this magma cools, it forms **igneous rocks**. You are probably most familiar with magma that flows across the surface of Earth as red-hot *lava* (Figure 12.6b) that erupts from volcanoes. This liquid rock ultimately cools to form **extrusive igneous rock**, which is so named because it results from magma that flows out of—or extrudes from—Earth.

In contrast to the magma that flows onto the Earth's surface and cools, the majority of magma cools within Earth (Figure 12.6a) to form **intrusive igneous rock**, so named because the magma intrudes into older rocks. This cooling process can occur in a variety of places beneath the Earth's surface, with any body of intrusive igneous rock referred to as a **pluton**. The largest plutons are referred to as *batholiths*, which are features of no known depth that are up to several hundred kilometers long and 100 km (62 mi) wide. A smaller, somewhat rounded pluton that is attached to a batholith is called a *stock*.

In some instances, magma will move upward from a magma chamber into relatively small fissures within the overlying rock. One such feature is a *dike*, which is a pluton that consists of a wall-like feature that develops when magma intrudes into a vertical rock fracture. Sometimes magma moves upward through such a fracture and then spreads horizontally within the crust along zones of rock weakness, forming a pluton called a *sill*. In other cases, upwardly moving magma spreads horizontally along a zone of rock weakness, while at the same time pushing upward in such a way that the surface rocks are warped upward in a dome-like feature. When this warping magma cools, it forms a *laccolith*.

Regardless of whether the igneous rock is intrusive or extrusive, it consists largely of silicate minerals that contain crystalline chemical compounds dominated by silicon and oxygen atoms. Although many different kinds of silicate minerals occur, seven are most common (Figure 12.7). Three of these minerals are *felsic*, which means (among other things) that they have (1) high silica content; (2) relatively low melting temperatures; and (3) lighter colors. The remaining four of the seven silicate minerals are *mafic*, which essentially have opposite characteristics to felsic minerals.

Figure 12.7 indicates how silicate minerals can combine in different ways to form various kinds of igneous rocks. A good example on the mafic side of the diagram is the intrusive rock *gabbro*, which forms when plagioclase feldspar mixes with minerals from the pyroxene group and olivine. If the same minerals combine in an extrusive environment, the mafic rock is called *basalt*. On the felsic side of the diagram, a good example is the intrusive rock *granite*, which forms when potash feldspar combines with plagioclase feldspar, biotite mica, and quartz. In an extrusive setting, these minerals combine to form *rhyolite*. Note the other kinds of combinations that can occur.

You may wonder, How can you tell the difference between extrusive and intrusive igneous rocks if they are made from the same materials? The answer is that extrusive and intrusive rocks differ fundamentally in their appearance. Intrusive igneous rocks have coarse crystalline grains that you can see with your unaided eye. These coarse textures evolve because magma cools relatively slowly, perhaps over several thousand years, when it is trapped within the Earth's crust, such as in a sill, dike, or laccolith (see Figure 12.6a). A good example

---

Igneous rocks *Rocks that form when magma rises from the mantle and cools, either within the Earth's crust or on the surface.*

Extrusive igneous rocks *Rocks that form when magma cools on the Earth's surface.*

Intrusive igneous rocks *Rocks that form when magma cools within the Earth's crust.*

Pluton *An extremely large mass of intrusive igneous rock that forms within the Earth's crust.*

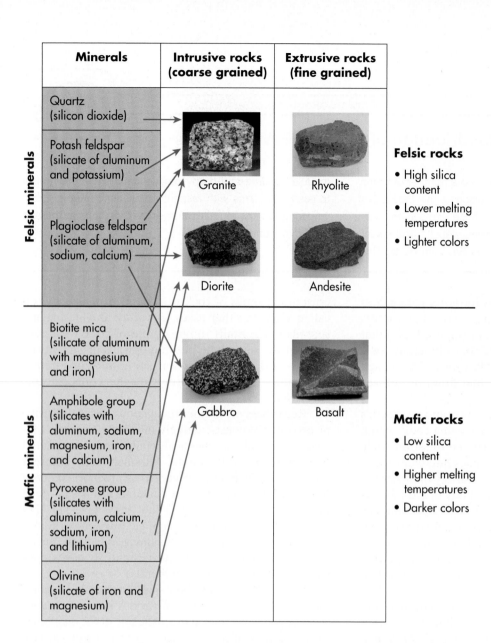

Minerals	Intrusive rocks (coarse grained)	Extrusive rocks (fine grained)

Felsic minerals

Quartz (silicon dioxide)	Granite	Rhyolite
Potash feldspar (silicate of aluminum and potassium)		
Plagioclase feldspar (silicate of aluminum, sodium, calcium)	Diorite	Andesite

Mafic minerals

Biotite mica (silicate of aluminum with magnesium and iron)	Gabbro	Basalt
Amphibole group (silicates with aluminum, sodium, magnesium, iron, and calcium)		
Pyroxene group (silicates with aluminum, calcium, sodium, iron, and lithium)		
Olivine (silicate of iron and magnesium)		

Felsic rocks
- High silica content
- Lower melting temperatures
- Lighter colors

Mafic rocks
- Low silica content
- Higher melting temperatures
- Darker colors

Figure 12.7 Silicate minerals and igneous rocks. This diagram shows the most important groups of silicate minerals, as well as some common igneous rocks. Notice that various combinations of minerals produce different rocks, both in extrusive and intrusive forms.

of a coarse-grained igneous rock is *granite* (Figure 12.7). Granite is a felsic rock that has distinctive dark grains consisting of the minerals *amphibole* and *biotite*, pinkish-tan grains of *feldspar*, and clear grains of *quartz*.

In contrast to intrusive igneous rocks, extrusive igneous rocks reach the surface and cool relatively quickly, sometimes in a matter of days. A good example of an extrusive igneous setting can be seen in Figure 12.6b, which shows a lava flow in Hawaii. Given that the lava cools relatively quickly in this setting, silicate minerals do not have time to congregate with one another, as they do in intrusive igneous rocks. The result is a fine-grained rock. A common kind of extrusive igneous rock is *basalt*, which is a dark mafic rock (Figure 12.7). Perhaps the most distinctive extrusive igneous rock is *obsidian* (Figure 12.8). Obsidian is also called *volcanic glass* and forms when nongaseous lava cools so fast that crystals cannot form.

Sedimentary Rocks

Now let's turn to the second major rock class: **sedimentary rocks**. Although some forms of sedimentary rock have organic origins, most consist of vast quantities of formerly loose minerals that collected in some kind of depositional setting, such as a river floodplain, shallow sea floor, interior valley, lake, or marsh (Figure 12.9). These minerals can arise from any of the three major rock groups—igneous, sedimentary, or metamorphic (to be discussed later)—but their ultimate source is usually some form of igneous rock that was weathered

---

Sedimentary rocks *Rocks that form through the deposition and lithification of small fragments or dissolved substances from other rocks or, in some cases, marine animals.*

in a way that liberated scores of individual particles such as clay (from feldspar) or sand (from quartz). These particles could then be transported as **sediment** by wind, water, or glaciers to a new place, where they subsequently accumulate as sedimentary deposits. Most sedimentary rocks are the remnants of sediments that accumulated in an oceanic environment. Many of these sediments were eroded from the continents in various ways and then carried to the ocean where they settled to the bottom.

After the sediment accumulates, all the grains gradually become cemented to one another in a gradual process called **lithification** to form rock. This process occurs because water is squeezed out of the lower layers of sediments when they are compacted by the cumulative weight of the deposits above them. As a result of compaction and associated water loss, minerals such as calcium and silica recrystallize in the sediments and essentially glue them together to form rock.

Within the overall category of sedimentary rocks, three major classes of sediments occur: clastic, chemical, and organic. Table 12.1 lists the basic characteristics of these rocks.

Sediment   *Solid fragments of rocks that are transported to some location and deposited by wind, water, or ice.*

Lithification   *The process whereby sediments are cemented through compaction to form rock.*

Figure 12.8 Obsidian. Obsidian is an extrusive igneous rock that develops when nongaseous lava cools very quickly, before crystals can develop. This rock was prized by prehistoric Native Americans because it has a rich, dark color and a smooth glassy surface.

Clastic Sedimentary Rocks   Perhaps the easiest type of sedimentary rock to understand is the category of clastic sedimentary rocks. These rocks contain mineral fragments derived from any of the three major rock groups that accumulate and lithify. The most common minerals in clastic sedimentary rocks are silicates, such as quartz and feldspar. Quartz is by far the most important clastic sediment, primarily because it is hard and very resistant to alteration. On the finer end of the

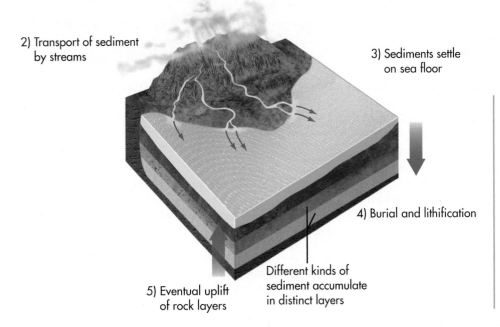

1) Sediments erode from continent

2) Transport of sediment by streams

3) Sediments settle on sea floor

4) Burial and lithification

5) Eventual uplift of rock layers

Different kinds of sediment accumulate in distinct layers

Figure 12.9 Marine sedimentary environments and rock formation. This generalized diagram illustrates how clastic marine sediments ultimately become rock: (1) sediments erode from continents; (2) sediments are transported by streams into the ocean; (3) sediments settle to the ocean bottom; (4) the deposited sediments are buried and slowly lithify to become rock; and (5) the lithified sediments eventually uplift as new rock layers.

TABLE 12.1	Some Common Types of Sedimentary Rocks	
Class	Rock Type	Composition
Clastic: Formed from rock or mineral fragments	Conglomerate	Coarse-grained sandstone that contains rocks and pebbles of different sizes
	Sandstone	Sand grains cemented by minerals
	Siltstone	Silt particles cemented by minerals
	Claystone	Clay particles cemented by minerals
	Shale	Fine-grained particles that often contain fossils; most abundant sedimentary rocks; forms layers along which rock can split
Chemical: Formed from mineral precipitates from seawater or salty lakes	Limestone	Calcium carbonate, including that from animals and microorganisms; forms rock layers
	Dolomite	Magnesium carbonates, mostly formed by chemical replacement in limestone
	Evaporites	Minerals (gypsum, rock salt, calcite) left in deposits after evaporation of water
	Chert or flint	Extremely fine grains of silica (quartz) formed in layers or nodules within limestone
Organic: Formed from carbon-based organic matter	Coal	Carbon-based materials lithified by heat and pressure
	Oil (petroleum)	Liquid hydrocarbon trapped in sedimentary deposits; important fossil fuel
	Natural gas	Gaseous hydrocarbon trapped above oil in sedimentary deposits; important fossil fuel

textural scale, clays can be weathered from feldspars and carried by wind and water to some point where they are deposited.

Several types of clastic sedimentary rocks occur on Earth. **Sandstone** is created when individual sand grains are deposited in thick layers by water along a beach or river, or by the wind within dunes. This collection of sand grains then lithifies, forming rock. When the sandstone also contains abundant rocks and pebbles of various sizes, it is called **conglomerate** (Figure 12.10).

An excellent place to see beautiful sandstone outcrops is the southwestern United States, with one of the best known rock bodies being the Navajo Sandstone (Figure 12.11). The sediments comprising this rock accumulated in sand dunes approximately 175 million years ago when the region was a desert and sand was blown and deposited by the wind. Subsequently, the climate became more humid, resulting in rivers that buried the dunes with stream sediments. Still later, the region was covered by an ocean that deposited tremendous quantities of marine sediments. As a result of the compression and cementation of sands in the old dunes, they gradually became rock.

Figure 12.10 Conglomerate rocks. Conglomerate is a type of rock consisting of large clasts contained within a matrix of fine-grained particles. Note in this photo the mix of coarse grains, cobbles, and fine sediments. Such a mix of particles likely accumulated in a stream environment.

In addition to the sedimentary rocks created from individual sand grains, rocks can also be composed of finer grained silts and clays. When fine grains dominate a sedimentary deposit, it is referred to as *mud*. Silts and

Sandstone *A sedimentary rock created when individual sand grains are deposited in thick layers by wind or water and lithify.*

Conglomerate *Sandstone that contains a wide variety of particle sizes.*

Figure 12.11 The Navajo Sandstone. Although it may be hard to believe, this landscape was once an active dune field in what is now Zion National Park, Utah. You can tell that the sands accumulated in dunes because of the distinct bedding lines exposed in the outcrop. Over a long period of time, the sand grains in the dunes lithified to form this rock body.

clays are easily suspended in and transported by moving water. Thus, mud typically accumulates in bodies of relatively still water, such as a coastal bay or lagoon (Figure 12.12a). After they are deposited in these settings, they lose a great deal of their original volume through compaction. The origin of these rocks and their associated names are logically based: *siltstone* is dominated by silt-sized particles and *claystone* consists largely of clays. The most common fine-grained sedimentary rock is **shale**, which is a claystone that easily breaks into small flakes and has a platy appearance (Figure 12.12b).

Chemically Precipitated Sedimentary Rocks When environmental conditions are favorable, thick deposits of mineral compounds can accumulate at the bottom of the ocean or inland lakes. In ocean environments, this process is associated with the precipitation and recrystallization of calcium carbonate from seawater. This deposition can occur for a variety of reasons, including increase in water temperature, intense evaporation, upwelling of water, and bacterial decay, to name a few.

Shale *Sedimentary rock consisting of lithified clay-sized sediments.*

(a)

(b)

Figure 12.12 Formation and character of shale. (a) Mud collectively consists of individual clay particles, along with some silts, which settle out of suspension in places where water is very calm. If these sediments remain in place, they will gradually lithify to form shale. (b) Close-up of shale. Note the flaky appearance of this rock type.

Clastic Rocks

Clastic rocks can be easily identified based on the size of sediment grains contained within them. In order to get a better feel for these rock types, go to the *GeoDiscoveries* website and access the module *Clastic Rocks*. This module is a simple interactivity that allows you to better visualize the character of clastic rocks. To test your understanding, be sure to answer the questions at the end of the module.

An excellent place to see the formation of chemically precipitated sedimentary rocks is the Bahama Islands, which lie off the eastern coast of Florida. Figure 12.13 is a satellite image that clearly shows the precipitation of calcium carbonate in this island chain. In this image, carbonate appears as the whitish plumes within the matrix of turquoise and bluish waters that lie between the islands. The turquoise and bluish colors appear because the water is shallow in the island complex and the satellite sensor can see the ocean bottom. Away from the islands, however, the water rapidly deepens, which is reflected by the dark ocean colors. Once the carbonates accumulate on the ocean floor, they subsequently lithify to form rock. The Bahama Islands, in fact, are made of calcium carbonate deposited in this manner.

In other situations, organisms rich in calcium carbonate such as clams, mussels, and oysters accumulate on the ocean bottom after they die. Subsequently, they become cemented when calcium carbonate precipitates and settles within the matrix of animal remains. The resulting rocks can contain abundant fossils (Figure 12.14), which enable geologists to reconstruct environmental conditions at the time the sediments were initially deposited. Regardless of the particular depositional process, whether it is entirely through carbonate precipitation or due to the accumulation

Figure 12.13 Carbonate depositional environments in the Bahama Islands. In this region, carbonate minerals are precipitating in the shallow waters of the island chain. They form the whitish haze that surrounds the islands.

Figure 12.14 Fossils in limestone. This rock formed tens of millions of years ago when dead ammonites accumulated on the ocean bottom and calcium carbonate subsequently precipitated within the matrix of shells. The combined materials subsequently lithified to form limestone.

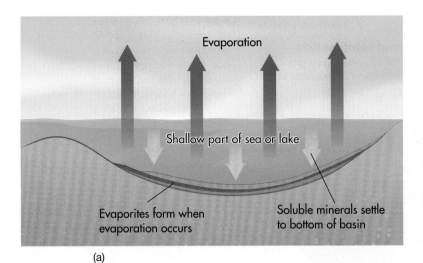

(a)

(b)

Figure 12.15 Formation of evaporites. (a) Evaporites form when minerals recrystallize on the exposed surface after waters evaporate from shallow oceans or inland lakes. (b) The whitish deposits in this image are salts and other highly soluble minerals that recrystallize as a crust when evaporation occurs in this extremely arid environment of Death Valley.

of marine organisms, the resulting rocks are typically classified as **limestone**. If the deposits are rich in magnesium, the resulting carbonate rock is called **dolomite**. Under more unusual conditions, silica can precipitate from solution at the bottom of the ocean to form *flint* or *chert*.

In addition to forming in marine environments, chemically precipitated rocks also develop within continental locations when minerals evaporate from concentrated solutions (Figure 12.15a). These minerals collect at the surface and are called **evaporites**. As you can imagine, evaporates are most likely to evolve in semi-arid to arid regions such as the southwestern United States (Figure 12.15b).

Organic Sedimentary Rocks As the name implies, organic sedimentary rocks consist of carbon-based materials that accumulate in thick deposits at the surface of Earth. An example of this process was shown in Chapter 11, specifically the accumulation of plant and organic matter as peat in cool and moist environments. Such thick accumulations of solid organic remains have occurred on a large scale over time, only to be buried by other sediments. When these organic

deposits are buried progressively deeper, they slowly compact under the weight of the overlying sediments. In addition, they begin to be heated because temperature slowly rises with increased depth of burial. Over the course of millions of years, the amount of water and oxygen in the carbon progressively lowers and the deposit gradually solidifies, resulting in **coal**. Sometimes these deposits are "cooked" so much that they liquefy to form **petroleum** (crude oil). **Natural gas** comes from the remains of microscopic plants that live in the surface waters of the ocean. When they die, they settle to the bottom of the ocean where they decompose and form gas. This form of gas is typically about 85% methane and 10% ethane, with smaller amounts of propane and butane.

In places where organic sedimentary rocks occur, they are usually found in distinct beds that are separated by layers of limestone, sandstone, or shale. These deposits frequently accumulate in domes that form when rocks are bent due to compression for one reason or another; this process will be discussed more thoroughly in Chapter 13. Deposits of coal, petroleum, and natural gas are very important as fossil fuels to modern human civilization.

Limestone *Sedimentary rock consisting of over 50% calcium carbonate ($CaCO_3$).*

Dolomite *Sedimentary rock consisting of over 50% calcium-magnesium carbonate ($CaMg[CO_3]_2$).*

Evaporites *Surface salt residue that collects through the evaporation of water and the crystallization of sodium.*

Coal *A solid fossil fuel that consists of carbonized plants and animals.*

Petroleum *Naturally occurring oily liquid that consists of ancient hydrocarbons.*

Natural gas *Naturally occurring mixture of hydrocarbons that often occurs in association with petroleum and is found in porous geologic formations.*

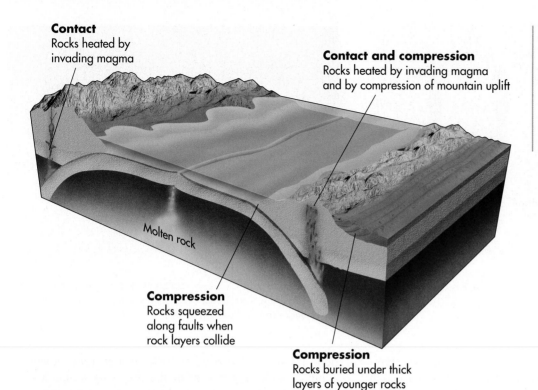

Contact
Rocks heated by
invading magma

Contact and compression
Rocks heated by invading magma
and by compression of mountain uplift

Molten rock

Compression
Rocks squeezed
along faults when
rock layers collide

Compression
Rocks buried under thick
layers of younger rocks

Figure 12.16 Formation of metamorphic rocks. Metamorphic rocks evolve due to intense pressure of overlying rocks or because rocks are heated dramatically when they come into contact with igneous bodies.

Metamorphic Rocks Igneous and sedimentary rocks can be altered after they develop. This alteration results in **metamorphic rock**, which occurs when a former igneous or sedimentary rock is subjected to intense heat or pressure within Earth over millions of years. Thus, the name of the rock comes from a Greek word meaning "to change form." (A good example of metamorphosis is when a caterpillar changes into a butterfly.) Figure 12.16 shows the various ways in which metamorphic rocks form. One way occurs when rocks are slowly compressed due to the thick accumulation of overlying sediments or because they are deeply buried by crustal processes. Another way that alteration occurs is through *contact metamorphism*, which occurs when rocks are heated dramatically when they come into contact with invading magma in batholiths, dikes, or sills such as those illustrated in Figure 12.6. Whatever kind of alteration takes place, the result is that mineral components within the rock are rearranged to form different mineral varieties, or the minerals recrystallize into new mineral forms.

Although many different kinds of metamorphic rocks occur, just a few will be described here. *Slate* forms when shale is heated and compressed, forming a grayish, smooth rock that is quite hard. You may be familiar with slate because it is used as roofing

shingles, patio flagstones, chalkboards, and the surface of the best pool tables. When slate is further heated and compressed, such that distinct planes and coarse texture develop, it becomes *schist* (Figure 12.17a). Limestone that has undergone metamorphism becomes *marble*, which has a distinctive appearance because mineral impurities show up as swirling bands. Marble is popular because it is frequently used in expensive tabletops and flooring. Finally, *gneiss* is a hard, banded metamorphic rock (Figure 12.17b) that forms when igneous or sedimentary rocks have been in close contact with intrusive magmas. The banding occurs because minerals become aligned during the metamorphic process. These kinds of rocks are described as being *foliated*, whereas those that lack banding are *nonfoliated* (Figure 12.17a).

The Rock Cycle

Recall from earlier chapters that water and carbon move from one reservoir on Earth to others as part of the hydrologic and carbon cycles, respectively. In a similar way, minerals contained within rocks move in a *rock cycle* (Figure 12.18). Although movement within this cycle is much slower than other cycles, often over millions of years, it nevertheless comprises a distinct series of steps where rock gradually moves from one physical state or location to another.

Let's examine the rock cycle from a holistic perspective by examining Figure 12.18 and starting with the formation of an extrusive igneous rock such as

---

Metamorphic rocks *Rocks that form when igneous, sedimentary, or other metamorphic rocks are subjected to intense heat and pressure.*

Figure 12.17 Some metamorphic rocks. (a) Schist forms when the metamorphic rock slate is further altered. (b) Gneiss evolves when igneous or sedimentary rocks are dramatically heated. Note the banding in this rock where minerals aligned under the intense heat.

basalt. This rock originates from magma that flows out of the ground, usually from a volcano, but perhaps just a fissure in the ground. This magma cools and crystallizes, forming basalt that lies at or close to ground level. With time, the basalt gradually weathers due to various erosion processes, causing individual grains of the former rock mass to be transported and deposited someplace else. As the deposit of sediments thickens, those at the bottom of the stack become compressed, causing lithification and formation of sedimentary rocks. If these sedimentary rocks become deeply buried over time, or come into contact with the intense heat of an intrusive magma body, they will be altered

to metamorphic rock because the original sediments will be recrystallized, recombined, or replaced. These metamorphic rocks, in turn, may be further heated such that they melt back to the magma state. If the sedimentary rocks remain near the surface, they too will be weathered, causing sediments to be loosened from the host rock mass. These sediments will then accumulate someplace else to form rock of a potentially different character. Remember that these are possible directions that the rock cycle might take. It is possible that sedimentary rocks may be re-melted directly to magma or that metamorphic rocks may be weathered to form new sedimentary rock.

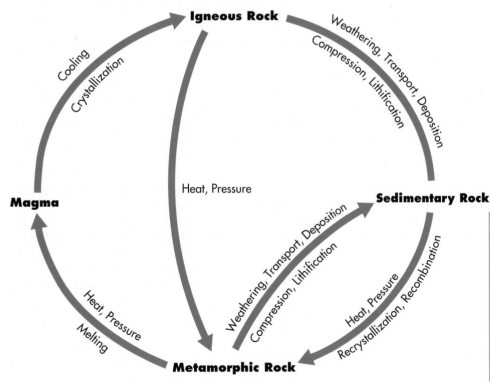

Figure 12.18 Simplified view of the rock cycle. Over time, minerals move from one rock phase to another through a variety of processes, such as heating, melting, weathering, and recrystallization. This illustration does not contain all forms of movement within the cycle. For example, igneous rocks can melt and re-form as igneous rocks. Can you think of other forms not shown here?

This image shows rock outcrop along a highway. The lower arrow points to a shale bed, whereas the upper arrow focuses on a limestone rock unit. Which one of the following choices best explains this combination of rocks?

a) They are igneous rocks.
b) They were modified by intense heat and pressure.
c) These units reflect fluctuating sea levels in the past.
d) They formed when sandy sediments accumulated along a beach.

 **Geo Discoveries**

www.wiley.com/college/arbogast

 WILEY PLUS

The Rock Cycle

Although the rock cycle moves very slowly, it is possible to animate the concept to capture its essence. Go to the *GeoDiscoveries* website and select the module *The Rock Cycle*. This module describes the ways in which rocks are created on Earth and how their minerals move from one reservoir to another. It begins with the formation of igneous rock and then describes how these rocks can be altered to form new kinds of rock, whether it be a new kind of igneous rock, sedimentary rock, or metamorphic rock. After you complete the animation, be sure to answer the questions at the end of the module to test your understanding of this concept.

A good example of how the rock cycle works is the carbonate environment of the Bahama Islands (see Figure 12.13). Recall that in this area recrystallized calcium carbonate accumulates on the shallow ocean floor. In the context of the rock cycle, this carbonate likely originated from continental limestones that were eroded, and the resulting calcium carbonate was carried in solution by rivers to the ocean. Some of this carbonate is currently recrystallizing in the Bahama Islands to form new rock. Some of that rock has been eroded and deposited on the shore to create the beautiful white, powdery beaches of the islands. Naturally, it has taken a long time for these combined processes to occur.

Geologic Time

In the context of the rock cycle, what does "a long time" mean? To answer this question requires considering the concept of **geologic time** (or *deep time*). In order to somehow grasp the enormity of deep time, it is essential to realize that it is radically different from the human timescale. Given our relatively short lifetimes, human beings tend to think that anything 100 years old, or older, is ancient. For example, the American Civil War seems like a very long time ago; does it not? After all, it occurred in the 1860s, long before your grandparents, or even great, great grandparents, lived. As far as the history of Earth is concerned, however, 100 years is a blink of an eye. Scientists accept that Earth is about 4.6 billion years old and that it lies within a much larger and older universe that is approximately 14 billion years old. This kind of timescale is naturally difficult to comprehend. It is important to understand the concept of deep time, however, because it provides a context within which we can study Earth. It also helps us realize the incredibly

Geologic time *The period of time that encompasses all of Earth history, from its formation to the present.*

rapid nature of environmental change associated with ongoing human impacts on the planet.

In an effort to make the concept of geologic time more understandable, geologists have devised a geologic timescale that is subdivided on the basis of major events in Earth history. Figure 12.19 shows this timescale and its various subdivisions. The timescale is categorized into (from largest time expanse to shortest) eons, eras, periods, and epochs. All time earlier than 540 million years (my) ago is lumped into the division of *Precambrian* time, which

includes both the *Archean* and *Proterozoic* Eons. This interval of time represents 88% of Earth history, during which very little life of any kind existed. Notable events during early Precambrian time include the initial stages of atmospheric development, the emergence of the first cyanobacteria, and the onset of photosynthesis. Toward the end of Precambrian time the modern atmosphere evolved.

Beyond Precambrian time is the *Phanerozoic* Eon, which is subdivided into three major *eras*, the *Paleozoic*, *Mesozoic*, and *Cenozoic*. These eras are further

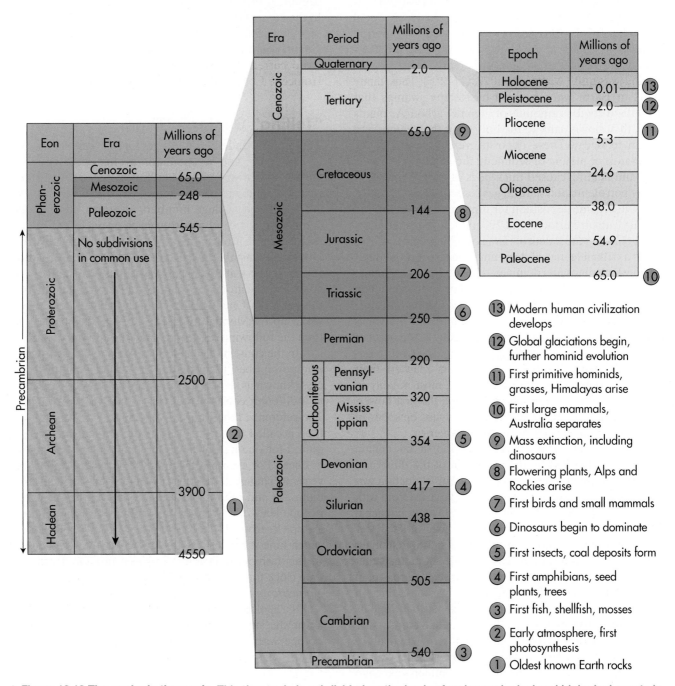

Figure 12.19 The geologic timescale. This timescale is subdivided on the basis of major geological and biological events in Earth history, such as the appearance of flowers or major extinctions.

subdivided into *periods*. The beginning of the Paleozoic Era (the *Cambrian* Period) is marked by the emergence of the first fish and shellfish about 540 million years ago, whereas the end of the Paleozoic Era is related to a major extinction during the *Permian* Period around 250 million years ago. Other significant events in the Paleozoic Era include the development of the first trees and amphibians at the beginning of the *Devonian* Period (~417 million years ago) and the first winged insects at the onset of the *Mississippian* Period (~354 million years ago).

The Mississippian Period is particularly noteworthy because organic deposits began to accumulate in sufficient quantity to ultimately form the extensive coal and other fossil fuel resources we use today. This period of massive organic deposition began because the average global temperature during the early Mississippian Period was quite warm, about 22°C (72°F). Thus, many parts of Earth were covered by steamy swamps that contained copious amounts of organic carbon. As plants died, they slowly sank to the bottom of the swamps and began to decay. These deposits were later buried by thousands of meters of sediment; the resulting compaction and heating caused thick beds of coal to form and many petroleum and natural gas reserves. This period of massive organic deposition terminated at the end of the Pennsylvanian Period about 290 million years ago.

Following the Paleozoic Era is the Mesozoic Era. This era is a definable unit of deep time because it is the interval in which dinosaurs dominated life on Earth, beginning around 250 million years ago and ending about 65 million years ago. *Isn't it amazing that dinosaurs existed for over 180 million years?* In addition, the first birds and mammals evolved during this era of time, specifically at the onset of the *Jurassic* Period 206 million years ago. The Cretaceous Period is significant because dinosaurs reached their maximum dominance during this time, most notably in the form of the *Tyrannosaurus Rex*, and the first flowering plants appeared. In addition, two major episodes of mountain building occurred in North America.

The end of the Cretaceous Period at 65 million years ago is marked by the extinction of the dinosaurs. Currently, most geologists believe that this extinction was caused by the catastrophic impact of a large asteroid with Earth. This impact is thought to have severely impacted Earth through a combination of intense fires and high levels of atmospheric dust that may have blocked sunlight for months. As a result, the vast majority of animals that depended directly on plants, either through eating them or by eating the animals that ate plants, became extinct. In contrast, it is thought that animals such as small rodents and insects survived because they fed on the corpses and organic waste of the larger animals that were decimated.

The Cenozoic Era is significant because it represents all of post-dinosaur time, beginning 65 million years ago and continuing until the present. During this period most current landscape features such as mountain chains and river valleys on Earth developed, as well as life as we know it. The Cenozoic Era is subdivided into two periods, the *Tertiary and Quaternary*. Given the relative youth of these two periods, plus the fact that most current landforms developed within this time interval, scientists know a great deal more about them than earlier periods. Thus, these periods are further subdivided into *epochs*. The *Paleocene* Epoch, for example, is marked by the emergence of the first large mammals following the extinction of the dinosaurs 65 million years ago. Subsequently, the first primitive hominids developed at the beginning of the *Pliocene* Epoch 5.3 million years ago. In the context of human habitation of Earth, the *Pleistocene* and *Holocene* Epochs are important because they are intervals of time during which most human evolution has occurred. As discussed in Chapter 9, we live in the Holocene Epoch, which spans the past 10,000 years.

"Telling" Geologic Time

You might wonder, How do geologists really know the age of Earth and when major events occurred? The answer lies in our understanding of the way in which rock elements radioactively decay through time. Rocks are composed of huge numbers of atoms, all of which contain protons and neutrons in their nuclei. Although many of these atoms remain stable indefinitely, some do not. These unstable atoms are called **radioactive isotopes**. In these atoms, particles within the nucleus break apart and the atom decays into a different element. For the purpose of analogy, imagine a pet dog that suddenly woke up on Monday morning and had become a cat, then the following Monday changed again into a mouse.

In the context of calculating rock age, isotopic decay is important because radiation is emitted when this process (known as *radioactivity*) occurs. Each isotope decays at a different constant rate that is known by geologists. The reference timeframe for the decay rate for any isotope is called its *half-life*; this is the amount of time required for one-half of the isotopes in any given sample to decay. For example, thorium-232 requires 14.1 billion years to change through its decay series (including radium-228 and radon-220, among others) into the stable isotope lead-208. The half-life of the radioactive carbon isotope (carbon-14), in contrast, is 5730 years, during which time it converts to the stable isotope nitrogen-14. In this fashion, radioactive isotopes provide a running time clock for the history of Earth and geologists use *radiometric dating* to calculate age.

If a geologist is interested in the age of a rock sample, he or she simply compares the amount of the original isotope with the amount of the decayed end product. Using radiocarbon dating, for example, a geologist can

---

Radioactive isotopes *Unstable isotopes that emit radioactivity as they decay from one element to another.*

Geologic Time

Understanding geologic time is difficult because it is hard to relate a diagram such as Figure 12.19 with the significant geologic events used to mark major intervals of time. Such events include major periods of mountain building, catastrophic extinctions, and the emergence of certain kinds of plants or animals in the fossil record. In an effort to improve this understanding, go to the *GeoDiscoveries* website and select the module *Geologic Time*. This module illustrates,

with images and text, some of the major geologic events associated with past time intervals. The module's simulation is interactive because you can explore various time periods on your own and view their unique qualities. After you visit all the time intervals within this simulation, be sure to answer the questions at the end of the module to test your understanding of the geologic timescale.

---

compare the amount of carbon-14 to the amount of nitrogen-14. If the ratio between the two is 1:1—that is, equal parts—then it means that one-half of the radioactive isotope has decayed. In other words, about 5730 years have passed since the carbon was deposited.

Putting Geologic Time in Perspective

Although understanding radiometric dating provides confidence in the ages reported in Earth history, it does not really aid with the comprehension of deep geologic

time. If you want to put geologic time in some kind of perspective, consider that virtually all scientists believe that anatomically modern humans—that is, people looking basically like you and me—have lived on Earth for about 150,000 years. Sounds like a long time, does it not?

Here is another analogy that helps explain deep time. Imagine that all geologic time is contained within a single calendar year, with time beginning on January 1 and extending until December 31, as shown in Table 12.2. In this time context, each day is equivalent to 12.6 million years, each hour is 525,000 years, each minute is 8750 years, and each second is 146 years. The year

TABLE 12.2	The Geologic Timescale Viewed Analogously Within a Calendar Year	
Event	**Age***	**Analogous Time in Calendar Year**
Earth formed	4.6 by	Jan. 1
First single-celled organism	3.2 by	mid-April
Oxygen in atmosphere	2.0 by	mid-July
First cell with nucleus	1.0 by	Oct. 12
First vertebrates	625 my Nov. 10	Nov. 10
First land plants	340 my Dec. 3	Dec. 3
First reptiles	220 my Dec. 13	Dec. 13
First mammals	155 my Dec. 18	Dec. 18
Grand Canyon downcutting	10 my	Dec. 31, 5 A.M.
Early hominids	3 my	Dec. 31, 6 P.M.
Your birth	~18	Dec. 31 (0.05 sec before midnight)

*by = billion years ago; my = million years ago.

begins with the formation of Earth. From that point until the middle of April, no life forms inhabit Earth. In other words, Earth is a lifeless planet for the first 4.5 months of the "year." *Nothing* except single-cell organisms such as amoebas and simple bacteria exist from early May until late November, when the first vertebrate animals (fish) develop. Subsequently, the first land plants emerge on December 3. Reptiles develop on about December 13, followed by the first mammals on December 18. Dinosaurs dominate Earth until late December, when they become extinct and mammals such as shrews and other small rodents emerge and more fully evolve. Early hominids do not evolve until 6 P.M on December 31, which means that anatomically modern people have existed for an extremely short period of time within the year. Your life occurs within the last 0.05 sec of the year. No wonder people have difficulty understanding the concept of geologic time!

A Holistic View of Geologic Time and the Rock Cycle: The Grand Canyon and the Spanish Peaks of Colorado

Although time analogies are fun and useful to consider, it is possible to sense the essence of deep time more directly when you travel. An excellent place to feel the incredible depth of time is the Grand Canyon in Arizona (Figure 12.20). The Grand Canyon is famous because it is one of the most scenic places on Earth. From a geologic perspective, the Grand Canyon is fascinating because it contains rocks that span much of Earth history (Figure 12.21). As a result, the rocks tell us a great deal about Earth's evolution and the environmental changes that have occurred over time in what is now northwestern Arizona.

The oldest rocks are Precambrian in age and are found in the inner gorge of the canyon. The best known of these oldest rocks is the Vishnu Schist, which is a metamorphic rock about 2.0 to 1.7 billion years old. This rock layer was the foundation of an ancient mountain chain that was subsequently eroded between about 1.7 and 1.25 billion years ago. Additional rocks of Precambrian age formed between approximately 1.25 billion and 825 million years ago.

After the last Precambrian sediments accumulated, another lengthy period of erosion occurred. This period lasted about 300 million years and is called the *Great Unconformity* because no rocks from this period exist. In fact, geologists believe that rocks were almost certainly there, but they were eroded by an advancing sea sometime late in that interval of time. The rocks above the Great Unconformity consist of alternating layers of sandstone, shale, and limestone that collectively accumulated between 550 and 250 million years ago. These rocks are testament to fluctuating sea levels that resulted in the deposition of sandstone when the area was above sea level, shale when the area was a coastal bay or lagoon, and limestone when the region was beneath a shallow sea. These rocks have been progressively exposed in the past 10 million years because the Colorado River has been downcutting. To put this period of downcutting in the context of the time analogy just discussed, consider that it began on December 31 at 5 A.M. (Table 12.2).

Another place to appreciate the concept of geologic time is a place called the Spanish Peaks in southern Colorado, because this area has a complex geologic history that spans a relatively recent interval of time. In addition, you can see how the rock cycle fits within the context of the geologic timescale here. The Spanish

Figure 12.20 The Grand Canyon. One of the reasons that the Grand Canyon is such a fascinating place is because it exposes rocks that span the past 2 billion years of Earth history.

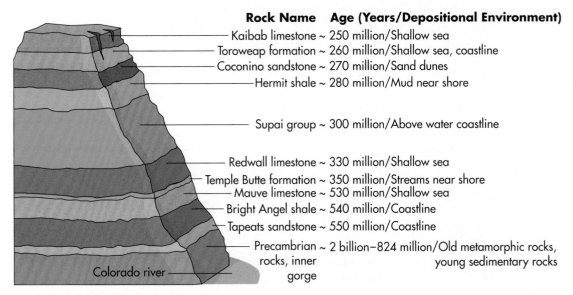

Rock Name	Age (Years/Depositional Environment)
Kaibab limestone	~ 250 million/Shallow sea
Toroweap formation	~ 260 million/Shallow sea, coastline
Coconino sandstone	~ 270 million/Sand dunes
Hermit shale	~ 280 million/Mud near shore
Supai group	~ 300 million/Above water coastline
Redwall limestone	~ 330 million/Shallow sea
Temple Butte formation	~ 350 million/Streams near shore
Mauve limestone	~ 530 million/Shallow sea
Bright Angel shale	~ 540 million/Coastline
Tapeats sandstone	~ 550 million/Coastline
Precambrian rocks, inner gorge	~ 2 billion–824 million/Old metamorphic rocks, young sedimentary rocks

Figure 12.21 Rock layers in the Grand Canyon. Many different kinds of rocks are exposed in the Grand Canyon, Arizona, reflecting several kinds of depositional environments that have occurred in the region over time. Note the relationship between the rock type and the kind of place in which the sediments originally collected.

Peaks lie on the eastern flank of the Sangre de Cristo mountain range (Figure 12.22a) and rise about 2134 m (7000 ft) above the Great Plains (Figure 12.22b), which lie immediately to the east. The evolution of this landscape began sometime before 75 million years ago when thick sedimentary deposits accumulated in a marine environment that fluctuated in depth. These deposits, consisting of alternating layers of sand, clay, and calcium carbonate, may have originated from other rocks and slowly compressed and lithified over millions of years to form sandstone, shale, and limestone, respectively, after the ocean receded.

(a)

(b)

Figure 12.22 The Spanish Peaks in southern Colorado. (a) This true color satellite image of Colorado and adjacent states shows the location of the Spanish Peaks. This image was acquired in September 2002 and shows the first snow of that year, which can be seen as white on the mountain crests. (b) Looking south at the Spanish Peaks. The crest of East Spanish Peak (left) is 3866 m (12,683 ft) high, whereas the top of West Spanish Peak (right) is 4153 m (13,626 ft) in elevation. Note the prominent radial dike that extends to West Spanish Peak. Several dikes like this one occur in the area.

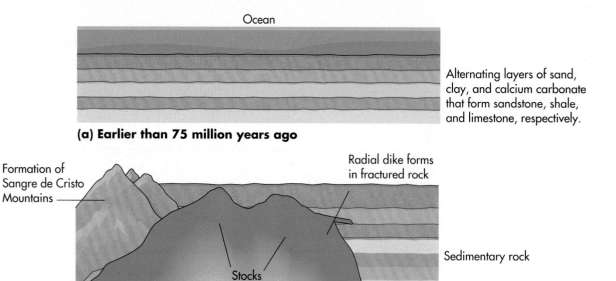

Ocean

Alternating layers of sand, clay, and calcium carbonate that form sandstone, shale, and limestone, respectively.

(a) Earlier than 75 million years ago

Formation of Sangre de Cristo Mountains

Radial dike forms in fractured rock

Stocks
Magma intrusion
about 25 million years ago

Sedimentary rock

(b) 75–25 million years ago

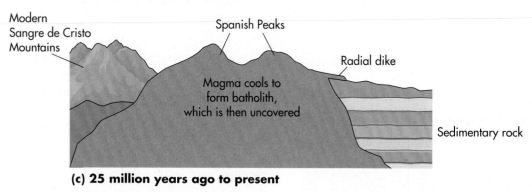

Modern Sangre de Cristo Mountains

Spanish Peaks

Radial dike

Magma cools to form batholith, which is then uncovered

Sedimentary rock

(c) 25 million years ago to present

Figure 12.23 Evolution of the landscape at the Spanish Peaks. (a) Before 75 million years ago, the region was covered by a fluctuating ocean, which deposited alternating layers of sediment that became sandstones, shales, and limestones. (b) Formation of the Sangre de Cristo Mountains occurred between about 75 and 25 million years ago. The force of this uplift bent and fractured the sedimentary rocks to the east. Between about 25 and 14 million years ago magma intruded into these fractures. The magma subsequently cooled, forming a batholith, two stocks, and a number of radial dikes. During this same interval of time, the Sangre de Cristo range began to intensively erode and wear down. (c) In the past 5 million years the sedimentary rocks overlying the igneous features have been eroded, which has exposed the stocks and radial dikes seen in Figure 12.22b.

Approximately 75 million years ago, during the late Cretaceous Period, the nearby Sangre de Cristo Mountains began to form (Figure 12.23). Over the course of the next 50 million years, until about 25 million years ago, the Sangre de Cristo Mountains slowly rose above the surrounding plains. The force of this uplift caused large fractures to appear in the preexisting layers of sedimentary rock. Between about 25 million and 14 million years ago, tremendous quantities of magma began to intrude into this portion of the Earth's crust, creating a giant chamber within the sedimentary rock body and filling smaller cracks in that same mass of rocks.

Sometime after 14 million years ago the intruded magma cooled to form a granite batholith with an associated pair of stocks at the top. Extending outward several kilometers from the stocks were **radial dikes** that also consisted of granite. In the past 5 million years,

the sedimentary rocks that were once present were completely eroded away from this part of Colorado, exposing the granite stocks and dikes on the surface of Earth. The reason why the igneous bodies remained intact is that these kinds of rocks are much harder and more resistant to erosion than sedimentary rocks. The end result is that the Spanish Peaks now tower over the landscape.

With the formation of the Spanish Peaks in mind, think about the broader questions related to geologic time and the rock cycle. In the context of the rock cycle, note that the sediments once contained within the layers of sandstone, limestone, and shale, which in turn covered the stocks and radial dikes, have been transported to other locations

---

Radial dikes *A long, wall-like feature of intrusive igneous rock that forms when magma is injected into thin cracks within older rock and cools.*

where new rock formations developed or are now developing. What is particularly impressive is that these layers of sedimentary rock must have collectively been greater than 2134 m (7000 ft) thick! Otherwise, the magma that formed the batholith, stocks, and radial dikes could not have been intruded within them. It is even more impressive that the process of erosion subsequently removed *all* the sedimentary rock above the stocks. In the time since the stocks and dikes were exposed, they were being slowly worn down by erosion. The resulting liberated sediments have also been transported elsewhere to form new rock.

In the context of deep time, the entire sequence of events that led to the development of what we now call the Spanish Peaks took over 75 million years, beginning with the deposition of more than 2134 m (7000 ft) of sand, calcium carbonate, and clay in a marine environment. These sediments slowly lithified to form rock. Next, the formation of the Sangre de Cristo Mountains lasted 50 million years and caused large fractures to develop in the sedimentary rocks. These fractures, in turn, filled with magma some time after 25 million years ago that intruded from a pair of developing stocks associated with an underlying batholith. This magma then cooled, forming granite. Then *all* the sedimentary rocks were eroded away, exposing the granite stocks and dikes, and leaving a pair of mountains. Imagine how much time *that* process took. And, perhaps the most impressive thing of all in this history is that the past 75 million years is encompassed within the last 2 weeks in the calendar analogy used to represent geologic time (see Table 12.2). In other words, the Spanish Peaks have existed for only a very short period of time within the geologic timescale. *Think about that!*

Human Interactions with the Rock Cycle: The Case of Petroleum

From a scientific perspective, geology is a fascinating discipline because it provides important insights into the evolution of Earth. It also has many practical applications that are highly significant to our industrial way of life. An excellent example of the relevance of geology to our lives is our dependency on organic sedimentary rocks (Figure 12.24). Organic sedimentary rocks are particularly important to humans because they are the source of fossil fuels, particularly coal and petroleum, which are used to power most of our industrial activities. If you use electrical devices of any kind, the "juice" used to power them was most likely derived from burning coal at your local power plant (Figure 12.25).

Petroleum is especially vital for many aspects of modern civilization. It is by far the most dominant energy source for automobiles in the form of refined gasoline. In addition, many consumer products, such as computers,

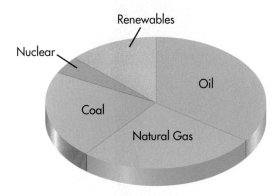

World Energy Consumption by Type

Figure 12.24 World energy consumption by type. Most of our energy comes from fossil fuels, with the majority of it derived from petroleum and coal. Natural gas is also important for energy production. Nuclear energy and renewable energy sources comprise a small part of global energy production. (*Source:* U.S. Department of Energy.)

televisions, cell phones, and plastic bottles, to name a very few things, are made with it. In short, the global economy in its present form is utterly dependent on petroleum.

Although humans have used petroleum in some way at least since the 4th century, when the upper classes in ancient Babylon burned it to provide light, the modern history of oil began in the mid-1800s when the process through which it could be distilled to kerosene was discovered in Canada. The first oil well in the United States was the Drake well in western Pennsylvania, which produced about 20 barrels of crude per day. Demand for oil grew rapidly throughout the late 1800s, primarily for kerosene use in oil lamps. With the advent of internal combustion engines in the early 20th century, demand skyrocketed. In the early years of industrialization, supply easily kept

Figure 12.25 A coal power plant. Most of the electricity generated in the United States is from power plants like this one. Note the huge mounds of coal, which are burned to produce electricity. These mounds are constantly replenished from faraway coal mines.

pace with demand. By 1906, for example, U.S. oil production was about 126.5 million barrels per year and extensive fields were discovered in Texas (Figure 12.26), Oklahoma, and California. During the 1950s petroleum supplanted coal as the world's dominant fuel source and the United States was the world's leading oil producer at approximately 4 billion barrels per year. At about this same time, huge reserves that dwarfed our own were discovered in the Middle East and Venezuela by oil companies like Shell and British Petroleum. These reserves were ultimately nationalized by the host nations, who then formed the Organization of the Petroleum Exporting Countries (OPEC) to coordinate petroleum policies.

The economic geography of petroleum began to change in the late 1960s as the United States slowly reached peak oil production, which occurred in 1971. Our demand had grown to the point where we had to import another 23% from foreign sources to meet our needs. Our vulnerable dependence on foreign supplies first became apparent in 1973 when exporting nations organized an oil embargo for political purposes. As a result of this disruption in supply, prices in the United States jumped and long gas lines were commonplace (Figure 12.27). In an effort to conserve gasoline, the speed limit on the nation's highways was lowered to 55 mph and automobile manufacturers focused on building fuel-efficient cars. Yet another embargo occurred in 1979, further disrupting supplies and increasing gas prices. Throughout most of the 1980s global oil supplies were tight and prices at the pump remained high.

In the late 1980s and 1990s global oil supply began to increase when better extraction techniques were developed and new reserves, such as those in Alaska's North Slope and in the North Sea of Europe, were systematically tapped. As a result, prices fell dramatically to about $15 per barrel and Americans generally moved away from

Figure 12.26 The Lucas gusher. In 1901 the Corsican oil field in southeastern Texas was tapped by Patillo Higgins, a self-taught geologist. The pressure in the well was so great that the oil (or "black gold") erupted to a height of about 50 m (150 ft) for 9 days after the initial strike. Approximately 1,000,000 barrels of oil were lost, which would be worth about $17,000,000 today.

Figure 12.27 Gas lines in the early 1970s. Waiting for gasoline like this was a common occurrence during the first oil embargo because service stations often ran out of gas. Will you see lines like this again in your lifetime?

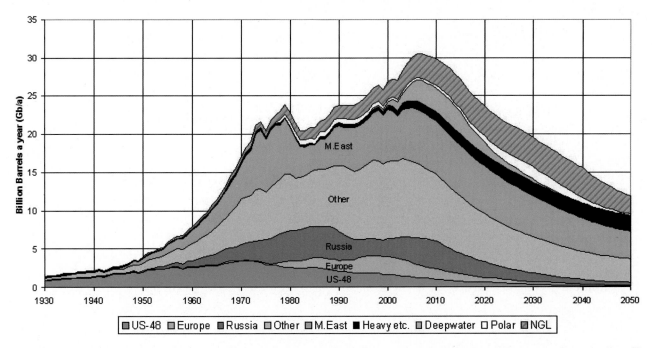

Figure 12.28 Peak oil curve. A growing number of geologists believe that we may be reaching global peak oil production. The various subdivisions of the graph represent the various contributions of specific geographical regions to the overall global oil supply over time. NGL refers to natural gas liquids, whereas heavy supplies are those that do not flow easily and have thus been previously ignored. If this scenario is accurate, note the progressive drop in global oil supplies in the coming decades. (*Source:* Association for the Study of Peak Oil and Gas.)

fuel-efficient cars to Sport Utility Vehicles (SUVs), which consume far more fuel. At the same time suburban development in the United States exploded and the length of daily work commutes increased. Global oil prices slowly began to rise again in 2004 and reached about $88 per barrel in 2006, in large part because of rising demand in China and India, which have rapidly growing populations. Prices rose quickly in late 2007 and early 2008 when future supply fears, speculation, and a weak dollar pushed the price of a barrel of oil to over $140. This rapid increase caused prices at the pump to exceed $4 per gallon in many places within the United States. Just as consumers were adjusting to this new reality, the world experienced a massive financial crisis in late 2008 and supply fears decreased due to a deepening global recession. As of this writing, the average price for a gallon of gasoline in the Midwest United States is about $2.70 and is expected to increase slightly in the coming months.

So what does the future hold for global oil prices? As long as the current economic uncertainty in the world exists, prices are likely to remain fairly stable due to decreased demand. The longer-term outlook is far less certain, however, due to further growth in global population, increased industrialization, and decreasing supply. The growing belief among a number of scientists is that we have reached (or soon will reach) peak global oil production (Figure 12.28). This projection is controversial, as some believe huge untapped reserves exist and are yet to be found. Nevertheless, signs of the peak oil scenario were beginning to emerge in early

2008, when global oil production (88 million barrels per day) barely exceeded global demand (86 million barrels of oil per day). According to some estimates, global oil demand will reach 130 million barrels per day in 2030.

Given that oil is a finite resource due to the way and lengthy time it takes to form, and all the easily accessible reserves have been found and heavily utilized, it is conceivable that gas prices will likely rise far beyond 2008 levels at some point in the relatively near future, especially when the economy recovers and oil demand increases. Americans are particularly vulnerable to these pressures because we easily consume more (25%) of the world's petroleum than any other nation, even though we comprise only 4% of the world's population. Furthermore, we now import over 60% of our oil supply, with much of it coming from politically unstable OPEC nations (Figure 12.29).

Another problem that contributes to price instability is the geography of our domestic oil production. In particular, the location of our own wells and refineries in the Gulf of Mexico is a problem because the region is periodically hit with destructive hurricanes that disrupt supply. The environmental risk of extracting oil from such remote sources was highlighted when an offshore well exploded in April 2010. This explosion ruptured the well, releasing an estimated 200 million gallons of oil into the Gulf until the leak was sealed in early August. This spill quickly dwarfed the 1989 Exxon-Valdez spill in Alaska to become America's worst oil spill. To combat the spill, British Petroleum (BP) employed hundreds of skimmers to sop up oil in the water. Beaches along the Gulf Coast

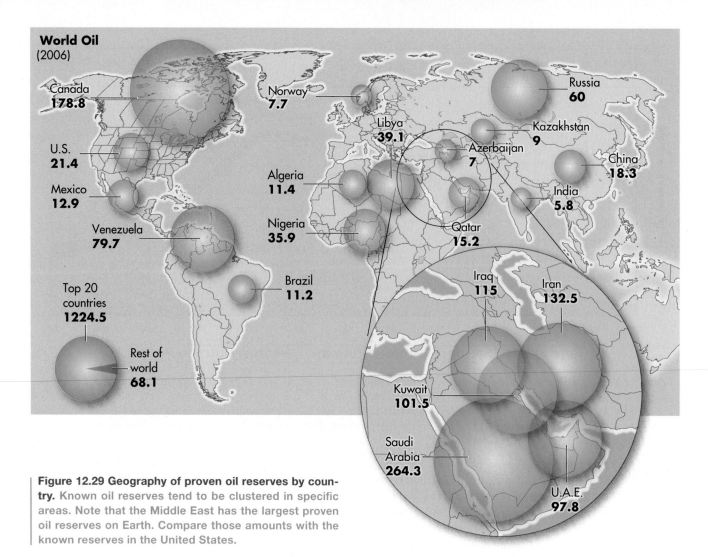

World Oil
(2006)

Canada
178.8

U.S.
21.4

Mexico
12.9

Venezuela
79.7

Top 20
countries
1224.5

Rest of
world
68.1

Brazil
11.2

Norway
7.7

Libya
39.1

Algeria
11.4

Nigeria
35.9

Azerbaijan
7

Qatar
15.2

Russia
60

Kazakhstan
9

China
18.3

India
5.8

Iraq
115

Iran
132.5

Kuwait
101.5

Saudi
Arabia
264.3

U.A.E.
97.8

Figure 12.29 Geography of proven oil reserves by country. Known oil reserves tend to be clustered in specific areas. Note that the Middle East has the largest proven oil reserves on Earth. Compare those amounts with the known reserves in the United States.

were repeatedly combed by teams hired by BP to rake tar balls from the sand. BP also applied over 1.5 million gallons of toxic dispersant in the water to thin the oil.

As of this writing, the ultimate environmental, economic, and psychological impact of the spill is unknown. Many millions of gallons of oil are unaccounted for and numerous fragile coastal marshes remain fouled with oil. No one knows the impact that the chemical dispersants will have on the Gulf ecosystem, including the fishing industry upon which so many Gulf residents depend. Tourism has suffered and many people are uncertain about the future.

The Gulf oil disaster reinvigorated our perpetual national debate about the need to develop "environmentally friendly" energy sources that reduce carbon footprints and reduce the likelihood of massive oil spills. This desire by many, coupled with overall concerns about future oil supplies, is spurring the development of highly fuel-efficient vehicles such as hybrids (Figure 12.30) or cars powered by alternative sources such as electricity, biofuels, or even hydrogen. No clear answer has yet been found to our reliance on petroleum, however, which is why you should expect to see vigorous debate about our oil dependency and a volatile oil market for years to come.

Figure 12.30 The futuristic Toyota 1/X. This concept vehicle, which is proposed to be made with a carbon-reinforced plastic frame and bio-plastic roof, is a tribrid that is propelled either by electricity, conventional gasoline, or ethanol (E85). Average fuel economy is expected to be 125 mpg.

1. Three kinds of basic rocks occur: igneous, sedimentary, and metamorphic.

2. Igneous rocks form when magma cools within Earth (forming intrusive igneous rocks) or on Earth's surface (forming extrusive igneous rocks).

3. Sedimentary rocks consist of large layers of formerly loose sediments that slowly compacted and lithified.

4. Metamorphic rocks form when igneous, sedimentary, or other metamorphic rocks are altered by intense heat and pressure.

5. The age of rocks can be confidently dated because radioactive isotopes decay at a known rate, called the half-life.

6. The geologic timescale involves a concept of deep time that transcends the normal confines of the human timescale.

7. Our current economy depends a great deal on fossil fuels for energy. Many scientists are concerned that we are reaching the peak in global oil production.

THE BIG PICTURE

Now that we have discussed basic geology and the internal structure of Earth, the next logical step is to examine how these concepts relate to landform evolution. In Chapter 13 we will focus on plate tectonics, continental drift, and the origin of mountain chains. These topics are directly related to Chapter 12 because they occur due to movements in the upper mantle. People experience these movements as earthquakes that rattle cities, and volcanoes, such as Mount Saint Helens, that eject material originating from deeper layers within Earth. We will discuss how these processes operate and the landforms that result when they occur.

1. Four major layers occur in the internal structure of Earth, including (from lowest to highest): (a) inner core, (b) outer core, (c) lower mantle, (d) upper mantle (including the asthenosphere, lithosphere, and crust). These layers are identified on the basis of their chemical composition and solid or molten character. Of particular interest is the oceanic and continental crust. Oceanic crust consists largely of silica and magnesium (*sima*) and is denser than continental crust, which is composed of silica and aluminum (*sial*).

2. Three basic kinds of rocks occur on Earth: igneous, sedimentary, and metamorphic. Igneous rocks form when magma cools within Earth (forming intrusive igneous rocks) or on the Earth's surface (forming extrusive igneous rocks). Sedimentary rocks consist of large layers of formerly loose sediments that slowly compacted and lithified. Metamorphic rocks form when igneous, sedimentary, or other metamorphic rocks are altered by intense heat and pressure.

3. The rock cycle describes the ways in which sediments move from one place to another on Earth over time to become new kinds of rock, or how they are transformed to become new kinds of rock. A key concept in this cycle is that rocks gradually break apart as time goes by. The liberated rock fragments are then transported to another place, where they are deposited and gradually lithify to form new rock.

4. Earth is about 4.6 billion years old. Geologic time refers to the length of time from the origin of Earth to the present and is hierarchically subdivided into (from longest to shortest time) Eons, Eras, Periods, and Epochs.

5. Fossil fuels are derived from organic remains that have been buried within Earth. We use these remains to power our economy.

CHECK YOUR UNDERSTANDING

1. Compare and contrast the four major layers of Earth.

2. Describe the concept of isostatic depression and rebound. Why do each of these processes occur and why does this concept reflect the relationship between the crust and asthenosphere?

3. Describe the nature of the Earth's inner core and why it has the characteristics that it does.

4. Describe the two types of crust found on Earth. What is the basic composition of each type of crust?

5. What is the difference between intrusive and extrusive igneous rock? What are some examples of each kind of rock and how do you tell the difference between them?

6. What are the various kinds of sedimentary rock and why do they occur? How does the process of lithification fit into the context of sedimentary rock?

7. What are the two ways that rocks become metamorphic rocks?

8. Describe the flows that occur within the rock cycle. Where does the cycle begin? How do igneous rocks become sedimentary rocks? How do sedimentary rocks become metamorphic rocks? How do the processes of erosion, deposition, and lithification fit within the rock cycle?

9. How is the geologic timescale configured? Using the terms Epoch, Era, Period, and Eon, arrange them into the hierarchy of geologic time.

10. How does the concept of radioactivity fit within the context of calculating the age of ancient rocks? Why are the terms *radioactive isotope* and *half-life* relevant?

11. Describe how the formation of the Grand Canyon is related to the geologic timescale and rock cycle.

12. How is our economy dependent on organic sedimentary rocks and why are we vulnerable as far as petroleum resources are concerned?

ANSWERS TO VISUAL CONCEPT CHECKS

VISUAL CONCEPT CHECK 12.1

Two types of crust are commonly present in an image like this—oceanic and continental crust. Oceanic crust, which is also called *sima*, is denser than continental crust because it consists largely of silica and magnesium. Continental crust, which is also called *sial*, is less dense because it consists largely of silica and aluminum.

VISUAL CONCEPT CHECK 12.2

The answer to this question is *c*. The rocks collectively indicate fluctuating sea levels in the past. Limestone reflects the accumulation of carbonates in a shallow sea, whereas shale reflects the deposition of mud in a very shallow coastal bay or lagoon. So, at this site, the environment was once a shallow coastal bay or lagoon, and then changed to a shallow sea when sea level rose. The many layers in this outcrop indicate that these fluctuations occurred several times.

CHAPTER THIRTEEN

TECTONIC PROCESSES AND LANDFORMS

The previous chapter discussed the internal structure of Earth, as well as the way rocks form and evolve in the lithosphere. Now our attention turns to the ways in which interactions among the crust, mantle, and asthenosphere shape the landscape to form mountains and volcanoes, as well as cause destructive earthquakes. These processes fall under the general framework of plate tectonics and are associated with the kinds of major geologic events frequently covered in the news. The next time you hear about a major earthquake or volcano, think about this chapter to understand what happened.

CHAPTER PREVIEW

Plate Tectonics
   On the Web: *GeoDiscoveries Continental Drift*

Types of Plate Movements
   On the Web: *GeoDiscoveries Folding*
   On the Web: *GeoDiscoveries Plate Tectonics*

Earthquakes
   On the Web: *GeoDiscoveries Types of Faults*

Mountainous landscapes are some of the most spectacular on Earth, as this image from the Southern Alps of New Zealand demonstrates. Mountains such as these pictured here form by tectonic processes, which consist of movements within the Earth's crust that deform rocks, build volcanoes, and cause earthquakes. This chapter focuses on those processes and the landforms that result.

Volcanoes

On the Web: *GeoDiscoveries Walk the Pacific Ring of Fire*

On the Web: *GeoDiscoveries Volcanoes*

The Big Picture

■ LEARNING OBJECTIVES

1. Describe the theory of plate tectonics and the evidence supporting the concept of continental drift.

2. Explain the various types of plate movements that occur as a result of continental drift.

3. Discuss how plate convergence promotes folding of rock, resulting in the formation of anticlines, synclines, and other structural features.

4. Explain the general development of the Appalachian Mountains and why segments of the Ridge and Valley Province may be structurally more complex than they appear.

5. Compare and contrast the various kinds of faults that occur on Earth.

6. Describe why earthquakes occur, the various features associated with them, and how their location can be precisely determined.

7. Compare and contrast the various types of volcanoes and why they occur.

8. Discuss the Pacific Ring of Fire and explain the geography of the Cascade Volcanic Arc.

Plate Tectonics

If you travel abundantly in the United States, or live in the eastern or western parts of the country, you are probably somewhat familiar with the Appalachian and Rocky Mountains. These regions contain prominent mountain chains that have formed over the course of geologic time. Although many of these mountains are partially composed of igneous rocks that represent some kind of volcanic origin, many contain marine sedimentary rocks that are now thousands of meters above the modern seafloor. How is that possible? You are also probably aware that earthquakes tend to occur in distinct places such as California, and that volcanoes are common in the Pacific Northwest of the United States, as well as other regions on Earth. Why is this so?

This chapter focuses on these questions by investigating the concept of **plate tectonics**, which is the now accepted theory that Earth's crust consists of individual pieces (plates) that move individually and collectively.

Plate tectonics *The theory that the Earth's crust is divided into a number of plates that move because they float on the asthenosphere.*

Understanding plate tectonics is important because it explains the geographic distribution of mountain chains and why earthquakes and volcanoes occur along plate boundaries. The first part of the chapter closely examines this theory and how it explains the formation and geography of mountain ranges. The second part of the chapter focuses on the specific processes and landforms associated with volcanism and earthquakes.

The Lithospheric Plates

A good place to begin a discussion of plate tectonics is the early 20th century, when a German geophysicist named Alfred Wegener noticed that many of the continents look as if they could have fit together at one time as a single landmass (Figure 13.1). Wegener noticed that the east coast of South America, for example, appears to fit into the southwestern edge of Africa. Wegener looked more deeply into the problem and discovered that portions of some continents share the same kind of rocks and fossils with other continents (Figure 13.1), which is precisely what one would expect if the continents were indeed connected at some point in the past. As a result of these studies, Wegener proposed in 1915 that a *supercontinent*

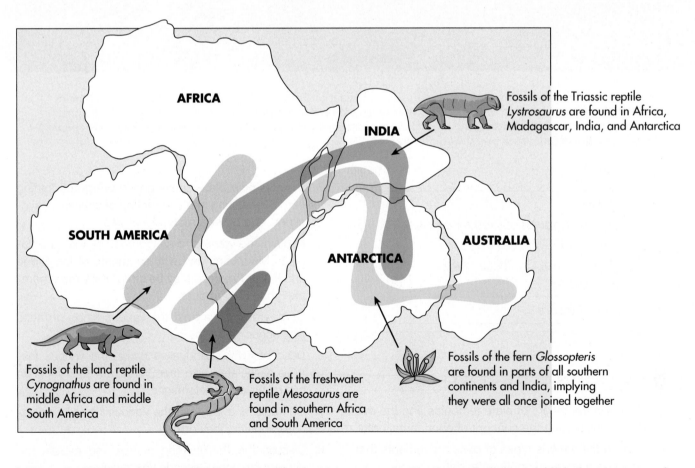

Figure 13.1 Fossils and continental drift. A strong line of evidence supporting the theory of continental drift is the occurrence of specific fossil plant and animal species on continents that are now far apart. (*Source:* U.S. Geological Survey.)

existed about 300 million years ago that he named **Pangaea** (meaning *whole Earth*). He further argued that the continents had drifted apart since that time in a process he called **continental drift.**

The reaction to Wegener's idea in the geologic community was openly hostile, as it directly challenged the long-standing notion that the continents had been connected by a variety of land bridges at one time. In addition, his explanation for how the continents moved, specifically that they "plowed" through the denser oceanic crust, was physically impossible. As a result, Wegener's ideas were largely ignored until the 1950s, when a systematic study began of the ocean basins around the world. In addition, geologists began to investigate the cause of earthquakes more closely. These combined studies demonstrated that the Earth's crust consists of a series of interconnected plates (Figure 13.2) that indeed move relative to one another. This new understanding was the foundation for the theory of plate tectonics that emerged in the 1960s.

Pangaea *The hypothetical supercontinent, composed of all the present continents, that existed between 300 and 200 million years ago.*

Continental drift *The theory that the continents move relative to one another in association with plate tectonics.*

This theory was revolutionary because it accurately explained a great deal of volcanic and earthquake activity on Earth, as you will see later in this chapter. It also validated Wegener's basic premise of continental drift because it provided the mechanism for such migration. As for Wegener himself, he unfortunately never got to enjoy this vindication of his ideas because he froze to death in 1930 on an expedition to Greenland, some 30 to 40 years before his hypothesis was accepted by the geological community.

Plate Movement

The next obvious question is *why* does continental drift occur; how is it possible that large landmasses move around Earth over time? At a fundamental level, it happens because tectonic plates float on the asthenosphere and are moved by convection loops driven by *geothermal* activity. Recall from Chapter 6 that convection occurs in the atmosphere when solar energy heats the ground surface and the warmed ground heats the overlying air. As the warm air rises, the cool air sinks. The same loop of rising and sinking currents also happens inside Earth. In contrast to atmospheric convection, which is driven by an outside source (the Sun), convection within Earth occurs because the radioactive decay of elements such as uranium, deep within Earth, creates very high temperatures between 1000°C and 2000°C (1800°F to 3600°F).

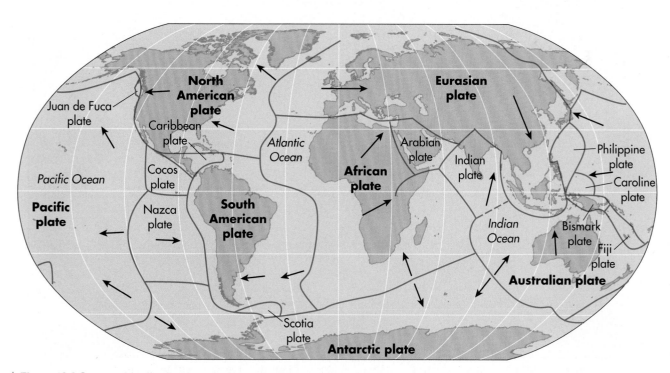

Figure 13.2 Geographic distribution of lithospheric plates. The seven major plates (boldface type) cover 94% of Earth; about a dozen smaller plates cover the remaining 6% of Earth. Notice the "jigsaw" appearance of the map and the way that the plates move (arrows) relative to one another. (The boundary between the Indian and Australian plates is uncertain; many sources refer instead to the "Indo-Australian" plate.)

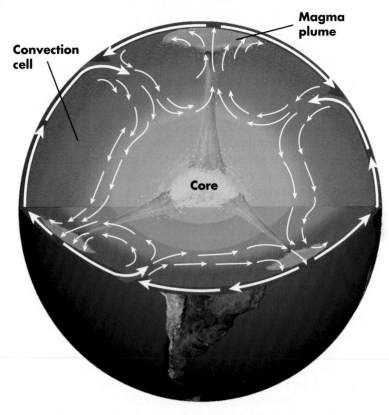

Figure 13.3 Convection plumes within Earth. Magma plumes rise from deep within Earth and spread apart at the crust. Rock created as the magma cools subsequently sinks back into the mantle, where it melts again.

These high temperatures cause plumes of magma to rise slowly by convection within the mantle and into the asthenosphere (Figure 13.3). As these magma plumes reach the base of the crust, they spread horizontally and cool, moving segments of the crust in the process. To visualize this movement, consider a mechanized people mover in the floor of an airport terminal or mall. This device works somewhat like a treadmill that continuously loops around. If you stand on it where the tread rises out of the floor, it carries you forward. The process of continental drift works in somewhat the same way, with plates carried forward by the moving magma beneath. Ultimately, the magma slowly cools, solidifies, and sinks back into the mantle, where it melts again. In this sense, a magma convection current is somewhat similar to a tropical Hadley cell in the atmosphere.

How do we know that the plates are moving? For one thing, scientists can now measure the motion through remote sensing and global positioning systems discussed in Chapter 2. In California, for example, sensors placed on either side of major faults actually move relative to one another over the course of time, indicating that the underlying plates are shifting. Evidence of plate movement also appears in the age of the seafloor. Convection in the mantle brings magma up through fractures in the crust, where it extrudes onto the seafloor, cools, and forms new oceanic crust. Figure 13.4

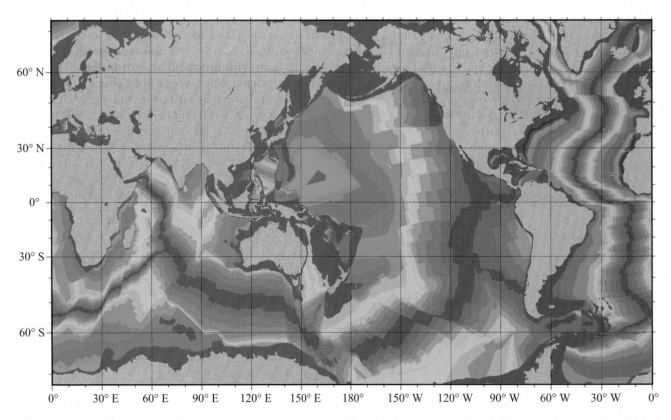

Figure 13.4 Seafloor ages. In this map colors are associated with seafloors, whereas the continents are gray. Notice how the youngest seafloor rocks (in red) occur in the zones of seafloor spreading, such as in the Atlantic Ocean, with progressively older rocks (yellows, greens, and blues, respectively) away from these regions.

Continental Drift

To see how the plates have drifted over time, go to the **GeoDiscoveries** website and access the module **Continental Drift**. This animation shows the position of continental landmasses during different intervals of geologic time. Use the controls in the animation to play it, to step forward or backward in time, or to rewind to the beginning of the animation. You can also put the individual continental positions into the context of the geologic timescale by accessing specific time periods such as the Triassic or Cretaceous. After you complete the animation, be sure to answer the questions at the end of the module to test your understanding of continental drift.

---

shows isochrons (lines of equal age) for the ocean floors around the world, and clearly indicates areas where oceanic plates have been spreading apart. Notice that the oldest part of the oceanic crust in the Atlantic Ocean basin is along the east coast of North America, where it is about 180 million years old. In contrast, the seafloor in the center of the Atlantic Ocean is less than 10 million years old. This age pattern supports the theory that new rocks are being constructed in the center of the ocean basin and are gradually spreading away from there. An additional piece of evidence supporting the theory of continental drift is that magnetic stripes run parallel to the Mid-Atlantic Ridge across the seafloor in a pattern similar to the isochrones shown in Figure 13.4. These stripes reflect the fact that the Earth's magnetic field has periodically reversed through time, with magnetic north becoming south, and vice versa. About 800,000 years have passed since the last such reversal. The ocean floor contains a record of magnetic reversals because, as new seafloor is progressively created, its magnetic orientation is aligned with the direction at that time. Overall, the isochrones and magnetic strips indicate that the rate of crust movement appears to range from 1.0 cm to 15.0 cm (0.4 in. to 5.9 in.) annually.

Given the overwhelming evidence for continental drift, it is possible to reconstruct the patterns of movement through time. Geologists generally agree that the continents began to coalesce to form Pangaea as early as 300 million years ago, during the Pennsylvanian Period. If you recall from Chapter 12, this is the period of time when Earth was much warmer and massive amounts of carbon began to accumulate in steamy swamps. Figure 13.5a shows how Pangaea probably looked around 275 million years ago. Notice that South America and Africa were indeed joined at this time, as were the eastern coast of North America and Africa. Approximately 200 million years ago, most of the continents began to slowly drift apart to reach their present positions. An important exception is the Indian landmass, which moved relatively quickly (as far as deep time is concerned) across what is now the Indian Ocean to collide with the southern side of the Asian continent. This collision created the Himalaya Mountains, which will be discussed in more detail later in the chapter.

Types of Plate Movements

With the concept of plate movement in mind, let's now look more closely at the various types of plate movements and the impact they have on the Earth's surface. For the sake of simplicity, consider that the range of potential plate movements includes (1) moving away from each other, (2) sliding past each other, and (3) colliding head on. This section discusses these types of plate movements and how they shape Earth's surface.

Passive Margins

The simplest kind of tectonic interaction occurs at **passive margins**. As the name implies, passive margins are relatively stable localities from a tectonic perspective. These regions frequently occur in places where the continental crust and the bordering oceanic crust are actually on the same tectonic plate. One such place is the eastern seaboard of North America, which has a passive margin with the oceanic crust in the western part of the Atlantic Ocean basin. Even though the continental crust that forms the North American landmass is a separate rock body from the adjoining oceanic crust, they are

---

Passive margin *A place where the continental crust and the oceanic crust are on the same tectonic plate and thus do not move relative to one another.*

both a part of the North American plate (Figure 13.6). As a result, they have a passive margin because it is not geologically active.

Transform Plate Margins

Another type of plate boundary is the **transform plate margin**. In contrast to passive plate margins, transform boundaries are places where plates slide horizontally past each other (Figure 13.7a). At transform margins, the plane of motion is along a nearly vertical break (or fault) that extends through much of the lithosphere. The San Andreas Fault in California and the Dead Sea Fault along the border of Israel and Jordan (Figure 13.7b) are excellent examples of transform faults. The San Andreas Fault will be discussed in more detail later on in this chapter in the context of earthquakes.

Transform plate margin *A plate boundary where opposing plates move horizontally relative to each other.*

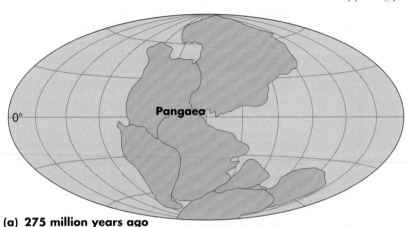

(a) 275 million years ago

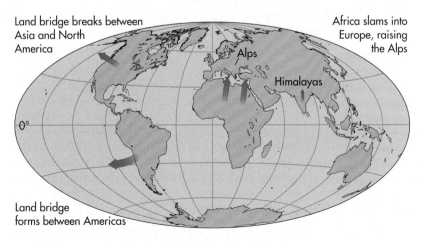

India slams into southeast Asia, raising the Himalayas

Australia breaks off from Antarctica

(b) 60 million years ago

Land bridge breaks between Asia and North America

Africa slams into Europe, raising the Alps

Land bridge forms between Americas

(c) Today

Figure 13.5 Continental drift in the past 275 million years. (a) Geography of the Pangaea supercontinent during the Permian Period about 275 million years ago. At this time, all the continents were connected to form one giant landmass. (b) Geography of continents at the beginning of the Cenozoic Era about 60 million years ago. Note the relative direction of continental movement. (c) Configuration of continents today. The continents continue to move around the surface of Earth and will eventually change the world's appearance again.

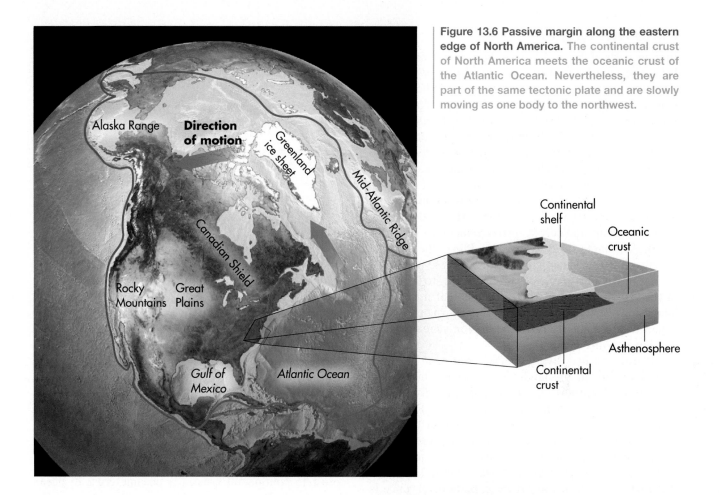

Figure 13.6 Passive margin along the eastern edge of North America. The continental crust of North America meets the oceanic crust of the Atlantic Ocean. Nevertheless, they are part of the same tectonic plate and are slowly moving as one body to the northwest.

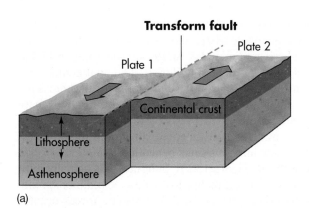

(a)

Figure 13.7 Transform plate margins. (a) Plates that meet at transform margins move horizontally past each other along transform faults. (b) Satellite image of the Dead Sea Fault. This transform fault occurs along the boundary of the Arabian and African plates and extends from the Gulf of Aqaba on the southern tip of Israel through the Dead Sea and north into Lebanon. The Arabian plate is generally moving to the north, whereas the African plate is moving in a southerly direction in this region.

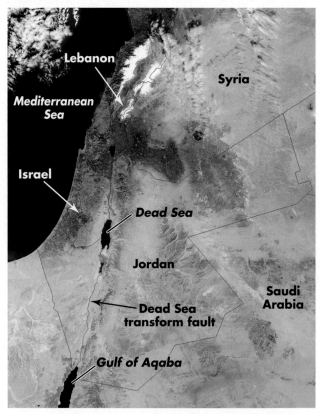

(b)

Plate Divergence

In some parts of the world, lithospheric plates move away from one another in a process called *plate divergence*. This type of movement occurs in regions where rising magma plumes within Earth move upward and outward between plate fractures, spreading Earth's plates apart in a process called **rifting** (or divergence). A great place to see the process of rifting is on the seafloor where magma extrudes through fractures in the oceanic crust (Figure 13.8). This extrusion of magma produces a ridge-like feature, appropriately called a **mid-oceanic ridge**, which lies parallel to the rift zone. The Mid-Atlantic Ridge is perhaps the best-known oceanic ridge and is located, as its name suggests, at the middle of the Atlantic Ocean along the Atlantic Rift Zone. If these plates are diverging, what do you suspect is happening to the width of the Atlantic Ocean? The answer is that it is slowly increasing every year. This process began when Pangaea began to break apart about 200 million years ago and continues to this day.

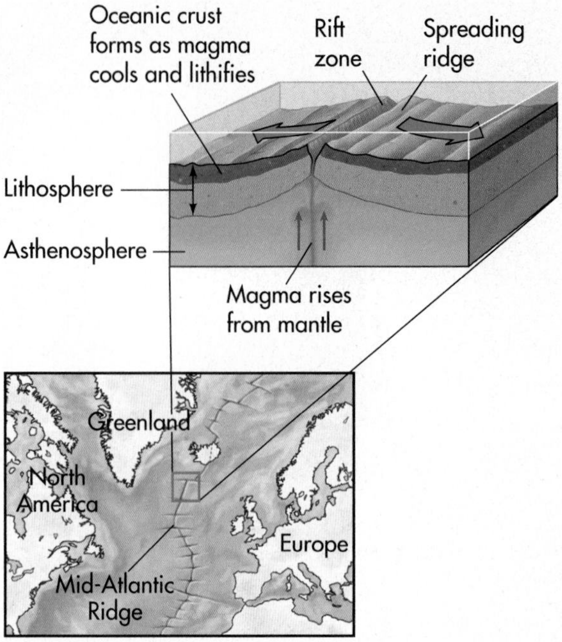

Plates can also diverge within continents, causing a gradual split in the landmass to occur. One of the best-known locations of continental plate divergence is in eastern Africa where the East African Rift is located (Figure 13.9a). Continental rifting in this area has produced a distinct valley landscape bordered by steep canyon walls (Figure 13.9b), as well as several large lakes such as Lake Victoria (Figure 13.9a). In the northern part of the rift zone, plate divergence is occurring in three places that merge at a single place known as a *triple junction* (Figures 13.9a and c). This junction is in close geographic association with the Arabian, African, and Indian plates. Two of the rifts are oceanic, with one opening the Red Sea and the other causing enlargement of the Gulf of Aden on the southern side of Saudi Arabia. On land, the third rift, the East African Rift, extends in a southerly direction from the juncture of the Red Sea and the Gulf of Aden. Along this rift, the eastern part of the African continent is slowly splitting in two, resulting in several large lakes, such as Lake Victoria.

Plate Convergence

Having just seen what happens when plates diverge from one another at rift zones, let's turn now to the places where two plates collide directly into each other in a process called *plate convergence*. Plate convergence occurs in three general settings, including (1) oceanic

Figure 13.8 Oceanic rifting, seafloor spreading, and plate divergence. In ocean basins, the seafloor spreads where plumes of magma interact with the crust, creating a rift. One of the best examples of seafloor spreading exists in the middle of the Atlantic Ocean.

crust to continental crust, (2) oceanic crust to ocean crust, and (3) continental crust to continent crust. The type of convergence associated with each of these settings results in specific processes that produce distinctive landforms on the Earth's surface.

Collision of Oceanic Crust with Continental Crust

The first type of plate collision discussed occurs when two plates collide where one consists of oceanic crust and the other is continental crust. Recall from Chapter 12 that oceanic crust is generally denser than continental crust because the former consists of basaltic rock, whereas the latter tends to be granitic. In convergence zones where this difference occurs, the oceanic crust can be viewed like a heavy lead that encounters the more spongy continental plate and thus sinks beneath it in a process called **subduction** (Figure 13.10). Subduction is actually initiated when oceanic plates diverge due to seafloor spreading. As noted previously, this spreading forces oceanic crust to move horizontally away from the spreading zone and toward continental margins. When the oceanic crust

---

Rifting   *The spreading apart of the Earth's crust by magma rising between fractures in the Earth's plates.*

Mid-oceanic ridge   *A ridge-like feature that develops along a rift zone in the ocean due to magma upwelling.*

Subduction   *The process by which one lithospheric plate is forced beneath another. This usually happens when oceanic crust descends beneath continental crust, but can happen where two plates of oceanic crust meet.*

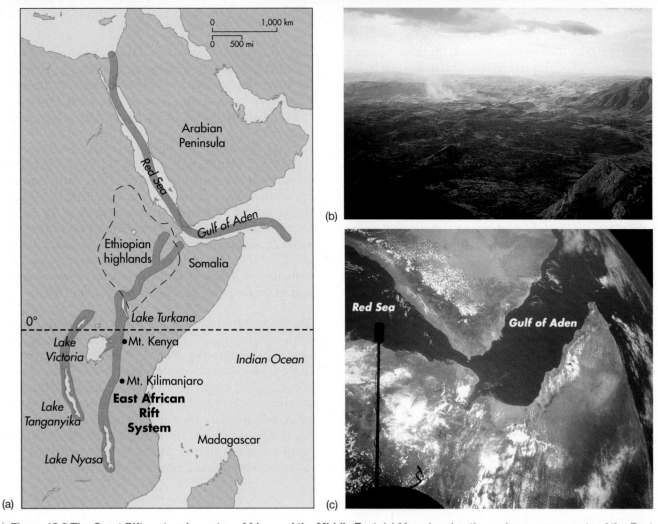

(a)

(b)

(c)

Figure 13.9 The Great Rift system in eastern Africa and the Middle East. (a) Map showing the various components of the East African Rift system. (b) The Suguta Valley in Kenya is part of the East African Rift; note the steep bluffs bordering the valley. (c) Satellite image of the triple junction with view to east/southeast. The Red Sea is to the upper left, the Gulf of Aden is to the upper right, and the East African Rift is in the foreground.

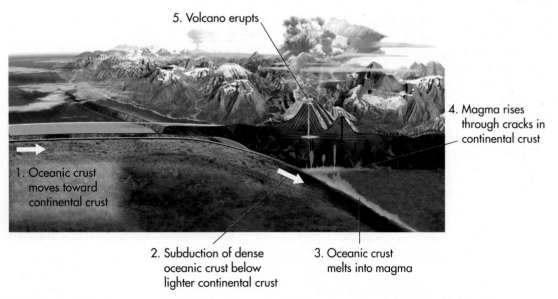

5. Volcano erupts

4. Magma rises through cracks in continental crust

1. Oceanic crust moves toward continental crust

2. Subduction of dense oceanic crust below lighter continental crust

3. Oceanic crust melts into magma

Figure 13.10 The subduction process. Subduction occurs where dense oceanic crust is forced beneath less dense continental crust. This process causes the oceanic crust to move down into the upper mantle and melt. When the resulting magma rises to the surface through cracks in the overlying crust, it frequently causes volcanoes.

converges with the continental crust, the denser oceanic crust material is slowly forced beneath the less dense continental crust and down into the upper mantle. As the slab is forced more deeply, the internal temperature gradually increases in a distinct *geothermal gradient*. When the oceanic crust ultimately encounters the upper mantle, it melts and is recycled into magma as part of the rock cycle.

It is important to note that the process of subduction is not continuous, but is instead episodic. In other words, the oceanic plate does not flow smoothly beneath the continental crust. Instead, the oceanic crust repeatedly stops and starts in its downward movement below the continental crust. It moves only when the buildup of stress behind it—which is ultimately derived from the zone of seafloor spreading—exceeds the force of friction holding it in place beneath the continental crust. This stress can cause earthquakes and significant deformation of the rocks above the subduction zone, which also results in the construction of mountain ranges and even volcanoes in the places where magma works its way to the surface. This relationship will be discussed in more detail later in the chapter.

Collision of Oceanic Crust with Oceanic Crust A second type of plate convergence occurs when two bodies of oceanic crust collide with one another. Although both plates have similar densities in these settings, one

plate will typically be subducted beneath the other, forming a deep *oceanic trench* (Figure 13.11a). As previously discussed, this subduction will produce stress in the overriding seafloor that may cause it to crumple upward to form mountains below the ocean surface. More often, subduction of one plate of oceanic crust beneath the other will result in the formation of volcanoes (due to melting of the crust in the upper mantle) that grow upward in the water. If these volcanoes enlarge sufficiently, they will breach the surface of the ocean to form distinctive island chains, such as Northern Marianas Islands of the Pacific Ocean (Figure 13.11b). Another great example of such an island chain formed by the collision of two plates of oceanic crust is the islands associated with Japan. Again, the formation of volcanoes will be discussed in more detail later in the chapter, including Japan's Mount Fuji.

Continent-to-Continent Collision The third type of plate collision is one that occurs when continents on separate plates collide with one another. This type of convergence is distinctive because both plates consist of continental crust that has similar densities. In these situations, one plate does not ride over or below the other, which means they smash together like two cars in a head-on collision. The crust, like the cars, crumples, causing folding of the formerly horizontal bedrock through compression (Figures 13.12a and b).

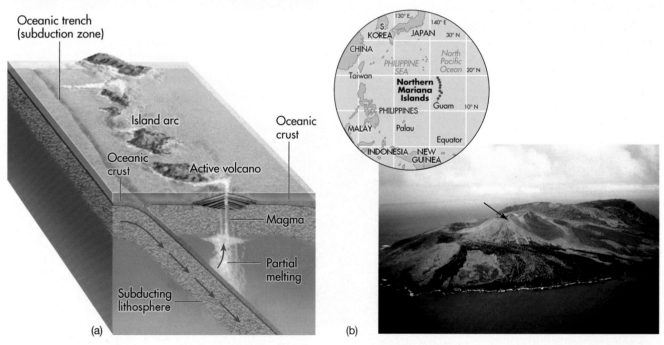

Figure 13.11 Collision of two plates of oceanic crust. (a) When two plates of oceanic crust collide, one of the plates will subduct beneath the other despite the fact they both have similar densities. An oceanic trench forms along the subduction zone. An island arc also develops due to volcanic activity along the plate boundary. (b) Guguan Island in the northern Marianas Islands in the Pacific Ocean. This island is a volcano (note the cone in the middle of the island) that formed due to subduction along the Marianas Trench. This trench is the deepest location on the surface of Earth's crust, with a maximum depth of 9,636 m (31,614 ft) below sea level.

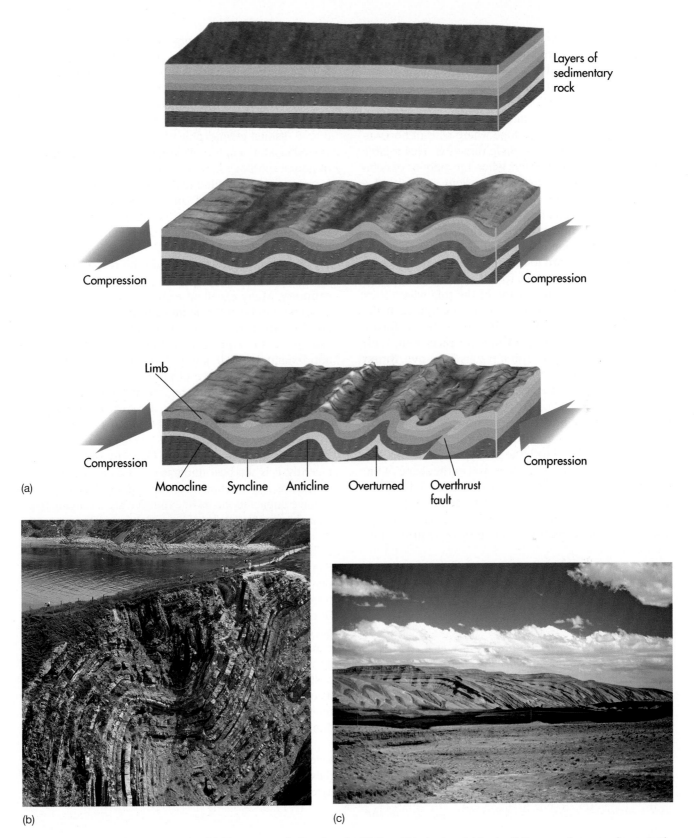

(a)

Layers of sedimentary rock

Compression

Compression

Limb

Compression

Compression

Monocline   Syncline   Anticline   Overturned   Overthrust fault

(b)

(c)

Figure 13.12 The folding process. (a) The vast majority of rocks lithify within horizontal beds. If these rocks are subsequently compressed by plate collision, they are deformed in some way. In many places, folded rocks are associated with mountains and valleys. (b) Folds in a rock outcrop in the United Kingdom. These rocks were once horizontal but have been bent in several different directions by compressional forces. (c) The Comb Ridge Monocline near Mexican Hat, Utah. Compressional forces produced a single fold here that connects two beds of relatively horizontal sedimentary rock. One of these beds of horizontal rock is on the top of the mountain, whereas the other underlies the landscape in the foreground.

You can see evidence for past compression and folding of rocks by looking at the orientation of rock layers or **rock structure**. Figure 13.12 illustrates how rock layers are folded and the various features that result. In general, a close correlation exists between the amount of compression and the nature of the fold. A **monocline** is a one-sided slope where beds of horizontal rock are inclined in a single direction over a large area. This kind of fold occurs when the amount of compression is relatively low. An excellent example of a monocline occurs in southeastern Utah, near the town of Mexican Hat (Figure 13.12c). More extensive compression causes anticlines and synclines to form (Figure 13.12a). An **anticline** is a portion of the fold where the rock layers arc upward to form a concave arch along the fold axis. One-half of such a fold is called a *limb*. In contrast, a **syncline** is the portion of the fold where rock layers dip downward to form a convex trough. If the collision is especially intense, the rocks can be folded so much that an **overturned fold** (or recumbent fold) results. Still more compression results in an **overthrust fold**, which occurs when one part of the rock mass is shoved up and over the other.

Geomorphology and the Evolution of the Appalachian Mountains Large-scale compression of rocks is frequently associated with the growth of mountains. The study of mountain formation, as well as the shape of the Earth's surface in general, falls within the subdiscipline of physical geography called **geomorphology**. Geomorphology is the study of the formation, shape, spatial distribution, and evolution of landforms on Earth—thus, the name *geo* (Earth) *morphology* (shape). In contrast to a *landscape*, which is the overall appearance of a place in terms of its vegetation, topography, or human modifications, a **landform** is a distinct geographic feature, such as a mountain, river valley, coastline, or sand dune, to name but a very few. Each of these landforms evolved due to distinct geomorphic processes, which will be discussed in later chapters. As you might imagine, the study of geomorphology is rooted deeply in geology because it requires a thorough understanding of how sediments are eroded, transported, and deposited. The discipline is also geographically based because geomorphologists are interested in the way landforms change across space and time.

As far as the growth of mountains is concerned, they typically form during a distinct period of time called an **orogeny**. Many excellent examples occur on Earth of how continental collision and compression cause orogenies in which rocks are intensively folded. One such example is the Appalachian Mountains, which extend approximately 2500 km (~1600 mi) along the eastern United States. Several phases of mountain building have occurred in this region, including the *Taconic Orogeny* (~450 million years ago), *Acadian Orogeny* (~375 million years ago), and *Alleghany Orogeny* (~290 to 248 million years ago).

Given that the Alleghany Orogeny occurred most recently, it is the best understood because the modern landforms are remnants from that time. It also appears to have impacted the most extensive area, ranging from what is now Pennsylvania to Alabama. This area was intensively folded when the northwest corner of Africa collided with the east coast of North America at the time Pangaea formed. During this period of time it is believed that rugged mountains formed that may have been between 5000 m and 8000 m (~16,400 ft and 26,200 ft) high. In other words, if you could go back in time, you would have seen mountains in what is now Virginia that looked much like the modern Rocky Mountains in Colorado, or maybe even the modern Himalayas in Tibet. Over the past 200 million years these once towering mountains have been eroded to well-rounded landforms, such as those seen in the Great Smoky Mountains in North Carolina, that are less than 2000 m (6550 ft) high.

The Appalachian Mountains are subdivided into several different subregions. Perhaps the best known of these regions is the Ridge and Valley Province, which extends along much of the length of the Appalachian

---

Rock structure *The internal arrangement of rock layers.*

Monocline *A geologic landform in which rock beds are inclined in a single direction over a large distance.*

Anticline *A convex fold in rock in which rock layers are bent upward into an arch.*

Syncline *A concave fold in rock in which rock layers are bent downward to form a trough.*

Overturned fold *A structural feature in which the fold limb is tilted beyond vertical, which results in both limbs inclined in the same direction, but not at the same angle.*

Overthrust fold *A structural feature where one part of the rock mass is shoved up and over the other.*

Geomorphology *The branch of physical geography that investigates the form and evolution of the Earth's surface.*

Landform *A natural feature, such as a hill or valley, on the surface of Earth.*

Orogeny *A period of mountain building, such as the Alleghany Orogeny.*

range. As the name implies, the landscape here consists of a series of prominent ridges and intervening valleys. Figure 13.13a shows a satellite view of this pattern. A characteristic ground-level view is seen in Figure 13.13b.

When you look at the Appalachian landscape, it is tempting to assume that the ridges are anticlines because they arch up and the valleys are synclines because they bend downward. In fact, the relationship of surface landforms and the underlying geologic structure is often much more complex than you might think. It is well beyond the scope of this text to examine every such complexity, but to see some simple examples, look at Figure 13.14. This diagram shows a general time sequence illustrating how portions of the landscape evolved. Part (a) shows what the topography might have looked like after it was initially folded during the Alleghany Orogeny. As you would expect, the landscape at that time consisted of well-developed ridges underlain by structural anticlines and prominent valleys associated with the synclines.

The expected relationship between the shape of the Earth's surface and the underlying structure began to change in some cases due to the behavior of streams. Notice in Figure 13.14a how a network of parallel streams developed on the landscape after it was compressed. See how the path of streams follows the structure? These streams began to vigorously erode the uplifted ridges, but did so at a variable rate that depended on the resistance (or hardness) of the rocks on which they flowed. The behavior of streams will be discussed in more detail in Chapter 16, but, for now, understand that streams erode some rocks (such as shale) more easily than others (such as limestone). As a result of this preferential erosion, some parts of the Appalachian landscape were eroded extensively, whereas others were not. In addition, stream erosion focused on the highest points, which were the crests of the original anticlines.

Because of the variability just described, the relationship between surface topography and underlying structure in the Appalachians became less clear over time. Look at the ridge marked *A* in Figure 13.14b, for example. This ridge was initially underlain by an anticline (Figure 13.14a). Erosion removed the crest of the ridge, however, forming a valley parallel to the underlying structure. The ridges on each side of the valley are remnants of the limbs of the anticline. Such a ridge is called a **hogback ridge** because one side of the ridge is steeper than the other. Also note that the valley formed

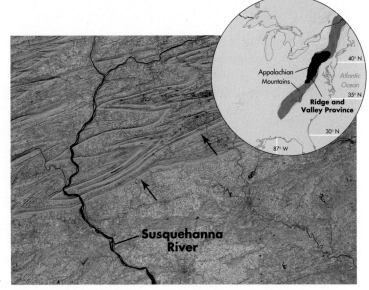

(a)

(b)

Figure 13.13 Appalachian Ridge and Valley. (a) Infrared satellite image of a portion of the Ridge and Valley Province in central Pennsylvania. The ridges (arrows) are covered by forest, which makes them stand out particularly well. (b) Oblique air photograph of the Ridge and Valley Province. Note the well-developed ridges and intervening valleys in this panoramic image.

between the two hogbacks at point *A*. In Figure 13.14b the ridge is underlain by the original anticline; that is, the rocks arch up there as you would expect. In other words, the stream eroded into the core of the anticline, forming a valley within it along its length. Such a landform is called an **anticlinal valley**. A **synclinal valley** lies immediately

---

Hogback ridge *A ridge underlain by gently tipped rock strata with a long, gradual slope on one side and a relatively steep scarp or cliff on the other.*

Anticlinal valley *A topographic valley that occurs along the axis of a structural anticline.*

Synclinal valley *A topographic valley that occurs along the axis of a structural syncline.*

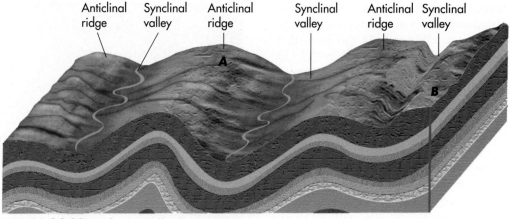

(a) Initial folding during Alleghany Orogeny

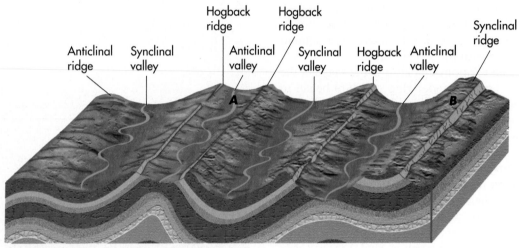

(b) After erosion of softer rock layers

Figure 13.14 General evolution of the Ridge and Valley Province. (a) The landscape shortly after folding. Anticlines form the high ground, whereas valleys occur in synclines. (b) As time progresses, erosion modifies the landscape. Anticlines at *A* and *B* are eroded by streams, forming valleys in upward-arching structures. Hogback ridges exist on both sides of the anticlinal valley at *A*, where limbs of the former anticline form high ground. To the right at point *B*, extensive erosion has inverted the topography so that the ridge is underlain by a syncline.

to the right of the hogback ridge-and-valley network at point *A* and is what you would expect since it lies below the adjacent ridge. To the right of that valley is another hogback ridge, which borders yet another anticlinal valley. See how the valley is cut into rocks that arch up?

Now look at point *B* in Figure 13.14b. A ridge exists at the surface here, but it is actually underlain by a syncline. See the U-shape of the rock layer that caps the ridge? This is the same rock body that forms the hogback ridge to the left and now is on top of the ridge here. Such a pattern is interesting because it means that the topography at point *B* has been *inverted*. In other words, what was once the valley floor (point *B*, upper image) is now the crest of a ridge (point *B*, lower image). Such complex relationships between the underlying geologic structure

and the way the landscape now appears demonstrate that landscapes are sometimes more complex than they appear. They also illustrate how much *time* must be involved for this kind of landscape to evolve. First, there had to be uplift and then *substantial* amounts of erosion. Overall, it took about 250 million years!

Although the Appalachian Mountains are a great example of how collision produced a landscape, it is certainly not the only example on Earth. Crustal collision continues to uplift mountains in several prominent places around the world, creating many of Earth's distinct mountain chains shown in Figure 13.15. The famous Swiss Alps, for example, formed by compression. This mountain range forms the border between Austria and Italy and began to develop during the

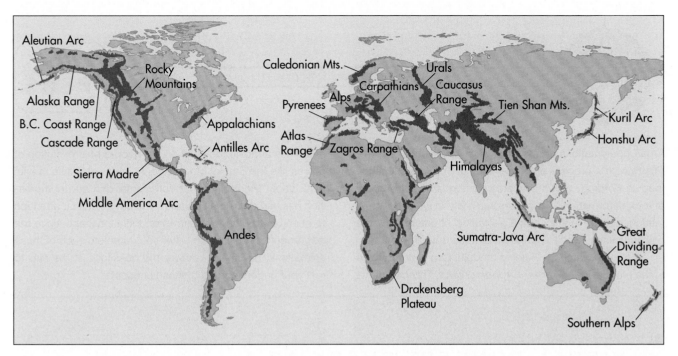

Figure 13.15 Location of major alpine chains on Earth. Many of the Earth's major mountain chains have resulted from compression along crustal plates. Others have developed due to other processes, such as volcanism. The Andes is the longest mountain chain, at 7200 km (4500 mi); the Rockies are not far behind, at 6000 km (3700 mi). However, the tallest mountains are in the Himalayas.

Cretaceous Period (144 my to 65 my ago) when the continental plate of Africa collided with the southern side of the European plate. A second orogeny occurred during the Tertiary Period (65 my to 2 my ago). An interesting aspect of the Alps geology is that part of the mountain range consists of rock material from the African plate. In other words, part of the European landmass is actually rock from Africa. Another mountain range that has formed largely through continental collision is the Himalaya Mountains (Figure 13.16), which border India and China. Most people probably

know that Mount Everest is the highest mountain in the world. What you may not have known is that this peak, as well as the entire Himalayan chain, has been uplifted because the northeastern part of the Indian plate has been colliding into the southern side of the Eurasian plate for about the past 30 million years. As a result of this collision, the mountains have shot out of the ground, at least as far as geologic time is concerned. In so doing, metamorphic and sedimentary rocks (including marine fossils) have been uplifted thousands of meters above sea level.

Figure 13.16 The Himalaya Mountains. Mount Everest is the tallest mountain in the world at 8850 m (29,035 ft). The 25 tallest mountains in the world are all in the Himalayas, including all 14 peaks that are over 8000 m (~26,250 ft). This mountain chain extends in an arc about 2400 km (1500 mi) long and forms the border between India and Tibet. The Himalayas formed due to collision of the Indian plate with the Asian landmass.

Folding

A pair of animations on the *GeoDiscoveries* website nicely illustrates the process of crustal folding. The first one is the module *Folding*, which animates the fundamental process of rock deformation by compression. As you watch this animation, be sure to think about geological structure and note how anticlines, synclines, and other folds develop.

After you watch the *Folding* module, turn your attention to the module *The Southern Appalachians*. This part of the

module is a movie focusing on the geomorphic evolution of places like the Smoky Mountains. As you watch this movie, think about the impact that continental drift and compression has had on this landscape. Also pay particular attention to changes that have occurred on this landscape since the last orogeny in the region. After you complete each of these animations, be sure to answer the questions at the end to test your understanding of these concepts.

VISUAL CONCEPT CHECK 13.1

Which one of the following statements is accurate with respect to the geologic structure in this roadcut?

a) The rocks are horizontal.
b) This is a nice example of a rift system.
c) The structure shows the limb of an anticline.
d) The structure shows the base of a syncline.

Plate Tectonics

To review the process of plate tectonics and the movements that occur along plate boundaries, go to a pair of animations on the *GeoDiscoveries* website. The first module is *Active and Passive Margins*, which compares the movements that occur on unstable and stable tectonic boundaries. The second module is *Tectonic Plate Boundary Relationships*, which focuses on subduction, transform, and colli-

sion boundaries. As you watch these animations, be sure to associate the boundary process with the type of landscapes discussed in the preceding text. After you complete each animation, be sure to answer the questions at the end of the media to test your understanding of plate tectonics and plate margins.

Earthquakes

The previous discussion described the various ways in which plate margins move relative to one another and how mountains can form in these areas. Now it is time to discuss more specific processes associated with plate boundaries—volcanism and earthquakes. The logical place to begin this discussion is with the concept of earthquakes because they occur both as independent events and frequently in association with volcanoes.

Seismic Processes

You have no doubt heard about a major earthquake that happened someplace in your lifetime, perhaps one that devastated a city and unfortunately killed hundreds or even thousands of people. Examples abound of recent earthquakes that have made the news in such a fashion. For example, an earthquake occurred in Java, Indonesia, in May 2006 that killed over 5000 people. In May 2008 a massive earthquake occurred in central China that killed over 65,000 people. A week before this section was written, a horrific earthquake struck Haiti, killing by some estimates over 200,000 people and causing chaos in its aftermath. It is quite possible another one will have occurred by the time you read this text. Clearly, earthquakes are a significant natural event that humans should understand.

An **earthquake** occurs when the sudden release of accumulated tectonic stress results in an instantaneous movement of the Earth's crust. Although this type of movement is usually associated with plate boundaries, it can also occur due to the rapid movement of magma within a volcano. Such movements produce shock waves that radiate through the lithosphere, causing the ground to shake in sometimes dramatic ways. The strongest earthquakes are usually most associated with plate boundaries because stress builds at these locations as plates grind against each other. This stress creates a fracture between adjoining plates called a **fault**, along which the adjacent rock bodies are displaced relative to each other. At some locations, faults extend deeply into the Earth's crust, whereas at other places they are very shallow. Faults and earthquakes can also occur in the middle of plates, but this association is less common. Although most large earthquakes occur along major plate boundaries, smaller earthquakes can take place whenever a small fault occurs within a rock body.

To see the relationship of earthquakes and plate boundaries, look at Figure 13.7a, which shows the plate movements along a transform margin. The important thing to remember is that these plates do not slide by one another gradually. Instead, these opposing masses of rock are locked by friction for long periods of time, causing elastic stress to build on both sides of the fault. This stress builds to a critical point when a rupture occurs along the fault plane. The stress released is somewhat analogous to the stress instantly released in a stretched rubber band when it is cut, causing the rock layers on either side of the fault to instantly adjust. The place within the lithosphere where the fault breaks is called the *focus* and usually occurs a few kilometers deep within the ground (Figure 13.17). In California, for example, the average focal depth is between 6.4 km and 9.5 km (4 mi to 6 mi). The point on the surface directly above the focus is known as the **epicenter**. When the rocks finally break free of each other, they do so in an abrupt movement and release powerful seismic waves into the surrounding crust.

Locating the Epicenter When an earthquake occurs, the first goal of geologists is to determine the location of the epicenter. This search is accomplished through

---

Earthquake *Shaking of the Earth's surface due to the instantaneous release of accumulated stress along a fault plane or from underground movements within a volcano.*

Fault *A crack in the Earth's crust that results in the displacement of one lithospheric plate or rock body relative to another.*

Epicenter *The point on the Earth's surface that lies directly over the focus of an earthquake.*

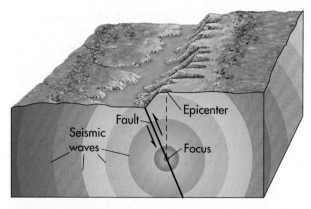

Figure 13.17 Schematic diagram of earthquake processes. Earthquakes occur when tension is abruptly released along a fault, causing opposing rock bodies or plates to move in opposite directions relative to one another. This movement generates seismic waves that radiate through the crust. Uplift along a fault can result in a distinct cliff or bluff along a fault, known as a fault scarp. The motion of plates that causes an earthquake can come from stress buildup associated with a transform fault, subduction, and collision.

the process of triangulation, in which the distance to the epicenter is compared among seismographs stationed at three separate locations. This methodology is based on the fact that earthquakes produce two kinds of seismic waves, or *body waves*, that radiate through the Earth's interior. These body waves consist of *P waves* (or *primary waves*) and *S waves* (or *secondary waves*) that move in different ways and travel at different speeds. P waves are

compressional waves that cause the Earth's crust to expand and contract rapidly in a horizontal manner as the waves radiate out from the epicenter. P waves typically move at about 1.5 km to 8 km (~1 mi to 5 mi) per second through the crust. S waves, in contrast, move about 60% to 70% slower than P waves and shake the ground in a vertical fashion that is similar to sending a wave pulse along a rope or string when you whip it up and down. In an earthquake, S waves cause the ground to vibrate in a rolling fashion that can be quite noticeable and frightening. The magnitude of both these waves is measured at observation stations with a device called a seismograph (Figure 13.18).

Given the known speed difference between P waves and S waves, it is possible to calculate the distance to an earthquake epicenter at any observation station by noting the time lag between the passing of each wave. When the passing of seismic waves is viewed in a regional context, it is logical to assume that a wave will require more time to reach an observation station that is farther away from the epicenter than one that is closer. With these time patterns in mind, geologists can determine the earthquake epicenter by triangulating the time lag between three separate observation stations. The epicenter is located at the point where the respective distance measurements overlap (Figure 13.19). Yet a third kind of wave is a surface wave, which moves only through the upper crust. Although *surface waves* move much more slowly than the P and S waves within Earth, and actually arrive after the two body waves, they are responsible for most of the damage associated with an earthquake.

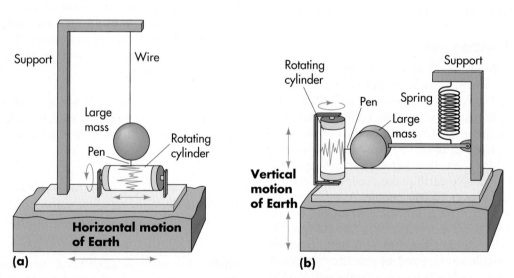

(a)
Horizontal motion of Earth

(b)
Vertical motion of Earth

Figure 13.18 Measuring seismic waves. Seismic waves associated with an earthquake are measured with a seismograph. When the Earth shakes, a platform moves back and forth beneath a stationary pen that records the movement on a continuous roll of paper. (a) Primary waves (P waves) consist of compressional waves that cause vibrations parallel to their travel direction. (b) In contrast, secondary waves (S waves) cause vertical vibrations to develop. P waves move more quickly than S waves and are thus the first to be felt.

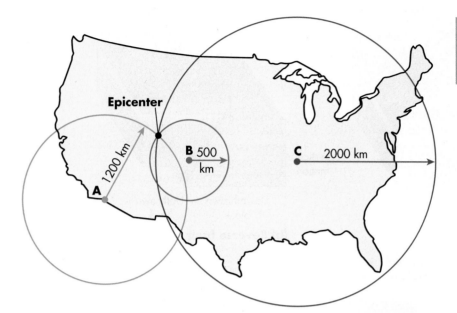

Measuring Earthquake Magnitude The strength of earthquakes is measured on the **Richter scale**, which is related to the amplitude of seismic waves moving through the Earth's crust as determined by a seismograph. The Richter scale is logarithmic and is represented by whole numbers and decimal fractions. An increase of one magnitude unit (such as 4 to 5) corresponds to 10 times greater ground motion; an increase of two magnitude units corresponds to 100 times greater ground motion; and so on (Figure 13.20). The magnitudes of earthquakes range between 0 (weakest) and 9 (strongest) on the Richter scale. The actual ground motion for, say, a magnitude 5 earthquake is about 0.04 mm at a distance of 100 km from the epicenter; it is 1.1 mm at a distance of 10 km from the epicenter.

Types of Faults In addition to the production of seismic waves, earthquakes cause deformation of the rocks both within the crust and on the Earth's surface. This deformation is usually associated with the various kinds of faults that occur when opposing rock bodies move relative to each other in association with plate tectonics (Figure 13.21).

One type of fault is a **normal fault**, which is a vertical fault in which one slab of the rock is displaced up and the other slab down (Figures 13.21a and 13.22). This kind of fault is created by tension forces acting in opposite directions. Where such faults occur, the opposing blocks are pulled away from one another by gravity, which causes

---

Richter scale *The logarithmic scale used to measure the strength of an earthquake.*

Normal fault *A steeply inclined fault in which the hanging rock block moves relatively downward.*

one of the fault blocks to slip up relative to the fault plane as an upthrown block (called a **horst**), while the other slips down as a downthrown block (called a **graben**). The exposed side of the upthrown block forms a cliff-like

---

Horst *An upthrown block of rock that lies between two steeply inclined fault blocks.*

Graben *A downthrown block of rock that lies between two steeply inclined fault blocks.*

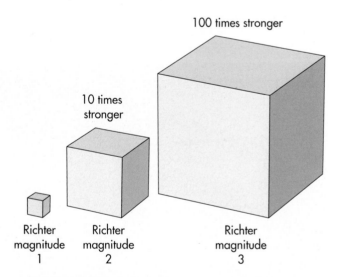

Figure 13.20 The Richter scale. These cubes show the logarithmic nature of the Richter scale, with the size of each cube representing relative power. For example, a magnitude 2 earthquake is 10 times stronger than a magnitude 1 earthquake. Similarly, a magnitude 3 earthquake is 100 times stronger than a magnitude 1 earthquake.

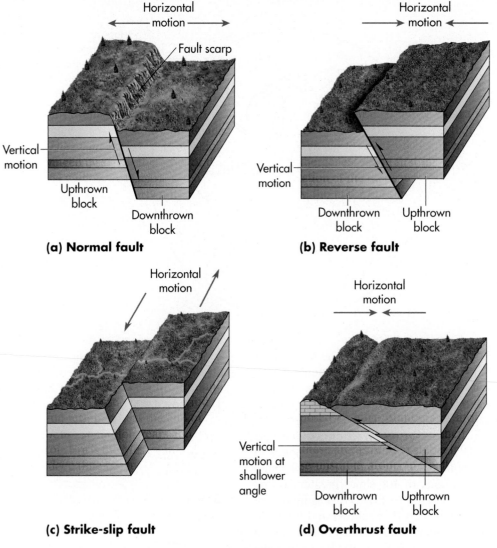

(a) Normal fault

Horizontal motion

Fault scarp

Vertical motion

Upthrown block

Downthrown block

(b) Reverse fault

Horizontal motion

Vertical motion

Downthrown block

Upthrown block

(c) Strike-slip fault

Horizontal motion

(d) Overthrust fault

Horizontal motion

Vertical motion at shallower angle

Downthrown block

Upthrown block

Figure 13.21 Types of faults. (a) Normal faults and (b) reverse faults result when one block moves up while the other moves down. (c) Strike-slip faults result when blocks move horizontally relative to one another. (d) Overthrust faults occur when the upthrown block also slides over the downthrown block.

Figure 13.22 A normal fault. This roadcut exposure shows a small normal fault in surface rocks near Death Valley, California. The fault plane is clearly visible as the diagonal contact between the two blocks of rock.

Figure 13.23 The Sierra Nevada. The Sierra Nevada forms the western boundary of the Basin and Range Province. Uplift and westward tilting of the horst has formed an approximately 2700-m (~9000-ft) escarpment into the Owens Valley graben (foreground).

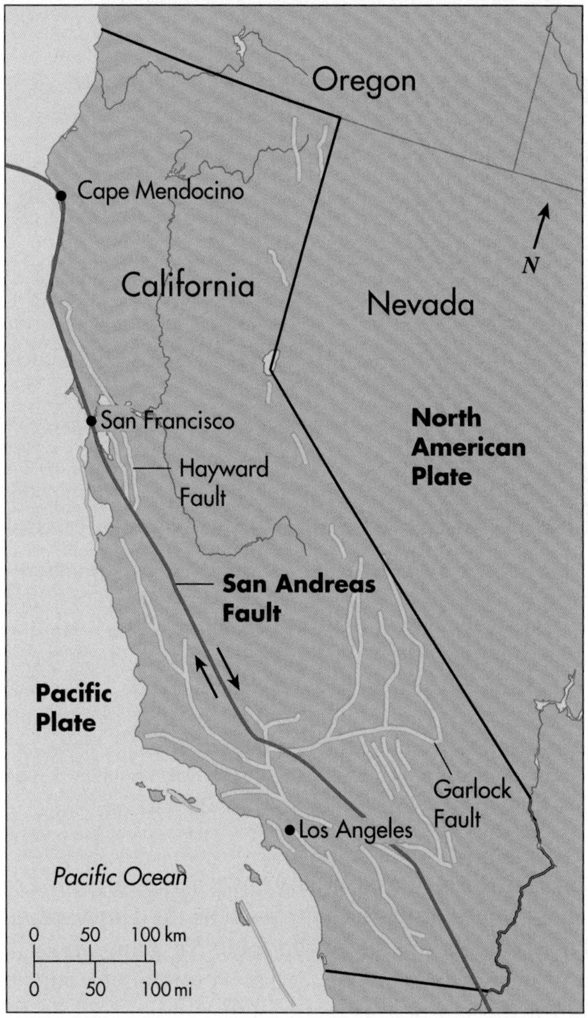

feature known as a **fault escarpment** (or *scarp* for short). A fantastic example of a landscape created by normal faulting is the awesome Sierra Nevada in eastern California. As you approach the mountain range from the east, the horst forms an abrupt escarpment that rises over 2700 m (~9000 ft) above Owens Valley, the graben to the east (Figure 13.23). Given that the Sierra horst is tilted to the west, the slope toward the west is much more gradual.

Another kind of fault is a **reverse fault** (Figure 13.21b). Although a reverse fault looks very similar to the normal fault, the cause and nature of the movement differ. Whereas a normal fault entails movement of blocks away from each other, a reverse fault results when rock blocks move toward each other, causing one block to ride up steeply over the other.

As noted previously, overthrust faults result when one rock body is thrust up and over another, usually in association with folding. These faults differ from normal and reverse faults because their fault planes are usually at a shallower angle in comparison.

Of the four types of faults illustrated in Figure 13.21, three are directly associated with some kind of uplift — specifically the normal, reverse, and overthrust faults. In contrast to these types of faults, the **strike-slip fault** entails purely horizontal movement of the two plates past each other. Strike-slip faults and transform faults

Figure 13.24 California fault systems. The San Andreas Fault is a transform fault that forms the boundary between the North American (to the east) and Pacific (to the west) plates. Arrows on either side of the fault reflect the relative movement of plates. Note the various subsidiary faults in the system.

are closely related because they share the same kinds of horizontal movement. In fact, the primary difference between transform and strike-slip faults is that transform faults are associated with large, tectonic plate boundaries, whereas strike-slip faults occur where small rock blocks move horizontally relative to one another.

As indicated previously, one of the best-known transform faults is the famous San Andreas Fault in California. This massive fault occurs along the boundary of the Pacific and North American plates, with the Pacific plate moving in a northwesterly direction relative to the North American plate at an average rate of about 5 cm/year (2 in./year). Figure 13.24 shows the fault geography,

---

Fault escarpment *A step-like feature on the Earth's surface created by fault slippage.*

Reverse fault *A steeply inclined fault in which the hanging rock block moves relatively upward.*

Strike-slip fault *A structural fault along which two lithospheric plates or rock blocks move horizontally in opposite directions and parallel to the fault line.*

Figure 13.25 The San Andreas Fault. In southern California, the Pacific plate (foreground) is moving toward the northwest (left) relative to the North American plate. Note the change in stream direction that occurs at the fault boundary. This distinct pattern evolved because the stream flows onto the Pacific plate from the North American plate. Because the Pacific plate has been gradually moving in a northwest direction relative to the North American plate, the stream has slowly changed direction as well.

including the major subsidiary faults in the San Francisco region. Many of these faults, such as the Hayward and Garlock faults, are strike-slip faults. In many areas, the San Andreas Fault can be seen as a prominent trough on the Earth's surface, as pictured in Figure 13.25. In this photograph, you can see the Pacific plate is in the foreground, whereas the North American plate is the hilly ground in the background. You can also see how incremental the movement of the fault has been by looking at the stream that flows from the North American plate onto the Pacific plate. Notice that the stream bends

sharply at the fault and then flows directly on it for a short distance before turning onto the Pacific plate. In other words, the Pacific plate has moved at a sufficiently slow speed to allow the stream to bend with it. Geologists say that the total accumulated displacement along this fault is at least 560 km (~350 mi) since it came into being about 15 to 20 million years ago.

Given the immense size of the San Andreas Fault, the earthquakes associated with it are sometimes quite violent and destructive. The largest earthquake on the San Andreas in recorded history is the 1906 San Francisco earthquake, which had a magnitude of 7.9 on the Richter scale and caused a rupture along 470 km (290 mi) of the fault. This extensive earthquake caused as much as 8.5 m (28 ft) of near instantaneous slip in some places and resulted in extensive damage in San Francisco and the surrounding area. Between the effects of the earthquake and the catastrophic fire that occurred afterward, 700 to 800 people were killed and about 28,000 buildings were destroyed.

Human Interactions with Earthquakes as Natural Hazards

As discussed at the beginning of this chapter, examples abound of recent earthquakes that have severely impacted people in many places around the world. The recent earthquake in Haiti is the best example of such a tragedy. This earthquake measured 7.0 on the Richter scale and occurred along the Enriquillo-Plantain Garden fault system. It was particularly horrific because the epicenter was near Port-au-Prince, which is a densely populated and impoverished city that is ill equipped to handle such an event. Such earthquakes have even occurred in the United States, as with the just-noted San Francisco earthquake in 1906. Another big earthquake

VISUAL CONCEPT CHECK 13.2

This view of the San Andreas Fault is in central California. Which one of the following choices explains why there are hills along each side of the fault?

a) There is subduction occurring at the fault.
b) The San Andreas Fault is a normal fault.
c) There is rifting occurring along the fault.
d) There is compression of the rocks on either side of the fault.

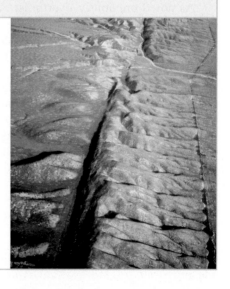

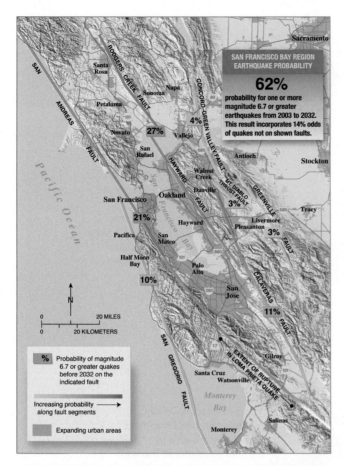

Figure 13.26 Earthquake probability map in the San Francisco Bay region. This map shows the probability of an earthquake with a magnitude 6.7 or greater on the San Andreas Fault and associated faults (such as the Hayward Fault) in the Bay region. Also shown is the extent of urban areas and the extent of the rupture that occurred along the San Andreas Fault during the Loma Prieta earthquake in 1989.

(magnitude 7.1), called the *Loma Prieta* earthquake, occurred along a branch of the San Andreas Fault south of San Francisco in 1989. You may be surprised to learn that an even more intense series of events occurred in 1811 and 1812 in the central part of the United States along the New Madrid Fault, which extends through southeastern Missouri and eastern Arkansas. At least four earthquakes occurred during this period of time that had reconstructed magnitudes greater than 8.0 on the Richter scale. These earthquakes were felt over a huge area, with even reports of church bells ringing 1800 km (~1100 mi) away in *Boston* due to the Earth shaking! Fortunately, the central United States was lightly populated at that time and the damage was thus minimal. If such an event occurred today, however, it would be absolutely devastating with significant impacts likely in cities such as Memphis, St. Louis, and even Chicago. It is not difficult to imagine that thousands of people could perish.

In this context, an earthquake is viewed as a **natural hazard** that has the potential of having a negative effect

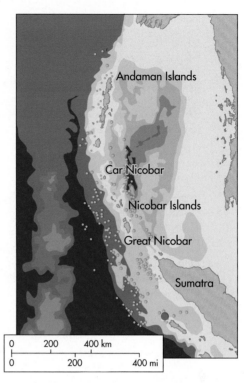

Figure 13.27 The epicenter of the Sumatra-Andaman earthquake in December 2004. The initial earthquake (at red dot) measured 9.1 on the Richter scale and was the second most powerful quake on record. As a result of this earthquake, a massive tsunami was created that killed over 200,000 people along the coasts of the Indian Ocean. The yellow dots show the location of aftershocks.

on people. Although many different kinds of natural hazards occur on Earth, such as hurricanes, tornadoes, landslides, and floods, to name a few, earthquakes are probably the most intense because they essentially occur without warning and can kill hundreds of thousands of people in a very short period of time. Given the devastating potential impact of earthquakes on people, it is no wonder that geologists are working vigorously to enhance their ability to predict when and where earthquakes will occur. Unfortunately, with current knowledge the best predictions are nothing more than statements of probability. In northern California, for example, geologists now believe a 62% chance exists that an earthquake with a magnitude greater than 6.7 or greater will occur in the San Francisco region by 2032 (Figure 13.26).

Perhaps the best example of the extreme hazard that earthquakes represent occurred in December 2004 when the second strongest earthquake ever recorded occurred near Sumatra, an Indonesian island in the eastern part of the Indian Ocean (Figure 13.27). This earthquake, known

Natural hazard *An actual or potentially occurring natural event, such as an earthquake or volcanic eruption, that has a negative effect on people or the environment.*

Earthquakes • 375

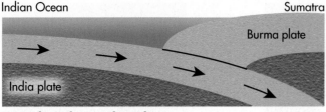

Indian Ocean   Sumatra

Burma plate

India plate

Stress gradually builds at
subduction zone along
India and Burma plate boundary.

(a) Before the earthquake

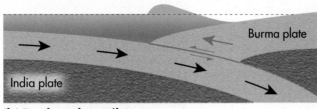

Burma plate

India plate

Earthquake causes Burma plate
to thrust up and over the India plate
several meters, causing massive
displacement of overlying water.

(b) Earthquake strikes

Tsunami spreads

West   East

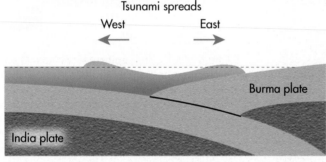

Burma plate

India plate

Tsunami spreads rapidly,
reaching the nearby coast of
Sumatra to the east in 15 minutes
and the more distant shore of
Somalia to the west in 7 hours.

(c) Tsunami radiates outward along the length of fault rupture

Figure 13.28 Evolution of the 2004 tsunami in the Indian Ocean. (a) Tectonic stress builds up where the India plate subducts beneath the Burma plate along the Sumatra Fault. (*Source:* Geoscience Australia.) (b) Pent-up stress is suddenly released in a massive earthquake. In addition to significant horizontal movement, the Burma plate is thrust up and over the India plate. The vertical movement displaces water in the overlying Indian Ocean. (c) After the water is initially displaced, seismic waves rapidly radiate outward in the water.

as the *Sumatra-Andaman* earthquake, occurred along the boundary between the Burma and Indian plates on the Sumatra Fault. The quake measured a staggering 9.1 on the Richter scale and continued for 10 min, which is a very long period of time when you consider that most earthquakes last no more than a few seconds. The quake was so strong that the *entire* Earth shook, with ground motions of at least 1.3 cm (0.5 in.) occurring *everywhere* on the globe as seismic waves spread from the epicenter. According to the United States Geological Survey, the ground motion associated with the quake caused the water level in a well in *Virginia* to suddenly increase 9 m (3 ft)! The length of the rupture along the Sumatra fault associated with this quake was 1200 km to 1300 km (720 mi to 780 mi), making it the longest such rupture ever observed.

In addition to the substantial lateral movement of the adjoining plates, the seafloor on the Burma plate was suddenly thrust up and over the India plate about 4 m to 5 m (13 ft to 16 ft; Figures 13.28a and b). This thrust instantly displaced 30 km³ of water, in a manner somewhat analogous to what happens if you flick your hand upward in water, resulting in another form of seismic wave known as a **tsunami** that rapidly radiated outward in the ocean along the entire length of the rupture (Figure 13.28c). Once in motion, the tsunami raced across the Indian Ocean at speeds ranging from 500 km/h to 1000 km/h

Tsunami   *A destructive sea wave caused by a disturbance within the ocean, such as an earthquake, volcanic eruption, or landslide.*

(310 mi/h to 620 mi/h). It reached the northwest coast of Sumatra to the east in about 15 min and the southern tip of South Africa, which is about 8500 km (5300 mi) to the southwest, in about 12 h (Figure 13.29). Coincidentally, a French radar satellite happened to pass over the Indian Ocean 2 h after the earthquake and measured the tsunami wave heights as they moved across deep water. These observations indicated that the maximum wave height in the open ocean was only 60 cm (2 ft), which would have scarcely been noticed by ships in the area. As the waves approached the shore, however, they grew rapidly due to the shallow water, reaching an estimated height of 24 m (80 ft) at Banda Aceh in the northwest coast of Sumatra. Some of the tsunami's energy spread into the Pacific Ocean, even creating a tsunami along the west coasts of North and South America that measured from 20 cm to 40 cm (7.9 in. to 15.7 in.) in height.

Amateur video of the tsunami in the hardest-hit areas of Indonesia and Sri Lanka illustrated the incredible power of moving water. These videos did not show a single wave, but instead a series of massive water surges that totally engulfed coastal communities. In some places the water surged 2 km (1.2 mi) inland. The energy released by the tsunami is estimated to have been equivalent to about 5 megatons of TNT, which is *more than all* of the explosive energy used during World War II (including the two atomic bombs). This tremendous energy wreaked havoc along the coast (Figure 13.30a), causing billions of dollars in damage and killing at least 265,000 people. Many thousands more are missing and may never be found. The devastation was so extensive and widespread that it can clearly be seen in satellite imagery that shows the landscape before (Figure 13.30b) and after (Figure 13.30c) the tsunami.

The severity of the 2004 tsunami underscores the need for a warning system in the Indian Ocean. Such a warning system was established in the Pacific Ocean in 1949 because tsunamis are relatively common in this part of the world due to extensive tectonic activity. The Japanese, in particular, are well acquainted with

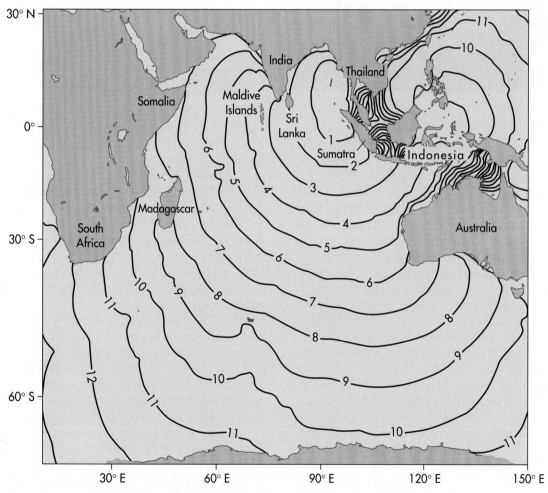

Figure 13.29 Travel time in hours for the 2004 tsunami in the Indian Ocean. The tsunami reached the shores of Indonesia and Thailand within minutes of the earthquake. In contrast, it reached the southern tip of South Africa in about 12 h (*Source:* NOAA.)

(a)

(b)

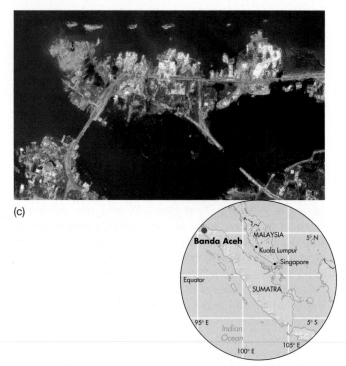

(c)

Figure 13.30 Devastation caused by the 2004 tsunami in the Indian Ocean. (a) Damage in coastal communities due to the surging water was catastrophic. (b) Satellite image of Banda Aceh, Indonesia, before the tsunami. (c) Satellite image of Banda Aceh after the tsunami. Note the change in the configuration of the shoreline and reduction in island size.

tsunamis (hence the name: *tsu* meaning "strong" and *nami* meaning "wave") because about 200 significant events have occurred in Japan alone in the past 1300 years. One of the most recent Pacific tsunamis was triggered by the massive 1964 Alaska earthquake, which created a wave that reached a height ranging from about 2 m to 6.3 m (7 ft to 21 ft) in northern California and killed eleven people. In fact, a significant tsunami is considered by the United States Geological Survey to be statistically more likely along the U.S. west coast than in the Indian Ocean. Because of this concern, an extensive network of sensors was deployed throughout the Pacific basin in the 1950s and 1960s that monitors seismic events and associated sea level fluctuations. Although this system is of little use during a sudden tsunami, it can provide warning within 15 min for events that originate far away. Such a system is now being considered for coastal communities along the Indian Ocean.

A side story of the tragic 2004 tsunami demonstrates why it is important to understand natural Earth processes. Although the tsunami happened quickly along the coastlines it struck, warning of its approach was evident very shortly before the catastrophic water surge arrived. In those brief moments before tragedy, the ocean was drawn rapidly away from the coast, as much as 2.5 km (1.6 mi) in some places, as water was pulled from the shore into the rapidly approaching wave. Video shows that many people were excited by this sudden expansion of the beach and even ran on to it to collect fish stranded on the exposed ocean bottom. Unfortunately, they were unaware that this sudden beach expansion meant a horrific wave was about to engulf them. Sadly, most of them perished. On the coast of Thailand, however, someone did know the warning signs of a tsunami, a 10-year old British girl who was on holiday there with her parents. Ironically, she had just studied tsunamis 2 weeks beforehand in school. She recognized that a sudden recession of water from the beach meant a tsunami was approaching and pleaded with her parents that they should run for safety. Her parents were understandably skeptical at first, but she was insistent. As they left the beach, the girl warned others nearby of the approaching disaster. In the end, the awareness and persistence of this little girl saved her life, as well as over 100 others.

A final thought about tsunamis illustrates the truly dynamic nature of Earth and the relevance of physical geography to ongoing events. As this chapter was being written in September 2009, a magnitude 8 earthquake occurred near the islands of Samoa and American Samoa in the south-central Pacific. This earthquake produced three successive tsunami waves, with the highest

Types of Faults

To review the various kinds of faults, go to the *GeoDiscoveries* website and access the module *Types of Faults*. This animation contains a video presentation that illustrates the fault types discussed in this chapter and will demonstrate the various ways that faults move and the stresses that cause this movement. As you watch the media, try to relate the various fault movements with the character of the landscape. After you have completed the media, be sure to answer the questions at the end of the module to test your understanding of fault types.

KEY CONCEPTS TO REMEMBER ABOUT FAULTS AND EARTHQUAKES

1. There are four major kinds of faults: normal, reverse, overthrust, and strike-slip faults.

2. Earthquakes occur when moving plates are stuck along fault planes, causing stress to increase until a rupture occurs along the fault.

3. The place within Earth where a fault breaks is called the focus. The location on the surface directly above the focus is called the epicenter.

4. Earthquakes are measured with a seismograph, which measures the intensity of P and S waves that radiate outward from the earthquake focus.

5. The strength of earthquakes is measured on the logarithmic Richter scale, which indicates the magnitude of seismic waves passing through the Earth's crust.

6. Earthquakes are perhaps the most intense natural hazard on Earth, often resulting in catastrophic damage and loss of human life. As a result, geologists are working to improve earthquake prediction.

about 1.5 m (5 ft). Although small, these waves struck the coastline with great force and devastated coastal communities much like the massive 2004 tsunami. A tsunami warning was generated following the quake, but was given only 18 min before the tsunami arrived. Over 200 people died in this event.

Volcanoes

After investigating the processes and geography of earthquakes, it is time to discuss volcanoes, which are the other major feature associated with plate tectonics. As

briefly discussed in Chapter 12, a **volcano** is a mountain or large hill containing a conduit that extends down into the upper mantle, through which magma, ash, and gases are periodically ejected onto the surface of Earth or into the atmosphere (Figure 13.31). In most cases, volcanoes are inactive for some time and erupt only when the pressure of material rising from the mantle becomes excessive. The length of inactivity varies dramatically between volcanoes. Some volcanoes lie dormant for hundreds or thousands of years before they erupt, whereas others are in a near-constant state of eruption.

Although no hard and fast classification scheme exists for volcanoes, most geologists distinguish three basic kinds: cinder cones, composite volcanoes, and shield volcanoes. This classification depends in large part on whether the eruption is explosive or fluid. The first part of the discussion will focus on the volcanoes associated with explosive eruptions.

Explosive Volcanoes

As the name implies, an explosive volcano is one that erupts very quickly and with great force. Of the two kinds of explosive volcanoes that occur on Earth, the easiest kind to understand is a **cinder-cone volcano**, which usually forms very quickly after a single eruption. These volcanoes are small relative to other types of volcanoes, have steep sides (~30°), and consist of solidified magma fragments, rock debris, and ash that are ejected from a central vent. A great example of

---

Volcano   *A mountain or large hill containing a conduit that extends down into the upper mantle, through which magma, ash, and gases are periodically ejected onto the surface of Earth or into the atmosphere.*

Cinder-cone volcano   *A small, steep-sided volcano that consists of solidified magma fragments and rock debris that may form in only one eruption.*

(a)

(b)

Figure 13.31 Volcano eruptions. (a) Ash and steam cloud from Augustine volcano in Alaska. (b) Magma eruption at the Pu'u 'O'o volcano in Hawaii.

a cinder-cone volcano can be seen in northeastern New Mexico at Mount Capulin (Figure 13.32). This volcano has erupted only once, about 62,000 years ago, and is approximately 300 m (1000 ft) high.

In contrast to cinder-cone volcanoes, **composite volcanoes** are volcanoes that build up and grow over the course of several eruptions. Composite volcanoes are typically inactive for long periods of time between eruptions, but when they do erupt, they tend to do so quite violently. Such an eruption occurs because the magma within composite volcanoes is rich in silicas, which are minerals containing silicon (Si), and therefore highly viscous (meaning sticky and slow flowing). As a result, gases are gradually trapped in the magma during the inactive phase and build up pressure within the volcano until it explodes. Such an eruption sends volcanic ash high into the atmosphere and thick layers of volcanic debris will accumulate on the slopes of the volcano, causing it to enlarge. This debris may consist of alternating layers of lava, which is magma flowing on the surface, and fragmented rock debris called **pyroclastic material** (or *tephra*), such as volcanic ash, cinders, and boulders (Figure 13.33a). Also called *stratovolcanoes* because they contain strata (or layers) of volcanic debris, these volcanoes typically have moderately steep cones with a semi-horizontal top containing the crater. Composite volcanoes are much larger than cinder cones, perhaps over 3000 m (10,000 ft) high. A beautiful example of a composite volcano is the famous Mount Fuji in Japan (Figure 13.33b).

Composite volcano *A large, steep-sided volcano that grows through progressive volcanic eruptions, which are usually explosive, and consists of layers of volcanic debris.*

Pyroclastic material *Fragmented rock materials resulting from a volcanic explosion or ejection from a volcanic vent.*

Sometimes the eruption of a composite volcano is so explosive that it literally blows the top off of a mountain, creating a large crater. After this massive kind of eruption occurs, the crater may partially fill with a **lava dome**, which forms slowly in the ensuing decades. Such a volcanic feature consists of a steep-sided mound built by highly viscous magma that gradually oozes from the central vent. Although these features sometimes occur alone and are thus classified as a particular type of volcano by many geologists, the vast majority of the recently active lava domes occur in association with composite volcanoes.

Lava dome *A steep-sided volcanic landform consisting of highly viscous lava that does not flow far from its point of origin before it solidifies.*

Figure 13.32 Cinder-cone volcanoes. Cinder-cone volcanoes are relatively small volcanoes that build up by the accumulation of solidified magma fragments, rock, and ash over a short period of time. Mount Capulin is a typical cinder-cone volcano and is about 305 m (1000 ft) high.

Figure 13.33 Composite volcanoes. (a) Composite volcanoes consist of layers of lava and pyroclastic material that pile up around one or more conduits to subterranean magma chambers. (b) Mount Fuji in Japan, reverently called *Fuji-san* by the Japanese people, is a composite volcano.

Volcanic Arcs at Plate Boundaries Composite volcanoes are most commonly associated with plate boundaries in places where subduction is occurring. This geographical relationship occurs because the subducting plate melts when it encounters the upper mantle and the molten material gradually moves upward through the crust. Given that subduction tends to occur along the length of certain plate boundaries, these zones are places where a chain of volcanoes is typically located at the surface. Such a chain is called a **volcanic arc** and is genetically similar to the island arc shown in Figure 13.11. A well-known tectonic feature associated with a number of volcanic arcs is the **Pacific Ring of Fire**, which follows most of the outline of the Pacific plate (Figure 13.34). Using New Zealand as an arbitrary starting point, the Pacific Ring of Fire extends northwestward to the eastern coast of Asia, and north across the Aleutian Islands of Alaska. From there, it continues south along the western coast of North and South America. This region contains about 75% of the world's volcanoes, including Mount Ruapehu in New Zealand, Mount Pinatubo in

the Philippines, Mount Fuji in Japan (Figure 13.33b), Kliuchevskoi Volcano on Russia's Kamchatka Peninsula, Augustine volcano in Alaska, Mount Garabaldi in Canada, and Cotapaxi in Chile, just to name a few. Most of these volcanoes are composite structures and have formed because subduction is occurring at many places along the Pacific plate. Given the tectonic stresses that occur in these areas, the plate boundary is known for earthquakes.

In the Pacific Northwest of North America, the Pacific Ring of Fire is associated with a chain of volcanoes known as the *Cascade Volcanic Arc* (Figure 13.35). This volcanic arc is present because the Juan de Fuca plate, which is located off the west coast of North America (see Figure 13.2), is subducting beneath the North American plate to the east. The best known of the Cascade volcanoes is Mount St. Helens, which lies in the southern part of the state of Washington. This volcano is so well known because the most recent major volcanic eruption in the mainland United States occurred here in May 1980. It has also been the most active volcano in the arc within the past 4000 years (Figure 13.35).

Prior to the eruption, Mount St. Helens had a classic conical shape (Figure 13.36a) and, as a result, it was frequently called the "Mount Fuji of North America" because it was such a picturesque volcano. Mount St. Helens had been inactive for 123 years when it slowly began to reactivate in the spring of 1980. Several small eruptions occurred throughout April and early May of

---

Volcanic arc *A chain of volcanoes created by rising magma derived from a subducting tectonic plate.*

Pacific Ring of Fire *The chain of volcanoes that occurs along the edge of the Pacific lithospheric plate.*

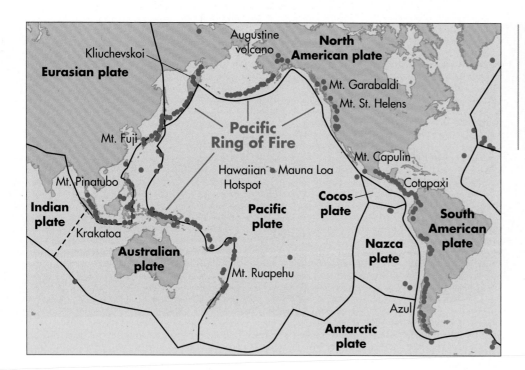

Figure 13.34 The Pacific Ring of Fire. Subduction along the Pacific plate and subsidiary plates (such as the Nazca and Cocos plates) has produced many prominent composite volcanoes. The locations of a few of them are shown here.

that year that opened a new crater and sent periodic clouds of volcanic debris onto the surrounding landscape. On May 18 the volcano erupted catastrophically (Figure 13.36b). Much of the volcanic cone of Mount St. Helens was blown off in this eruption and the surrounding landscape was devastated (Figure 13.36c). Subsequently, a nice example of a lava dome formed in the center of the crater. Evidence for this eruption, such as the new pattern of streams and the partial filling of nearby Spirit Lake with volcanic debris, can still be seen on satellite imagery.

Mount St. Helens provided a recent reminder that it remains the most active volcano in the Cascade Vol-

canic Arc, apparently because it continues to be more consistently fed with magma from the subducting Juan de Fuca plate than the other volcanoes in the region. Following a short period of increased seismic activity, Mount St. Helens erupted again in October 2004, creating a moderate ash plume. Volcanic activity continued sporadically into May 2005, including low levels of seismic activity, emissions of steam and volcanic gases, and minor productions of ash. None of this activity was catastrophic in nature and was merely associated with enlargement of the lava dome.

Given the active history of Mount St. Helens in the past 4000 years (Figure 13.35), this volcano continues

Cascade eruptions during the past 4,000 years

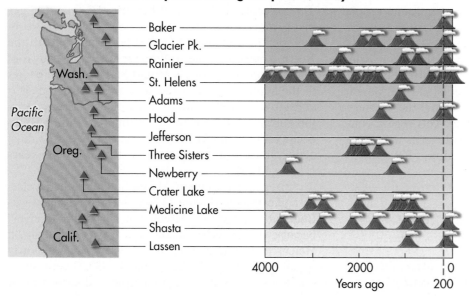

Figure 13.35 The Cascade Volcanic Arc. This diagram shows the geography and eruption history of volcanoes within the Cascade Volcanic Arc in the United States. The arc also extends for a short distance into British Columbia, Canada, where 5 volcanoes occur. Mount St. Helens in Washington has clearly been the most active volcano in the arc, with 15 eruptions in the past 4,000 years. Mount Jefferson, in contrast, has not erupted during this timeframe. The Canadian volcanoes have been largely inactive as well. (*Source:* United States Geological Survey.)

(a)

(b)

(c)

Figure 13.36 Eruption of Mount St. Helens. (a) The near-perfect conical shape of Mount St. Helens before the May 1980 eruption. (b) The catastrophic eruption on May 18, 1980. (c) Mount St. Helens after the eruption. The blast was so explosive that much of the peak was blown off, reducing the height of the mountain from 2950 m (9677 ft) to about 2550 m (8364 ft) and leaving a wide crater. Note the lava dome (arrow) in the center of the crater.

to be closely monitored by the United States Geologic Survey, as are other large volcanoes such as Mount Rainier and Mount Adams. It is possible that one of these volcanoes, or another one in the Cascade Volcanic Arc, will violently erupt in your lifetime. If such an eruption were to occur on Mount Rainier, for example, it would truly be a catastrophic event. In addition to the destruction caused by the blast, the eruption would cause ancient glaciers on the mountain summit suddenly to melt, resulting in a massive flood of volcanic debris and water (called a *lahar*) that would sweep toward the populated areas near Tacoma, Washington. This flood would likely cause numerous deaths and massive property loss.

 www.wiley.com/college/arbogast

Walk the Pacific Ring of Fire

A good way to appreciate the Pacific Ring of Fire is to experience it on a virtual tour. To do so, go to the *GeoDiscoveries* website and access the module *Walk the Pacific Ring of Fire*. This module allows you to work your way around the margin of the Pacific plate and observe the nature of its boundary with adjacent plates. It will also provide a sampling of notable tectonic events that have occurred at individual places along the plate boundary. Once you complete this tour, be sure to answer the questions at the end of the module to test your understanding of this concept.

CRATER LAKE

Oregon's Crater Lake is the second deepest lake in North America, measuring 389 m (1934 ft) deep. It is roughly 14.5 km (9 mi) wide and is almost perfectly circular in shape. At first glance it seems like a typical lake. If you look closely at the island in the middle of the lake, however, and the steep rock walls surrounding it, you begin to realize an even more remarkable fact— Crater Lake is the site of a dormant volcano.

Crater Lake in Oregon is the site of a dormant volcano, known as Mount Mazama, that exploded over 6800 years ago. Note Wizard Island, which formed during the most recent volcanic activity a little over 4000 years ago.

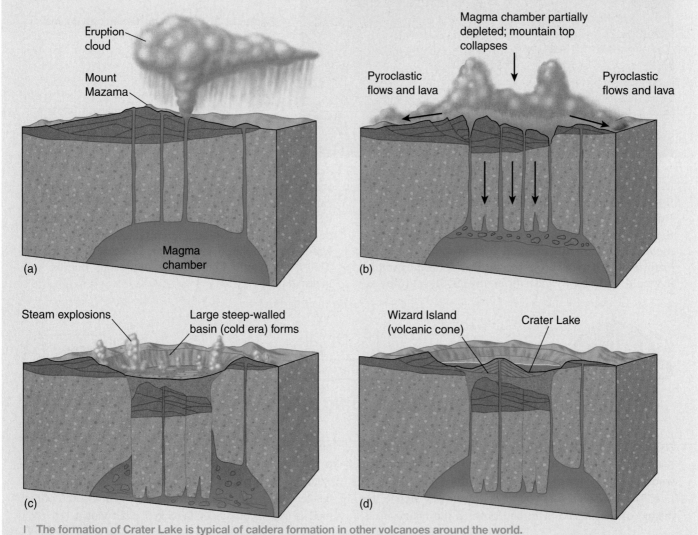

I   The formation of Crater Lake is typical of caldera formation in other volcanoes around the world.

Geologists have reconstructed how the crater formed when the top of the volcano, known as Mount Mazama, collapsed into the partially empty magma chamber after eruption. The resulting crater is a *caldera* and is particularly deep, with sheer rock walls. Mount Mazama began to form about 420,000 years ago and last erupted only 6850 years ago, in one of the biggest explosions of the past 10,000 years. Ash from Mount Mazama has been traced as far as Alberta, Canada, and pyroclastic flows extend up to 40 mi from the central vent. Approximately 4000 years ago, a relatively small eruption created Wizard Island, a small volcanic cone that protrudes out of the water. At some point following the major eruption, the caldera filled with water, forming Crater Lake. Another interesting fact about Crater Lake is that the water is incredibly blue. This occurs because there is no stream supplying water to the lake, so there is very little sediment washing into it.

Fluid Volcanoes

In addition to the volcanoes that erupt explosively, a number of prominent volcanoes have very fluid eruptions with flowing rivers of lava. In contrast to the highly viscous magma associated with explosive volcanoes, the magma that results in fluid eruptions contains far less silica and is thus less viscous. As a result, the lava associated with such an eruption flows across the surface until it cools to form basalt. Such eruptions can occur for weeks and months at a time and generally take place more frequently than explosive eruptions.

Although fluid eruptions are occasionally associated with composite volcanoes, such as Mount Nyaragongo in the Democratic Republic of the Congo in eastern Africa, they usually result in the formation of a **shield volcano** that has shallow, sloping sides (Figure 13.37). This type of volcano develops because successive eruptions of fluid magma cause the volcano to gradually build upward over a much broader area than composite volcanoes. A number of prominent shield volcanoes occur on Earth in association with subduction zones and rift zones. One of the best known is Skjaldbreiður,

---

Shield volcano *A very broad volcano with shallow slopes that forms in association with nonviscous lava flows.*

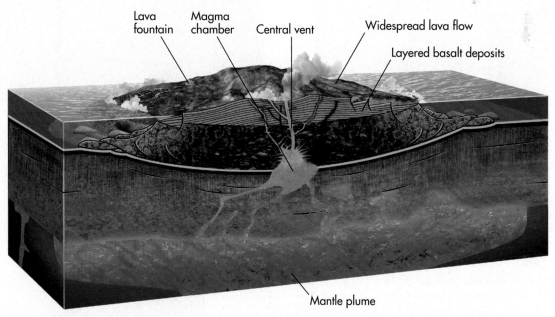

Figure 13.37 Shield volcanoes. Shield volcanoes are broad, low features that surround a central magma chamber. Magma associated with shield volcanoes typically has a low viscosity, which allows it to flow freely in broad streams of molten lava.

which straddles the Mid-Atlantic Ridge in Iceland. The name of the volcano is derived from the Icelandic term for *broad shield* and is the source of the more general term that is applied to similar volcanoes around the world. In Alaska, a prominent shield volcano is Mount Sanford, which is part of the Wrangell Volcanic Field in the eastern part of the state. This volcano began to form about 900,000 years ago and was most recently active approximately 320,000 years ago. This shield volcano is unusual because one of the slopes is very steep, rising 2400 m (8000 ft) over the course of 1.6 km (1 mi).

Hotspots

The previous discussion focused on volcanoes that form along subduction zones and rift zones. In some cases, however, a volcano is associated with a stationary zone in the asthenosphere where upwelling magma from a mantle plume is released at the surface. Such a geologic feature is called a **hotspot**. Although hotspots occur more randomly than subduction zones and rift zones, which can be traced for great distances, they are associated with some of the most intensive volcanic activity on the planet. These eruptions can be of either a fluid or explosive nature.

A great example of fluid eruptions associated with a hotspot can be seen in the state of Hawaii, which lies in the middle of the Pacific tectonic plate where the *Hawaiian hotspot* occurs. The hotspot currently underlies the island of Hawaii, which is the easternmost island in the chain (Figure 13.38a). This hotspot is releasing highly fluid magma that has built a variety of shield volcanoes on the island, such as Mauna Loa (Figure 13.38b). Mauna Loa is the largest volcano on Earth and is one of five that collectively comprise the island (Figure 13.38c).

Evidence indicates that the Hawaiian hotspot has been active for a long time and has a fascinating history related to continental drift. Figure 13.39 shows a trail of islands and submerged highlands called *seamounts* that extend to the northwest from the island of Hawaii. This trail exists because the oceanic crust in the Pacific basin has rotated over the Hawaiian hotspot for the past 70 million years. Do you see the sharp turn in the island/seamount chain? The features north of that

---

Hotspot *A stationary zone of magma upwelling that is associated with volcanism within the interior of a crustal plate.*

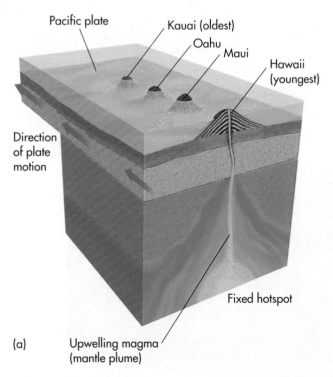

(b)

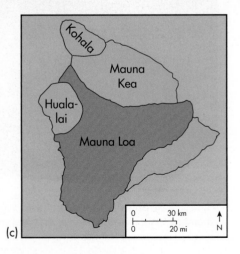

(c)

Figure 13.38 The Hawaiian Hotspot. (a) This hotspot is fixed because it is at the top of a mantle plume. The Hawaiian island chain has evolved through progressive movement of the Pacific plate over the hotspot. (b) Mauna Loa is the largest volcano on Earth. (c) The island of Hawaii consists of five separate shield volcanoes.

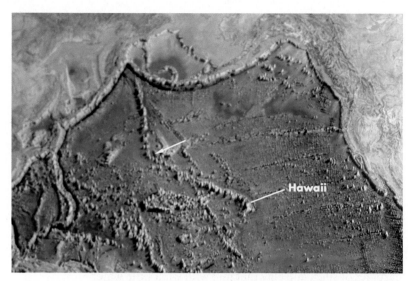

Figure 13.39 Migration of the Hawaiian hotspot over time. In addition to creating the modern Hawaiian Islands, the apparent migration of the hotspot formed the Emperor Seamount Chain. Note that the chain contains a sharp turn (arrow) that reflects a shift in the drift of the Pacific plate.

point, beginning with the *Emperor Seamount Chain*, must have formed when the plate was moving generally to the north as it flowed over the hotspot. Shortly after the Emperor Seamount Chain formed, approximately 43 million years ago, the plate began moving more toward the northwest. This movement has continued to the present day and explains why the state of Hawaii

includes a string of islands northwest of the island of Hawaii. Each of these islands was formed when that particular part of the Pacific plate was over the hotspot. With an eye to the future, where would you expect the next island of the Hawaiian Islands to develop—to the northwest of Hawaii or to the southeast? Why?

Whereas the Hawaiian hotspot provides an excellent example of a fluid system that lies within the middle of the ocean, hotspots are also associated with explosive volcanic eruptions that occur within the interior of continents. Such a continental example is the Yellowstone hotspot, which lies beneath Yellowstone National Park in northwestern Wyoming. Similar to the Hawaiian hotspot, the Yellowstone hotspot is a fixed zone of upwelling magma. In this case, however, the hotspot lies beneath the North American plate and the magma is highly viscous. Geological studies indicate that the apparent location of the Yellowstone hotspot has moved over time (Figure 13.40a), reflecting the west/southwestern migration of the North American plate. About 16 million years ago the hotspot was located in what is now the southeastern corner of Oregon. Between 12 million and about 2 million years ago, it migrated across what is now southern Idaho.

The Yellowstone hotspot has been in its present location for approximately the past 2 million years, with three major eruptions occurring during this time at 2.2 million years ago, 1.3 million years ago, and about 630,000 years ago. Each of these eruptions was cataclysmic, resulting in the formation of a giant caldera (Figure 13.40b). The

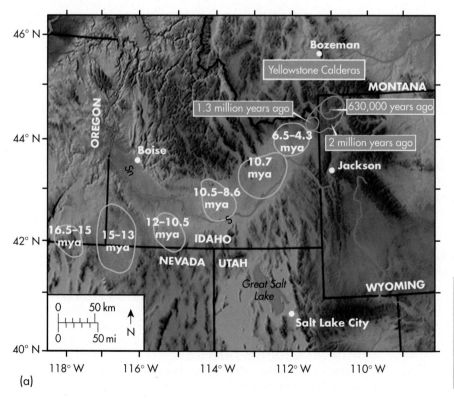

(a)

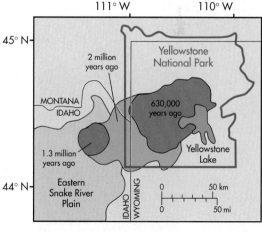

(b)

Figure 13.40 The Yellowstone Hotspot. (a) Migration of the Yellowstone hotspot between 16 million and 600,000 years ago. Although it appears that the hotspot has moved, it is the North American plate, in fact, that has migrated toward the southwest. (b) Calderas in the vicinity of Yellowstone National Park in northwest Wyoming. Each of these calderas formed during individual eruptions.

Yellowstone caldera—that is, the crater from the most recent eruption—is over 50 km (~32 mi) wide. As you can imagine, eruptions of this magnitude must have ejected enormous amounts of volcanic debris. The first eruption in the area, which is the largest of the three, ejected 2500 km³ (~1550 mi³) of material. By comparison, only 1 km³ to 2 km³ (~0.6 mi³ to 1.2 mi³) of material was ejected by Mount St. Helens in 1980. Some more recent eruptions have occurred at Yellowstone, such as a lava flow around 70,000 years ago and a steam explosion about 14,000 years ago, but these events were relatively small and isolated.

Imagine what the large Yellowstone eruptions must have been like! If a similar eruption happened in the present time it would impact the United States in a truly catastrophic way. It is conceivable that the eastern half of the country would *shut down* as long as the eruption occurred and volcanic ash fell across the landscape. A glimpse into the nature of this impact was seen in April 2010 when an erupting volcano in Iceland grounded air travel in Europe for about a week due to high concentrations of ash in the atmosphere. Given the truly gargantuan size of a potential Yellowstone eruption, and the fact that it would likely last for weeks, it would have devastating consequences for the nation's economy and human health downwind. Global climate would also be impacted because the extensive ash plume would reflect solar radiation, potentially causing a series of very cool years. Fortunately, the current best guess of geologists is that the next major eruption of Yellowstone is thousands of years away.

Nevertheless, if you have been to Yellowstone you know that the underlying hotspot indeed remains active today. Geothermal features such as geysers, mudpots, and fumeroles are very common within the park. A **geyser** is a superheated fountain of water that suddenly sprays into the air on a periodic basis (Figure 13.41). This process occurs because boiling water beneath Earth is constricted as it rises through a subterranean passageway. When the pressure builds sufficiently, the water bursts into the sky. When the geyser has used all the water in its reservoir, it dies away until the water is replenished from the surface or underground springs. A **mudpot** consists of a bubbling mixture of gaseous mud and water. These systems form where hot water is limited and hydrogen sulfide gas is present, creating sulfuric acid. This acid dissolves the surrounding rock into fine particles of silica and clay that mix with what little water there is to form the mudpot. A **fumerole** is a steam vent that results because underlying groundwater is boiled away before reaching the surface.

Geyser   *A superheated fountain of water that suddenly sprays into the air on a periodic basis.*

Mudpot   *A bubbling mixture of gaseous mud and water at the Earth's surface that is associated with geothermal activity.*

Fumerole   *A steam vent that results because underlying groundwater is boiled away before reaching the surface.*

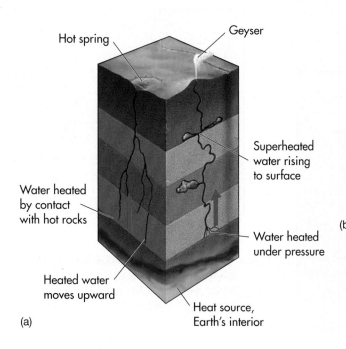

(a)

(b)

Figure 13.41 Geothermal features at Yellowstone National Park. (a) Schematic illustration of geyser processes. Water below the Earth's surface is superheated and rises in small cracks in the overlying rock. If a constriction exists in this rock, the heated water backs up until the growing pressure causes it to blast toward the surface where it erupts. **(b)** Old Faithful Geyser is probably the most famous system of its kind in the world. It derives its name because it erupts on average every 91 minutes.

Volcanoes

Now that we have discussed volcanoes thoroughly, you can see their associated processes in animated form. Go to the *GeoDiscoveries* website and access the module *Volcanoes*. This module contains a video that compares shield and composite volcanoes and shows various examples of each kind. As you watch the video, note how the concept of plate tectonics fits into the evolution of volcanoes, where they occur, and how they erupt. After you complete the animation, be sure to answer the questions at the end of the module to test your understanding about volcanoes.

VISUAL CONCEPT CHECK 13.3

Given your understanding of volcanoes, describe the volcano pictured here. Is it a cinder-cone, composite, or shield volcano? How can you tell? What kind of eruption will most likely occur when it does erupt?

KEY CONCEPTS TO REMEMBER ABOUT VOLCANOES

1. Volcanoes occur most frequently on plate boundaries. In particular, most of the world's volcanoes are found along the Pacific Ring of Fire.

2. Volcanoes are most frequently associated with the process of subduction because oceanic crust melts and then rises through cracks in overlying continental crust. The Cascade volcanic arc is an example of this relationship.

3. Volcanic eruptions can be broadly classified as explosive or fluid. Explosive eruptions occur when the magma is viscous, forming composite volcanoes (also called stratovolcanoes) such as those in the Cascades. When the magma is not viscous, it flows freely, resulting in broad rivers of lava that collectively form shield volcanoes such as those in the Hawaiian Islands. Cinder-cone volcanoes evolve through the accumulation of solidified magma fragments, rock debris, and ash that are ejected from a central vent.

4. Some volcanoes form over hotspots, which are places where upwelling magma reaches the surface. Examples of hotspot activity are the Hawaiian Islands and Yellowstone National Park.

Now that we have described how rocks form and mountains grow, we can next examine how these features are modified over time. For instance, this chapter discussed how the Appalachian Mountains have been extensively eroded from much higher peaks, resulting in today's well-rounded mountains. One of the ways in which this erosion occurred was through the process of rocks falling under the force of gravity, such as those pictured here. This image illustrates one of the many different ways in which rocks and landforms are modified after they form. The next chapter begins this investigation by focusing on weathering, which is the process through which rocks and landscapes are altered incrementally or suddenly by water and gravity.

■ SUMMARY OF KEY CONCEPTS

1. The Earth's crust is not a single sheet, but rather a jigsaw puzzle of interconnected plates. A variety of evidence (location of fossil remains, apparent magnetic polar wander, and actual measurement of motion) strongly supports the theory of continental drift, indicating that plates have shifted through geologic time.

2. Plate margins can be passive, transform, converging, or diverging. At diverging plate margins, magma plumes from the asthenosphere cause rifting, either on continents or on the seafloor. At converging plate margins, plates collide or subduct, causing geomorphic features such as alpine chains and volcanoes.

3. Folding is the process by which rocks are compressed and deformed. Several kinds of geologic structure are associated with folded topography, including (a) monoclines, (b) synclines, (c) anticlines, (d) overturned folds, and (e) overthrust faults.

4. Four major kinds of faults occur on Earth: (a) normal, (b) reverse, (c) overthrust, and (d) strike-slip faults. Earthquakes occur when moving plates are stuck along a fault, which increases stress until a rupture occurs along the fault plane. The strength of earthquakes is measured on the logarithmic Richter scale, which indicates the magnitude of seismic waves passing through the Earth's crust.

5. A volcano is a mountain that erupts periodically when the buildup of pressure beneath it passes a critical threshold. Most volcanoes occur along subduction zones. Three primary kinds of volcanoes occur on Earth: (a) cinder-cone volcanoes, (b) composite volcanoes, and (c) shield volcanoes. The type of volcano depends on the character of the magma associated with the system.

CHECK *YOUR* UNDERSTANDING

1. How does fossil evidence support the theory of continental drift?

2. How does the age of the Atlantic seafloor support the theory that the mid-Atlantic plate boundary is a rift?

3. What is the difference between a converging and diverging plate boundary? Why do diverging plate boundaries occur?

4. How does continental collision cause mountains to grow? What are the two processes through which alpine chains evolve?

5. Describe the geologic structure of the Ridge and Valley Province in the Appalachian Mountains. How can it be that some mountains are underlain by structural synclines, whereas others are underlain by anticlines?

6. Compare and contrast the various kinds of faults on Earth.

7. Discuss earthquake processes, including why they happen, the nature of seismic waves, and how they are measured.

8. Why are earthquakes typically associated with plate boundaries?

9. How much stronger is a magnitude 5 earthquake than a magnitude 2 earthquake?

10. Why are volcanoes associated with subduction zones?

11. What are the differences between composite volcanoes and shield volcanoes?

12. What is the Pacific Ring of Fire and why does it have that name?

13. What tectonic feature is associated with Yellowstone National Park and what unique features does it contain? In the context of your answer, explain how the migration of the Yellowstone hotspot fits into the picture.

ANSWERS TO VISUAL CONCEPT CHECKS

VISUAL CONCEPT CHECK 13.1

The answer is *c*. The structure shows the limb of an anticline. You can recognize this as an anticline because the rock structure rises rather steeply.

VISUAL CONCEPT CHECK 13.2

The answer is *d*. There is compression of the rocks on either side of the fault. Movement along the fault produces incredible strain that is causing deformation of the rocks along the fault boundary.

VISUAL CONCEPT CHECK 13.3

This volcano is a composite volcano, specifically Mount Rainier in Washington. You can tell the nature of this volcano because it is much too large to be a cinder-cone volcano. It is not a shield volcano because it is tall with moderately steep sides. Mount Rainier will probably erupt violently because the magma is highly viscous; thus, a great deal of pressure will build before the eruption occurs.

| CHAPTER FOURTEEN | **WEATHERING AND MASS MOVEMENT** |

The rest of this text focuses on the various geomorphic processes that shape the Earth's surface on a much smaller scale than the grand mountain ranges described in Chapter 13. These smaller-scale processes produce the rolling hills, valleys, and floodplains that are common throughout the world. Before these landscapes can be investigated, however, we first need to examine the mechanisms that break apart rock and loosen sediment so that they can be subsequently eroded, transported elsewhere, and deposited.

It is useful to think of the various geomorphic processes described in the rest of the text as part of a continuum of landscape evolution that begins with tectonic processes, which tend to elevate large blocks of land above sea level and provide topographic relief. Once landforms have been uplifted, other forces begin to reduce the overall relief and ultimately smooth the landscape over a long period. Two major components of this process are weathering and mass movement.

Look at this amazing rock outcrop. It is located in Arches National Park in Utah and is one of the many beautiful places in the southwestern United States. This landform is a remnant of a solid rock mass that once covered the area. Year after year, piece by piece, rock fragments broke off from this unit and were carried away by erosion to create this stunning landscape. This chapter focuses on the way that rocks break apart and wear down through the process of weathering.

CHAPTER PREVIEW

Weathering

On the Web: *GeoDiscoveries*
Weathering

Mass Wasting

On the Web: *GeoDiscoveries*
Weathering and Mass Movements

The Big Picture

LEARNING OBJECTIVES

1. Describe the process of weathering and the role of rock resistance.

2. Compare and contrast mechanical and chemical weathering.

3. Discuss the various weathering processes.

4. Explain the concept of acid rain in the United States.

5. Define the concept of mass wasting.

6. Compare and contrast the various kinds of mass wasting.

Weathering

The logical place to begin a discussion of landscape evolution is with the concept of weathering because it is often the initial force that changes rocks into sediment. Recall from Chapter 11 that **weathering** is the process by which rocks break down or decay into smaller fragments. Two primary kinds of weathering processes take place: *mechanical* (also called *physical*) and *chemical*. Although each of these weathering processes can operate alone, they often work together to break rocks apart. Figure 14.1 shows the relationship between these weathering processes and climate. Notice that physical weathering is more pronounced in colder/drier environments, whereas chemical weathering is related more to warmer/wetter climates. The reasons for these environmental relationships will become clear as you work through the chapter.

An important factor that influences the amount or pattern of weathering is the relative resistance of different kinds of rock at an outcrop. Some rocks are simply harder than others and are thus more resistant to weathering and erosion. Limestone, for example, is generally more resistant to physical weathering than shale. Within the limestone rocks, the dolomite units are more resistant than chalky rock units, which sometimes dissolve easily due to the chemical interactions discussed in Chapter 15. It is also possible that certain parts of an individual rock unit are more resistant to weathering and erosion than others within the same unit due to variations in the firmness of the cementing materials. This kind of differential cementation is particularly common in sandstone. In places where this kind of variability exists within a rock outcrop or within an individual rock unit, long-term weathering can have a noticeable impact and can even result in very unusual landforms (Figure 14.2). Such a weathering pattern is called *differential weathering* because rock units weather at different rates. If a more resistant rock layer overlies a less resistant layer, it protects the underlying rock somewhat from erosion and is thus known as **caprock**. Keep these variations in mind as you work your way through this chapter.

Weathering   *Physical or chemical modification of rock or sediment that occurs over time.*

Caprock   *A unit of relatively resistant rock that caps the top of a landform and thus protects underlying rocks from erosion.*

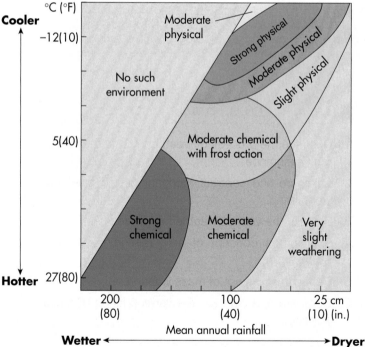

Figure 14.1 The geography of weathering. Physical weathering dominates in colder, drier environments, whereas chemical weathering tends to prevail in warmer, wetter regions (although some overlap occurs).

Figure 14.2 The effects of differential weathering. Differential weathering occurs when rock resistance varies between or within individual rock units. This form of weathering often leaves unusual rock forms, such as this *hoodoo* in Arches National Park in Utah. In this particular example, the caprock is a sandstone that is more resistant to weathering than the underlying rock layer.

Figure 14.3 Joints in rocks. Joints are fissures that develop along horizontal or vertical planes within rock. These columnar joints formed in basalts now exposed at Devil's Postpile National Monument in California.

Mechanical Weathering

Mechanical weathering, also called *physical weathering*, involves the destruction of rocks through physical stresses. When this kind of weathering occurs, rocks do not change their chemical composition. Instead, they simply break into progressively smaller pieces of the same kind of rock. As they do so, the overall surface area of the rock progressively increases, which facilitates even more weathering because more area within the rock mass is available for additional weathering to occur. This portion of the chapter describes the various kinds of mechanical weathering.

Frost Wedging The most common type of mechanical weathering is **frost wedging**. It begins when water works its way into fractures in the rock. A good example of such a rock fissure is a **joint** (Figure 14.3). Joints can form when rock repeatedly expands and contracts with heating and cooling, respectively, causing fractures to develop along horizontal and vertical planes that essentially break the rock into large blocks. Joints can also form due to regional tectonic forces that produce stress within the rock mass. Although joint features

---

Mechanical weathering *The breakup of a body of rock into smaller rocks of the same type.*

Frost wedging *Expansion and contraction of water in rock cracks due to freezing and thawing.*

Joint *A crack or fissure along horizontal or vertical planes in a rock mass that divides the rock into large blocks.*

Figure 14.4 The effects of frost wedging. Frost wedging occurs when water gets into cracks and joints within rocks and subsequently expands and contracts upon freezing and thawing, respectively. This process causes cracks to widen over time. Frost wedging can cause even large boulders to fracture in two, such as this specimen in California's Sierra Nevada.

may look like faults, they are not because no relative vertical or horizontal movement is occurring on either side of the fracture.

Once joints develop, they allow water to flow deeper into the rock mass at different points. If the rock body lies in a region that has distinct winter and summer seasons, the water will freeze and thaw, respectively, over the course of the year. When water freezes, it expands its volume as much as 9%. You can see this in the way that water in an ice cube tray freezes into cubes that are slightly larger than the liquid water level. If water subsequently thaws back to liquid form, it then contracts. (By the way, water is one of the very few substances that act this way; most materials are larger in volume as a liquid than as a solid.) Over the course of time, the stress produced by this freeze-thaw cycle causes the joints progressively to enlarge so that boulders slowly break away from each other (Figure 14.4). In places where frost action is most active, such as high mountain peaks or arctic environments, large fields of fractured boulders form (Figure 14.5); they are called *felsenmeer* (German for "rock sea").

THE LONG-TERM EFFECTS OF WEATHERING AND EROSION ON MOUNTAINS

The relative age (which feature is older or younger) of landforms often can be determined by looking at the impact that weathering and erosion have had on them over time. This kind of comparison is particularly valid in mountain ranges. To see such an example, compare this typical scene in Morocco's Atlas Mountains with a landscape from the Andes Mountains, which extend the length of the west coast of South America. Note that the Atlas Mountains are well rounded, whereas the Andes are very angular with sharp jagged peaks. Which one of the mountain ranges looks older to you? Why?

Based on the relative appearance of the two mountain ranges alone, it is reasonable to assume that the Atlas Mountains are probably older than the Andes Mountains and that they have gradually been worn down through a variety of weathering and erosion processes. In fact, the Atlas Mountains were created between 290 and 248 million years ago during the same Alleghany orogeny that formed the Appalachian Mountains in North America. The Andes Mountains, in contrast, are only about 7 million years old. This dramatic age difference means that weathering and erosion have been active for at least 240 million years longer in the Atlas Mountains than in the Andes Mountains.

The Atlas Mountains are well-rounded mountains in northwest Africa.

The Andes Mountains are jagged mountains that extend along the west coast of South America.

In addition to its effects on rock, the freezing and thawing of water can also cause modifications in soil. These modifications are most pronounced where soils are fine-textured with horizontal planes between silt and clay particles. Under these conditions, water collects in the planes and then expands upward upon freezing, causing the soil to rise upward in a process called **frost heaving**. If you happen to live in the northern part of the United States or in Canada, you can often see evidence of frost heaving in farm fields. In these areas, individual stones and cobbles (large, rounded rock fragments) can be slowly brought to the surface over time when water collects beneath them and freezes. After each winter, farmers pull any freshly surfaced rocks

---

Frost heaving   *Upthrust of sediment or soil due to the freezing of wet soil beneath.*

Figure 14.5 A rock sea. This rock sea of boulders (or *felsenmeer*) in the Colorado Rockies was created through the long-term process of frost action at high altitudes.

out of their fields to make plowing easier and stack them in a pile at the edge of the lot. You can also see the effects of frost heaving on northern paved roads when cracks develop in them because they are uplifted somewhat during the winter. Building contractors are acutely aware of the frost-heaving process. In fact, to avoid problems with structures such as decks settling in uneven ways, builders are required to install support footings below the depth of the frost line. The depth of this line is progressively greater with higher latitude or altitude.

Two other weathering processes have the same effect as frost wedging in that they break down rocks into progressively smaller fragments. One is related to the impact of plants when their roots grow into cracks and joints in rock. As the plants and associated roots grow, they can literally cause rock to split farther apart (Figure 14.6) in a process called *root wedging*. The second process takes place in association with temperature changes that occur over the course of seasons, between day and night, or due to changes in the direction the Sun faces a rock over the course of the day. These temperature fluctuations can cause rock to expand and contract when heated and cooled, respectively. Over the course of countless such cycles, the rock will gradually break into progressively smaller pieces. This kind of mechanical weathering is very prominent in arid regions (Figure 14.7) or at high elevations where intense solar radiation occurs during the day and radiational cooling prevails at night.

Salt-Crystal Growth Another kind of physical weathering process is **salt-crystal growth**, which involves the buildup of salts on rock surfaces. Salt is a dispersing agent and its concentration subsequently causes the minerals that cement the sediment grains within the rock to weaken their bonds; this, in turn, loosens sediment so that it can then be eroded. This process is similar to frost action in that it requires the presence of water moving within rock.

Figure 14.6 Plant roots and mechanical weathering. Plants can slowly break apart rocks when they grow in cracks such as this one and enlarge over time.

Figure 14.7 Temperature changes and mechanical weathering. This rock in the Mojave Desert has split into four pieces due to thermal expansion and contraction.

Salt-crystal growth *A weathering process involving the buildup of salts on rock surfaces through evaporation. These salts weaken the cement that bonds rock particles, allowing them later to be washed or blown away.*

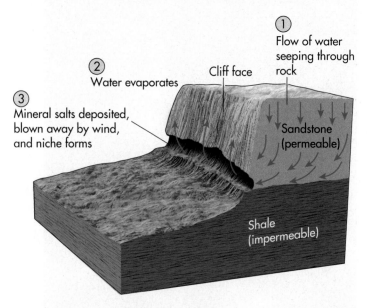

① Flow of water seeping through rock

② Water evaporates

Cliff face

③ Mineral salts deposited, blown away by wind, and niche forms

Sandstone (permeable)

Shale (impermeable)

Figure 14.8 Salt-crystal growth and niche formation. Salt-crystal growth is most prominent where permeable sandstones overlie relatively impermeable shales. As a result, water seeps out at the contact between the two rock types, and salt-crystal growth in the sandstone creates a niche.

An excellent place to see the effects of salt-crystal growth is in the steep cliffs of the southwestern United States, where permeable sandstones often overlie relatively impermeable shales (Figure 14.8). Water stored within the sandstones tends to move downward fairly easily because these rocks have large pore spaces. This moving water easily dissolves salts such as halite (sodium chloride), calcite (calcium carbonate), or gypsum (calcium sulfate). When this downward-moving water meets the shale, it can no longer move vertically and thus flows horizontally toward the cliff wall, where it seeps out and rapidly evaporates.

As a result of this evaporation, salts build up on the rock surface at the contact between the sandstone and shale. This high concentration of salts dissolves the cementing mineral bonds, which, in turn, causes the sand grains within the rock to separate slowly from another. These loosened grains are then washed or blown away by the water or wind, respectively, forming a niche at the contact between the two rock types. Although salt-crystal growth is a small-scale process, it has contributed greatly to the evolution of landscapes in the canyons of the desert Southwest by creating large niches and cave-like recesses in cliff walls. From an archaeological perspective, these niches are interesting because many contain the remains of ancient villages constructed by the Anasazi people about 1000 years ago. A great place to see such ruins is Mesa Verde National Park.

Exfoliation Another way in which rock physically weathers is when it expands through the process of *unloading*. Unloading occurs when deep rocks are slowly uncovered because overlying rock gradually erodes off of them. As the overlying rock mass is gradually removed, the amount of compression on the deeper rocks progressively lessens. If unloading progresses to the point where the formerly deep rocks become surface rocks, they may expand so much as the pressure lessens that curved joints develop parallel to the surface. These joints are called *sheeting structure* and result in the flaking of loosened rock from the main rock mass in a process called **exfoliation** (Figure 14.9). Although unloading may be the dominant cause of exfoliation, the process is probably enhanced by daily and seasonal temperature cycles that cause the rock to expand and contract.

Exfoliation is most active in igneous and metamorphic rocks, but also occurs in some sedimentary rocks. A great place to see exfoliation on a large scale is Yosemite National Park in California, where several large exfoliation domes occur in granite monoliths. The most famous of these landforms is *Half Dome*, which towers over the adjacent Yosemite Valley (Figure 14.9b). A close look at this peak reveals the curved sheeting structure associated with exfoliation.

Chemical Weathering

Another way in which rocks are altered is by **chemical weathering**. In contrast to physical weathering, which causes rock simply to break apart into smaller fragments, chemical weathering decays the rock by changing its chemical composition. This decomposition occurs through the interaction of air, water, biological processes, and rock minerals in various ways. In particular, water is very important as a weathering agent because of its ability to dissolve minerals. As a result, chemical weathering is typically more active in warm, humid environments (see Figure 14.1) where abundant water is present and various chemical reactions freely occur. Several chemical weathering processes are common, each having a unique combination of environmental variables that cause rock to decompose.

Exfoliation *Form of physical weathering in which sheets of rock flake away due to seasonal temperature changes or by the expansion of the rock due to unloading.*

Chemical weathering *The decomposition and alteration of rocks due to chemical actions of natural physical and biological processes.*

(a)

(b)

Figure 14.9 The impact of exfoliation. Exfoliation occurs when overlying rocks are removed through unloading and when temperature changes cause expansion and contraction of surface rocks. When these fluctuations occur, sheeting structure develops and rock flakes off the surface. (a) Exfoliation on Enchanted Rock, an exposed batholith in central Texas. Note how flaky the surface appears. (b) The peak of Half Dome mountain in Yosemite National Park is also composed of granite. Its unique shape gives rise to its name. The peak shows the sheeting structure and jointing associated with the process of exfoliation. These sheets gradually flake off, revealing underlying rock that will then itself expand.

Hydrolysis **Hydrolysis** is a weathering process that decomposes silicate minerals within rocks. This form of weathering occurs when silicate molecules are split after the addition of hydrogen and hydroxyl ions derived from water. In granitic rocks, for example, hydrolysis causes feldspars to weather into clays, such as kaolinite and silica, which may be subsequently washed or blown away and deposited somewhere in a soil or stream system. If a sufficient amount of clay is removed in this fashion, more resistant quartz grains are exposed and, with further weathering, liberated from the host rock, to be transported by a stream or perhaps blown into sand dunes.

You can see the effects of hydrolysis in several different ways. The most prominent impact of hydrolysis on the landscape is that the sharp edges and corners of rocks become rounded in a process called *spheroidal weathering* (Figure 14.10). This process

---

Hydrolysis *A chemical weathering process that results in the decomposition of silicate molecules within rock through the reaction of hydrogen and hydroxyl ions in water.*

Figure 14.10 Spheroidal weathering. (a) Corners and edges of rocks offer more surfaces on which water can act, so they are worn away faster than the rest of the rock. (b) Rounded rocks such as these in Joshua Tree National Park in California usually result from spheroidal weathering. Note the small trees for scale.

Figure 14.11 The effects of oxidation. Oxidized sandstones in the southwest United States. Scenes like this at the Fiery Furnace in Arches National Park are very common in this part of the country because iron-bearing rocks crop out in many places due to the arid climate and lack of vegetation.

Figure 14.12 Effects of carbonation in limestone. When carbon dioxide in the air dissolves in rainwater it forms a weak carbonic acid that can erode limestone. This limstone bed in Switzerland was once a solid rock layer that has been extensively worn by acidic rainwater flowing over it.

has the greatest impact where numerous joints occur in rocks, providing more corners and edges on which decomposition can occur.

Oxidation Another form of chemical weathering is called **oxidation**. In contrast to hydrolysis, which requires water, oxidation occurs when oxygen is added to chemical compounds and causes electrons within the compounds to be lost. In everyday life, we are most familiar with oxidation when tools and household equipment rust when left outside for a few days. Rusting occurs because oxygen reacts with iron to produce iron oxides, which are reddish-colored and flake off with time. The iron underneath is then exposed to the air and forms rust in its turn. The same process happens in rocks that contain abundant iron, such as those in the southwestern United States. These rocks have been heavily oxidized, resulting in their orange appearance (Figure 14.11).

Carbonation A third kind of chemical weathering occurs when water containing carbon dioxide causes minerals to dissolve and wash away in solution. This form of weathering is called **carbonation** (because carbon is reacting with minerals) and occurs because carbon dioxide is easily dissolved by atmospheric water vapor, forming carbonic acid. As a result, precipitation contains this carbonic acid, which is sufficiently acidic to dissolve many minerals, especially limestone (calcium carbonate). When rainfall collects on the rock surface, it dissolves the limestone and forms pits and rounded surfaces (Figure 14.12). If you are interested in your local weathering rates, visit a cemetery and compare the decomposition in limestone versus marble or slate tombstones. Since marble and slate are much more resistant than limestone to weathering, the difference can be dramatic.

Oxidation *A form of chemical weathering in which oxygen chemically combines with metallic iron to form iron oxides, resulting in the loss of electrons.*

Carbonation *A type of chemical weathering caused by rainwater that has absorbed atmospheric carbon dioxide and formed a weak carbonic acid that slowly dissolves rock.*

Human Interactions: Acid Rain

As just described, carbonation is the natural process through which rocks are chemically weathered when they come into contact with acidic water. In a similar way, by-products of human industrial activity cause chemical weathering of rocks through the process of acid rain. **Acid rain** is the broad term applied to the way in which industrial acids fall out of the atmosphere in association with precipitation.

In North America, acid rain is a particular problem in the northeastern part of the United States and eastern Canada. This area lies downwind of large industrial cities, such as Chicago, Detroit, Cleveland, and Pittsburgh, where factories emit chemical residue into the atmosphere. In particular, sulfur dioxide and nitrogen oxides are the primary acids linked to acid rain. According to the United States Environmental Protection Agency, about two-thirds of all sulfur dioxide and one-fourth of all nitrogen oxides in the atmosphere come from electric power generation plants that rely on burning fossil fuels such as coal.

After these acids are emitted from factories in the industrial Midwest, they are carried aloft and eastward by the prevailing westerly winds. Subsequently, they are deposited on the surface of Earth through precipitation or within fog. As a result, the pH of rainfall in the eastern part of the United States is about 4, whereas it is generally above 5 in the western half of the country. Similar patterns have been identified in other parts of the world, particularly Eastern Europe.

The impact of acid rain on the landscape and as a weathering agent has been significant. With respect to chemical weathering, acid rain dramatically accelerates the weathering process. Although some of this weathering would have naturally occurred by carbonation, it has certainly been enhanced by acid rain. In addition to the acceleration of chemical weathering, acid rain is also associated with several negative environmental impacts especially upon forests and lakes. Forests that lie within the belt of acid precipitation are often stressed (Figure 14.13). This stress occurs in part because supporting soils are acidified, which accelerates the leaching of nutrient cations before the trees absorb them. Evidence also exists that acid deposition on the needle leaves of conifers reduces their tolerance to cold. Similarly, the pH of lakes and ponds can be lowered due to acid rain, causing stress to the plants and animals that live within those water bodies. Many lakes in New England, for example, do not support fish because they contain toxic levels of inorganic acids.

Although acid rain is a serious environmental problem, recent studies indicate that environmental regulations imposed in 1970 are having a positive effect, at least as far as the United States and Canada are concerned. Prior to the major amendments to the Clean Air Act in 1970, the amount of industrial acids being emitted into the air had been essentially unchecked since the onset of the Industrial Revolution in the 19th century. These emissions peaked at 28.8 million metric tons in 1973 in the United States alone. With the amending of the Clean Air Act in 1970, which specified a variety of industrial regulations designed to curb acid emissions, these rates in the United States decreased 32% to 19.6 million metric tons in 1998. Although nitrogen oxide emissions remain unchanged, we appear to have at least turned the corner with respect to acid rain in North America. Major problems remain in the former Soviet bloc of Eastern Europe, however, where virtually no environmental regulations were in effect during the Communist era. As a result, the relatively new democratic governments are struggling with how to cope with the serious environmental degradation in these areas.

Figure 14.13 Forest stress due to acid rain. Acidic rainfall is concentrated in the eastern part of the United States. Forests in these areas, such as this stand of Fraser Firs in the Appalachian Mountains, are affected because soils are acidified, which accelerates leaching of nutrient cations before the trees can absorb them. In addition, trees become less tolerant to cold weather.

Acid rain *The precipitation by rain, fog, or snow of strong mineral acids, primarily sulfur dioxide and nitrogen oxides, that originated in factories.*

This image is of a rock in the southwestern United States. Given what you see, and the location of the rock, which one of the following weathering processes produced the rock pattern visible here?

a) Hydrolysis
b) Carbonation
c) Freeze-thaw activity
d) Thermal expansion
e) Oxidation

www.wiley.com/college/arbogast

Weathering

Now is a good time to review the various weathering processes discussed so far in this chapter. Given they involve water movement and the gradual wearing down of the landscape, it is useful to observe them in an animated format. Go to the *GeoDiscoveries* website and open the module *Weathering*. This module shows, for example, how frost wedging gradually causes rock to split apart and how the surface area of a rock mass increases with progressive mechanical weathering.

The animation also reviews the various ways that rock decomposition occurs as a result of chemical weathering processes such as hydrolysis and carbonation. As you watch this animation, think about how weathering contributes to the wearing down of rock masses and the time involved for this erosion to occur on a noticeable scale. After you complete the animation, be sure to answer the questions at the end of the module to test your understanding of weathering.

KEY CONCEPTS TO REMEMBER ABOUT WEATHERING

1. Two general kinds of weathering processes take place: mechanical (physical) and chemical. Mechanical weathering causes rock bodies to break into smaller rock fragments, whereas chemical weathering causes the rock to change chemical composition.

2. Although mechanical weathering and chemical weathering can occur anywhere, most mechanical weathering occurs in cold environments, whereas chemical weathering is more pervasive in warm/humid climates.

3. The three major kinds of mechanical weathering are frost wedging, salt-crystal growth, and exfoliation. Frost wedging involves water that expands and contracts upon freezing and thawing in joints and cracks in rock. Salt-crystal growth occurs when water seeps out of rock in arid environments. Exfoliation happens when rock expands because of unloading and temperature changes.

4. The three major kinds of chemical weathering are hydrolysis, oxidation, and carbonation. Hydrolysis occurs when water interacts with rocks, causing chemical reactions to occur. Oxidation occurs when oxygen combines with metals to form oxides. Carbonation takes place when carbonic acid in precipitation accumulates on rocks and causes them slowly to dissolve.

5. Humans have increased the amount of chemical weathering downwind of industrial cities through the process of acid rain. Most of these acids come in the form of sulfur dioxide and nitrogen oxides derived from coal-burning power plants.

Mass Wasting

Once sediment is liberated from rock in some fashion as a result of the weathering process, it can be moved and deposited by gravity, wind, waves, or flowing water and glaciers. In this way, the Earth's landscape is modified and shaped to form features such as hills, valleys, beaches, and sand dunes. The remainder of this chapter will focus on the effect of gravity and the role it plays in landscape evolution by moving large volumes of sediment through the process of **mass wasting**.

From a geographical perspective, the place to consider mass wasting is on hillslopes, which are land surfaces inclined at various angles from the horizontal. This process takes place when the downward force of gravity overcomes the resisting force of the slope material's shearing strength—in other words, its cohesiveness and internal friction. The point at which sediment becomes unstable is called the *threshold point* and depends a great deal on the material's *angle of repose*. This angle is the natural maximum slope that any deposit of a particular kind of sediment—for example, sand—can achieve without moving under the force of gravity. The angle of repose ranges between about 25° for fine sands and approximately 40° for angular pebbles and cobbles. You can see these relationships when you see piles of sediment deposited at construction sites or sand and gravel quarries. After the sediment is dumped, its slope naturally moves to the angle of repose (Figure 14.14).

Generally speaking, most landscapes contain a variety of hillslopes that grade downward to a stream of some size (Figure 14.15). The *local relief* of the area—that is, the vertical distance between the top of the hills and the stream downslope—may have been created either by uplift of the landscape or through downcutting by the stream. In some cases, both processes occur.

Regardless of how the relief was created, the resulting hillslopes contain many distinctive components. In all probability, the core of the slope is some form of bedrock that dates to either an igneous or sedimentary process that occurred tens of millions or hundreds of

millions of years ago. In relatively humid areas, the solid core is mantled by regolith, which, if you recall, is formed when rock is weathered. Also recall that regolith is further modified through various pedogenic processes at the surface to form soil. Depending on the steepness of the hillslope and the regional environmental conditions, the soil will be either well developed and thick, or poorly developed and thin. Most mass-wasting processes occur within the mantling regolith and soil, although where relief is particularly steep, they may also impact outcropping bedrock. Sediment that moves down the hillslope due to some mass-wasting process is called **colluvium**. Collovium tends to be deposited at the base of hillslopes because the gradient becomes less at those locations and the energy to transport sediment is lost.

Several different kinds of mass wasting occur on Earth, as shown in Figure 14.16. Although these variations tend to grade into one another, an association exists between specific environmental settings and individual mass-wasting processes. These processes are subdivided on the basis of:

1. The kinds of Earth materials being moved, such as rock or regolith, and their association with the presence of water.
2. Their physical properties, including whether they are hard, plastic, or fluid.
3. The kind of motion, for example, whether the material is falling or flowing.
4. How quickly the process occurs.

The following discussion focuses on the processes outlined in Figure 14.16.

Rockfall

The easiest kind of mass-wasting process to envision is **rockfall.** As the name implies, rockfall occurs when rocks fall quickly down a hillslope under the force of gravity. Rockfall is usually most prominent in places

---

Colluvium   *Unconsolidated sediment that accumulates at the base of a slope.*

Rockfall   *A mass-wasting process in which rocks break free from cliff faces and rapidly tumble into the valley below.*

Mass wasting   *Movement of rock, sediment, and soil downslope due to the force of gravity.*

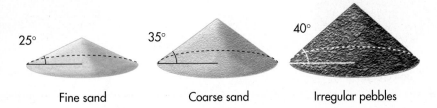

Figure 14.14 The angle of repose. The angle of repose is the natural maximum angle at which sediments of a given size can rest before they begin to slide under the force of gravity. In general, coarse sediments have a steeper angle of repose than finer sediments.

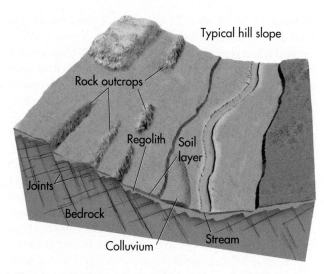

Typical hill slope

Rock outcrops

Regolith   Soil
              layer

Joints

Bedrock

Colluvium   Stream

Figure 14.15 Generalized view of a typical hillslope. Hillslopes typically have distinctive characteristics that are easily visible. Note the relationships of bedrock, regolith, soils, and valley stream in the image, which are common in many hillslopes.

where extensive rock outcrops occur on steep hillslopes or canyon walls (Figure 14.17a). Frost wedging and large temperature changes contribute significantly to rockfall because these processes gradually split outcropping rocks apart from one another. Eventually, the gravity threshold is passed and stones and boulders break free to fall onto the lower part of the hillslope. Rockfall also occurs in association with differential weathering in places where relatively resistant rocks such as limestone and sandstone overlie softer rocks like shale. In these situations, the shale surfaces gradually retreat into the cliff through a combination of weathering and erosion, leaving the resistant rock as a ledge that juts out over the shale. Sooner or later the force of gravity exceeds the strength of the rock and it collapses. This process is very common in places like the Grand Canyon.

Extensive deposits of fallen rock, called **talus** (or *scree*), often accumulate through rockfall. These

Talus   *A pile of rock fragments and boulders that accumulates below a cliff due to rockfall.*

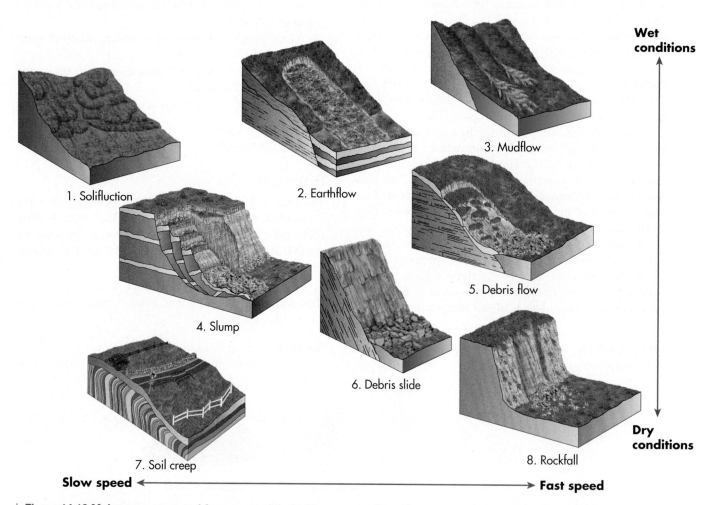

Wet conditions

1. Solifluction

2. Earthflow

3. Mudflow

4. Slump

5. Debris flow

6. Debris slide

7. Soil creep

8. Rockfall

Dry conditions

Slow speed ← → Fast speed

Figure 14.16 Major processes and forms associated with mass wasting. Movements to the right of this figure are relatively fast, whereas those to the left are relatively slow. Movements near the top of this figure are triggered by wet conditions, such as heavy rain; those near the bottom of the figure can occur under dry conditions.

(a)

(b)

Figure 14.17 Rockfall. (a) A typical rockfall in a steep-walled canyon. Rockfall occurs when rocks in canyon walls are gradually loosened from the host rock mass by frost wedging and/or large temperature changes and thus quickly fall under the force of gravity. (b)Talus cones, such as these in Banff National Park in Canada, form along canyon walls and mountain escarpments where ravines and gullies funnel boulders into the adjacent valley.

deposits can collect in relatively small amounts along the base of individual, small hillslopes or in larger quantities along high cliffs. Typically, talus slopes have a relatively high angle of repose, ranging between 35° and 40°, because boulders become lodged together and thus support one another. In most cases, talus accumulates in cone-shaped bodies called *talus cones* (Figure 14.17b). This pattern occurs because most canyon walls are interspersed with ravines and gullies that funnel boulders down where they accumulate. In places where numerous talus cones lie adjacent to each other, the resulting landform is called a *talus apron*.

Soil Creep

The slowest mass-wasting process is called **soil creep.** Soil creep occurs on virtually every soil-covered slope, as the perpetual force of gravity slowly pulls surface particles downhill. In this fashion, the soil behaves as a plastic substance that is gradually molded over time. This process is enhanced by cycles of wetting, drying, freezing, and thawing in the soil, as well as from disturbances caused by burrowing animals and even people. The result of soil creep is that the surface deposits of a hillslope, and even artificial features such as fence posts and telephone poles, tend to shift slightly downhill (Figure 14.18). Although this process does not produce any distinctive landforms, it causes the angles of hillslopes and the elevation of hilltops to slowly decrease. Although soil creep

Soil creep *The gradual downhill movement of soil, trees, and rocks due to the force of gravity.*

Figure 14.18 Evidence of soil creep. Soil creep occurs when hillslopes gradually move downhill under the force of gravity. Notice here the tilted wooden columns, which is a good sign that soil creep is occurring.

Figure 14.19 The effects of solifluction. The distinctive lobes on this hillslope in Kyrgyzstan have evolved because unfrozen soil becomes saturated in the summer and sags downhill over underlying permafrost.

is a process that occurs virtually everywhere hillslopes exist, a distinctive kind of creep called **solifluction** (meaning "soil flowage") occurs above the tree line in tundra landscapes. This process is very straightforward and involves the impact of seasonal temperature changes on the landscape. Most of the year, the soil is frozen from the surface to some depth. During the short summer months, the upper part of the soil thaws. Water within this part of the soil (the active layer) cannot percolate very deeply, however, because the sediment beneath it is continuously frozen as permafrost. As a result, the overlying (unfrozen) soil becomes saturated and slowly sags downhill in a very uneven way, producing distinct lobes that overlap one another (Figure 14.19).

Landslides

Another form of mass-wasting process is called a **landslide**. As the name implies, a landslide is a mass of rock, regolith, and soil that suddenly moves downhill. This movement is initiated when an event such as an earthquake or period of heavy rain causes the downward force of gravity to overcome the sediment's cohesiveness and internal friction. Slope-shear strength depends on the frictional resistance and overall cohesion of the underlying sediment. When slope shear strength is reduced for some reason, a portion of a hillslope can move downhill at a rapid rate, either as a debris slide or a slump.

A **debris slide** occurs when slope failure occurs along a plane that is roughly parallel to the surface. A famous example of a debris slide is the *Madison slide*, which occurred in Montana near Yellowstone National Park on August 17, 1959, when a magnitude 7.1 earthquake rocked the region late in the evening for about 30 sec. This earthquake initiated a massive landslide (Figure 14.20) on a mountain that borders the Madison River, which flows through the area. Approximately 28 million m³ (37 million yd³) of rock was instantly detached from the mountain in an area that measured about 600 m (about 2000 ft) in length and 300 m (about 980 ft) in thickness. During the ensuing landslide, the rock mass slid down the mountain slope, crossed the Madison River, and, in the process, killed 28 people who had been camping in the vicinity. The landslide was so large that it dammed the Madison River, creating a new lake, now known as Earthquake Lake, which is nearly 100 m (330 ft) deep.

In contrast to debris slides, which flow parallel to the surface, a **slump** occurs when rock and sediment rotate and move down the slope along a concave plane relative to the surface. Slumps typically occur when impermeable clay-rich deposits such as shale underlie more porous rock or sediment. Under such circumstances, water will flow through the upper deposits and then parallel to the clay surface upon encountering it. This water lubricates the contact between the rock layers and thus reduces frictional force, causing the rock and sediment mass to rotate as it slides down the hill (Figure 14.21a).

At the top of the slump, a wall-like escarpment is produced where the slump block breaks away, leaving a steep face. The sediment in the middle of the slump usually moves more quickly than near the scarp, resulting in a series of step-like features called *slump terraces*. At the base of the slump, the sediment flows more slowly and bulges out from the middle section of the flow as a distinctive *toe*. Although slumps happen virtually everywhere, they are by far more common in humid regions because the sediments are more frequently saturated there. They usually occur in isolated places, such as in Figure 14.21b, with very little impact on humans. Occasionally, however, a large slump occurs in a residential zone and causes severe damage to personal property.

Solifluction *A form of soil creep that occurs in arctic environments where freeze-thaw processes result in lobes of soil moving gradually downslope.*

Landslide *An instantaneous movement of soil and bedrock down a steep slope in response to gravity, triggered by an event such as an earthquake.*

Debris slide *A mass-wasting process in which slope failure occurs along a plane that is roughly parallel to the surface.*

Slump *A mass-wasting process in which rock and sediment rotate and move down the slope along a concave plane relative to the surface.*

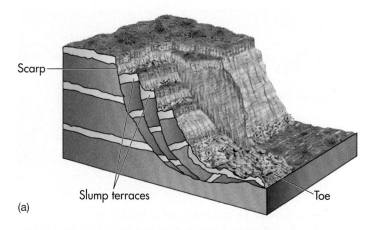

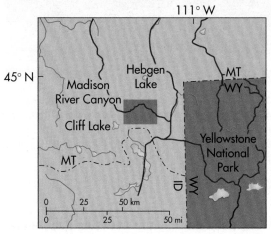

Figure 14.20 The Madison slide. Slopes can instantly fail when the threshold for cohesiveness is passed, causing tons of rock and sediment to move rapidly down the hill. The massive Madison slide, for example, was prompted by a large earthquake in the region. This slide blocked the Madison River, forming Earthquake Lake (arrow).

Figure 14.21 Slump processes. (a) A typical slump includes a scarp, slump terraces, and a toe. (b) Slumps like this frequently occur in north-central Kansas because a thin shale (and associated regolith) at the base of hills becomes lubricated during heavy spring rains. Repeated slumps like this one have gradually caused the hillslopes to retreat throughout the area, forming these small hills.

Flows

As the name implies, flows are mass-wasting events that involve sediments that are very wet. Such wet conditions occur when rainfall is high for a period of time, or when humans have stripped hillslopes of protective vegetation. These processes are catastrophic forms of mass wasting and are a serious environmental hazard in many parts of the world. Flows are common in southern California, for example, and are especially dangerous in suburban areas where homes are built on steep hillslopes. In fact, every 2 or 3 years some form of major flow occurs in this area that causes extensive property loss and even the occasional human death. Three basic kinds of flow occur: (1) earthflows, (2) mudflows, and (3) debris flows.

An **earthflow** is a slow-to-rapid type of mass movement that involves soil and other loose sediments, some of which may be coarse. This kind of flow may move downhill across a fairly broad surface. Although a **mudflow** is very similar to an earthflow, it differs because it consists of fine-textured sediments and is extremely fluid; thus, it moves more quickly. Mudflows are most frequently associated with steep canyons in mountainous regions and are initiated in humid zones when heavy rains fall on already saturated slopes. In arid zones, they occur when heavy rains fall on hillslopes that are unprotected by vegetation. Regardless of the environment, mudflows start when large volumes of water begin to run directly off the surface down the canyon. As the water flows, it picks up increasing amounts of mud from the underlying soil,

Earthflow *A slow-to-rapid type of mass movement that involves soil and other loose sediments, some of which may be coarse.*

Mudflow *A well-saturated and highly fluid mass of fine-textured sediment.*

Figure 14.22 Mudflow deposits. Mudflows occur when a significant amount of saturated fine sediments flow quickly. This mudflow deposit is located in the Never-Summer Range in Colorado. Note the lack of coarse fragments such as rocks and boulders.

Figure 14.23 The La Conchita debris flow. This massive debris flow along the coast of southern California was triggered by heavy rains over the course of 3 days. Thick deposits of mud, trees, and other debris swept into the community of La Conchita and unfortunately killed several people.

resulting in a thick, viscous flow of water-laden sediment that has tremendous power. In large mudflows, the power is so great that large boulders are frequently transported great distances and human structures such as homes and bridges can be destroyed. Once the mudflow reaches the valley, it often spreads out to form an apron-like landform (Figures 14.16 and 14.22).

Another kind of flow associated with heavy rainfall is a **debris flow**. Debris flows differ from mudflows because, in addition to mud, they also contain a mix of boulders, trees, and even buildings as they flow rapidly downhill. These rapidly moving and powerful flows can be triggered by extremely heavy rains in mountainous regions or by a volcanic eruption that instantly melts summit glaciers.

Debris flows are very serious environmental hazards that can impact people catastrophically. An unfortunate example was the massive debris flow that swept down a steep hillslope into the coastal community of La Conchita, California, in January 2005 (Figure 14.23). This debris flow was triggered by a 3-day storm that produced nearly 25 cm (10 in.) of rain in the region.

The slopes above La Conchita consist of ancient debris flow and landslide deposits that had failed to a smaller extent in 1995 when a large slump destroyed nine homes and blocked a county road. When these deposits became saturated again in early 2005, they failed on a much larger scale, releasing approximately 14,000 m³ (18,311 yd³) of material that flowed more than 100 m (330 ft) into the community. This debris flow destroyed 13 homes and killed 10 people. Given that slope failures frequently occur in California during winter, this disaster demonstrates why it is important to understand how

mass-wasting processes operate and how they can be triggered by passing weather systems.

Avalanches

Another mass-wasting process that you have probably heard of is an **avalanche**, which is a large mass of snow or rock that suddenly slides down a mountainside (Figure 14.24a). Avalanches are most commonly associated with snow and typically occur when thick deposits of snow accumulate at steep angles on mountain slopes. These steep-angled snow deposits can become unstable when a subsequent large storm produces another layer of snow of lighter density than the snow it buries. Under these conditions, an avalanche will occur because the two layers of snow are not firmly bound together to form a uniform mass. When the uppermost layer is somehow destabilized, a section will break free as a *slab avalanche*. Such an avalanche may release over 225 million m³ (~300 million yd³) of snow, which is equivalent to about 20 football fields covered with about 3 m (10 ft) of snow.

Avalanches have three major components (Figure 14.24b). They begin in a place appropriately called the *starting zone*, which is usually located on the steepest slopes of a mountain where deep snows lie at precarious

Debris flow *A rapidly flowing and extremely powerful mass of water, rocks, sediment, boulders, and trees.*

Avalanche *A large mass of snow or rock that suddenly slides down a mountainside.*

(a)

(b)

starting zone

track

runout zone

Figure 14.24 Avalanches. (a) Avalanches are usually associated with the sudden failure of enormous quantities of snow that cascade down steep mountain slopes. (b) Avalanches begin at the starting zone, travel down the track, and end at the runout.

angles. After the avalanche begins, it typically follows a chute or ravine system on the mountain slope called a *track*. The track is where most destruction takes place, either through trees being knocked down or through the movement of boulders and other debris. As you travel through the mountains in summer, you can often see the track of a recent avalanche by looking for avalanche scars, which are treeless zones extending down the slope of a mountain. Ultimately, an avalanche loses its energy when it reaches the base of the mountain and the slope lessens. This area is called the *runout* and is the place where snow and debris are deposited.

Although most avalanches happen in remote places that have no impact on people, they can be a worrisome natural hazard at ski resorts and even along mountain highways. In 2009 more than 25 skiers died (and many more injured) around the world, for example, after being swept away in avalanches. Avalanche hazards were a particular problem in the Rocky Mountains and Pacific Northwest, but also caused fatalities in the European Alps and Southern Alps in New Zealand. People who live and work in mountainous areas are keenly aware that some areas are particularly prone to avalanches. In an effort to control the hazard, avalanches are often purposefully started when conditions appear ripe. To minimize the risk, explosives are dropped from helicopters onto the unstable snow pack. The resulting explosion causes an avalanche to develop under somewhat controlled conditions.

VISUAL CONCEPT CHECK 14.2

What type of mass-wasting process is pictured here?

a) Debris flow
b) Mudslide
c) Rockfall
d) Slumping
e) Avalanche

Weathering and Mass Movements

To better understand the process of mass movements, go to the *GeoDiscoveries* website and access the module *Weathering and Mass Movements*. This module allows you to visualize the various mass-wasting processes in action. As you work through this module, observe how these processes operate, how they differ from each other, and how they shape the landscape. In addition, you will better comprehend the impact that these hazards have on people and the structures in which they work and live. Once you complete the module, be sure to answer the questions at the end to test your understanding of this concept.

KEY CONCEPTS TO REMEMBER ABOUT MASS WASTING

1. Mass wasting refers to the downslope movement of sediment, soil, and rock under the force of gravity.

2. Mass-wasting processes are differentiated on the basis of how quickly they move and the relative saturation of sediments.

3. Rockfall, creep, solifluction, and landslides are mass-wasting processes that occur under the force of gravity. Rockfall and landslides occur when rocks and sediments break free for some reason, such as lubrication of underlying clays or earthquakes.

4. Flows occur when sediments become saturated and then move under the force of gravity. Flow processes include earthflows, mudflows, and debris flows.

5. Avalanches occur in mountainous areas when steep snow packs break free.

THE BIG PICTURE

Now that the various mass-wasting processes have been investigated, it is time to examine some other geomorphic processes that produce landforms. The next chapter focuses on the processes and landscapes associated with groundwater, which, as the name implies, is water stored within the ground. Much of Chapter 15 focuses on karst landscapes, which are created when rocks dissolve. In this sense, the next chapter logically follows Chapter 14 because chemical weathering and the formation of karst landscapes are related processes. A cave, such as the one illustrated here, is an example of a karst feature. If you have been to a cave, you know that karst landscapes can be some of the more mysterious features on Earth. The next chapter will describe how these features form, as well as the ways in which water moves and is stored in the ground.

SUMMARY OF KEY CONCEPTS

1. Weathering refers to the decay and breakup of rocks on the surface of Earth. Two general kinds of weathering processes take place: mechanical (physical) and chemical. Mechanical weathering causes rock bodies to break into smaller rock fragments, whereas chemical weathering causes the rock to change chemical composition. These processes are closely associated with environmental conditions and may vary locally due to the relative resistance of different rock units.

2. The three major kinds of mechanical weathering are frost wedging, salt-crystal growth, and exfoliation. The three major kinds of chemical weathering are hydrolysis, oxidation, and carbonation.

3. Mass wasting refers to the downslope movement of sediment, soil, and rock under the force of gravity. Mass-wasting processes are differentiated on the basis of how quickly they move and the relative saturation of sediments.

4. Rockfall and soil creep are mass-wasting processes that are associated with unsaturated sediments.

5. Flows occur when sediments become saturated and then move under the force of gravity. Flow processes include solifluction, earthflows, mudflows, and debris flows.

CHECK YOUR UNDERSTANDING

1. Why is the term *weathering* appropriate for the process it describes?

2. What does the term *differential weathering* mean and why does it occur?

3. Why is chemical weathering more likely to occur in warm/wet environments, whereas frost action mostly occurs in colder regions?

4. Why does frost wedging cause physical or mechanical weathering? How is it similar to the effect that plant roots and large temperature changes have on rock? Why do these processes increase the likelihood of further weathering of a rock mass?

5. How does the process of unloading lead to exfoliation?

6. How are the processes of hydrolysis and carbonation similar? How are they different?

7. What is acid rain and why is it most closely associated with the northeastern part of the United States and eastern Canada?

8. Define the concept of mass wasting.

9. Why is the process of soil creep considered to be a mass-wasting process?

10. Describe the process of solifluction and its environmental context.

11. Why is slumping more likely to occur in humid regions than arid regions? What particular rock type—sandstone or shale—is most commonly associated with slumping? Why?

12. What is the difference between a mudflow and a debris flow? How would the process of human-induced deforestation make a mudflow or debris flow more likely?

ANSWERS TO VISUAL CONCEPT CHECKS

VISUAL CONCEPT CHECK 14.1

This weathering pattern was most likely produced by thermal expansion. It is certainly the result of mechanical weathering because the rock is merely breaking up into smaller fragments. Freeze-thaw is a mechanical process, but is unlikely to have much of an effect in southern Texas.

VISUAL CONCEPT CHECK 14.2

The mass-wasting process pictured here is rockfall. The rocks at the base of the cliff broke free from the exposure above and fell to their current location.

CHAPTER FIFTEEN

GROUNDWATER AND KARST LANDSCAPES

Have you ever considered the significance of freshwater in your daily life? It is used in many ways, including for drinking, cooking, and washing, as well as for farming and recreation. Water is critical to life on Earth in many different ways, yet people often take it for granted. How often do you think about where freshwater comes from and the various places where it is stored? Do you ever even consider what it would be like to run out of water? If you happen to live in the humid eastern half of the United States, you may never think about these questions. In many other parts of the world, however, freshwater is a precious commodity that people think about every day.

Recall from Chapter 6 that water covers about 71% of the Earth's surface. In subsequent chapters, we examined how water moves in the air and within the ocean. Now we focus on the very small portion of the water on Earth that is stored just below the surface as groundwater. This form of water may very well be the dominant supply of freshwater where you live.

Groundwater is water that is stored below Earth's surface. Understanding groundwater is important because much of our water supplies come from this source. Groundwater also shapes the ground below you, forming fascinating subterranean chambers such as this cave in France. This chapter focuses on the characteristics of groundwater and the impact that it can have on the landscape, both above and below the Earth's surface.

CHAPTER PREVIEW

Movement and Storage of Groundwater

Human Interactions with Groundwater
On the Web: *GeoDiscoveries*
Hydrologic Cycle and Groundwater

Karst Landforms and Landscapes

The Big Picture

LEARNING OBJECTIVES

1. Describe the way that groundwater is absorbed into the ground and stored within it.

2. Compare and contrast the various components of the groundwater model.

3. Explain the geography and formation of the High Plains Aquifer, and human interactions with it.

4. Understand the concept of groundwater contamination.

5. Discuss the different processes that form caves, and how these contribute to the development of karst topography.

Movement and Storage of Groundwater

The logical place to begin a discussion of groundwater is with a brief review of the hydrologic cycle presented in Chapter 7. If you recall, the hydrologic cycle refers to the way that water is stored in various reservoirs (such as rivers) and how it moves to other places (like the oceans). Earth's groundwater system is also part of the hydrologic cycle because it is a place where water can move to, be stored for a period of time, and flow to some other reservoir.

In this context, it is time to examine the various parts of the groundwater system and establish some fundamental terminology. Start first by considering what happens to water after it falls to Earth as some form of precipitation. In most circumstances, this water is absorbed by the soil, where it is stored in the *soil–water belt* (Figure 15.1). Remember from Chapter 11 that water is absorbed into the soil by infiltration, which occurs through pore spaces between sediment grains (Figures 11.3 and 11.4), along pathways associated with the soil structure, or down plant roots. When the soil is very dry, water is held tightly to sediment grains as **hygroscopic water** that is largely

unavailable for plant absorption (Figure 15.2). Recall that the *wilting point* is the threshold at which, for any given soil, water is no longer available for plant uptake. With increasing precipitation, the pore spaces begin to fill with *capillary water*, which is water held to soil particles by molecular attraction. Some of this water is lost to the atmosphere through evaporation, but plants absorb a great deal of it before the water is returned to the air through the process of transpiration (see Figure 15.1). The combined processes of evaporation and transpiration are often referred to as *evapotranspiration*.

If precipitation continues to the point where soil pore spaces become completely filled with water, the soil is considered to be at its **field capacity**. Any excess water that does not run off the surface is then free to move more deeply down into the sediment or rock, under the force of gravity, as gravitational water. As this water moves downward, it first passes through an area called the **unsaturated zone** (see Figure 15.1 again), which is the portion of the rock or sediment mass where pore spaces may be partially filled with water, but also contain air (i.e., oxygen and other gases).

As water continues to percolate downward, it will eventually reach a depth where pore spaces are entirely filled with water and will stop moving. This part of

---

Hygroscopic water   *Soil water held so tightly by sediment grains that it is unavailable for plant use.*

Field capacity   *The maximum amount of water the soil can hold after gravitational water has moved away.*

Unsaturated zone   *The area between the soil–water belt and the water table where pore spaces are not saturated with water.*

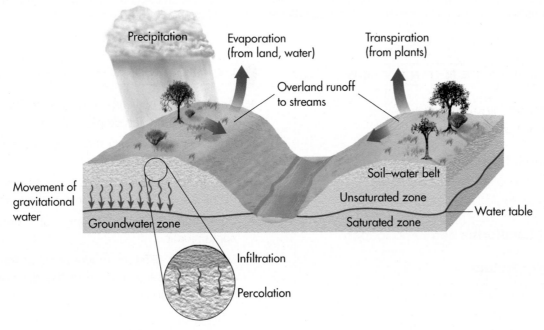

Figure 15.1 The groundwater model. Precipitation is initially stored in the soil–water belt until it becomes saturated. After that occurs, water can either run off the surface in streams or percolate through the unsaturated zone to the saturated zone.

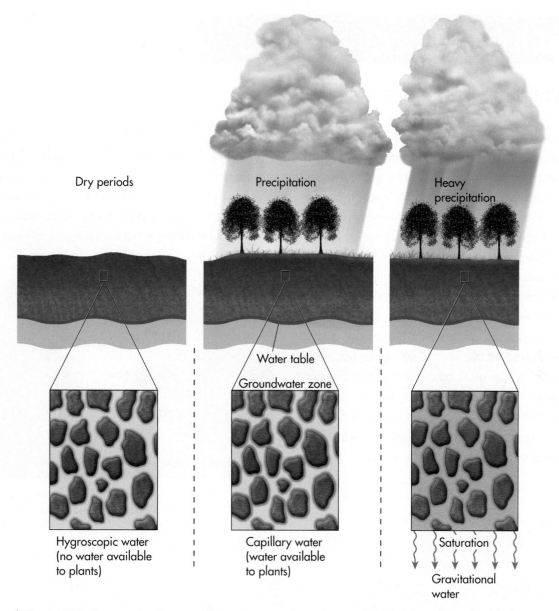

Figure 15.2 Stages of soil water. During dry periods, soil grains tightly hold onto hygroscopic water. With increasing rainfall, the pores begin to fill with capillary water. Once field capacity is reached, pores are full and excess water can run off the surface and flow downward as gravitational water.

the groundwater system is called the **saturated zone** to reflect these conditions. Immediately below the saturated zone is a feature called an **aquiclude** that consists of some kind of impermeable rock mass, such as shale or granite, which keeps water from moving further downward. The top of the saturated zone is called the **water table** and marks the boundary between the groundwater below and the unsaturated zone above. In most places, the water table is not visible because it is beneath the ground. Sometimes, however, topographic

---

Saturated zone *The zone of rock below and including the water table where pore spaces are completely filled with water.*

Aquiclude *An impermeable body of rock that may contain water but does not allow transmission of water through it.*

Water table *The top of the saturated zone.*

NATURAL LAKES

Have you ever visited a region containing many natural lakes? Such places are fascinating ecosystems and are prime vacation spots for people. They are also favored zones of development, specifically for second homes and weekend cabins. Perhaps your family has such a cabin. Although natural lakes are very attractive places to be, few understand their origin.

An excellent example of a region with numerous natural lakes is the state of Minnesota, which is famous as the "Land of 10,000 Lakes." The majority of these lakes formed in topographic depressions created by glacial ice between about 18,000 and 10,000 years ago. This landscape, as well as other similar places in the upper Midwest, looks this way due to an intimate and complex relationship between topography and local and regional water tables. A geographer or hydrologist can tell by looking at the map of Douglas County, for example, that the

elevation of the water table must be high to create so many lakes. Examination of an air photo of a typical Minnesota landscape further reinforces this impression. The next time you are in Minnesota, or a place where lots of natural lakes occur, try to look at the landscape in a more holistic way and think about the local groundwater supply.

(b)

A geographer knows these lakes in Minnesota mean that the regional water table is sufficiently high to fill depressions with water.

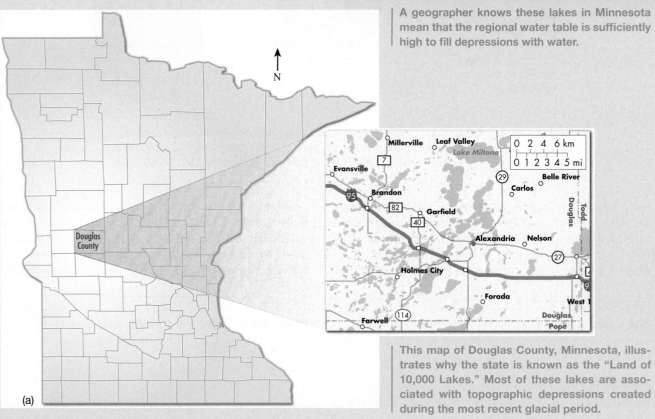
(a)

This map of Douglas County, Minnesota, illustrates why the state is known as the "Land of 10,000 Lakes." Most of these lakes are associated with topographic depressions created during the most recent glacial period.

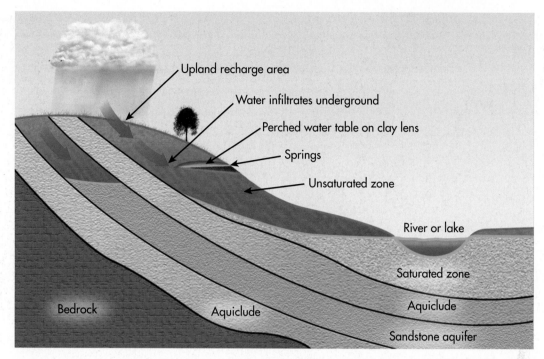

Figure 15.3 Water tables and aquifers. A water table is the top of the saturated zone of rock and soil. Depending on the kind of rock deeper in the ground, water may collect in saturated zones called aquifers, enclosed by layers of impermeable rock called aquicludes. An aquifer is a saturated zone that holds enough water to serve as a source for irrigation or drinking water.

depressions or stream channels dip below the level of the water table, as you can see in Figure 15.3. Where such topographic relationships occur, a stream or lake will be present.

In some places, the groundwater system is simple and consists of a single saturated zone of broad regional extent that overlies an aquiclude. At other places, however, the situation may be more complex with a layered assemblage of saturated zones, aquicludes, and unsaturated zones (Figure 15.3). Where such complexity occurs, saturated zones deep within the system may be slowly *recharged* by water gradually moving from the saturated zones higher in the column or from sources of water far away. In still other settings, a saturated zone may be highly localized because it occurs in association with an isolated aquiclude such as a body of clay-rich sediment deposited by an ancient glacier or stream. Such a localized saturated zone is called a **perched water table** because it is located above the primary groundwater body. Regardless of the setting of a specific saturated zone, or whether it is regionally extensive or more isolated, if it is sufficiently large to be a significant source of water for people, it is called an **aquifer**.

Human Interactions with Groundwater

The High Plains Aquifer

The most extensive aquifers occur in regions where abundant sandy sediments or sandstone rocks are found across a broad geographical area. Aquifers are one of the most important natural resources on Earth. Perhaps the best-known and most heavily utilized aquifer in North America is the *High Plains Aquifer*. Also known as the *Ogallala Aquifer*, this groundwater reservoir is found in the Great Plains region of the United States and underlies portions of eight states, ranging from South Dakota to Texas (Figure 15.4). The High Plains Aquifer is largely formed in the thick, porous deposits of unlithified sand and gravel of the Ogallala Formation, which consists of sediments that weathered and washed out from the Rocky Mountains between 19 and 5 million years ago. These sediments covered the western part of the Great Plains and slowly filled with water from rainfall and snowmelt during the ensuing

---

Perched water table *A localized area of saturated sediment that is elevated above the regional water table by a zone of impermeable rock or sediment.*

Aquifer *A geological formation that contains a suitable amount of water to be accessed for human use.*

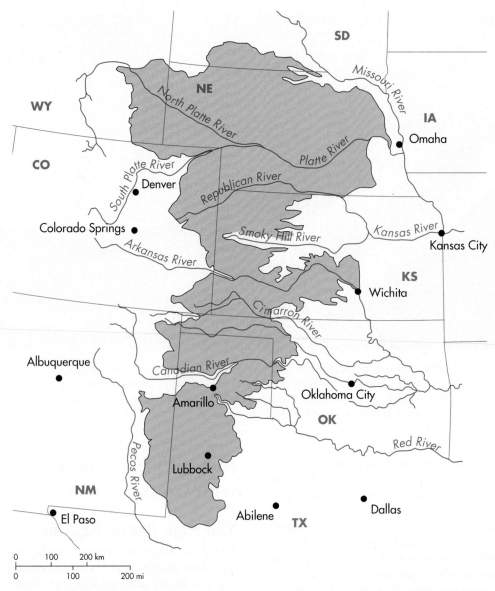

Figure 15.4 Distribution of the High Plains Aquifer. This vast underground reservoir of water underlies eight states in the central United States.

millennia. Much of this groundwater accumulated during the ice ages, between about 1.6 million and 10,000 years ago, when the regional climate was much wetter than it is today.

As you can see in Figure 15.4 and Table 15.1, the High Plains Aquifer is huge. It covers an area of about 280,000 km² (174,000 mi²) and contains 1,315,000 hectare feet (3,250,000 acre feet) of water. In other words, the aquifer contains sufficient water to cover over 1,000,000 hectares (3,000,000 acres) to a depth of about 30 cm (12 in.). This volume is approximately equivalent to one of the larger Great Lakes, such as Lake Huron. As you can see from Table 15.1, the thickness of the saturated zone varies dramatically across the region. In Nebraska, for example,

the average saturated thickness is currently 104 m (342 ft). In Colorado, on the other hand, the average saturated thickness of the aquifer is only 24 m (79 ft). This variation exists because the sediments are sandy all the way to the surface in Nebraska, resulting in rapid infiltration of water, whereas in Colorado they are covered in most places by relatively impermeable deposits.

The Great Plains is one of the most important agricultural regions on Earth, producing much of the world's wheat, corn, and soybeans. One of the primary reasons for this high productivity is that the soils are mostly Mollisols, which, as we saw in Chapter 11, have thick, organic-rich A horizons and B horizons laden with nutrients such as calcium and magnesium. Despite

TABLE 15.1 **Characteristics of the High Plains Aquifer***

Characteristic	Unit	Total	CO	KS	NE	NM	OK	SD	TX	WY
Area underlain by aquifer	mi²	174,050	14,900	30,500	63,650	9,450	7,350	4,750	35,450	8,000
% of total aquifer area	%	100	8.6	17.5	36.6	5.4	4.2	2.7	20.4	4.6
% of each state underlain by aquifer	%	—	14	38	83	8	11	7	13	8
Average area-weighted saturated thickness in 1980	ft	190	79	101	342	51	130	207	110	182
Volume of drainable water in storage in millions 1980	acre-ft	3250	120	320	2130	50	110	60	390	70

* Modified from E. D. Gutentag et al., 1984, *Geohydrology of the High Plains Aquifer in Parts of Colorado, Kansas, Nebraska, New Mexico, Oklahoma, South Dakota, Texas, and Wyoming*, U.S. Geological Survey Professional Paper 1400-B, p. 63.

the semi-arid environment, these fertile soils attracted large numbers of farmers to the region in the late 1800s and early 1900s during the period of western settlement. These farmers quickly learned, however, that the region could be an unforgiving place for agriculture because precipitation could be inconsistent and droughts often decimated crops.

The early settlers saw abundant evidence for a large groundwater supply. First, many artesian wells were scattered throughout the region. An **artesian well** occurs when the pressure applied by incoming water elsewhere in the groundwater system causes water to move up to the surface. This pressure can be enhanced by a variety of factors, such as if a saturated zone is sandwiched between two aquicludes, or because of faulting (Figure 15.5a). Second, settlers also noticed that many streams had a fairly consistent supply of water in them, even during periods of relative drought. This consistent flow resulted because stream channels intersected with the aquifer's water table, providing a steady source of water. These patterns are common not just to the High Plains Aquifer, but to any place where the water table is high and intersects depressions and streams.

For the first half of the 20th century, access to the High Plains Aquifer was limited and largely confined to artesian and hand-excavated wells. However, the access to water changed dramatically in the early 1950s with the advent of a new technology called *center-pivot irrigation*. These systems consist of motorized pumps that tap into the aquifer and suck up water like drinking from a straw. They then distribute the water to farm fields through an elevated sprinkler system that extends to the edge of the field. This sprinkler system slowly rotates around the pump on a set of wheels so the field is consistently supplied with water. Most such systems cover a circular area approximately 64 hectares (160 acres) in size, although some are larger.

Development of center-pivot systems was a major advancement for the agriculture-based economy of the region, enabling farmers to control the semi-arid environment by providing a reliable supply of water to high-yield crops such as corn that would otherwise not consistently grow there. In fact, these systems have been so wildly successful that tens of thousands of them are currently in operation around the region, forming a patchwork of circles that are easily seen from the air

---

Artesian well   *A well in which water from a confined aquifer rises to the surface through natural pressure.*

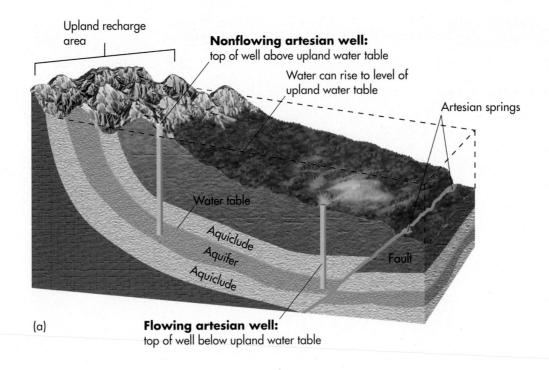

(a)

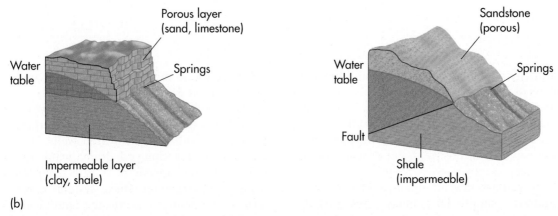

(b)

Figure 15.5 Artesian wells and springs. (a) When aquifers are confined between aquicludes, water may rise to the surface in an artesian well due to pressure applied by incoming water elsewhere in the aquifer. Where depressions or channels intersect the water table—for instance, at a fault—lakes and streams develop. (b) A spring is a place where water naturally seeps out of the ground. This may occur because the groundwater lies between aquicludes, as in an aquifer, or water may flow horizontally along the contact between overlying coarse-textured sediments and an aquiclude.

(Figure 15.6). The next time you fly across the country, be sure to notice this important part of the nation's cultural landscape and consider how it supplies much of your food.

Although the rapid expansion of center-pivot irrigation has enabled farmers to be successful in marginal regions such as the High Plains, the long-term environmental consequences of extensive groundwater mining in the region are significant. The most obvious impact of extensive groundwater extraction is that the water table has dropped in many places. This decline can be seen at the local level with any individual well as a **cone of depression**

in which water is drawn down in a circular pattern from the aquifer surrounding the well (Figure 15.7a). If pumping from the well ceases, the cone will fill back in, both by recharge from the surface and by water seeping into the empty pores from the surrounding saturated zone.

In many regions, however, groundwater extraction is increasing because more wells are being installed and

---

Cone of depression *The cone-shaped depression of the water table that occurs around a well.*

Figure 15.6 Aerial view of center-pivot irrigation. Center-pivot pumping systems such as these in Nebraska were developed in the mid-20th century and are almost ubiquitous now in parts of the Great Plains. Look for them when you fly over the country.

water is used year-round for farming and ranching. As a result, the drain on underground water has become serious at a regional level in many parts of the world. Figure 15.7b shows the impact that two wells—a deep well and a shallow, seasonal well—may have on the groundwater supply in a particular area. In the Great Plains of North America, such drawdown has occurred in thousands of places, causing a significant decline in the elevation of the water table in many locations. In areas within southwestern Kansas and the Texas Panhandle,

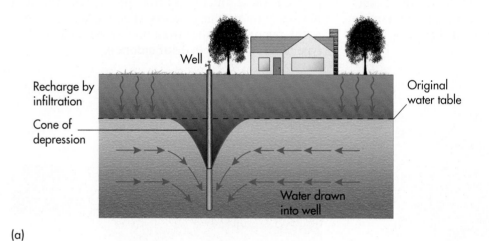

(a)

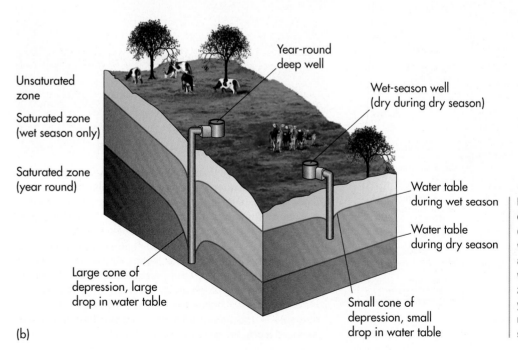

(b)

Figure 15.7 Groundwater depletion. (a) When water is drawn from an aquifer, the water table is depressed in a conical shape. (b) Deep wells draw water from zones that are saturated year round, draining far more water than shallower, seasonal wells.

for example, the average drop in the level of the water table was more than 12 m (40 ft) between 1980 and 1996 (Figure 15.8a). This rapid decline has occurred in these regions because the withdrawal rate is as much as 22 times greater than the recharge rate, which averages only 0.5 cm/year (0.2 in./year). Signs of this drop in the water table can be seen throughout the region. For example, many of the streams have dried up (Figure 15.8b) because the water table is now below the base of the channel.

As a result of the heavy reliance on groundwater mining, the High Plains Aquifer may disappear in many places by the middle of the 21st century. This estimate is based on the amount of groundwater extracted from the aquifer compared to its recharge rate due to infiltration of precipitation. Estimates are similar for the depletion of other aquifers around the world. In the Sahara Desert, for example, estimates suggest that aquifers there may last 50 years. Put another way, natural resources that took hundreds of thousands of years to evolve will disappear essentially within a timeframe of 200 years (from the mid-1800s to 2050) due to human impact.

In response to this important environmental issue, many farmers in the Great Plains are shifting to dryland agriculture, which is a crop-rotation system designed to conserve soil moisture by leaving fields fallow (or crop-free) every other year. Other farmers are attempting to maximize short-term profits by mining the aquifer until it disappears in their region, at which time they will shift to dry-land agriculture or leave farming entirely.

Subsidence

The High Plains Aquifer demonstrates what happens to the water table elevation when groundwater is tapped excessively. In addition to these changes, which occur deep within the ground, adjustments sometimes take place on the surface when large amounts of groundwater are removed, either for irrigation or to supply drinking water in heavily populated regions. One kind of impact that can occur is that sediments become compacted because water is lost from pore spaces. The most obvious surface change occurs when the land sinks due to the removal of water from underlying aquifers. This type of response is called **subsidence**.

Subsidence *The settling or sinking of a surface as a result of the loss of support from underlying water, soils, or strata.*

(b)

Figure 15.8 Drawdown of the High Plains Aquifer. (a) Due to excess groundwater mining, the water table has dropped significantly in an area that ranges from southwestern Kansas to the Texas panhandle. (*Source:* U. S. Geological Survey.) (b) The effect of aquifer depletion has been significant on many streams. For example, prior to aggressive groundwater extraction in the region, Rattlesnake Creek in south-central Kansas was a perennial, spring-fed stream. Now, it is dry most of the time, as shown here, because the water table is far below the level of the stream bed. The students in the foreground are studying the sand dunes exposed along the old stream.

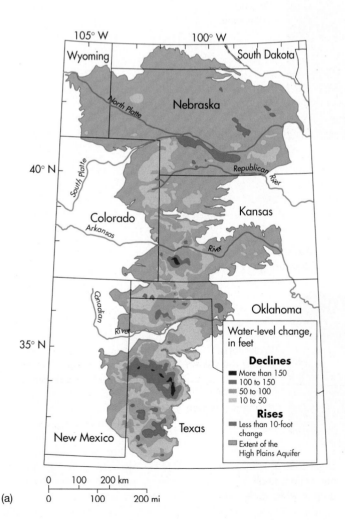
(a)

Subsidence is a particularly significant problem in and around large cities. A great example of this kind of impact can be seen at Venice, Italy. This beautiful city is approximately 1400 years old and was constructed on a landscape consisting of over 100 low-lying islands in a lagoon in the northern part of the Adriatic Sea. Underlying the islands are about 1000 m (about 3300 ft) of weakly cemented deposits of sand, gravel, clay, and silt. The coarse sands and gravels of these deposits contain fresh groundwater resources that have been extensively exploited for human use. The associated subsidence of the landscape has been so great that many Venetian homes are now at the shoreline and flooding is common. In addition, salt water from the Adriatic is beginning to seep into the city's water supply. Similar problems have also recently occurred in China and California. In the San Joaquin Valley of California, for example, groundwater removal for irrigation has caused the ground to subside as much as 4.8 m (16 ft) in places over a 35-year period. This issue will be discussed more thoroughly in Chapter 20.

Groundwater Contamination

Another major environmental problem associated with the relationship between humans and groundwater is contamination by pollutants. Our industrial society produces enormous amounts of solid and liquid waste. Until the latter half of the 20th century, no regulations governed the disposal of this waste. As a result, much of it was simply carted off to large holes in the ground, some excavated and some natural,

where it was perhaps burned and buried. Infiltrating rainwater would interact with chemical compounds in the waste and carry pollutants into the saturated zone, where they would then flow along the various groundwater paths to wells and streams (Figure 15.9). Given that drinking water is often obtained from groundwater, contamination can be a significant health risk.

Since the establishment of the Environmental Protection Agency (EPA) in the mid-1970s, a more systematic effort has been made to control the disposal of solid waste in the United States. One important way in which solid waste disposal is intensely regulated by the EPA occurs in highly engineered landfills. In contrast to the traditional "hole-in-the-ground" approach used previously, modern engineered landfills are designed to eliminate the flow of toxic chemicals and other contaminants into the groundwater supply. For example, landfill engineers can line the base and sides of the landfill with impermeable material so that groundwater contamination is minimized. Any toxic chemicals and contaminants produced in the landfill are pumped to a settling pond where the waste is oxidized and neutralized. In spite of these measures, landfills and other hazardous waste facilities often remain very controversial.

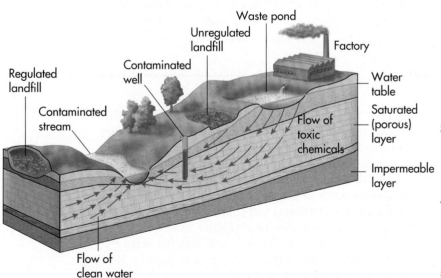

Figure 15.9 Movement of groundwater contaminants. Infiltrating water at solid waste disposal sites, such as waste ponds and unregulated landfills, carries toxic chemicals into the groundwater, which flows along natural pathways to pollute wells and streams.

This photograph shows the shoreline of a natural lake. Given what you see here, which one of the following statements is accurate?

a) The shoreline represents the place where the ground level meets the local water table.
b) The water table lies below the level of the lake bottom.
c) The local water table is very low.
d) A tremendous amount of groundwater mining has occurred in the area.

 **GeoDiscoveries**   www.wiley.com/college/arbogast    **WILEY PLUS**

Hydrologic Cycle and Groundwater

Now that the concept of groundwater has been discussed, go to the *GeoDiscoveries* website and open the module *Hydrologic Cycle and Groundwater*. This module reviews the processes that have been described so far in this chapter. It begins with the hydrologic cycle and demonstrates how water moves from one reservoir to another. The module also animates the relationship between evaporation, precipitation, and the way that groundwater is stored and moves. Review how the water table fluctuates and how groundwater depletion impacts the water table. A final exercise tests your understanding of groundwater terminology. Once you complete the module, be sure to answer the questions at the end to assess your knowledge of this topic.

Karst Landforms and Landscapes

Have you ever been to a famous cave system, such as Mammoth Cave in Kentucky or Carlsbad Caverns in New Mexico? If so, you explored how groundwater can shape the subterranean landscape over a long period of time. These effects are visible not just beneath the ground, but can also be seen at the surface in the form of depressions and, sometimes, even hills.

A region that contains a lot of caves and other soluble rock features is said to be a landscape with **karst topography**. Karst topography is most closely associated with extensive, thick deposits of limestone. As discussed in Chapter 14, these rocks dissolve easily through the weathering process of carbonation. This erosion produces a complex system of joints that allows water to flow from the surface into the underlying rock. Vegetation also plays a role because it provides organic acids that enhance the solution process. When these conditions occur in humid regions, such as the eastern part of the United States and southeastern China, karst landscapes can be widespread (Figure 15.10). They also occur in some regions that are now arid but were humid at some point in the geologic past, such as New Mexico and northwestern Africa.

Karst topography *Terrain that is underlain by soluble rocks, such as limestone and dolomite, where the landscape evolves largely through the dissolution of rock.*

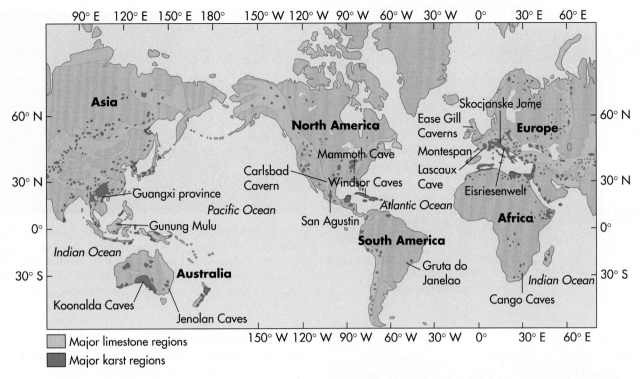

Figure 15.10 Location of major limestone and karst regions on Earth. Large areas of limestone and karst occur on every continent.

Caves and Caverns

Perhaps the best-known karst landforms are **caves**, which consist of underground voids in rock that are sufficiently large for people to enter and explore. Although most caves are small, some cave systems are enormous and contain numerous passageways and large chambers. Such a large cave system is called a *cavern*. To give you an idea of how large chambers can be, consider that The Big Room at Carlsbad Caverns is about 110,000 m^2 (1,184,030 ft^2) in area.

People have been aware of caves and caverns for a long time. In Europe and Asia, for example, caves provided shelter for prehistoric people during the Pleistocene Epoch when gargantuan glaciers covered much of the landscape. Evidence for these occupations can be seen in Europe in the number of beautiful paintings (known as *cave art*) that adorn the cave walls. These paintings, as well as burials deep within the caves, have given archaeologists important clues about the way people lived and the animals that they hunted.

Caves and caverns form due to the complex interaction of climate, groundwater, and local streams over time. Figure 15.11 illustrates the evolution of these landscapes. Cave formation begins during a time (stage 1) when the water table is high because the stream has

Cave *A cavity in rock, produced by the dissolution of calcium carbonate, that is large enough for someone to enter.*

not yet downcut a deep valley by erosion. This concept of downcutting will be discussed in more detail in Chapter 16, but for now, suffice it to say that one of the things that streams do as they evolve is carve out deep valleys. In the context of karst topography, the important relationship is that the water table mirrors the behavior of the stream because groundwater flows to the lowest point on the landscape—in other words, the river channel.

During stage 1, the process of carbonation is concentrated in the deposits of limestone that occur just below the water table. The resulting dissolved sediments are then carried by underground streams to the surface channel. These kinds of streams evolve because water concentrates along joints in the rocks where it can move freely. These cracks slowly enlarge through dissolution and the water can thus flow more quickly as an underground stream. Over time, the process of carbonation and associated groundwater flow creates myriad passageways and caverns at the water table boundary. In stage 2, the stream has cut a valley into the rock, lowering the water table. Groundwater cuts large channels and caverns into the rock as it flows through to the stream. At this time, these features are submerged in groundwater. Finally, in stage 3, as the stream continues to downcut and deepens its valley, the water table also lowers still farther. As a result of this lower water table, the formerly submerged passageways of stage 1 are now in the unsaturated zone and thus become open caves and caverns.

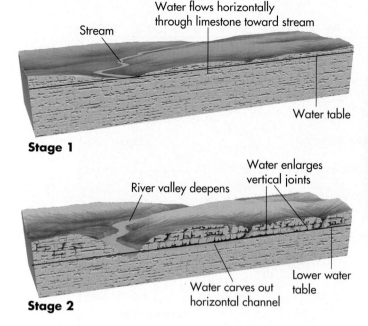

Stage 1

Stream

Water flows horizontally through limestone toward stream

Water table

River valley deepens

Water enlarges vertical joints

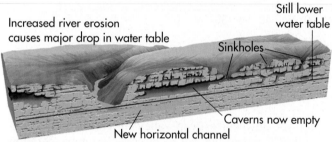

Lower water table

Water carves out horizontal channel

Stage 2

Increased river erosion causes major drop in water table

Still lower water table

Sinkholes

Caverns now empty

New horizontal channel

Stage 3

Figure 15.11 Cave and cavern evolution. In Stage 1, limestone below the water table dissolves, forming submerged caverns. As the stream downcuts in Stage 2, the water table drops in response, causing water to carve out channels in the limestone as the water flows to the river. In Stage 3, the water table drops farther, enough so that the formerly submerged caverns are now located in the unsaturated zone. Surface water percolates into these open features, resulting in the formation of stalagmites and stalactites.

Once stage 3 is complete, the cave or cavern has formed and the dominant processes shift to those that occur within the chamber. In order for the cavern to be preserved, some kind of more resistant cap rock must be present to minimize the amount of water percolating into it and keep the roof from collapsing. If such preservation occurs, features such as stalactites (hanging rods), stalagmites (upward-pointing rods), columns, and drip curtains can begin to form within the cave or cavern (Figure 15.12). These unusual formations evolve in caves where large quantities of surface water percolate downward into the roof of the open cavern. As this water moves through the overlying rock, the solution becomes enriched in dissolved carbonates before it slowly drips into the cave. Stalactites in the roof of a cave form in much the same way that an icicle lengthens downward during a winter thaw. Stalagmites, in contrast,

Figure 15.12 Drip formations at Carlsbad Caverns in New Mexico. Here, dripping water enriched in calcium carbonate has formed numerous stalactites, stalagmites, and columns in the Papoose Room. This form of deposition occurs when percolating groundwater dissolves limestone and drips, leaving calcium carbonate behind.

develop when water drips to the cave floor. This water evaporates, leaving the calcium carbonate behind in a deposit that slowly grows upward over time.

Karst Topography

Karst landforms can also evolve at the surface. The most common surface landform is a **sinkhole**, which is a depression that occurs in regions of cavernous limestone (Figure 15.11). Sinkholes form when caves enlarge so much that their ceilings collapse due to the force of gravity, causing the surface rocks and sediment to sink, often suddenly. They can also form due to excessive groundwater extraction because caves that were previously filled with water become empty, causing the weight of the overlying land to increase to a critical point. Sinkholes frequently enlarge for a period of time after their initial formation because surface water tends to funnel into them. Sometimes this surface flow becomes a **disappearing stream**, which flows down into a sinkhole from the surface and then enters underground channels. Often, this downward-flowing stream causes a pipe to evolve that connects the sinkhole with the caverns and passageways beneath.

At other times sinkholes remain detached from the underlying cavern and simply fill with water, appearing as ponds on the landscape and in topographic maps (Figure 15.13). In regions where karst activity occurs in densely populated regions, sinkholes are a significant geological hazard because they can unexpectedly

---

Sinkhole   *A topographic depression that forms when underlying rock dissolves, causing the surface to collapse.*

Disappearing stream   *A surface river or stream that flows into a sinkhole and subsequently moves into an underground river system.*

(a)

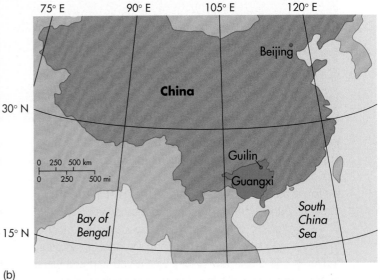

(b)

Figure 15.13 Sinkhole topography. (a) Sometimes sinkholes develop that extend below the water table, forming ponds. **(b)** Sinkholes show up clearly on topographic maps as ponds, where they are filled with water, and as depressions, where contour lines make concentric circles.

swallow homes or portions of highways. A spectacular form of karst topography occurs in humid subtropical areas where thick beds of limestone and high water tables exist. This kind of karst is called *tower* (or *haystack*) *karst* and results in spectacular landscapes. In order for tower karst to evolve, karst processes must have operated for a long period of time, during which a variety of subterranean caves and passageways developed. Over time, these features gradually enlarge so that sinkholes and various other collapse structures form. If enough of these karst features exist, the effect is to produce a number of towering hills that are the unaffected remnants of the block of solid limestone that once existed. Streams also play a role in the development of tower karst because they cut steep valleys into the regional rock mass.

An excellent place to see fantastic tower karst is in southeastern China (Figure 15.14), near the city of Guilin in the northern part of Guangxi province. This area lies within the humid subtropical climate region and is affected by the southeastern monsoon. The Guilin area is covered by carbonate rocks that may reach a thickness of 4600 m (~15,000 ft). Two major surface karst landforms occur in the region: (1) the peak-cluster depressions (Fengcong) and (2) the peak-forest plain (Fenglin). These areas collectively cover an area 2429 km² (938 mi²) in size. The Fengcong landscape consists of a group of peaks separated from each other by collapse depressions, whereas the Fenglin areas are a group of isolated peaks separated by flat ground created by streams. Taken together, these hills literally tower over the landscape, reaching heights of 200 m (660 ft) above the surrounding lowlands and giving the region a beautifully surreal quality. Close to the mainland United States, the only place to see tower karst is the northern part of Puerto Rico.

(a)

Figure 15.14 Tower karst in China. (a) This surreal landscape of isolated peaks near Guilin, China, formed when thick limestones were dissolved and collapse structures formed on a massive scale. **(b)** The Guangxi province in southeastern China is especially known for its tower karst.

(b)

Given what you see in this photograph, which one of the following statements is accurate?

a) The local water table intersects with the ground level at this place.

b) This photograph shows the formation of tower karst.

c) The top of a cave collapsed, forming a sinkhole.

d) The local water table was much higher at some point in the past.

KEY CONCEPTS TO REMEMBER ABOUT KARST LANDSCAPES

1. Karst landforms and landscapes form in association with groundwater and include features such as caves and sinkholes.

2. Caves begin to form during a period of high water table when flowing groundwater dissolves carbonate-rich limestones through the process of carbonation. Caves become open when the water table falls, usually in association with stream downcutting or climate change.

3. Once caves are opened, drip processes form features such as stalagmites and stalactites.

4. Sinkholes are surface depressions that result when underlying caves collapse, causing subsidence.

5. Tower karst forms in humid areas when numerous sinkholes and collapse structures develop, leaving prominent hills rising above the landscape.

THE BIG PICTURE

Water is stored and moves around Earth in a variety of ways. This chapter focused on the different ways in which water moves and is stored within the ground. In the next chapter, we focus on how water is stored and flows at the surface in streams. Although this form of water comprises a tiny part of the hydrologic cycle, rivers are responsible for most of the sculpting that occurs on the Earth's surface. In addition, a river is one of the few places where you can see water on the surface actually flow as part of the hydrologic cycle.

Over the course of your lifetime, water issues will become increasingly more important as human demands on water increase. Not only will this demand impact groundwater resources, as you have seen, but it will also impact rivers and streams. To fully comprehend some of the critical events that occur in association with stream systems, or thoroughly enjoy a canoe ride down a nice river, you must recognize how streams function and what landforms they produce. The next chapter will provide you with that background.

SUMMARY OF KEY CONCEPTS

1. Groundwater refers to water that is stored within the ground. Some water is stored in soil pore spaces to form the soil–water belt. If the soil and associated pores are saturated, water is free to move downward under the force of gravity. When water percolates downward, it first encounters the unsaturated zone.

2. At some depth the sediment pore spaces are saturated with water, forming the saturated zone. The sediments in this zone usually consist of sand and gravel that are underlain by an aquiclude of impermeable clay or shale. The top of the saturated zone is called the water table. In places where the groundwater supply is sufficiently large for people to use for irrigation or drinking water, the saturated zone is called an aquifer.

3. Groundwater can become contaminated when chemicals from old landfills or dumps slowly move into the system.

4. Karst landforms and landscapes form in association with groundwater and include features such as caves and sinkholes.

5. Caves begin to form during a period of high water table when flowing groundwater dissolves carbonate-rich limestones through the process of carbonation. Caves become open when the water table falls, usually in association with stream downcutting or, perhaps, climate change.

6. Tower (or haystack) karst topography forms when numerous caverns and subterranean passageways collapse, leaving individual high hills.

CHECK YOUR UNDERSTANDING

1. What is the soil–water belt and how do the terms hygroscopic water, capillary water, and field capacity relate to this term?

2. What is the unsaturated zone and where is it located in the groundwater model?

3. What role does an aquiclude play in the formation of a saturated zone?

4. Why do sandy and gravelly deposits make good aquifers?

5. Describe the relationship among topography, the water table, and the development of natural lakes.

6. Describe the physical geography of the High Plains Aquifer.

7. What is a cone of depression and how is it related to declining water tables in places like the Great Plains?

8. Why are many streams going dry in places where aggressive extraction of groundwater occurs?

9. Why are modern landfills more environmentally safe than unregulated disposal sites of years past?

10. What is karst topography and how do caves evolve?

11. How does the formation of a sinkhole relate to the process of land subsidence and the formation of tower karst?

ANSWERS TO VISUAL CONCEPT CHECKS

VISUAL CONCEPT CHECK 15.1

The answer to this question is *a*. The shoreline represents the place where the ground level meets the local water table. Such a natural lake exists when depressions dip below the level of the water table.

VISUAL CONCEPT CHECK 15.2

The answer to this question is *d*. The local water table was much higher at some point in the past. This cave formed when carbonate rocks dissolved when the water table was higher. After the water table fell, the cave was left behind.

CHAPTER SIXTEEN

FLUVIAL SYSTEMS AND LANDFORMS

This chapter investigates water that flows across the surface of Earth in streams. Whether these flowing bodies of water occur in the largest rivers or smallest creeks and brooks, they are collectively viewed as *fluvial systems*. This chapter discusses the input of water into these systems, how it behaves within them, and the predictable outputs such as landforms that result.

In many cases, streams are a visible reflection of the more distant geological past. In humid regions, streams have likely been flowing along the same basic course for thousands of years, shaping the landscape gradually through erosion and deposition. Where streams occur in semi-arid to arid regions, in contrast, they may flow only during the wet season, or only every few years when intense rains fall. When they do flow,

however, they may dramatically alter the landscape because they have incredible erosive power. If you like watching streams, or relish canoeing on them, you will enjoy this chapter.

CHAPTER PREVIEW

Overland Flow and Drainage Basins
  On the Web: *GeoDiscoveries The Rhine River*

Hydraulic Geometry and Channel Flow

Fluvial Processes and Landforms
  On the Web: *GeoDiscoveries Bedload Transport and Braided Streams*

430

Flowing water is the most important geomorphic agent on Earth, shaping the landscape in many interesting ways. Humans also interact with streams on a daily basis, as this image of London's Thames River shows, using them for transportation arteries, supplies of drinking water, and energy production. This chapter focuses on the way that water flows across the Earth's surface in streams, the landforms that are produced by these processes, and some of the ways that people interact with them.

On the Web: *GeoDiscoveries* *The Graded Stream*

On the Web: *GeoDiscoveries* *Stream Meandering*

On the Web: *GeoDiscoveries* *Fluvial Geomorphology and Stream Processes*

Human Interactions with Streams

The Big Picture

LEARNING OBJECTIVES

1. Describe the concept of a drainage basin (or watershed) and how such basins are arranged from a hierarchical perspective in the United States.

2. Discuss the various components of hydraulic geometry and the way that water flows in channels.

3. Compare and contrast the concepts of base flow, bankfull discharge, and flood discharge in the context of recurrence interval.

4. Explain how hillslope erosion processes operate and the formation of rills and gullies.

5. Define the concept of a graded stream and how the longitudinal profile of such a stream changes over time.

6. Discuss why base level is an important variable in the behavior of a stream.

7. Compare and contrast the character and development of braided streams with meandering streams.

8. Explain the concept of a stream meander, how water flows within it, and how the landscape is shaped at this location.

9. Describe the various ways that humans impact streams and how they have both positive and negative outcomes.

431

Overland Flow and Drainage Basins

Streams are important features for a variety of reasons. In the context of the hydrologic cycle, they connect the interior of continents with oceans. Water that falls as precipitation in high mountainous areas, for example, will ultimately flow back to the sea. Flowing water is also the most powerful geomorphic agent on Earth, shaping the landscape in both subtle and profound ways. Streams are important to humans because they are sources of water, both for direct consumption and agriculture. In addition, people have long used streams as transportation networks to move economic goods from one place to another. As a result, rivers such as the Rhine in Europe, the Nile in Africa, the Mekong in Southeast Asia, the Amazon in South America, and the Hudson and Mississippi in the United States, have played critical roles in the evolution of human societies. Figure 16.1 shows the location of major rivers around the world.

Origin of Streams

A good place to begin considering streams is their origin and the sources of water within them. Some streams flow all year long and are called *perennial streams*, whereas others contain water for only a short time and are thus *ephemeral (or intermittent) streams*. Although you might think stream water comes directly from precipitation that falls into the channel, it is only a negligible contribution. In most places, a more important source of stream water is water that seeps out of the ground where stream channels intersect the water table, as was briefly discussed in Chapter 15 (see Figure 15.3). In some areas, such as high mountain areas, stream water comes from melting snow and glaciers.

Still another important source for streams is water that is stored in lakes and ponds, which are fed both by precipitation and groundwater. If the water level in such a lake reaches a sufficient depth, it may spill over at a low spot on the landscape, called an *outlet*, and become the source of a river. Many rivers start at an outlet from a major lake. An excellent example is Lake Victoria in east central Africa, which is the source of the Nile River. The lake straddles the borders of Uganda, Kenya, and Tanzania (Figure 16.2a) and lies in a depression created by the East Africa Rift system (Figure 13.9). Approximately 67,850 km² (26,197 mi²) in size, Lake Victoria is the second largest freshwater lake in area on Earth (behind Lake Superior in North America). Excess water spills over an outlet at the north end of Lake Victoria (Figure 16.2b) and marks the beginning of the Nile River. The Nile River is about 10,730 km (6670 mi) long, making it the longest river in the world, and flows northward through Sudan and Egypt to the Mediterranean Sea. The best-known river in North America, the Mississippi, also begins at a lake outlet, Lake Itasca in Minnesota. At this place, you can literally walk across the Mississippi River!

Another significant source of stream water is surface runoff. Runoff occurs most commonly in association with wet periods when soils are saturated and pore

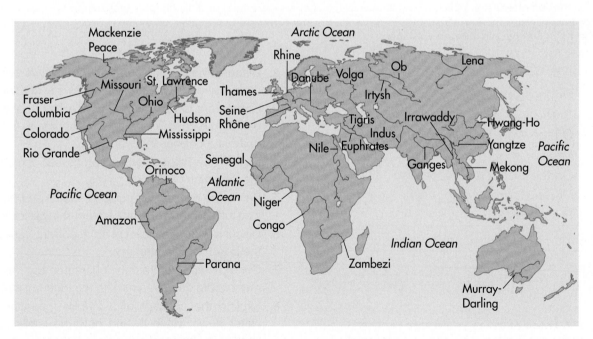

Figure 16.1 Important rivers on Earth. Rivers are channelized bodies of flowing water that connect the continental interior with the ocean. As a result, these rivers, and streams like them, have been used as transportation systems over the course of human history.

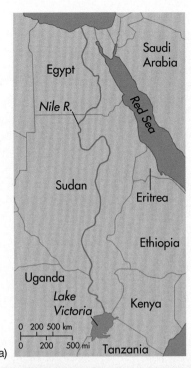

(a)

(b)

Figure 16.2 Source of the Nile River. (a) The Nile River begins at the north end of Lake Victoria and flows northward for about 10,730 km (6670 mi) before it ultimately reaches the Mediterranean Sea in Egypt. (b) This image shows Bujagali Falls in Uganda. This waterfall is located just downstream of Lake Victoria, which is the source of the Nile River.

spaces can no longer absorb additional precipitation. As discussed in Chapter 15, these periods are times of groundwater recharge because water is free to move from the soil downward through the force of gravity. At the same time that water is moving downward through the soil, water runs off the surface and toward the

stream as overland flow (Figure 16.3). You can see the process of runoff clearly in any parking lot during and shortly after a strong storm. The asphalt keeps water from soaking into the soil beneath; as a result, the water flows across the surface in sheets toward a drain of some kind. This type of runoff, called *sheet runoff*, also occurs in areas where slopes are very steep and rainfall simply does not have a chance to soak into the ground.

Drainage Basins

The next step in our discussion of stream systems is to consider how they are spatially organized. A good place to begin is to consider the concept of a **drainage basin**, which is the geographical area that contributes groundwater and runoff to any particular stream. Another term frequently used to define the same area is *watershed*.

To begin this discussion, some basic terminology must first be established. Streams in any two watersheds are separated by a topographic feature called a **drainage divide**, which is an area of elevated terrain that forms a kind of high rim around any given basin (Figure 16.4). Below this rim, the topography slopes downward into the core of the basin. As a result, all runoff and groundwater flow into a network of streams that collectively funnel into the **trunk stream**, which is the largest stream in the drainage basin. Each stream that flows into the trunk stream, or a stream that flows into another stream, is called a **tributary**. Tributaries are

Drainage basin *The geographical area that contributes groundwater and runoff to any particular stream.*

Drainage divide *An area of raised land that forms a bordering rim between two adjacent drainage basins.*

Trunk stream *The primary stream of a drainage basin.*

Tributary *A stream or river that flows into a larger stream or river.*

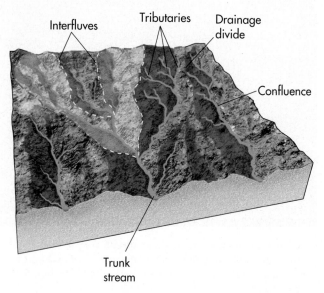

Figure 16.4 The concept of drainage basins. Also called watersheds, drainage basins are separated from one another by high points on the surrounding landscape called divides. Each large drainage basin contains numerous tributaries, which themselves are separated from each other by interfluves.

Figure 16.5 A typical river confluence. The Smoky Hill River flows into this image from the upper left, whereas the Republican River enters from the upper right. They join at a confluence in the center of the photograph to form the Kansas River, which is the trunk stream of the Kansas River basin, and flows toward the bottom (east) of the image.

themselves separated from one another by relatively low topographic rims called **interfluves**. The point where a tributary joins the trunk stream, or any stream for that matter, is referred to as a **confluence** (Figure 16.4).

Drainage basins range from several hectares to thousands of square kilometers in size, with numerous watersheds contained within larger ones in a nested fashion, as you can see in Figure 16.4. Notice, for example, that the small drainage basin portrayed in light orange is nested within the drainage basin highlighted in yellow. In other words, water that drains into the orange basin ultimately flows into the larger stream contained within the yellow basin at a confluence. This stream ultimately flows into the trunk stream at another confluence, like the one pictured in Figure 16.5. This image shows the confluence of the two major tributaries of the Kansas River, which is the trunk stream of the Kansas River basin in the central United States. To see the geographic relationship of these streams, refer back to the map in Figure 15.4. In this map note that the Republican River drains the northern portion of Kansas and southern Nebraska, whereas the Smoky Hill River drains western Kansas. These two streams meet in eastern Kansas to form the Kansas River.

Interfluves *Topographic high points in a drainage basin that separate one tributary from another.*

Confluence *The place where two streams join together.*

Now put this discussion of drainage basins into the context of major watersheds in the United States. The map in Figure 16.6 not only shows the geographic position and size of the largest drainage basins, but also indicates into which ocean the trunk stream of any particular basin flows. The largest watershed in North America is the Mississippi River basin (bold outline), which drains most of the United States (and a small part of Canada) and funnels water to the Gulf of Mexico. The Mississippi watershed contains several other large basins, including the Ohio, Arkansas, and Missouri, which are named after their trunk streams. Another way to describe these trunk streams is that they are major tributaries of the Mississippi River. Each of these trunk streams, in turn, contains a number of smaller tributaries. A major tributary of the Missouri River, for example, is the Kansas River shown in Figure 16.5. In other words, one can accurately say that the Smoky Hill and Republican Rivers are not only tributaries of the Kansas, but also of the Missouri River and, ultimately, the Mississippi River.

The western divide of the Mississippi basin is the Rocky Mountains. This divide is the famous *continental divide* because it separates the Mississippi watershed and its drainage to the Gulf of Mexico from streams that deliver runoff into the Pacific Ocean, such as the Colorado and Columbia Rivers. The eastern divide of the Mississippi basin is the Appalachian Mountains, which is also a continental divide because it separates Atlantic-flowing drainages such as the Delaware and Hudson basins from the Gulf-flowing waters of the continental interior.

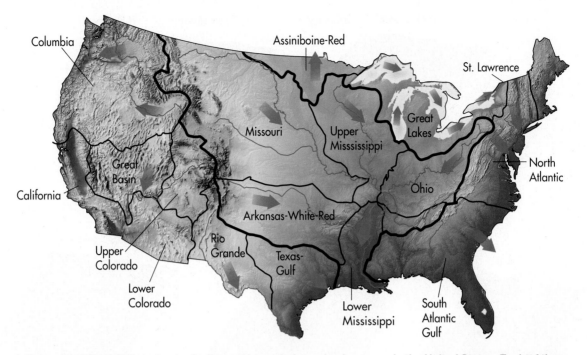

Figure 16.6 Major U.S. watersheds. Several large drainage basins occur in the United States. Each of these basins contains an array of nested drainage basins like those shown in Figure 16.4. Note the direction of flow for each trunk stream. The Mississippi River is the largest drainage basin in the country and is outlined in bold.

Drainage Patterns, Density, and Stream Ordering

In the context of individual drainage basins, it is often useful to characterize the spatial configuration of the stream network. Such an assessment sheds light on the evolution of the basin or the nature of the rocks and sediments that lie within it. One method is to look at the drainage pattern, which is the way that the various streams within a drainage basin are spatially organized. The five primary drainage patterns are shown in Figure 16.7.

1. **Dendritic**—A branching, tree-like drainage pattern evolves in areas of uniform rock resistance and structure, with little distortion by folding or faulting. Notice in Figure 16.7a that the streams develop a random branching network similar to a tree.

2. **Rectangular**—A rectangular drainage pattern occurs when joints and faults steer streams at right angles to one another. This pattern occurs because water flows preferentially to these zones of weakened bedrock where the water more freely erodes. Notice in Figure 16.7b the angular nature of the drainage pattern.

3. **Trellis**—A trellis drainage pattern (Figure 16.7c) is a system of streams that develops in areas such as the Ridge and Valley Province in the Appalachian Mountains where rocks are folded. In this area, major streams tend to flow parallel to one another in

adjacent valleys within the folded mountain belts. Minor tributaries flow into larger streams at right angles.

4. **Radial**—In a radial drainage pattern (Figure 16.7d), streams radiate outward from a central point, forming a spoke-like pattern of rivers. This kind of pattern tends to evolve where streams flow away from rounded upland areas such as a volcano.

5. **Deranged**—A distinctively chaotic drainage pattern (Figure 16.7e), characterized by irregular direction of stream flow, few tributaries, swamps, and many lakes, that develops in recently glaciated terrain.

Another way to characterize a drainage basin is by calculating the drainage density of a watershed. **Drainage density** refers to the relative density of natural drainage channels in a given area. This value is calculated by dividing the total length of all streams in the basin by the area of the basin:

$$\text{Drainage density} = \frac{\text{Total length of all streams}}{\text{Area of drainage basin}}$$

Figure 16.8 shows an example of differences in drainage density. This diagram shows two watersheds (A and B) of the same size that clearly contain a variable number of streams. Which area, basin A or B, has

---

Drainage density *The measure of stream channel length per unit area of drainage basin.*

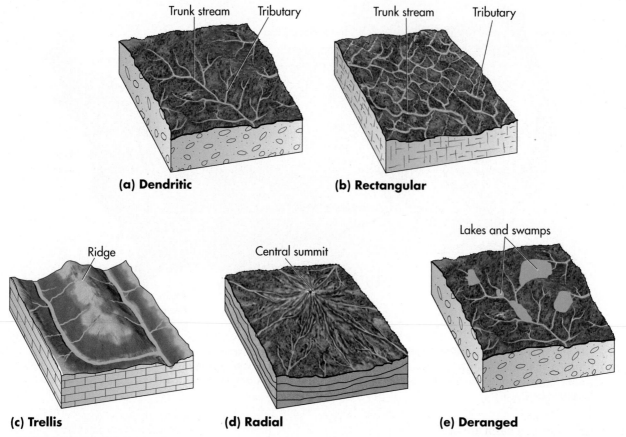

Figure 16.7 Drainage patterns. Stream networks are configured in five primary drainage patterns: (a) dendritic, (b) rectangular, (c) trellis, (d) radial, and (e) deranged.

the higher drainage density? The answer is basin B. Why might such a difference occur? One reason might be that basin B occurs in a more humid climate, so more streams exist in the area to handle the extra runoff of this region. A second possibility is that basin A occurs on a landscape lying over porous sands that allow water to seep rapidly into the ground, whereas watershed B occurs in a region of relatively impermeable clay that causes water

to flow overland. The measurement of drainage density provides a hydrologist or geomorphologist with a useful numerical measure of runoff potential and how much streams have shaped the landscape.

A third way to characterize a drainage network is through the process of *stream ordering*. Streams within any drainage network can be arranged into a hierarchy based on their size. Stream ordering is useful because it

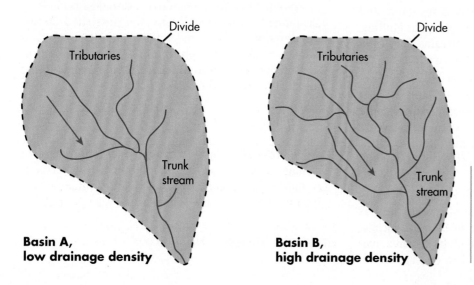

Figure 16.8 Drainage density differences. Basin A has a lower drainage density than basin B. This may result from differences in climate or type of underlying rock or sediment. The arrows indicate the general direction of flow in each watershed.

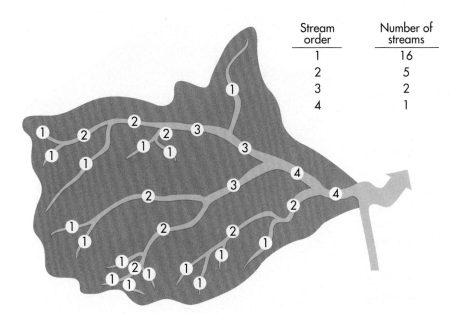

Stream order	Number of streams
1	16
2	5
3	2
4	1

Figure 16.9 Stream ordering. Stream ordering classifies the hierarchy of channels in a drainage basin. Stream order changes only where two streams of the same order merge. A positive relationship exists between stream size and stream order, with streams of progressively higher order portrayed by progressively thicker lines on the diagram.

provides a good relative measure of a stream's place in a basin hierarchy, which is typically a function of how many tributaries occur in any given watershed. Ordering the streams of any basin is a simple task, as shown in Figure 16.9. An important rule to remember when calculating stream order is that it changes whenever two streams of the *same order* join at their confluence. The smallest tributaries of a basin are first-order streams. At the confluence of the first-order tributaries, a second-order stream forms. Subsequently, two second-order streams join to form a third-order stream.

It is tempting to think that the ordering of a stream changes *whenever* it meets another stream, regardless of the second stream's order. For example, Figure 16.9 shows several places where first-order streams flow into streams of a higher order, such as a second- or third-order stream. You can also see a confluence where a second-order stream meets a third-order stream. Remember that stream order changes *only* at the conflu-

ence of streams of the same order. For example, just because a first-order stream meets a second-order stream that does not mean the second-order stream becomes a third-order stream. It does not; rather, it remains a second-order stream. The second-order stream becomes a third-order stream *only* where it meets with another second-order stream.

A last point to make about stream ordering is that a positive relationship exists between increased stream order and stream size. In other words, you can expect a second-order stream in a drainage basin to be larger than a first-order stream. Similarly, a third-order stream is bigger than a second-order stream. This increase occurs because a stream of a higher order contains the combined flow of the lower ones. Thus, higher-order streams logically contain more water than lower-order streams. In Figure 16.9, this geographical relationship is portrayed by using progressively thicker lines to represent streams of progressively higher order.

 www.wiley.com/college/arbogast

The Rhine River

The next section of the chapter focuses on hydraulic geometry and channel flow. You can preview this material by going to the *GeoDiscoveries* website and accessing the module *The Rhine River*. This video follows the course of the famous Rhine River in Europe from its origin in the Swiss Alps to where it meets the Atlantic Ocean in Holland. As you watch the video, notice the various landscapes through

which the Rhine flows. Listen for the technical terms, such as *graded stream*, *rapids*, and *valley*, and try to understand their meaning as they will be discussed in this text. As you continue through the chapter, keep this video in mind as an excellent example of a stream. Once you complete the module, be sure to answer the questions at the end of the module to test your understanding of this concept.

1. Streams consist of channels in which water flows downhill. Runoff refers to water that runs off land surfaces into streams.

2. Streams are fed by runoff or groundwater, or may have a large lake as their source.

3. A drainage basin is the area that contributes runoff or groundwater to any given stream. Drainage basins are separated from one another by divides, which are topographic rims consisting of relatively high ridges and hills.

4. Drainage density refers to the relative density of streams in comparable areas and is influenced by climate, underlying bedrock, or both.

5. Streams are classified by their order, which ranges from low (such as 1 or 2) to high (such as 5 or 6). Stream order changes only at the confluence of two streams of the same order.

Hydraulic Geometry and Channel Flow

Now that the organization of stream networks has been discussed, let's look at it more closely as well as the character of flowing water within channels. This assessment falls within the framework of *stream hydrology* and focuses on the geometrical attributes (or *hydraulic geometry*) of river channels, including the measurable width, depth, velocity, slope, and discharge (Figure 16.10). These concepts are respectively defined as:

- w = channel width (how wide the channel is in which the stream is flowing).

- d = depth (how deep the channel is from the water surface to the bed).

- v = velocity (how fast the water is moving in the channel; velocity is greatest in the middle of a channel where water flows freely and is slower on the channel bottom and sides due to frictional forces).

- s = slope (how steep the slope [also called *gradient*] is on which the stream is flowing).

- Q = discharge (how much water is flowing in the channel).

As you might imagine, these variables are intimately related to one another in a holistic way. For example, the velocity of a stream closely correlates to the channel slope, with streams on steep gradients moving

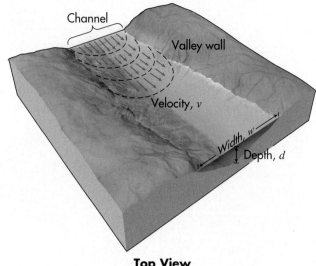

Top View

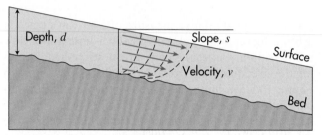

Side View

Figure 16.10 Hydraulic variables. Streams have a measurable width, depth, velocity, slope (or gradient) down which they flow, and discharge. Velocity is slowest on the sides and bottom of the channel due to the effects of friction.

more quickly than those on shallower slopes. Increased discharge naturally leads to greater channel width and depth to accommodate higher flow.

Fluctuations in Stream Discharge

As far as human interactions are concerned, the most important hydraulic variable is stream discharge Q, which refers to the amount of water flowing in the channel. This value is calculated by the simple equation

$$Q = w \times d \times v$$

Discharge is measured with a device called a *stream gauge* that can either be portable or fixed in one place in a gauging station. The United States Geological Survey has a network of stream-gauging stations to monitor discharge in all the major streams in the country. To see this network, examine Figure 1.4 in Chapter 1. Stream discharge is typically expressed in cubic meters per second (m³/sec) or cubic feet per second (ft³/sec). With increases in width, depth, and velocity, stream discharge becomes greater.

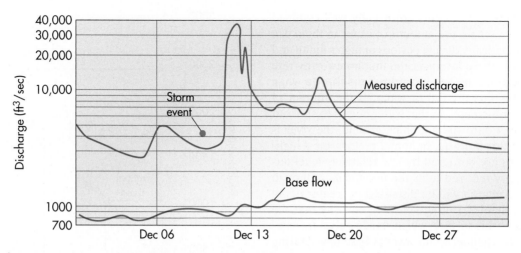

Figure 16.11 Rappahannock River hydrograph. This hydrograph illustrates how discharge varied during December 2003 (blue line) in the Rappahannock River near Fredricksburg, Virginia. Base flow (red line) indicates the rate of flow maintained by groundwater influx. Note the peak that occurs in association with increased runoff caused by a storm event. (*Source:* U.S. Geological Survey.)

In most streams, discharge varies over the course of the year, depending on cycles of rainfall and relative drought. The amount of discharge that serves as the baseline level for any stream at any given place and time is **base flow**, which is the flow rate that is sustained solely by groundwater influx. During periods of heavy precipitation, the amount of runoff to the stream increases, causing stream discharge to increase above the base flow amount. Figure 16.11 is a typical **stream hydrograph** that shows how stream discharge varies over the course of time due to changes in groundwater influx and runoff. This particular hydrograph is from the Rappahannock River in Virginia and was collected over the course of a month from a measuring station near the city of Fredricksburg. Notice the relationship between base flow and the actual amount of discharge. Also observe the fluctuations in discharge that occurred over the month. In particular, note that stream discharge does not immediately increase during a storm event. This *time lag* between the period of heaviest rainfall and maximum stream discharge is typical of most streams because it takes a period of time for the water to flow overland and reach the channel. This concept will be discussed more thoroughly later in the chapter.

Base flow *The amount of stream discharge at any given place and time that is solely the product of groundwater seepage.*

Stream hydrograph *The graphical representation of stream discharge over a period of time.*

VISUAL CONCEPT CHECK 16.1

The famous Danube River, shown here, is the second longest river in Europe (the Volga is the longest), flowing from Germany to the Black Sea. Which one of the following statements is correct and can be determined accurately?

a) The Danube is a high-order stream.
b) The discharge of the Danube is low.
c) The watershed of the Danube is small.
d) The slope of the Danube is very steep.

Flooding Although streams are usually well confined within their channels, the amount of water in them can increase dramatically during periods of high rainfall and runoff. During these intervals, the first noteworthy rate of heightened stream flow is **bankfull discharge** when the channel is literally full of water. If the amount of runoff further increases so that water begins to spill out of the channel onto the adjoining ground, the river is said to be at **flood stage**. In humid regions, most high-order streams are bordered by low, relatively level terrain commonly referred to as the *floodplain*. Although floodplain formation will be discussed in greater detail later in this chapter, it is sufficient to now say that the stream creates this surface, which stands only a few meters above the channel water level at base flow. During the wet season, the combination of significant surface runoff and groundwater influx can cause the floodplain to become inundated by excess discharge. In most humid locations floods are annual to semi-annual events and are part of the natural evolution of river systems. In addition, they are important in the maintenance of the flora and fauna that live along a river (which, if you recall from Chapter 10, is called the *riparian zone*).

Occasionally, periods of extremely heavy rains occur in a watershed, causing the streams to rise to spectacular levels that result in great damage to human property and even sometimes significant loss of life. Fortunately, these kinds of floods are rare, with a close statistical relationship between the *return period*—in other words, the frequency of the event—and its magnitude. To put it more simply, more extreme floods happen less frequently than do lower-magnitude floods. Hydrologists who monitor flood frequency may describe a particular large flood as being, for example, a 30-year or 50-year event; in other words, it is expected to statistically occur once every 30 or 50 years. They also refer to the probability of a particular discharge being exceeded by an even larger one.

To see how flood magnitude, return period, and statistical probability can be used to predict stream behavior, examine Figure 16.12, which shows data from the Skykomish River at Goldbar, Washington. The Skykomish has a base flow of about 200 m³/sec (about 7060 ft³/sec), which statistically occurs every year and has a 99% probability of being equaled or exceeded on an annual basis. According to the graph, the 10-year peak discharge in this stream is about 1700 m³/sec (~60,000 ft³/sec) and this flow has a 20% chance of being equaled or exceeded each year. A more unlikely event is the 50-year flood, which has a discharge of approximately 2500 m³/sec

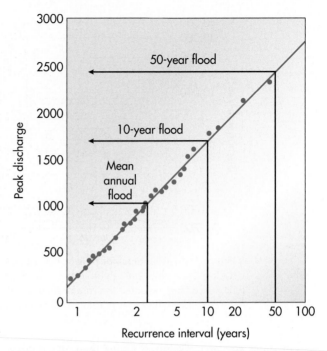

Figure 16.12 Flood frequency data for the Skykomish River at Goldbar, Washington. Each data point is the measured maximum discharge over a 53-year period.

(about 88,300 ft³/sec). This discharge has only a 2% probability of being equaled or exceeded on a yearly basis. This is not to say that a 50-year flood or even a larger one could not happen in successive years—it could. It is just not very likely from a statistical perspective.

1993 Mississippi Basin Floods Perhaps the best example of a major flood in the fairly recent past within the United States is the Mississippi River flood of 1993, which affected a large part of the upper Midwest. This flood was at least a 100-year event and has been classified by some hydrologists as a 500-year flood. It dominated the national media during June and July of that year because of its catastrophic impact on people living along the Mississippi River and its tributaries. The causes of this flood and the basin response provide an excellent example of how environmental variables are interrelated in a holistic way.

The stage for this massive flood was actually set during the preceding winter when unusually heavy rain and snow saturated the soils in the region. Following the wet winter, an unusual weather pattern developed during June and July in which a strong subtropical high-pressure system formed and stalled over the southeastern United States (Figure 16.13). Recall from Chapter 8 that atmospheric pressure systems in the midlatitudes rarely stay in one place but instead typically migrate from west to east in association with the westerly winds. The year 1993 was unusual because a high over the southeastern United States did not move and, in fact, remained stationary for about 2 months. This stalled high caused two significant

---

Bankfull discharge *The amount of discharge at which the stream channel is full.*

Flood stage *The level at which stream discharge begins to spill out of the channel into the surrounding area.*

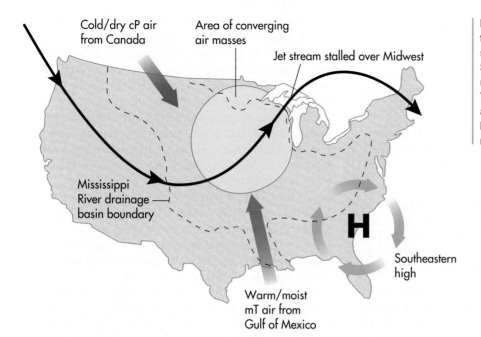

Cold/dry cP air from Canada

Area of converging air masses

Jet stream stalled over Midwest

Mississippi River drainage basin boundary

H

Southeastern high

Warm/moist mT air from Gulf of Mexico

Figure 16.13 Meteorology of 1993 flood. A high-pressure system stalled over the southeastern United States and persistently pumped humid mT air into the upper Midwest. This water vapor condensed and fell as rainfall along the stalled frontal boundary in the region. This intense rain caused widespread flooding.

things to occur. First, the midlatitude jet stream became "stuck" in a sense over the Midwest because it was blocked from moving to the east by the southeastern high. Second, water vapor from the Gulf was continuously pumped into the Midwest by the clockwise rotation of the southeastern high. Given that cool/dry cP air was steadily colliding with warm/moist mT air along the stalled jet stream in the upper Midwest, prodigious rains fell during June and July. During July 4 and 5, for example, between 10 cm and 15 cm (4 in. and 6 in.) of rain fell in a broad swath of southeastern Iowa. This kind of rain event was a common occurrence that summer as the region received over *200%* of normal precipitation for that time of year!

Given that the soils of the region were already saturated from the wet winter, most of the rainfall could not be absorbed. As a result, flooding was widespread, affecting most of the major streams and tributaries in Iowa, southeastern Wisconsin, western Illinois, northern Missouri, eastern Nebraska, and northeastern Kansas. Since all these streams progressively merge and ultimately flow into the Mississippi River, flooding along this stream was intense (Figure 16.14). Overall, this flood caused $15 billion of damage and was responsible for at least 50 deaths. Another interval of heavy flooding occurred in the upper Mississippi basin in 2001, although it was not as widespread.

(a)

(b)

Figure 16.14 Flooding along the Mississippi River in 1993. (a) Aerial photograph of flooding along the Mississippi River in 1993. (b) Landsat images of the confluence of the Missouri and Mississippi Rivers north of St. Louis during a normal year (left) and during 1993 (right). In these images, vegetation and urban areas appear in green and pink tones, respectively. The extensive flooding in 1993 appears as blue and black areas, indicating submerged floodplains.

1. Stream flow in a channel is defined by its hydraulic variables, including width, depth, velocity, slope, and discharge.

2. Hydraulic variables are interrelated with one another. For example, stream discharge is calculated by the equation $Q = w \times d \times v$. Increased slope results in faster-flowing streams.

3. Base flow is the average stream discharge in a given stream at any particular time of the year. Bankfull discharge occurs when the channel is full. Floods occur when discharge is sufficiently high that water spills out of the river channel.

4. Prediction of flood discharge is based on statistical probability. The return period refers to the frequency with which a particular discharge is expected to occur. High-magnitude floods occur less frequently than less intense discharges.

Fluvial Processes and Landforms

Now that the nature of stream channel flow has been discussed, let's next examine the landforms that streams produce. Landforms created by fluvial processes are called *fluvial landforms*. Although processes such as glaciation, wind, and ocean waves also act to shape the Earth's surface (and will be covered in subsequent chapters), running water is the single most important geomorphic agent on Earth. This process is driven by gravity, which is the most important geomorphic force on Earth.

Erosion and Deposition

Discussion of stream landforms requires an understanding of erosion and deposition as geomorphic processes because they act in concert to shape the landscape. Simply put, erosion occurs when sediment is removed from one place by a geomorphic process such as slumping, stream flow, or glaciation. Erosion is usually more intense in areas of high relief because the force of gravity enhances the ability of geomorphic processes to do work by increasing their power. Other factors that influence the process of erosion are vegetation and climate. Erosion tends to be less in areas of dense vegetation because the plants protect hillslopes from the effects of running water. Stream erosion is also limited in the driest deserts because very little running water occurs there. In fact, the most intense erosion tends to occur in semi-arid to subhumid zones where plant cover is relatively sparse but heavy rains that can move sediment sometimes occur.

In contrast to erosion, the process of deposition occurs when sediments that are being transported (after being eroded from someplace else) stop moving and are dropped. Deposition occurs when the transporting agent (such as wind or running water) simply loses the power to carry the sediment, which can happen for a variety of reasons. Although erosion and deposition often work in tandem to shape a landscape, one process or the other tends to dominate in any particular place at specific periods of time. Thus, it is possible to classify landforms generally as being either erosional or depositional in their nature. Erosional landforms are created when sediment, soil, or rock is stripped away by some geomorphic process. Depositional landforms, in contrast, form when sediment accumulates after being dropped.

Figure 16.15 shows a simplified example of these categories. Here, the mountain slopes have largely been shaped by erosion due to the high energy created by the steep relief. This process creates some distinctive landforms. The most prominent feature is a *peak*, which is the highest point on any given mountain. Peaks are typically separated by a lower landform called a *saddle*. As streams cut into the mountain slopes, they first create a shallow *gully*, which can enlarge to become a *ravine* and then, if sufficient time and erosion later occur, a deep and broad *canyon*. These features are separated from one another by a relatively high ridge called a *spur*, which is, in effect, a drainage divide. Over time, the eroded hillslope sediments are transported into the valley below, where they may be deposited on more level terrain within an *alluvial fan* or *river floodplain*. Here the relief lessens and geomorphic processes lose their power. These landforms will be described in more detail later in the chapter.

Fluvial Erosion on Hillslopes The logical place to begin a discussion of stream erosion is by focusing on hillslopes, which are the part of the landscape that is most intensely eroded by running water. Hillslopes are the most active zones of fluvial erosion because, as indicated before, the force of gravity is greatest in areas of high relief, which, in turn, causes running water to flow more quickly and thus with more energy. Sediment transport begins on hillslopes as soil erosion, initiated when the fluid drag associated with overland flow picks up sediment. In humid regions, dense vegetation protects slopes, and erosion occurs sufficiently slowly that the soil maintains all its horizons. However, the process of slope erosion accelerates in more arid regions, deforested areas, and agricultural fields because raindrops fall directly on bare soil. When this pounding occurs, sediment is loosened, lifted, and dropped into new positions through a process called *splash erosion* (Figure 16.16). It subsequently begins to flow down the slope as sheet runoff, mentioned earlier.

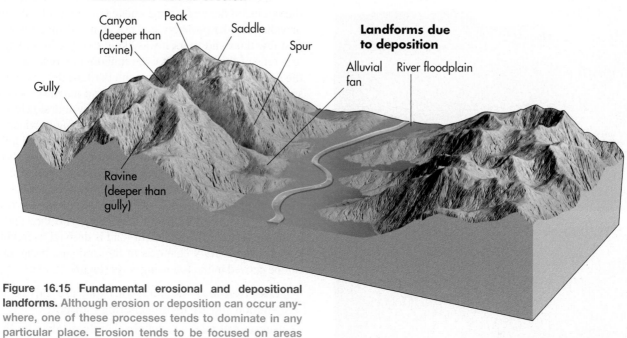

Landforms due to erosion

Canyon (deeper than ravine)

Peak

Saddle

Spur

Gully

Ravine (deeper than gully)

Landforms due to deposition

Alluvial fan

River floodplain

Figure 16.15 Fundamental erosional and depositional landforms. Although erosion or deposition can occur anywhere, one of these processes tends to dominate in any particular place. Erosion tends to be focused on areas of high relief, whereas deposition typically dominates in more level terrain.

As sheet flow moves across the surface, the total water volume and degree of turbulence increases down the slope. This turbulent water loosens sediment and results in the formation of tiny furrows in the ground called **rills** that look like parallel seams on the slope (Figure 16.17a). These rills, in turn, begin to merge as stream power and erosion intensifies, forming larger and deeper ditch-like features called *gullies* (Figure 16.17b).

Rills *Small drainage channels that are cut into hillslopes by running water.*

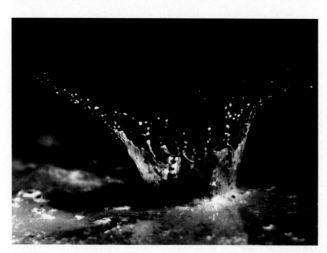

Figure 16.16 The process of splash erosion. Splash erosion occurs because large raindrops create a small crater when they impact on bare soil.

Sediments eroded from hillslopes by overland flow are either deposited quickly or carried to a stream where they can be transported great distances. Deposition of eroded slope sediments often takes place at the base of the hill where the slope becomes gentler, the effect of gravity lessens, and less power is available to carry sediment. Recall from Chapter 14 that sediment deposited in this manner is called *colluvium* and can form an apron-like landform that may be difficult to distinguish in humid regions of dense vegetation.

KEY CONCEPTS TO REMEMBER ABOUT FLUVIAL EROSION AND DEPOSITION

1. Erosional landforms are created where flowing water carries sediment away, as on a steep hillslope.

2. Depositional landforms are constructed through accumulation of sediment when the power to carry sediment lessens, as on a level floodplain.

3. Hillslope erosion is most active on unvegetated slopes and begins with a process called *splash erosion*, caused by raindrops.

4. Flowing water first begins to erode hillslopes in rills. As rills coalesce and enlarge, they form trench-like features called gullies, which eventually broaden into canyons.

(a)

(b)

Figure 16.17 Formation of rills and gullies. (a) Rills are small channels that evolve when heavy rains fall on hillslopes. (b) This gully formed due to the enlargement and merging of rills as a result of rapid runoff.

Stream Gradation

Let's now examine how landscapes evolve along larger, high-order streams. Although these streams do not vigorously erode like rills and gullies, they nevertheless shape the landscape in subtle ways that have profound effects over a long period of time. In fact, in most higher-order streams the combined processes of runoff, erosion, and deposition have been occurring for thousands of years.

Think of a stream in your neighborhood or city. Although its fundamental appearance (e.g., its width, depth, or valley width) probably has not changed much in the past few thousand years, it was likely a very different system tens of thousands (or even millions) of years ago in the early stages of its development. Perhaps this ancestral stream had a much steeper gradient and narrower valley than it currently does. Through time, the system slowly adjusted internally to the environmental setting in which it occurs—specifically, the amount of precipitation that falls, the size of the drainage basin, the density of vegetation on the landscape, and the rock and sediment types found there. Another way to look at this change is that streams ideally evolve in a way that allows them to ultimately balance the amount of sediment they carry with the capacity of the stream. This sediment load is derived from hillslopes and tributary channels in the drainage basin and can be carried in the following ways (Figure 16.18):

1. **Dissolved load**—Mineral ions that are carried in solution and are invisible during transport.

2. **Suspended load**—Sediment that floats along in the stream. This load usually consists of clays and silts that are held up by turbulent flow within the water.

3. **Bed load**—Larger particles such as sand and gravel that roll, slide, or bounce along the channel bed in a process called *saltation*. This form of transport dominates in mountain streams where slopes are steep and stream velocity is high.

It is usually possible to determine the dominant kind of sediment a stream is carrying by noting its fundamental channel characteristics. If a stream is transporting mostly suspended sediment, it will typically have a single, deep channel that gradually curves from side to side across the landscape (Figure 16.19a). Such a stream is called a **meandering stream** and often contains muddy-looking water due to the high concentration of suspended

Meandering stream   *A river or small stream that curves back and forth across its valley.*

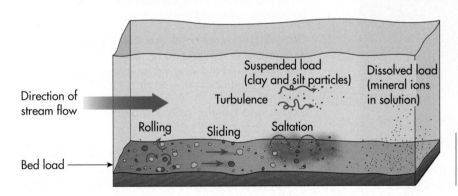

Figure 16.18 The sediment load of streams. Streams carry their sediment load as dissolved, suspended, and bed load.

Bedload Transport and Braided Streams

Braided streams are interesting fluvial systems that develop when the sediment they carry is predominantly bedload of sand and gravel. To better visualize the characteristics of these streams, go to the *GeoDiscoveries* website and select the module *Bedload Transport and Braided Streams*. This module contains three videos related to braided streams. The first video shows the nature of bedload transport in flowing water. Subsequently, a second video demonstrates the evolution of a braided system in an experimental setting. The final clip shows the Platte River in the central United States. This stream is one of the best-known braided systems east of the Rocky Mountains. Once you complete the module, answer the questions at the end to test your understanding of this concept.

sediment that was eroded from deposits containing abundant silt and clay. In the Mississippi River, for example, approximately 90% of the sediment carried within the river is suspended load. No wonder this river is frequently called *Big Muddy*! If a stream's dominant load is bed load, then it will have a **braided stream** pattern (Figure 16.19b), which consists of a maze of interconnected, wide, shallow channels that look like the braids in someone's hair. These streams have this form because the channel banks consist of easily eroded deposits of sand, gravel, and even cobbles that allow the stream to develop a wide, shallow form. In many cases, these coarse-textured deposits are derived from melting glaciers somewhere upstream. This relationship will be examined more thoroughly in Chapter 17.

The amount and kind of sediment a stream can carry at a given point are determined by a variety of complex factors within the system that are related to the concept of *stream gradation*. This concept is based on the fact that streams are capable of internal adjustments that allow them to maintain an equilibrium state with the surrounding environmental variables. In this context, it is essential to view streams as holistic systems that can respond in many different ways. Imagine, for example, if the vegetation on hillslopes were suddenly removed, as in a forest fire—how would the stream respond? Rill and gully erosion would increase and more sediment would be delivered to streams because vegetation would no longer be present to protect the hillslopes from heavy rain. In an effort to adjust to this higher sediment supply, much of

(a)

(b)

Braided stream   *A network of converging and diverging stream channels within an individual stream system that are separated from each other by deposits of sand and gravel.*

Figure 16.19 Stream load and channel characteristics. (a) A meandering stream usually consists of a single channel that winds back and forth across the landscape. (b) Braided streams such as this one in Alaska consist of a maze of wide, shallow channels because the dominant sediment load is sand, gravel, and even cobbles.

the sediment would initially be deposited on the channel bottom (because the stream could not carry it) in such a way as to increase the overall stream gradient. This kind of deposition, which raises the elevation of the channel bed, is called **aggradation**. The increased gradient caused by aggradation would result in higher flow velocity and power, which would, in turn, enable the stream to carry the increased sediment load. If hillslope vegetation was reestablished and erosion slowed through time, the stream would require less energy to carry the reduced sediment load. As a result, the stream would gradually erode its channel bed in a process called **degradation** (or *downcutting*) until a new gradient was established that enabled the stream to be able to carry just the lessened supply of sediment. A stream in this equilibrium condition is referred to as a **graded stream** because a balance exists among erosion, sediment load, deposition, and the capacity of flowing water.

Now examine more closely how a stream might evolve to reach a graded condition. Figure 16.20a shows several stages in the development of a hypothetical landscape. The diagram traces the evolution of a stream's **longitudinal profile** (or simply "profile"), which shows the change in the stream's gradient along its length. The beginning point for this developmental sequence could be a period of rapid uplift that elevated the upper part of the basin. Conversely, this sequence could be triggered by a major climate shift or a major environmental change such as the uncovering of the landscape following deglaciation. For simplicity, imagine that the sequence begins with some kind of uplift. Notice in Figure 16.20 that several profiles can be seen in the side of the illustration. Each one of these profile lines represents the gradient of the channel bed at a distinct period of time in the evolution of the system. For reference, let's say that the entire sequence of events takes about 1 million years to complete.

The top profile line represents the stream gradient following the initial uplift of the landscape above sea level. Soon after the landscape was raised, a stream would begin to develop to handle the runoff associated

---

Aggradation *The progressive accumulation of sediment along or within a stream.*

Degradation *The topographic lowering of a stream channel by stream erosion.*

Graded stream *An idealized stream that has achieved a balance between sediment transport and stream capacity.*

Longitudinal profile *A graph that illustrates the change in stream gradient in cross section along a stream from its source to its mouth.*

with heavy rains and groundwater flow. Given the nature of streams, runoff on this landscape would naturally flow downhill toward the sea and would follow the path of least resistance. The gradient would naturally be very steep in some places and relatively shallow in others. The steepest gradients would most likely be associated with resistant rock layers because stream erosion would be slow in these locations compared to the areas of less resistant rock. As a result, the stream gradient is higher where resistant rocks occur and less in areas of softer rock. A place where the stream gradient is significantly steeper than other places within the stream is called a *knickpoint*. Knickpoints are visible within streams as rapids and, where gradients are especially steep, as waterfalls. Where the gradient is low, overland flow would collect in lakes. Assuming that climate and vegetation cover remained essentially the same, the stream would spend perhaps its first 300,000 to 400,000 years adjusting to the variations in gradient that occur across the landscape. Note, however, that the overall stream gradient becomes smoother between the top two profiles.

One reason why stream gradients smooth is that knickpoints slowly retreat upstream as they wear down by stream erosion. A good example of how this process occurs is at a great waterfall like Niagara Falls at the Ontario–New York border (Figure 16.21). Niagara Falls is on the Niagara River, which flows from Lake Erie into Lake Ontario over a resistant layer of limestone that overlies a layer of easily eroded shale. This limestone was scraped clean and covered by a continental glacier until about 12,000 years ago, at which time the ice receded into Canada and an ungraded landscape remained. Since the ice melted, the falls have retreated more than 11 km (6.8 mi) at a rate of about 1.3 m/year (4.3 ft/year). The primary reason for this retreat is that the force of the water striking the plunge pool erodes the shale at the base of the falls. As the shale erodes headward over time, the overlying limestone collapses under its own weight and the falls retreat. This cycle has been repeated countless times in the past 12,000 years and continues to the present time. Today about 195,000 feet^3 of water pours over the falls every second. Most of the water flows over the Canadian (or Horseshoe) Falls, which measure about 790 m (2592 ft) in length and are about 49 m (160 ft) high. Although the American Falls are the same height, they are only approximately 305 m (1000 ft) long and thus have less water flowing over them.

Returning to the concept of a smoothing stream profile, let's examine our hypothetical stream about 500,000 years into its history when it becomes graded. In other words, it can carry the average sediment load in the basin. At this time, the stream gradient has the ideal longitudinal profile (see Figure 16.20a), with a steeper

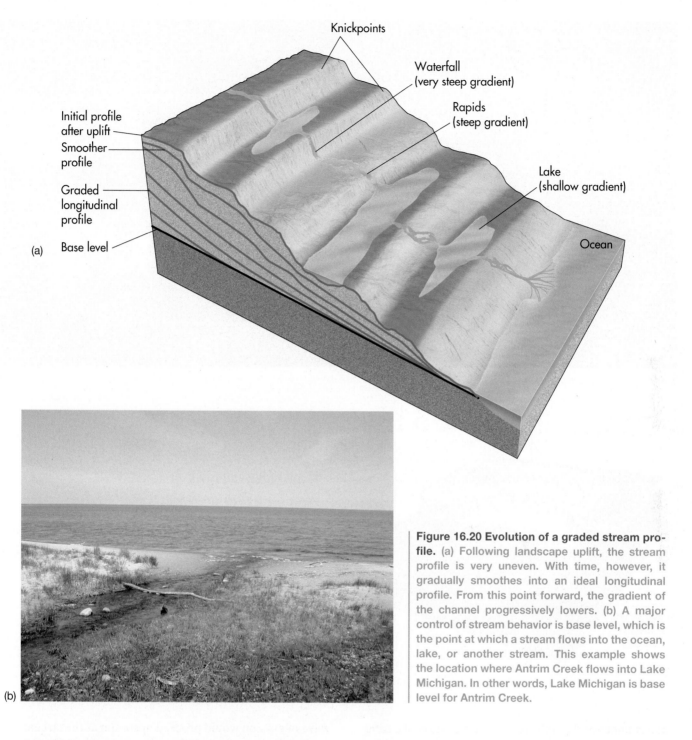

Figure 16.20 Evolution of a graded stream profile. (a) Following landscape uplift, the stream profile is very uneven. With time, however, it gradually smoothes into an ideal longitudinal profile. From this point forward, the gradient of the channel progressively lowers. (b) A major control of stream behavior is base level, which is the point at which a stream flows into the ocean, lake, or another stream. This example shows the location where Antrim Creek flows into Lake Michigan. In other words, Lake Michigan is base level for Antrim Creek.

slope in the upper reaches of the basin and a shallower slope downstream. From this point forward, this graded profile is maintained and gradually lowered as the landscape is further eroded.

Now let's turn our attention to the place where the stream enters the ocean. This place is called the stream's **base level** because it represents the lowest elevation at which a stream can erode its channel bed. This concept is easy to visualize as where a stream meets the ocean or large lake because the stream cannot erode its bed below the level of the water body in which it flows (Figure 16.20b). If it did, it would be flowing uphill to meet the ocean or lake, which is physically impossible. The concept of base level also applies on a more local level, for example, where one stream flows into another

***

Base level *The lowest level at which a stream can no longer lower its bed, because it flows into the ocean, a lake, or another stream.*

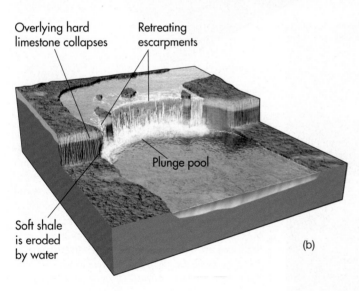

Figure 16.21 Niagara Falls. (a) Niagara Falls is a 51-m (167-ft) waterfall on the Niagara River. It actually consists of two waterfalls, the Horseshoe Falls on the Canadian side (to the right) and the American Falls (to the left). The eastern side of Lake Erie lies in the background. (b) The Niagara River flows over a resistant layer of limestone that overlies relatively soft shale. This shale is eroded headward in the plunge pool, which results in collapse of the overlying limestone and retreat of the falls. Niagara Falls has retreated more than 11 km (6.8 mi) in the past 12,000 years.

at a confluence. As at the ocean, the elevation of a tributary's stream bed must be the same as the channel bed of the stream in which it flows.

Return to Figure 16.20a to see how base level relates to stream behavior. In this particular example the elevation of base level did not change throughout the evolution of the stream profile. In reality, however, sea level—in other words, the stream's base level elevation—would probably change frequently over such a period of time and would thus be a major control in the behavior of the stream. If sea level fell, then a wave of erosion would proceed upstream as the stream eroded down to the new base level elevation. On the other hand, if sea level rose, then the stream would begin to deposit sediment and aggrade its bed as it rose to the new base level elevation. The stream would also begin to erode if uplift occurred in the source area of the stream, which would cause the channel slope to increase and result in more energy. If such a scenario occurred, the stream system would be said to be *rejuvenated*; that is, it would begin the developmental cycle all over again.

The Graded Stream

Now that the graded stream concept has been discussed, you can review it by viewing an animation. Go to the *Geo-Discoveries* website and access the module *The Graded Stream*. This animation nicely illustrates the idealized process of stream evolution and gradient adjustment. As you watch it, you will first see a longitudinal profile similar to the uppermost one in Figure 16.20a. Notice how this profile smoothes as time progresses through the combined process of knickpoint reduction and filling of lakes and ponds with sediment. Ultimately, the stream will obtain a slope that enables it just to carry the sediment that is provided through hillslope erosion. Once you complete the animation, answer the questions at the end of the module to test your understanding of this concept.

KEY CONCEPTS TO REMEMBER ABOUT STREAM GRADATION

1. Streams carry sediment as dissolved load, suspended load, or bed load.

2. A graded stream is a stream that has evolved to the point where the amount of sediment it carries is balanced with the capacity of the stream.

3. An example of an ungraded landscape is one that was recently uplifted. This uplift creates a very steep gradient. As a result, a stream is able to carry more sediment than is provided to it from eroded hillslopes. This stream will downcut its bed through a process called *degradation*.

4. Stream aggradation occurs when the sediment load is greater than the stream's capacity to carry it.

5. Base level is the point where a stream flows into the ocean and is a major control of a stream's behavior. When base level rises, streams deposit sediment to match that elevation. In contrast, when base level falls, streams erode their channel beds to cut down to sea level.

Evolution of Stream Valleys and Floodplains

The next step in this discussion of stream systems is to investigate the landforms that evolve in association with the long-term process of stream gradation. As you work your way through this section of the chapter, refer to Figure 16.22. This diagram shows a simplified four-stage evolution of a stream and its associated valley.

For simplicity, let's begin by imagining that a landscape was uplifted or deglaciated at some point in the distant past. With this uplift in mind, compare Figure 16.22 with the top stream profile in Figure 16.20a. Lakes and waterfalls are present in both illustrations. Recall that lakes occur in areas where the slope is shallow, whereas waterfalls are associated with knickpoints where the rocks are resistant and the gradient is steep. In this early stage of gradation, the stream has a capacity to carry more sediment than is being delivered to it. As a result, the stream begins to downcut vigorously in an attempt to reach a state of equilibrium. In so doing, the stream carves out a deep V-shaped valley (Figure 16.22b) that is consistent with the Grand Canyon of the Yellowstone River (Figure 16.23).

As time progresses (Figures 16.22b and c), the lake disappears through a combination of infilling by eroded upstream sediments and draining because the river cut back into it after eroding through its downstream knickpoint. During this interval of time, the gradient becomes smoother, as in the second curve in Figure 16.20a. Turning back to Figures 16.22b and c, the stream begins to extend its valley farther upstream (or *headward*) because the gradient is steepest in this part of the system. Because of this steep gradient, the stream naturally flows at a greater speed and thus has more energy to erode sediments. In Figure 16.22c, you can see that tributaries begin to develop and also extend their valleys upstream by headward erosion. At the same time as this headward erosion is occurring, the canyon continues to be more deeply incised.

Look at Figure 16.22d. The stream valley looks different from its appearance in the previous three images. Now, the valley is no longer V-shaped, but is quite wide. This change is evidence that the stream has reached a graded condition and has the graded longitudinal profile seen in Figure 16.20a. Recall that during the early stages of stream

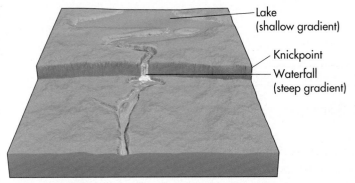

Lake
(shallow gradient)

Knickpoint

Waterfall
(steep gradient)

(a) Initial uplift

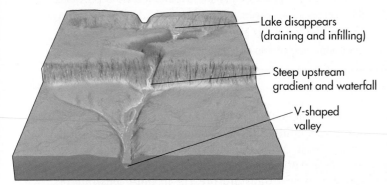

Lake disappears
(draining and infilling)

Steep upstream
gradient and waterfall

V-shaped
valley

(b) Vigorous downcutting

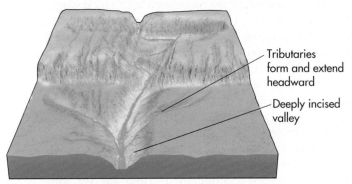

Tributaries
form and extend
headward

Deeply incised
valley

(c) Reaching equilibrium

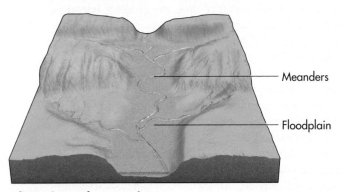

Meanders

Floodplain

(d) Horizontal expansion

Figure 16.22 Evolution of a graded stream and associated valley.
During the first stages of valley evolution, the stream downcuts vigorously because it has a steep gradient and excess energy. When the stream reaches a graded state, it begins to migrate horizontally, resulting in a wide valley.

Figure 16.23 The Grand Canyon of the Yellowstone River.
Here, the Yellowstone has yet to reach a graded condition, as indicated by the deep V-shaped canyon and waterfall. Compare this photograph with Figure 16.22b.

valley evolution the stream erodes vertically because it has proportionately more energy than sediment load. When a stream becomes graded, however, it no longer needs to downcut throughout the entire system because it has reached a gradient on which it can transport the average load that is provided from the hillslopes in the upper reaches of the system. Although downcutting may continue in the upper reaches of a watershed where the gradient is steep, the stream begins to expend its energy in a horizontal fashion in the lower reaches where the gradient is less.

Stream Meandering As noted previously, streams migrate horizontally through a process called meandering, in which numerous pronounced arcs and curving bends develop in the valley (Figure 16.22d). Stream meanders are fascinating places because both erosion and deposition occur simultaneously. As a result, stream meanders move back and forth across the valley over time. Figure 16.24 shows the processes associated with a typical stream meander. As the river flows around the meander bend, the centrifugal force associated with its current carries the main line of flow of the channel, or *thalweg*, to the outside of the curve. Thus, the channel is

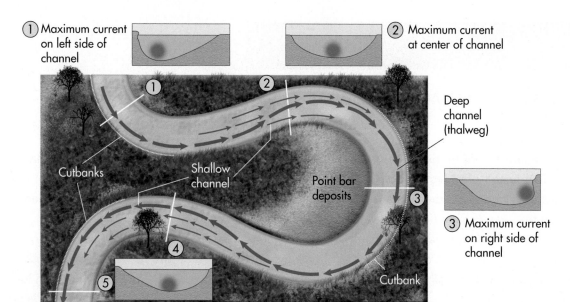

① Maximum current on left side of channel

② Maximum current at center of channel

Deep channel (thalweg)

③ Maximum current on right side of channel

Cutbanks

Shallow channel

Point bar deposits

Cutbank

④ Maximum current at center of channel

⑤ Maximum current on left side of channel

(a)

Cutbank

Point bar

(b)

Figure 16.24 Features of stream meanders.
(a) Deposition of sediment, forming a point bar, occurs on the inside of a meander because stream velocity is low. In contrast, the stream erodes the cutbank on the outside of the meander because velocity is high. (b) Note the distinct point bar and cutbank in this stream meander along the Kansas River.

GeoDiscoveries

www.wiley.com/college/arbogast

WILEY PLUS

Stream Meandering

To view an animation on the meandering process, go to the **GeoDiscoveries** website and access the module **Stream Meandering**. This animation illustrates how meander bends form over time in an initially straight river and also shows the formation of an oxbow lake. As you watch this animation, relate it back to the evolution of a stream valley. Once you complete the animation, be sure to answer the questions at the end of the module to test your understanding of this concept.

MISSISSIPPI RIVER MEANDERS

You can learn much about a river by examining satellite images of it. This image provides an excellent example. It is a synthetic aperture radar image of the Mississippi River that was acquired on the space shuttle in October 1994. The image is centered at about 32.75° N, 90.5° W and covers an area of about 23 km by 40 km (14.2 mi by 24.8 mi). North is toward the upper right of the image and water appears black. The accompanying diagram shows the location of the oxbow lakes and some of the meander scars. See if you can find some more.

As you can see, the Mississippi River is actively meandering in this area, with numerous oxbow lakes and meander scars visible. By looking at such an image, you can discover where the fastest and slowest velocities are within the river. Note that two such places are identified on the diagram. In addition, one can determine the areas that are prone to flooding. For example, the elevation between the current channel and the modern floodplain boundary is probably low, which means that the area is likely to flood during high discharge. The numerous signs of meandering also mean that the Mississippi River must be close to base level in this area and that the stream is in, or close to, a graded condition.

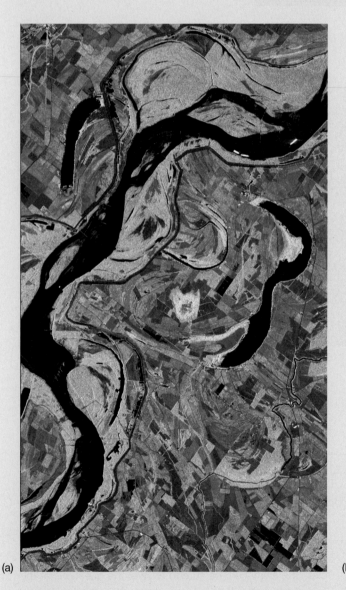

(a)

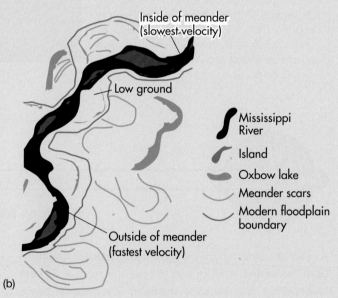

(b)

(a) The meandering Mississippi along the borders of Mississippi, Arkansas, and Louisiana. (b) Note the numerous oxbow lakes and meander scars highlighted in the diagram.

The chapter thus far has focused on fluvial processes and the nature of stream channels. This point marks a good time to review these concepts with two interactivities. Go to the *GeoDiscoveries* website and first select the module *Fluvial Geomorphology and Stream Processes*. This module reviews the concepts of channel geometry, stream flow, and the way water erodes and transports sediment. As you work your way through the module, consider how streams are viewed as holistic systems that shape Earth's surface. A second module is *Identification of Stream Landforms,* which is a drag-and-drop exercise that assesses your ability to identify features within streams. Once you complete the exercises, answer the questions at the end to test your understanding of these concepts.

---

deepest and the flow velocity is highest in this part of the stream. Given the high stream power that exists on the outside of the meander, this part of the arc is a zone of aggressive erosion. As a result, the surface leading away from the river is undercut, which causes the overlying sediment to collapse into the stream, forming a steep slope called a *cutbank*. Erosion also scours the channel bottom at these locations, forming deep *pools*. At the same time that erosion shapes the cutbank, deposition occurs on the inside of the meander bend because the channel is shallow in this part and energy is low. As a result, alluvium accumulates as a long, curving deposit of sediment called a *point bar* that is only about 1 m (about 3.3 ft) above the level of the stream.

If you are looking for clues as to whether a stream is actively meandering, a good piece of evidence is the presence of oxbow lakes in a stream valley. An **oxbow lake** is a water-filled former meander bend that was detached from the main channel in a stream valley. This lake develops when the stream erodes the cutbank as water migrates downstream. At the same time that cutbank erosion is occurring, deposition of sediment takes place at the corresponding point bar. Given the combined impact of these processes, the neck between meander loops gradually becomes narrower. Ultimately, one or more meander loops can be entirely cut through when the stream floods. This neck cutting is part of the stream's continuous process of gradient adjustment and results in a temporary lake (Figure 16.25). With time, an oxbow will usually fill with sediment derived from major floods along the active channel. In the first stages of sediment infilling, the oxbow will become a swamp that contains plants such as reeds and cattails. As sediment continues slowly to accumulate with successive floods, the swamp will ultimately fill completely with alluvium. Nevertheless, evidence that an oxbow lake was once present will remain in the form of a *meander scar*.

Valley Widening and Floodplain Formation In the context of stream meandering, remember that the processes of cutbank erosion and point bar deposition occur simultaneously. These combined processes cause stream meanders to migrate across the valley over time. This lateral migration, in turn, is important because meanders slowly cut into the side slopes, or *bluffs*, that flank the channel, causing stream valleys gradually to widen, as you can see in Figure 16.22d. Note in this diagram that the stream valley is now broad, essentially flat, and bordered by steep, rocky bluffs.

A stream valley usually contains thick deposits of alluvium that were deposited by the combined processes of point bar development and overbank deposition during large floods (Figure 16.26). During floods, the coarsest sediments (usually sand grains) are deposited along the channel margin because they cannot be carried far away from the stream due to their size. This near-channel deposition results in a ridge-like landform along the river called a **natural levee** that may be 1 m to 2 m (3.3 ft to 6.6 ft) high. Finer silt and clay-sized particles are carried farther away from the stream during floods and, when deposited, form low **backswamps** because they are poorly drained, for reasons outlined in Chapter 11.

Oxbow lake *A portion of an abandoned stream channel that is cut off from the rest of the stream by the meandering process and filled with stagnant water.*

Natural levee *A small ridge that develops along the channel of a stream through the deposition of relatively coarse sediment when flooding occurs.*

Backswamps *Marshy floodplain landforms that develop behind natural levees in which fine-grained sediments settle after a flood.*

(a)

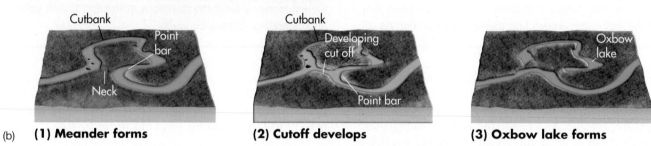

(b) **(1) Meander forms**   **(2) Cutoff develops**   **(3) Oxbow lake forms**

Overall, however, the combined processes of meander development, point-bar formation, and overbank deposition produce the low relief of the stream valley. This part of the landscape is typically called the river *floodplain* because it can be submerged during large floods, even though such an event may be rare. The river floodplain differs slightly from the *active floodplain*, which is the low, frequently flooded surface that results

mostly from point-bar formation on the inside of meander bends.

Entrenched Meanders As previously discussed, a meandering stream develops when a stream reaches a graded condition. These conditions do not last forever, however, as streams periodically adjust to new environmental conditions such as tectonic uplift, a drop in base level, or

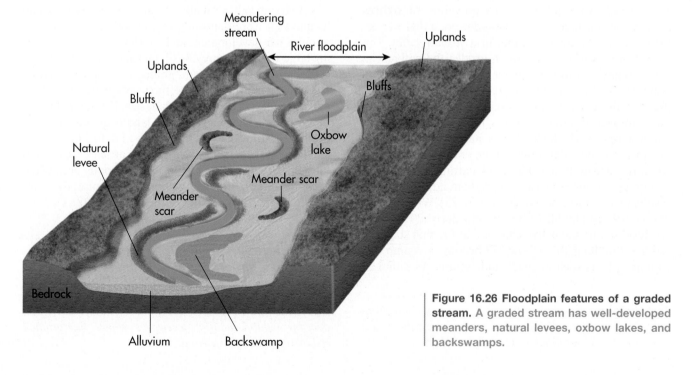

Figure 16.26 Floodplain features of a graded stream. A graded stream has well-developed meanders, natural levees, oxbow lakes, and backswamps.

This image shows a stream in an alpine setting. Given what you see here, which one of the following statements accurately reflects the stream in this landscape?

a) The stream has a wide valley.
b) The stream is actively meandering.
c) The stream has high discharge.
d) The stream is not graded.
e) The stream is a high-order stream.

1. When a landscape is recently uplifted, the resulting ungraded streams degrade by cutting V-shaped valleys that probably contain numerous knickpoints.

2. Waterfalls occur in places where rock is resistant to erosion and progressively erode headward until the gradient is smoothed.

3. As streams approach a graded state, they begin to erode horizontally through the process of meandering. This process produces a wide valley because bordering bluffs are eroded.

4. In the process of meandering, erosion occurs on the outside of the meander at the cutbank, whereas deposition is focused at the point bar on the inside of the meander.

5. You can tell that a stream is actively meandering (in present time) if it has well-developed oxbow lakes and meander scars.

climate change. The landscape response to these changes is usually preserved in stream valleys as distinct landforms.

For example, when a graded, meandering stream responds to a new period of landscape uplift associated with a major tectonic event or base level drop, the gradient of the stream is dramatically increased. Such a stream once again has a capacity to carry more sediment than is delivered to it from the hillslopes. As a result, the rejuvenated stream will begin a new period of downcutting and channel-bed erosion.

When a graded stream is rejuvenated, interesting landforms called *entrenched meanders* can develop, such as the well-known Goosenecks of the San Juan River in Utah (Figure 16.27). Prior to uplift of the Colorado Plateau region, the San Juan River was a meandering stream and the channel bed was somewhere near what is now the lip of the gorge. In other words, the elevation of the stream channel was much higher at that point in time and thus no gorge was present. In the past 10 million years, the entire Colorado Plateau has been uplifted about 1.5 km (0.9 mi). This period of uplift caused the San Juan River to begin eroding its channel bed while preserving its meandering pattern (because it was no longer eroding horizontally). This same period of uplift also initiated the downcutting by the Colorado River that created the Grand Canyon in Arizona.

Alluvial Terraces Entrenched meanders illustrate the kind of landscape evolution that can occur when extensive steepening of a stream's slope happens due to large-scale regional uplift. Most of the time, however, the environmental adjustments within a drainage basin are more subtle, consisting of relatively minor fluctuations in base level or in the regional climate. Although these adjustments do not result in dramatic canyons and gorges, such as the Goosenecks of the San Juan River, they nevertheless create distinctive landforms.

The most prominent landform created by smaller-scale fluvial adjustments is an alluvial terrace. An **alluvial terrace** is a broad, flat surface that occurs

---

Alluvial terrace *A level, step-like landform that forms when a stream erodes its bed so that a horizontal surface is raised relative to the channel.*

Figure 16.27 Goosenecks of the San Juan River in Utah. These entrenched meanders formed as a result of extensive uplift of the Colorado Plateau in the past 10 million years, which increased the stream gradient of the San Juan River. The resulting gorge is about 370 m (1400 ft) deep.

within the stream valley and is elevated with respect to the river so that it floods much less frequently than the topographically lower point bars. Although an alluvial terrace is usually considered to be part of the floodplain, it is also useful to think of the surface as being an abandoned floodplain because it is no longer frequently flooded.

Floodplain abandonment can occur in one of two ways. One way is that, through repeated flooding, thick deposits of alluvial sediment (or **alluvium**) accumulate on the active floodplain in a process called *alluviation*. As these sediments gradually thicken with repeated flooding, the active floodplain slowly becomes elevated with respect to the stream so that progressively higher discharges are required to flood the surface.

Another way in which floodplain abandonment and terrace formation occur is when a stream downcuts following a period of meandering. This downcutting may be associated with either a subtle period of regional uplift, base level drop, or regional climate change. When such an event occurs, the stream responds to the new ungraded conditions by downcutting through its own alluvium. This downcutting elevates the former active floodplain, creating terraces on either side of the river that are at the same elevation. Such terraces are called *paired terraces*. If the stream is able to continue meandering as it downcuts, it may remove a portion of the alluvium on the outside of meander bends through cutbank erosion, leaving surfaces of unequal eleva-

Alluvium   *Sediment deposited by a stream.*

tion called *unpaired terraces* on either side of the river. Many streams have numerous terraces that progressively step down to the river (Figure 16.28). Whenever you see such a sequence of surfaces, it indicates that the stream has had a very complex history of flooding and downcutting.

A Holistic View of the Graded Stream Model At this point in the chapter, the concepts of stream order, hydraulic variables, stream behavior, and fluvial landform evolution have been discussed. With these concepts in mind, it is time to integrate them into a holistic framework that links them within the graded stream model. Remember that once a stream reaches a graded condition, it has the idealized longitudinal profile seen in Figure 16.20a. Once this condition is reached, a very close and predictable relationship arises among stream order, hydraulic variables, and the type of landforms you would see along the stream's profile (Figure 16.29). Stream gradient is highest in the upper reaches of the system, which translates to greater velocity in these areas and higher *stream power*. It also has high *stream competence* because it is capable of carrying large particles, perhaps even boulders, at high flows. As a result, the stream expends its energy vertically, and well-developed gullies and canyons appear in this part of the watershed due to extensive erosion. Given that the streams here are low-order systems, however, their width and depth are relatively small because they contain little water compared to higher-order streams. This means that discharge is also relatively low in streams in this part of the drainage basin.

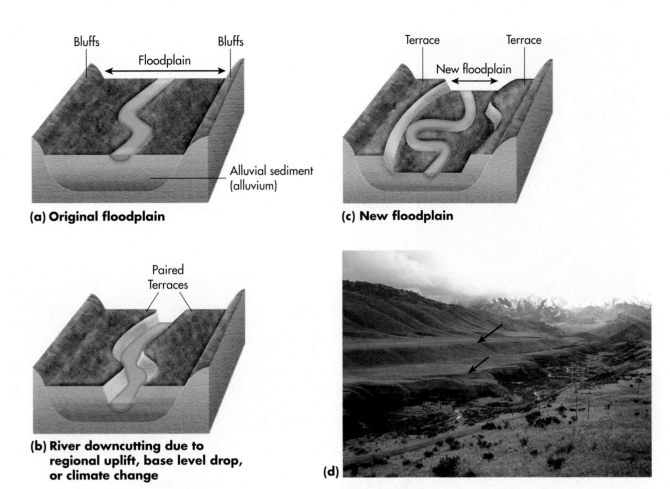

(a) Original floodplain

Bluffs   Floodplain   Bluffs

Alluvial sediment (alluvium)

(b) River downcutting due to regional uplift, base level drop, or climate change

Paired Terraces

(c) New floodplain

Terrace   New floodplain   Terrace

(d)

Figure 16.28 Formation of alluvial terraces. Alluvial terraces form through a combination of meandering, flooding, and downcutting in a river valley. (a) As the stream meanders, it creates an active floodplain that is underlain by alluvial sediment. (b) During a period of downcutting, the formerly active floodplain is elevated with respect to the river; thus, it rarely floods and a terrace forms. (c) A new active floodplain forms at a lower elevation when the stream begins to meander again. (d) Alluvial terraces along the Cave River on the South Island of New Zealand. Notice that two terraces (arrows) occur along this valley, which means that at least two periods of downcutting have occurred.

As we move farther downstream into the middle reaches of the profile, notice that slope decreases, which causes stream velocity to drop correspondingly. As a result, the stream has less power and thus begins to meander rather than downcut. This shift results in valley widening in this part of the system. Width, depth, and discharge also increase because streams in this reach have a higher order due to the progressive funneling of low-order streams into the system. With higher discharge the potential of flooding increases, which will result in deposition of alluvium on the active floodplain because the valley is wider than it is in the higher reaches. Deposition of alluvium means that stream terraces may also occur in this part of the profile, especially if some waves of downcutting progressed through the area in response to a drop in base level. The size of the sediment load is usually lower in this part of the system compared to the higher reaches due to a loss in competence.

Moving into the lowest reaches of the profile, the pattern seen in the middle part of the system becomes even more pronounced. Here, the trunk stream is fully developed, which naturally has a relatively high order because it contains virtually all the water that lower-order tributaries can provide. Stream gradient is very low here, resulting in low stream power and the formation of well-developed meanders and a correspondingly wide valley. Discharge is high, however, resulting in progressively greater width and depth. Given that all the runoff in the drainage basin now flows through this central conduit, the potential exists for major flooding in this part of the system. Consequently, the alluvium is probably very thick and well-developed terraces are likely present. These landforms may be very prominent if significant base-level fluctuations have occurred through time.

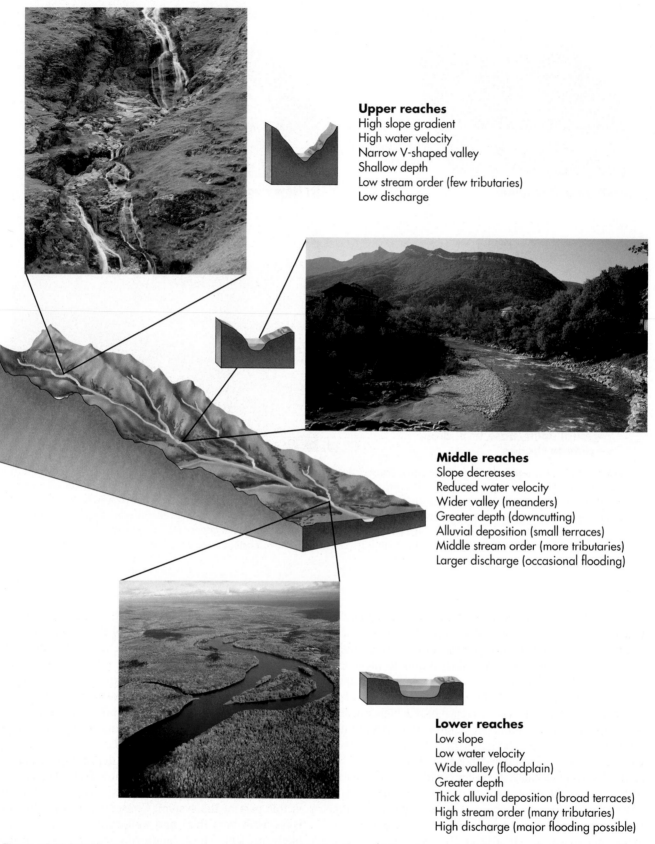

Upper reaches
High slope gradient
High water velocity
Narrow V-shaped valley
Shallow depth
Low stream order (few tributaries)
Low discharge

Middle reaches
Slope decreases
Reduced water velocity
Wider valley (meanders)
Greater depth (downcutting)
Alluvial deposition (small terraces)
Middle stream order (more tributaries)
Larger discharge (occasional flooding)

Lower reaches
Low slope
Low water velocity
Wide valley (floodplain)
Greater depth
Thick alluvial deposition (broad terraces)
High stream order (many tributaries)
High discharge (major flooding possible)

Figure 16.29 Relationship of stream order, valley form, and hydraulic variables along a graded stream. Streams in the upper reaches of the profile typically have the highest slopes and velocity, but the lowest widths, depths, and discharge. With distance along the profile, stream order and discharge increase, and gradient and velocity correspondingly decrease.

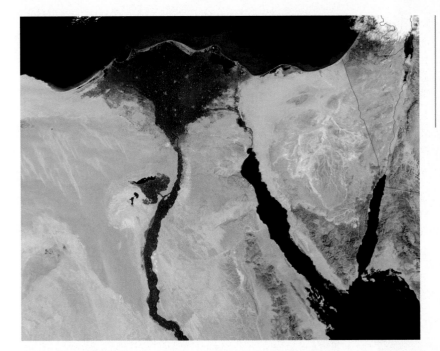

Figure 16.30 The Nile River Delta. This true-color image shows the triangular shape of the Nile River Delta where the Nile meets the Mediterranean Sea. The color green represents dense foliage and crops that are being supplied with water and nutrients from the fertile soils of the Nile River Valley and Delta. Note the surrounding desert.

Deltas The graded stream model considers the changes in stream variables and behavior over the course of an idealized longitudunal profile. Now let's examine what happens geomorphically at the location where some streams terminate. In most cases, streams are tributaries of larger streams and simply flow into them at a confluence such as shown in Figure 16.4. In other cases, however, a stream flows directly into a lake or ocean. Such a place is often called the *river mouth* and is a zone of intense sediment dispersal and deposition. This shift in stream behavior takes places because stream velocity slows dramatically when it enters the relatively still waters of the larger water body. Think of what happens to your own velocity when you run into a lake at the beach. It slows and you then fall. When a stream slows in a similar setting, it drops its sediment.

The landform created by sediment deposition at a river mouth is called a **delta** because it usually has a triangular shape similar to the Greek letter delta (Δ). An example of this shape is the Nile River Delta (Figure 16.30). A delta consists of a level plain of sediment that is sorted geographically by size. In other words, the coarsest sediments, such as sands, are deposited at the point where the stream begins to lose energy. As stream energy further weakens away from the river

---

Delta *A low, level plain that develops where a stream flows into a relatively still body of water so that its velocity decreases and alluvial deposition occurs.*

mouth, the silt-sized fraction of the sediment load is then deposited. The clay fraction is carried the farthest in suspension and ultimately accumulates far from the river mouth.

Another excellent example of a river delta is the Mississippi River Delta in Louisiana, where the Mississippi River meets the Gulf of Mexico. This delta is a classic *birdfoot delta* because it appears to contain a number of individual toes that can easily be seen in the satellite image (Figure 16.31). Each one of these "toes" is, in fact, a river channel that is part of a well-defined distributary network (Figures 16.31b and c). Unlike the tributary system in the upper part of the drainage basin, which funnels water into the trunk stream, the distributary network systematically drains water away from the main channel into the ocean.

Such a network develops because the delta gradient is very low and new channels form wherever a breach occurs in the natural levee along the stream. Wherever these breaks occur, freshwater pours into the distributary bay, which subsequently fills with alluvial sediment. When the bay ultimately fills, it seals off and a levee breach occurs at another place in the delta. This cycle occurs simultaneously at several places, which is why the delta has the distinct distributary network. The entire delta is underlain by enormous quantities of sediment (Figure 16.31b) that originated in places ranging from the Appalachian Mountains to the Rocky Mountains. Deltas will be further discussed in Chapter 19 when we investigate how coastal processes shape them.

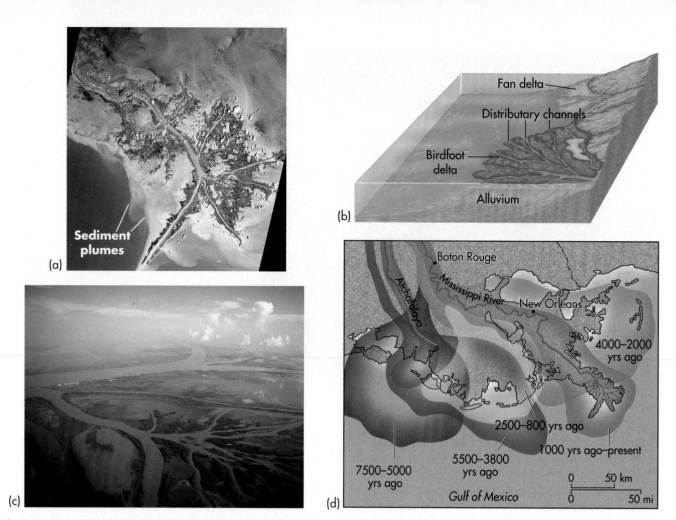

Figure 16.31 The Mississippi River Delta. (a) This satellite image beautifully shows the Mississippi River's birdfoot delta. The delta consists of a distributary network that funnels sediment (whitish hues) into the Gulf of Mexico. (b) The Mississippi River Delta is a complex system of distributary channels that is underlain by copious amounts of sediment derived from the continental interior. (c) The Mississippi Delta is a landscape of very low relief, consisting of numerous swamps and marshes that lie within a network of distributary stream channels. (d) Migration of the delta locations over the past 7500 years.

VISUAL CONCEPT CHECK 16.3

This satellite image shows the deltas of the Krishna (bottom) and Godarvi (top) Rivers in India. The whitish areas in the Indian Ocean are plumes of sediment from these rivers. Why are they present?

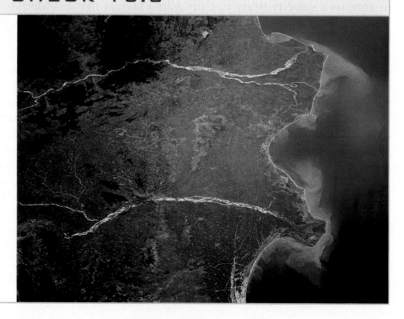

1. Entrenched meanders form when the slope of a graded, meandering stream is increased through extensive regional uplift, base-level drop, or climate change.

2. Alluvial terraces are abandoned floodplains that have been elevated through the process of alluviation or stream downcutting so that they are no longer frequently flooded.

3. A delta is a low-lying plain that forms at the mouth of rivers where they meet the ocean. They develop because stream power is lost when the stream flows into the relatively still ocean water. This change in velocity causes massive sedimentation. The sediment contained within a delta is well sorted, with coarse deposits accumulating at the river mouth and progressively finer sediments transported to the delta front.

Human Interactions with Streams

This section of the chapter focuses on some of the ways that humans interact with streams. As indicated previously, rivers and streams have long been important for human societies. They function as transportation net-

works, sources of water for direct human consumption and irrigation, and power that drives factories and even cities. Although this section does not describe every way that people interact with rivers and streams, it provides a good feel for the way we impact streams, and how they can impact us. The need to understand these relationships was tragically demonstrated in August 2010 when severe monsoon flooding in Pakistan killed over 1,500 people and affected another 10 million due to the low-lying nature of the country.

Urbanization

People have a big impact on fluvial systems through the process of urbanization. In the early part of this chapter, an example of storm runoff was water draining off an impervious parking lot. The reality is that with increasing urbanization, parking lots and other urban structures are covering greater amounts of ground around the world. For example, think about the sizes of the parking lots associated with the malls and shopping centers where you live. In many cases, entire drainage basins now lie within urban areas where the ground cover is no longer vegetation and soil, but mostly asphalt and concrete.

The impact of urbanization on stream behavior is significant, especially following a heavy rainfall. Figure 16.32 shows a hydrograph that illustrates the theoretical response of a stream to a heavy storm prior to urbanization (green line) and after urbanization (blue line). Before the city was constructed, the lag time between the storm peak (when it was raining hardest) and peak stream discharge was slow. This long lag time occurred because the soil–water belt first had to be saturated before the stream could be supplied with runoff or water from a recharged aquifer.

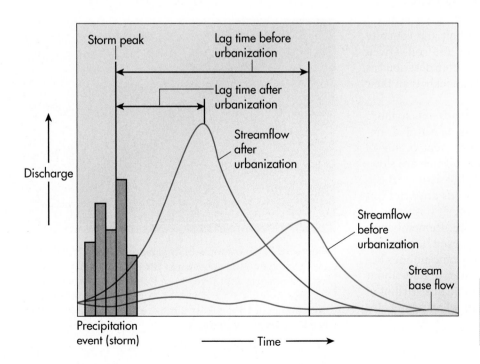

Figure 16.32 Stream hydrograph before and after urbanization. The response of the urbanized stream is much faster and larger than during pre-urban conditions.

Figure 16.33 An example of flash flooding. A flash flood is a brief but intense flood that occurs more frequently in urban areas where stream response is rapid. About half of flood-related fatalities occur when people try to drive across flooded roadways. This particular flash flood occurred in Brisbane, Australia.

After urbanization, however, the response of stream discharge is markedly different in two specific ways. Can you see them in the stream hydrograph? First, the stream response is much faster; in other words, the lag time between the storm peak and peak discharge is less. This decreased lag time occurs because rainwater is no longer absorbed first into the soil, but runs off the urban parking lots and other concrete surfaces directly into the stream. Second, the peak discharge in an urbanized setting is *greater* after a given storm event than it was following a storm of similar magnitude during pre-urban times. Very simply, this increased discharge occurs in the urbanized stream because more water runs off into the stream from the various asphalt and concrete areas rather than being absorbed by the soil. Given these relationships, urban streams are often described as being *flashy* because they react more quickly than their rural counterparts.

These conditions make *flash floods* much more likely in urbanized watersheds than in rural drainage basins if the same type of storm event occurs in both places. In contrast to river floods, which are slow-evolving events such as the 1993 Mississippi Basin floods that lasted for weeks, flash floods are brief but intense (Figure 16.33), and can result in both significant property damage and loss of life. Remember to *never* drive across a flooded roadway, even if the water appears to be shallow. After all, many automobiles become buoyant and can be carried away in only about 0.6 m (2 ft) of water. Of all the human deaths attributed to flooding, approximately half occur when people are in their vehicles.

Artificial Levees

Given the increasing number of people living on floodplains around the world, major river and flash floods can be economically disastrous. As noted previously, for example, the 1993 floods in the Mississippi Basin caused $15 billion in damage. In the context of this kind of cost, humans have engineered a variety of control structures to mitigate the effects of flooding. One kind of flood control structure is an artificial levee.

Recall that a natural levee is a small ridge that forms along the stream channel where coarse sediments accumulate during a flood. This ridge increases the height of the stream bank so that the next flood must have a greater discharge, and thus be deeper, in order to reach flood stage. An **artificial levee** is based on the same principle as a natural levee, with the difference that an artificial levee is designed by engineers and is usually built to a height such that a given city or valuable farmland is protected from a major event such as a 50-year or 100-year flood (Figure 16.34). In the United States, most of the large artificial levees along major rivers such as the Mississippi, Ohio, and Missouri are built and maintained by the United States Army Corps of Engineers. For example, in the Memphis district of the Corps of Engineers, levees are built to a height of about 7.6 m (25 ft) above the natural elevation of the Mississippi River bank.

Artificial levee *An engineered structure along a river that effectively raises the height of the river bank and thus confines flood discharge.*

Figure 16.34 An example of engineered levees. This artificial levee along the Mississippi River in Guttenberg, Iowa, kept the Mississippi River from flooding the town during a minor flood in 2001. The U.S. Army Corps of Engineers built and maintains this structure, as well as many others like it, along the Mississippi.

Figure 16.35 Hoover Dam on the Colorado River. This dam was constructed in the early 1930s to provide flood control and hydroelectricity for the region. Lake Mead is the artificial lake upstream of Hoover Dam.

Although levees certainly provide a societal benefit with respect to flood protection, they also have negative side effects. One problem with levees is that they keep the river from replenishing the sensitive wetland ecosystem that borders most rivers. This ecosystem is intimately related to the frequent but low-level floods that naturally occur along the stream and bring a fresh supply of sediment and nutrients into the wetlands. This relationship is lost, however, along a stream that is heavily protected by levees. Another problem with levees is that they may actually increase the intensity of flooding when major floods do occur. Under natural flood conditions, a stream will slowly spill out of its banks and spread across its floodplain. Although this slow spread can certainly be damaging, the gradual nature of the flood decreases the stream power and has a cushioning effect on locations downstream. In a levied section of river, however, the water cannot slowly spread horizontally because it is confined between the levees. As a result, the depth of the water increases beyond what it naturally would and the associated stream power grows. Much of this energy is transferred to the levee system, which may then fail during especially large floods like the 1993 and 2001 Mississippi River floods. If such a failure occurs, water pours catastrophically through the breach in the levee, resulting in tremendous damage to farmland and human structures where the break occurs.

Dams and Reservoirs

Another way that humans interact with rivers is through the construction of dams, which are among the largest human structures in the world. A **dam** is an engineered obstruction that is built across a river to control its flow. When you were young, you may have built dams by piling dirt across small streams to control them. In large dams, this control is obtained through hydraulic gates in the dam that can be adjusted to allow a specific discharge of water to proceed downstream. Dams typically have behind them a large lake or reservoir that evolves through the process of impounding stream water that would otherwise flow downstream. A well-known example of a dam is Hoover Dam, which straddles the Colorado River on the border of Arizona and Nevada (Figure 16.35). Formed behind this dam is Lake Mead, which is one of the largest artificially created bodies of water in the world, with an area of 603 km^2 (233 mi^2). Lake Mead will be examined again in Chapter 20.

Historically, dams have been built for three primary reasons: (1) to provide hydroelectric energy, (2) to provide flood control, and (3) to enhance river navigation. The reservoirs behind dams are also extensively used for recreation in many places. Dams improve navigation because the amount of water in the river downstream of a dam can be kept at a constant level by controlling the amount of water released from the reservoir, except during the most extreme droughts or floods. For example, Figure 16.36 is a hydrograph from the Chattahoochee River at Narcross, Georgia, which lies downstream of Buford Dam. Note how the discharge of the stream was held much more constant following dam construction than would have occurred naturally before. As a result of this controlling effect, rivers such as the Ohio, Mississippi, and Missouri are important transportation corridors because water depth is maintained at more or less the same level.

With respect to hydroelectricity, this form of energy is produced when water from the upstream side of the dam (from the reservoir) flows down and through tunnels in the dam to its outlet into the channel. As the water flows through the dam, it spins turbines within generators that produce electricity. The hydroelectric generators at Hoover Dam, for example, are capable of supplying nearly 1.5 million kW of power and provide electricity to Arizona, Nevada, and southern California.

The third major function of dams is to provide flood protection for areas downstream of the structures. Very simply, dams reduce the chances of a flood

---

Dam *A barrier that blocks or restricts the downstream movement of a stream.*

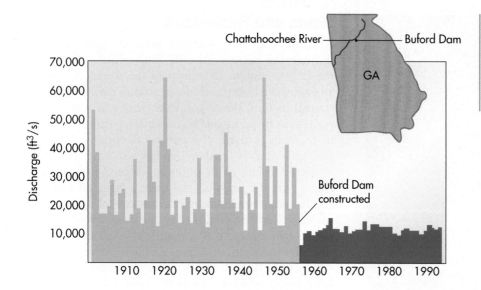

Figure 16.36 Annual flows for the Chattahoochee River downstream of Buford Dam in Georgia. Notice the change in the variability of annual discharge before and after the construction of Buford Dam in 1956. (*Source:* U.S. Geological Survey.)

because they can be used to store excess runoff in the reservoir by controlling the outflow of water at the dam during periods of heavy rainfall. In theory, the gates of a dam can be shut completely during a flood, which would cause all the stream discharge upstream of the dam to be trapped in the reservoir. Upstream of the dam, the lake would become larger and progressively fill the trunk and tributary valleys upstream. Below the dam, however, the stream channel would be dry until it began to receive tributary discharge someplace downstream. This kind of extreme scenario rarely occurs, however.

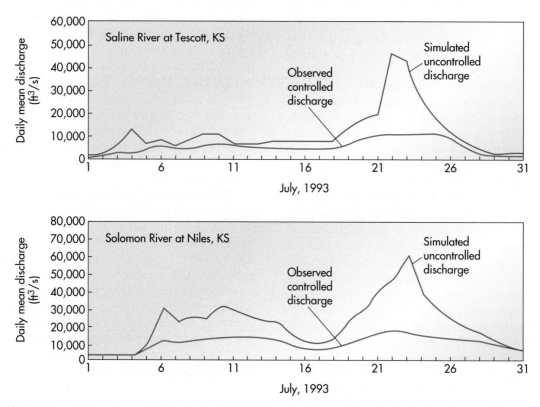

Figure 16.37 River hydrographs from Tescott and Niles, Kansas, during the 1993 flood in the Mississippi watershed. Note the difference between simulated peak discharge (blue line) and actual peak discharge (red line) resulting from controlled outflow from the dams upstream of this pair of towns. (*Source:* U.S. Geological Survey.)

Figure 16.37 shows hydrographs during the 1993 flood from Tescott and Niles, Kansas, which lie downstream of dams on the Saline and Solomon Rivers, respectively. In both graphs the blue line represents the simulated discharge at each town if the dams had not been present, whereas the red line shows the actual discharge. Although actual peak discharge certainly increased during the flood interval, it was about 80% less than what it is simulated to have been. As a result of these two dams, the small towns of Niles and Tescott were protected from the flood.

Negative Environmental Effects of Dams Although dams have a clear benefit to society because they control flooding, produce electricity, and enhance shipping, they also have a variety of societal and environmental costs associated with them. The most basic societal cost is that dams are extremely expensive. One example of extraordinary dam cost is the Itaipu Dam on the border of Paraguay and Brazil. This structure is currently the largest hydroelectric dam in the world, producing up to 12,600 mW of electricity, and provides 78% of the electricity used in Paraguay and 26% of that consumed in Brazil. Construction of this dam began in 1975 and was finished in 1983. It cost between US$20 billion and US$25 billion to complete and was such a large financial drain that it contributed to the downturn of the Brazilian economy in the 1990s. Similar costs occurred during construction of the Three Gorges Dam on the Yangtze River in China. This controversial dam cost US$30 billion.

Another negative effect associated with dams is specifically related to the reservoirs they impound. In the course of their creation, reservoirs flood large areas upstream of the dam site. If this flooding occurs in heavily populated areas, many people need to be relocated. Upstream of the Three Gorges Dam, approximately 632 km^2 (244 mi^2) of land will ultimately be inundated. As a result, up to 1 million people have had to move to avoid being flooded. In a related vein, ancient cultural artifacts are sometimes lost in this way. Upstream of the Aswan Dam in Egypt, which was constructed across the Nile River, impoundment of Lake Nasser resulted in the loss of numerous classical Egyptian sites and artifacts.

In addition to the societal costs of reservoirs, they also have negative environmental impacts. These impacts are most closely related to the loss of ecosystems and spectacular scenery. Prior to construction of the Itaipu Dam, for example, the Parana River flowed through a spectacular gorge that contained a variety of beautiful waterfalls. Many of these waterfalls, along with much of the habitat unique to the plants and animals of the region, are now submerged beneath the reservoir. Similar scenery and habitat loss occurred in the canyon land region of the southwestern United States when dams such as Glen Canyon Dam and Hoover Dam (Figure 16.35) were constructed.

In the context of ecosystem impacts, dams also have negative environmental costs because of the way they regulate the discharge of streams, when, in fact, stream flow is naturally highly variable over the course of the year. The riparian ecosystem is adapted to this variable discharge and can be pressured because stream flow is largely held constant to meet the various societal needs. A variety of animal species, for example, are currently threatened by dams in the United States. In the Pacific Northwest, dams impact salmon because they often form a barrier that stops fish from migrating upstream to spawning sites. Along the Missouri River, the pallid sturgeon is threatened because it requires high spring flows to trigger its reproductive cycle; these flows are moderated significantly by the numerous dams along the river. At this point in time, environmentalists and government agencies are working toward reducing the environmental impact of dams.

KEY CONCEPTS TO REMEMBER ABOUT HUMAN IMPACTS ON STREAMS

1. Most streams in the world have been impacted by humans in some way.

2. Asphalt and concrete structures in cities increase storm runoff. As a result, the lag time of a stream decreases, peak discharge increases, and flash floods are more likely.

3. Artificial levees are ridges built along stream channels to increase the height of the stream bank so that protected areas are rarely flooded. These systems work, but increase the overall power of the stream at high discharge by confining it to a narrow channel rather than allowing gradual flooding.

4. Dams are used to control downstream discharge on rivers by impounding water (in reservoirs) and releasing it slowly. Dams are used for flood control and hydroelectric production, and to facilitate channel navigation.

5. Although dams provide a definite societal need, they have significant environmental impact because the natural rhythm of seasonally variable stream flow is lost. In addition, they are a barrier for migrating river species.

THE BIG PICTURE

In this chapter, we focused on the way that flowing water moves and shapes the landscape. Next we turn to the way that frozen water flows across the Earth's surface in the form of glaciers. Remember that most freshwater on Earth is stored in glaciers and polar ice sheets. The next chapter focuses on the way that glaciers evolve, how they behave, and the landforms they produce. Our discussion will incorporate concepts related to climate, the hydrologic cycle, and Earth–Sun geometry, and should give you a much greater appreciation for the role that ice has played on Earth.

SUMMARY OF KEY CONCEPTS

1. Streams consist of channels in which water flows downhill under the force of gravity. They are fed by runoff or groundwater and may have a large lake as their source. Drainage basins are separated topographically from one another by bordering high ground called divides and interfluves.

2. Channel flow is defined by its hydraulic variables, including width, depth, velocity, slope, and discharge. These variables are interrelated with one another. For example, stream discharge is calculated by the equation $Q = w \times d \times v$.

3. In the course of their evolution, streams gradually sculpt the landscape so that the slope of the stream is sufficiently steep to carry the average discharge in the river. During this shaping process, the slope of the stream decreases. Once the stream reaches this stage of evolution, it is said to be "graded." The behavior of the stream is controlled in large part by base level and climate.

4. As streams approach a graded state, they begin to erode horizontally through meandering. This process produces a wide valley because bordering bluffs are eroded.

5. Streams produce a variety of erosional and depositional landforms, including gullies, cutbanks, point bars, deltas, and terraces.

6. Given the importance of streams to human societies, they are heavily managed.

CHECK YOUR UNDERSTANDING

1. Describe the concept of runoff.

2. Define the concept of *drainage basin* and describe one of the major watersheds in North America. What are the boundaries of this drainage basin and what are some of the major tributaries?

3. Imagine that two drainage basins are found in a humid environment. One of the watersheds has high drainage density, whereas the other one does not. Assuming that each drainage basin lies within a region of similar relief, what environmental variable most likely accounts for the difference in drainage density? Why?

4. What is the minimum number of tributaries that a fourth-order stream has? How many stream confluences are there in this watershed?

5. Name four primary hydraulic variables. Provide an example of how one variable influences another.

6. Compare and contrast the concept of *base flow* and *base level*.

7. Which landscape is more likely to contain graded streams: one that has recently been uplifted or one that has been stable for a long period of time with a consistent climate? Why?

8. Name and describe three landforms associated with a stream that is striving to reach a graded condition.

9. Name and describe three landforms associated with a graded stream.

10. How can deposition and erosion both occur along a stream meander?

11. Imagine that you encountered a stream with well-developed meanders that are deeply entrenched. Describe two reasons why this landscape would evolve.

12. Why is a terrace an abandoned floodplain? Describe the two ways that terraces form.

13. Describe the formation of river deltas.

14. Why is peak discharge greater in streams contained within urban areas than those in the surrounding countryside?

15. Describe how a dam provides flood control for a city that lies downstream of the structure.

ANSWERS TO VISUAL CONCEPT CHECKS

VISUAL CONCEPT CHECK 16.1

The correct answer is *a*. The Danube is a high-order stream. You can tell that the Danube is a high-order stream because it is large, with high discharge. This appearance results from the numerous lower-order tributaries that feed into the river.

VISUAL CONCEPT CHECK 16.2

The correct answer is *d*. The stream is not graded. Remember that graded streams have shallow slopes that allow a stream to migrate horizontally. This stream has a very steep slope, which means that most of its energy will be used to erode vertically.

VISUAL CONCEPT CHECK 16.3

The whitish plumes of sediment in the Indian Ocean from the Godarvi and Krishna rivers are present because the ocean is relatively still compared to the flowing streams. As a result, stream velocity slows where they flow into the ocean, causing sediment to be deposited. The whitish plumes are fine-grained sediments such as silts and clays that are floating in suspension.

CHAPTER SEVENTEEN

GLACIAL GEOMORPHOLOGY: PROCESSES AND LANDFORMS

The preceding chapter focused on the processes and landforms associated with water that flows across the surface in rivers and streams. Now it is time to examine the processes related to flowing *ice* and the landforms that result from this movement. About 77% of the freshwater on land is stored as ice. Although such may not seem possible, this form of water moves if it becomes sufficiently thick to flow under its own weight. Such a river of flowing ice is called a glacier and can shape the landscape in predictable ways through erosion and deposition.

At the present time, ice covers about 11% of the Earth's land surface. As recently as about 18,000 years ago, however, glaciers covered about 30% of the landmasses, including much of North America and Europe. As the climate warmed in the late Pleistocene and early Holocene, these giant ice sheets melted and new landscapes created by glacial processes were uncovered. If you live in Canada or any of the northern states such as Wisconsin, Michigan, New York, Washington, or Illinois, your home may very well be built on deposits left by glaciers.

Flowing ice has had a huge impact on Earth throughout history. Although glaciers such as these in the Swiss Alps move slowly, they have tremendous power and are capable of eroding and depositing prodigious quantities of sediment. Much of the landscape seen here has been shaped in this way. This chapter focuses on glacial processes and the landforms that result from flowing ice.

CHAPTER PREVIEW

Development of a Glacier
On the Web: *GeoDiscoveries The Glacial Mass Budget*

Types of Glaciers
On the Web: *GeoDiscoveries Glaciers in the Cascade Mountains*

Glacial Landforms
On the Web: *GeoDiscoveries Depositional Glacial Landforms*

History of Glaciation on Earth

Probable Human Impact on Glaciers
On the Web: *GeoDiscoveries Glaciers and Climate Change*

Periglacial Processes and Landscapes

The Big Picture

LEARNING OBJECTIVES

1. Describe how glaciers develop and the environmental conditions that are required.

2. Discuss the various kinds of glaciers that occur and the types of landscapes in which they are found.

3. Compare and contrast the various erosional and depositional glacial landforms.

4. Explain the history of glaciation on Earth, particularly the number of Pleistocene glaciations that occurred.

5. Describe the impact that ongoing climate change is having on glaciers around the world. What is the evidence for that change?

6. Discuss periglacial processes, the environments in which they occur, and the impact they have on the landscape.

Development of a Glacier

A **glacier** is a slowly moving mass of dense ice (Figure 17.1) formed by the gradual thickening, compaction, and recrystallization of snow and water over time. For these conditions to take place, glaciers develop in regions where heavy snows fall in the winter and then do not melt completely during the subsequent summer. In other words, a key factor in glacial formation is that summer temperatures must be sufficiently cool that some snow remains when the next winter arrives. Only then can snow depth slowly increase so that it can be transformed to a river of flowing ice that shapes the landscape.

The Metamorphosis of Snow to Glacial Ice

How is it possible that snow can be transformed into a flowing glacier? Recall the discussion in Chapter 11 about certain kinds of metamorphic rocks and how they form because of high pressure due to compaction from overlying sediments. The process of snow transformation to glacial ice is very similar. The first thing to remember is that this snow-to-ice metamorphosis takes a great deal of time, often hundreds or even thousands of years. Over this course of time, snow accumulates annually in a way that slowly increases the overall thickness of the snow mass. These annual accumulations can be seen as distinct layers that are somewhat similar to sedimentary deposits (Figure 17.2), which is why some glaciologists think of glacial ice as being a form of sedimentary rock.

Glacier *A slow-moving mass of dense ice that flows under its own weight.*

Figure 17.1 A typical glacier. Valley glaciers like this one in Alaska's Chugach National Park are rivers of moving ice that slowly flow down the mountains from their snowy source areas.

Figure 17.2 Annual snow layers in a glacier. A key factor in glacial formation is summer temperatures. If the snow from the preceding winter does not melt, then it can accumulate in layers like these exposed in the Quelccaya Ice Cap in Peru. The average thickness of these layers is about .75 m (2.5 ft). Note the similarity between these layers and those found in nondeformed sedimentary rocks.

It is useful to think of snow as being the raw material for a developing glacier. When snow first falls, it is approximately 80% air. With time, this snow is progressively buried by new annual layers of snow. As a result, it gradually compacts and recrystallizes due to the weight of snow above it. In addition, some of the snow might melt and refreeze. Water may percolate downward into the compacting snow from summertime rain that falls on the surface of the snow pack and from the limited snowmelt that takes place within it. As a result of these combined processes, the lowest deposits of snow are slowly transformed into a compact, granular substance called **firn** (Figure 17.3). Firn is somewhat similar to the crunchy substance that forms within the top of snow after it melts a bit during the daytime and then refreezes at night. With additional compaction, pockets of air in the firn further decrease in size or are squeezed out entirely, and the ice crystals become aligned with one another. At this time, the mass is 90% solid and begins to flow slowly under its own weight, much like maple syrup across a pancake. This point in the transformation of snow marks the time when the ice mass is a glacier. In most mountain glaciers, the full transformation to active glacier begins when the snow and ice mass reach a depth of about 40 m (130 ft).

Firn *The compact, granular substance that is the transition stage between snow and glacial ice.*

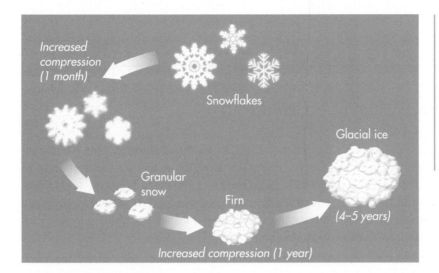

Figure 17.3 The transformation of snow to glacial ice. When snow first falls, it is 80% air. As snow compacts, its density increases to the point that it becomes granular. When these grains of snow become more compact, they collectively form firn. With additional compression, the firn becomes glacial ice. In general, the most dense and compressed layers of ice are at the bottom of a glacier, whereas the top layers reflect relatively recent snowfalls.

The Glacial Mass Budget

Like streams, glaciers are natural systems that depend on inputs and outputs to function. In the case of streams, the input is liquid water in the form of precipitation, runoff, and groundwater. The outputs are stream discharge and water vapor (through evaporation). The input to the glacial system is snow that accumulates at the upper end of the glacier (Figure 17.4). This area is fittingly called the **zone of accumulation** and is where the addition of annual snows exceeds what is lost through melting, evaporation, and sublimation. Glacial outputs occur primarily down the valley toward the front of the glacier, where the temperatures are higher. This area is called the **zone of ablation** (or *wastage*) and is the place where more annual melting and evaporation of ice take place than snow accumulation. Somewhere between the ablation and accumulation zones is the **equilibrium line**, which marks the place where system inputs and outputs are in balance.

The overall budget of the glacial system is continually in flux, with variable amounts of input and output over the course of time. During periods when

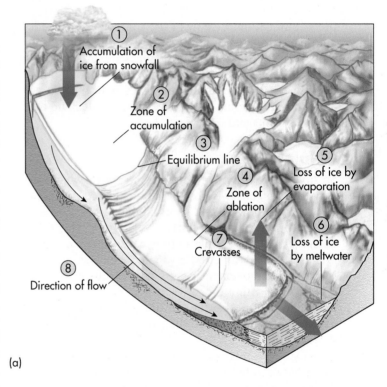

(a)

Zone of accumulation *The geographical area where snow accumulates and feeds the growth of a glacier.*

Zone of ablation *The part of a glacier where melting exceeds snow accumulation.*

Equilibrium line *The place on a glacier where snow accumulation and melting are in balance.*

Figure 17.4 The glacial mass budget. (a) The glacial mass budget is the balance between snowfall in the zone of accumulation and melting and evaporation at the zone of ablation. (b) Side view of the glacial mass budget. Note the geographical relationship of the zones of accumulation and ablation and the way in which ice flows.

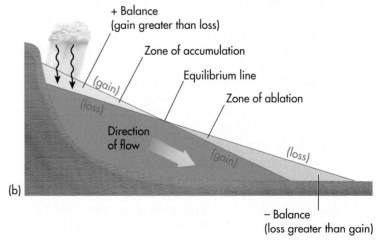

(b)

The Glacial Mass Budget

Glacier development depends on the complex interaction of snow accumulation in a source area and melting in the zone of ablation. The advance or retreat of a glacier is governed by the balance between the input of snow or output due to melting. In order to better visualize this process, go to the *GeoDiscoveries* website and select the module *The Glacial Mass Budget*. This module allows you to adjust temperature and the input of snow to influence the advance or retreat of a glacier. It also provides links to studies of glacier flow. After you complete the module, answer the questions at the end to test your understanding.

the accumulation of snow in the source area exceeds melting at the zone of ablation, the equilibrium line moves down the valley and the glacier grows and advances. Conversely, when melting in the zone of ablation exceeds the accumulation of snow in the source area, the equilibrium line moves up the valley and the glacier shrinks and retreats.

Glacial Movement

So how does a glacier move? Contrary to what you may think, a glacier is not just a block of ice that happens to slide down a hill. After all, the largest glaciers on Earth moved across fairly level terrain. This type of movement is possible only because these massive bodies of ice are able to flow under their own weight. Most of this movement occurs as a result of *internal deformation* within the ice. This internal movement is related to the fact that the interior of a glacier behaves like a plastic that can be shaped and manipulated to some degree.

Given the relatively high amount of friction that exists between the ice and the rock on which it flows, the ice moves most quickly in the core of the glacier (Figure 17.5). Thus, ice-flow dynamics are quite similar to those observed in a stream system. However, in contrast to streams, which flow quickly, average glacial flow is on the order of 10 cm to 30 cm (about 4 in. to 12 in.) per day. This internal deformation continues even if the glacier is melting, shrinking, and retreating. In other words, the interior of the glacier continues to move forward *even though* the front of the glacier may actually be retreating up the mountain valley as a result of fluctuations within the overall mass budget (Figure 17.6).

Not all the ice is plastic, however. In fact, the upper part of the ice is quite brittle and breaks easily. When this part of the glacier fractures, prominent **crevasses** can develop that extend from the surface of the ice to some

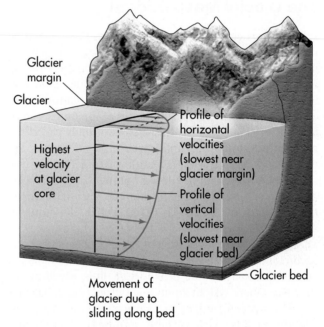

Figure 17.5 Internal flow in a glacier. Glacial ice is a plastic substance that flows under its own weight. Because of friction, the ice flows more slowly on its edge and base where it is in contact with bedrock. Velocity gradually increases toward the core of the glacier because it can move more freely.

depth within it. Crevasses usually form when the ice flows over a small ridge in the bedrock below (Figure 17.7). This ridge causes compression to build in the ice on the up-valley side of the ridge. On the downstream side of the ridge, the ice stretches, causing tension to build and crevasses to form. Crevasses are most closely associated with the equilibrium line because that is the place where

Crevasse *A deep crack in a glacier.*

they become visible as you move down the glacier (see Figure 17.4). These giant cracks in the ice are visible in the zone of ablation because they are not covered by fresh snow. In the zone of accumulation, however, they are not as visible because fresh snow tends to hide them.

In addition to the flow associated with internal deformation, glacier movement also transpires through a process called *basal slip*. This process results when high amounts of meltwater accumulate at the base of the glacier, thereby providing a lubricant between the underlying rock and the ice mass. If conditions are right, this lubricant can suddenly cause a glacier to move rapidly through a process called *surging*. In Alaska and the Himalayas, for example, some glaciers have been observed to have surged up to 100 m (333 ft) in a single day!

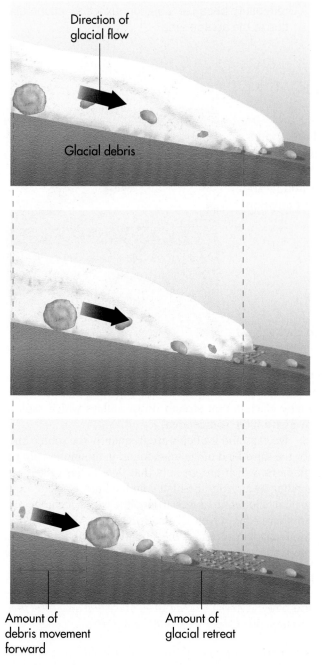

Figure 17.6 Internal movement of a glacier. The interior of a glacier continues to flow forward, carrying large rocks and other debris, even though the position of the glacier's front may retreat due to fluctuations in the glacial mass budget.

Figure 17.7 Formation of a crevasse. (a) A crevasse is a deep fissure that extends from the surface to some depth within the ice. Crevasses form when ice is compressed on the upstream side of an underlying bedrock ridge. On the downstream side of the ridge, the ice stretches, tension is high, and crevasses develop. (b) A typical crevasse can be 27 m (~90 ft) deep or more, causing great danger to skiers or hikers attempting to cross a glacier.

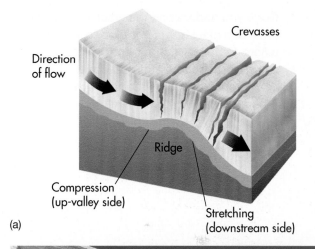

(a)

(b)

1. Glaciers form when unmelted annual snows reach a sufficient thickness to promote the metamorphosis of snow to firn, then to glacial ice.

2. Glaciers originate from source areas where snow accumulation is greatest. These source areas are typically located at high altitudes or latitudes where snow does not melt completely in summer because of cool temperatures that season.

3. The glacial mass budget refers to the balance between snow accumulation in the source area and the melting of ice within the zone of ablation.

4. Glaciers behave as a plastic that flows most quickly in the core of the system.

5. The interior of a glacier is always moving forward, even though the actual front of the glacier may be retreating because ablation exceeds accumulation in the source area.

VISUAL CONCEPT CHECK 17.1

Once you understand the glacial mass budget and the appearance of a glacier, it is possible to look at a glacier and interpret where features such as the equilibrium line and zones of ablation and accumulation are located. Look at this image and see if you can identify the zones of accumulation and ablation. Where is the equilibrium line? How can you tell?

Types of Glaciers

Now that the processes associated with glacier movement have been discussed, let's discuss the different kinds of glaciers by examining the environments in which they form.

Glaciers in Mountainous Regions

Perhaps the most obvious place that glaciers form is in high mountainous environments that have long winters and short, cool summers. Such places are excellent locations for glacier development because heavy orographic snows often accumulate there. Due to the short summers much of this snow is able to remain into the next winter, forming the foundation of an evolving glacier.

Three fundamental kinds of ice masses occur in such mountainous areas (Figure 17.8). An **ice cap** is a dome-shaped sheet of ice that entirely covers less than 50,000 km² (~19,000 mi²) of land area. In contrast, an **ice field** is more constrained by topography and forms in large basins or on top of plateaus. Many ice fields are located in places where tall mountains, or *nunataks*

("lonely stones"), occasionally rise above the ice. Ice fields are similar in size to ice caps and can also have many glaciers that stream down valleys which radiate away from the source area.

Ice caps and ice fields are frequently the source area for the third kind of ice mass found in mountains, **alpine glaciers**, which are glaciers that flow down valleys extending away from the high country. Such glaciers occur in many mountain ranges around the world, including the Southern Alps of New Zealand, the Swiss Alps in Europe, the Canadian Rockies, the Himalayan Mountains, and the Andes Mountains in South America. Several alpine

---

Ice cap *A dome-shaped sheet of ice that covers an area less than 50,000 km² (~19,000 mi²) in size.*

Ice field *A topographically constrained sheet of ice in mountainous areas that frequently has glaciers streaming away from it.*

Alpine glacier *A glacier in mountainous regions that flows down preexisting valleys.*

Figure 17.8 Ice caps and ice fields.
(a) The Vatnajökull ice cap in south-eastern Iceland. (b) The Patagonia ice field in the Andes Mountains of southern Chile. This long ice mass has numerous glaciers flowing out of it. (c) Nunataks rise above the ice in the Juneau ice field in Alaska.

glaciers occur in the United States, with the majority in Washington and Alaska (Figure 17.9). As you can see in that figure, most of these glaciers are quite small.

In areas where such glaciers flow in steep-sided valleys, they are called *valley glaciers* (Figure 17.10) because they occupy landscapes originally carved by streams. If a valley glacier extends entirely out of a valley into the lowland beyond the mountain front, it is called a *piedmont glacier*. Some valley glaciers even terminate at the ocean. These glaciers are called *tidewater glaciers* and are particularly common along the coast of Alaska. Such landscapes are attractive tourist destinations because they mark a beautiful interaction of ice and ocean. One of the particularly interesting aspects of tidewater glaciers is that large blocks of ice break off the front of the glacier in a process called *calving*. The calved blocks then fall into the ocean where they become *icebergs*. Such an iceberg, of course, was involved in the tragic sinking of the Titanic in 1912.

In addition to ice caps and ice fields, another source area for an alpine glacier is a **cirque**, which is a small, bowl-like depression on a mountain flank formed by

Cirque *A bowl-like depression that serves as a source area for some alpine glaciers.*

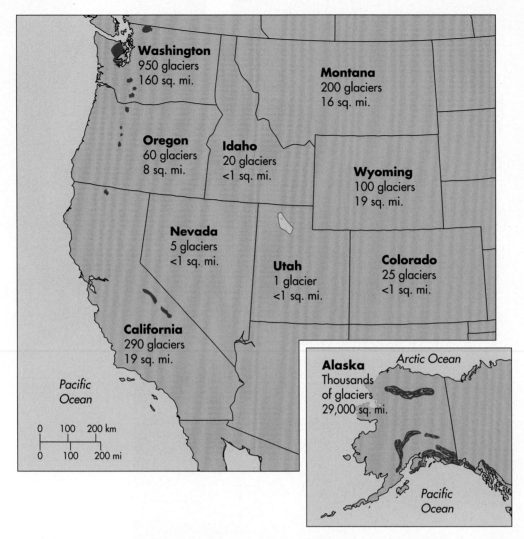

Figure 17.9 Alpine glaciers in the Pacific Northwest and Alaska. Most glaciers in the continental United States are relatively small, as indicated by the total area covered in each state. The largest and most impressive glaciers are in Alaska.

glacial erosion. A good example of an environment where numerous cirque glaciers have formed is the Cascade Mountains in the Pacific Northwest. Recall from Chapter 7 that this region receives abundant orographic precipitation from moisture-laden air flowing off the Pacific Ocean to the west. This precipitation is especially high on the numerous volcanoes that arc along the range. Given that the peaks of the volcanoes

www.wiley.com/college/arbogast

Glaciers in the Cascade Mountains

Although alpine glaciers are now rare in the contiguous United States, a few nice ones can be found in the Cascade Mountains in the Pacific Northwest. The best developed of these glaciers occur on the flanks of the large volcanoes in the range, such as Mount Rainier and Glacier Peak. In order to explore these ice masses more thoroughly, go to the *Geo-Discoveries* website and select the module *Glaciers in the Cascade Mountains*. After you complete the interactivity, answer the questions at the end of the module to test your understanding of Cascade glaciers.

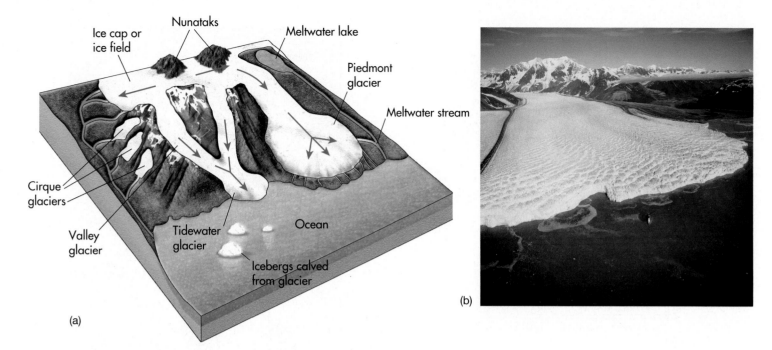

Figure 17.10 Valley glaciers. (a) Valley glaciers frequently originate at ice fields and ice caps and flow down terrain initially carved by streams. They become piedmont glaciers if they expand beyond the valley front. (b) Tidewater glaciers occur where glacier ice meets the sea. In these situations, such as seen here at the terminus of the Hubbard glacier in Alaska, ice calves off the front of the glacier and falls into the water to form icebergs. Note the cruise ship in front of the glacier for scale.

are usually greater than 3000 m (9800 ft) in elevation, summer temperatures near the mountain crests are typically very cool.

Figure 17.11a shows the effect that these cool temperatures have on snow cover. Note the extent of the snow cover on the flanks of Mount Hood in Oregon during late summer. Such snow cover is the foundation

for several cirque glaciers on this mountain, as well as on other volcanoes in the range. On Mount Hood, a variety of glaciers flank the mountain, extending from near the crest at about 3400 m (about 11,200 ft) down to an elevation of about 1800 m (6000 ft). These glaciers ring the mountain on all sides, as shown in Figure 17.11b.

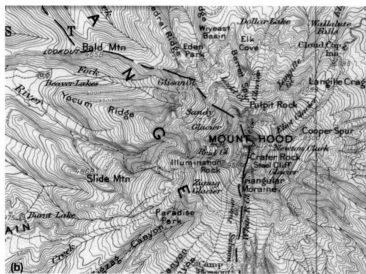

Figure 17.11 Cascade alpine glaciers. (a) Mount Hood is a composite volcano that reaches an elevation of over 3400 m (11,200 ft) and thus is an excellent place for the development of alpine glaciers. (b) This topographic map illustrates the various glaciers that ring the flanks of Mount Hood.

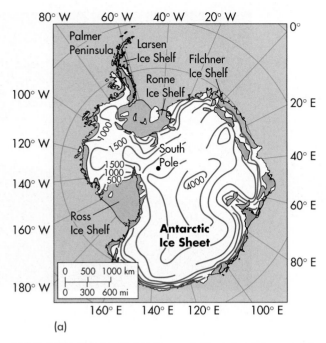

(a)

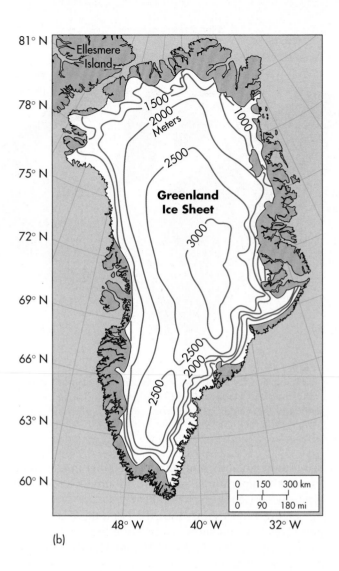

(b)

Figure 17.12 The Antarctic and Greenland ice sheets.
These ice sheets collectively contain over 90% of the world's glacial ice. The Antarctic Ice Sheet (a) is over 4000 m (13,000 ft) thick in places, whereas the Greenland Ice Sheet (b) is over 3000 m (9800 ft) thick in the center.

Continental Glaciers

In addition to mountain glaciers, another form of glacier is called a continental glacier. As the name implies, a **continental glacier** is a huge ice mass that covers a large part of a continent or large island. Also called *ice sheets*, continental glaciers once covered much of North America and Europe. However, they are now largely confined to the island of Greenland and the continent of Antarctica (Figure 17.12) because summer temperatures are cold at these high latitudes and abundant snowfall rarely melts.

Modern continental glaciers contain an enormous volume of ice that is more than 3000 m (10,000 ft) deep in places. The Antarctic Ice Sheet covers 90% of the continent and alone contains 91% of all the glacial ice on Earth (amounting to 13.9 million km³ or 3.3 million mi³). Similarly, the Greenland Ice Sheet covers 80% of that island and is about 1.8 million km³ (0.43 million mi³) in size. In both Greenland and Antarctica, the tremendous weight of these extensive ice sheets is so great that the underlying lithosphere has been pushed down into the upper part of the asthenosphere to some degree by a process called *isostatic depression* (see Figure 12.3).

Continental glacier   *An enormous body of flowing ice that covers a significant part of a large landmass.*

Glacial Landforms

Like stream systems, glaciers create landforms through the processes of erosion and deposition. Put simply, as the glacier advances, it erodes the surface by picking up rocks and other debris that it will later deposit at some other place. In this section, glacial landforms are divided into two groups: (1) landforms created mainly by erosion and (2) landforms produced mainly by deposition. The actual processes associated with glacial erosion and deposition are discussed in some detail.

Landforms Made by Glacial Erosion

Erosion by glaciers occurs mainly through the processes of abrasion and plucking. **Glacial abrasion** takes place because the moving ice contains particles of sand or rock fragments that are frozen to the bottom of the ice. As the ice moves, these fragments scratch and gouge the underlying bedrock. Such abrasion can result in **glacial striations** or scratches that indicate the direction of glacial movement (Figure 17.13). When the scratches are particularly deep and well expressed, they are called **glacial grooves**.

Another form of glacial erosion is **glacial plucking**, which happens when a glacier pulls large rocks or boulders from the ground as it moves. These boulders are subsequently carried within the ice or are frozen to the bottom of the glacier, where they act as tools for abrasion. When the glacier melts and retreats, it drops the boulders in what appear to be very random locations. These isolated boulders are known as **glacial erratics** because they appear to be out of place since they are now far from their natural outcrops. A great example of glacial plucking and long-distance transport can be seen in the Sioux Quartzite in the midwestern United States. This rock unit is a Precambrian quartzite that crops out in southwestern Minnesota at several places,

Figure 17.13 Glacial striations. Glacial striations such as these in Minnesota are linear scratches in bedrock produced when rocks at the bottom of a glacier grind across the bedrock.

including Pipestone National Monument. This area was overridden and smoothed during the ice ages when a massive continental glacier stretched at times from northeastern Canada to southern Illinois. (This history will be discussed in more detail later in the chapter.) One such advance of the ice plucked rocks from the Pipestone region and carried them to southeastern Nebraska and northeastern Kansas, where they were dropped as glacial erratics (Figure 17.14).

The combined processes of glacial abrasion and plucking scrape and round the hills. The most distinctive landform produced by these combined processes is the roche moutonnée. A **roche moutonnée** is a rounded bedrock hill that has a gradual slope on one side, called the stoss side, and a steep slope on the other, the lee side (Figure 17.15). The stoss side faces the direction from which the glacier came and was smoothed by abrasion as the ice moved over it. The lee side, in contrast, is steeper and more irregular with respect to its surface because the ice plucked rock from this side of the hill.

Glacial abrasion *An erosional process caused by the grinding action of a glacier on rock.*

Glacial striations *Scratches in rock produced by glacial abrasion.*

Glacial grooves *Deep furrows in rock produced by glacial abrasion.*

Glacial plucking *An erosional process by which rocks are pulled out of the ground by a glacier.*

Glacial erratics *Large boulders that have been plucked and transported a great distance before they are deposited.*

Roche moutonnée *A landform produced by glacial abrasion and plucking that has a shallow slope on one side and a steep slope on the other side.*

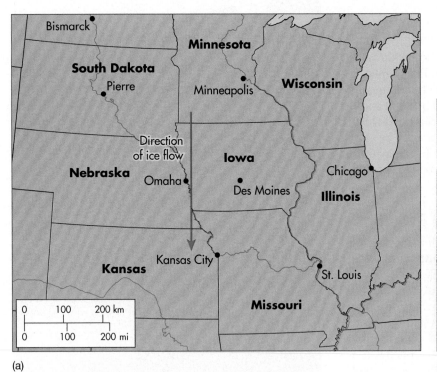

(a)

(b)

Figure 17.14 Plucking and erratics of Sioux Quartzite. (a) The Sioux Quartzite is Precambrian bedrock that crops out in southwestern Minnesota. Rocks from this outcrop were plucked by glacial ice about 600,000 years ago and carried as far south as northeastern Kansas. (b) This Sioux Quartzite erratic lies in northeastern Kansas. It is highly resistant to erosion, which is why it remains a prominent feature even though it has been resting there for about 600,000 years.

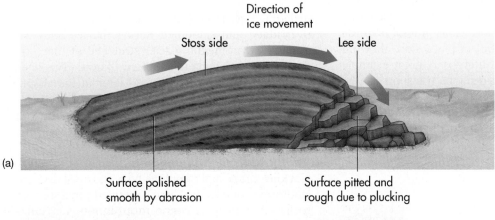

(a)

(b)

Figure 17.15 Roche moutonnée. (a) A roche moutonnée is created through the combined processes of abrasion on the stoss side and plucking on the lee side. (b) A roche moutonnée formed in the Manhattan Schist in New York City's Central Park. The stoss side is on the right-hand side of the image.

Although glacial erosion occurs in both alpine and continental environments, its effects are typically more visible in mountainous regions, resulting in very dramatic landscapes that are easy to interpret. In general, alpine glaciers modify a landscape by rounding parts of it at the same time that other aspects are sharpened. To see how a pre-glacial mountainous landscape is transformed by glaciation, refer to Figure 17.16 during the following discussion.

The most obvious landform created by glacial erosion in alpine regions is a cirque, which, as described earlier, is a broad amphitheater that forms on the flanks of a mountain. Cirques typically have very steep side and head walls and a floor that is flat to shallow sloping

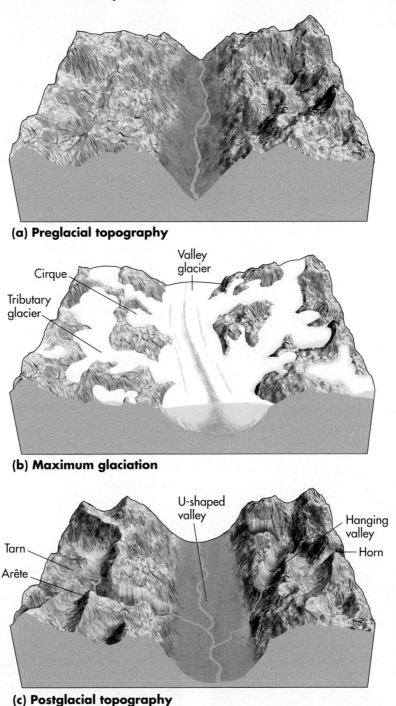

(a) Preglacial topography

(b) Maximum glaciation

(c) Postglacial topography

Figure 17.16 Modification of mountainous landscapes by glaciation. (a) The preglacial topography of alpine regions consists of rounded mountain slopes and a V-shaped stream valley. **(b)** During maximum glaciation, well-defined cirque glaciers shape mountain slopes and a thick valley glacier fills the former stream valley. The valley glacier is fed by tributary glaciers flowing into it. **(c)** The postglacial topography is much sharper on the mountain slopes and the former V-shaped stream valley is rounded to a U-shaped valley.

Glacial Landforms • 481

(a) (b)

Figure 17.17 Glacial cirques and tarns. (a) Glacial cirques, such as this one in the Canadian Rockies, are bowl-shaped depressions created by the scouring action of glacial ice. (b) A tarn is a meltwater lake that fills the depression within a cirque. This one is located in the North Cascades of Washington State.

(Figure 17.17a). As noted previously, cirques are source areas for alpine glaciation and are enlarged through the combined processes of glacier plucking and mass wasting of the adjacent walls. Sometimes the ice in a cirque melts completely away, leaving water in the quarried depression. Such a small lake is called a **tarn** (Figure 17.17b).

In many cases, cirques exist on several sides of an individual mountain. Over time, these cirques may progressively enlarge and steepen their respective side and headwalls so that the ridges between them narrow and become quite steep and thin, with serrated crests (Figure 17.18). Such a thin, steep ridge is called an **arête**, the French word for "fishbone." If a mountain has three or more arêtes resulting from intersecting cirques, then

the mountain is called a **horn**. The most famous example of a horn is the Matterhorn in the Swiss Alps.

Glacial erosion not only happens on high alpine peaks, but also within the valleys that slope away from the accumulation zones of snow. Sometimes glaciers never leave their cirque source areas, presumably because snow accumulation is not sufficient to cause glacial advance. If snow accumulation is high, however, a glacier will advance out of the cirque basin and move down the adjoining valley. When this type of advance takes place, the glacier modifies the preexisting V-shaped valley into a U-shaped valley called a **glacial trough** (Figure 17.19). These landforms are *very distinctive* and easy to identify the next time you are in

---

Tarn   *A small lake that forms within a glacial cirque.*

Arête   *A sharp ridge that forms between two glacial cirques.*

Horn   *A mountain with three or more arêtes on its flanks.*

Glacial trough   *A deep, U-shaped valley carved by an alpine glacier.*

Figure 17.18 The Matterhorn in Switzerland. This mountain is 4478 m (14,692 ft) tall and has been dramatically sculpted by alpine glaciations. Most of this shaping occurred due to erosion on the sides of the mountain.

Figure 17.19 A glacial trough. Extensive alpine glacial erosion creates deep glacial troughs. This well-developed U-shaped trough in Montana was carved by a trunk valley glacier.

VISUAL CONCEPT CHECK 17.2

With an understanding of glacial processes, it is possible to visualize how a landscape has evolved through time. Given what we have discussed in this chapter so far, can you describe how glaciation has modified this landscape? What is the evidence for that modification?

KEY CONCEPTS TO REMEMBER ABOUT GLACIAL EROSION AND RESULTING LANDFORMS

1. Glaciers erode through the processes of abrasion (grinding rock) and plucking (pulling rock from the ground).

2. Glacial striations and grooves are produced through abrasion.

3. A roche moutonnée is a streamlined bedrock landform produced by the combined processes of abrasion (on the upstream side) and plucking (on the lee side).

4. Extensive glacial erosion occurs in alpine regions. In general, a preglacial landscape is converted to one that is more angular, with sharper ridges, after glaciation. Resulting landforms include cirques, tarns, arêtes, horns, troughs, hanging valleys, and waterfalls.

Deposition of Glacial Drift and Resulting Landforms

Now that we have examined the landforms created by glacial erosion, let's consider those constructed through glacial deposition. As you study this section of the text, refer frequently to Figure 17.20, which shows most of the landforms discussed as well as their fundamental geographical relationships.

Glacial Till Sediment deposited by a glacier is called **glacial drift** and is found in two general forms: glacial till and glacial outwash. **Glacial till** is sediment that is deposited directly by a glacier. This form of glacial drift typically contains many different particle sizes, ranging from clay to sand and even cobbles and small boulders. As a result, it is considered to be an unsorted deposit because it is a mix of many different sediment sizes.

One form of glacial till is *basal till* (or *lodgement till*), which is deposited at the base of the glacier in way that is somewhat analogous to the way peanut butter is smeared onto bread with a knife. A glacier deposits basal till in somewhat the same way and can do so either as the glacier advances or retreats. Because of the intense pressure at the base of the glacier, due to the weight of the ice, basal till is usually quite dense and relatively fine-grained because it has been ground and crushed.

Another form of glacial till is *ablation till*, which is sediment that is carried within or on top of the ice and deposited when the glacier melts. Ablation till usually contains coarser fragments such as boulders because they are not crushed at the base of the glacier. As a result, ablation till tends to be even less well sorted than basal till.

Deposition of glacial till can result in very distinctive landforms. A notable type of till landform is a **moraine**, which is a winding ridge-like feature that forms

the mountains. The deepest and best-developed glacial troughs are formed by the trunk glacier in a particular mountain system (see Figure 17.16 again). Smaller, tributary glaciers also carve U-shaped troughs, but not to the same depth as the trunk valleys into which they flow because they contain less ice. As a result, a distinctive **hanging valley** is produced at the intersection of the tributary and trunk valleys when the glaciers fully recede. A hanging valley is often a place where a waterfall develops, such as the famous Yosemite Falls, as the tributary stream flows toward the trunk stream.

Hanging valley *An elevated U-shaped valley (with respect to a glacial trough) formed by a tributary alpine glacier.*

Glacial drift *Sediment deposited indirectly or directly by a glacier.*

Glacial till *Sediment deposited directly by a glacier.*

Moraine *A winding ridge-like feature that forms at the front or side of a glacier or between two glaciers.*

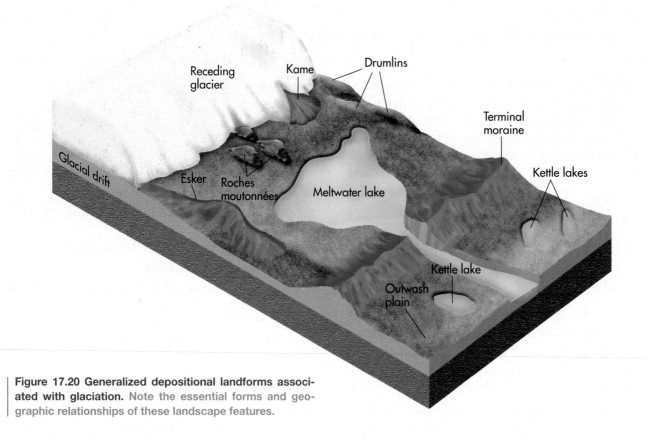

Figure 17.20 Generalized depositional landforms associated with glaciation. Note the essential forms and geographic relationships of these landscape features.

at the front or side of a glacier, or between two glaciers. A glacial moraine is a very poorly sorted deposit, ranging from clays to boulders, that appears when sediments are brought forward by the conveyor action within a glacier until they are dropped out of it (Figure 17.21). It is important to note that a well-developed moraine forms only when the ice front or side is stationary for a relatively long period of time. Only then can the persistent internal motion of the ice continually bring forth enough sediment for a visible ridge to grow.

The specific classification of any given moraine depends on its spatial relationship to the glacier that deposited its till. Figure 17.22 shows examples of the different kinds of moraines. A *lateral moraine* forms on the edge of a glacier and is usually best expressed along the sides of alpine glacial troughs. When a lateral moraine is sandwiched between two glaciers, it is called a *medial moraine*. A *terminal moraine* (also called an *end moraine*) forms at the front of the ice and marks the long-term stationary position at the farthest advance of a glacier. Another type of moraine that forms at the ice front is a *recessional moraine*. Although a recessional moraine may look exactly the same as a terminal moraine, it marks a pause in the overall retreat of the ice, rather than the farthest advance. A final kind of moraine is called *ground moraine*. This kind of moraine forms when the retreat of the ice is slow but steady, allowing an irregular pattern of deposition

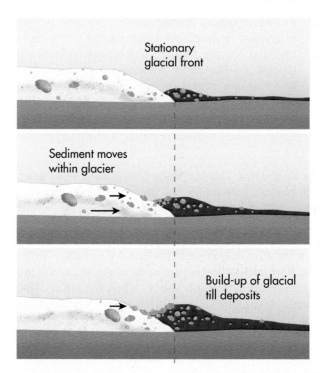

Figure 17.21 Formation of a moraine. A moraine forms when the position of the ice front is stationary for a long period of time. This stationary position allows till to be deposited in one place due to the conveyor-like action within the interior of a glacier.

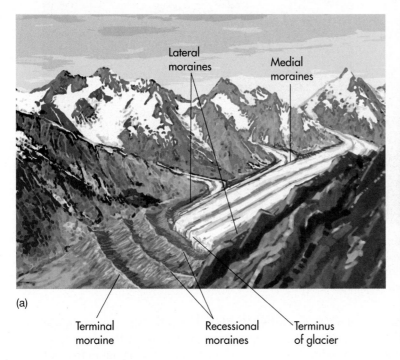

(a)

Lateral moraines

Medial moraines

Terminal moraine

Recessional moraines

Terminus of glacier

(b)

Figure 17.22 Types of moraines. (a) Lateral, medial, and end moraines in an alpine glacier system. (b) The hill in the background is a portion of the Port Huron moraine in northwestern Lower Michigan. This moraine can be traced around much of the northern part of the state's southern peninsula.

that creates a hummocky landscape consisting of small hills and depressions (Figure 17.23).

After till is deposited, it can subsequently be modified by direct and indirect glacial processes, again leaving distinctive landforms. One such landform is a **drumlin**, which is till or other soft sediment that has been streamlined in the direction of ice flow (Figure 17.24). A drumlin is somewhat similar to a roche moutonnée, but, in this

---

Drumlin *A streamlined landform created when a glacier deforms previously deposited till.*

---

case, the blunt end faces the direction from which the ice advanced. The formation of drumlins is not fully understood. Some may form because a glacier flows over previously deposited till that has uneven resistance, perhaps due to scattered boulder deposits or because some parts of the till have a frozen core. As the glacier flows over the resistant till, it may streamline less resistant sediment downstream of the core. Also, meltwater at the base of the glacier may have a role in the formation of some drumlins by streamlining sediments when periodic dramatic discharge bursts occur. Drumlins can range in length from 100 m to 5000 m (300 ft to over 3 mi) and

Figure 17.23 Ground moraine on the Coteau du Missouri in northwestern North Dakota. This rolling landscape was created because the glacier was retreating slowly at the same time it was depositing sediment in front and beneath it.

Figure 17.24 A drumlin field in Wisconsin. These prominent hills were created by the streamlining effect of a glacier on the sediment beneath it.

Figure 17.25 Outwash plain at Jackson Hole, Wyoming. Note the broad, horizontal surface that dominates the center of this image. This landform evolved because meltwater streams deposited the sediment in front of a valley glacier.

Figure 17.26 An esker. Eskers form when meltwater streams flow in winding tunnels beneath a glacier, depositing sediment that forms a ridge when the ice melts. This esker forms a prominent ridge in Blue Lake, Minnesota.

can be as much as 200 m (650 ft) high. They often occur in "swarms" and are particularly common in areas of Wisconsin, Michigan, and New York where continental glaciation was intense during the most recent ice age.

Glacial Outwash In contrast to glacial till, which is sediment that accumulates while in direct contact with an ice mass, we now turn to sediment that is carried and deposited by water that flows out and underneath a glacier as it melts. This kind of sediment is called **glacial outwash** and results from *glaciofluvial processes*, which are the combined effects of glaciers and streams.

As a deposit, glacial outwash differs fundamentally from glacial till in that it is well sorted, which means that most of the sediments are roughly the same size, namely sand and gravel. This higher degree of sorting results from the fact that flowing water naturally segregates sediments on the basis of their size (see Chapter 16). For example, flooding stream systems deposit sand near the channel (resulting in natural levees) and clays toward the valley edge (forming wetlands and swamps). When glaciers melt, they produce prodigious amounts of meltwater that flows underneath and out in front of the ice. These flowing bodies behave according to hydraulic principles and therefore sort sediments in the same way as "normal" rivers do. These sediments are also stratified, which means that distinctive layers of sand, silt, or gravel can be traced horizontally for some distance.

Deposition of glacial outwash creates some distinctive landforms. The most common and extensive glaciofluvial landform is an outwash plain. As the name

implies, an **outwash plain** is a relatively flat landscape (Figure 17.25) that was created through the deposition of sediments carried by meltwater flowing out in front of a glacier. The associated meltwater streams are usually choked with coarse sediment and are therefore braided. Recall that braided streams are associated with aggradation, which, in the case of outwash plains, can result in very thick deposits of water-laid sediment in front of the ice.

Another type of landform associated with glacial meltwater is a **kame**, which is a large mound of sediment deposited by a melting glacier (see Figure 17.20). The formation of a kame begins when sediment accumulates in a depression on the surface of a retreating glacier. The sediment is subsequently deposited on the ground when the ice melts. A kame can form as an alluvial fan at the ice margin or as a deltaic deposit if the ice borders a lake. Kames usually contain stratified deposits of sand and gravel that progressively bury older deposits of a similar kind. Kame landscapes typically consist of rolling and irregular hills that are quite large, so large, in fact, that many ski resorts in the upper Midwest are built on these landforms.

Yet another landform created by flowing water associated with a glacier is an **esker**, which is a winding ridge (Figure 17.26) of coarse sediment deposited by a stream flowing *under* the ice. In this instance, the cross section of a subglacial stream channel is inverted with respect to a normal river channel; that is, the channel looks like an upside-down U in cross section. The flowing water is confined by the ice at the top and sides of the tunnel and by the ground beneath it, and so is the deposition of

Glacial outwash *Sediment deposited by meltwater streams emanating from a glacier.*

Outwash plain *A broad landscape of limited relief created by the deposition of glacial outwash.*

Kame *A large mound of sediment deposited along the front of a slowly melting or stationary glacier.*

Esker *A winding ridge formed by a stream that flows beneath a glacier.*

Figure 17.27 Kettle lakes. Kettle lakes form when ice blocks break off the front of a glacier, are subsequently buried by glacial sediment, and then melt, forming a depression and a lake if the water table is high. This air photo shows two beautiful kettle lakes in Wisconsin.

sediment. When the ice eventually melts, a meandering ridge of well-sorted and stratified glacial sediment is left behind. Most eskers are discontinuous because the conditions required for subglacial fluvial deposition are difficult to maintain over long distances. Nevertheless, eskers can be over 30 m (100 ft) high and tens of kilometers long.

Still another kind of landform associated with glacial outwash is called a **kettle lake**. A kettle lake forms when a large block of ice falls off the front of a glacier and is subsequently buried by glacial outwash as the ice retreats. At some point, long after the glacier is gone, the buried block of ice melts. In the process of melting, the block creates its own lake, first by forming the depression that results from the subsiding sediments, and second through providing the water to fill the depression. This lake can subsequently be maintained at its original size if the depression intersects the water table. Kettle lakes are very common in heavily glaciated terrain where they are scattered on the landscape (Figure 17.27). As discussed

---

Kettle lake *A lake that forms when a block of ice falls off the glacial front, is buried by glacial drift, and then melts, forming a depression that fills with water.*

in Chapter 15, the state of Minnesota is known as the "Land of 10,000 Lakes" because it contains thousands of kettle lakes that intersect the water table. The next time you camp at a small lake in this area, or any similar place in the northern tier of the United States, remember that the lake probably evolved because an ancient ice block was buried and then melted. Amazing.

 www.wiley.com/college/arbogast

Depositional Glacial Landforms

Like many other Earth processes, the formation of glacial landforms can be better understood when visualized in animated form. To do so, go to the *GeoDiscoveries* website and select the module *Depositional Glacial Landforms*. This module illustrates the way that moving and melting ice shapes the landscape. It also contains a nice video of glaciers in Peru. After you complete the interactivity, be sure to answer the questions at the end of the module to test your understanding.

Glaciers construct landforms in many different ways, both in direct contact with the ice and in close proximity to it. Given your understanding of landscape evolution and glaciers, what kind of landform is being constructed in this image? How can you tell? What will the relief of this landform be like when it is fully developed?

History of Glaciation on Earth

The many examples of landforms created by glaciers show that ice sheets have been an important and widespread feature around the world in the past. The presence of drumlins in places like New York and Michigan, for example, indicates that ice sheets once existed where they no longer do. You might wonder if it is possible for such conditions to return at some point in the future. This is a great question and provides the context for a discussion of Earth's glacial history.

In order to provide the proper context about the glacial past, some fundamental terminology must be established. Periods of glacial advance, when glaciation is a dominant worldwide process, are called *glacials*. During glacials, temperatures are cooler on average, ice sheets expand and become larger, and sea level is lower because more water is stored in the ice sheets. Periods of glacial retreat, such as we are in now and have been for about the past 10,000 years, are called *interglacials*. During interglacials, average global temperatures are warmer, ice sheets and alpine glaciers are smaller, and sea level is higher because meltwater from the glaciers has returned to the sea.

Although the evidence is extremely vague, the first major glacial period on Earth appears to have occurred during Precambrian time about 2.3 billion years ago. This glaciation was in response to a major fluctuation in the composition of the early atmosphere. Prior to this period, the atmosphere was composed mostly of methane, an important greenhouse gas. Sometime approximately 2.3 billion years ago the amount of oxygen in the atmosphere increased dramatically, which in turn cooled the planet such that much of it was covered by ice. In fact, ice was so extensive at this time the period is sometimes called the *snowball Earth*. During this glacial interval, many of the world's oceans were frozen to a depth of about 0.8 km (0.5 mi). Similar "snowball" glacials occurred later in Precambrian time, about 750 million and 600 million years ago, and are most closely associated with decreases in carbon dioxide, another greenhouse gas.

Although much of Earth was heavily glaciated during these so-called snowball glaciations, the most recent and best-understood glacial periods are those associated with the Pleistocene Epoch between about 1.6 million to 10,000 years ago. During this epoch of time, alpine and continental glaciation occurred on such a large scale that 30% of the Earth's landmasses were covered by glacial ice at its maximum extent. For perspective, remember that contemporary glaciers now extend over only 11% of the Earth's land surface.

Until the latter part of the 20th century, glacial geomorphologists believed that four major advances of glacier ice occurred during the Pleistocene Epoch. This history was suggested through the reconstruction and dating of stratified tills in Europe and North America. In North America, these "classic" advances were named after the locations in which the ice advanced to its farthest point and where prominent evidence was found in the form of distinctive glacial drift. These glacial periods were given the following names (and approximate ages):

1. Nebraskan—1 million years ago
2. Kansan—600,000 years ago
3. Illinoisan—300,000 years ago
4. Wisconsin—35,000 to 10,000 years ago

In each of these advances, a thick continental glacier extended from the Arctic to a point deep within the American Midwest. East of the Rocky Mountains, this glacier is typically referred to as the **Laurentide Ice Sheet** because it covered the Laurentian Mountains in eastern Canada. Thick ice caps also covered much of

---

Laurentide Ice Sheet *The continental glacier that covered eastern Canada and parts of the northeastern United States during the Pleistocene Epoch.*

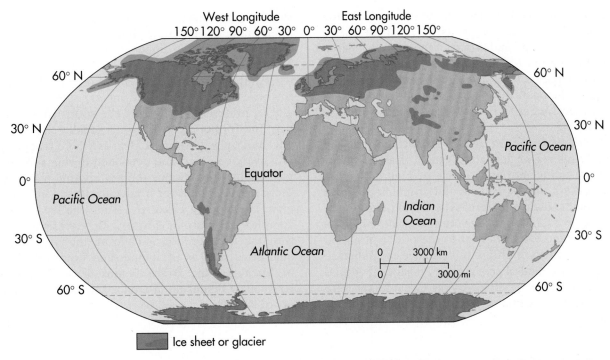

West Longitude   East Longitude
150° 120° 90° 60° 30° 0° 30° 60° 90° 120° 150°

■ Ice sheet or glacier

Figure 17.28 Maximum extent of glaciation during the Pleistocene Epoch. During this period of time, glaciers covered 30% of the Earth's land surface, compared to 11% today.

the Rockies and extended west to the Pacific coast and south into Washington. Collectively, these ice caps are referred to as the **Cordilleran Ice Sheet**, which, at its maximum extent, merged with the Laurentide Ice Sheet in central Canada to form a single ice mass across much of North America. According to classical theory, the Laurentide Ice Sheet reached its southernmost point during the so-called Kansan glaciation about 600,000 years ago (Figure 17.28). This is the glacial advance that stopped at present-day northeastern Kansas and deposited the glacial erratic shown in Figure 17.14.

As discussed in Chapter 9, however, reconstruction of oxygen-isotope records derived from glaciers and ocean sediments indicates that more than 20 such glacial periods occurred during the Pleistocene Epoch. These numerous cycles are apparently related to Earth's orbital fluctuations associated with the Milankovitch theory, which was also discussed in Chapter 9. Recall that this theory describes how Earth's geometric relationship with the Sun slowly fluctuates over time, with changes in axial tilt, precession of the equinoxes, and orbital shape at distinct cyclic intervals. In this context, the most intense glaciations would be theoretically associated with periods when (1) Earth was farthest away from the Sun in summer, (2) axial tilt is less than present, and (3) equinox occurred during the sum-

mer and winter months. As these geometric conditions slowly aligned over tens of thousands of years, the amount of solar radiation striking the high latitudes of the Northern Hemisphere in summer would gradually decrease, resulting in progressively cooler temperatures and the growth of ice sheets. Because of these new understandings, the only names maintained from the traditional glacial model in North America are the Illinoisan and Wisconsin advances because the deposits associated with these most recent glaciations are easily recognizable. All other glaciations are lumped into one category called *Pre-Illinoisan*.

The Wisconsin Glaciation and Evolution of the Great Lakes

Given that the Wisconsin glacial advance is the most recent, it is easily the most studied of all Pleistocene glaciations because its effects can still be seen clearly at the surface. Although strong evidence for this glaciation is found in northern Europe and western Canada, a great place to examine its impact on the landscape is the Great Lakes region of North America. Following the post-Illinoisan interglacial period, the Laurentide Ice Sheet began to expand again about 110,000 years ago. For the next approximately 90,000 years ice advanced south, reaching its most southerly position in central Ohio, Indiana, and Illinois by about 20,000 years ago. At this time, *all* of what is now known as the Great Lakes and the state of Michigan were buried beneath a sheet of ice (Figure 17.29a) that was probably

Cordilleran Ice Sheet   *The ice cap that covered much of the mountains in the northwestern part of North America during the Pleistocene Epoch.*

hundreds of meters thick. This ice subsequently melted from the Great Lakes region between about 16,000 and 10,000 years ago and is part of the overall collapse of the Laurentide Ice Sheet at this time. The glacier melted in large part because Earth's axial tilt had increased to the point that sufficiently more solar radiation was received at high latitudes in summer to cause rapid melting.

Although the most recent deglaciation of the Great Lakes region left a multitude of glacial landforms, including moraines (such as Figure 17.22b), kames, eskers, outwash plains, and drumlins, the most prominent landscape feature is the Great Lakes themselves (Figure 17.29b). The evolution of these lakes is fascinating and is a testament to the power of glacial ice and how rocks of different resistance can influence the shape of a landscape. Prior to Pleistocene glaciation, the areas now occupied by the Great Lakes were large river valleys that were cut into soft deposits of shale. Very resistant granites, limestones, and sandstones underlie the areas that are now regional landmasses, such as the upper and lower peninsulas of Michigan.

As the Laurentide Ice Sheet oscillated back and forth across the region during the Pleistocene, the glacier preferentially eroded the softer rocks. In this way, the former river valleys gradually changed into enormous glacial troughs. Because these troughs progressively enlarged and deepened during subsequent glaciations, they became the preferential pathways of the ice front whenever it advanced back into the region. The modern Great Lakes are the legacy of that complex geomorphic history. More resistant rocks were much less eroded and became the foundation for landforms such as the Upper and Lower Peninsulas of Michigan, the Bruce Peninsula in Ontario, the Keweenaw Peninsula in Upper Michigan, and a variety of islands.

DISCOVER...

THE CHANNELED SCABLANDS IN EASTERN WASHINGTON

The Channeled Scablands in eastern Washington are known for the numerous and dry canyons carved into the landscape. In the early 20th century, the geologist J. Harlan Bretz hypothesized that these landforms were created by catastrophic floods caused by melting glaciers. He was ridiculed for most of his life about this theory because a majority of geologists then believed that Earth-changing processes due to glaciers had to be slow and gradual. However, by the time Bretz died in 1981, his hypothesis had been accepted. In fact, he received geology's highest honor, the Penrose Medal, in 1979.

The horizontal lines visible in this hillside in Missoula, Montana, are old shorelines created by wave action of glacial Lake Missoula during the most recent ice age. To imagine the depth of the water, note the school building in the foreground for scale.

Geologists now believe that a lobe of the Cordilleran Ice Sheet—the Purcell Trench Lobe—advanced far enough into Montana that it blocked a valley through which glacial meltwater flowed west. This blockage created an ice dam that allowed a deep lake to form behind it, spreading back into the mountain valleys to the east. This lake is known to geomorphologists as *Glacial Lake Missoula* because it formed shorelines that are still visible around Missoula, Montana. These shorelines indicate that the lake was as big as today's Lake Ontario and Lake Erie combined. As the lake deepened, it eventually began to flow over the top of the ice dam, finally causing the dam to break catastrophically.

Torrential floodwaters, up to 300 m (1000 ft) deep, subsequently burst onto the plains of what is now eastern Washington, flowing at a rate 10 times the total of all the rivers on Earth and emptying Lake Missoula in perhaps as little as 48 h. This type of catastrophic discharge may have happened up to 100 times and created the enormous valleys carved into the landscape. The network of now-dry channels is so well defined that it can be seen on satellite imagery.

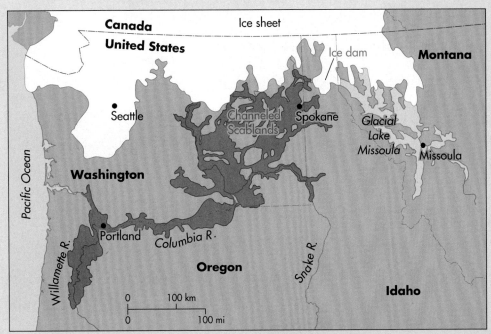

The Channeled Scablands cover a large area that includes most of southeastern Washington.

The Channeled Scablands show up beautifully as light brown channel forms (arrows) in this satellite image that focuses on eastern Washington, northeast Oregon, and western Idaho.

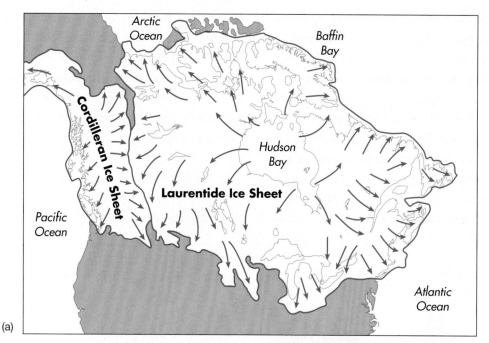

(a)

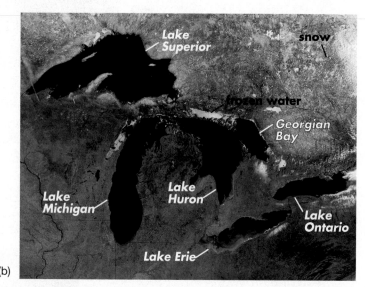

(b)

Figure 17.29 Extent and impact of the Wisconsin glaciation. (a) During the most recent ice age, the Laurentide Ice Sheet and Cordilleran Ice Sheet collectively covered most of northern North America. (b) Satellite image of the Great Lakes acquired in April 2005. These lakes were formed when the advancing ice further enlarged and deepened preexisting glacial troughs that had been scoured out of relatively soft rocks. Note the extent of snow cover, frozen lakes, and the suspended sediment in Lake Erie.

KEY CONCEPTS TO REMEMBER ABOUT THE HISTORY OF GLACIATION ON EARTH

1. Several periods of intense glaciation occurred during Precambrian time in which most of the planet was covered by ice. Such a period is referred to as a snowball Earth.

2. In the context of geologic time, the most recent period of major continental glaciations is the Pleistocene Epoch, which lasted from approximately 1.6 million to 10,000 years ago.

3. Until recently, it was believed that four major glaciations occurred during the Pleistocene Epoch.

Oxygen isotope records derived from ice cores and marine sediments indicate that more than 20 major glacial periods occurred during the Pleistocene. The best understood of these glaciations are the Illinoisan and Wisconsin periods, which occurred most recently. All other glacial periods are lumped into one category called the Pre-Illinoisan.

4. Most glacial landforms in the upper Midwest, the northeastern United States, and Canada are products of the Wisconsin glaciation, which occurred between about 110,000 and 10,000 years ago. The most distinctive large-scale landforms are the Great Lakes, which are glacial troughs cut into soft bedrock.

Probable Human Impact on Glaciers

With an understanding of past glaciations in mind, a key question to now consider is the status of contemporary glaciers with respect to the issue of ongoing climate change. Although this period of warming may be part of a natural cycle, most climatologists believe that it is associated with increased levels of atmospheric carbon dioxide (and other greenhouse gases) caused by human industrial activity. Regardless of the cause, Earth appears to be in a distinct warming phase that has been linked to the increased incidence of drought, species migration and loss, and rise in sea level. Global warming is also having a major impact on glaciers around the world.

As previously discussed in this chapter, glacial advance and retreat are controlled by fluctuations in the glacial mass budget (see Figure 17.4). When snowfall in the source area exceeds melting in the zone of ablation, glaciers thicken and expand, either down valleys in alpine settings or across landmasses in continental glaciers. Glaciers naturally recede during periods when the rate of melting in the zone of ablation exceeds the accumulation of snow in the source area. During the Pleistocene Epoch, the budget of global ice masses was controlled largely by fluctuations in Earth–Sun geometry as described in the Milankovitch theory. The most recent noteworthy episode of glacial advance occurred during the Little Ice Age, which was a time of global cooling that occurred between about 1500 and 1850 A.D. Since then, glaciers have generally been retreating around the world—in many instances at rates faster than previously recorded. This rapid rate of glacial retreat has been attributed by many scientists to human-induced global warming.

Although some of the world's glaciers have actually advanced in the past few decades, such as the Perito Moreno Glacier in Argentina, the vast majority are melting at rapid rates. Glacier retreat has been very noticeable in most alpine environments. In the Swiss Alps, for example, the Rhône glacier has retreated about 2.5 km (1.5 mi) in approximately the past 140 years. In Peru, the Ururashraju Glacier retreated 500 m (1640 ft) between 1986 and 1999 alone! Another excellent example of rapid alpine glacial retreat is Mount Kilimanjaro, which is the largest mountain in Africa at an elevation of 5895 m (19,336 ft). Although it lies near the Equator, this volcano on the East African Rift has had a substantial summit glacier for more than 11,000 years. This glacier, which is seen in Figure 17.30a, inspired the famous Hemingway novel, *The Snows of Kilimanjaro*. In this past century the extent of the ice mass at Mount Kilimanjaro has shrunk about 80%. Extensive glacial melting was observed in Landsat imagery between 1993 and 2000 alone (Figures 17.30b and c).

(a)

(b)

(c)

Figure 17.30 The melting snows of Kilimanjaro. (a) Although Mount Kilimanjaro lies near the Equator, it has long had a substantial ice cap due to its high elevation. (b) Landsat image of the ice cover on Mount Kilimanjaro in February 1993. (c) Landsat image of Mount Kilimanjaro in February 2000. Notice the extent of melting that occurred over only 7 years.

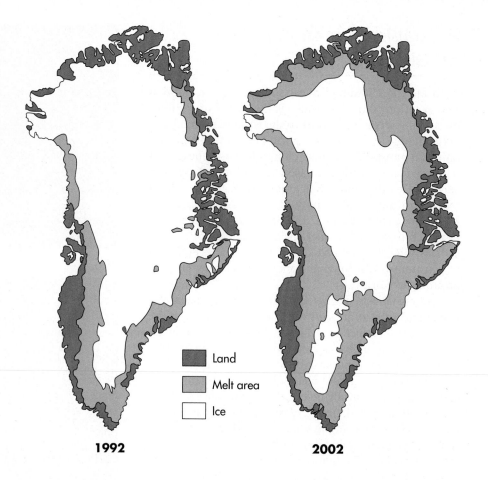

Figure 17.31 Melting of the Greenland Ice Sheet between 1992 and 2002. The Greenland Ice Sheet is melting at a rapid rate. Notice, for example, how the area of melting ice in 2002 is much greater than it was in 1992. (*Source:* NASA.)

Land

Melt area

Ice

1992   **2002**

If the rate of melting continues, it is estimated that the entire glacier will disappear by 2015.

Rapid melting of glaciers has also been observed on the Greenland and Antarctic ice sheets. On the Greenland Ice Sheet, the area on the edge of the glacier affected by melting increased by 16% between 1979 and 2002. You can see how the area affected by melting changed between 1992 and 2002 in Figure 17.31. On the Antarctic Ice Sheet, the rate of glacial melting is even

more impressive. An example of this melting is the collapse of the Larsen B Ice Shelf that occurred over a 35-day period beginning January 31, 2002. An ice shelf is a plate of glacial ice that extends from the landmass into the water; many of these features extend off the Antarctic landmass. During the Antarctic summer of 2002, about 3500 km² of the Larsen B Ice Shelf disintegrated. For scale, consider that the entire state of Rhode Island is 2717 km² (1049 mi²) in size. Figure 17.32 shows

www.wiley.com/college/arbogast

Glaciers and Climate Change

Evidence strongly indicates that glaciers around the world are melting at rapid rates due to climate change. These changes have been documented at many glaciers through repeat photography and time-lapse imagery. To see examples of such changes, go to the *GeoDiscoveries* website and select the module *Glaciers and Climate Change*. This

module contains an excellent discussion of changes at Glacier National Park in Montana. It also contains links to sites where glaciers have been closely monitored in the past decade. Once you complete the module, be sure to answer the questions at the end to test your understanding.

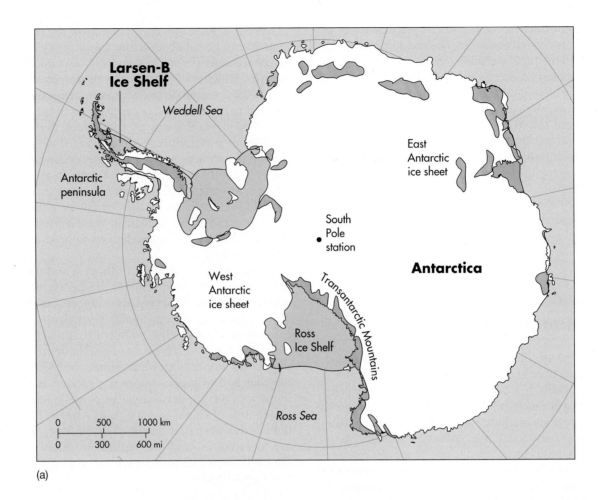

(a)

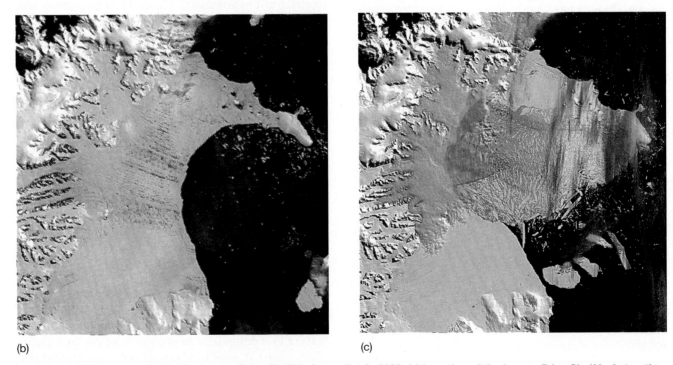

(b)

(c)

Figure 17.32 Disintegration of the Larsen B Ice Shelf in Antarctica in 2002. (a) Location of the Larsen B Ice Shelf in Antarctica. (b) The ice shelf when breakup began on January 31, 2002. The dark spots on the shelf are patches of meltwater. (c) Status of breakup on March 5, 2002. The largest ice blocks are hundreds of square kilometers in size.

the location of the ice shelf and the progression of its breakup in two satellite images. The breakup is believed to be due to the rapid warming (about 0.5°C [0.9°F] per decade) that has occurred since the 1940s. This warming has had a profound impact, causing the extent of seven ice shelves on the Antarctic peninsula to decline by a total of about 13,500 km² (5212 mi²) since 1974.

As evidence that these processes are still active, a chunk of ice about 16 km (10 mi) wide and 80 km (50 mi) long rapidly broke off the Greenland ice cap in late July 2010, right before this text was published. This break raised concerns that the ice cap may be reaching an important threshold as far as its future stability is concerned.

Periglacial Processes and Landscapes

Although we are clearly living in an interglacial period, one that may even become significantly warmer if global warming models prove accurate, much of Earth remains very cold over much or all of the year. Naturally, these very cold regions are located at high altitudes and latitudes and are associated with polar-type climates. Associated with these climates is a range of specific geomorphic processes referred to as **periglacial processes** that are closely associated with substantial amounts of frost action, specifically the effects that continual freezing and thawing have on the landscape. These processes are closely related to physical weathering, mass movement, and soil modification and affect approximately 20% of the Earth's surface.

Permafrost

The most common periglacial processes are associated with **permafrost**, which, as the name implies, is ground that is permanently frozen. These conditions typically develop when soil or rock temperatures remain below 0°C (32°F) for at least 2 years. Permafrost may develop because the area is covered by a glacier or it may be in close proximity to the ice margin and thus periglacial.

Periglacial processes *The suite of processes involving frost action, permafrost, and ground ice that occurs in arctic environments or along the margins of ice sheets.*

Permafrost *Ground that is permanently frozen.*

From a geographical perspective, regions of permafrost can be subdivided into two primary categories: (1) continuous permafrost and (2) discontinuous permafrost. Figure 17.33 shows the geographical distribution of these regions in the Northern Hemisphere. Large areas of permafrost also occur in Antarctica but are not described here. Regions of continuous permafrost in the Northern Hemisphere are associated with the most frigid temperatures, which essentially occur poleward of the –7°C (19°F) mean annual isotherm. In areas of continuous permafrost, all surfaces are frozen except those insulated beneath deep lakes or rivers. Although the average depth of permafrost in these regions is about 400 m (1300 ft), it can reach the impressive depth of 1000 m (3300 ft).

South of the region of continuous and deeply frozen ground, the mean annual temperatures become warmer and permafrost is discontinuous and shallower. This region of discontinuous permafrost generally extends in a broad area bounded by the –7°C (19°F) and the –1°C (30.2°F) isotherms. Permafrost in this region is discontinuous where slopes face south toward the Sun and in places where the ground is insulated by snow. Overall, permafrost of some kind occurs in about 50% of Canada and 80% of Alaska. To see how the depth and spatial extent of permafrost vary across the continuous and discontinuous zones, examine Figure 17.34, which shows a typical cross section across the high latitudes.

Permafrost Processes Figure 17.34 also shows how permafrost impacts the landscape. For example, notice the areas denoted as *active layer*. The active layer is the part of the soil that melts and refreezes on a daily or seasonal basis. This zone lies above the subsurface permafrost layer and predictably varies with latitude, ranging in thickness from up to 2 m (6.6 ft) in areas of discontinuous permafrost to only 10 cm (4 in.) in northern areas of continuous permafrost.

As you can imagine, the active layer is sensitive to climate change. Although the response is slow, colder temperatures gradually cause the thickness of the active layer to decrease, whereas warmer temperatures result in permafrost melting and an increased thickness of the active layer. In this context, abundant evidence indicates that Arctic permafrost is rapidly thawing in Earth's current warming phase. This thawing is releasing tons of CO_2 into the atmosphere, which should contribute to more warming in the future.

In addition to the active layer, another important feature in a permafrost landscape is a talik (see Figure 17.34 again). A *talik* is a body of unfrozen ground that is found within a zone of discontinous permafrost or

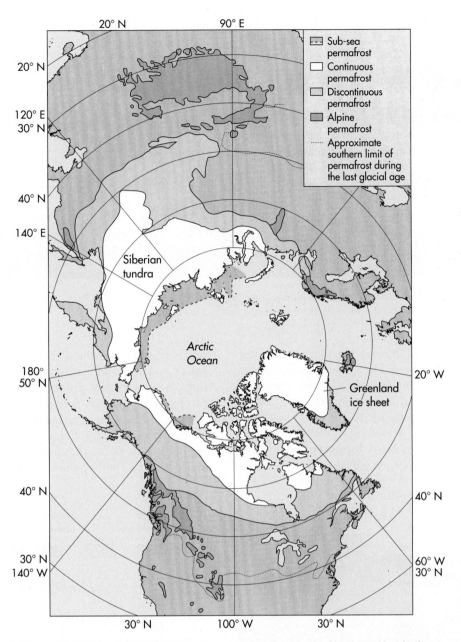

Figure 17.33 Geographical distribution of permafrost in the Northern Hemisphere.
An extensive part of the Northern Hemisphere contains permafrost. Note that it can occur in the oceans as well as on land.

beneath a lake or river in the continuous region. Taliks are especially important in discontinuous permafrost zones because they are a link between the active layer and groundwater, forming a kind of conduit through which water can move freely.

Ground Ice and Associated Landforms

Within permafrost regions, distinct zones of frozen water occur within the ground. These areas are referred to as **ground ice** and contain highly variable amounts of water, ranging from near zero to completely saturated conditions. Areas of ground ice expand and contract along freezing fronts, which are the boundaries between frozen and unfrozen ground. As these areas expand and contract, they cause physical weathering of the landscape through the frost action processes

---

Ground ice   *Distinct zones of frozen water that occur in permafrost regions.*

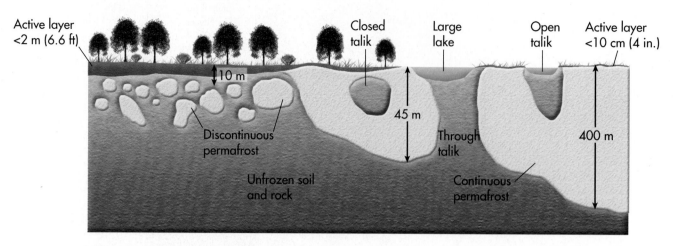

SOUTH   NORTH

Active layer <2 m (6.6 ft)

10 m

Closed talik

Large lake

Open talik

Active layer <10 cm (4 in.)

Discontinuous permafrost

45 m

400 m

Through talik

Unfrozen soil and rock

Continuous permafrost

Figure 17.34 Typical periglacial landscapes. Note the differences in the spatial extent and depth of permafrost across this hypothetical cross section (not to scale).

described in Chapter 14. Remember that when water freezes, it expands by about 9%. This expansion causes rocks to break along joint lines. In addition, water expansion can cause soil to move vertically or horizontally through the processes of *frost heaving* and *frost thrusting*, respectively.

Ground ice comes in many forms. The most basic kind of ground ice is *pore ice* where water freezes in soil pore spaces. Ground ice also occurs as *horizontal lenses* and veins of ice that extend in random directions. A third kind of ground ice is an *ice wedge*, which results when water enters a crack in the ground (Figure 17.35). *Segregated ice* is ground ice that is buried but grows when additional water is added. Finally, *intrusive ice* is ground ice that forms when water is injected under pressure, perhaps because it lies between a pair of

(a)

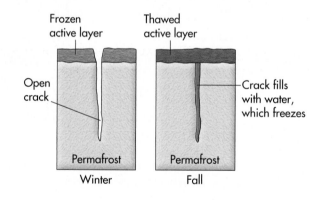

Frozen active layer   Thawed active layer

Open crack

Crack fills with water, which freezes

Permafrost   Permafrost

Winter   Fall

After 500 years

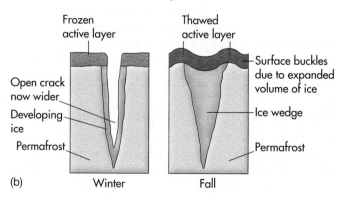

Frozen active layer   Thawed active layer

Open crack now wider

Developing ice

Permafrost

Winter

Surface buckles due to expanded volume of ice

Ice wedge

Permafrost

Fall

Figure 17.35 Formation of an ice wedge. (a) An ice wedge in northern Canada. (b) Sequence of steps in the formation of an ice wedge. These features form when water penetrates an open crack in the ground and freezes. Over time, repeated freezing and thawing slowly cause the crack to expand, forming a wedge. (*Source:* Modified from Lachenbruch, 1962, "Mechanics of Thermal Contraction and Ice-Wedge Polygons in Permafrost," Geological Society of America Special Paper 70.)

(b)

Figure 17.36 A pingo. A pingo forms when water is injected into the ground under some form of hydraulic pressure. This pressure forces the ground to bulge upward.

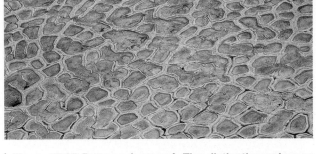

Figure 17.37 Patterned ground. The distinctive polygons on this landscape form when frost action preferentially brings large stones and rocks to the surface.

aquicludes. Where intrusive ice is found, the surface of the ground can bulge up to 60 m (200 ft) upward, forming a landform called a *pingo* (Figure 17.36).

In addition to pingos, frost action can result in a variety of other landforms. A *palsa* (Spanish for "elliptical") is a rounded or elliptical mound of peat that is similar to a pingo in that it is thrust upward. It differs from a pingo, however, because it is only 1 m to 10 m (3 ft to 30 ft) high and contains ice lenses rather than a solid core. Another distinctive periglacial landform is *patterned ground*. As the name implies, patterned ground is a landscape that has evolved such that it contains distinctive shapes. These shapes are usually polygons (Figure 17.37) that form when frost action preferentially brings coarser materials—stones and boulders—to the surface.

THE BIG PICTURE

The previous two chapters have focused on the processes and landforms associated with flowing water and ice. In the next chapter, we turn to another kind of flow, moving air. You will discover that wind can have a dramatic impact on the landscape through the erosion and transport of wind-blown sediment. These processes shape the landscape in predictable ways that produce landforms such as sand dunes and vast plains covered by wind-blown silt. Eolian processes are most closely associated with dry places, so the chapter will begin with the geomorphology of arid environments and will integrate the elements of climate, weathering, and rock structure discussed earlier.

SUMMARY OF KEY CONCEPTS

1. Glaciers form when snow accumulates annually and does not melt during the summer months. In this way, the snow mass gradually thickens with each passing year and gradually changes into glacial ice through compression.

2. A growing ice mass becomes a glacier when it begins to flow under its own weight. Whether a glacier advances or retreats depends on the glacial mass budget, which is the balance between snow accumulation in the source area and melting at the zone of ablation. Melting and accumulation are in balance along an equilibrium line.

3. Ice fields and ice caps are large ice masses that spread horizontally across mountainous landscapes. Valley glaciers flow down preexisting stream valleys in alpine settings. Continental glaciers, such as the Antarctic Ice Sheet, are thousands of meters thick and spread across continental landmasses.

4. Glaciers erode the ground beneath them through the combined processes of abrasion and plucking. In alpine settings, these processes change the landscape from generally rounded to much more angular, with sharp features such as arêtes, horns, and cirques.

5. Glaciers deposit sediment either directly as till or indirectly from meltwater streams as outwash. These processes produce landforms such as moraines, kames, and outwash plains.

6. During the Pleistocene Epoch (between about 1.6 million and 10,000 years ago), glacial ice covered up to 30% of the Earth's land surface including extensive parts of North America, Europe, and Asia. The most recent period of massive glacial advance within the Pleistocene was the Wisconsin glaciation, which ended approximately 10,000 years ago. Glaciers now cover about 11% of Earth's landmasses. These tremendous fluctuations in ice volume are intimately related to gradual changes that occur with respect to Earth–Sun geometry.

7. Evidence for the Pleistocene ice ages is found in many places in the northern part of the United States and Canada. The Great Lakes, for example, were created because large ice lobes enlarged preexisting stream valleys.

8. Permafrost refers to ground that is perpetually frozen, forming features such as ice wedges and pingos. Areas of current permafrost conditions occur at high latitudes and altitudes.

CHECK YOUR UNDERSTANDING

1. Why are summertime temperatures a critical variable with respect to the formation of a glacier?

2. Describe the steps involved in the metamorphosis of snow to glacier ice.

3. How does a glacier flow? Where does the ice move the fastest and slowest? Why?

4. Compare and contrast the various kinds of glaciers on Earth.

5. Describe the difference between glacial abrasion and glacial plucking. What landform is produced by a combination of these two processes?

6. How can you tell if a mountainous landscape has been glaciated? What are some of the characteristic landforms that should be present if the landscape was glaciated?

7. How does a hanging valley form?

8. In what two regions of Earth are large continental glaciers currently found?

9. What is the difference between glacial till and glacial outwash? What are the diagnostic characteristics of each deposit?

10. How does the formation of a recessional moraine differ from that of a terminal moraine?

11. Explain the history of glaciations on Earth.

12. What is the impact of ongoing climate change on most glaciers around the world?

13. What is the difference between continuous and discontinuous permafrost? What is their geographical distribution, and why does it make sense given your understanding of permafrost processes?

14. What is ground ice? Describe two landforms created by this phenomenon.

ANSWERS TO VISUAL CONCEPT CHECKS

VISUAL CONCEPT CHECK 17.1

The glacial mass budget refers to the balance between snow accumulation in the source area and melting in the zone of ablation. The point where these two processes are in balance is known as the equilibrium line. Below the equilibrium line, crevasses are visible, whereas above the equilibrium line they are covered with snow.

VISUAL CONCEPT CHECK 17.2

Glaciers modify alpine landscapes in a very intense way. Prior to glaciation, alpine landscapes are generally rounded, with streams flowing in V-shaped valleys. Through the combined processes of abrasion and plucking, alpine glaciers cause mountainous regions to become more angular, forming features such as cirques, arêtes, and horns. Valley glaciers enlarge and deepen former stream valleys, changing them to U-shaped valleys.

VISUAL CONCEPT CHECK 17.3

This photograph of a stream system in the Southern Alps of New Zealand is a nice example of how outwash plains form. Several lines of evidence point to this conclusion. The snow-capped peaks in the background suggest that alpine glaciers are present, which must be delivering coarse sediment to the stream in glacial meltwater. You can tell that the sediment is coarse because the stream is braided. Given that fluvial deposition is occurring, you can surmise that the landscape will have very limited relief and will thus be consistent with an outwash plain.

CHAPTER EIGHTEEN

ARID LANDSCAPES AND EOLIAN PROCESSES

This chapter focuses on the geomorphology of arid regions and the way that flowing air shapes the landscape. Wind processes are typically referred to as *eolian processes*, which produce eolian landforms. (The term *eolian* is derived from Aeolus, the Greek god of the winds.) Although the ability of wind to shape the landscape is low relative to moving water and ice, a fascinating array of erosional and depositional landforms can nevertheless be produced by flowing air. It is important to understand the role that wind plays with respect to landform evolution because more than one-third of the Earth's land surface is located within arid or semi-arid climate zones. In addition, many of these landscapes are agriculturally important and can be adversely affected by the wind.

Some of the most beautiful landforms on Earth are in arid regions, such as these sand dunes in Death Valley, California. This chapter focuses on these areas, including the way rock structure influences topography and how streams shape the landscape. It also examines the behavior of wind and how it creates landforms such as those pictured here.

CHAPTER PREVIEW

Arid Landscapes

Eolian Erosion and Transport

Eolian Deposition and Landforms
    On the Web: *GeoDiscoveries Eolian Processes and Landforms*

Human Interactions with Eolian Processes

The Big Picture

LEARNING OBJECTIVES

1. Describe the physical geography of arid landscapes.
2. Discuss how rock structure influences the landscapes of the southwestern United States.
3. Explain the formation of an alluvial fan.
4. Understand the character of eolian processes and how they move sediment.
5. Describe the formation of a sand dune.
6. Compare and contrast the various forms of sand dunes.
7. Describe the character of loess and where it is found on Earth.
8. Discuss human interactions with eolian processes in the African Sahel and Great Plains.

Arid Landscapes

A good place to begin investigating eolian processes is by looking more closely at the geography and character of arid environments because these areas are often most closely associated with the influence of flowing air. This discussion incorporates many of the topics covered earlier in the text—atmospheric circulation, plant geography, soils, geology, and fluvial processes. A goal of this part of the chapter is to provide you with a good understanding of desert environments and their geography.

Figure 18.1 shows the geography of arid and semi-arid regions on Earth. Such climates occur in both warm and cold settings and collectively form the largest climate region on the planet, one that covers perhaps as much as 35% of Earth's land surface. Within the Köppen climate system (Figure 9.2), these climate regions fall within the arid desert (BW) and semi-arid steppe (BS) climate categories. The spatial distribution of these dry regions is based on one of three factors:

1. *Dominance of Subtropical High-Pressure Systems*— In areas located between about 15° and 35° N and S latitude, strong high-pressure systems dominate the weather for all or significant portions of the year. These descending air masses are components of the Hadley cells (shown in Figures 6.16 and 6.19) that circulate within the atmosphere in the tropical latitudes. The subtropical high-pressure systems are very closely associated with the largest deserts on Earth, such as the Sahara and Kalahari in Africa, the Simpson Desert in Australia, and the Chihuahuan Desert in Mexico (Figure 18.1).

2. *Rain Shadow Effect*—Another environmental factor associated with deserts is the rain shadow effect. If you recall from Chapter 7, a rain shadow occurs on the lee side of mountain ranges where descending air warms adiabatically. This process was outlined in Figure 7.22 and is responsible for the large zones of arid and semi-arid climates in the western United States. It also is related to the presence of the Atacama Desert along the west coast of South America.

3. *Distance from Large Water Bodies*—The last environmental variable that is related to dry regions is the proximity to a water body such as an ocean. Locations deep within continents tend to receive relatively small amounts of moisture-laden air and can therefore be quite dry. An excellent example of this kind of situation is the Taklimakan and Gobi Deserts in Asia (Figure 18.1), as well as the other large areas of arid and semi-arid lands that lie deep within that continent.

Desert Geomorphology

A key element that all deserts and semi-arid regions share is a relative lack of vegetation when compared to more humid areas. As a result, descriptive terms such as *sparse* and *barren* are frequently used to describe dry regions. This relationship of climate and vegetation was thoroughly explored in Chapters 9 and 10 and can be seen in numerous photographs of desert environments, such as Figures 9.5 and 10.17.

Although the barren nature of these environments may at first seem unattractive or even foreboding, desert landscapes are beautiful in their own way, with

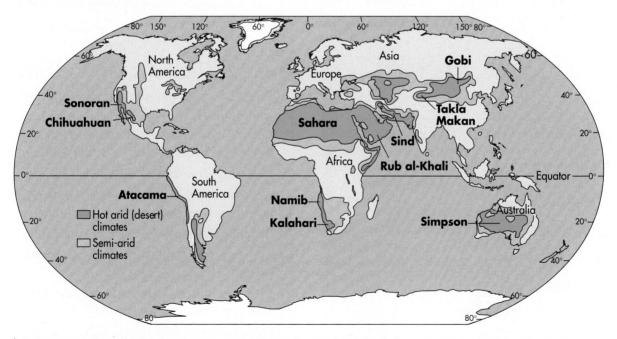

Figure 18.1 Deserts of the world. Arid and semi-arid deserts collectively cover about one-third of the Earth's land surface. The largest deserts are associated with the belt of subtropical high-pressure systems.

spectacular rock outcrops and panoramic vistas. In the context of geomorphology, deserts are great places because you can actually *see* important elements like rock structure, differential weathering, and the entire outline of landforms clearly. A superb example of this visual effect can be seen at *Uluru* (formerly known as Ayer's Rock) in central Australia (Figure 18.2). Uluru is the best example of an *inselberg* (island mountain) in the world. Composed of hard sandstone, this prominent landform towers over the surrounding landscape because differential weathering over millions of years removed the rocks that once covered it. In this desert environment, it is possible to visualize this process. Although landforms in humid regions are being thoroughly investigated in many places, it can be somewhat more complicated because thick stands of vegetation often mar the view.

The Southwestern United States An excellent place to see desert geomorphology where you may one day visit is the southwestern United States. This region lies in the rain shadow of the Sierra Nevada (Figure 13.23) and encompasses a number of previously discussed arid places such as the Sonoran Desert (Figure 10.16) and the Grand Canyon (Figure 12.20a). Perhaps the most obvious variable, other than climate, that strongly influences landforms in the West is rock structure. A spectacular example of the effects of normal faulting (Figure 13.22), for example, is the geologically famous Basin and Range Province in the southwestern United States. This region extends from the Wasatch Mountains in eastern Utah to the Sierra Nevada along the border of California and Utah (Figure 18.3a). It is characterized by numerous mountain ranges and intervening valleys that are generally aligned from north to south (Figure 18.3b). The relief between the crests of mountain ranges and the adjacent basins can be as much as about 3050 m (10,000 ft), resulting in a distinctive view from the air (Figure 18.3c).

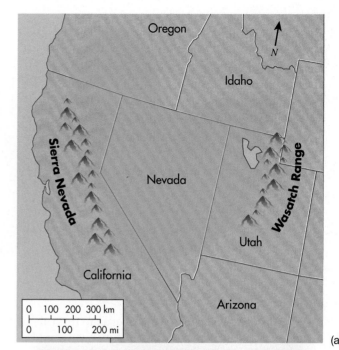
(a)

(b)

(c)

Figure 18.2 Uluru in central Australia. Uluru is the best example of an inselberg, or island mountain, in the world. This iconic landform is composed of hard sandstone that was once buried by other rocks that have been removed by erosion. The beauty of Uluru is magnified in the arid environment of central Australia.

Figure 18.3 The Basin and Range Province. (a) The Basin and Range Province extends over much of the American Southwest. (b) Satellite image of the Basin and Range Province. Notice the semi-linear network of mountain ranges (darker shades) and intervening basins that lie between the Wasatch Range and Sierra Nevada. (c) Panoramic view of a portion of the Basin and Range Province. Note the distinct mountain ranges and the intervening basins.

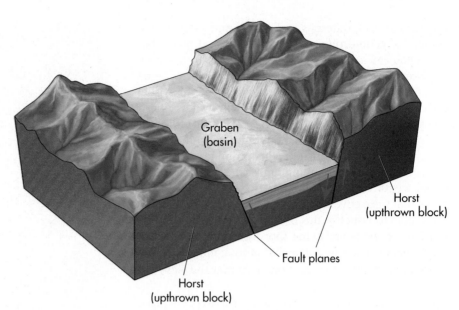

Figure 18.4 Formation of the Basin and Range. The Basin and Range Province consists of a series of normal faults produced by stretching of the continental crust over the past 20 million years. After uplift, erosion attacks the mountain ranges (horsts). The resulting sediments are then deposited in the intervening basins (grabens), slowly filling them with sediment.

Graben
(basin)

Horst
(upthrown block)

Fault planes

Horst
(upthrown block)

The Basin and Range Province began to form about 20 million years ago when the North American crust began to stretch in this part of the continent due to upward pressure applied by an underlying magma plume. As a result of this stretching, the crust thinned, cracked, and pulled apart, creating numerous normal faults that have planes inclined at about a 60° angle (Figure 18.4). This faulting created the series of alternating basins and ranges that characterize the area.

The importance of rock structure can also be seen where the rocks are horizontal and have not been deformed (Figure 18.5). In these areas, which are very common in the West, the rocks usually consist of alternating layers of limestone, sandstone, and shale that contribute to a distinctive pattern of differential weathering. The limestones and sandstones are generally the most resistant rocks to erosion, whereas the shales are relatively soft and thus easily eroded. Where these patterns occur, the hard rocks usually form a resistant caprock that protects the underlying shale to some extent from erosion. Notice in Figure 18.5 that shale tends to form shallow slopes, whereas the more resistant rocks

are associated with steep slopes. Shale tends to erode backward gradually due to slope wash when it does rain. This gradual slope retreat ultimately undermines the caprock above and causes it to collapse due to rockfall and the cycle begins again.

The net effect of this kind of landscape evolution is that large parts of the western United States are extensively dissected (appear to be cut up), with upright landforms of various sizes that rise above the surrounding lower ground. To the novice these landforms seem to have some relationship to one another, but it is difficult to imagine how. In fact, this landscape really represents a slow progression of erosion over millions of years. Figure 18.5 illustrates the landforms associated with this progressive evolution. It is interesting to notice how each of the rock layers can be visually traced from one landform to another. This linkage occurs because the rock was once a uniform mass with broad regional extent. Through a combination of stream erosion, differential weathering, and mass wasting, the landscape was slowly dissected, leaving a succession of flat-topped remnants that tower throughout the area.

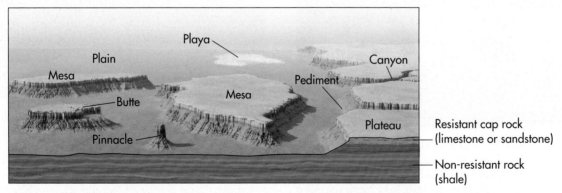

Playa

Plain

Mesa

Butte

Pinnacle

Mesa

Pediment

Canyon

Plateau

Resistant cap rock
(limestone or sandstone)

Non-resistant rock
(shale)

Figure 18.5 Prominent desert landforms associated with horizontal rock structure. Over time, a plateau is dissected into progressively smaller landforms that give the region a distinct appearance.

Figure 18.6 A typical canyon in the western United States. Canyons such as this one are very common in the region and evolved due to vigorous stream incision. Exposed in the canyon walls are layers of sedimentary rock.

Figure 18.7 Landforms in Monument Valley, Utah. This photograph nicely illustrates the concept that the mesa, buttes, and pinnacles were once part of a uniform rock mass that was extensively eroded over a long time.

The landforms produced by these long-term processes are easy to see if you travel in the Southwest. A **plateau** is a broad platform tens of kilometers across that is elevated up to perhaps 300 m (about 980 ft) above the surrounding terrain. Dissection of a plateau occurs when a stream downcuts vigorously into the underlying rock strata to form a **canyon**, which is a deep, narrow valley with very steep sides (Figure 18.6). The most spectacular canyon, of course, is the Grand Canyon, but numerous smaller canyons carve the landscape. This system of canyons is largely associated with the uplift of the Colorado Plateau, which began about 30 million years ago. This uplift increased the slope of streams dramatically. Since that time, streams and their tributaries have been vigorously downcutting to reach a new graded state (see Figure 16.20a).

With the system of canyons in place, further stream erosion and mass-wasting processes begin to attack the slopes of the dissected plateau. These combined processes gradually leave isolated remnants of the plateau standing alone in progressively smaller and smaller landforms. You can easily visualize this progression in Figure 18.5. The largest of these plateau fragments is a **mesa**, which can be up to several kilometers across. A somewhat smaller remnant is a **butte**, which gradually wears down to form a tower-like landform called a **pinnacle**. A great place to see these kinds of landforms in close proximity to one another is Monument Valley in Utah (Figure 18.7).

The net effect of the progressive dissection and shrinking of plateaus, mesas, buttes, and pinnacles is that the landscape is gradually worn down into a series of rocky surfaces. Where such a surface extends away from an escarpment of the Basin and Range or cliff slope of a mesa or butte, it is called a *pediment*. In other places, the intervening plains and basins are covered with alluvium derived from long-term erosion of hillslopes. A common feature in these areas is a dried lake bed called a **playa** (Figures 18.5 and 18.8). Playas are associated with closed topographic depressions that may temporarily fill with water during wet periods. They

---

Plateau *A very broad, horizontal surface that is upheld by resistant caprock.*

Canyon *A very steep-sided valley that is cut into bedrock.*

Mesa *A broad horizontal surface, smaller than a plateau, that is upheld by resistant caprock.*

Butte *A steep-sided hill or peak that is often a remnant of a plateau or mesa.*

Pinnacle *A steep-sided, narrow tower that is the final remnant of a plateau, mesa, or butte.*

Playa *A dried lake bed that forms when runoff collects in closed topographic depressions in arid regions.*

Figure 18.8 Playa in Death Valley, California. Lakes form in basins and on plains in the American West because they collect storm runoff. They subsequently dry out during drought periods, leaving this distinctive landscape behind.

Figure 18.9 Arroyos in the Southwest. Arroyos are steep-sided gullies cut into alluvium. Flowing water within this arroyo suggests that a heavy storm recently passed through the area.

subsequently dry out during extended droughts, leaving salty evaporites on the surface. In many cases, these old lake beds contain a lengthy record of climate change and sedimentation in the region.

In this discussion of southwestern geomorphology, you may wonder how streams can play such an important geomorphic role in such an arid place. Although it might seem that the influence of water would be minimal in dry lands where most streams are ephemeral, streams are actually very important geomorphic agents in these landscapes. One reason for this impact is that when it does rain in arid regions, it often rains very hard for a short period of time, resulting in flash flooding. This high discharge has a disproportionate impact in desert regions because vegetation cover is minimal and hillslopes are not protected by plants. As a result, the landscape can be shaped dramatically when streams do flow. Also remember that these processes have been ongoing for many millions of years and you are seeing the *net effect* of long-term change.

Along with the overall dissected terrain of the Southwest, ephemeral streams also produce some distinctive landforms. One such feature is an **arroyo**, which is a steep-sided gully cut into alluvial sediments (Figure 18.9). Arroyo formation appears to be linked to wet/dry climate cycles related to El Niño and La Niña conditions, respectively, in the Pacific Ocean. Research shows that episodic cutting of arroyos has been especially intense the past 4000 years.

Perhaps the most interesting landform created by streams in the Southwest and many other arid regions is an alluvial fan. An **alluvial fan** is a depositional landform created where bedload-dominated streams flow out of mountainous or hilly areas onto an adjacent plain. Very simply, as the stream travels down the steep gradient, it can carry the coarse sediment derived from the mountain or hill. When the stream reaches the plain at the base of the hill or mountain, however, it loses energy because the gradient there is much lower. As a result of this reduced gradient, stream power is lost and aggradation occurs. Sediment deposition begins at the fan apex, which is the point where the stream leaves the mountain. Below the apex, the stream sweeps back and forth through time, and in so doing creates a semicircular landform that looks much like a hand-held fan from above (Figure 18.10a).

Although alluvial fans can occur virtually anywhere that streams flow abruptly from areas of high relief to surfaces that have lower slopes, even in humid areas, some of the most impressive fans in the world are those in the southwestern United States. Here, alluvial fans occur along many of the mountain fronts within the region and are sometimes many kilometers wide and have steep gradients. These fans are so large that they often overlap with one another along the mountain front to form a *bajada*. Many of these fans are complex in that they contain interbedded deposits of mud, sand, gravel, cobbles, and even boulders. The sand and gravel

Arroyo *A deep, steep-sided gully that is cut into alluvium.*

Alluvial fan *A fan-shaped landform of low relief that forms where a stream flows out of an area of high relief into a broad, open plane where the gradient is less and deposition thus occurs.*

(a)

(b)

Figure 18.10 Alluvial fans in the southwestern United States. (a) This typical alluvial fan in Death Valley, California, has a fan-like shape where sediment is deposited by water periodically flowing out of the mountains. (b) Surface of an alluvial fan in Death Valley, California. Note the steep slope and coarse texture of this surface.

layers indicate deposition by a bedload stream, whereas the very coarse rocks indicate prior debris flows. The fine sediments accumulate during major mudflows that periodically occur. Often, the surface of fans consists of a mix of gravels and cobbles (Figure 18.10b).

Eolian Erosion and Transport

In addition to the progression of erosional landforms that occurs in many deserts, another prominent geomorphic agent in these areas is **eolian processes**, which are those involving the wind. Although such processes can occur in subhumid and even humid regions, they are most active in desert regions because (1) strong winds are common, (2) a large supply of sand and silt is often available that can be blown, and (3) vegetation cover is minimal and the wind is thus free to erode sediment. This section of the chapter focuses on the way that moving air behaves and the way it shapes the landscape.

The Fluid Behavior of Wind and Sediment Transport

Like water, wind behaves according to specific physical laws associated with fluids. From this perspective, the primary difference between wind and water is that water is much denser than wind. For example, think of how much more buoyant you are in water than in air. Because of this variation in buoyancy, winds must have speeds almost 30 times greater than the speed of water currents to move the same particle.

Both water and wind behave as a fluid with a predictable velocity gradient in the vertical dimension. When air flows over a bare, unvegetated surface, a very thin boundary layer is present where the wind velocity is zero. Above this layer, to a height of about 3 cm, velocity initially increases rapidly, but then slows with additional height.

The ability of wind to erode sediment is similar to that of water in that sediment particles begin to move

Eolian processes *Geomorphic processes associated with the way that wind erodes, transports, and deposits sediment.*

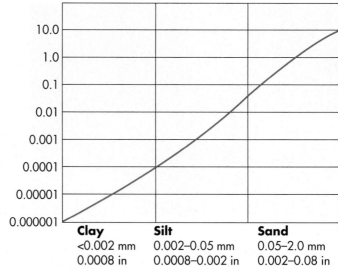

Figure 18.11 Threshold velocity required for wind to carry various-sized particles. The largest sand particles require a wind velocity of 10 m/sec (~33 ft/sec) to be moved by the wind.

Threshold velocity (m/sec)

Clay	**Silt**	**Sand**
<0.002 mm	0.002–0.05 mm	0.05–2.0 mm
0.0008 in	0.0008–0.002 in	0.002–0.08 in

once a critical threshold velocity is reached. In most circumstances, the mobilization of sand grains begins when wind velocity is greater than about 5 m/sec (~16 ft/sec) at a height of approximately 1 m (3.3 ft) above the ground. This threshold varies depending on the size of any given particle, with small grains more easily moved than larger ones (Figure 18.11). To see how wind moves sediment, consider Figure 18.12. Focus for a minute on a single particle of sediment and how it begins to move. Initial movement of the particle begins when the threshold velocity is crossed, which causes the sediment to oscillate at the surface beneath the wind. If the particle is less than 0.06 mm—in other words, classified as silt or clay—it will suddenly fly nearly vertically into the air and be carried in suspension for great distances. These fine grains are kept in the air by turbulence (Figure 18.13) and may travel hundreds of kilometers before they settle back to Earth.

In contrast to silts and clays, which are carried in suspension, larger particles such as sand either bounce across the ground through the process of *saltation* or roll along the surface in a process called *creep* (Figure 18.12). Saltation begins when a sand grain flies into the air. Because this grain is large relative to silts and clays, it is carried downwind only a short distance and then falls back to the surface under the force of gravity. When the grain strikes the ground, it transfers momentum to other sand grains that, in turn, fly into the air to repeat the process farther downwind. Creeping grains are larger, such as small pebbles, and consistently maintain contact with the surface. Like saltating grains, however, they cause movement in other grains through momentum transfer.

Eolian Erosional Landforms

Let's examine the process of eolian erosion on a larger scale and the landforms that result. Wind erosion transpires by either one of two processes: deflation or abrasion. When **deflation** occurs, turbulent air blows away loose soil particles. If this process persists over time across a broad surface, all fine particles are ultimately removed, leaving a concentration of coarser pebbles and cobbles that collectively form a surface called **desert pavement** (Figure 18.14). These surfaces are usually a mix of dark colors, collectively referred to as *desert varnish*, a coating of fine clays and bacteria that forms on rock surfaces over many decades. Although desert pavement protects underlying layers of fine particles from further deflation by capping them, the surface can easily be disturbed by human activities. When deflation is more localized, a landform called a deflation hollow may form. As the name implies, a **deflation hollow** is a basin formed by the removal of fine material. Although most deflation hollows are small, some may exceed 1.6 km (1 mi) in diameter.

The second way in which eolian erosion occurs is through **abrasion**. You can visualize the process of eolian abrasion by comparing it with the sandblasting conducted by work crews. As you have probably seen, sandblasting is a method used to clean buildings and bridges by blowing sand in a compressed stream of air. The process of eolian abrasion works much slower than mechanical sandblasting and is confined to the space immediately above the ground, but nevertheless has the same effect over a long period of time. Many factors influence

---

Deflation *Removal of sediment from a surface by wind action.*

Desert pavement *A resistant, pavement-like surface created when fine particles blow away and coarse sediment such as pebbles and gravel are left behind.*

Deflation hollow *A depression created by wind erosion.*

Abrasion *Erosion that occurs when particles grind against each other.*

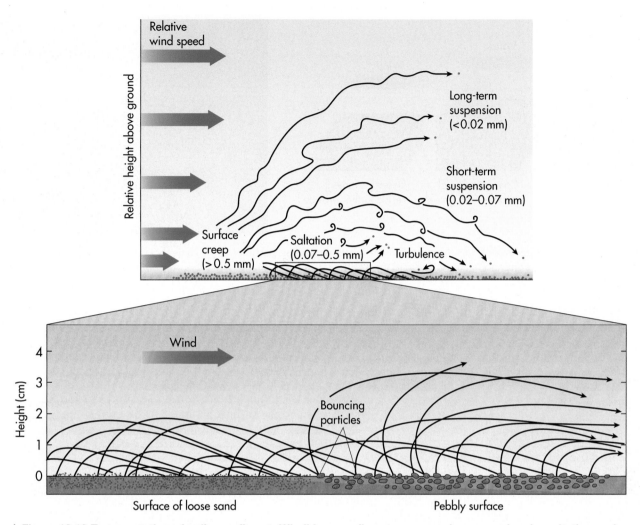

Figure 18.12 Transportation of eolian sediment. Windblown sediment can move in suspension, by saltation, or by surface creep. Note the flow lines to visualize how these processes operate.

how rapidly abrasion occurs, including the strength and persistence of the wind, the hardness and angularity of the blowing sand grains, and the resistance of the rock being abraded.

Individual rocks that have been influenced by abrasion are called **ventifacts** ("artifacts of the wind") and tend to be pitted, grooved, or polished (Figures 18.15a and b). In addition, they are typically aerodynamically shaped in the direction of the prevailing winds. Sometimes an entire rock outcrop or series of rock outcrops is streamlined by abrading winds. When this larger-scale abrasion occurs, elongated, wind-sculpted ridges called **yardangs** form (Figure 18.15c). Yardangs are similar to

Ventifact *An individual rock that is pitted, grooved, or streamlined through wind abrasion.*

Yardangs *Ridges that are sculpted and streamlined by wind abrasion and deflation.*

Figure 18.13 Eolian dust in suspension. Silt and clay particles can be carried in the air by turbulent flow, as demonstrated here in north central Iowa.

Eolian Erosion and Transport • 511

Figure 18.14 Desert pavement. (a) Desert pavement forms when fine particles progressively blow away, concentrating the coarse particles in a continuous gravel surface. (b) A typical desert pavement. Note the very coarse texture of this surface.

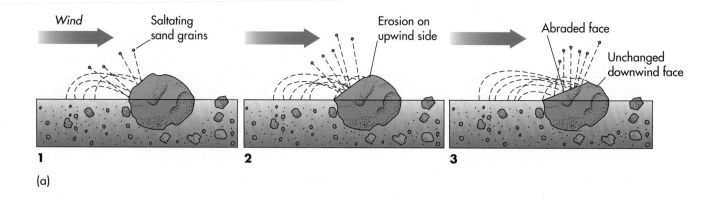

Figure 18.15 Erosional features created by the wind. (a) Ventifacts are individual rocks that have been streamlined by the wind. (b) Ventifacts such as these in Death Valley form when wind sandblasts rocks. (c) These yardangs in southeastern Iran are up to 80 m (~263 ft) high and are some of the largest on Earth.

Imagine that you are on a vacation in the Sonoran Desert in the southwestern United States. As you travel around this region, you frequently see landscapes such as this one, covered with small stones and pebbles. How have these landscapes formed?

glacial roches moutonées in that they form as a result of two distinct processes, with abrasion dominating on the windward side of the feature and deflation on the leeward side. Yardangs range from a few meters to several kilometers in length and are very prominent in desert regions such as those in Egypt and Iran. These landforms have also been identified on Mars, indicating that this planet also has an active eolian landscape.

Eolian Deposition and Landforms

The preceding section discussed how wind behaves as a fluid and how it erodes and transports sediments. These processes produce distinct erosional landforms, such as desert pavement and yardangs. However, the sediment removed from one place must accumulate someplace else, forming depositional landforms. This part of the chapter focuses on the landforms produced by eolian deposition.

Airflow and the Formation of Sand Dunes

Of all desert and wind phenomena, sand dunes have probably received the most scientific attention because they are beautiful and exotic landforms that are good indicators of environmental change. In some cases, even a slight decrease in rainfall over a short period of time, or increased wind speed, can result in the formation and movement of sand dunes in sandy areas. Given this sensitivity, dunes are fun to study because they can literally change shape overnight, whereas most landforms associated with rivers and glaciers usually require years, decades, or even centuries to change.

The formation of dunes is dependent on the interactions of sand supply, wind speed and direction, and vegetation, as these factors relate to the erosion and deposition of sand. Given that these relationships are complex and not always easily explained, let's look at a simplistic example that is easy to visualize. Imagine first that a bare, sandy patch is exposed to wind flowing over it and is thus a source for a dune that is about to form (Figure 18.16a). This patch might be present because a lot of sand accumulated along a beach or because drought

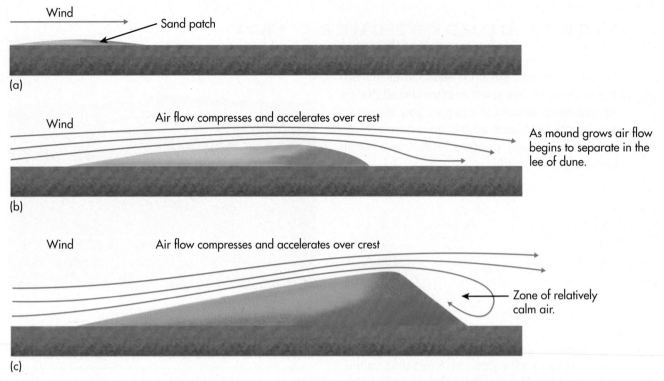

Figure 18.16 Formation of a sand dune. (a) Sand dune formation begins when a bare patch of sand sediment is eroded by the wind. (b) Moving sand encounters an obstruction or change in surface texture that traps sediment, causing a mound of sand to grow. Airflow is compressed and accelerates on the windward side of the mound. At the same time, airflow on the opposite side of the mound begins to separate and slow. As a result, the mound slowly grows and begins to move downwind. (c) The mound grows further, resulting in full separation of flow and the development of a relatively calm zone in the lee of the dune. Note also that the dune has migrated further downwind.

reduced vegetation cover. During the initial phase of sand movement, the airflow over the patch is essentially laminar, which means that air molecules are moving parallel to one another downwind with no turbulence. Now imagine that the moving sand grains encounter some obstruction, such as closely spaced clumps of grass, that results in the deposition of sand and the formation of a small mound. Such accumulation of sand can also be caused by a change in surface texture that slows air speed.

As deposition increases, wind speed at the top of the mound accelerates because the vertical space in which the air is flowing compresses (Figure 18.16b). At the same time, airflow slows and begins to separate in the lee of the mound because the wind can no longer follow the downwind slope. Given this reduced wind speed, sand deposition begins to occur downwind of the growing mound. This separation of airflow increases as the mound further grows, creating an area of erratic air movement in the lee of the mound and even reversal of flow (Figure 18.16c). Although gusts of wind can occur in this area, this zone is relatively calm compared to the flowing air above.

Once these relationships between airflow and topography are fully developed, the mound of sand transitions into a fully active sand dune with distinct patterns of erosion and deposition that result in down-

wind migration of the landform. Figure 18.17a illustrates how these processes collectively operate. The best way to visualize these patterns is to begin on the windward slope of the dune where the compressed and accelerated airflow shown in Figure 18.16 causes erosion. Notice in Figure 18.17a that this slope is called the **backslope** and has the shallowest slope of the feature, ranging between about 10° and 15°. Once wind begins to move sand grains on this surface, they bounce up the slope by saltation. When they reach the top of the dune, which is called the **crest**, they fly into the zone of calm air in the lee of the dune. As they fall, they are deposited on an evolving slope that lies at the angle of repose. Remember from Chapter 14 that this angle for sand stands between about 30° and 34°. This slope on a dune is called the **slip face** because sand on

---

Backslope *The gradual slope of a dune that faces the prevailing winds.*

Crest *The highest point of a dune.*

Slip face *The steep slope that lies on the leeward side of a sand dune at the angle of repose.*

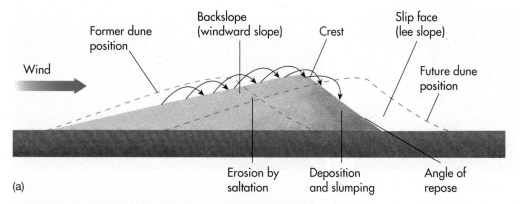

Figure 18.17 Dune components and migration. (a) When viewed in cross section, a freely active dune has three primary components: the backslope, crest, and slip face. Dunes migrate downwind through the combined processes of sand saltation on the backslope and slumping on the slip face. (b) This large sand dune in central Brazil has a beautiful example of a slip face, which slopes down toward the river. Airflow is left to right across the image. Note the pinnacle in the background and the person (arrow) for scale.

this relatively steep slope sometimes slides downhill under the force of gravity after it is deposited. Due to the combined effects of erosion on the backslope and deposition on the slip face, the dune migrates downwind. This migration is depicted in Figure 18.17a by the dashed red lines that represent the former and future position of the dune.

The extent of sand dune migration depends on the balance between erosion on the backslope and deposition on the lee slope. As a result, maintenance of the dune form depends on the forward movement of the entire feature because the amount of backslope erosion is balanced by the amount of deposition on the lee slope. In reality, variations in fluid dynamics and other factors result in complex cross-sectional characteristics in most dunes. Overall, the most important factors that influence the shapes of dunes are wind speed, the amount of stabilizing vegetation, and the sand supply—that is, how much sand is present for the wind to move. The amount and kind of vegetation is a critical variable because plant roots hold deposits of eolian sand together. If a sand dune is well vegetated, it tends to remain stable—that is, it stays in one place—unless it is significantly disturbed through human impact, drought, or a major storm. If such an event occurs, bare sand can be exposed and the dune becomes active and begins to move.

Classification of Sand Dunes and Related Landforms

Although dunes acquire a distinctive cross-sectional shape through the combined processes of erosion and deposition, their overall form can vary in a way that promotes classification. On a very broad scale, the largest depositional features related to eolian sand are *sand sheets* and *sand seas*. Sand sheets are horizontal to semi-horizontal bodies of sand that exhibit little or no surface topography. In contrast, sand seas are vast regions where enormous quantities of sand result in a wide variety of dune types. The best-known sand sea is in the Sahara Desert in Africa (Figure 18.18a). Sand seas occur elsewhere in the world, however, such as the Namib sand sea in the Namib Desert on the southwest coast of Africa (Figure 18.1). The largest sand sea in North America is the Nebraska Sand Hills, which is over 32,000 km^2 (about 12,350 mi^2) in size. This sand sea is largely stabilized at present due to an extensive cover of grass (Figure 18.18b). Research indicates, however, that it has been very active at various times within the Holocene Epoch (the past 10,000 years) when intensive droughts reduced the vegetation cover.

On a much more local level, it is possible to classify individual dunes based on a variety of factors that

(a)   (b)

Figure 18.18 Sand seas. (a) The Sahara Desert is best known for the massive sand dunes that have formed in this very dry environment, which is dominated by the Subtropical High pressure system. (b) The Nebraska Sand Hills is the largest sand sea in the Western Hemisphere and contains a wide variety of dune forms. Although isolated areas of blowing sand occur in the area, the dunes are mostly stabilized by an extensive cover of grass. Do you see the similarity in form with the Sahara?

depend on the complex interaction of sand supply, the amount of vegetation present, and wind speed and prevailing direction. The variability of these interactions results in at least eight major classes of dunes that can be seen in Figure 18.19. For simplicity, this discussion will be initially subdivided on the basis of dunes that develop in landscapes that are poorly vegetated as opposed to those where plants are common. Another way to consider this relationship is that dunes in poorly vegetated areas are "free," whereas those in vegetated areas are "anchored." As you work your way through this classification system, refer to Figure 18.19.

Beginning with free dunes, let's first examine those that form where sand supply is limited. The first of such dunes to consider is a *dome dune*, which is an individual mound of sand that has an oval form (Figure 18.20a) with no slip face. Dome dunes are temporary features with low relief and are thought to form because the wind has not blown from one direction long enough for a slip face to form, or because wind direction frequently changes. If sufficient time elapses, or winds become unidirectional, dome dunes can be modified into *barchan dunes*. These dunes look like a crescent from above and have a gently inclined windward slope with steep lee side, around which tapering cusps of the dune project downwind (Figure 18.20b). Usually, barchans are isolated dunes that form in areas of strong wind and relatively small amounts of sand. Where winds are weak, however, and massive quantities of sand are present, *barchanoid ridges* may develop (Figure 18.19). These dunes are sinuous, asymmetrical landforms that are transitional to yet another kind of dune called a *transverse ridge*. Transverse ridges are primarily a consequence of unidirectional winds that result in a linear deposit of sand perpendicular to the prevailing wind direction.

Another kind of dune that forms in barren landscapes is a *longitudinal dune* (Figure 18.21a). These dunes have the same linear form as transverse ridges,

but differ because they develop along the axis of two prevailing wind directions. In areas of multidirectional winds with very little vegetation, *star dunes* may form (Figure 18.21b). Star dunes can grow to be very large. In the Sahara, for example, star dunes may be several hundred meters in height and many kilometers in diameter. Rather than migrate, star dunes simply grow in height. A transitional dune between star dunes and transverse ridges is the *reversing dune*, which forms where two winds from nearly opposite directions are balanced with respect to strength and duration. As a result, a second slip face periodically develops.

In contrast to free dunes, which actively migrate, anchored dunes are somewhat more stable landforms. The best known anchored dunes are *blowout* and *parabolic* dunes (Figure 18.19). Both of these dune types are essentially deflation features with slip faces that slope in many directions. A typical blowout is a semicircular to circular deflation basin with lateral erosional walls (Figure 18.22). Sand eroded from the deflation basin is moved immediately downwind to form a depositional lobe. If the blowout continues to enlarge and elongate downwind, a parabolic dune will form with U- or V-shaped arms that point upwind (Figure 18.19). These arms are usually stabilized by vegetation, which causes the dune to further elongate as it migrates downwind.

Loess

Another form of eolian deposit is called **loess** (a German word that, in the United States, is usually pronounced "luss"), which consists of fine-grained, windblown silt. Loess is typically highly calcareous (as much as ~40% $CaCO_3$) and yellowish-tan in color (buff-colored). Often, loess originates within the same region as the sand that

---

Loess   *Windblown silt.*

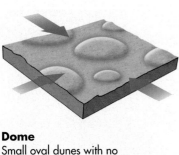

Dome
Small oval dunes with no slip face

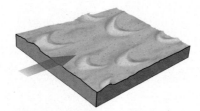

Barchan
Strong wind in one direction, limited amounts of sand

Barchanoid ridge
Formed from rows of merged barchans

Transverse
Weak wind in one direction, large amounts of sand deposited perpendicular to wind

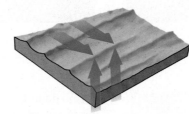

Longitudinal
Opposing winds flowing in similar directions, two slip faces, aligned with wind

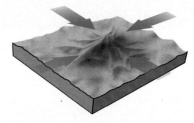

Star
Multidirectional winds, largest dunes grow tall instead of moving

Reversing
Two winds in opposite directions, changes shape periodically

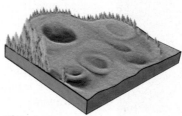

Blowouts
Localized deflation basins in partially vegetated landscape

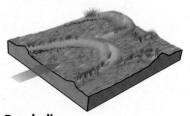

Parabolic
Arms stabilized upwind by vegetation, often start as semi-circular blowouts that elongate downwind

Figure 18.19 Sand dune classification. Eight major varieties of sand dunes occur, with form depending on wind direction, sand supply, and the amount of vegetation cover. Arrows represent the direction of prevailing winds.

(a)

(b)

Figure 18.20 Dome dunes and barchan dunes. (a) A typical dome dune (foreground). Note the oval shape, lack of a slip face, and low relief of this dune. (b) A well-developed barchan dune. Air flow is from left to right across the image. Note the crescent form, lee slope, and tapering cusps.

(a)

(b)

Figure 18.21 Logitudinal dunes and star dunes. (a) A nicely developed longitudinal dune. Such a dune forms as a result of two dominant wind directions. (b) Star dunes such as these in the Sahara desert form when winds are multidirectional.

Figure 18.22 A typical blowout. Blowouts like these in the Nebraska Sand Hills form when vegetation thins and strong winds deflate the core of a sandy deposit.

forms dune fields. However, because silt can be carried far in the air by suspension, it is sorted out from the sand and subsequently deposited great distances from its original source, leaving the dune fields behind (Figure 18.23). Loess also originates in stream valleys after silt is deposited on floodplains during floods. A great deal of loess also owes its origins to the thick deposits of glacial till and outwash deposited during the Ice Age. Following the deposition of these sediments by the glacier, the ice retreated, exposing surfaces that were unvegetated and easily deflated by strong winds. These winds picked up the silts contained within these deposits and carried them downwind.

Loess is a distinctive deposit for two primary reasons. One is that deposits of loess often occur as thick blankets of sediment that are regional in their extent, with approximately 10% of the Earth's land surfaces

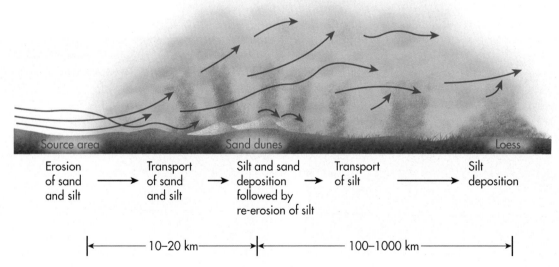

| Erosion of sand and silt | → | Transport of sand and silt | → | Silt and sand deposition followed by re-erosion of silt | → | Transport of silt | → | Silt deposition |

|←—— 10–20 km ——→|←——— 100–1000 km ———→|

Figure 18.23 The process of eolian sorting. After sediments are deflated from a source area, silts and clays are transported great distances in suspension. Coarse sands, in contrast, travel a relatively short distance, forming dune fields.

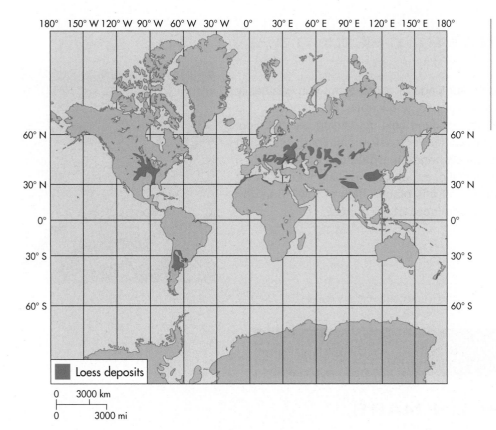

60° N

30° N

0°

30° S

60° S

■ Loess deposits

0   3000 km

0   3000 mi

covered in this way. For example, extensive deposits of loess cover much of the central United States, Argentina, Russia, and China (Figure 18.24). These areas are some of the most productive agricultural regions on

Earth because the loess is so calcareous and can absorb a lot of water. A second distinctive aspect of loess is that it is capable of maintaining nearly vertical faces upon exposure (Figure 18.25). This durability occurs because the silt grains are strongly attracted to one another and are bound together by the calcium carbonate contained within it.

Figure 18.25 Loess deposits in Nebraska. All the sediment visible in the exposure is loess that accumulated during the Illinoisan and Wisconsin glacial periods. Note the vertical nature of the pit wall, which is held together by calcium carbonate. The arrows point to buried soils, which represent intervals of time when loess was not accumulating and pedogenic processes dominated. Note the people at the top of the exposure for scale.

KEY CONCEPTS TO REMEMBER ABOUT EOLIAN LANDFORMS

1. The best-developed depositional eolian landforms are sand dunes. Sand dunes occur in a variety of forms, determined by wind direction and strength, sand supply, and the amount of stabilizing vegetation.

2. The wind is an efficient sorting mechanism, which means that sediments are separated by size. Clay-sized particles travel the farthest from the source area. Silts move an intermediate distance, and pebbles and gravels remain in place because they are too large to move by wind.

3. Another eolian deposit is loess, which is windblown silt that is carried great distances in suspension before it settles to the ground. Loess covers much of the Earth and is very fertile.

If you understand how different kinds of dunes form, you can easily interpret the dune formations in arid landscapes. For example, this photo shows a complex type of dune field containing barchanoidal transverse dunes near the Skeleton Coast in Namibia. Assuming that north is at the top of the photograph, what does the orientation of the dunes tell you about the direction of prevailing winds? What must the sand supply be like at this place—high or low? Can you identify the backslopes and slip faces of the dunes?

DISCOVER...

SAND DUNES ON MARS

You may know that much research is being conducted to learn about the environment and geologic history of Mars. The reason for this intensive research program is that Mars and Earth are about the same size, are relatively close to one another, and are about the same age. Evidence suggests that flowing water once existed on Mars, which means that the planet may have had an atmosphere somewhat similar to Earth's at one time. It also means that some form of primitive life may have once existed there, or may still exist in some obscure place.

One of the ways that Mars is being studied is with satellite remote sensing. This program is revealing much about the character and history of Mars, including what you can see in this image. Do you recognize these features? They are barchan dunes. These dunes mean that wind blows sand on the planet, just like on Earth. It also means that the amount of sand available to blow is somewhat limited. Of particular interest on these dunes are the randomly spaced dark patches. If these dunes were on Earth, you might assume that these spots are patches of vegetation. But, of course, vegetation does not exist on Mars. So what are they? The best guess so far is that the dunes are covered with a light frost in the Martian winter and that the spots represent places where the frost sublimates (turns to vapor) as summer approaches. Fascinating, yes?

Now that we have discussed the basic components of eolian processes and associated landforms, let's visualize these concepts in an animated format. To do so, go to the *GeoDiscoveries* website and open the module *Eolian Processes and Landforms*. This module describes the fluid flow of air and how it shapes the landscape. In this animation, you will see how abrasion shapes rock outcrops; how the variables of sand supply, wind direction, and vegetation influence the formation of dunes; and how loess develops. After you complete the animation, be sure to answer the questions at the end of the module to test your understanding of this concept.

Human Interactions with Eolian Processes

In Chapter 16 we examined how humans interact with stream systems. Humans can also impact the eolian landscape in many ways. In several examples around the world, people have contributed to the degradation of a landscape through the process of **desertification**; that is, transforming a formerly vegetated landscape into one that is relatively barren and highly susceptible to wind erosion. This process is enhanced in marginal, semi-arid landscapes where the vegetation density is already low. Figure 18.26 shows extensive regions of Earth that are susceptible to desertification. Until recent times, people have generally avoided these kinds of landscapes because rainfall to sustain crops and animal herds is unpredictable. However, with increasing population pressure, more and more people have been moving into these formerly uninhabited places in search of land that can be cleared for farmland. In the process of this development, the impact on the landscape has been enormous. As you will see in the following discussion, human-induced desertification can have catastrophic consequences.

Desertification in the African Sahel

One place where human-induced or human-enhanced desertification has occurred is the Sahel region of Africa. The Sahel is a narrow band of latitude that lies between about 15° and 18° N and extends from the country of Senegal on the west coast to the country of Sudan on the east coast (Figure 18.27a). Recall from Chapter 9 that this part of Africa has a distinct wet/dry precipitation cycle that depends on the seasonal passage of the Intertropical Convergence Zone (ITCZ). Consequently, the wet season occurs from late June to mid-September when the ITCZ is in the region. The remainder of the year, in contrast, is quite dry because of the dominance of the Subtropical High (STH) Pressure System. In the context of vegetation, the region lies between the Sahara Desert to the north and the tropical savanna (grass and open forest) to the south. In other words, this region is an **ecotone** because it lies between two distinct ecosystems and serves as the transition from one to the other. The photograph in Figure 18.27b gives you a feel for the character of the place.

Prior to the 19th century, the Sahel was largely home to nomadic pastoralists and small-scale sedentary farmers. The nomads lived mostly in the driest (northern) part of the region, which bounds the Sahara Desert to the north, whereas the farmers lived in the humid southern part of the region. The nomads maintained small herds of cattle and goats and followed a practice of moving their herds to find grass that their cattle could graze. Given the seasonal movement of the ITCZ and associated rainfall, their lifestyle evolved into a pattern of annual north–south migrations across the region; in other words, they moved their cattle to follow the rain. The farmers, in contrast, tended to stay in one place and had evolved a subsistence strategy that allowed them to be successful in this marginal environment. These strategies included staggering their plantings to have a continuous source of food, growing grains that mature quickly, and farming near a source of water such as river valleys.

Desertification *The process through which a formerly vegetated landscape gradually becomes desert-like.*

Ecotone *The transition between two distinct ecosystems that contains species of each area.*

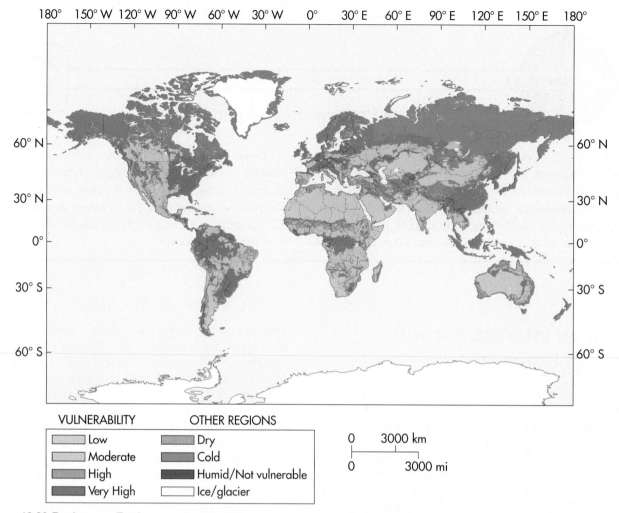

Figure 18.26 Regions on Earth prone to desertification. Much of the Earth's land surface lies in semi-arid to arid climate zones where desertification is a potential hazard.

VULNERABILITY
- Low
- Moderate
- High
- Very High

OTHER REGIONS
- Dry
- Cold
- Humid/Not vulnerable
- Ice/glacier

0   3000 km
0   3000 mi

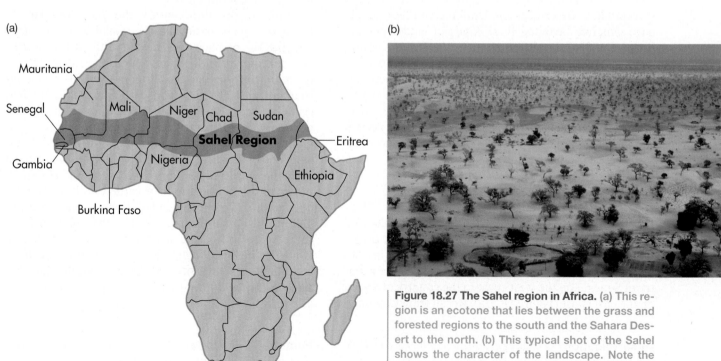

(a)

Mauritania
Senegal
Gambia
Mali
Niger
Chad
Sudan
Sahel Region
Eritrea
Nigeria
Burkina Faso
Ethiopia

(b)

Figure 18.27 The Sahel region in Africa. (a) This region is an ecotone that lies between the grass and forested regions to the south and the Sahara Desert to the north. **(b)** This typical shot of the Sahel shows the character of the landscape. Note the widely spaced trees and intervening barren areas.

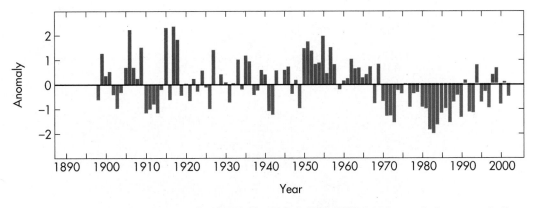

Figure 18.28 Annual rainfall from June through October in the Sahel, 1898–2002. In this graph, 0 represents the mean amount of rainfall during this period. Positive numbers above the mean are annual amounts of rainfall that are above the mean. Conversely, negative values below the mean represent below-average amounts of precipitation. Until the late 1960s, the region was relatively wet. Drought conditions have dominated since that time. (*Source:* World Meteorological Surface Station.)

Overall, the people in the Sahel had evolved a lifestyle that was compatible with the carrying capacity of the landscape; that is, the number of people living there did not exceed what the landscape could support without degradation. This sensitive relationship began to change in the late 1800s when Great Britain and France began to organize the region into distinct colonial entities. With this organization, boundaries arose that became barriers to the annual cycle of nomadic migration. In addition, the economic focus changed from living within the regional carrying capacity to exploitation of raw materials for consumption in Europe. Although the colonial era ended in the mid-20th century with the advent of African nationalism, the political boundaries, and their effect on human behavior, remained in place. People became increasingly sedentary, which resulted in the cultivation of fragile soils, overgrazing, and the removal of trees for firewood. In addition, overall population increased because people from more humid regions in the south migrated to the Sahel in search of places to cultivate.

As a result of the combined factors of boundary establishment and increasing population, the landscape of the Sahel began to degrade in the early 20th century because the carrying capacity was greatly exceeded. Landscape degradation and desertification of the Sahel reached a catastrophic level during the late 1960s and early 1970s. As you can see in Figure 18.28, this period coincides with a major period of drought that has basically lasted to the present time. Coupled with the pre-existing human-induced degradation that had already occurred on the landscape, this drought resulted in widespread desertification in the region.

In association with this drought, hundreds of thousands of people and countless numbers of livestock died of starvation; in fact, it has been estimated that approximately 100,000 people died in 1973 alone. The amount of wind erosion has also increased over the region, frequently resulting in the transportation of large amounts of atmospheric dust (Figure 18.29). Some researchers believe that this increased eolian dust is a direct result of the human impact on the landscape. Others believe, however, that the dust is a result of increased occurrence and intensity of tropical waves (low-pressure systems) in the region.

Desertification in the Great Plains of the United States: The Dust Bowl

People in the United States sometimes think that catastrophic environmental impacts such as those in the Sahel occur only in poor countries where most people have little education and economic opportunity. In fact, within the so-called developed world many examples have occurred of human impacts on the landscape resulting in dire environmental consequences. An excellent

Figure 18.29 Atmospheric dust off the northwest coast of Africa. The combination of drought and human environmental impacts in the Sahel may have increased the incidence of dust blowing off of the African continent.

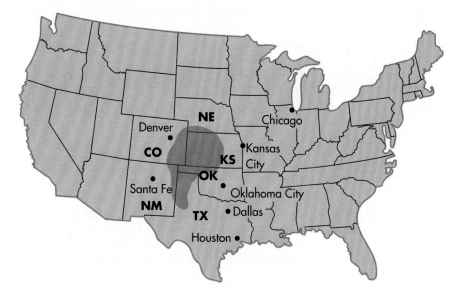

Figure 18.30 The American Dust Bowl. Most native grasslands in the High Plains were plowed in the late 19th and early 20th centuries during a settlement boom. Subsequently, a severe drought in the 1930s impacted a very large area that included parts of Nebraska, Kansas, Colorado, Oklahoma, New Mexico, and Texas.

example in the United States of human-induced desertification on a large scale is the region known as the *Dust Bowl*, which became part of the national consciousness in the 1930s.

The Dust Bowl region is located in the western Great Plains and includes parts of Colorado, Kansas, Nebraska, New Mexico, Oklahoma, and Texas (Figure 18.30). As discussed previously, the climate of this region is semi-arid due to the rain shadow of the Rocky Mountains. In response to this climate, the natural vegetation is predominantly short-grass prairie.

Until the latter part of the 1800s, this region was largely uninhabited, except for small numbers of nomadic groups of Native Americans such as the Comanche and Pawnee. Given the hunting and gathering subsistence of these groups, they had very little impact on the landscape because they moved frequently from one place to another. However, beginning in the 1870s, large numbers of European Americans from the eastern United States began to settle in the western Great Plains in association with the Homestead Act of 1862. In this context, each family of homesteaders was given a free quarter section, or about 65 hectares (160 acres), of land with the understanding that it would be developed in some way. Given the high fertility of the Mollisols found in the region, the vast majority of this land was plowed and subsequently farmed. In this way, most of the native grassland in the region was destroyed over the next 50 years.

Although the region is known for recurring drought because of its rain shadow location, the first few decades of European settlement were sufficiently wet to give farmers reason for hope regarding their economic future in the area. A drought in the early 1890s caused some concern, but was not overly extreme. In the middle 1930s, however, a period of especially intense drought occurred that decimated the landscape and thus the regional farm economy. Undoubtedly, this drought would have severely impacted the landscape even if it had occurred during the pre-settlement period when the ground was covered with grass. However, since most of the grass had been plowed under in the preceding few decades, the soils were exposed and easily deflated by the very high winds that accompanied the drought.

One of the lasting images from the Dust Bowl era is the massive dust storms that developed when tons of topsoil were blown away. Figure 18.31 shows an immense wall of windblown dust from one of these storms. They were so intense that they were called *black blizzards* and resulted in the deposition of fine sediment as far away as Chicago. In addition, the dust was so thick that as far east as Kansas City, streetlights were frequently left on during daylight hours. To give you a feel for just how bad it was in the region during this period, consider that from 1933 through the first nine months of 1937 there were 352 dust storms like the one pictured in Figure 18.31. Can you imagine what it must have been like to live there?

The overall result of the drought was to desertify the landscape in much of the region. From a cultural and economic standpoint, the impact was devastating, leaving many families literally on the verge of starvation. Although many people chose to stay and hope for rain, large numbers migrated to California in search of a better life. This migration is best depicted in John Steinbeck's famous novel, *The Grapes of Wrath*.

Figure 18.31 Wall of windblown dust.
Dust storms in the Great Plains, such as this one in Oklahoma, were common during the Dust Bowl era. These massive storms resulted from the combined impacts of drought, high winds, and plowed soils.

For those who stayed in the Dust Bowl region or moved into the area at a later time, the drought and resulting desertification demonstrated that any agricultural strategy must include conservation measures to mitigate future topsoil loss. As discussed in Chapter 15, one of the major changes that occurred in the region was the development of circle-pivot irrigation. This technology not only maintains soil wetness during drought conditions but also leaves the portion of the field outside the cultivated circle available for stabilizing grass. Another soil conservation measure is a *windbreak* (Figure 18.32), which is a row of tall trees planted along a field that blocks the wind from blowing directly across the soil. Still another soil-conservation strategy is *conservation tillage*. In this method, the soil is not plowed as deeply or as thoroughly, leaving at least 30% of the previous year's crop residue on the ground. This residue increases the surface roughness of the ground, which helps protect the soil from wind erosion and conserve water in the soil.

Figure 18.32 A conservation windbreak. Windbreaks such as this one are rows of tall trees planted along fields that block the wind from blowing directly across the ground.

KEY CONCEPTS TO REMEMBER ABOUT DESERTIFICATION OF SEMI-ARID LANDS AND SAND DUNES AS INDICATORS OF CLIMATE CHANGE

1. Large areas of Earth's surface are prone to desertification, which is the transformation of formerly stable (by vegetation) landscapes to desert-like conditions.

2. The Sahel region of Africa is an ecotone that lies between the Sahara Desert and the tropical rainforest. This region has undergone extensive desertification in the past 30 years due to human impacts and extensive drought.

3. The American Dust Bowl is a large area in the central United States that was extensively desertified during the 1930s due to a combination of human settlement patterns and extensive drought.

4. Soils in semi-arid lands can be protected to some extent through such measures as windbreaks and conservation tillage.

In the next chapter, we come close to finishing our tour of the various geomorphic processes and associated landforms by discussing coastal landscapes. The coastal zone is a dynamic interface between the atmosphere, large water bodies, and landmasses. A variety of geomorphic processes occur along coastlines that are quite specific to a very narrow portion of the shore. This part of the Earth is shaped, often in dramatic ways, because waves contain a tremendous amount of energy that is continuously dispersed along the coast over time. If you happen to live by the shore or enjoy traveling to it, you will learn a lot in Chapter 19.

SUMMARY OF KEY CONCEPTS

1. Arid and semi-arid environments collectively comprise over 30% of the Earth's land surface. Arid landscapes are excellent places to study geomorphology because elements such as rock structure are easily seen. In places where the structure is horizontal, desert landscapes often contain a progression of erosional landforms, including plateaus, mesas, buttes, and pinnacles.

2. Eolian processes involve the shaping of the Earth due to the wind. The wind moves sediment in one of three ways: (1) fine grains such as silts and clays are carried great distances in suspension; (2) sand grains bounce along the ground in the process of saltation; and (3) heavier particles roll along the ground by creep.

3. The wind shapes the landscape through the combined processes of erosion and deposition. The formation of yardangs is an example of wind erosion. In contrast, sand dunes form when windblown sand is deposited and shaped.

4. Sand dunes come in many different shapes and sizes. The shapes of dunes depend largely on the supply of sand, the amount of vegetation, and prevailing wind speed and direction.

5. Loess is windblown silt carried a great distance by suspension. Thick deposits of loess occur in many places around the world, with most accumulating during glacial cycles.

6. Human-induced desertification is a major environmental issue facing the world today. Desertification occurs when the stabilizing vegetation on the landscape is reduced through human activity. This process exposes soils to the wind, which causes severe soil erosion and the transport of eolian sediment.

CHECK YOUR UNDERSTANDING

1. What are the three large-scale environmental variables that contribute to the presence of deserts? What are the processes associated with each of these factors and why do they produce arid environments?

2. Discuss how normal faulting is responsible for the dominant landforms of the Basin and Range Province.

3. Beginning with a plateau, describe the sequence of landform evolution where horizontal rocks of differing resistance influence the regional geomorphology.

4. What is a playa and where is a good place to find one?

5. Describe the formation of an alluvial fan.

6. How does flowing air compare to flowing water? How is it similar and how is it different?

7. In the context of eolian processes, describe suspension, saltation, and creep. How do these various processes function and what kind of sediment is associated with each one?

8. What is the difference between deflation and abrasion?

9. Describe how ventifacts and yardangs form.

10. Discuss the early stages of dune formation and the nature of airflow during this time.

11. Sketch and describe a typical dune cross section and identify the three important components. How does sand move in these places and why do these processes collectively result in dune migration?

12. How do the factors of sand supply, wind (speed and direction), and vegetation influence the behavior and form of sand dunes? Provide at least three examples.

13. What is loess, what is its source, and how is it transported?

14. Describe the process of desertification. Why are marginal landscapes most likely to experience human-induced desertification?

15. Name and describe two conservation techniques used to protect soils from wind erosion.

ANSWERS TO VISUAL CONCEPT CHECKS

VISUAL CONCEPT CHECK 18.1

This landscape contains an example of desert pavement, which consists mostly of pebbles and gravel. Desert pavement forms when smaller sediment particles, such as sand and silt, are carried away by the wind, leaving the coarser particles behind. Over time the gravel pebbles and gravel coalesce to form a surface that resembles pavement.

VISUAL CONCEPT CHECK 18.2

Since north is at the top of the photograph, the orientation of these barchanoidal transverse dunes indicates that prevailing winds are northeasterly, that is, from the north and east. These winds are moving the dunes in a somewhat southwesterly direction toward the lower left of the photograph. The backslopes face the direction from which the winds are coming, whereas the slip faces lie in the shadows. The presence of these dunes indicates that the available sand supply is relatively low, causing distinct dunes to form rather than a sand sea.

CHAPTER NINETEEN

COASTAL PROCESSES AND LANDFORMS

The previous three chapters investigated fluvial, glacial, and eolian processes and their distinctive landforms. Now we turn to the various processes and landforms that occur along coastlines. Although these processes affect a very small part of the Earth's surface, they create some of the most distinctive and popular landscapes you see.

Why should you be interested in coastal processes and landforms? One very important reason is that you may very well live along or very near a coastline, as do a slim majority of people in the United States (and indeed, the world). People are attracted to coastlines for a variety of economic, recreational, and lifestyle reasons and are affected by the geomorphic processes that occur along these narrow strips of land. In turn, people influence the behavior of coastlines in dramatic ways through development. Thus, understanding the geomorphology of coastlines is important, not only for gaining a better appreciation of this landscape, but also for helping you become a good steward for these areas.

Are you attracted to this landscape? If so, you are certainly not alone. Coastlines are perhaps the most exotic landscapes on Earth, and people love to live near or visit them. This chapter focuses on the oceans of the world and the way they shape the landscape along coastlines.

CHAPTER PREVIEW

Oceans and Seas on Earth

The Nature of Coastlines: Intersection of Earth's Spheres
On the Web: *GeoDiscoveries Tides*

Coastal Landforms
On the Web: *GeoDiscoveries Waves and Coastal Erosion*
On the Web: *GeoDiscoveries Longshore Processes and Depositional Coastlines*

Human Interactions with Coastlines
On the Web: *GeoDiscoveries Evolution of the Louisiana Coastline*

The Big Picture

LEARNING OBJECTIVES

1. Distinguish among oceans, seas, and gulfs on Earth.

2. Describe the nature of eustatic sea-level changes the past 150,000 years.

3. Discuss how waves form and the ways in which they interact with coastlines.

4. Explain how tides form and the variables that influence them.

5. Compare and contrast the various erosional coastal landforms.

6. Describe the character of coastal depositional landforms and how they form.

7. Discuss the various ways that humans interact with (and manage) coastlines.

Oceans and Seas on Earth

Before we begin discussing various coastal processes, let's take a closer look at the character of the world's oceans and seas because it provides the context in which to view the shore. Figure 19.1 shows the names and locations of most of the notable large water bodies on Earth. The term **ocean** refers to the enormous body of salty water that covers about 71% of Earth. Five geographic divisions of the world's ocean are recognized. The largest is the Pacific Ocean, which encompasses a third of Earth's surface. This ocean is significantly larger than Earth's entire landmass, having an area of 179.7 million km^2 (69.4 million mi^2). The next largest ocean is the Atlantic, followed by the Indian Ocean, then the recently recognized (in 2000) Southern Ocean, and the Arctic Ocean. The Arctic Ocean covers an area of about 14,090,000 km^2 (5,440,000 mi^2), which is slightly less than 1.5 times the size of the United States. The Southern Ocean was recognized because it encircles Antarctica and has distinctive circulatory patterns.

We use the term *ocean* to refer to these large water bodies. The terminology for smaller areas of water more closely associated with land is inconsistently applied. In general, the next largest bodies of water are considered to be **seas**, which are subdivisions of oceans and partially enclosed by land. For example, the Mediterranean Sea (Figure 19.1) is mostly surrounded by the European, Middle Eastern, and African landmasses. In fact, it opens to the Atlantic only in its far western end.

After seas the next largest body of water is generally considered to be a **gulf**, which is a smaller arm of an ocean or sea that is also partially enclosed by land. Unfortunately, the use of the terms *gulf* or *sea* can be misleading because some named gulfs are actually

---

Ocean *The entire body of saltwater that covers about 71% of Earth.*

Sea *A subdivision of an ocean that is partially enclosed by land.*

Gulf *A relatively small body of saltwater that is surrounded by land on three sides and opens to a sea or ocean.*

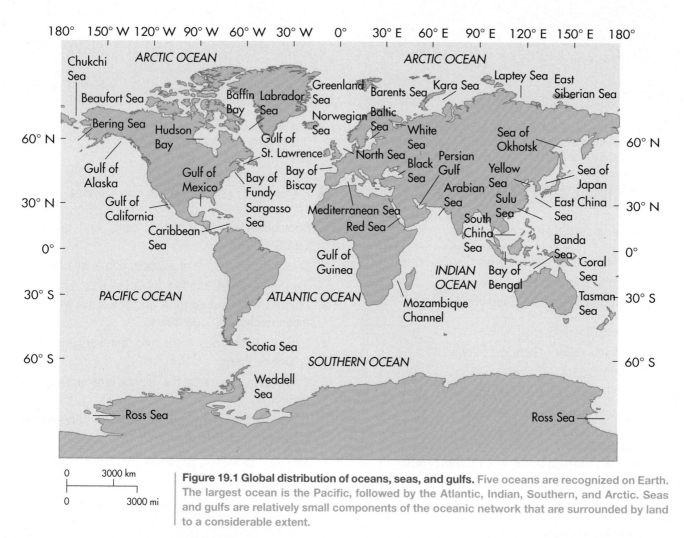

Figure 19.1 Global distribution of oceans, seas, and gulfs. Five oceans are recognized on Earth. The largest ocean is the Pacific, followed by the Atlantic, Indian, Southern, and Arctic. Seas and gulfs are relatively small components of the oceanic network that are surrounded by land to a considerable extent.

larger than some seas. The Gulf of Mexico, for example, is much larger at 1.6 million km² (615,000 mi²) than the Black Sea in Eastern Europe, which is 422,000 km² (163,000 mi²). A distinctive characteristic of a gulf is that it usually has a recessed shoreline that opens outward to a larger body of water. Notice in Figure 19.1 how the shoreline of the Gulf of Mexico curves northward and that the Gulf opens to the Caribbean Sea east of Central America. On a still smaller level, a **bay** is a relatively small indentation on the coast that is directly connected to an ocean, sea, or gulf.

Although oceans, seas, and gulfs differ by size and their relationship to land, they are essentially interconnected and therefore have the same basic chemical composition. This chemistry is a result of complex interactions among several factors, including the atmosphere, seawater, minerals, sediments, and the myriad organisms living in the water. Given that water is an excellent solvent, the world's oceans and seas contain a wide variety of dissolved solids, such as chlorine, sodium, and magnesium ions, among others. The term **salinity** refers to the concentration of dissolved solids in seawater and is most commonly expressed in parts per thousand. Overall, global salinity varies between 34‰ and 37‰. Water that exceeds a salinity of 35‰ is considered **brine**, whereas water with salinity less than 35‰ is called **brackish**.

The Nature of Coastlines: Intersection of Earth's Spheres

The primary distinguishing characteristic of coastlines is that they are places where landmasses and large water bodies intersect (Figure 19.2), often for hundreds or even thousands of kilometers. Although coastlines are usually associated with oceans and seas, they also occur around large lakes such as the Great Lakes of North America. Coastlines are especially noteworthy because they represent the intersection of three major Earth spheres, namely the hydrosphere, lithosphere, and atmosphere.

Coastlines come in all shapes and sizes. Some coasts have wide, sandy beaches, whereas others feature rocky bluffs and cliffs. Along some coastlines, mountains extend straight down to the water. Because of the wide range of environments in which coastlines occur, some are passive and change slowly, whereas others erode vigorously and are in a near-constant state of adjustment.

Figure 19.2 A typical coastline in northern California. Coastlines such as this one at Big Sur are where the hydrosphere, lithosphere, and atmosphere meet in a very big way.

Processes That Shape the Coastline

A variety of physical processes influence the shape of a coastline, with most directly related to how water moves along the shore. This movement occurs at several different temporal and spatial scales—some movements occur very slowly over centuries, whereas others are related to daily oscillations and flow patterns. This section discusses the primary ways in which water moves along a coastline.

Fluctuations in Water Level Over long periods of time, water levels in oceans and large lakes change for a variety of reasons. For instance, changes can occur when tectonic forces cause a landmass to be uplifted or sink. Another way in which water levels fluctuate is when the amount of water in the ocean or lake varies due to adjustments in the hydrologic cycle. This form of water-level fluctuation is called **eustatic change**.

The best example of eustatic change in recent Earth history is the great change in sea level that occurred at the end of the Pleistocene Epoch, when the massive continental glaciers melted and returned water back to the oceans. Figure 19.3 shows that sea level changed dramatically between the Illinoisan glaciation (called the *Riss glaciation* in Europe) around 130,000 years ago and the Holocene period (the past 10,000 years). Major periods of low ocean water level occurred during both the

---

Bay *An indentation in the shoreline that is generally associated with an ocean, sea, or gulf.*

Salinity *Concentration of dissolved solids in water that is measured in parts per thousand (‰).*

Brine *Water that has salinity greater than 35‰.*

Brackish *Water that has salinity less than 35‰.*

Eustatic change *Fluctuations in sea level associated with adjustments in the hydrologic cycle.*

Figure 19.3 Eustatic sea-level changes during the late Quaternary Period at New Guinea. Dramatic low sea levels occurred during the Illinoisan glaciation (called the *Riss glaciation* in Europe) and Last Glacial Maximum, with increases following these glaciations. Sea-level rise the past 10,000 years is often called the *Holocene Transgression*. (*Source:* J. Chappell and N. J. Shackelton, 1986, "Oxygen Isotopes and Sea Level," *Nature* 324: 137–140.)

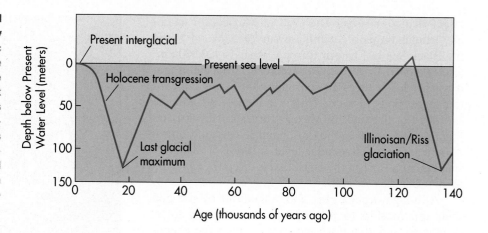

Illinoisan/Riss glaciation and Last Glacial Maximum (the Wisconsin glaciation discussed in Chapter 17), when enormous volumes of water were stored in the immense ice sheets that covered much of North America and Europe. When the ice subsequently melted, the sea level rose rapidly because water poured back into the ocean.

One of the important outcomes of the Pleistocene eustatic changes was that the amount of continental land sub-aerially exposed—that is, above water—varied a great deal over time. This variation can be seen in Figure 19.4, which shows the configuration of the North American coastline today, at peak glacial time, and at a peak interglacial if all global ice sheets melted. As you can see, the amount of land above sea level during the peak glacial period was significantly greater than today.

Referring back to Figure 19.3, note that water levels during the most recent glacial periods were about 125 m (410 ft) lower than today. At these times, large parts of the now-submerged continental shelf were sub-aerially exposed (see Figure 19.4 again) and rivers extended onto that platform. The evidence for these rivers can be seen in the form of canyons cut into the shelf deposits during the major low ocean levels. An excellent example of such a canyon is the Hudson Canyon, which was eroded into the continental shelf by the ancestral Hudson River. Similar features also exist on the floor of Chesapeake Bay (Figure 19.5a), which is essentially a large river valley, or **ria**, that was flooded when the sea level rose during the Holocene. Rias are common along many coastlines, with some of the most beautiful found in New Zealand (Figure 19.5b). The glacial counterpart of a ria is a **fjord**, which is an ice-formed trough that becomes flooded when sea level rises. Excellent examples of fjords exist in places such as Alaska, Norway, and the south island of New Zealand.

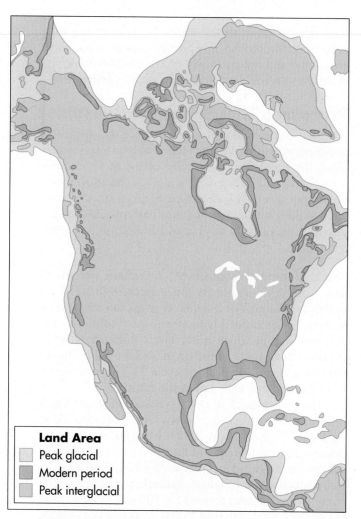

Land Area

- Peak glacial
- Modern period
- Peak interglacial

Figure 19.4 Configuration of the North American coastline during peak glacial, peak interglacial, and modern periods. Over time sea level has fluctuated greatly due to the amount of water tied up in ice sheets. Note the configuration of the coastline if all ice sheets melted.

Ria *A former river valley along the coast that is flooded by rising sea level.*

Fjord *A former glaciated valley along the coast that is flooded by rising sea level.*

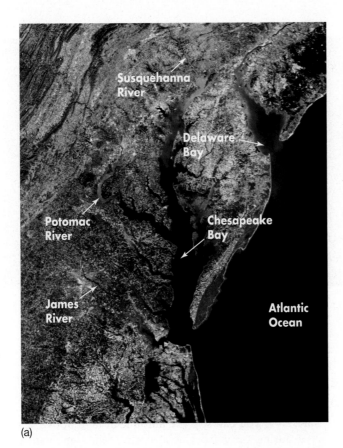

(a)

Figure 19.6 Tidal fluctuations in Chesapeake Bay. This image shows a boat dock at low tide. At high tide, the water reaches approximately half way up the support beams of the dock (arrow).

Tides Tides are regular and predictable oscillations that occur with respect to the level of the world's oceans (Figure 19.6). These oscillations are related to Newton's law of gravitation, which states that every particle in the universe attracts every other particle. According to this law, both the Moon and Sun exert a gravitational force on the Earth. Although it would seem that the Sun has the biggest influence on tides due to its immense size, the Moon's gravitational effect is actually greater. This is because the strength of the force is inversely proportional to the distance: the closer the two bodies, the stronger the pull. Recall from Chapter 3 that the average distance from Earth to the Sun is approximately 149,000,000 km (93,000,000 mi). In contrast, the Moon is only about 386,000 km (240,000 mi) away from Earth. As a result, the Moon's gravitational pull is responsible for about 56% of the daily tides, whereas about 44% is related to the Sun's effect. Other factors influencing the nature and timing of tides are related to the geometric relationship of Earth to the Moon and Sun and how it varies on a daily and monthly basis.

The simplest tide to explain is the one created on the side of Earth that faces the Moon. Think of this tide as a bulge of water that forms when ocean water is drawn up and along by gravity as the Moon passes over. A similar bulge also forms on the opposite side of Earth, however, which may seem odd because this side of the planet faces away from the Moon. How can this tide form? The reason why tidal bulges appear on both sides of Earth is because of the combined effects of the Moon's gravitational pull *and* the fact that the Earth and Moon rotate around each other on a common axis.

(b)

Figure 19.5 Flooded valleys on the continental margin. (a) Chesapeake Bay is fundamentally a large river valley, created by joining rivers, including the Susquehanna, Potomac, and James. The valley was later flooded by post-glacial rising sea levels. Delaware Bay, shown in the upper right of the image, is also a ria that was created when a portion of the ancestral Delaware River valley was flooded because of eustatic change. (b) Queen Charlotte Sound, New Zealand, evolved when rising sea levels flooded this valley on the northern tip of New Zealand's south island.

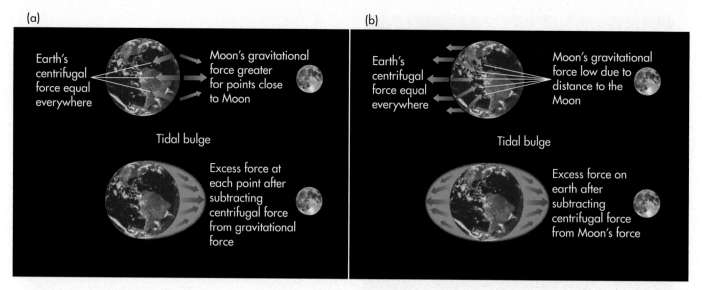

Figure 19.7 Formation of tides. (a) Tide-raising force on the side of Earth facing the Moon. On this side of Earth, a tidal bulge forms because the Moon's attractive forces exceed Earth's centrifugal force. The bulge is most pronounced where Earth is closest to the Moon due to the curvature of both bodies. (b) Tide-raising force on the side of Earth opposite the Moon. A tidal bulge forms on this side of Earth because the centrifugal force associated with the combined rotation of Earth and Moon is greater than the Moon's gravitational force. The height of the observed tides in the figure is significantly exaggerated to illustrate the forcing variables and effects.

Examine Figure 19.7 to see how these relationships work. If the Earth and Moon did not rotate around each other, they would fly through space side by side and the tidal bulge would always be on the side of Earth facing the Moon. Because they rotate around a common axis, however, a centrifugal force exists that is equal to the gravitational forces between the two objects. As a result, the Earth and Moon are always the same distance apart and each particle of Earth experiences both forces. A tidal bulge forms on the side of Earth most directly facing the Moon because the Moon's gravitational pull there is slightly greater than the centrifugal force toward the other side of Earth (Figure 19.7a). The high tide associated with this gravitational force is called a *direct tide*. A tidal bulge also forms on the other side of Earth because the centrifugal force (of the rotating bodies) on this side of the planet is slightly greater than the gravitational pull of the Moon (Figure 19.7b). The high tide associated with this centrifugal force is called an *indirect tide*. Each of these tides has a corresponding low tide.

To fully understand the timing of the direct and indirect tides at any given point, you must first consider the daily (24-h) rotation of Earth on its axis. If the Moon were stationary relative to Earth, any coastal location on the planet would rotate under the two tidal bulges (the direct and indirect) each day. Similarly, each coastal locality would also experience two low tides between the bulges. If the Moon were stationary, the timing of

these high and low tides each day would always be the same. The Moon is not stationary, of course, but instead slowly orbits Earth approximately every 28 days. As a result, the timing of the high tides is about an hour later each day.

Tides are even more complicated when the role of the Sun is factored into their timing and character (Figure 19.8). Although the gravitational force of the Sun is less than that of the Moon, it is nevertheless important. When the Sun, Moon, and Earth are aligned, the tide-raising force is maximized, resulting in the very highest tides called *spring tides*. In this case, "spring" refers to upwelling, rather than the season. When the Sun and Moon are at right angles to Earth, however, somewhat lower high tides result; these tides are called *neap tides*.

In addition to the tidal variations that occur with respect to the geometric relationship of the Earth, Moon, and Sun, the overall tidal range can vary dramatically between places. For example, as you can see in Figure 19.9, the average tidal range at San Francisco, California, is about 2.5 m (8.2 ft). The largest tidal range on Earth occurs in the Bay of Fundy, which lies between the Canadian provinces of New Brunswick and Nova Scotia in eastern North America (Figure 19.1). The shape of the coastline here has a funneling effect that produces a tidal range which varies from about 6 m (~20 ft) at the mouth of the bay to 16 m (53 ft) at the head of the bay!

Tides

The formation of tides involves the complex geometric interaction of the Earth, Moon, and Sun. These interactions cause gravitational forces that pull on the Earth's oceans in such a way as to cause predictable oscillations in local sea level. To see how these forces appear in animated form, go to the *GeoDiscoveries* website and select the module *Tides*. This module will allow you to better visualize the various forces acting on the Earth's oceans and how they move water. Once you complete the module, be sure to answer the questions at the end to test your understanding of this concept.

---

the wave base and the ocean bottom intersect. From this point forward, the vertical space in which the rotating water has to move becomes progressively restricted, causing the overall wave height to increase and the overall speed of the waves to slow. At the same time that the forward speed of the waves is decreasing, from the point where the interaction of the wave base and ocean bottom begins, additional waves continue to advance into the area, resulting in compressed wavelengths as more waves crowd into a smaller space. Ultimately, the height of each individual wave exceeds its vertical stability and the wave crashes in a feature called a **breaker**. This point typically occurs when the wave height is about seven times greater than the wavelength. From the time the wave breaks onward, the forward momentum of the wave causes water to rush onto the beach as *surf*, which is the only part of the wave system where water is actually moving forward, instead of up and down. In places that are typically sheltered from waves in some fashion, or have a short fetch, a sudden increase in wave energy can cause erosion and significant modification of the coast. On the other hand, many other coastlines are consistently influenced by strong waves and are thus in equilibrium with no net erosion or deposition of sediment due to high wave energy.

Littoral Processes and Sediment Transport   Let's now take a closer look at wave/shore interactions and how they affect the way in which sediment is transported along the coast. These interactions depend on many factors, including the shape of the coastline, the direction and seasonal variation of prevailing winds, intensity of waves, tidal range, and the nature of coastal sediment, to name but a few. In some places sediment moves freely along straight coasts for great distances, whereas

in others it can be trapped in nooks and crannies along the shore after moving only a short way. As you might imagine, the various combinations of these factors can result in myriad interactions, most of which are far beyond the scope of this text.

Perhaps the most simplistic and commonly considered way to view these interactions is to observe what happens on a relatively straight coast when waves approach at some kind of angle (Figure 19.11), as they do in many places. Because of this oblique approach, the force of the water is deflected downwind when it interacts with the shore. This deflection causes the formation of a **longshore current** that flows parallel to the shore,

---

Longshore current   *The current that develops parallel to a coast when waves approach the coast obliquely and forward momentum is deflected.*

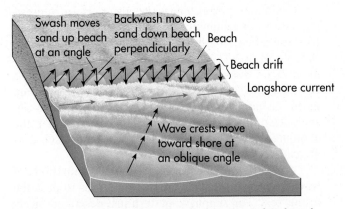

Figure 19.11 Littoral drift. Longshore current develops in places where waves strike the coast obliquely. Sediment subsequently moves down the shore as littoral drift. Littoral drift includes beach drift, which is the transport of sediment down shore due to the zig-zag pattern of swash and backwash, and longshore drift, which is the movement of sediment by the longshore current.

Breaker   *A wave that rises and crashes when the forward momentum of oscillations cannot be maintained.*

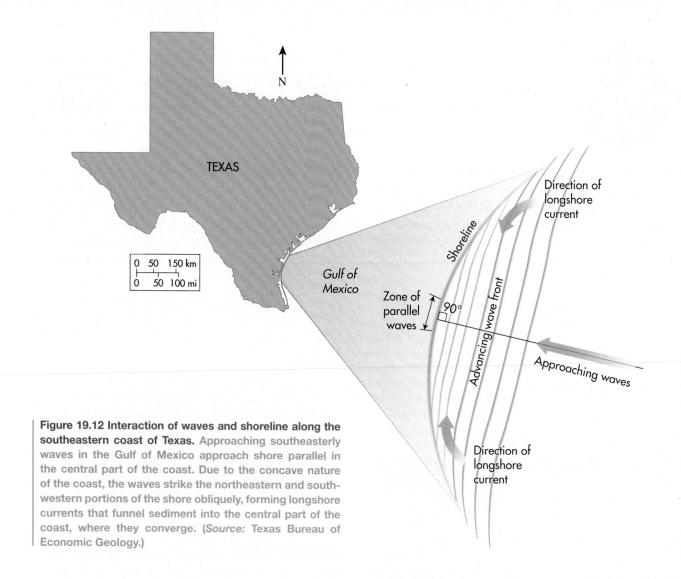

Figure 19.12 Interaction of waves and shoreline along the southeastern coast of Texas. Approaching southeasterly waves in the Gulf of Mexico approach shore parallel in the central part of the coast. Due to the concave nature of the coast, the waves strike the northeastern and southwestern portions of the shore obliquely, forming longshore currents that funnel sediment into the central part of the coast, where they converge. (*Source:* Texas Bureau of Economic Geology.)

one that is capable of moving sediment down the coast in a process called **longshore drift**.

Obliquely approaching waves also cause an interesting pattern of sediment transport on the beach proper. In these situations the surf also flows onto the beach (known as *swash*) at an angle. After the swash reaches its apex on the beach, it flows back to the ocean at a right angle to shore as *backwash*. Sediment is picked up in this fashion and transported down the coast in a zig-zag pattern known as **beach drift**. Taken together, the combined processes of longshore drift and beach drift are called **littoral drift**.

As noted previously, the various combinations of wave/beach interactions are too numerous to describe in this text. The simplistic pattern just described nicely explains the pattern of littoral drift along the coast of southern California where the strong, persistent longshore current freely moves sediment from northwest to southeast. Along the eastern shore of Lake Michigan, however, seasonal prevailing winds shift from northwesterly in winter to southwesterly in summer. This

shift in wind direction causes sediment to move up and down the coast to some degree. Still another pattern is observed along the concave southeastern coast of Texas where waves approach from the southeast (Figure 19.12). In the central part of this portion of the Texas coast, the waves approach in a perpendicular fashion. On the flanks of this section of coast, however, the curvature of the shore causes the waves to approach obliquely. This interaction results in opposing longshore currents that converge in the central part of the coast.

Longshore drift *Transport of sediment by the longshore current.*

Beach drift *Sediment that is transported in the surf zone by swash and backwash, which form due to the oblique approach of waves.*

Littoral drift *Sediment that is transported through the combined processes of longshore drift and beach drift.*

With your understanding of littoral processes in mind, examine this image from Queensland, Australia. Assume the axis of the coast is aligned north and south and see if you can determine:

a) The direction of wave-forming winds
b) Which way the longshore current is flowing
c) Which way the beach drift is moving

Hint: North is at the bottom of the photo.

KEY CONCEPTS TO REMEMBER ABOUT WATER MOVEMENT ALONG COASTLINES

1. Coastlines are the places on Earth where the hydrosphere, lithosphere, and atmosphere intersect over a large spatial extent.

2. The amount of water in the oceans fluctuates with time due to adjustments in the hydrologic cycle. These changes are called eustatic changes. Many coastlines show evidence of long-term eustatic change in the form of rias and fjords.

3. Tides are daily adjustments that occur with respect to the water level along any given coastline. Tidal processes are driven largely by the gravitational pull of the Moon and, to a lesser extent, the Sun. Other important factors are the centrifugal force of the combined Earth–Moon rotation and the geometric relationship of the Sun and Moon.

4. Waves are the dominant factor that shapes coastlines. Waves are disturbances in bodies of water and form as a result of frictional forces related to wind. They generate rotating bodies of water that migrate obliquely toward the shore. At the point where the wave base interacts with the ocean bottom in shallow water, the oscillations become deformed, ultimately causing waves to crash.

5. Littoral processes refer to the way that sediment is transported and deposited in the shore zone. An important feature is a longshore current that develops along a coastline due to the deflected force of oblique waves striking the shore. In a related process, water runs up the beach and back into the ocean or sea in processes called swash and backwash, respectively.

Coastal Landforms

Let's now turn to the various landforms created along shorelines. Recall that fluvial, glacial, and eolian landforms can be subdivided into erosional and depositional categories. Coastal landforms can be viewed in the same way. This discussion begins with erosional coastlines.

Erosional Coastlines

Most coastal erosion is accomplished through the unceasing pounding of the shoreline by waves. Waves have tremendous power after they break, with the associated spray moving as fast as 113 km/h (70 mi/h). Given that water has a very high density, the rapidly moving water has a lot of power and is responsible for the majority of coastal erosional landforms.

Although wave erosion can occur along any part of the coast, it tends to be focused on a particular part of the landscape called a headland. A **headland** is a promontory that juts into the ocean or sea and thus is surrounded on three sides by water (Figure 19.13). Headlands tend to form along shorelines where bands of rock with alternating resistance run perpendicular to the coast. The headlands are associated with the more resistant rock such as limestone or granite. Where the rock is relatively soft, perhaps shale or sandstone, waves can erode sediment more effectively and a bay forms.

Once the pattern of headlands and intervening bays develops, the erosive power of waves begins to concentrate on the headlands. This change occurs because approaching waves initially begin to slow down in front of the headlands, where they first encounter shallow

---

Headland   *A portion of the coast that extends outward into a large body of water.*

Figure 19.13 Headlands along the southern coast of Tasmania in Australia. Headlands are prominent coastal features that project out into the water. Note that at least three headlands appear in this image.

water. This friction causes the waves to pivot around the headlands in a process called **wave refraction**. As you can see in Figures 19.14a and b, waves bend around a headland, which results in the energy of the waves being expended on three sides of the landform. In this way, headlands are more vigorously eroded than the

intervening bays. The bays become zones of deposition, forming features called *pocket beaches*, because sediment is funneled into them from the eroded headlands (Figure 19.14c).

The basic coastal erosional landform is the wave-cut bluff, which is a vertical face that commonly forms along rocky shorelines. Excellent examples of large wave-cut bluffs occur on many parts of the coast of Scotland (Figure 19.15), as well as the coasts of England, Australia, and California. Wave-cut bluffs evolve in a predictable sequence of events as shown in Figure 19.16. This sequence begins with waves that are pounding against a noneroded shoreline (Figure 19.16a). With time (Figure 19.16b), the waves cut a vertical face into the rock outcrop, and at the same time they plane a more horizontal surface called a *wave-cut platform* into the rocks in front of it. During higher water stages or strong storms, waves run up and cut a notch into the lower rocks of the bluff outcrop and further plane the nearshore platform (Figure 19.16c). If the notch is cut sufficiently deep, the rock overhang falls under the influence of gravity and the bluff retreats farther inland, leaving a longer wave-cut platform in front of it (Figure 19.16d). As the coastline retreats due to erosion, it is said to be undergoing **retrogradation**.

Wave refraction   *The process through which waves are focused and bent around headlands.*

Retrogradation   *The process through which a shoreline retreats through erosion.*

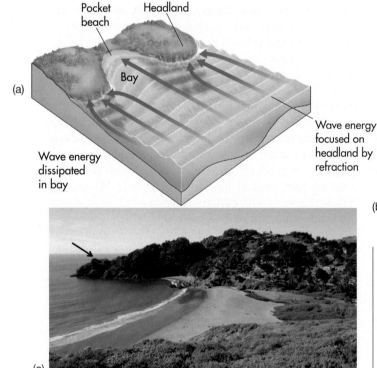

Figure 19.14 Headland erosion. (a) A model showing how waves refract around headlands, causing the formation of a pocket beach in the intervening bay. (b) Wave refraction along the coast of California. Notice how the waves are bending around this headland? (c) Pocket beach on the Pacific Coast north of San Francisco. This beach formed when sandy sediment eroded from headland in the background (arrow) and behind (south) the view was funneled into this small embayment.

Figure 19.15 Wave-cut cliff of Kilt Rock, Scotland. This massive wave-cut bluff was eroded into basalt cliffs over time. Note the waterfall in the foreground.

As described earlier, one way that water moves along coastlines is through eustatic changes in sea level. In addition to the submerged river channels that occur on the continental shelf, one of the best lines of evidence for prehistoric eustatic sea-level changes is the presence of abandoned bluffs and marine terraces. Similar to a fluvial terrace, a marine terrace represents the former position of the active shoreline on the landscape. Figure 19.17 shows how this kind of landscape evolves. In Figure 19.17a, waves create a wave-cut bluff and platform similar to that shown in Figure 19.16. Imagine what would happen if the landscape were subsequently uplifted or if sea level fell. If such a eustatic change occurred, a new bluff and platform would form below the abandoned one (Figure 19.17b), leaving a stair-stepped coastal landscape. Such landscapes are common along the active margin of the west coast of the United States (Figure 19.18).

In addition to wave-cut bluffs and terraces, additional evidence for coastal retrogradation comes in a variety of forms, which evolve in a continuum as the shoreline recedes. *Sea caves* are semicircular notches cut into the base of rock bluffs. If the rock is entirely cut through so that water passes through the notch, a *sea arch* results

(a) Noneroded rocky shoreline

(b) Formation of vertical bluff

Eroded vertical face
Wave-cut platform

(c) Waves cut notch into base of cliff

Rock overhang
Notch

(d) Overhang collapses, bluff retreats

Direction of retrogradation

New wave-cut platform

Figure 19.16 Bluff formation and retreat. (a) Waves attack a noneroded rock outcrop. (b) With time, the waves cut a vertical bluff into the rock. (c) During strong storms or relatively high water, the waves cut a notch into the lowermost rocks of the bluff. (d) If the notch is cut sufficiently deep, the overhanging rocks in the bluff collapse, causing bluff retreat.

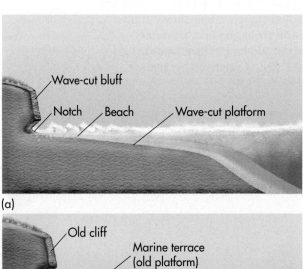

Wave-cut bluff
Notch   Beach   Wave-cut platform

(a)

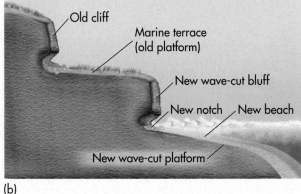

Old cliff
Marine terrace (old platform)
New wave-cut bluff
New notch   New beach
New wave-cut platform

(b)

Figure 19.17 Evolution of a marine terrace. (a) Waves cut a bluff and platform. (b) Tectonic uplift or sea-level fall raises the previous wave-cut features relative to the position of current erosion, resulting in a marine terrace.

Figure 19.18 Marine terrace in California. This marine terrace formed sometime during the late Pleistocene, perhaps between 60,000 and 100,000 years ago, in association with tectonic uplift following a period of high sea level. Uplift rates in this area are about 40 cm (16 in.) every 1000 years. Note the prominent bluff that stands between the terrace and the beach.

(Figure 19.19). With time and additional erosion, the top of the arch may collapse, leaving a detached fragment of the original bluff, called a *sea stack*, isolated in the ocean. These features are particularly common along the rocky coast of the Pacific Northwest. The net effect of shoreline retrogradation in places like this is that the coastline slowly transforms from one that has a number of prominent headlands to one that is relatively straight (Figure 19.20).

Figure 19.19 Erosional landforms along the coast of Wales. Sea arches develop when sea caves are entirely cut through, allowing free water passage beneath the arch. If the top of the arch collapses, an isolated sea stack remains, such as the one to the left. Note the tilted rock strata.

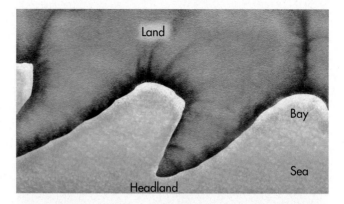

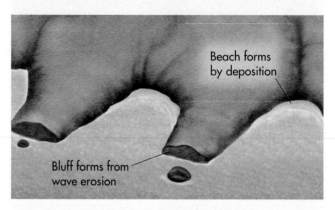

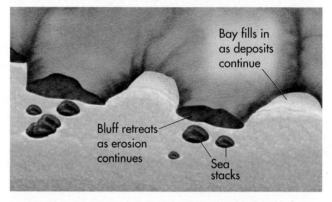

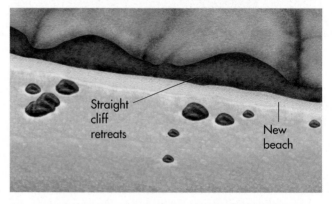

Figure 19.20 Evolution of a rocky coastline. Through the combined processes of erosion and deposition over time, rocky coastlines evolve in a predictable way that wears down headlands to form relatively straight beaches.

Imagine that you are driving down the Pacific Coast in Oregon and see the beautiful landscape pictured here. What conclusions can you reach about the history of landscape evolution at this location? What were the processes involved? In short, how did this landscape form?

DISCOVER...

MARINE TERRACES ON SAN CLEMENTE ISLAND, CALIFORNIA

The elevation of marine terraces can provide a great deal of information about past ocean levels and tectonic uplift. This image shows marine terraces on San Clemente Island, which is 115 km (~72 mi) west of the southern coast of California. Each of these terraces formed during a prehistoric high sea level associated with an interglacial period, with the highest surface representing the oldest high-level sea stand. Sea level subsequently dropped during glacial periods when enormous volumes of water were stored as ice. At the same time, the old wave-cut bluff was slowly uplifted due to tectonic activity. With each successive interglacial, sea level rose again, creating new wave-cut bluffs that were progressively lower than the previous ones. Given that the rate of uplift is known, it is possible to reconstruct the height of the former sea-level high stands.

GeoDiscoveries   www.wiley.com/college/arbogast

Waves and Coastal Erosion

Waves are fascinating features that often shape the coastline in dramatic ways. Aside from witnessing these processes yourself, the best way to visualize them is to see video of them in action. To do so, go to the **GeoDiscoveries** website and open the module **Waves and Coastal Erosion**. The first part of this module is a video that demonstrates wave processes and their interaction with the shore. Following this video is another film clip that shows coastal erosion along the California coast in 2008. This video was filmed during a strong storm that produced strong waves that battered coastal bluffs. Watch how the waves crashed against the base of the bluff and the way the upper part of the cliff was destabilized in the manner discussed in this chapter.

KEY CONCEPTS TO REMEMBER ABOUT EROSIONAL COASTLINES

1. A coastline that is essentially retreating through erosion is considered to be retrograding.

2. Erosional coastlines are shaped primarily by strong waves. These coastlines tend to be rocky with high bluffs.

3. Headlands are prominent landforms that protrude into the ocean. These landforms are eroded more vigorously than other places along the shore because waves are refracted around them and erode on three sides.

4. Marine terraces represent ancient shorelines. The landforms indicate either variable water levels or uplift due to tectonic activity.

5. Other indicators of coastline recession include caves, sea arches, and sea stacks. These erosional features evolve in a continuum as a coastline erodes.

Depositional Coastlines

We have just seen how coastlines are shaped by erosion and some of the diagnostic landforms that result from this process. Now, let's turn to the coastal landforms created when sediment is deposited in various ways and places. You were briefly introduced to this concept with the formation of a pocket beach between two eroding headlands (Figure 19.14c). Coastlines in the process of extending outward into the water through deposition are said to be undergoing **progradation**.

Beaches The part of the coastal landscape people are probably most familiar with is the beach. Beaches are dynamic places where sediment is deposited through the combination of waves, beach drift, and wind. In some places, beaches are supplied with alluvial sands derived from land far inland. The beach is a transition between the water and the landmass and consists of exposed, unconsolidated sediments that usually range from sand to cobbles. Although finer silt and clay-sized particles are sometimes contained within a beach, they are usually carried away in suspension by the longshore current. Most beach growth occurs during the summer months in the Northern Hemisphere when the weather is relatively calm. In winter, however, beaches can be significantly eroded due to large waves created by strong storms. If you happen to live on or near a beach, notice how the shape of the beach can change between seasons.

Figure 19.21 shows how beach components are differentiated on the basis of their relative position to the water. The **offshore** is permanently submerged and is the zone where waves break and the surf is most active. Moving landward, the **foreshore** is the part of the beach that is influenced by the rise and fall of the tides; that is, it is regularly exposed and submerged. Separating the offshore and foreshore is a submerged ridge called an **offshore bar**,

---

Progradation *Outward extension of the shoreline through deposition of sediment.*

Offshore *The nearshore zone that is permanently submerged and where waves break.*

Foreshore *The nearshore zone that is regularly exposed and submerged through the tidal fluctuations and movement of surf.*

Offshore bar *A small ridge on the bottom of the ocean that separates the offshore and foreshore.*

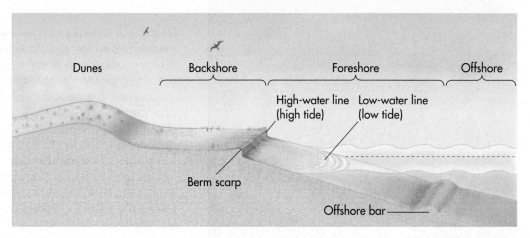

Figure 19.21 Generalized beach cross section. Beaches are divided into several regions, each associated with a different water level or the action of wind.

Dunes   Backshore   Foreshore   Offshore

High-water line (high tide)   Low-water line (low tide)

Berm scarp

Offshore bar

which forms as a result of the interaction between breakers and the ocean bottom in nearshore environments, where wave velocity is relatively low. Another noticeable feature on many coasts is a **berm scarp**, which is cut at the high water line of the beach. Landward of the high water line is the **backshore**, which is the relatively flat part of the beach that is covered by water only during severe storms.

Spits and Baymouth Bars   As discussed earlier, one of the ways in which water moves along the coast is in association with the longshore current (see Figure 19.11). This current is capable of eroding a great deal of fine sediment

from the foreshore and offshore environments and transporting it down the beach. As long as the beach continues in an uninterrupted fashion along the shore, the current and its sediment load will flow progressively down the coast because of the interaction of waves and the landmass. Once the current encounters the mouth of a deep bay, however, its speed reduces in a manner consistent with a stream's loss of velocity when it meets the ocean. As with river deltas, sediment carried by the longshore current is deposited in a predictable fashion that produces distinctive landforms in bay areas (Figure 19.22).

The variety of landforms created by the deposition of longshore drift in bays forms a geomorphic continuum that evolves with time. This continuum begins with the formation of a **spit**, which is a linear bank of

---

Berm scarp   *A miniature cliff created by wave erosion that fronts a beach berm.*

Backshore   *The part of the beach that lies between the berm scarp and foredune and is covered by water only during strong storms.*

Spit   *A linear bank of land that extends into a bay made by the deposition of longshore sediment.*

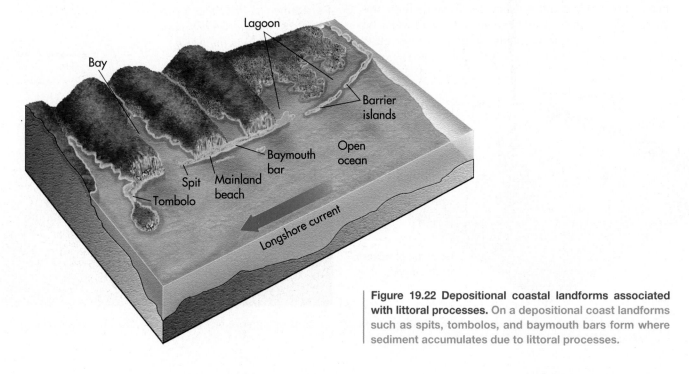

Lagoon

Bay

Barrier islands

Open ocean

Spit   Baymouth bar

Mainland beach

Tombolo

Longshore current

Figure 19.22 Depositional coastal landforms associated with littoral processes. On a depositional coast landforms such as spits, tombolos, and baymouth bars form where sediment accumulates due to littoral processes.

Figure 19.23 Cape Cod, Massachusetts. This satellite image shows a beautiful example of a curved spit, one that formed because the prevailing longshore drift flows to the north.

land that extends into a bay when longshore sediment is deposited (Figure 19.22). An important distinction to make here is that a spit is connected to land on the up-current side of the bay and extends into open water on the other side. A spit is usually straight for most of its length, but typically curves landward at its tip, forming a hook when viewed from the air above. This form develops because at the same time the longshore sediment is being deposited as the spit extends, incoming waves are moving the tip shoreward. An excellent example of a curved spit occurs on Cape Cod, Massachusetts (Figure 19.23). With sufficient time, the spit could extend entirely across the bay, closing it off with a landform called a **baymouth bar**. If and when this closing occurs, the bay will no longer be directly connected to the ocean and will then be called a **lagoon**. The water in a lagoon is typically brackish relative to the open ocean and thus is an important ecosystem. Sometimes longshore sediments converge from two different directions, causing a spit to grow perpendicular to the shore. If this kind of spit grows so that it reaches a large sea stack or island, it is called a **tombolo** (Figure 19.24).

Coastal Dunes Another type of depositional landform associated with sandy coasts is sand dunes. In contrast to desert environments where dunes form in poorly vegetated, sandy areas, coastal dunes develop because the supply of sand along beaches can be very high. A common type of dune form associated with beaches is a **foredune**. Foredunes range in height from 1 m to perhaps 10 m

---

Baymouth bar   *A spit that entirely encloses a bay.*

Lagoon   *A brackish body of water that lies behind a baymouth bar.*

Tombolo   *A spit or sandbar that connects an island to the mainland.*

Foredune   *A dune that forms parallel to the shore when sand blows inland from the beach.*

Figure 19.24 Tombolos. (a) Tombolos form when longshore drift extends to a preexisting island. (b) A beautiful example of a tombolo is Mont-Saint-Michel in France.

(3.3 ft to 33 ft) and are parallel to the shore (Figure 19.25a). These dunes develop due to shore vegetation trapping sand blown by the wind off the beach and gradually forming a linear (alongshore) ridge or dune. Other kinds of dunes commonly seen in coastal settings are blowouts and parabolic dunes, discussed in Chapter 18 (Figures 18.19 and 18.22). In places where the supply of sand is especially large and the winds are consistently strong, massive *transgressive dune fields* form. These dune fields are so named because they transgress (advance) inland and bury older surfaces (Figure 19.25b).

Coastal dunes are found all over the world. Spectacular dune fields occur along the coast of Namibia in Africa, Australia, Brazil, India, many shores in Europe, Mexico, and numerous other places. Coastal dune fields also line much of the U.S. shore. Perhaps the best example is the fantastic dune fields along much of the coast of Oregon and Washington (Figure 19.25c), where they are supplied by sand driven by strong westerly winds blowing off the Pacific Ocean. Dune fields also occur along much of the Great Lakes coastline, with the dunes along the western shore of Lower Michigan the most impressive due to prevailing westerly winds and high sand supply (Figure 19.25a). Some of these dunes reach heights of about 60 m (200 ft)!

Barrier Islands Another type of depositional landform that occurs along coastlines is a **barrier island**, which is an accumulation of sediment (sand, gravel, boulders, and/or shells) parallel to the shore in the ocean that is formed by waves, tides, and winds (Figure 19.26). These islands tend to form on broad, gently sloping continental shelves. Although no firm agreement exists about how these islands develop, a leading hypothesis is that they are linked to rising sea levels following the most recent ice age. In this model, sediment deposited on the continental shelf by rivers during the Wisconsin glacial period was reworked and molded when sea level began to rise as the ice sheets melted. As the water level slowly rose, the shelf sediments were shaped by waves, tides, and the wind into the linear islands. At the same time as these islands began to form, they were gradually nudged toward their present position by the slowly rising water and storms crashing into them.

A key factor in the overall topography of a barrier island is the supply of sand fed to the system. In barriers with low sand supply, the landform will have very low relief and lie only a meter or two above sea level. Where barriers have high sand supply, dunes typically form and provide additional relief to the landscape. Barrier islands lie in various proximities to the shoreline, with some being only several hundred meters away and others many kilometers from the coast. An excellent place to see barrier islands

---

Barrier island *An elongated bar of sand that forms parallel to the shore for some distance.*

(a)

(b)

(c)

Figure 19.25 Coastal sand dunes. (a) Foredune (arrow) along the east coast of Lake Michigan. The dune formed because sand blew up from the beach, which can be seen to the left. Note the person for scale standing in a swale in the lee of the dune and the way the dune axis mirrors the beach. (b) Transgressive dune field on North Head, New Zealand. Here, eolian sand has spread inland due to very high sand supply and strong winds blowing off the Tasman Sea (left). (c) Coastal dunes on the Oregon coast. Westerly winds here have produced this beautiful dune field. Do you see the sea stacks in the background?

Figure 19.26 A typical barrier island. Barrier islands are essentially long sand bars, with some capping sand dunes, that form parallel to the coast. These landforms may have formed when rising sea levels at the end of the Wisconsin glacial period reworked sediments on the continental shelf into these linear features. Note the dunes on this island due to high sand supply.

is along the Atlantic and Gulf Coasts of the southeastern United States, where approximately 280 barrier islands are found. Figure 19.27a shows a satellite image of a reach of barrier islands along the Gulf Coast from Louisiana (left) to Alabama (right). The Chandeleur Islands are part of the complex, lying east of the Mississippi delta. These islands have very low relief due to limited sand supply and are thus pounded when large waves from strong storms and hurricanes wash over. As you can see in Figure 19.27b, hurricanes Rita (2004) and Katrina (2005) dramatically altered the shape of these islands, producing numerous gaps and reducing island size.

Like baymouth bars, brackish lagoons form behind barrier islands that are close to the shore because they are not fully in contact with the ocean and thus have lower salinity. Once these lagoons become established, they become the focus of sediment deposition from three sources: (1) streams flowing off the mainland, (2) eolian sand blowing landward from the island, and (3) tides. Taken together, this deposition produces new land bodies called mud-flats (Figure 19.28a), which, if vegetated, become salt marshes (Figure 19.28b). If plant colonization continues unchecked, due to the lack of strong storms or an outlet to the ocean, the lagoon is slowly transformed into a continuous landmass connected to the barrier island.

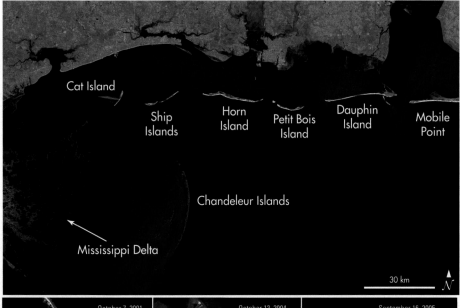

Cat Island

Ship Islands

Horn Island

Petit Bois Island

Dauphin Island

Mobile Point

Chandeleur Islands

Mississippi Delta

30 km

N

(a)

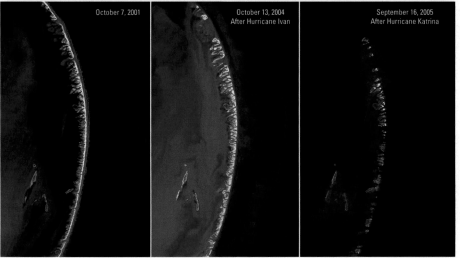

October 7, 2001

October 13, 2004
After Hurricane Ivan

September 16, 2005
After Hurricane Katrina

(b)

Figure 19.27 Barrier islands along the Gulf Coast. (a) Satellite image of a complex of barrier islands east of the Mississippi Delta along the coast of Mississippi and Alabama. (*Source:* NASA.) (b) Time sequence of the Chandeleur Islands from 2001, 2004, and 2005. Notice the profound impact hurricanes Rita and Katrina had on these inlands. (*Source:* USGS.)

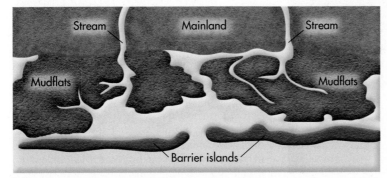

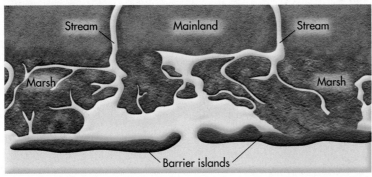

Figure 19.28 Infilling of a lagoon. (a) Because lagoons are water bodies of low energy, they are the focus of sediment deposition, forming mudflats. (b) With time, the lagoon will be transformed into a vegetated salt marsh that connects the mainland with the barrier island.

VISUAL CONCEPT CHECK 19.3

Look at this satellite image of a landform on the coast of Japan. Assuming that north is at the top of the image, what way is the longshore current flowing? Why is the curve present at the end of the feature?

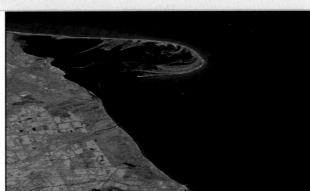

www.wiley.com/college/arbogast

WILEY PLUS

Longshore Processes and Depositional Coastlines

Now that we have discussed the basic processes associated with longshore drift, we can view them in an animated format. To do so, go to the *GeoDiscoveries* website and access the module *Longshore Processes and Depositional Coastlines*. This module describes the way that water moves along a shoreline and the landforms that result from this flow. Watch how oblique-approaching waves cause a current to develop parallel to the shore. Also see how this current moves sediment until it reaches bays where it is deposited to first form spits and then baymouth bars. Most importantly, watch how the coastal landscape is transformed through time and process. After you complete the animation, be sure to answer the questions at the end of the module to test your understanding of this concept.

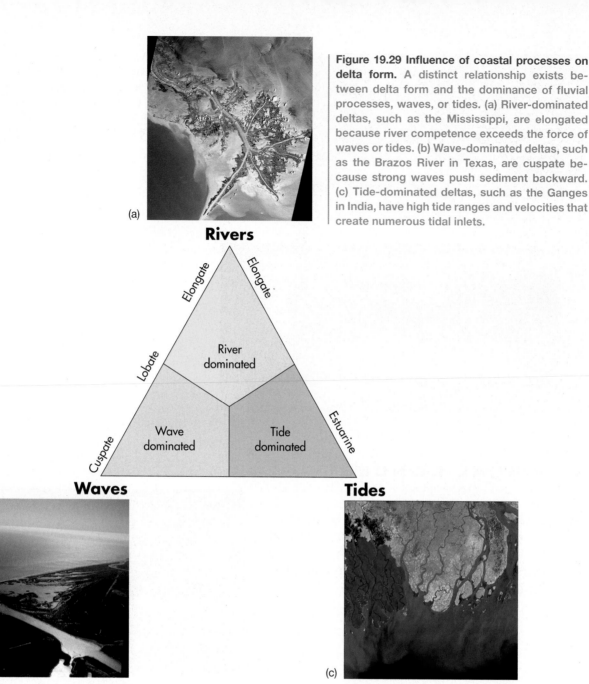

(a)

Rivers

Elongate

Elongate

Lobate

River
dominated

Cuspate

Wave
dominated

Tide
dominated

Estuarine

Waves

(b)

Tides

(c)

Figure 19.29 Influence of coastal processes on delta form. A distinct relationship exists between delta form and the dominance of fluvial processes, waves, or tides. (a) River-dominated deltas, such as the Mississippi, are elongated because river competence exceeds the force of waves or tides. (b) Wave-dominated deltas, such as the Brazos River in Texas, are cuspate because strong waves push sediment backward. (c) Tide-dominated deltas, such as the Ganges in India, have high tide ranges and velocities that create numerous tidal inlets.

Impact of Coastal Processes on Delta Form As discussed in Chapter 16, one of the most interesting landscapes on Earth is the place where a river meets the ocean to form a delta. Deltas are noteworthy purely in the context of fluvial processes because they mark the location where sediment from the continental interior is deposited due to very low stream gradients and reduced velocity. These places are also fascinating from the standpoint of various coastal processes because their overall form is directly related to the influence (or lack thereof) they have on the delta as it develops.

The most common way to examine delta form is to consider whether waves, tides, or fluvial processes dominate the coastline at any given place. Although many deltas are influenced equally by both waves and tides, or fluvial processes and waves, it is instructive to consider their shape as it relates to the singular dominance of any particular process. This triangular relationship is shown in Figure 19.29. As you examine the corners of this diagram, remember that a continuum of delta forms actually exists and that you are observing the extreme examples of each kind at this level.

Let's begin this investigation by considering the shape of deltas dominated by rivers. This kind of delta occurs along a coastline where wave energy and tidal range are typically low and stream energy is relatively

Figure 19.30 Pillar coral in the Cayman Islands. Coral comes in a wide variety of colors, shapes, and species, but they all produce calcium carbonate external skeletons.

high. In these situations, such as the Mississippi Delta, the delta is elongated because the stream is able to extend into the ocean through progradation because waves and tides are unable to blunt this growth. Wave-dominated deltas are those that form where the energy

of incoming waves is high. These deltas tend to be blunted and even cuspate because strong waves push sediments back toward the mainland. A good example of such a cuspate delta can be seen where the Brazos River meets the Gulf of Mexico along the Texas coast. Finally, tide-dominated deltas form along coasts that have a high tidal range. In these settings, such as the Ganges River Delta in India, the velocity of incoming and outgoing tides exceeds river velocity except during floods. As a result, the delta takes on the form of an estuary with numerous tidal inlets.

Coral Reefs   Thus far our discussion of depositional coastlines has focused on landforms that develop due to the accumulation of sediment. Another way that coastal landforms evolve is through the growth of **coral reefs** (Figure 19.30). Coral reefs form due to the symbiotic interaction between coral polyps and microscopic algae (called *zooxanthellae*) that give the corals their color. When the polyps die, they leave an exoskeleton behind composed of calcium carbonate. These deposits grow larger with time because, when one generation of corals dies, a new one grows on top of it. Coral reefs grow best in tropical waters between about 30° N and 25° S that are warmer than 20°C (68°F), free of suspended sediment, and well aerated. Figure 19.31 shows the distribution of coral reefs on Earth.

Coral reefs form in three distinct types of settings. A *fringing reef* forms on a shallow platform attached to the

---

Coral reefs   *Resistant marine ridges or mounds consisting largely of compacted coral together with algal material and biochemically deposited calcium carbonates.*

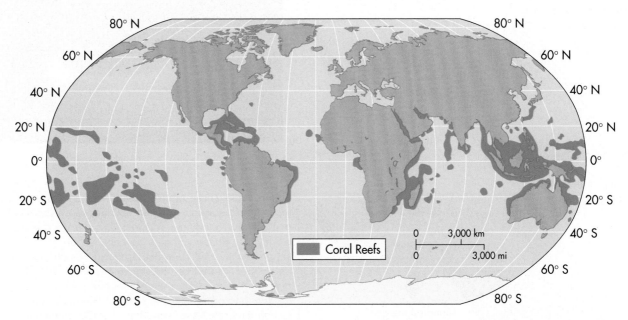

| Figure 19.31 Global distribution of coral reefs. Coral reefs are very common in tropical areas.

Figure 19.32 Coral reefs. A fringing reef along the island of Moorea in Tahiti, South Pacific Ocean.

shore or island, such as the one shown in Figure 19.32. A second kind of reef is a *barrier reef*, which is a line of coral that is roughly parallel to the shore. Such reefs are separated from the coast either by a deep and wide lagoon or, in some cases, by several kilometers of open

ocean that is much deeper than the water above the coral. The best-known barrier reef is the Great Barrier Reef in Australia. This spectacular reef is actually a complex of more than 2000 individual reefs that lie about 50 km (30 mi) from the northeast Australian coast. This complex of reefs extends for about 2000 km (1250 mi), making it the longest reef system in the world.

A third type of coral reef is an atoll. *Atolls* are semicircular reefs that are most closely associated with degradation of volcanic islands. These reefs evolve in a predictable pattern. In the first stage of development (Figure 19.33a), a fringing reef grows on the nearshore platform surrounding a relatively new volcanic island. In the second stage of development (Figure 19.33b), the volcano is dormant and begins to wear down and become somewhat submerged due to erosion and subsidence. During this phase of reef development, the reef

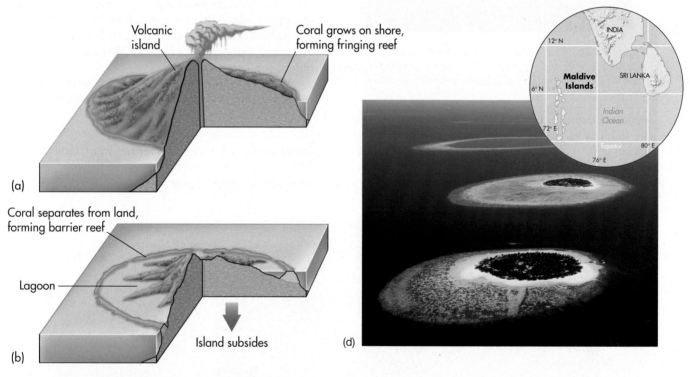

Figure 19.33 Three-stage evolution of an atoll. (a) During the first stage, a fringing reef grows on the shallow platform surrounding a volcanic island. (b) In the second stage, the volcano begins to collapse on itself, which separates the reef from the landform, resulting in a barrier reef. (c) With time, the volcanic island is entirely submerged, leaving only the coral above water as an atoll. (d) This photograph of the Maldive Islands in the Indian Ocean illustrates the evolution of atolls, beginning with the fringing reef in the foreground.

Figure 19.34 Coral reefs in the South Pacific Ocean. These shallow reefs in the Solomon Islands are home to a fascinating array of life. Such a colorful view is common in most coral reefs around the world.

becomes more like a barrier reef because it is separated from the volcanic island. With the passage of additional time, the volcano disappears entirely (Figure 19.33c), leaving an atoll reef ring isolated in the ocean. Atolls are particularly common in the South Pacific Ocean and, like reefs everywhere, are home to a fascinating and colorful variety of ocean animal species (Figure 19.34).

KEY CONCEPTS TO REMEMBER ABOUT DEPOSITIONAL COASTLINES

1. Depositional coastlines form through the process of sediment accumulation. A coastline that is extending outward into a body of water is said to be prograding.

2. The most familiar coastal depositional feature is the beach. Sediment is supplied to beaches through longshore deposition, swash/backwash, and wind. This landform is subdivided into three regions (offshore, foreshore, backshore) on the basis of water depth and frequency and nature of submergence.

3. Sand dunes commonly occur in the lee of beaches where the supply of sand is high and strong winds frequently blow.

4. A range of depositional features is associated with the decreased velocity of the longshore current, including spits, baymouth bars, and tombolos.

5. A coral reef evolves through the growth of small marine organisms and can be categorized as a fringing reef, barrier reef, or atoll.

Human Interactions with Coastlines

As noted previously, coastlines are a favored place for human interactions related to economic development, trade, housing, and recreation. Of the 6 billion people that live on Earth, approximately 37% (2 billion) live within 100 km (60 mi) of a major coastline. Nearly 50% live within 200 km (120 mi) of a shore. To see how these figures compare to the United States, examine Table 19.1. Overall, nearly 144 million people, (~53% of the population) live near a coastline. Almost 63 million people live along the Atlantic Coast in cities such as New York, Boston, Philadelphia, and Miami. Another 17.3 million people live along the Gulf Coast in cities such as Tampa, New Orleans, and Houston. The Great Lakes region is home to almost 27 million people who live in cities such as Cleveland, Detroit, and Chicago. Still another 37 million people reside along the Pacific Coast, with most living in cities such as Seattle, Portland, San Francisco, and Los Angeles.

Given that most people in the United States (and the world) live close to coastlines, you should understand some of the human impacts in these geomorphically sensitive regions. Many of these impacts involve stabilizing beaches in order to maintain attractive tourist destinations for economic purposes. Others are potential future outcomes related to global climate change and the effects rising sea level will have on people living near the shore. Still other human interactions are based on the desire to keep coastlines as environmentally healthy as possible.

In this latter context, the nation was recently focused on the catastrophic oil leak from a ruptured well in the Gulf of Mexico. Estimates indicated that over 757,000 L (200,000 gal) of oil were streaming into the Gulf every day, causing grave concerns about the impact on beaches and marine life in the region. Geographers were involved in mapping the extent of the spill and the coastal zones potentially harmed. This section will discuss some of the ways humans impact coastlines.

Coastal Engineering

One of the most obvious ways that people impact the coastline is through engineered structures. Although these structures come in a wide variety of forms, they are generally built with the idea of (1) protecting the shore and shorefront homes from coastal hazards such as erosion and storm surges, (2) stabilizing and nourishing beaches, or (3) maintaining or improving the flow of trade or traffic into ports.

Shoreline Protection Of the many people who live near U.S. coastlines, a large number choose to reside or

Total U.S.	Atlantic Coast	Gulf of Mexico	Great Lakes	Pacific Coast	Total Coastal	Balance of U.S.
Population in Millions						
272.7	62.7	17.3	26.8	37.0	143.9	128.8
Percent of Total U.S. Population						
100.0	23.0	6.3	9.8	13.6	52.8	47.2

Source: U.S. Department of Commerce, Bureau of the Census.

work directly on the shore (Figure 19.35). The reasons for this choice are probably obvious, as the shore is a pleasurable landscape in which to live and work, with spectacular views and numerous recreational opportunities. However, coastlines are one of the more sensitive landscapes on Earth, with changes occurring on a daily basis and major adjustments taking place when strong storms such as hurricanes strike.

Although extreme events such as hurricanes are indeed dangerous, they are statistically rare along any particular stretch of coast. As a result, people tend to be much more concerned about the erosion associated with less severe storms, waves, and high water level. Unchecked, the effects of shoreline erosion on coastal communities can be disastrous (Figure 19.36). In an effort to control these effects, people have devised a variety of protective structures.

Perhaps the most common way people attempt to mitigate coastal erosion is to build *sea walls* and *revetments* that armor the shore. As the name implies, a sea wall is a vertical, wall-like structure built along the coast, usually at the landward edge of a beach, where a bluff or dune may

occur. Figure 19.37 shows sea wall at Palm Beach, Florida. Many protective structures are less elaborate, consisting of large rocks and sometimes even cement blocks from old roads. This kind of protection is called *rip-rap* and forms a revetment because it is usually applied to a preexisting slope, rather than being built vertically. Although sea walls and revetments protect the property they front from erosion, they cause more extensive erosion up and down shore than would otherwise occur naturally because the erosive power of waves is transferred and concentrated on those locations (Figure 19.38).

Beach Nourishment   One of the primary reasons so many people like coastlines is that they are attracted to the beach and all the recreational and leisure opportunities it provides. Much economic activity is directly tied to beaches, through either tourism or the many people who live there. Given these important relationships, people have a strong desire to maintain the physical integrity of their particular beach, whether it fronts a luxury hotel, condominiums, or a row of homes.

Figure 19.35 Coastal development on Padre Island, Texas. Much of this barrier island, which is shown in the satellite image in Figure 19.29, is heavily developed. This pattern of development is very common along the Gulf and Atlantic coasts of the United States.

Figure 19.36 Effects of coastal erosion on personal property. Severe bluff erosion along this stretch of the California coastline can be a problem for those who choose to live really close to shore.

Figure 19.37 Sea walls. This sea wall (arrow) at Palm Beach, Florida, protects a hotel complex from coastal erosion. Structures like this armor much of the shore in southeastern Florida.

However, this desire to maintain an attractive beach is frequently at odds with natural coastal processes because beaches often consist of unconsolidated sands that are easily eroded. In an effort to offset coastal erosion, many communities have organized beach nourishment programs that bring in more sand to replace that which is lost. Although effective, these programs are very costly. In the very small state of Delaware, for example, over 2,300,000 m³ (3,008,286 yd³) of sand has been used to nourish beaches since the early 1960s at a cost in 2010 dollars of over $60,000,000. In the tourist-dominated state of Florida, the cost to nourish beaches over the same length of time has been over $432,000,000 in 2010 dollars.

One way that people nourish beaches is simply to bring fresh deposits of sand into a beach system to replenish sand lost through erosion. In this manner of beach maintenance, which usually occurs at the end of the winter storm season, large earth-moving equipment

Figure 19.39 Groins and shoreline processes. Groins such as these along Lake Michigan in Chicago are walls built perpendicular to the shore. These walls intercept the longshore current, causing deposition of sand on the up-current side and erosion on the down-current side of the structure. In Lake Michigan, the prevailing longshore current flows from north (at the top of the photo) to south. Note how sand has accumulated on the north side of these groins and has washed away on the south side.

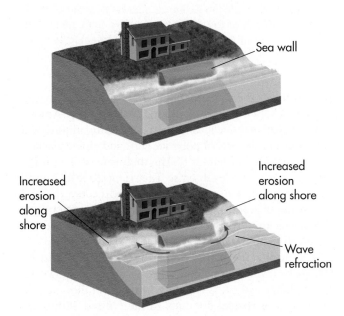

Figure 19.38 Beach walls and coastal erosion. Although sea walls protect the property they front from erosion, they cause extensive erosion up and down shore because wave power is deflected.

transports sand from another source to a heavily eroded beach. Another way that people try to maintain a functional beach is to limit the loss of sand to erosion. The most common method used to stabilize sand supply is building a simple structure called a **groin**, which is a low wall built at a right angle from the shore a short distance into the water. This wall intercepts and slows the longshore current on the up-current side of the structure, causing deposition of sand at that locality (Figure 19.39).

Although this process results in beach widening on one side of the groin, the portion of the beach immediately down-current of the structure becomes a zone of erosion because the longshore current is regenerated there due to unaltered waves striking the shore obliquely. Thus, it is possible for one landowner, say, landowner A, to build a wide beach in front of her property by installing a groin that, in turn, causes erosion and beach loss at the adjacent property of landowner B down the coast. To avoid this

Groin   *A short, low wall built at a right angle to the coast that intercepts the longshore current.*

Figure 19.40 A typical jetty. This jetty on the Texas coast provides a reliable passageway into the ocean for boats. Given the variable beach thickness on either side of the jetty, can you tell which way the longshore current is flowing? Note the sequence of foredunes on the updrift (arrow) side of the jetty. These dunes formed due to high sand supply because sand was trapped.

loss, landowner B might also install a groin to intercept the sand eroded on his property caused by the groin on the property immediately up the coast. Given this downshore ripple effect, it is quite common to see a series of groins along a coast in distinct *groin fields*.

Jetties   Another purpose for protecting beaches and coastlines is to maintain a channel between the open water and interior water bodies such as lagoons, lakes, and rivers. This access is maintained with **jetties**, which are long stone or concrete structures used to create a permanent channel by reinforcing both sides of the passageway (Figure 19.40). From the shore, jetties project several hundred meters out into the ocean and provide a zone of relatively quiet water for a boat to approach the channel safely. Like groins, jetties cause deposition on the up-current side of the structure and erosion on the down-current side. To offset beach loss on the down-current side of jetties, beaches are often nourished either through direct deposition of sand or with groins.

Global Climate Change and the Impact on Coastlines

Recall from Chapter 9 that one of the major environmental issues facing people on Earth today is global climate change. As noted in that chapter, evidence is increasing

---

Jetties   *Long stone or concrete structures used to create a permanent opening for a channel by reinforcing both sides of the passageway.*

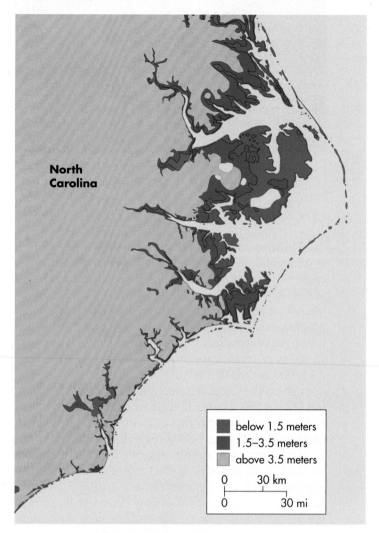

North Carolina

- ■ below 1.5 meters
- ■ 1.5–3.5 meters
- □ above 3.5 meters

0   30 km

0   30 mi

Figure 19.41 Predicted extent of coastal flooding in North Carolina. Much of the coastline lies less than 1.5 m (5 ft) above sea level and thus will likely be impacted to some extent by coastal flooding associated with global warming.

that average global temperature may increase between 2°C and 4°C (4°F to 7°F) by the end of this century. Among other negative effects, this increased temperature will cause further melting of polar ice caps and alpine glaciers, as we discussed in Chapter 17. This meltwater will return to oceans, causing sea level to rise. The rising sea level caused by more ocean water will be magnified by the thermal expansion of water due to increased global temperature. According to current predictions, sea level will rise between 15 cm and 90 cm (5.11 in. and 35.4 in.) by 2100.

Naturally, the places that will be affected most by rising sea levels will be coastal zones around the world. On a global scale, it is estimated that 92 million people are at risk of coastal flooding. In the United States, significant regions are susceptible to rising water levels. For example, Figure 19.41 shows the areas vulnerable to rising sea level along the coast of North Carolina. This map is color-coded: areas in red are coastal zones less than

Evolution of the Louisiana Coastline

Rising sea levels threaten many coastlines around the world. Perhaps the most sensitive coastal zone in the United States is the Gulf shore from Texas to Florida. The coast of Louisiana is particularly susceptible to rising sea levels because of the close relationship it has with the Mississippi River and its delta. Large areas of coastal loss have already occurred in the past few decades. To see the complex history of coastal zone, and the rates of projected sea-level change in the future, go to the *GeoDiscoveries* website and select the module *Evolution of the Louisiana Coastline*. Once you complete the module, be sure to answer the questions at the end to test your understanding.

1.5 m (5 ft) above sea level, whereas those in blue lie between 1.5 m (5 ft) and 3.5 m (11.5 ft) above sea level. As you can see, much of the eastern part of the state is very close to sea level. Of these regions, the extensive zone less than 1.5 m (5 ft) will likely be impacted somehow from coastal flooding associated with rising sea level. The likely response of environmental planners and government officials to this threat will be more funds devoted to coastal engineering and beach stabilization.

An area that will be particularly hard hit by rising sea levels is the Pacific Island region, which contains hundreds of low-lying atolls and volcanic islands. Countries such as the Marshall Islands, which consists of an archipelago of 34 atolls and coral reefs with a maximum elevation of less than 3 m (10 ft) above sea level, are already being affected by rising sea level. Several islands have been lost to erosion throughout the Pacific Island region, and plans are being developed in many places to relocate coastal inhabitants to more interior locations within the islands. In response to this regional crisis, people of the various island nations have formed the *Alliance of Small Island States* to lobby the United Nations about their particular plight, which, they argue, is impacting them disproportionately given that they produce only 0.06% of all carbon emissions.

Coral-Reef Bleaching Another by-product of global climate change is the impact that warmer ocean waters are having on coral-reef systems around the world. These systems are being affected by a process called *coral bleaching*, which occurs when coral reefs are stressed by warmer-than-normal waters. Coral bleaching can occur naturally as a result of short-term temperature changes, such as those that occurred in association with the El Niño in the early 1980s. Coral bleaching also occurs due to overexploitation of coral resources by people and other forms of local environmental degradation. Since the 1980s, however, growing stress caused by warmer oceans has been increasingly linked to coral bleaching. This stress loosens the algae that help feed the coral-building organisms from the reef. Because the algae give the corals their color, the starved corals look pale or bleached. Ultimately, continued bleaching kills corals.

In the past decade, significant amounts of bleaching associated with above-normal ocean temperatures have been reported in a variety of Pacific islands, the Caribbean Sea, Indian Ocean, and along the coast of Australia. In August 2010, for example, a major bleaching event occurred in the corals of Sumatra in Indonesia. Over 60% of corals in the area were bleached, with over 80% of some species lost. This event occurred because ocean temperatures were over 32°C (90°F).

KEY CONCEPTS TO REMEMBER ABOUT HUMAN IMPACTS ON COASTLINES

1. Coastlines are favored places for people to work, live, and play. Given the inherent sensitivity of the coastal landscape, human impacts are significant and growing. Most human impacts along coastlines are associated with beach stabilization, protection, and access (to port) maintenance.

2. Beaches are stabilized predominantly in two ways: beach nourishment and groin construction. Beach nourishment is an active process involving the physical transport of sand to the beach, whereas groins take advantage of the natural behavior of the longshore current.

3. Beach and bluff protection is mostly accomplished through the installation of sea walls and revetments, which armor the shore. Although these features work in the sense that they protect the portion of shore they front, they transfer wave energy to adjacent localities where it is concentrated as an erosive force.

(continued)

4. Jetties are long walls built along river channels and inlets to maintain a passageway from the open ocean to the interior. These features project some distance into the water to provide a safe coastal approach for ships.

5. Global warming appears to be causing a rise in sea level. In this context, significant parts of global coastlines are susceptible to increased erosion and, in some extreme cases, ultimate submergence. Warmer oceans are also contributing to coral bleaching.

THE BIG PICTURE

Up to this point, the text has focused largely on the physical processes and outcomes associated with various Earth systems and their spatial variability. Where appropriate, the text briefly examined the way that humans interact with these processes. In this chapter, for example, you investigated some of the ways that humans interact with coastlines. The next chapter expands on this theme by focusing on the relationship of physical geography to many of the complex environmental issues facing people today around the world. Three case studies are presented, including water use in Las Vegas, soil salinization in the United States and Australia, and Giant Panda conservation in China. Each of these case studies is presented in the context of human population, which has increased rapidly in the past century and is predicted to further grow in your lifetime.

SUMMARY OF KEY CONCEPTS

1. A coastline is a narrow zone where the hydrosphere, lithosphere, and atmosphere interact on a very large scale. These interactions result in very distinctive processes and landforms. Most coastlines are associated with the world's oceans, seas, and gulfs, but some prominent coastlines occur along very large lakes such as the North American Great Lakes.

2. The world's oceans collectively constitute the largest component of the hydrological cycle. Smaller bodies of water associated with oceans are seas and gulfs.

3. The most important agents of coastal change include water fluctuations, tides, waves, and littoral processes. The most extensive water fluctuations occurred when continental glaciers advanced and retreated during the Pleistocene. Tides are daily fluctuations caused by the gravitational pull of the Moon (mostly), and to a lesser degree the Sun. Waves form when wind blows across the water, causing energy within the water to roll forward due to friction. Waves often approach coastlines obliquely, resulting in the littoral processes of (1) longshore current in the water along the shore and (2) swash/backwash of water moving up and down, respectively, on the beach.

4. Erosional coastlines evolve when sediment is removed by coastal processes. Prominent erosional landforms along coastlines are headlands, bluffs, sea stacks, and marine terraces. Depositional coastlines develop where sediment accumulates. These landforms include beaches, spits, baymouth bars, tombolos, and barrier islands.

5. A slight majority of people on Earth live near coastlines. Given this population distribution, coastlines are impacted greatly by human behavior. Most of this impact occurs in the form of engineered structures, such as seawalls and jetties, that protect the shore and maintain the location of shipping channels.

6. It appears that global warming is having a major impact on coastlines around the world. These impacts include bleaching of coral reefs and rising sea levels, which are threatening coastal communities.

CHECK YOUR UNDERSTANDING

1. What is the difference between an ocean and a sea? Name the oceans of the world and at least five major seas.

2. What does the term *salinity* mean and how is it related to the chemical composition of ocean water?

3. In the past approximately 150,000 years, what has been the primary cause of eustatic sea-level change? Why have these changes occurred and what are some prominent landscapes that indicate such fluctuations?

4. With respect to tides, what is the difference between centrifugal and gravitational forces and how do they produce tidal bulges?

5. Compare and contrast spring and neap tides.

6. Describe the formation and movement of waves. What happens to waves when they approach the shore?

7. Describe the processes associated with littoral drift. How are they related to the approach of waves to the shore?

8. How does a marine terrace indicate that (1) a period of coastal erosion occurred, and (2) sea level fell or tectonic activity transpired?

9. How are sea caves, sea arches, and sea stacks part of a geomorphic continuum that indicates shoreline erosion is occurring?

10. What is the difference between an erosional and a depositional coastline?

11. Name and describe the depositional landforms associated with the longshore current.

12. Why are coastlines a good place for dune fields to evolve?

13. How is the shape of river deltas influenced by coastal processes?

14. How do coral reefs form and why is this process fundamentally different from other geomorphic processes?

15. What are the various ways that people attempt to protect beaches and bluffs from erosion? Why do these approaches also cause problems in nearby parts of the coast?

16. How does a groin take advantage of the longshore current to cause beach growth? Why are groin fields common?

17. Name and describe the two ways in which global warming may cause sea level to rise.

ANSWERS TO VISUAL CONCEPT CHECKS

VISUAL CONCEPT CHECK 19.1

Assuming that the axis of this part of Australia's coast aligns north and south, the wave-forming winds must be out of the southeast. These winds are causing swells to move slightly toward the northwest before they break as waves and strike the coast. This oblique approach causes the force of the water to be deflected toward the north, resulting in the formation of a longshore current flowing in that direction. In association with this current and swash/backwash of the surf, littoral drift is toward the north along the shore.

VISUAL CONCEPT CHECK 19.2

Sea stacks and arches are indicative of a retrograding coastline. These landforms are bodies of relatively resistant rock that have, until now, withstood the force of erosion caused by waves.

VISUAL CONCEPT CHECK 19.3

Assuming north is at the top of the photograph, this landform indicates that the longshore current is flowing slightly from northwest to southeast. The spit has formed because the longshore current encountered the relatively still water in the open ocean, causing deposition of sediment and progradation of the landform. The spit is curved because waves are moving sediment landward at the same time that it is being deposited at the tip of the feature.

CHAPTER TWENTY

RELEVANCE OF PHYSICAL GEOGRAPHY TO ENVIRONMENTAL ISSUES

This text has focused on the physical geography of Earth. The vast majority of topics dealt with the fundamental natural geographic patterns of the planet, such as seasonal variations in solar radiation, climate variability, distribution of vegetation and soils, location of tectonic plates, and the configuration of coastlines, to name a few. In addition, an underlying theme has been some interactions people have with the physical environment. In Chapter 9, for example, we investigated the potential role people may have in terms of the Earth's climate. We also addressed some of the ways that humans interact with streams, soils, eolian processes, and coastlines, and the ways that humans harness energy from the Sun, wind, and fossil fuels. This closing chapter will sharpen the focus on this theme by looking at some key issues that you will likely hear about in the future—issues that relate to the natural environment and the way in which humans interact with it. The purpose of this discussion is to demonstrate how an understanding of physical geography is extremely relevant to many societal issues.

560

Harvesting wheat in the Great Plains. Wheat production requires an excellent understanding of the holistic relationship between soils, climate, and vegetation. In a world of rapidly growing human population, a good knowledge of physical geography is essential to create a sustainable food supply. This chapter focuses on some of the various ways that physical geography relates to important human/environmental interactions on Earth.

CHAPTER PREVIEW

A Short History of Human Population

Case 1: Water Issues in the Arid American Southwest

Case 2: Soil Salinization in Arid and Semi-Arid Lands

Case 3: Giant Panda Conservation in China

The Future Picture

LEARNING OBJECTIVES

1. Describe how an understanding of physical geography is essential to a consideration of contemporary human/environmental issues.

2. Discuss the history of human population and the factors that have fueled its growth over time.

3. Compare and contrast the physical setting of Las Vegas, Nevada, the history of population growth, and ongoing water issues in the region.

4. Describe the various ways that soil salinization occurs and why it is a major problem in California and Australia.

5. Discuss the nature of panda habitat and how it has been progressively fragmented over time. Explain how geographical principles are being used in an effort to stabilize panda populations in the wild.

561

A Short History of Human Population

Before beginning the investigation of human/environmental interactions, we should first consider the general history of human population on Earth and how it is expected to change over the course of your lifetime. This topic is relevant because it is directly related to resource consumption and the ability of people to live in a particular place in a *sustainable* way—in other words, in a fashion that can continue for future generations. The **carrying capacity** of any given landscape refers to the number of individuals who can live in it without degrading the natural, social, cultural, and economic environments beyond sustainable limits. A number of variables directly associated with physical geography affect the carrying capacity of the land, including climate, vegetation, and soils.

Technological Development and Population Growth

Let's consider the changes in human population since the end of the last ice age approximately 10,000 years ago. As discussed in Chapter 9, the beginning of the Holocene Epoch marks the onset of what climatologists generally consider to be "modern climate" (Figure 20.1). Warming in the early Holocene was rapid. By 9000 years ago the average Earth temperature was about 4°C (~7°F) warmer than it had been just a thousand years before. Although distinct warmer/drier or cooler/wetter periods have

---

Carrying capacity *The maximum number of organisms that can live in a given area of habitat without degrading it and causing social stresses that result in population decline.*

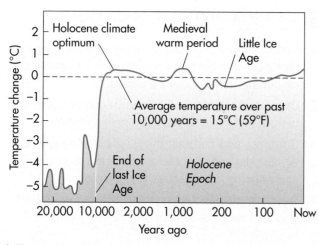

Figure 20.1 Global temperature changes the past 20,000 years. Note the rapid warming that occurred at the end of the Pleistocene and beginning of Holocene. Average temperatures for this time span are compared to the present.

occurred within the Holocene, as shown in Chapter 9, the overall climate has generally been much more hospitable to humans the past 10,000 years than it was at the end of the Pleistocene when glacial ice covered much of the northern latitudes. Partially in response to this more favorable climate, human societies gradually became more complex. People moved into newer areas and technological innovation increased. A key result of this cultural evolution was the beginning of agriculture about 8000 years ago in several widely separated places on Earth. Prior to that time, people were essentially hunters and gatherers who spent much of their lives foraging for food.

The onset of agriculture was critical in human history because people could now grow their own food rather than search for it. In so doing, humans gradually raised the carrying capacity of many landscapes in such a way that larger numbers of people could congregate in centralized locations that ultimately became villages, towns, and even cities. With food more easily accessible, people were freer to devote time to improving agricultural techniques and inventing other things that made life still easier. Ultimately, these efforts led to the Industrial Revolution in the late 18th and early 19th centuries, during which numerous technological advances led to mechanized production and increased supply of goods. The social and economic changes brought on by the Industrial Revolution ushered in the modern world as we know it.

Historical growth in human population during the past 10,000 years closely mirrors the evolution of human technology and society. At the end of the most recent ice age, probably fewer than 500,000 people lived on Earth (Figure 20.2) and average life expectancy may have been as short as 20 years. With the advent of agriculture and enhanced food production, human population slowly increased to approximately 500 million about 2000 years ago. The population essentially held steady for another 1000 years and then gradually increased to about 600 million around the year 1700, with an average life expectancy of perhaps 40 years at that time.

The impact of early industrialization on human population was dramatic. By 1800 the global population had reached about 1 billion. Much of this increase was driven by further technological developments, many of which were associated with sanitation, as well as improvements in medicine that dramatically enhanced human health. In response to these advances, average life expectancy in the United States approached 50 by 1900. By 2000 the human population on Earth reached over 6 billion and average life expectancy for those born in the United States was over 77 years. In August 2009 the global population was about 6.8 billion. According to the United Nations, world population is projected to reach 7 billion by 2013, 8 billion by 2028, and 9 billion by 2050. In other words, by the time a current college senior is about 60 years old, 2 billion more people will probably inhabit Earth than do now.

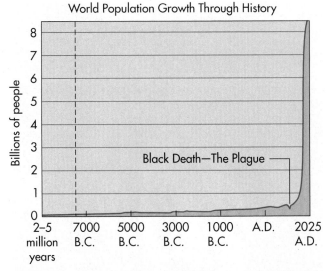

World Population Growth Through History

Figure 20.2 Global population change throughout history. Note the dramatic increase the past 200 years. It is estimated that human population will be 9 billion by 2050. (*Source:* Population Reference Bureau, *World Population: Toward the Next Century*, 1994.)

The Impact of Growing Human Population on the Natural Environment

The impact of this rapidly growing human population on the Earth's environment is important in many ways. At the most fundamental level a growing population means that more food is required. The need for more food, in turn, means that more forests are cleared to increase the amount of farmland. As discussed in Chapter 11, soils will also be more heavily utilized and thus degraded as farmers attempt to sustain or improve crop yields. A growing population also means that more environmental pollution occurs, often in ways that the average person cannot imagine.

An excellent example of a growing but little known environmental impact is the huge expanse of trash in the northern Pacific Ocean where the North Pacific gyre circulates (Figure 20.3a). Many of the coastal countries along the Pacific Rim, such as Japan, the Philippines, and the United States, are densely populated and produce prodigious amounts of garbage. Some of this trash, unfortunately, winds up in the ocean and circulates around the basin in the gyre before it washes up on shore (Figure 20.3b). The amount of trash in the northern Pacific is now so high that the area has been referred to as the *North Pacific garbage patch*. Most of this trash is plastic, some of which is being consumed by a variety of animals with negative results (Figure 20.3c). Countless examples of this kind of impact are occurring around Earth at the present time and will continue into the future.

The environmental impacts of a growing population are enormous. Some parts of the world have industrialized and reached relatively stable population levels, but others are in the process of developing economically and are rapidly growing. An example of an industrialized (or "developed") country is the United States, which has long been a global leader in technological innovation. As a result, the quality of life for the vast majority of Americans is generally high, certainly at least with respect to most other countries. Developing countries such as India and China, in contrast, are in the process of industrializing, which means that large numbers of people are reaching the middle class for the first time and thus now have the income to purchase consumer goods. Still other countries, such as a number in Africa, have just begun to develop economically and are populated mostly by people living on less than $1 per day. Many people in these countries live at a level of poverty that the average American simply cannot imagine.

The distinction between the developed and developing countries is important because their population growth rates are very different. Population in the industrialized world reached about 1 billion in 2000 and has essentially stabilized since that time (Figure 20.4). Future projections suggest that it may even slowly decrease for the next century due to lower birth rates. In stark contrast, population in the developing world began to grow rapidly in the late 20th century when technological and health advances reached those areas. Leading the way have been China and India, which now each contain more than 1 billion people. According to current projections, population in the developing world will continue to increase rapidly for the foreseeable future, with perhaps 7 billion people in these areas by 2050.

The rapid population growth rates in the developing world are very important as far as sustainable development and environmental change are concerned. For one thing, increased population pressure in regions with a moderate climate means that more people will move out of those places into marginal areas with limited carrying capacity, such as the African Sahel or the Middle East. As far as global warming is concerned, the Intergovernmental Panel on Climate Change (IPCC) believes that the combination of increasing population and rapid economic growth in the developing world makes further temperature increases more likely because this growth will largely depend on fossil fuels that produce greenhouse gases. For example, to meet the rapidly increasing demand for energy in the growing Chinese economy, the government there plans to build about 500 new coal power plants in the next decade. In a similar vein, an Indian automaker is now producing an automobile for the country's growing middle class that sells for less than $3,000. Although this car is highly efficient (>40 mi/gal or mpg), its development and potential

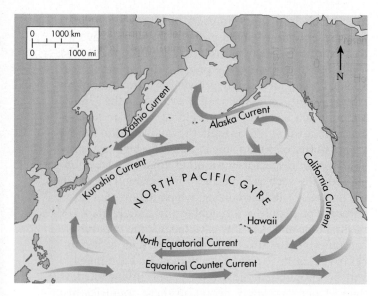

Figure 20.3(a) The North Pacific gyre. Note the counterclockwise circulation in the northern Pacific Ocean. This current moves trash in a circular motion around the Pacific Ocean.

Figure 20.3(c) Dead albatross on Midway Atoll, Hawaii. This bird died because it consumed plastic from the Pacific garbage patch. Numerous albatross chicks have died in this manner because plastic was brought back to them by their parents to eat.

Figure 20.3(b) Coastal trash on the Pacific Rim. This garbage circulated in the North Pacific gyre for an unknown period of time before it washed ashore on this beach.

high sales make increased greenhouse gas production a certainty because tens of thousands of Indians will now own a car for the very first time.

These issues of economic development, growing human population, and environmental impacts are happening all around the world. In dealing with current and future environmental concerns, an understanding of physical geography is important to identify, manage, and, where possible, solve issues associated with sustainable development. The remainder of this chapter presents three case studies that examine the impact that humans have on the environment in a particular place, and vice versa. As you read about them, be sure to notice how these issues are distinctly related to physical geography and look for the topics that were covered in earlier chapters.

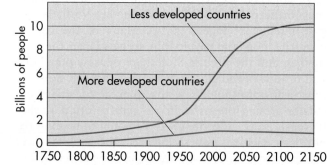

Figure 20.4 World population growth in the industrialized and nonindustrialized world from 1800 to 2100. Note the rapid growth in the developing world that is currently taking place, whereas population in the developed world has been stable for some time. (*Source:* The United Nations.)

URBAN SPRAWL

Urban sprawl is the expansion of cities outward from the original central core into formerly rural land. It is usually associated with increased suburbanization and expanding traffic networks, such as the example shown here at Phoenix, Arizona. Geographers like to visualize the spatial patterns of sprawl by mapping changes at specific intervals over time. For example, look at this map to see the patterns of urban sprawl in the Detroit metropolitan area in southeast Michigan since 1890. It is color-coded to illustrate the progression of urban expansion over time. The impacts of suburbanization on the physical environment are extensive, including deforestation, the loss of farmland, increased storm runoff because more surfaces are covered with concrete, a greater possibility for the urban heat island effect, and enhanced atmospheric CO_2 concentrations because more people are commuting farther to and from work.

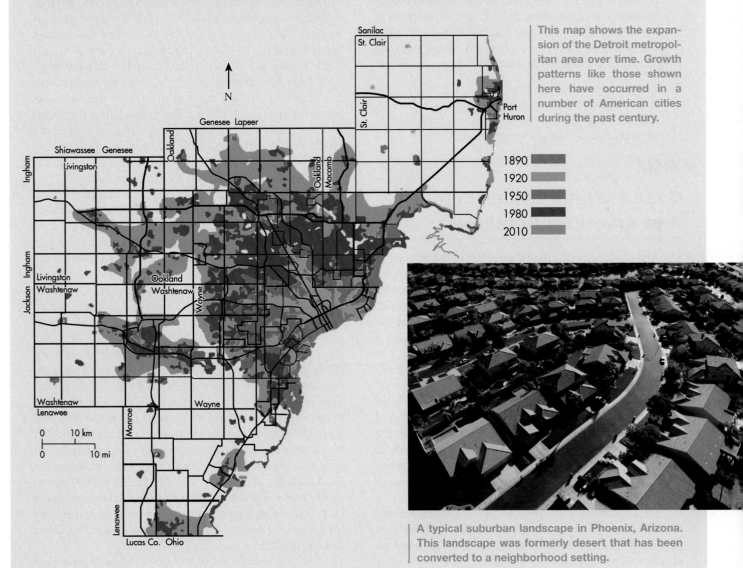

This map shows the expansion of the Detroit metropolitan area over time. Growth patterns like those shown here have occurred in a number of American cities during the past century.

1890
1920
1950
1980
2010

A typical suburban landscape in Phoenix, Arizona. This landscape was formerly desert that has been converted to a neighborhood setting.

1. Carrying capacity refers to the number of individuals who can be supported within a landscape in a way that does not degrade the natural, social, cultural, and economic environment beyond sustainable limits.

2. The onset of the Holocene Epoch is an important event in the development of human societies because the climate warmed, making life easier.

3. The development of agriculture about 8000 years ago set the stage for later population growth because it raised the carrying capacity of landscapes and allowed more people to live in smaller geographical areas.

4. The Industrial Revolution in the 19th century allowed human population to grow rapidly because mechanization, increased sanitation, and developments in medicine led to longer and better lives. In 1800 the global population was about 1 billion, whereas in 1900 it was approximately 1.7 billion.

5. Global population is now about 6.8 billion and is projected to reach approximately 9 billion by 2050. This large number of people is having huge impacts on the natural environment.

Case 1: Water Issues in the Arid American Southwest

The first case study examines water use in the American Southwest. It focuses on the Las Vegas metropolitan area, which has been one of the fastest-growing urban zones in America since the mid-20th century. According to the U.S. Census Bureau, the population of the Las Vegas metropolitan area increased from about 1.4 million in 2000 to about 2 million today. Ironically, Las Vegas is located within the low-latitude desert climate (*BWh*) in the Köppen system, which is extremely arid. In other words, hundreds of thousands of people have recently moved into a region where it rarely rains and natural water supplies are scarce. Although the current economic downturn has slowed population growth in the area, the effect at this time has been to essentially stabilize the number of people rather than to cause a steep decline.

The competing pressures of high population and the potential lack of freshwater pose challenging questions for the current and future residents of Las Vegas, as well as other large cities in the Southwest such as Phoenix and Tucson, Arizona, and Albuquerque, New Mexico.

Are these cities sustainable places that can continue to grow, or will they one day decline if and when water supplies give out? Regardless of the ultimate outcome, to understand the issues and appreciate the various solutions proposed require a deep understanding of physical geography.

As you probably know, Las Vegas is a world-famous resort city that draws millions of tourists every year. Located in the southern part of Nevada in Clark County (Figure 20.5), the metropolitan area lies within the Las Vegas Valley, which is part of the Basin and Range physiographic province discussed in Chapter 13. The Las Vegas Valley is bordered to the immediate west by the Spring Mountains and to the east by Frenchman Mountain. Las Vegas is located about 48 km (30 mi) northwest of Hoover Dam, which impounds Lake Mead.

The Las Vegas metropolitan area is located in the central part of the Mojave Desert and is thus one of the hottest cities in the United States. The average daily high temperature is >38°C (100°F) approximately 90 days out of the year (Figure 20.6). The record high temperature is 47°C (117°F), which occurred in July 2005. Winter months are typically mild, with an average daily high temperature of 13°C (57°F) in January. The Las Vegas climate is influenced strongly by the rain shadow effect associated with the Sierra Nevada in California to the west. Average annual precipitation is only about 10 cm (4 in.) and relative humidity is usually very low. As a result, Las Vegas is one of the sunniest cities in the country, with over 300 days of clear skies per year.

The development of Las Vegas is directly tied to its distinctive physical geography. The Las Vegas Valley was "discovered" in 1829 when a scouting party associated with a Mexican trading caravan traveling to Los Angeles left the main trail in search of water. While doing so, the scouts entered an unnamed valley and found an oasis fed by numerous artesian springs. As a result of this discovery, the site was named *Las Vegas*, which is Spanish for "The Meadows," and became an important place to replenish water supplies, first for Mexican traders and later for American explorers. The area was subsequently annexed by the United States following the Mexican-American war in 1848 and was soon occupied by Mormon missionaries, who built a fort near present-day downtown Las Vegas in the mid-1850s. The fort was soon abandoned and later reoccupied in 1865 by Octavius Gass. After constructing an irrigation system drawing water from the springs, Gass developed a farm complex that was about 243 hectares (600 acres) in size and renamed the area *Las Vegas Rancho*. Gass's farm became an important rest stop for people traveling on the nearby Mormon trail from Salt Lake City, Utah, to Los Angeles.

The population of Clark County began to grow very slowly after Nevada became a state in 1864. A major

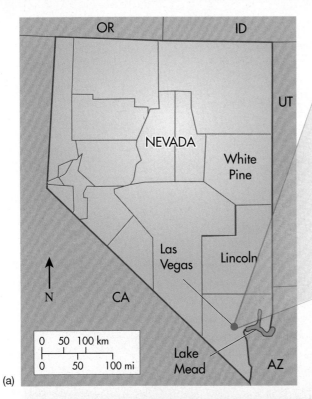

(a)

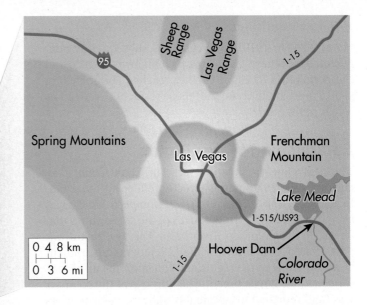

Figure 20.5 Las Vegas, Nevada. (a) Las Vegas is located within Clark County in the southern part of Nevada. (b) Skyline of Las Vegas. Las Vegas is a world-famous resort city that lies within the arid Mojave Desert.

(b)

Figure 20.6 Climograph for Las Vegas, Nevada. Average monthly temperatures range from mild in the winter to blistering hot in the summer. Total annual precipitation is about 10 cm (~4 in.).

factor in the early growth of the area was the State Land Act of 1885, which allowed homesteaders to purchase land for $1.25/acre ($309/km²). The lure of cheap land drew many people to the area, most of whom were farmers. As a result, agriculture dominated the local economy until the early 1900s and was supported by an increasingly elaborate irrigation system that drew water from the wells. Some of this water was directed into the town proper and Las Vegas became firmly established as a stop for travelers moving to and from Los Angeles and places to the east such as Albuquerque, New Mexico.

Despite the irrigation improvements in the Las Vegas Valley in the late 19th century, only about 1100 people lived in Clark County in 1900. The first surge in population occurred after the San Pedro, Los Angeles & Salt Lake Railroad linked the area to other places in the West in 1905. The city of Las Vegas was founded that same year and the population of Clark County grew to about 3300 in 1910. Las Vegas was subsequently connected to southern California by a highway in 1926 and the population of Clark County reached about 8500 in 1930. Approximately 60% of those residents lived in Las Vegas at that time.

Establishing Water Rights

The next major population surge in Clark County occurred in the 1930s and is directly related to the establishment of the *Colorado River Compact*, which was signed in 1922 by representatives of the seven states that lie within the Colorado River basin. This agreement was part of an overall effort by the U.S. federal government to draw people to the region and "reclaim" the arid lands for economic development. The purpose of the compact was (and remains) to sustainably allocate water to users in the basin in a fair and equitable manner by establishing a system of **water rights** defined by law. It does so by dividing the watershed into an upper basin (including Wyoming, Colorado, New Mexico, and Utah) and a lower basin (including Nevada, Arizona, and California; see Figure 20.7). The key stipulation of the compact is that both the upper basin and lower basin are allowed 7.5 million acre-feet of water per year, in other words, the amount of water required to cover 7.5 million acres in 1 foot of water. Given that the Colorado River flows for a short distance in Mexico before it reaches the Gulf of California, 1.5 million acre-feet of water is allocated to Mexico per year. This stipulation means that the upper basin states cannot use all the water they desire; instead, they must allow enough water to flow downstream so both the lower basin and Mexico receive their share. The lower basin, in turn, must assure that Mexico receives the amount of water required by the agreement.

The Colorado River Compact set the stage for the construction of the Boulder Dam (later renamed the Hoover Dam), which began near Las Vegas in 1931. In association with the Colorado River Compact, Hoover Dam was designed for two purposes: (1) impounding water (in Lake Mead) for irrigation and water conservation and (2) the production of hydroelectricity for southwestern states. During the course of construction, the population of Clark County swelled to about 25,000. It was at this time that the first of the now-famous casinos were established in Las Vegas to entertain the construction workers. After the dam was completed in 1935, the dam and Lake Mead became tourist attractions that further propelled the growth of Las Vegas, along with an influx of organized crime. By 1950 the population of Clark County was nearly 50,000.

Water rights *Legally protected rights to take possession of water occurring in a natural waterway and to divert that water for beneficial purposes.*

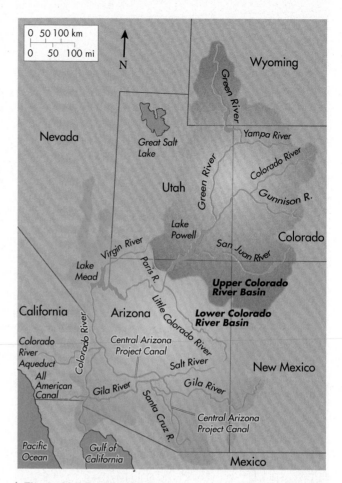

Figure 20.7 Regulation of water in the Colorado River basin. The Colorado River Compact divided the watershed into an upper and lower basin. Water in each of these basins is strictly regulated, with a set amount allowed for each region. The Central Arizona Project and Colorado River Aqueduct divert water from the Colorado River to southern Arizona and southern California, respectively, for human purposes.

Since the advent of air conditioning in the 1950s, the population in the Las Vegas metropolitan area has grown at a rapid rate to about 2 million (Figure 20.8a) and was expected to reach 3 million by 2020 prior to the current economic downturn. It may still reach that figure, but such projections are a bit uncertain as of 2010. Regardless, the city is visited by about *40 million* tourists every year who are attracted to the casinos, entertainment, and area's natural beauty. This rapid population increase mirrors the rate of change in other hot southwestern cities such as Phoenix, Tucson, and Albuquerque, and is part of a general migration of Americans during the past 50 years to the southern part of the country or *Sun Belt*. In 2007, for example, over 60,000 new housing permits were approved in Phoenix alone, resulting in extensive urban sprawl. A comparison of satellite images between 1973 and 2006 (Figure 20.8b) illustrates

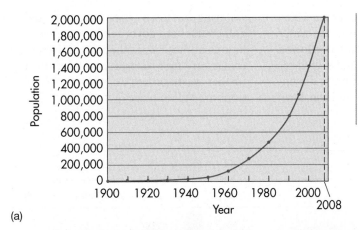

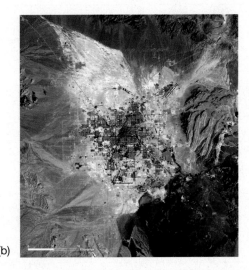

(a)

(b)

Figure 20.8 Growth in Las Vegas population. (a) Graph showing the growth of Las Vegas since 1900. Note the rapid increase after 1950. (b) Satellite images showing the expansion of Las Vegas between 1973 (left) and 2006 (right). The 1973 image shows an area about 25 km (~16 mi) across. (*Source:* NASA.)

the impact that rapid population growth in Las Vegas has had on the size of the city. Las Vegas is now a major American city with wide boulevards lined with luxury hotels, waterfalls, and fountains that spray millions of gallons of water into the dry desert air.

Will Water Supplies Disappear?

As the growth of Las Vegas accelerated in the latter part of the 20th century, many people began to become concerned about the supply of water in the arid desert. Was there enough water available for sustainable growth? Was it possible that water supplies would disappear? These concerns resulted in the establishment of the *Southern Nevada Water Authority (SNWA)* in 1991. The purpose of this organization has been to develop and manage local water resources, theoretically in a sustainable way. At the present time 90% of water used in Las Vegas comes from Lake Mead on the Colorado, whereas the remaining 10% is derived from groundwater. In an effort to manage these water supplies, the SNWA established a four-phase hierarchy of drought status, including (1) no drought, (2) drought watch, (3) drought alert, and (4) drought critical, to keep people informed of current environmental conditions and the various kinds of water use that are permitted at each level.

The concerns about water have magnified since the late 1990s, when the southwestern United States entered into a major drought that has only recently eased somewhat. Average annual precipitation in the region has been about 70% to 75% of normal, with a significant decrease in the amount of snowfall each winter in the Rocky Mountains. The amount of snow is important because it provides spring runoff to the Colorado River and its tributaries, which is the major source of water for Las Vegas and many other southwestern cities. As you can see in Figure 20.9a, the average spring runoff in the Colorado was generally almost half of the normal amount between 2000 and 2007. This decrease in runoff has caused water levels in Lake Mead to drop to the point where it now contains only about 50% of its total capacity (Figure 20.9b). The extent of this decrease is visible on the bleached canyon walls on the edge of the lake (Figure 20.9c).

The impact of drought on Las Vegas and other southwestern cities is potentially significant, especially given the recent influx of people into the region. As a result of the drought, the Southern Nevada Water

Authority moved the drought status from *drought watch* to *drought alert* and imposed a series of tougher water restrictions. For example, no new grass turf is allowed in residential yards and is limited in rear and side yards to 50% of the area, with a 3 m (10 ft) minimum dimension and a 464 m² (5000 ft²) maximum. In addition, lawn watering is allowed only 3 assigned days per week and golf courses are governed by strict water budgets. If violations of these and other restrictions are observed, customers are first warned, and then subsequently fined if further waste occurs. As of October 2010, these restrictions are still in place.

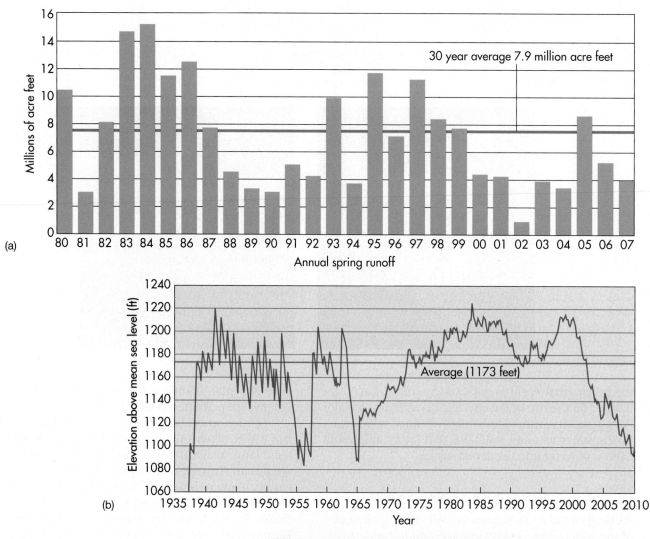

(a)

(b)

Figure 20.9 Colorado River runoff and Lake Mead. (a) Average spring runoff in the Colorado River for this decade has generally been about half the normal amount. (*Source:* U.S. Bureau of Reclamation.) (b) Due to reduced spring runoff, water levels in Lake Mead have dropped significantly this decade (*Source:* U.S. Bureau of Reclamation.) (c) Water level in Lake Mead, spring 2009. Since 2000, the lake level has dropped about 30 m (100 ft), exposing a ring of bleached rock around the edge of the lake. (c)

A Drought-Prone Future?

In the context of the recent drought, one cannot help but wonder what future climate conditions may be like in the Southwest. To answer this question it is useful to look at the past. Recent tree-ring studies, like those discussed in Chapter 9, indicate that eight droughts with an intensity similar to the recent one have occurred in the region within the past 500 years, with perhaps two or three being even more severe. These studies show that arid conditions are the norm and that extreme droughts are fairly common in the American Southwest. These studies also reveal that the Colorado River Compact was signed during a relatively wet phase in the regional climate, one resulting in more runoff into the Colorado than normal. In other words, the current distribution of water in the upper basin, lower basin, and Mexico is based on ideal conditions that may, in fact, be quite rare.

Unfortunately, it appears that ideal climate conditions in the Southwest may become even more rare in the future. This assessment is based on climate models that predict future environmental conditions. These models demonstrate that the southwestern climate indeed became more arid in the late 20th century and is continuing the trend into this century. The prevailing belief is that ongoing global climate change is altering the movement of storms and moisture in the atmosphere and that the drying trend will continue at least until 2100. In short, climate researchers are concerned that the drought conditions of the past decade will become the new norm for the region.

If this assessment of future climate conditions is true, it has important implications for water resources in southwestern cities such as Las Vegas and Phoenix, especially if recent growth rates pick up again. An increasing number of scientists argue that planners should acknowledge that water resources are limited and that shortfalls will occur more frequently in the future. Some even suggest that recent growth rates are unsustainable and that limits should actually be placed on the number of people who can live in the region. In response to these concerns, a 2007 planning study argued that limiting growth in Clark County was an unrealistic method of drought control because the economy of Nevada would suffer. Instead, the focus should be to increase the efficiency of water usage and to continue an aggressive search for other water sources to supplement the current supply from the Colorado River and local groundwater. Efficiency has indeed improved dramatically in the recent past, as Las Vegas residents now stretch their water supplies further through the restrictions described earlier. Unfortunately, these gains are offset by the increased overall demand due to growing population. As a result, a systematic effort is currently ongoing to acquire water from groundwater supplies in Lincoln and White Pine Counties, which lie north of Las Vegas (Figure 20.5a). The extent of these supplies is currently being assessed, with the goal of building a pipeline that will deliver water to Las Vegas in the near future.

Although these counties are lightly populated (~14,000 total people), many residents are fighting to keep what they consider to be *their* water. They believe that groundwater extraction from the counties will degrade the environment and destroy their traditional ranching lifestyle. They recall the negative impacts that water diversion from Owens Valley, California, to Los Angeles in the early 20th century had on the environment. Much of this water was diverted from

1. The Las Vegas Valley lies within the rain shadow of the Sierra Nevada Mountains and thus receives about 10 cm (4 in.) of precipitation per year. Average annual temperatures range from warm in the winter to exceptionally hot in summer.

2. A major factor in the growth of Las Vegas was the Colorado River Compact, which was signed in 1922. This document legally subdivided the Colorado River basin into separate basins with distinct water rights.

3. Rapid growth of Las Vegas began in the 1950s, in part due to the advent of air conditioning. Population has grown from about 50,000 in 1950 to over 2 million today.

4. Increased population has raised serious concerns about the sustainability of water supplies in the region. As a result, water managers in Las Vegas can employ strict use restrictions and are aggressively searching for supplemental sources of water from neighboring areas.

5. Regional climate models suggest that the southwestern United States will become more drought-prone in the coming decades, potentially making water concerns more acute.

Mono Lake, which is about 385 km (240 mi) northeast of Los Angeles, to fuel the rapid growth of southern California. As a result of this diversion, the level of Mono Lake fell about 14 m (45 ft), approximately 7,500 hectares (18,500 acres) of shoreline were consequently exposed to blowing wind, and the salinity of the lake doubled by the early 1980s. The concerns of residents in Lincoln and White Pine Counties are shared by officials in neighboring Utah because groundwater extraction in Nevada may impact groundwater resources there. They are particularly concerned with the viability of Fish Springs, which is a lake that is home to a fish called the least chub. Environmentalists are concerned that the least chub will become endangered if the lake level declines as a result of groundwater extraction in Nevada.

As you can see, a full understanding of population growth and water use in marginal environments such as the southwestern United States requires a holistic assessment of many factors directly related to physical geography. This particular case study has included several topics discussed in earlier chapters, such as Köppen climates, the Basin and Range province, the rain shadow

effect, groundwater, drainage basins, runoff, and global climate change. All these factors are intimately related to create a distinctive environment in the southwestern United States, one that happens to be the center of a recent population boom.

The societal tensions resulting from issues like this one are loosely referred to as *water wars* and are particularly intense in places where different groups of people compete for the same water supply. According to current estimates, about 1 billion people on Earth lack access to reliable freshwater supplies. This number will undoubtedly increase as global population continues to grow. In many other places, access to water is being used as a political weapon to influence the behavior of neighboring countries. An excellent example is Turkey, where dams have been built across the Tigris and Euphrates Rivers that reduce stream flow into neighboring Iraq. Examples such as these abound around the world and will be political flashpoints for the rest of your life. To fully understand them, and their potential solutions, requires an excellent understanding of physical geography.

Case 2: Soil Salinization in Arid and Semi-Arid Lands

As we discussed earlier in this chapter, the development of agriculture increased the carrying capacity of landscapes and gradually led to increased population. During the previous hunter/gatherer phase of human evolution, the global population was probably less than 5 million people. Today, modern agricultural techniques support a global population of about 6.8 billion, with 9 billion (perhaps more) projected by 2050.

The ability of agriculture to feed everyone on a sustainable basis poses enormous challenges. The production of cereal grains such as wheat and sorghum, for example, will have to *double* between now and 2050, not only to meet the basic needs of so many people, but also because more people will consume grain-fed beef as their standards of living improve. To meet this growing demand, further increases in agricultural output are essential for global political and social stability and equity. An even greater challenge is to meet this demand in a sustainable way that maintains the health of environmental systems.

One of the primary factors complicating the task of increasing global food production is simply the limited supply of **arable land** on Earth (Figure 20.10). Recall from the discussion of the hydrological cycle in Chapter 7

Arable land *Land that has the high potential to be cultivated for crop production.*

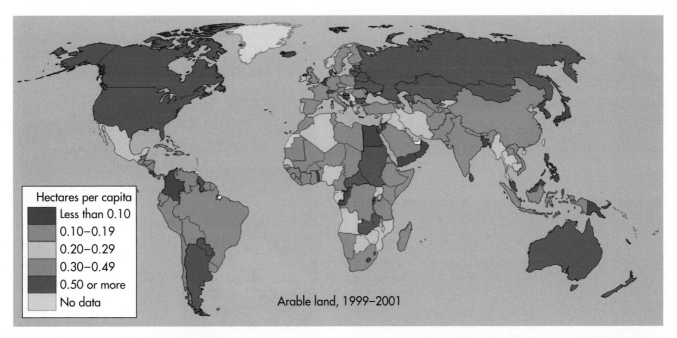

Figure 20.10 Map of global arable lands. This chloropleth map shows the countries that contain arable land; it is somewhat misleading. Much of the western United States, for example, as well as most of Australia, contains very few arable lands. Regardless, note that this map indicates extensive parts of the world contain few lands suitable for agriculture.

that about 71% of the Earth's surface is covered with water, mostly contained in oceans. Of the remaining 29% of the Earth's surface, significant parts of the land area are simply unsuitable for large-scale agriculture on a sustainable basis. Such places include the world's arctic and subarctic climate zones, the major desert regions, and large parts of the tropical rainforests. In fact, only 22% of the Earth's land area is potentially productive for agriculture. Of these areas, only a few such as the American Midwest, southeastern Asia, parts of South America, and Eastern Europe are ideal for intensive agriculture because they have fertile soils and abundant precipitation.

With the increasing demand for food, more and more agriculture is being practiced in marginal lands where the potential for major environmental degradation is high. In Chapter 18, two such examples of human impact were the overplowing of soils that created the American Dust Bowl of the 1930s and changing land-use policies that led to massive desertification in the Sahel region of Africa. This second case study focuses on another environmental impact of expanding agriculture in arid and semi-arid regions. This impact is **soil salinization**, which is the accumulation of soluble salts such as sulfates of calcium and potassium in soil. Salt buildup in soils is toxic to most plants because salts decrease the osmotic potential of the soil, that is, the ability of soil water to enter permeable

---

Soil salinization *The process in which soils become enriched in soluble salts such as sulfates from calcium and potassium.*

plant roots. As soils become increasingly salty, they have greater concentrations of solute than the plant roots do, so plants cannot absorb water from the soil.

Salinization of soils can occur under natural conditions and is most commonly associated with topographic depressions (Figure 20.11). In these micro-environments dissolved salts are washed by overland flow into the low area, or alternatively, the base of the depression lies just above the water table. The source of these salts can either be the weathering of minerals on the slopes above the depression, ions precipitated in the rainwater, or solutes leached from saline rocks such as gypsum. If the salts are washed into the depression from the nearby slopes, they will recrystallize at the surface when the water evaporates. If the water table is shallow, salts can be brought up from deeper horizons by the process of capillary action (or "wicking") described in Chapter 15. When excess salts accumulate, soil structure breaks down because clays no longer bind together. Recall from Chapter 11 that soils with a moderate amount of clay typically have good structure, which allows water to flow through the soil rapidly. When this binding ceases, the clay particles act individually to plug soil pores. As a result, the soil becomes increasingly impermeable to water.

Soil salinization can also occur when agricultural practices change the hydrological and chemical aspects of soils. This change can happen in two ways. One way is related to rising water tables associated with the clearing of native vegetation (Figure 20.12). In arid and semi-arid areas where natural vegetation occurs, rainwater typically leaches salt from the permeable soil deep into the ground

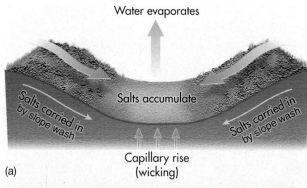

Water evaporates

Salts carried in by slope wash

Salts accumulate

Salts carried in by slope wash

Capillary rise (wicking)

(a)

(b)

Figure 20.11 Natural salinization of soils. (a) Salt crusts naturally form in topographic depressions within arid regions because slopewash evaporates and leaves a salt crust behind. (b) Salt crust at Bad Water in Death Valley, California. This playa contained water during the Wisconsin ice age that evaporated during the Holocene, leaving mineral salts behind. Today, occasional rains promote runoff into the playa. Some salts dissolve and then recrystallize when the water evaporates. The hexagonal pattern results from repeated freeze/thaw and evaporation cycles that push the crust into these forms.

Before land clearing, transpiration from trees helps to keep the water table down

Water table

Saline groundwater

(a)

After land clearing, discharge through transpiration decreases, causing the water table to rise

Dead trees

Barren land

Saline water flows to the surface

Water table

Saline groundwater

(b)

Figure 20.12 Salt crust formation after native vegetation is cleared. (a) Native vegetation keeps the water table down due to transpiration. (b) Once the native plants are removed, more groundwater recharge occurs because agricultural crops absorb less water. The water table then rises, causing salinization in topographic depressions.

and into the groundwater. Even though the salinity of the groundwater is high in these circumstances, native plants do fine as long as the water table is below about 4 m (13 ft) below the surface. In places where the water table is shallower, some native plants such as salt grass can cope with higher levels of groundwater salinity. When native vegetation is cleared in these areas, the water table typically rises because less precipitation is intercepted by plants and less evapotranspiration occurs. If the water table rises to within about 2 m (~6.5 ft) of the ground, saline water can begin to seep into topographic depressions and salt crusts can form through evaporation. In addition, saline water can begin to rise to the surface by capillary action. In this way, soils become laden with salts.

Another way that agriculture causes soil salinization is through irrigation (Figure 20.13). In irrigated fields the water applied to the soil can raise the water table. Evidence from around the world indicates that irrigation can cause the local water table to rise rapidly, with increases up to 3 m (~10 ft) reported in some places. As with the clearing of native vegetation, such a rise can lead to wicking of salt in the soil and formation of salt crusts as the water evaporates. Especially in warm and dry areas, if the applied water is already concentrated with salts, the irrigated water will rapidly evaporate from the surface, leaving higher levels of salt in the soil. Human-induced soil salinization has been a problem since the dawn of civilization. One of the best examples of the impact that increased soil salinity has on people is ancient Sumeria in Mesopotamia.

Ancient Sumeria

The Sumerians lived between about 4000 and 2300 B.C. in the fertile alluvial plain of the lower Tigris and Euphrates river valleys in what is now southern Iraq. This region lies within the southeastern part of the famed *Fertile Crescent* (Figure 20.14a), which is one of the key places where early human civilization developed. Although this region lies within the hot low-latitude desert climate region (*BWh*), the Sumerians thrived for centuries because they developed a complex irrigation network of canals and aqueducts to divert water from the rivers to outlying farm fields. It is

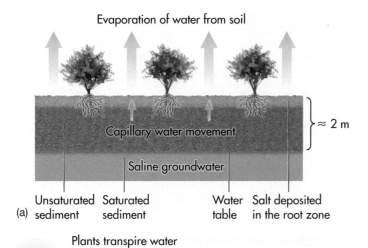

Evaporation of water from soil

Capillary water movement

≈ 2 m

Saline groundwater

(a) Unsaturated sediment | Saturated sediment | Water table | Salt deposited in the root zone

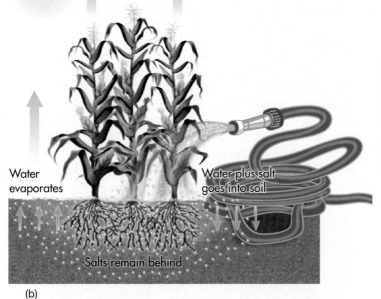

Plants transpire water

Water evaporates

Water plus salt goes into soil

Salts remain behind

(b)

Figure 20.13 Soil salinization due to irrigation. (a) Soil salinization can occur in irrigated fields because application of water causes the water table to rise close to the root zone. Once this occurs, water and salts move to the surface by wicking. The water dissolves, leaving salt behind. (b) Soil salinization in farm fields can also occur if saline water is applied to fields. Once the water evaporates, salt crusts form at the surface.

believed that wheat was first grown as a domestic crop during this time. As carrying capacity increased, the first city-states (Figure 20.14b) evolved in this part of the world and the first form of systematic writing began (Figure 20.14c).

At the same time that Sumerian society became increasingly sophisticated, its foundation was apparently slowly collapsing as soil salinity increased due to rising water tables and wicking. Indirect evidence for increased salinity can be seen in the counts of grain impressions in excavated pottery, which indicate that by 2500 B.C. wheat production had dropped by 83% and been replaced by

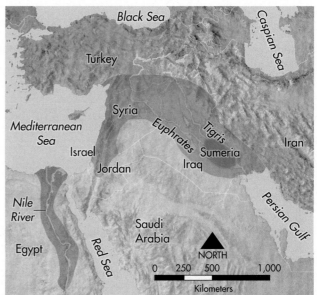

(a)

(b)

(c)

Figure 20.14 The ancient Sumerians. (a) The Sumerians lived in the lower Tigris and Euphrates valleys in what is now southern Iraq. This area was part of the Fertile Crescent. (b) Artist's rendering of a Sumerian temple. The development of agriculture raised the carrying capacity of the landscape and allowed people to congregate in smaller places and diversify tasks. (c) The Sumerians were the first people to begin writing, such as seen on this tablet.

TABLE 20.1   Extent of Salinized Soils Throughout the World

Region	Area in Millions of Hectares	Area in Millions of Acres
Australia	84.7	209.3
Africa	69.5	171.7
Latin America	59.4	146.8
Near and Middle East	53.1	131.2
Europe	20.7	51.2
Asia and Far East	19.5	48.2
North America	16.0	39.5

Source: FAO/UNESCO.

barley. This switch was apparently made because wheat is one of the least salt-tolerant crops, whereas barley does somewhat better. Between 2400 and 1700 B.C., however, barley yields dropped substantially as salinity increased further and massive crop failures began soon thereafter. Archaeologists believe that these stresses led the various city-states in the region to compete for the remaining arable land and to a near-constant state of conflict that ultimately caused Sumerian society to collapse. The legacy of salinity is still keenly felt in modern Iraq today, where at least 20% of the potentially arable land is permanently destroyed.

The Current Global Extent of Salinization

Soil salinization is currently a problem in many parts of the world (Table 20.1). Today, it affects between 20 and 30 million hectares (50 and 75 million acres) of the world's current 260 million hectares (642 million acres) of irrigated land and has the potential to severely limit world food production. Every inhabited continent has areas that are being affected by salinization, with Australia leading the way at over 84 million hectares (209 million acres) affected. In the United States, salinization may be lowering crop yields on as much as 25% to 30% of the nation's irrigated lands. In Mexico, salinization is estimated to be reducing grain yields by about 1 million tons/year, which is enough to feed approximately 1 million people. The remainder of this case study examines two examples of salinized soils—the San Joaquin Valley of California and Australia.

The San Joaquin Valley in California

Perhaps the best example of the impact of soil salinity in the United States today is the San Joaquin Valley of California, which encompasses the southern half of California's central valley (Figure 20.15). In contrast to much of the rest of California, the valley terrain has

very limited relief due to the complex geological history of the area. The geological foundation of the valley was created when subduction along the North American coastline during the Mesozoic Era established a structural basin that lay below sea level. This basin was later filled with about 7600 m (25,000 ft) of marine sediments. These deposits were then buried by as much as 730 m (2400 ft) of alluvial sands and gravels derived from the rising coast ranges to the west and the Sierra Nevada to the east. Taken together, this depositional sequence created an extremely flat landscape that generally lies less than 122 m (400 ft) above sea level. Given the higher

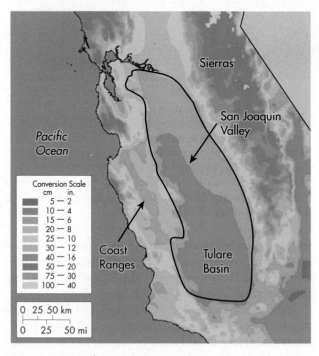

Figure 20.15 The San Joaquin Valley. Map of annual precipitation within the San Joaquin Valley and surrounding area. The dry area on the west side of the valley lies within the rain shadow of the coast ranges.

elevations of the coast ranges to the west, the San Joaquin Valley lies within a rain shadow and thus receives less than 38 cm (15 in.) of rainfall per year. The climate of the region is generally classified as hot low-latitude steppe (*BShs*) and even low-latitude desert (*BWh*) in places, with hot and dry summers and cool, damp winters. The northern part of the region is drained by the San Joaquin River. Streams in the southern end of the valley flow into Tulare Lake in the Tulare Basin.

In spite of the semi-arid climate of the region, the San Joaquin Valley is one of the most important agriculture zones in the United States, with 5 of the top 10 farming counties in the country. The valley has often been referred to as the "nation's salad bowl" for the amazing array of fruits and vegetables grown here. If you happen to like asparagus, grapes, raisins, almonds, and pistachios, they are likely grown in the San Joaquin Valley. In addition, the largest single cotton farm in the country is found here. The reason agriculture has been so successful in the valley is because a complex irrigation network was developed that delivers water to fields whenever it is needed (Figure 20.16).

Large-scale diversion of water began when the Bureau of Reclamation began the *Central Valley Project (CVP)* in 1935. Like the hydrological projects we viewed previously around Las Vegas, the purpose of the CVP (and later, the State Water Project) has been to manage water resources in central California and, among other things, relocate water with a system of dams and canals from the humid northern part of the Central Valley to the more arid south. In addition, the underlying Central Valley aquifer has been systematically tapped, much like the High Plains aquifer discussed in Chapter 15. Taken together, these water projects have turned what was previously a semi-arid desert into productive farmland. As a result, the area is now considered by some to be the engine of the California economy. This significance is not to be underestimated, as the California economy provides 13% of all goods and services in the United States, and would be the 10th largest in the world if the state was its own country.

Consequences of Irrigation   Nevertheless, the impact of irrigation on the natural environment in the valley has been extensive. In some places, withdrawal of water from the Central Valley aquifer has caused the land surface to subside, with millions of dollars of damage to buildings, aqueducts, bridges, and highways. Irrigation has also caused extensive salinization of the soils in the valley (Figure 20.17), the effects of which have been magnified by the hydro-geologic relationships in the area. To water their fields, farmers divert water from irrigation canals or draw fresh groundwater from the Central Valley aquifer (Figure 20.16). Early on, these irrigation strategies were beneficial to the soil because they flushed salts from the sediment,

(a)

(b)

(c)

Figure 20.16 Irrigation in the San Joaquin Valley. (a) Water used in spray irrigation is derived from the Central Valley aquifer. (b) Irrigation canals such as this one divert water from the streams to specific farm areas. (c) Row irrigation of lettuce. Water from irrigation canals is delivered to a key part of your next salad in this way.

which naturally occurred given the marine geologic history of the region. Unfortunately, a fine-grained marine deposit, called Corcoran clay, lies at a shallow depth below the surface in the central and western parts of the valley. This clay layer is impermeable, which causes percolating irrigation water to perch on top of it and the water table to rise. Because this part of the groundwater system does not reach the freshwater in the main aquifer, the groundwater perched on the clay becomes perpetually more saline, which, in turn, causes the salinization of soils in a manner entirely consistent with what the Sumerians experienced thousands of years ago.

In an effort to mitigate the impact of increased soil salinity, the Bureau of Reclamation devised a way in the 1960s to drain away the shallow groundwater to stabilize the depth of the water table below the root zone of plants and to facilitate leaching of surface salts. This diversion was accomplished by the construction of the San Luis Drain, which flowed 134 km from the southern end of the valley to Kesterson Reservoir (Figure 20.18), a man-made impoundment containing 12 evaporation ponds within the Kesterson National Wildlife Refuge. The drainage system was completed in 1971 and appeared to work well until it was discovered in the early 1980s that water in the impoundment contained elevated levels of selenium, which is toxic to wildlife in high concentrations. Although selenium occurs naturally in the area soils, especially the eastern side of the valley near outcrops of marine rocks within the coast ranges, it became concentrated in the groundwater because percolating irrigation water flushed the selenium from the surface and it ultimately drained into Kesterson Reservoir. The impact of this selenium enrichment in the groundwater became apparent when many over-wintering migratory birds and newborn chicks at the reservoir developed deformities or became so sick that

Figure 20.17 Soil salinization in the San Joaquin Valley. This photo nicely illustrates the effects of salinization in the San Joaquin Valley. Notice how the salty patch (large white area) is very poorly vegetated.

they drowned. In addition, ranchers near the drain and reservoir reported numerous livestock deformities and deaths. As a result of this environmental problem, the San Luis Drain was closed in 1985. Amid much public outcry, it was reopened in 1996 to receive limited amounts of drain water.

Possible Solutions Given the environmental problems associated with draining saline groundwater from the San Joaquin Valley, environmental managers continue to look for a way to reduce soil salinity in the region. In the mid-1980s the United States Geological Survey (USGS) took the lead to search for new and better ways to rid the valley of excess groundwater. A number of potential solutions continue to be discussed. One idea is to pay farmers hundreds of millions of dollars to take about 79,000 hectares (~195,000 acres) of

VISUAL CONCEPT CHECK 20.2

Imagine you are flying over the San Joaquin Valley in California and see this landscape when you look out the window. What is the best explanation for the whitish area visible in the center of the image?

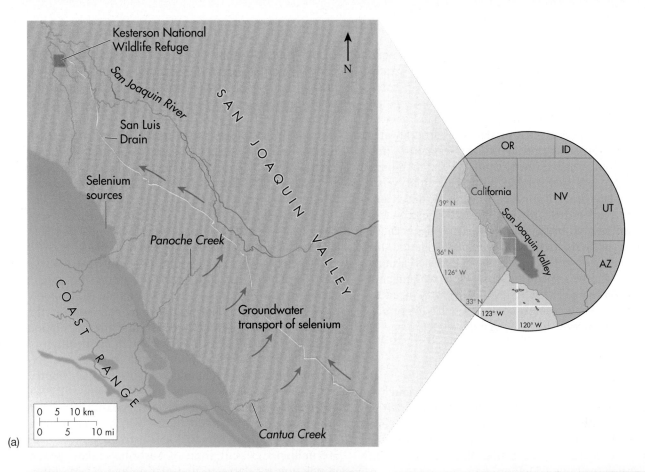

(a)

(b)

(c)

Figure 20.18 Drain system in the San Joaquin Valley. (a) Map of the San Luis Drain and Kesterson Reservoir. This system was used to drain excess groundwater from farm fields in the valley, some of which contained high levels of selenium derived from the coast ranges. **(b)** Photograph of Kesterson Reservoir. **(c)** Deformities in a bird chick linked to high levels of selenium in the water at Kesterson Reservoir.

the most sensitive farmland out of production. Another idea is to build more than 2000 artificial evaporation ponds into which groundwater would drain and evaporate. This method would concentrate salts at specific locations, which could then perhaps be transported to the ocean and dumped. It would require removing land from productivity, however, to create room for the ponds. In the western part of the valley, for example, it is believed that one pond is required for every nine fields. Still other ways of coping are to treat the water with expensive technology and to grow more salt-tolerant crops such as cotton.

At this point in time, no systematic plan has been implemented to deal with soil salinity in the San Joaquin Valley. Regardless of the method(s) employed, some kind of soil salt balance must be achieved for irrigated agriculture to be sustainable in the long term in this critical agricultural zone. Recent modeling of groundwater salt concentrations suggests that in the near term some kind of soil salt balance can be attained. The long-term prospects are not as favorable, however, primarily because it appears that salt concentrations in the deep Central Valley aquifer are slowly rising. These rising concentrations will be a problem in the coming decades because this water supply is used extensively for irrigation. Increasing salt levels in this aquifer mean that increasingly salty water will be applied to crops, thereby damaging them and raising soil salinity levels. Given the need to produce an ever-increasing amount of food in the future, this potential outcome has serious implications.

Australia

Australia is a fascinating country in many ways and has long been a draw to people throughout the world. It is the largest island on Earth, while at the same time it is a continent about the size of the contiguous United States (its lower 48 states). Some of the oldest rocks in the world, dating to about 3.5 billion years ago, are found in Australia. The Australian continent has been essentially inert from a geologic perspective for some 60 million years, which means that some of the oldest and most weathered soils on the planet occur there. Given its long-term isolation, the flora and fauna are exotic, with 80% of Australian plants and animals like the kangaroo, platypus, and koala found nowhere else on Earth. Some of the most lethal animals on Earth are indigenous to Australia, including the salt-water crocodile, trap-door spider, and box jellyfish. The largest living organism in the world, the Great Barrier Reef, is on the fringe of the northeast coast. Among many other unusual things, Australia has the most salinized soils in the world. Most of the country is semi-arid to arid, receiving less than 50 cm (20 in.) of precipitation a year (Figure 20.19). The combined effect of low precipitation and ancient soils poor in nutrients produced one of the most desolate places on Earth, the famed Australian outback.

Understanding the complexity of Australia's physical geography is essential to appreciating the impact humans are having there. The environment is extremely sensitive, and the potential environmental issues in the future are thus critical. Consider first the population of the country. As of February 2010, the population of Australia was about 22 million, with most people living on the relatively humid margins of the country, especially the continent's eastern side (Figure 20.20a). To put that number in perspective, consider that the population of the United States, which is fundamentally the same physical size as Australia, is now over 300 million. Certainly, some of Australia's relatively low population is related to its isolation and a very different immigrant history compared to the United States. Another more important factor is that the comparatively sterile physical environment of the country simply cannot support more people on a sustainable basis. In fact, some believe that the country is *already* overpopulated

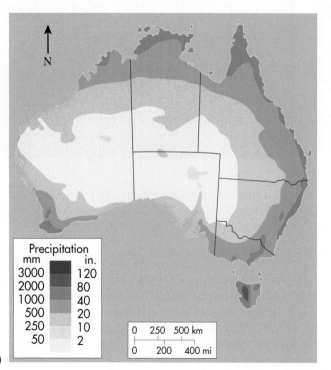

Precipitation

mm	in.
3000	120
2000	80
1000	40
500	20
250	10
50	2

0   250   500 km

0   200   400 mi

(a)

(b)

Figure 20.19 Australia precipitation. (a) Map of average annual precipitation in Australia. Note that most of the country is semi-arid to arid. (b) Typical scene in the Australian outback. This landscape is the result of sparse precipitation and ancient soils that are low in nutrients.

and that the sustainable population may be as low as 8 million people.

Given potential population issues in the country, it is essential that Australian farmers maximize food production. Unfortunately, a key variable limiting this goal is the simple fact that the vast majority of the country is unsuitable for agriculture, as you can see in the land-use map in Figure 20.20b. As a result of these opposing realities, many farmers have focused on marginal drylands on the fringe of the outback to produce food, particularly wheat. The primary wheat areas in Australia are the Murray-Darling River basin in the southeastern part of the country, and the wheat belt region in the extreme southwest (Figure 20.20c). The wheat belt region, for example, produces about 20 to 25 million tons of wheat per year.

As you have seen in ancient Sumeria and the San Joaquin Valley, dryland farming can increase soil salinity. Such soil salinization is a serious problem in the wheat areas of Australia. These regions are particularly susceptible to salt buildup because Australia is naturally a very salty place. This natural salinization of the continent has occurred due to the interaction of several factors related to physical geography. First, much of the continent is essentially a closed basin where very little water flows to the sea. As a result, when rain does fall, it tends to settle in specific places rather than flow away. Another factor is that any rain which does fall contains microscopic salt particles derived from the saline marine waters that surround the continent. Over millions of years these salts have gradually built up in Australian soils, especially since water rapidly evaporates in the semi-arid to arid climate. A third factor is that much of the groundwater in Australia is saline—in some cases, as much as the ocean that surrounds the continent. This natural groundwater salinity is a result of the gradual salinization of Australia over tens of millions of years and the slow, persistent downward movement of salt into underlying aquifers.

Given the natural salinity of Australia, the potential for soil salinization due to agriculture is extremely high. Significant areas of the country are already being affected

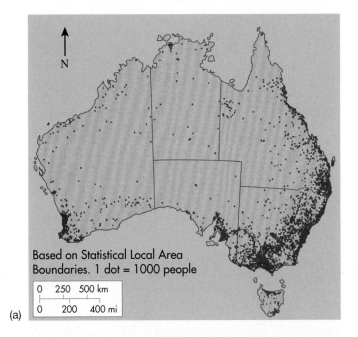

Based on Statistical Local Area Boundaries. 1 dot = 1000 people

| 0 | 250 | 500 km |
| 0 | 200 | 400 mi |

(a)

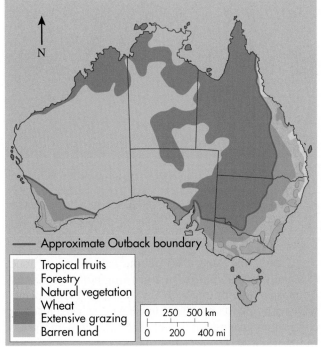

—— Approximate Outback boundary

Tropical fruits
Forestry
Natural vegetation
Wheat
Extensive grazing
Barren land

| 0 | 250 | 500 km |
| 0 | 200 | 400 mi |

(b)

(c)

Figure 20.20 Australian population and land use. (a) Population geography of Australia. Note the relationship of dense population clusters and the precipitation map in Figure 20.19. (b) Map of land use in Australia. Again, note the geographical relationship of land-use patterns and precipitation shown in Figure 20.19. (c) Photograph of the wheat belt region in southwestern Australia.

Case 2: Soil Salinization in Arid and Semi-Arid Lands • 581

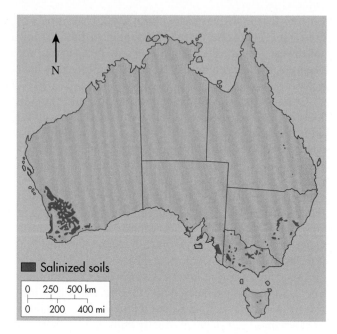

Figure 20.21 Australia soil salinization. Map of areas in Australia where soil salinization is a problem. (*Source:* Commonwealth of Australia.)

(a)

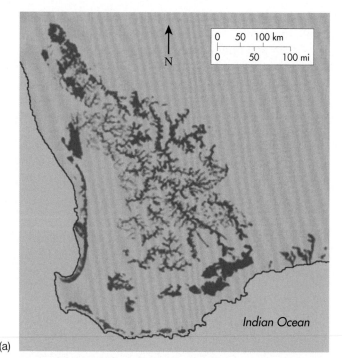

(b)

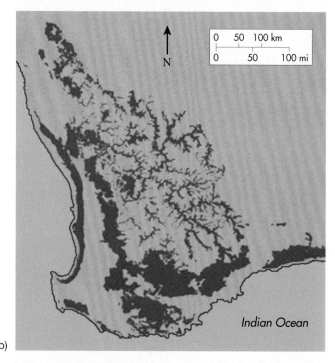

Figure 20.22 Dryland salinity in southwestern Australia. (a) Map showing the extent of dryland salinity in 2000. (b) Map showing area expected to be at risk for salinization in 2050. (*Source:* Commonwealth of Australia.)

by the problem (Figure 20.21). In the Murray-Darling River basin, salinization occurs because irrigation is increasing soil salinity in a manner very similar to what we discussed in the San Joaquin Valley. The largest areas of dryland salinity are in the wheat belt of Western Australia, where salinization is largely related to the clearing of native vegetation and associated rise of the water table. In this context, salts are either accumulating in depressions or are wicked up from the shallow water table.

The real and potential costs of soil salinization in Australia are staggering. About 5.7 million hectares (14 million acres), or approximately 16% of the agricultural area, are currently at risk. The reduced or lost crop yields from this large area have cost farmers about $1.3 billion Australian dollars (~$800 million U.S.) in lost income. Soil salinization is also posing risks to urban and rural infrastructure (because it deteriorates iron and concrete), as well as the health of stream ecosystems and wetlands.

The problem of soil salinization has a very high profile in Australia and is a major source of concern. Unfortunately, the situation is expected to become even worse in the future. A good place to see the likely future outcome is in the wheat belt in Western Australia. At present, about 3.5 million hectares (~8.6 million acres) of the region are at risk of eventual soil salinization. At current rates of salinization, the amount of land that will be affected is expected to double by 2050 (Figure 20.22). In the country as a whole, the amount of land salinized may reach 17 million hectares (42 million hectares).

Clearly, the Australian people face a critical issue, especially in the context of a growing population. Australians are attempting to deal with the problem, specifically with efforts to restore water and salt balances by controlling groundwater recharge rates in some areas. In other places, farmers and ranchers are attempting to reduce the effects of salinization by planting salt-tolerant trees, such as Saltgrow (a hybrid Eucalyptus tree). These efforts will take a substantial amount of time, however, which means that the problem of salinization in Australia is going to be chronic for the foreseeable future, and will likely get worse before it gets better. The potential negative impact that salinization has on wheat production is magnified if global climate change scenarios are included in the equation, because most models suggest that Australia will become even warmer and drier in the future. If this is true, then the challenge to produce enough wheat in Australia is obvious. Stay tuned.

The Challenge of Sustainable Agriculture in Semi-Arid and Arid Regions

This case study focused on the issue of soil salinization in semi-arid and arid regions of the world. The factors that influence the salinization of soils are clearly related to physical geography and the ways that humans interact with the environment. Salinization is prone to occur in marginal lands where people are attempting to farm in an effort to produce more food. These environments are extremely sensitive to disturbance and modification of the landscape to fit agricultural needs.

KEY CONCEPTS TO REMEMBER ABOUT SOIL SALINIZATION

1. Soil salinization refers to the buildup of salts in soil and usually occurs in semi-arid and arid regions. It can occur naturally when salty slopewash accumulates in a depression and evaporates, leaving a salt crust behind.

2. Soil salinization can also occur due to agricultural practices in marginal lands. It can happen when clearing of native vegetation causes local water tables to rise sufficiently close to the surface that wicking can occur. This form of salinization is magnified when irrigation causes the water table to rise further. Soil salinization can also occur when salty irrigated water evaporates after it is applied to fields, leaving behind salt deposits.

3. A major area of soil salinization in the United States is the San Joaquin Valley in California, which is one of the most important agricultural regions in the nation.

4. Extensive soil salinization is also occurring in the agricultural belts of Australia. Soil degradation in these areas has the high potential to significantly reduce crop yields such as wheat.

Understanding the subsequent salinization of soils that follows such disturbance requires an appreciation of soils, groundwater, and evaporation rates. Archaeological evidence indicates that soil salinization has been a problem since the dawn of human civilization, resulting in the collapse of past human societies. The problem remains with us today and will likely intensify in the future, barring some yet undeveloped way to deal with soil salts systematically. Given the expected growth in human population in the coming decades, the likely loss of farmland or reduced crop yields in salinized areas will increase the challenge of producing sustainable levels of food for the world.

Case 3: Giant Panda Conservation in China

In the third and final case study, we will investigate the effort to conserve giant panda habitat in China. Giant pandas (Ailuropoda melanoleuca) are among the most popular animals on Earth because of their distinctive markings and docile demeanor (Figure 20.23a). These animals are also a prominent cultural symbol in China. Nevertheless, pandas are highly threatened due to human land-use practices and thus are officially an endangered species. Only about 1600 of them live in the wild. Given these very small numbers and the potential for extinction, aggressive efforts are ongoing by the Chinese government, international conservation agencies, and leading scientists to protect remaining habitat so pandas can maintain a sustainable population in the wild. Physical geography helps us understand the panda's landscape, the impact that growing human population has on the panda environment, and the conservation efforts required to save them in the wild. In contrast to the potentially bleak case studies investigated so far, the panda case demonstrates how a potentially disastrous environmental issue may be stabilized and perhaps reversed in the future by using holistic geographical principles.

Giant pandas are solitary members of the bear family (Ursidae). They live in old-growth forests that have a healthy bamboo understory and forest clearings (Figure 20.23b). Although pandas are classified as carnivores and have been known to eat honey, bulbs, fish, and eggs, their diet consists almost entirely (>95%) of bamboo shoots. Given the low nutrient composition of bamboo, and the pandas' inability to digest cellulose efficiently, they derive little energy from their food and thus must consume up to 13.5 kg (~30 lb) of shoots daily. To meet this need, they spend 10 to 12 h a day feeding and conserve energy by moving slowly. Pandas neither hibernate nor make permanent dens like many other bears and use hollow trees and caves for shelter.

(a)

(b)

Figure 20.23 Giant pandas. (a) Pandas are members of the bear family and are known for their distinctive markings and docile demeanor. (b) Panda habitat consists of old-growth forests with a prominent bamboo understory and forest clearings.

In prehistoric times, the pandas' natural range once extended throughout the forests of southeastern China, Myanmar (formerly Burma), and even into India (Figure 20.24). This area lies largely within the humid subtropical hot-summer climate (*Cfa, Cwa*) region of southeastern Asia, which, if you recall from Chapter 9, is dominated by monsoon atmospheric circulation. During the 18th century the extent of panda habitat progressively shrank because lowland subtropical forests were cleared for agriculture and wood fuel for cooking and home heating. This expansion of agriculture and loss of panda habitat occurred during a major growth phase in Chinese population from about 140 million people in 1700 to approximately 400 million in 1900 (Figure 20.25). As a result, by 1900, panda habitat was confined to the Qinling Mountains in the north-central part of their historical range and the Minshan, Xiangling, Qionglai, and the Liangshan Mountains in the west-central part of their ancestral range (Figure 20.24).

During the 20th century panda habitat was further reduced as agriculture expanded along the major river valleys within the region. This growth of farming occurred at the same time that Chinese population grew rapidly from about 500 million people in 1950 to over 1.3 billion presently (Figure 20.25). Pandas were also hunted for pelts and because it was believed that their body parts had medicinal qualities valued in traditional Chinese culture. Today, pandas are found only in about

20 isolated parts of the six mountain ranges mentioned previously and live in dense stands of bamboo that lie between about 1500 m and 3500 m (~4900 ft and 11,500 ft) above sea level. Their optimal altitudinal range is between 2500 m and 3000 m (~8200 ft to 9800 ft) above sea level. Given their need to conserve energy, they prefer level terrain (<15° slopes) within the mountains so they can move about easily.

As population in China increased in the latter half of the 20th century, concerns about the viability of wild panda populations began to grow. In the early 1960s the first four panda reserves were established by the Chinese government and hunting was outlawed. Additional reserves were established in the 1970s, including a permanent research station at the Wolong Nature Reserve. In the 1980s pandas were given even more legal protection and hunters began to be prosecuted vigorously for killing them. At about this same time, however, geographical surveys indicated that the extent of panda habitat dropped from 29,500 km² (11,390 mi²) in 1975 to only 13,000 km² (5,019 mi²) by 1985 due to extensive logging. Results of the third national panda survey in 2004 indicated that China's wild giant panda population included approximately 1600 animals, which is about 40% more than previous estimates. Although hailed as good news, it was believed that this increase is probably related to more accurate survey methods rather than an actual increase in the number of pandas.

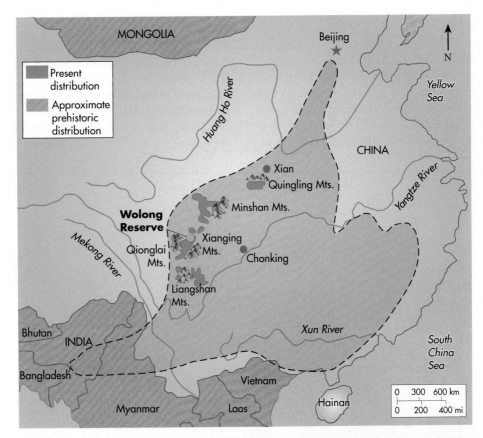

Figure 20.24 Prehistoric and current range of giant pandas in southeastern Asia. Note the extensive decline in panda range that has occurred over time. This loss of habitat has generally coincided with two major periods of population growth in China.

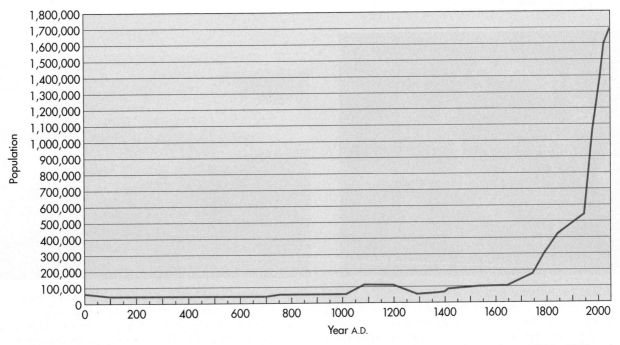

Figure 20.25 Graph of Chinese population the past 2000 years. Note the rapid growth phases from 1650 to 1900 and between 1950 and the present. Over 1 billion people now live in the country.

The Wolong Nature Reserve

In the past decade, a systematic effort has been made to fully document the extent of panda habitat and to monitor its health and distribution. Much of this effort has focused on the Wolong Nature Reserve, which is located in the Qionglai Mountains in the west-central part of the modern panda range (Figures 20.24 and 20.26a). The reserve was established in 1963 and is about 200,000 hectares (~494,200 acres) in size. It contains numerous high mountains and deep valleys, with overall elevation ranging from 1200 m to 6250 m (~3900 ft to 20,500 ft) above sea level. The vegetation in the reserve has a distinct vertical zonation, such as that discussed in Chapter 10, ranging from evergreen and deciduous broadleaf trees at the lower elevations and to subalpine coniferous forests at higher elevations. Given this broad ecological range, the reserve

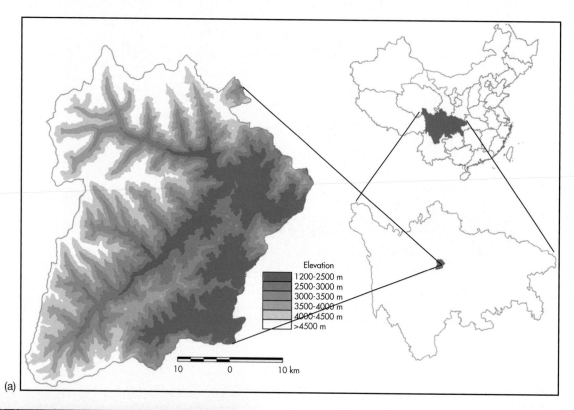

Elevation
1200-2500 m
2500-3000 m
3000-3500 m
3500-4000 m
4000-4500 m
>4500 m

10   0   10 km

(a)

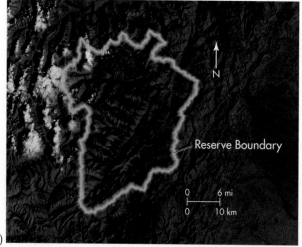

N

Reserve Boundary

0   6 mi
0   10 km

(b)

(c)

Figure 20.26 The Wolong Nature Reserve. (a) Map showing the location of the Wolong Nature Reserve in Sichuan Province and elevation in the area. (*Source:* A. S. Vin et al., "Temporal Changes in Giant Panda Habitat Connectivity across Boundaries of Wolong Nature Reserve, China," *Ecological Applications* 17(4), 2007, pp. 1019–1030.) (b) Satellite image of the Wolong Nature Reserve in China. Note the rugged terrain in the western part of the reserve and how it correlates with the elevation map in (a). (*Source:* NASA.) The best panda habitat is in the less rugged terrain on the east side of the reserve. (c) Qionglai Mountains in the Wolong Nature Reserve. This oblique image provides a great sense of the terrain in this part of China.

contains over 6000 different plant and animal species, including more than 150 giant pandas. The reserve also includes the Wolong Giant Panda Reserve Center, which was established in the 1980s and is thus one of the oldest research centers devoted to pandas in China. In an effort to support the wild panda population, much of the focus of this research center is devoted to giant panda breeding and the study of bamboo ecology.

A great deal of scientific research has been conducted in the Wolong Nature Reserve to monitor the health of panda habitat and the impact that humans have on it. Much of this research uses the remote sensing technology discussed in Chapter 2. Figure 20.26b is a good example of a satellite image of the Wolong Nature Reserve. This image is a composite digital elevation model (DEM) that provides a sense of the topographic relationships in the reserve. This image reveals that the most rugged terrain is on the western side of the reserve and is unsuitable for pandas (Figure 20.26c), whereas the best panda habitat is to the east.

A major effort at the Wolong Reserve is to document changes in the distribution of panda habitat over the past few decades and to predict future patterns. Much of this research assesses the changes in **land cover**, such as roads and forests, on the landscape over time. Satellite imagery from the mid-1960s to 2001 at the

---

Land cover *The various things that cover the landscape, such as forests, roads, and water bodies, at any point in time.*

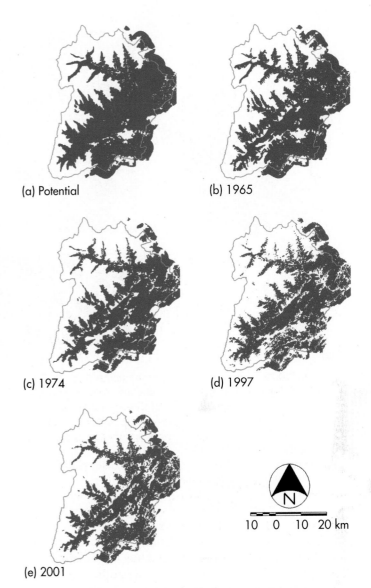

(a) Potential   (b) 1965

(c) 1974   (d) 1997

(e) 2001

Figure 20.27 Deforestation of the Wolong Nature Reserve between 1965 and 2001. Note the change in forest cover that occurred during this interval of time. The "potential" map refers to the expected amount of forest if no deforestation had occurred in the area. (*Source:* A. S. Vin et al., "Temporal Changes in Giant Panda Habitat Connectivity across Boundaries of Wolong Nature Reserve, China," *Ecological Applications* 17(4), 2007, pp. 1019–1030.)

Wolong Nature Reserve, for example, reveals that a significant amount of deforestation occurred throughout the reserve during that period of time (Figure 20.27). In 1965 about 43% of the nature reserve was forested, whereas by 2001 forests had decreased to approximately 36% of the area. This modification in vegetation was most closely associated with the change from forest to cropland, grassland, and shrub land, particularly along the rivers and the main roads. Although the area is technically a nature reserve, ~5000 people live within the boundaries of the park. About 90% of them are farmers

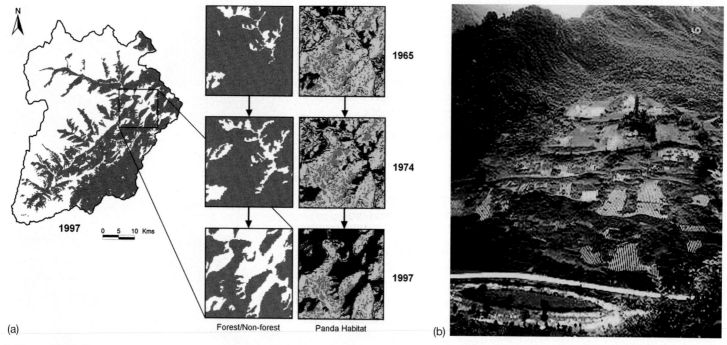

Figure 20.28 Fragmentation of panda habitat from 1965 to 1997. Map on left shows forest extent in Wolong Nature Reserve in 1997. Sequence of gray and white larger-scale maps in the center illustrate change from forest (gray) to nonforest (white) during study interval. Sequence of three color-coded maps on right shows changes in areas of highly suitable (red), suitable (yellow), marginally suitable (green), and unsuitable (black) panda habitats from 1965 to 1997. (*Source:* Used with permission of the American Association for the Advancement of Science; Liu et al., "Ecological Degradation in Protected Areas: The Case of Wolong Nature Reserve for Giant Pandas," *Science* 292, 2001, pp. 98–101.) (b) Deforested hillslopes in the Wolong Nature Reserve. Such a deforested area fragments the panda habitat and makes it difficult for pandas to cross to more suitable areas. (*Source:* Photo courtesy Jianguo Liu.)

who engage in agriculture or livestock breeding in cleared areas. Perhaps the most important cause of forest loss was the collection of wood fuel by peasant farmers who lacked the income to purchase other sources of fuel for cooking and home heating.

Aside from the extensive loss of forest cover, a major outcome of deforestation between 1965 and 2001 was that remaining panda habitat was intensively fragmented. In other words, patches of healthy bamboo in the understory were cut off from one another. You can see the nature and extent of habitat fragmentation by examining Figure 20.28, which again shows a time series of forest loss, this time between 1965 and 1997. This degree of fragmentation is potentially a serious problem because it reduces the quality and quantity of panda habitat and decreases the carrying capacity for the animals. It also limits the ability of pandas to move from one area of suitable habitat to another, which increases the potential for the animals to become genetically isolated from one another. If this trend were to continue, it would dramatically increase the probability of extinction through inbreeding. Unchecked, this type of human impact is projected to result in the loss of up to 16% of all existing panda habitat by 2035.

As the geography of forest loss in the Wolong Nature Reserve and other panda habitats became more apparent in the latter part of the 20th century, the Chinese government responded by outlawing logging in protected panda areas in 1998. In return, farmers were offered government payments to allow cropland to convert back to forest. At about the same time, conservation efforts by such international organizations as the World Wildlife Fund (WWF) began to have significant positive impacts. In 2002, for example, an agreement between the WWF and the Shaanxi Forestry Department was signed to create 13 new panda reserves and develop the first corridors in the Qinling Mountains to connect fragmented habitats. A further positive development was that rising wages associated with regional economic development reduced pressure on forests because more people gained access to electricity and thus did not need wood fuel for heating. These combined factors, among others, have the potential to stabilize the ongoing loss of panda habitat and to slowly turn things around in the coming decades. Panda habitat may begin to recover after about 40 years of forest regeneration. If the logging ban is maintained, previously forested tracts may regenerate entirely and the odds for long-term survival of pandas will increase.

THE FUTURE PICTURE

Our first two case studies dealt with water use in the American Southwest and soil salinization in marginal lands. These case studies showed that it is necessary to understand such variables as the nature and cause of regional climate characteristics, groundwater, runoff, and soils in order to place human/environment interactions in a proper context. In each of these cases, the current trends in water and land use are probably unsustainable in the long term and the potential for significant environmental and social impacts is very high. One cannot help but wonder what people will do in Las Vegas and other southwestern cities if and when water supplies decrease drastically should global warming scenarios actually come to pass in the coming decades. Similarly, what does the future hold for farmers in California and Australia if soil salinization continues? More importantly for you, perhaps, is what happens to the cost of food in the coming decades as global population continues to grow and

more farmland is potentially lost to salinization? These are tough questions indeed and the ultimate outcome is uncertain.

Our last case on panda habitat also requires a good knowledge of physical geography to fully comprehend the issue and possible outcomes. To understand panda conservation, it is necessary to know the geographical characteristics of their habitat, such as the distribution of bamboo, elevation, and relief. It also requires an understanding of human population trends in the area and how expansion of agriculture and wood fuel collection has impacted the habitat. This case study also demonstrates how scientists have used geographical techniques to reconstruct the changes in habitat over time. Their studies have helped shape conservation efforts within China and by various international organizations in a positive way that may very well have stabilized the giant panda population in the wild and saved them from extinction.

SUMMARY OF KEY CONCEPTS

1. An understanding of physical geography is essential to understanding human environmental interactions around the world. Many of the topics covered in this text, such as solar radiation, atmospheric circulation, soils, and plant geography, are directly related to all kinds of major societal issues.

2. Human population has grown dramatically in the past 200 years, from about 1 billion people in 1800 to approximately 6.7 billion today. It is projected that the population will reach over 9 billion by 2050. This rapid growth is tied directly to industrialization and improvements in human health. It poses profound questions regarding the impact on environmental systems around the world.

3. An important ongoing and future issue is the availability of freshwater for human consumption in many places around the world. As population grows, more people are moving to marginal lands where water supplies are scarce. This scarcity in places such as

Las Vegas, Nevada, poses major challenges regarding sustainable economic development in semi-arid and arid regions in the future.

4. With the growing need for food, agriculture is being practiced on a massive scale in many places, such as the San Joaquin Valley in California and in several parts of Australia. The clearing of native vegetation and use of irrigation have caused intensive soil salinization to occur in many parts of these landscapes. This process significantly harms soils and reduces crop yields at a time when more food is required.

5. A growing population has resulted in massive deforestation around the world. Such forest loss has threatened the habitat of countless plant and animal species, such as the giant panda in China. Conservation efforts have the potential to stabilize these losses and even reverse them in the coming decades.

CHECK YOUR UNDERSTANDING

1. Describe the overall history of human population on Earth. When were there major growth spurts and why did they occur? What is the projected growth of population in the future?

2. What is the difference between the developed and developing world, and what are the population trends in each of these areas?

3. Why does population growth in the developing world make it more likely that concentrations of atmospheric CO_2 will increase in the future?

4. Characterize the climate of Las Vegas and describe why this pattern occurs.

5. What is the Colorado River Compact and how did it influence the population of Las Vegas?

6. How do the concepts of carrying capacity and sustainability relate to the issue of water use and population in Las Vegas?

7. Describe the process of soil salinization and how it can occur naturally.

8. How can agricultural practices cause soil salinization to occur? What impact does increased soil salinity have on crops?

9. In the context of soil salinization, how are the San Joaquin Valley and ancient Sumeria similar? Why are these places particularly susceptible to soil salinization?

10. In the context of carrying capacity, why is soil salinization in Australia an important environmental issue?

11. Describe the nature of giant panda habitat, and how and/or why its geographical distribution has changed over time.

12. What do studies of land-cover change in the late 20th century indicate about panda habitat at the Wolong Nature Reserve? How have these findings influenced Chinese environmental policy?

ANSWERS TO VISUAL CONCEPT CHECKS

VISUAL CONCEPT CHECK 20.1

Water in this fountain is most likely derived from Lake Mead, which is a reservoir upstream of Hoover Dam on the Colorado River. Water in Lake Mead originates mostly as runoff from melting snow in the Rocky Mountains, which are the headwaters of the Colorado River.

VISUAL CONCEPT CHECK 20.2

The salt crust in the San Joaquin Valley formed because water tables in the basin have increased in the past few decades from irrigation. As a result, water and salts have moved to the surface by wicking. After the water evaporates, salts are left behind on the surface.

Abrasion   Erosion that occurs when particles grind against each other.   510

Absorption   The assimilation and conversion of solar radiation into another form of energy by a medium such as water vapor. In this process, the temperature of the absorbing medium is raised.   76

Acid rain   The precipitation by rain, fog, or snow of strong mineral acids, primarily sulfur dioxide and nitrogen oxides, that originated in factories.   401

Actual evapotranspiration (actual ET)   The quantity of water actually removed from a given land area by evaporation and transpiration.   213

Adiabatic processes   Changes in temperature that occur due to variations in air pressure.   160

Adret slope   The slope that faces the Sun most directly.   266

Advection   The horizontal transfer of air.   117

Advection fog   Fog that develops when warm air flows over cooler air.   164

Aerial photographs   Photographs taken of the Earth's surface from the air.   33

Aggradation   The progressive accumulation of sediment along or within a stream.   446

A horizon   The soil horizon that is enriched with humus.   296

Air mass   A large body of air in the lower atmosphere that has distinct temperature and humidity characteristics.   180

Air pressure   The force that air molecules exert on a surface due to their weight.   114

Albedo   The reflectivity of features on the Earth's surface or in the atmosphere.   76

Albic horizon   A diagnostic horizon of podzolization from which clay and free iron oxides have been removed, resulting in a light-colored E horizon.   305

Alfisols   Soils generally found in seasonal midlatitude regions that formed through podzolization and have an alkaline argillic horizon.   304

Alluvial fan   A fan-shaped landform of low relief that forms where a stream flows out of an area of high relief into a broad, open plane where the gradient is less and deposition thus occurs.   508

Alluvial terrace   A level, step-like landform that forms when a stream erodes its bed so that a horizontal surface is raised relative to the channel.   455

Alluvium   Sediment deposited by a stream.   456

Alpine glacier   A glacier in mountainous regions that flows down preexisting valleys.   474

Andisols   Soils formed in parent material that is at least 50% volcanic ash.   315

Angle of incidence   The angle at which the Sun strikes Earth at any given place and time.   79

Anthropogenic   Environmental changes caused by humans.   239

Anticlinal valley   A topographic valley that occurs along the axis of a structural anticline.   365

Anticline   A convex fold in rock in which rock layers are bent upward into an arch.   364

Anticyclones   High-pressure systems.   116

Aphelion   The point of the Earth's orbit where the distance between the Earth and Sun is greatest (~152 million km or 94.5 million mi).   48

Aquiclude   An impermeable body of rock that may contain water but does not allow transmission of water through it.   415

Aquifer   A geological formation that contains a suitable amount of water to be accessed for human use.   417

Arable land   Land that has the high potential to be cultivated for crop production.   572

Arête   A sharp ridge that forms between two glacial cirques.   482

Argillic horizon   A B horizon that is enriched in eluviated clay.   298

Aridisols   Mineral soils that form in arid environments and thus are poorly developed.   310

Arroyo   A deep, steep-sided gully that is cut into alluvium.   508

Artesian well   A well in which water from a confined aquifer rises to the surface through natural pressure.   419

Artificial levee   An engineered stucture along a river that effectively raises the height of the river bank and thus confines flood discharge.   462

Asthenosphere   The layer of very soft rock that occurs in the upper part of the upper mantle. This region is about 40 km to 250 km (25 mi to 105 mi) below the surface of Earth. The soft character of this rock allows isostatic adjustments to occur.   324

Atmosphere   The gaseous shell that surrounds Earth.   8

Autotrophs   Organisms that synthesize their own food using heat or light as the source of energy.   250

Avalanche   A large mass of snow or rock that suddenly slides down a mountainside.   408

Axis   The line around which the Earth rotates, extending through the poles.   48

Backshore   The part of the beach that lies between the berm scarp and foredune and is covered by water only during strong storms.   545

Backslope   The gradual slope of a dune that faces the prevailing winds.   514

Backswamps   Marshy floodplain landforms that develop behind natural levees in which fine-grained sediments settle after a flood.   453

Bankfull discharge   The amount of discharge at which the stream channel is full.   440

Barrier island   An elongated bar of sand that forms parallel to the shore for some distance.   547

Base flow   The amount of stream discharge at any given place and time that is solely the product of groundwater seepage.   439

Base level   The lowest level at which a stream can no longer lower its bed, because it flows into the ocean, a lake, or another stream.   447

Bay   An indentation in the shoreline that is generally associated with an ocean, sea, or gulf.   531

Baymouth bar   A spit that entirely encloses a bay.   546

Beach drift   Sediment that is transported in the surf zone by swash and backwash, which form due to the oblique approach of waves.   538

Berm scarp   A miniature cliff created by wave erosion that fronts a beach berm.   545

B horizon   The soil horizon that forms below the E horizon because translocated minerals recrystallize.   296

Big Bang theory   The theory that the Universe originated about 14 billion years ago when all matter and energy erupted from a singular mass of extremely high density and temperature.   46

Biomass   The amount of living matter in an area, including plants, animals, and insects.   253

Biome   The complex of living communities maintained by the climate of a region and characterized by a distinctive type of vegetation.   255

Biosphere   The portion of Earth and its atmosphere that supports life.   8

Bioturbation   The mixing of soil by plants or animals.   287

Brackish   Water that has salinity less than 35‰.   531

Braided stream   A network of converging and diverging stream channels within an individual stream system that are separated from each other by deposits of sand and gravel.   445

Breaker   A wave that rises and crashes when the forward momentum of oscillations cannot be maintained.   537

Brine   Water that has salinity greater than 35‰.   531

Butte   A steep-sided hill or peak that is often a remnant of a plateau or mesa.   507

Calcic horizon   A diagnostic soil horizon of calcification enriched in illuviated calcium carbonate.   309

Calcification   A regional soil-forming process in which calcium carbonate is cycled within the soil.   308

Canyon   A very steep-sided valley that is cut into bedrock.   507

Capillary action   The process through which water is able to move upward against the force of gravity.   148

Capillary action   The force that causes water to rise in the small tubular conduits within the soil.   281

Caprock   A unit of relatively resistant rock that caps the top of a landform and thus protects underlying rocks from erosion.   394

Carbonation   A type of chemical weathering caused by rainwater that has absorbed atmospheric carbon dioxide and formed a weak carbonic acid that slowly dissolves rock.   400

Carrying capacity   The maximum number of organisms that can live in a given area of habitat without degrading it and causing social stresses that result in population decline.   562

Cartography   The design and production of maps.   22

Cation exchange capacity (CEC)   The total amount of exchangeable cations that a soil can absorb.   294

Cations   Positively charged ions, such as sodium, potassium, calcium, and magnesium.   294

Cave   A cavity in rock, produced by the dissolution of calcium carbonate, that is large enough for someone to enter.   425

Celestial dome   A sphere that shows the Sun's arc, relative to the Earth, in the sky.   56

Chemical weathering   The decomposition and alteration of rocks due to chemical actions of natural physical and biological processes.   398

Chinook winds   Downslope airflow that results when a zone of high air pressure exists on one side of a mountain range and a zone of low pressure exists on the other.   138

C horizon   Unaltered soil parent material.   296

Cinder-cone volcano   A small, steep-sided volcano that consists of solidified magma fragments and rock debris that may form in only one eruption.   379

Circle of illumination   The great circle on Earth that is the border between night and day.   48

Cirque   A bowl-like depression that serves as a source area for some alpine glaciers.   475

Cirrus clouds   Thin, wispy clouds that develop high in the troposphere.   164

Climate   Average precipitation and temperature characteristics for a region that are based on long-term records.   212

Climax vegetation   The vegetation within an area that has reached its ultimate complexity.   268

Climograph   A graphical representation of climate that shows average annual precipitation and temperature characteristics by month.   218

Coal   A solid fossil fuel that consists of carbonized plants and animals.   335

Cold front   A frontal boundary where cold air is advancing into relatively warm air. This front is typically associated with intense rain of short duration.   182

Colloids   Very small (10 nanometers to 1 micrometer), evenly divided solids that do not settle in solution.   294

Colluvium   Unconsolidated sediment that accumulates at the base of a slope.   403

Composite volcano   A large, steep-sided volcano that grows through progressive volcanic eruptions, which are usually explosive, and consists of layers of volcanic debris.   380

Condensation   The process through which water changes from the vapor to liquid phase.   150

Condensation nuclei   Microscopic dust particles around which atmospheric water coalesces to form raindrops.   163

Conduction   The transfer of heat energy from one substance to another by direct physical contact.   74

Cone of depression   The cone-shaped depression of the water table that occurs around a well.   420

Confluence   The place where two streams join together.   434

Conformal projection   A map that maintains the correct shape of features on the Earth but distorts their relative size to one another.   24

Conglomerate   Sandstone that contains a wide variety of particle sizes.   332

Constant gases   Atmospheric gases such as nitrogen, oxygen, and argon that maintain relatively consistent levels in space and time.   67

Continental   A place that is surrounded by a large body of land and that experiences a large annual range of temperature.   101

Continental crust   Granitic part of the Earth's crust that makes up the continents. Continental crust averages about 40 km (25 mi) in thickness and is also called *sial* because it consists largely of silica and aluminum.   324

Continental drift   The theory that the continents move relative to one another in association with plate tectonics.   355

Continental glacier   An enormous body of flowing ice that covers a significant part of a large landmass.   478

Contours   Isolines that connect points of equal elevation.   31

Convection   A circular cell of moving matter that contains warm material moving up and cooler matter moving down.   75

Convectional uplift   Uplift of air that occurs when bubbles of warm air rise within an unstable body of air.   169

Convergent uplift   Uplift of air that occurs when large bodies of air meet in a central location.   169

Coral reefs   Resistant marine ridges or mounds consisting largely of compacted coral together with algal material and biochemically deposited calcium carbonates.   551

Cordilleran Ice Sheet   The ice cap that covered much of the mountains in the northwestern part of North America during the Pleistocene Epoch.   489

Coriolis force   The force created by the Earth's rotation that causes winds to be deflected to the right in the Northern Hemisphere and to the left in the Southern Hemisphere.   122

Counterradiation   Longwave radiation that is emitted toward the Earth's surface from the atmosphere.   68

Crest   The highest point of a dune.   514

Crevasse   A deep crack in a glacier.   472

Cumulus clouds   Individual puffy clouds that develop due to convection.   164

Cyclogenesis   The sequence of atmospheric events along the polar jet stream that produces midlatitude cyclones.   184

Cyclones   Low-pressure systems.   116

Dam   A barrier that blocks or restricts the downstream movement of a stream.   463

Debris flow   A rapidly flowing and extremely powerful mass of water, rocks, sediment, boulders, and trees.   408

Debris slide   A mass-wasting process in which slope failure occurs along a plane that is roughly parallel to the surface.   406

Decomposers   Organisms that consume dead or decaying organic substances for nutrition.   250

Deflation   Removal of sediment from a surface by wind action.   510

Deflation hollow   A depression created by wind erosion.   510

Deforestation   The removal of trees for economic or agricultural purposes.   270

Degradation   The topographic lowering of a stream channel by stream erosion.   446

Delta   A low, level plain that develops where a stream flows into a relatively still body of water so that its velocity decreases and alluvial deposition occurs.   459

Dendrochronology   The dating of past events and variations in the environment and climate by studying the annual growth rates of trees.   231

Deposition   The process by which water vapor changes directly to ice.   151

Desertification   The process through which a formerly vegetated landscape gradually becomes desert-like.   521

Desert pavement   A resistant, pavement-like surface created when fine particles blow away and coarse sediment such as pebbles and gravel are left behind.   510

Dew-point temperature   The temperature at which condensation occurs in a definable body of air.   156

Direct radiation   Solar radiation that flows directly to the surface of Earth and is absorbed.   76

Disappearing stream   A surface river or stream that flows into a sinkhole and subsequently moves into an underground river system.   426

Diurnal cycle   A 24-hour cycle.   56

Dolomite   Sedimentary rock consisting of over 50% calcium-magnesium carbonate ($CaMg[CO_3]_2$).   335

Downdraft   A rapidly moving current of cool air that flows downward in a thunderstorm.   193

Downwelling current   A current that sinks to great depths within the ocean because water temperature drops and salinity increases.   141

Drainage basin   The geographical area that contributes groundwater and runoff to any particular stream.   433

Drainage density   The measure of stream channel length per unit area of drainage basin.   435

Drainage divide   An area of raised land that forms a bordering rim between two adjacent drainage basins.   433

Drumlin   A streamlined landform created when a glacier deforms previously deposited till.   485

Dry adiabatic lapse rate (DAR)   The rate at which an unsaturated body of air cools while lifting or warms while descending. This rate is 10°C/1000 m (5.5°F/1000 ft).   160

Earthflow   A slow-to-rapid type of mass movement that involves soil and other loose sediments, some of which may be coarse.   407

Earthquake   Shaking of the Earth's surface due to the instantaneous release of accumulated stress along a fault plane or from underground movements within a volcano.   369

Easterly wave   A slow-moving trough of low pressure that develops within the tropical latitudes.   201

Ecosystems   A community of plants, animals, and microorganisms linked by energy and nutrient flows.   250

Ecotone   The transition area where two or more ecosystems merge.   255

Ecotone   The transition between two distinct ecosystems that contains species of each area.   521

E horizon   The soil horizon that is progressively depleted of minerals through dissolution and translocation.   296

Electromagnetic spectrum   The radiant energy produced by the Sun that is measured in progressive wavelengths.   64

Eluviation   The dissolution and downward mobilization of minerals by water in the soil.   284

Emissivity   The amount of electromagnetic energy released by some aspect of the Earth's surface.   35

Entisols   Soils that are very weakly developed and thus have no distinct horizonation.   313

Environmental lapse rate   The decrease in temperature that generally occurs with respect to altitude in the troposphere. This rate is 6.4°C per kilometer or 3.5°F per 1000 ft (a negative lapse rate).   93

Eolian processes   Geomorphic processes associated with the way that wind erodes, transports, and deposits sediment.   509

Epicenter   The point on the Earth's surface that lies directly over the focus of an earthquake.   369

Epipedon   The uppermost horizon or horizons of a soil that is used for diagnostic purposes.   299

Equator   The great circle that lies halfway between the North and South Poles.   17

Equatorial trough   Core of low-pressure zone associated with the Intertropical Convergence Zone.   126

Equilibrium line   The place on a glacier where snow accumulation and melting are in balance.   471

Equivalent projection   A map projection that accurately portrays size features throughout the map.   26

Esker   A winding ridge formed by a stream that flows beneath a glacier.   486

Eustatic change   Fluctuations in sea level associated with adjustments in the hydrologic cycle.   531

Evaporation   The process by which atoms and molecules of liquid water gain sufficient energy to enter the gaseous phase.   79

Evaporation   The process through which water changes from the liquid to vapor phase.   150

Evaporites   Surface salt residue that collects through the evaporation of water and the crystallization of sodium.   335

Evapotranspiration   The combined processes of evaporation and transpiration.   159

Exfoliation   Form of physical weathering in which sheets of rock flake away due to seasonal temperature changes or by the expansion of the rock due to unloading.   398

Extrusive igneous rocks   Rocks that form when magma cools on the Earth's surface.   329

Fall Equinox   Assuming a Northern Hemisphere seasonal reference, the Fall (or autumnal) Equinox occurs on September 22 or 23, when the subsolar point is located at the Equator (0°).   53

Fault   A crack in the Earth's crust that results in the displacement of one lithospheric plate or rock body relative to another.   369

Fault escarpment   A step-like feature on the Earth's surface created by fault slippage.   373

Field capacity   The amount of water remaining in the soil after the soil is completely drained of gravitational water.   282

Field capacity   The maximum amount of water the soil can hold after gravitational water has moved away.   414

Firn   The compact, granular substance that is the transition stage between snow and glacial ice.   470

Fjord   A former glaciated valley along the coast that is flooded by rising sea level.   532

Flood stage   The level at which stream discharge begins to spill out of the channel into the surrounding area.   440

Foredune   A dune that forms parallel to the shore when sand blows inland from the beach.   546

Foreshore   The nearshore zone that is regularly exposed and submerged through the tidal fluctuations and movement of surf.   544

Fossil fuels   Carbon-based energy sources, such as gasoline and coal, that are derived from ancient organisms.   240

Freezing   The process through which water changes from the liquid to solid phase.   150

Frontal uplift   Uplift of air that occurs along the boundary of contrasting bodies of air.   169

Frost heaving   Upthrust of sediment or soil due to the freezing of wet soil beneath.   396

Frost wedging   Expansion and contraction of water in rock cracks due to freezing and thawing.   395

Fumerole   A steam vent that results because underlying groundwater is boiled away before reaching the surface.   388

Gelisols   Soils in subarctic and arctic environments that contain permafrost within 2 m of the surface.   307

Geologic time   The period of time that encompasses all of Earth history, from its formation to the present.   338

Geomorphology   The branch of physical geography that investigates the form and evolution of the Earth's surface.   364

Geostationary orbit   An orbit where satellites remain over the same place on Earth every day. This orbit is achieved because the satellite is placed very high above Earth and travels at the same speed as the Earth's rotation.   35

Geostrophic winds   Airflow that moves parallel to isobars because of the combined effect of the pressure gradient force and Coriolis force.   123

Geyser   A superheated fountain of water that suddenly sprays into the air on a periodic basis.   388

Glacial abrasion   An erosional process caused by the grinding action of a glacier on rock.   479

Glacial drift   Sediment deposited indirectly or directly by a glacier.   483

Glacial erratics   Large boulders that have been plucked and transported a great distance before they are deposited.   479

Glacial grooves   Deep furrows in rock produced by glacial abrasion.   479

Glacial outwash   Sediment deposited by meltwater streams emanating from a glacier.   486

Glacial plucking   An erosional process by which rocks are pulled out of the ground by a glacier.   479

Glacial striations   Scratches in rock produced by glacial abrasion.   479

Glacial till   Sediment deposited directly by a glacier.   483

Glacial trough   A deep, U-shaped valley carved by an alpine glacier.   482

Glacier   A slow-moving mass of dense ice that flows under its own weight.   470

Graben   A downthrown block of rock that lies between two steeply inclined fault blocks.   371

Graded stream   An idealized stream that has achieved a balance between sediment transport and stream capacity.   446

Great circles   Circles that pass through the center of the Earth and that divide the planet into equal halves.   16

Greenhouse effect   The process through which the lower part of the atmosphere is warmed because longwave radiation from Earth is trapped by carbon dioxide (CO_2) and other greenhouse gases.   68

Groin   A short, low wall built at a right angle to the coast that intercepts the longshore current.   555

Ground ice   Distinct zones of frozen water that occur in permafrost regions.   497

Gulf   A relatively small body of saltwater that is surrounded by land on three sides and opens to a sea or ocean.   530

Gyres   Large oceanic circulatory systems that form because currents are deflected by landmasses.   140

Hadley cell   Large-scale convection loop in the tropical latitudes that connects the Intertropical Convergence Zone (ITCZ) and the Subtropical High (STH).   129

Hanging valley   An elevated U-shaped valley (with respect to a glacial trough) formed by a tributary alpine glacier.   483

Headland   A portion of the coast that extends outward into a large body of water.   539

Heterotrophs   Organisms that consume complex organic substances for food.   250

High latitudes   The zone of latitude that lies from about 55° to 90° in both hemispheres.   19

High-pressure ridge   An elongated area of elevated air pressure in the upper atmosphere that is typically associated with sunny skies and calm winds.   186

High-pressure system   A rotating column of air that descends toward the surface of Earth where it diverges.   116

Histosols   Organic soils that form in cool, wet environments where organic carbon decomposes very slowly.   306

Hogback ridge   A ridge underlain by gently tipped rock strata with a long, gradual slope on one side and a relatively steep scarp or cliff on the other.   365

Hook echo   The diagnostic feature in Doppler radar indicating strong rotation is occurring within a thunderstorm and tornado development is thus possible.   201

Horn   A mountain with three or more arêtes on its flanks.   482

Horst   An upthrown block of rock that lies between two steeply inclined fault blocks.   371

Hotspot   A stationary zone of magma upwelling that is associated with volcanism within the interior of a crustal plate.   386

Humidity   A measure of how much water vapor is in the air. The ability of air to hold water vapor is dependent on temperature.   153

Humus   Decomposed organic matter, typically dark, that is contained within the soil.   281

Hurricane   A tropical circulatory system with maximum sustained winds greater than 63 knots (73 mph).   202

Hydrologic cycle   A model that illustrates the way that water is stored and moves on Earth from one reservoir to another.   151

Hydrolysis   A chemical weathering process that results in the decomposition of silicate molecules within rock through the reaction of hydrogen and hydroxyl ions in water.   399

Hydrosphere   The part of Earth where water, in all its forms, flows and is stored.   8

Hydrosphere   The water realm on Earth.   151

Hygroscopic water   Soil water held so tightly by sediment grains that it is unavailable for plant use.   414

Ice cap   A dome-shaped sheet of ice that covers an area less than 50,000 km^2 (~19,000 mi^2) in size.   474

Ice field   A topographically constrained sheet of ice in mountainous areas that frequently has glaciers streaming away from it.   474

Igneous rocks   Rocks that form when magma rises from the mantle and cools, either within the Earth's crust or on the surface.   329

Illuviation   The recrystallization of minerals that occurs directly below the zone of eluviation.   284

Inceptisols   Soils that have one or more weakly developed horizons due to some alteration and removal of soluble minerals.   314

Indirect radiation   Radiation that reaches Earth after it has been scattered or reflected.   77

Inner core   The inner part of the Earth's core. This area is about 1220 km (760 mi) thick and consists of solid iron and nickel.   323

Insolation   Amount of solar radiation measured in watts per square meter (W/m^2) that strikes a surface perpendicular to the Sun's incoming rays.   74

Interfluves   Topographic high points in a drainage basin that separate one tributary from another.   434

International Date Line   This line generally occurs at 180° longitude, with some variations due to political boundaries, and marks the transition from one day to another on Earth.   50

Intertropical Convergence Zone (ITCZ)   Band of low pressure, calm winds, and clouds in tropical latitudes where air converges from the Southern and Northern Hemispheres.   126

Intrusive igneous rocks   Rocks that form when magma cools within the Earth's crust.   329

Isobars   Isolines that connect points of equal atmospheric pressure.   31

Isohyets   Isolines that connect points of equal precipitation.   31

Isolines   Lines on a map that connect data points of equal value.   28

Isopachs   Isolines that connect points of equal sediment or rock thickness.   31

Isotherms   Isolines that connect points of equal temperature.   31

Jetties   Long stone or concrete structures used to create a permanent opening for a channel by reinforcing both sides of the passageway.   556

Joint   A crack or fissure along horizontal or vertical planes in a rock mass that divides the rock into large blocks.   395

Kame   A large mound of sediment deposited along the front of a slowly melting or stationary glacier.   486

Karst topography   Terrain that is underlain by soluble rocks, such as limestone and dolomite, where the landscape evolves largely through the dissolution of rock.   424

Katabatic winds   Downslope airflow that evolves when pools of cold air develop over ice caps and subsequently descend into valleys.   138

Kettle lake   A lake that forms when a block of ice falls off the glacial front, is buried by glacial drift, and then melts, forming a depression that fills with water.   487

Kinetic energy   The energy of motion in a body, measured as temperature, that is derived from movement of molecules within the body.   94

Lagoon   A brackish body of water that lies behind a baymouth bar.   546

Land breeze   Nighttime circulatory system along coasts where winds from a zone of high pressure over land flow to a zone of relatively low pressure over water.   138

Land cover   The various things that cover the landscape, such as forests, roads, and water bodies, at any point in time.   587

Landform   A natural feature, such as a hill or valley, on the surface of Earth.   364

Landslide   An instantaneous movement of soil and bedrock down a steep slope in response to gravity, triggered by an event such as an earthquake.   406

Large-scale map   A map that shows a relatively small geographic area with a relatively high level of detail.   28

Latent heat   Heat stored in molecular bonds that cannot be measured.   78

Laterization   A regional soil-forming process in tropical and subtropical environments that results in extensive eluviation of minerals except for iron and aluminum.   301

Latitude   The part of the Earth's grid system that determines location north and south of the Equator.   17

Laurentide Ice Sheet   The continental glacier that covered eastern Canada and parts of the northeastern United States during the Pleistocene Epoch.   488

Lava dome   A steep-sided volcanic landform consisting of highly viscous lava that does not flow far from its point of origin before it solidifies.   380

Leeward side   The side of a mountain range that faces away from prevailing winds.   139

Level of condensation   The altitude at which water changes from the vapor to liquid phase.   161

Limestone   Sedimentary rock consisting of over 50% calcium carbonate ($CaCO_3$).   335

Liquid outer core   The outer part of the Earth's core. This area is about 2250 km (1400 mi) thick and consists of molten iron and nickel.   323

Lithification   The process whereby sediments are cemented through compaction to form rock.   331

Lithosphere   A layer of solid, brittle rock that comprises the outer 70 km (44 mi) of Earth.   8

Lithosphere   The outer, solid part of Earth that is about 70 km (44 mi) thick and includes the uppermost part of the asthenosphere and the crust.   324

Littoral drift   Sediment that is transported through the combined processes of longshore drift and beach drift.   538

Loess   Windblown silt.   516

Longitude   The part of the Earth's grid system that determines location east and west of the Prime Meridian.   20

Longitudinal profile   A graph that illustrates the change in stream gradient in cross section along a stream from its source to its mouth.   446

Longshore current   The current that develops parallel to a coast when waves approach the coast obliquely and forward momentum is deflected.   537

Longshore drift   Transport of sediment by the longshore current.   538

Longwave radiation   The portion of the electromagnetic spectrum that includes thermal infrared radiation.   64

Low latitudes   The zone of latitude that lies between about 35° N and 35° S.   19

Low-pressure system   A rotating column of air where air converges at the surface and subsequently lifts.   116

Low-pressure trough   An elongated area of depressed air pressure in the upper atmosphere that is typically associated with cloudy skies and rain.   186

Magma   Molten rock beneath the surface of Earth.   327

Mantle   The layer of the Earth's interior that lies between the liquid outer core and the crust. This area is about 2900 km (1800 mi) thick and consists largely of silicate rock.   323

Map projection   The representation of the three-dimensional Earth on a two-dimensional surface.   22

Map scale   The distance ratio that exists between features on a map and the real world.   28

Maritime   A place that is close to a large body of water that moderates temperature.   101

Maritime vs. continental effect   The difference in annual and daily temperature that exists between coastal locations and those that are surrounded by large bodies of land.   101

Mass wasting   Movement of rock, sediment, and soil downslope due to the force of gravity.   403

Maximum humidity   The maximum amount of water vapor that a definable body of air can hold at a given temperature.   153

Meandering stream   A river or small stream that curves back and forth across its valley.   444

Mechanical weathering   The breakup of a body of rock into smaller rocks of the same type.   395

Meridians   Lines of longitude.   20

Meridional flow   Jet stream pattern that develops when strong Rossby waves exist and the polar front jet stream flows parallel to the meridians in many places.   131

Mesa   A broad horizontal surface, smaller than a plateau, that is upheld by resistant caprock.   507

Mesocyclones   Strong updrafts that rotate within a supercell thunderstorm.   195

Mesopause   The upper boundary of the mesosphere where temperature reaches its lowest point.   94

Mesosphere   A layer of decreasing temperature in the atmosphere that occurs from about 50 km to 80 km (~30 mi to 50 mi) in altitude.   94

Metamorphic rocks   Rocks that form when igneous, sedimentary, or other metamorphic rocks are subjected to intense heat and pressure.   336

Microclimate   Average temperature and precipitation characteristics within a small area within a larger climate region.   266

Midlatitude cyclone   A well-organized low-pressure system in the midlatitudes that contains warm and cold fronts.   184

Midlatitudes   The zone of latitude that lies between about 35° and 55° in both hemispheres.   19

Mid-oceanic ridge   A ridge-like feature that develops along a rift zone in the ocean due to magma upwelling.   360

Milankovitch theory   The theory that best explains Pleistocene glacial/interglacial cycles through long-term variations in Earth's orbital eccentricity, tilt, and axial precession.   236

Minerals   Naturally occurring substances with distinctive chemical configurations that usually manifest themselves in some kind of crystalline form.   326

Mohorovicic discontinuity   The boundary between the Earth's crust and the upper part of the asthenosphere; seismic waves change speed at this boundary.   324

Mollisols   Soils that form through calcification and have a mollic epipedon that overlies mineral matter which is more than 50% saturated with base ions.   308

Monocline   A geologic landform in which rock beds are inclined in a single direction over a large distance.   364

Monsoon   The seasonal change in wind direction that occurs in subtropical locations due to the migration of the Intertropical Convergence Zone (ITCZ) and Subtropical High (STH) Pressure System.   133

Moraine   A winding ridge-like feature that forms at the front or side of a glacier or between two glaciers.   483

Mountain breeze   Downslope airflow that develops when mountain slopes cool off at night and relatively low pressure exists in valleys.   138

Mudflow   A well-saturated and highly fluid mass of fine-textured sediment.   407

Mudpot   A bubbling mixture of gaseous mud and water at the Earth's surface that is associated with geothermal activity.   388

Multipath error   Disruption of the GPS signal from satellites due to obstructions such as trees and buildings.   40

Natural gas   Naturally occurring mixture of hydrocarbons that often occurs in association with petroleum and is found in porous geologic formations.   335

Natural hazard   An actual or potentially occurring natural event, such as an earthquake or volcanic eruption, that has a negative effect on people or the environment.   375

Natural levee   A small ridge that develops along the channel of a stream through the deposition of relatively coarse sediment when flooding occurs.   453

Net radiation   The difference between incoming and outgoing flows of radiation.   81

Normal fault   A steeply inclined fault in which the hanging rock block moves relatively downward.   371

Northern Hemisphere   The half of Earth that lies north of the Equator.   17

Occluded front   The area where a cold front begins to overtake a warm front and thus lift warm surface air aloft.   189

Ocean   The entire body of saltwater that covers about 71% of Earth.   530

Oceanic crust   Basaltic part of the Earth's crust that makes up the ocean basins. Oceanic crust is about 8 km (5 mi) thick and is also called *sima* because it consists largely of silica and magnesium.   324

Offshore   The nearshore zone that is permanently submerged and where waves break.   544

Offshore bar   A small ridge on the bottom of the ocean that separates the offshore and foreshore.   544

O horizon   The uppermost soil horizon that consists of undecomposed plant litter.   296

Orogeny   A period of mountain building, such as the Alleghany Orogeny.   364

Orographic uplift   Uplift that occurs when a flowing body of air encounters a mountain range.   169

Outwash plain   A broad landscape of limited relief created by the deposition of glacial outwash.   486

Overthrust fold   A structural feature where one part of the rock mass is shoved up and over the other.   364

Overturned fold   A structural feature in which the fold limb is tilted beyond vertical, which results in both limbs inclined in the same direction, but not at the same angle.   364

Oxbow lake   A portion of an abandoned stream channel that is cut off from the rest of the stream by the meandering process and filled with stagnant water.   453

Oxic horizon   A diagnostic soil horizon in tropical and subtropical environments rich in iron oxides.   302

Oxidation   A form of chemical weathering in which oxygen chemically combines with metallic iron to form iron oxides, resulting in the loss of electrons.   400

Oxisols   Mineral soils in tropical and subtropical environments that form through laterization and thus have an oxic horizon within 2 m of the surface.   301

Oxygen isotope stages   Periods of time that have distinct O-18/O-16 ratios, which are used to reconstruct prehistoric climate change.   234

Ozone hole   The decrease in stratospheric ozone observed on a seasonal basis over Antarctica, and to a lesser extent over the Arctic.   72

Ozone layer   The layer of the atmosphere that contains high concentrations of ozone, which protect the Earth from ultraviolet (UV) radiation.   71

Pacific Ring of Fire   The chain of volcanoes that occurs along the edge of the Pacific lithospheric plate.   381

Pangaea   The hypothetical supercontinent, composed of all the present continents, that existed between 300 and 200 million years ago.   355

Parallels   Lines of latitude.   18

Parent material   The mineral or organic material in which soil forms.   285

Passive margin   A place where the continental crust and the oceanic crust are on the same tectonic plate and thus do not move relative to one another.   357

Pedogenic processes   The natural processes of soil formation that involve additions, translocations, transformations, and losses.   283

Perched water table   A localized area of saturated sediment that is elevated above the regional water table by a zone of impermeable rock or sediment.   417

Periglacial processes   The suite of processes involving frost action, permafrost, and ground ice that occurs in arctic environments or along the margins of ice sheets.   496

Perihelion   The point of the Earth's orbit where the distance between the Earth and Sun is least (~147 million km or 91.5 million mi).   48

Permafrost   Ground that is permanently frozen.   496

Petroleum   Naturally occurring oily liquid that consists of ancient hydrocarbons.   335

pH   The measure of acidity or alkalinity of a solution, ranging from 0 to 14, based on the activity of hydrogen ions (H^+).   293

Photosynthesis   The conversion of solar radiation into chemical energy. Sugars and starches are produced from carbon dioxide and water, through the interaction of light and chlorophyll in green plants. The process releases oxygen into the atmosphere.   251

Physical geography   Spatial analysis of the physical components and natural processes that combine to form the environment.   6

Pinnacle   A steep-sided, narrow tower that is the final remnant of a plateau, mesa, or butte.   507

Pixel   The smallest definable area of detail on an image; short for pixel element.   34

Plane of the ecliptic   The flat plane on which the Earth travels as it revolves around the Sun.   47

Plant succession   The predictable vegetation transition that occurs within a landscape over a period of time.   268

Plateau   A very broad, horizontal surface that is upheld by resistant caprock.   507

Plate tectonics   The theory that the Earth's crust is divided into a number of plates that move because they float on the asthenosphere.   354

Playa   A dried lake bed that forms when runoff collects in closed topographic depressions in arid regions.   507

Pluton   An extremely large mass of intrusive igneous rock that forms within the Earth's crust.   329

Podzolization   A regional soil-forming process in cool, humid environments that results in the eluviation of iron, aluminum, and organic acids to form well-developed E and Bs horizons.   305

Polar easterlies   Band of easterly winds at high latitudes.   132

Polar front   The contact in the midlatitudes between warm, tropical air and colder polar air.   129

Polar front jet stream   River of high-speed air in the upper atmosphere that flows along the polar front.   130

Polar High   Zone of high atmospheric pressure at high latitudes.   132

Potential evapotranspiration (potential ET)   A measure of the maximum possible water loss from a given land area assuming sufficient water is available.   213

Potential natural vegetation   The vegetation that would occur naturally within a specific area if no human influence occurred.   255

Pressure gradient force   The difference in barometric pressure that exists between adjacent zones of low and high pressure that results in airflow.   120

Prime Meridian   The arbitrary reference point for longitude that passes through Greenwich, England.   20

Process   A naturally occurring series of events or reactions that can be measured and that result in predictable outcomes.   6

Prograformation   Outward extension of the shoreline through deposition of sediment.   544

Proxy data   Indirect evidence of an event. For example, fossil pollen is a proxy indicator of climate change because vegetation reflects climate.   230

Pyroclastic material   Fragmented rock materials resulting from a volcanic explosion or ejection from a volcanic vent.   380

Radial dikes   A long, wall-like feature of intrusive igneous rock that forms when magma is injected into thin cracks within older rock and cools.   344

Radiation   Energy that is transmitted in the form of rays or waves.   74

Radiation budget   The overall balance between incoming and outgoing radiation on Earth.   81

Radiation fog   Fog that develops at night when a temperature inversion exists.   164

Radioactive isotopes   Unstable isotopes that emit radioactivity as they decay from one element to another.   340

Rain shadow   The body of land on the leeward side of a mountain range that is relatively dry and hot (compared to the windward side) due to adiabatic warming and drying.   171

Reflection   The process through which solar radiation is returned directly to space without being absorbed by Earth.   76

Regolith   The fragmented and weathered rock material that overlies solid bedrock.   285

Relative humidity   The ratio between the specific and maximum humidity of a definable body of air.   154

Relief   The difference between the high and low elevation of an area.   288

Remote sensing   The method through which information is gathered about the Earth from a distance.   33

Residual parent material   Parent material that forms by the weathering of bedrock directly beneath it.   285

Retrogradation   The process through which a shoreline retreats through erosion.   540

Reverse fault   A steeply inclined fault in which the hanging rock block moves relatively upward.   373

R horizon   Unweathered bedrock that underlies soil.   296

Ria   A former river valley along the coast that is flooded by rising sea level.   532

Richter scale   The logarithmic scale used to measure the strength of an earthquake.   371

Rifting   The spreading apart of the Earth's crust by magma rising between fractures in the Earth's plates.   360

Rills   Small drainage channels that are cut into hillslopes by running water.   443

Riparian zone   The strip of land, which borders a body of water, that supports plants and animals adapted to water systems.   269

Roche moutonnée   A landform produced by glacial abrasion and plucking that has a shallow slope on one side and a steep slope on the other side.   479

Rock   An amorphous mass of consolidated mineral matter.   326

Rockfall   A mass-wasting process in which rocks break free from cliff faces and rapidly tumble into the valley below.   403

Rock outcrop   A place where rocks are exposed at the surface of Earth.   327

Rock structure   The internal arrangement of rock layers.   364

Rossby waves   Undulations that develop in the polar front jet stream when significant temperature differences exist between tropical and polar air masses.   130

Salic horizon   The diagnostic horizon of salinization that forms due to the recrystallization of secondary salts.   311

Salinity   Concentration of dissolved solids in water that is measured in parts per thousand (‰).   531

Salinization   A regional soil-forming process in which soluble salts are cycled within the soil.   310

Salt-crystal growth   A weathering process involving the buildup of salts on rock surfaces through evaporation. These salts weaken the cement that bonds rock particles, allowing them later to be washed or blown away.   397

Sandstone   A sedimentary rock created when individual sand grains are deposited in thick layers by wind or water and lithify.   332

Saturated zone   The zone of rock below and including the water table where pore spaces are completely filled with water.   415

Scattering   The redirection and deflection of solar radiation by atmospheric gases or particulates.   77

Sea   A subdivision of an ocean that is partially enclosed by land.   530

Sea breeze   Daytime circulatory system along coasts where winds flow from a zone of high pressure over water to a zone of relatively low pressure over land.   137

Sea fog   Fog that develops when cool, marine air comes into direct contact with colder ocean water.   164

Sediment   Solid fragments of rocks that are transported to some location and deposited by wind, water, or ice.   331

Sedimentary rocks   Rocks that form through the deposition and lithification of small fragments or dissolved substances from other rocks or, in some cases, marine animals.   330

Seismic waves   Vibrations that travel through the Earth when stress is released in an earthquake.   322

Sensible heat   Heat that can be felt and measured with a thermometer.   78

Shale   Sedimentary rock consisting of lithified clay-sized sediments.   333

Shield volcano   A very broad volcano with shallow slopes that forms in association with nonviscous lava flows.   385

Shortwave radiation   The portion of the electromagnetic spectrum that includes gamma rays, X-rays, ultraviolet radiation, visible light, and near-infrared radiation.   64

Sinkhole   A topographic depression that forms when underlying rock dissolves, causing the surface to collapse.   426

Slip face   The steep slope that lies on the leeward side of a sand dune at the angle of repose.   514

Slope   The degree of steepness of a portion of the landscape.   266

Slump   A mass-wasting process in which rock and sediment rotate and move down the slope along a concave plane relative to the surface.   406

Small circles   Circles that intersect the Earth's surface and that do not pass through the center of the planet.   17

Small-scale map   A map that shows a relatively large geographic area with a relatively low level of detail.   28

Soil   The uppermost layer of the Earth's surface that forms by the influence of parent material, climate, relief, and chemical and biological agents.   280

Soil creep   The gradual downhill movement of soil, trees, and rocks due to the force of gravity.   405

Soil-forming factors   The variables of climate, organisms, relief, parent material, and time that collectively influence the development of soil.   285

Soil horizons   The distinct layers within a soil that result from pedogenesis.   295

Soil order   A group of 12 distinctive soils differentiated at the most general level.   298

Soil profile   A vertical exposure in which all soil components can be seen.   295

Soil salinization   The process in which soils become enriched in soluble salts such as sulfates from calcium and potassium.   573

Soil science   The study of soil as a natural resource through understanding of its physical, chemical, and biological properties.   298

Soil structure   The way soil aggregates clump to form distinct physical characteristics.   292

Soil taxonomy   The method of soil classification that is based on the physical, chemical, and biological properties of the soil.   298

Soil–water budget   The balance of soil water that involves the amount of precipitation, evapotranspiration, and water storage and loss.   282

Solar constant   The average amount of solar radiation (~1370 W/m^2) received at the top of the atmosphere.   65

Solar noon   The time of day when the Sun angle reaches its highest point as the Sun arcs across the sky.   56

Solifluction   A form of soil creep that occurs in arctic environments where freeze-thaw processes result in lobes of soil moving gradually downslope.   406

Solum   The A, E, and B horizons of a soil, which form through pedogenic processes.   296

Southern Hemisphere   The half of Earth that lies south of the Equator.   17

Spatial analysis   A method of analyzing data that specifically includes information about the location of places and their defining characteristics.   5

Spatial resolution   The area on the ground that can be viewed with detail from the air or space.   34

Specific humidity   The measurable amount of water vapor that is in a definable body of air.   154

Spit   A linear bank of land that extends into a bay made by the deposition of longshore sediment.   545

Spodic horizon   A mineral soil horizon characterized by the illuvial accumulation of aluminum, iron, and organic carbon.   305

Spodosols   Soils in cool, humid regions that form through podzolization and contain a spodic horizon enriched in eluviated iron, aluminum, and organic carbon.   305

Spring Equinox   Assuming a Northern Hemisphere seasonal reference, the Spring (or vernal) Equinox occurs on March 20 or 21, when the subsolar point is located at the Equator (0°).   53

Stable air   A body of air that has a relatively low environmental lapse rate compared to potential uplifting air; thus, strong convection cannot occur.   173

Stationary front   A boundary where contrasting air masses are flowing parallel to one another.   181

Stratopause   The upper boundary of the stratosphere where temperature reaches its highest point.   93

Stratosphere   The layer of the atmosphere, between the troposphere and mesosphere, that ranges between about 12 km and 50 km (between ~7.5 mi and 31 mi) in altitude.   93

Stratus clouds   Layered sheets of clouds that have a thick and dark appearance.   164

Stream hydrograph   The graphical representation of stream discharge over a period of time.   439

Strike-slip fault   A structural fault along which two lithospheric plates or rock blocks move horizontally in opposite directions and parallel to the fault line.   373

Subduction   The process by which one lithospheric plate is forced beneath another. This usually happens when oceanic crust descends beneath continental crust,

but can happen where two plates of oceanic crust meet.   360

Sublimation   The process through which water changes directly from ice to the vapor phase.   150

Subsidence   The settling or sinking of a surface as a result of the loss of support from underlying water, soils, or strata.   422

Subsolar point   The point on Earth where the Sun angle is 90° and solar radiation strikes the surface most directly at any given point in time.   47

Subtropical High (STH) Pressure System   Band of high air pressure, calm winds, and clear skies that exists at about 25° to 30° N and S latitude.   128

Summer Solstice   Assuming a Northern Hemisphere seasonal reference, the Summer Solstice occurs on June 20 or 21, when the subsolar point is located at the Tropic of Cancer (23.5° N).   53

Sun angle   The angle at which the Sun's rays strike the Earth's surface at any given point and time. This angle is high at low latitudes and is progressively less at higher latitudes.   47

Sun-synchronous orbit   A slightly inclined polar orbit that keeps pace with the Sun's westward progress as Earth rotates, resulting in regular return intervals over every location on Earth.   34

Supercell thunderstorms   Large thunderstorms that contain winds moving in opposing directions and are associated with strong winds, lightning, thunder, and sometimes hail and tornadoes.   195

Surface tension   The contracting force that occurs when the water surface meets the air and acts like an elastic skin.   282

Synclinal valley   A topographic valley that occurs along the axis of a structural syncline.   365

Syncline   A concave fold in rock in which rock layers are bent downward to form a trough.   364

Systems theory   The examination of interactions involving energy inputs and outputs that result in predictable outcomes.   6

Talus   A pile of rock fragments and boulders that accumulates below a cliff due to rockfall.   404

Tarn   A small lake that forms within a glacial cirque.   482

Temperature inversion   A layer of the atmosphere in which the air temperature increases, rather than cools, with altitude.   164

Temporal lag   The difference in time between two events, such as when peak insolation and temperature occur.   100

Thermohaline circulation   The global oceanic circulatory system that is driven by differences in salinity.   140

Thermosphere   The upper layer of the atmosphere, which occurs between about 80 km and 480 km (~50 mi and 300 mi) in altitude.   94

Thunderstorm   A brief, but intense storm that contains strong winds, lightning, thunder, and perhaps hail.   183

Tombolo   A spit or sandbar that connects an island to the mainland.   546

Topographic map   A map that displays elevation data regarding the Earth's surface.   31

Topography   The shape and configuration of the Earth's surface.   31

Trade winds   The primary wind system in the tropics that flows toward the Intertropical Convergence Zone on the equatorial side of the Subtropical High-Pressure System. These winds flow to the southwest in the Northern Hemisphere and to the northwest in the Southern Hemisphere.   126

Transform plate margin   A plate boundary where opposing plates move horizontally relative to each other.   358

Transpiration   The passage of water from leaf pores to the atmosphere.   159

Transported parent material   Parent material such as glacial or stream sediments that has recently been deposited and in which soil forms.   286

Tree line   The line that represents the upper limit in mountains and high latitudes where environmental conditions support the growth of trees.   267

Tributary   A stream or river that flows into a larger stream or river.   433

Tropical depression   A tropical low-pressure system with central sustained winds ranging between 20 knots and 34 knots (23 mph and 39 mph).   202

Tropical easterlies   Band of easterly winds that exists where northern and southern trade winds converge.   126

Tropical storm   A tropical low-pressure system with maximum sustained winds between 35 knots and 63 knots (39 mph and 73 mph).   202

Tropic of Cancer   The line of latitude at 23.5° N where the subsolar point is located (and Sun angle is thus 90°) on the Summer Solstice in the Northern Hemisphere.   53

Tropic of Capricorn   The line of latitude at 23.5° S where the subsolar point is located on the Winter Solstice (and Sun angle is thus 90°) in the Northern Hemisphere.   53

Tropopause   The top part of the troposphere, which is identified by where the air temperature is –57°C (–70°F).   93

Troposphere   The lowermost layer of the atmosphere, which lies between the Earth's surface and an altitude of about 12 km (~7.5 mi).   92

Trunk stream   The primary stream of a drainage basin.   433

Tsunami   A destructive sea wave caused by a disturbance within the ocean, such as an earthquake, volcanic eruption, or landslide.   376

Ubac slope   The slope that faces away from the Sun.   266

Ultisols   Mineral soils in subtropical environments that formed through laterization and thus are depleted of calcium and have an argillic horizon.   302

Unsaturated zone   The area between the soil–water belt and the water table where pore spaces are not saturated with water.   414

Unstable air   A body of air that has a relatively high environmental lapse rate compared to uplifting air within it; thus, strong convection can occur.   173

Updrafts   An area of rapidly flowing air that is moving upward within a thunderstorm.   193

Upwelling current   A current that ascends to the surface of the ocean because water temperature warms and salinity decreases.   141

Urban heat island   The relatively warm temperatures associated with cities that occur because paved surfaces and urban structures absorb and release radiation differently than the surrounding countryside.   103

Valley breeze   Upslope airflow that develops when mountain slopes heat up due to re-radiation and conduction over the course of the day.   138

Variable gases   Atmospheric gases such as carbon dioxide, water vapor, and ozone that vary in concentration in space and time.   67

Ventifact   An individual rock that is pitted, grooved, or streamlined through wind abrasion.   511

Vertical zonation The change in environmental characteristics that occurs with respect to altitude. 267

Vertisols Soils that contain an abundance of expandable clay and thus swell and shrink during wet and dry cycles, respectively. 303

Volcanic arc A chain of volcanoes created by rising magma derived from a subducting tectonic plate. 381

Volcano A mountain or large hill containing a conduit that extends down into the upper mantle, through which magma, ash, and gases are periodically ejected onto the surface of Earth or into the atmosphere. 379

Warm front A frontal boundary where warm air is advancing into relatively cool air. This front is typically associated with slow, steady precipitation. 182

Water rights Legally protected rights to take possession of water occurring in a natural waterway and to divert that water for beneficial purposes. 568

Water table The top of the saturated zone. 415

Wave amplitude The overall height of any given wave as measured from the wave trough to the wave crest. 64

Wavelength The distance between adjacent wave crests or wave troughs. 64

Wave refraction The process through which waves are focused and bent around headlands. 540

Waves Oscillations in a body of water that form mostly due to the frictional force generated by wind blowing across the surface of the water. 536

Weather Day-to-day changes that occur with respect to temperature and precipitation. 212

Weathering Physical or chemical modification of rock or sediment that occurs over time. 394

Westerlies Midlatitude winds that generally flow from west to east. 130

Wet adiabatic lapse rate (WAR) The rate at which a saturated body of air cools as it lifts. The average rate is about $5°C/1000$ m ($2.7°F/1000$ ft). 162

Wilting point The threshold amount of soil water below which plants can no longer transpire water. 282

Windward side The side of a mountain range that faces oncoming winds. 138

Winter Solstice Assuming a Northern Hemisphere seasonal reference, the Winter Solstice occurs on December 21 or 22, when the subsolar point is at the Tropic of Capricorn ($23.5°$ S). 53

Xerophytic plants Plants in very dry places that have a number of survival mechanisms in response to prolonged periods of drought. 264

Yardangs Ridges that are sculpted and streamlined by wind abrasion and deflation. 511

Zonal flow Jet stream pattern that is tightly confined to the high latitudes and is thus circular to semicircular in polar view. 130

Zone of ablation The part of a glacier where melting exceeds snow accumulation. 471

Zone of accumulation The geographical area where snow accumulates and feeds the growth of a glacier. 471

PHOTO CREDITS

CHAPTER 1

CO1 Josiah Davidson/Photographer's Choice/Getty Images, Inc.; **1.2a** Eric Nguyen/Jim Reed Photography/Photo Researchers, Inc.; **11.2b** Nick Gunderson/Corbis; **1.2d** James Randklev/Stone/Getty Images; **1.2c** Frans Lemmens/Iconica/Getty Images, Inc.; **1.3b** Science Photo Library/Photo Researchers; **1.3c** Alan Arbogast; **1.3d** David Gowans/Alamy; **1.3a** Roger Harris/Photo Researchers, Inc.; **1.4** USGS; **1.5a** Alan Arbogast; **1.5b** Alan Arbogast; **1.5c** Alan Arbogast; **1.5d** Alan Arbogast; **1.6a** NASA Media Services; **1.6b** Alan Arbogast; **1.6c** Nick Cobbing/Alamy; **1.6d** PhotoDisc, Inc./Getty Images; **p. 12** NASA/GSFC/MITI/ERSDAC//JAROS, and U.S./Japan ASTER Science Team.

CHAPTER 2

CO2 NASA/JPL/NIMA; **2.6** ©National Maritime Museum, London/The Image Works; **2.10a** Age Fotostock America, Inc.; **2.10b** Nasa Jet Propulsion Laboratory; **2.12** Alan Arbogast; **2.17c** USGS, Eros Data Center, National Air and Space Museum; **2.19a** USGS; **2.19b** Richard Price/Taxi/Getty Images; **2.20** David Zimmerman/Corbis Images; **2.22** NASA GSFC; **2.23a** Landsat.org/TRFIC NASA ESIP, Michigan State University; **2.24a** NASA/GSFC/MITI/ERSDAC/JAROS, and U.S./Japan ASTER Science Team; **2.24b** NASA; **2.26** Nasa Jet Propulsion Laboratory; **2.28** Floris Leeuwenberg/Corbis Images; **2.30** Landsat 7, GSFC, NASA; **p. 22 (top)** Michael Busselle/Corbis Images; **p. 22 (bottom left)** Aurora Photos Inc.; **p. 22 (bottom right)** Hubert Stadler/Corbis Images; **p. 38** NASA GSFC, MITI, ERSDAC, JAROS, and U.S./Japan ASTER Science Team; **p. 41** Landsat/GSFC/NASA.

CHAPTER 3

CO3 Paul Nicklen/National Geographic/Getty Images, Inc.; **p. 59** Andrea N. Hahmann, Ph.D. Research Applications Laboratory, National Center for Atmospheric Research; **p. 60** Alan F. Arbogast; **p. 66** SOHO's Extreme Ultraviolet Imaging Telescope/Science Photo Library/Photo Researchers; **p. 70** Raymond Gehman/Corbis Images; **p. 75** Corbis Images.

CHAPTER 4

CO4 Science Source/Photo Researchers; **4.4** NASA Johnson Space Center; **4.6** NOAA; **4.11** Dean Conger/Corbis Images; **4.13b** Courtesy NASA; **4.13a** NASA/Goddard Space Flight Center Scientific Visualization Studio; **4.14** ©AP/Wide World Photos; **4.18a** R. Spoenlein/Corbis Images;

4.18c Onne van der Wal/Corbis Images; **4.20** Andrea N. Hahmann, PhD Research Applications Laboratory Natl. Center for Atmospheric Research; **4.22** Andrea N. Hahmann, Ph.D./Research Applications Laboratory/Natl. Center for Atmospheric Research; **4.23** Andrea N. Hahmann, Ph.D./Research Applications Laboratory/Natl. Center for Atmospheric Research; **4.25a** Andrea N. Hahmann, Ph.D./Research Applications Laboratory/Natl. Center for Atmospheric Research; **4.25b** Andrea N. Hahmann, Ph.D./Research Applications Laboratory/Natl. Center for Atmospheric Research; **4.29** NREL/U.S. Dept. of Energy; **4.30** ©EJ West/Age Fotostock America, Inc.; **p. 87 (bottom)** Suomi Virtual Museum; **p. 88 (left)** Frans Lemmens/Getty Images, Inc.; **p. 88 (right)** Frans Lanting/Minden Pictures, Inc.

CHAPTER 5

CO5 Science Source/Photo Researchers; **5.3b** Science Photo Library/Chris Priest/Photo Researchers, Inc.; **5.3c** David Joel/Stone/Getty Images; **5.3d** Jody Doyle/The Image Bank/Getty Images, Inc.; **5.3e** Digital Vision/Getty Images; **5.7 (top)** PODAAC; **5.7 (bottom)** PODAAC; **5.9a** Digital Vision/Getty Images; **5.9b** Pete Seaward/Stone/Getty Images; **5.10** Courtesy NASA/EPA. PRovided by Dr. Dale Quattrochi, Marshall Space Flight Plan; **p. 94** George Hall/Corbis Images; **p. 97 (top)** Kevin Schafer/The Image Bank/Getty Images; **p. 97 (bottom)** NASA/Science Photo Library/Photo Researchers, Inc.; **p. 105 (top)** North Light Images/Age Fotostock America, Inc.; **p. 109** Patrick J. Bartlein/Department of Geography, University of Oregon; **p. 110** Science Photo Library/Photo Researchers, Inc.

CHAPTER 6

CO6 Andrew McConnell/Robert Harding World Imagery/Getty Images, Inc.; **6.1** MODIS/NASA; **6.8** Science Photo Library/Photo Researchers, Inc.; **6.17** Image Courtesy GOES Project Science Office; **6.18a** Image courtesy MODIS Land Group/Vegetation Indices, Alfredo Huete, Principal Investigator, and Kamel Didan, University of Arizona; **6.18b** Martin Zwick/WWI/Peter Arnold, Inc.; **6.20b** Science Photo Library/NASA/Photo Researchers, Inc.; **6.25** Emilio Suetone/Photolibrary; **6.27** Alamy; **6.32** NASA, Courtesy of Otis B. Brown, Robert Evans, and M. Carl, University of Miami, Rosentiel School of Marine and Atmospheric Science; **p. 116** Age Fotostock America, Inc.; **p. 119** Provided by the SeaWiFS Project, NASA/Goddard Space Flight Center, and ORBIMAGE; **p. 121** NOAA/NESDIS/NCDC; **p. 129** GOES; **p. 143** Digital Vision/Getty Images.

CHAPTER 7

CO7 Gordon Wiltsie/National Geographic/Getty Images, Inc.; **7.2a** Corbis Images; **7.2b** ImageState/Age Fotostock America, Inc.; **7.2c** Digital Vision/Getty Images; **7.8** Joseph M. Moran, American Meteorological Society; **7.9** Dr. Timothy Miller Deputy Chief, Earth and Planetary Science Brach NASA Marshall Space Flight; **7.9** Dr. Timothy Miller Deputy Chief, Earth and Planetary Science Brach NASA Marshall Space Flight; **7.11** Corbis Images; **7.17a** NOAA; **7.17b** NOAA; **7.18c** S. J. Krasemann/Peter Arnold, Inc.; **7.18d** Jim Reed/SPL//Photo Researchers, Inc.; **7.18e** MedioImages/Getty Images, Inc.; **7.18f** John Meade/Science Photo Library/Photo Researchers, Inc.; **7.18g** John Howard/SPL/Photo Researchers; **7.18h** Punchstock; **7.18i** Jim Wark/Peter Arnold, Inc.; **7.18j** John Sanford/Science Photo Library/Photo Researchers, Inc.; **7.18k** David Parker/Science Photo Library/Photo Researchers, Inc.; **7.18l** Sheila Terry/SPL/Photo Researchers, Inc.; **7.19a** Alan Arbogast; **7.19b** Paul Harris/Stone/Getty Images; **7.23a** Alan Arbogast; **7.23b** Alan Arbogast; **7.24c** Jeff Schmaltz,MODIS Rapid Response Team, NASA/GSFC; **7.25b** Alan Arbogast; **p. 152** Ric Ergenbright/Corbis Images; **p. 163 (bottom)** Anthony Jay West/Corbis Images; **p. 165 (top)** Art Wolfe/The Image Bank/Getty Images; **p. 165 (bottom)** Pekka Parviainen/Science Photo Library/Photo Researchers, Inc.; **p. 170** NOAA; **p. 176** Lyle Leduc/Getty Images, Inc.

CHAPTER 8

CO8 NOAA/Science Source/Photo Researchers; **8.1** GOES/NOAA; **8.4** NOAA; **8.12d** Jonathan Smith; Cordaiy Photo Library Ltd./Corbis Images; **8.13e** Lyle Leduc/Getty Images, Inc.; **8.15b** Charles O'Rear/Corbis Images; **8.16d** Courtesy NOAA; **8.18** Topeka Capital-Journal Newspaper; **8.19a** Photo by Greg Henshall/FEMA; **8.19b&c** Kansas Division of Emergency Management; **8.26a** Courtesy NOAA; **8.26b** ©AP/Wide World Photos; **8.26c** ©AP/Wide World Photos; **p. 195 (top)** ©AP/Wide World Photos; **p. 198 (top)** Clint Farlinger/Visuals Unlimited; **p. 198 (bottom)** Gene & Karen Rhoden/Visuals Unlimited; **p. 205** Data from NOAA GOES satellite. Image produced by Hal Pierce, Laboratory for Atmospheres, NASA Goddard Space Flight Center; **p. 207** NASA Goddard Space Flight Center Image by Reto Stockli.

CHAPTER 9

CO9 Charles Kogod/National Geographic/Getty Images, Inc.; **9.3** Wolfgang Kaehler/Corbis Images; **9.4** Bruce Burkhardt/Corbis Images; **9.5a** Yann Arthus-Bertrand/Corbis Images; **9.5b** Tony Craddock/Stone/Getty Images, Inc.; **9.6a** David Muench/Corbis Images; **9.6b** Bob Krist/©Corbis; **9.7** ©Paul Rezendes; **9.8** Animals Animals/Earth Scenes; **9.9** Terry W. Eggers/Corbis Images; **9.10** Adam Woolfitt/Corbis Images; **9.11** Jim Wark/Index Stock Imagery/Photolibrary; **9.12** Raymond Gehman/National Geographic/Getty Images;

9.13a George F. Mobley/National Geographic/Getty Images; **9.13b** Doug Allan/Tartan Dragon Ltd./Getty Images, Inc.; **9.14** David Scharf/Science Photo Library/Photo Researchers, Inc.; **9.15** Eric Grimm, Illinois State Museum; **9.16a** Andrew Brown; Ecoscene/Corbis; **9.16b** Walter Hodges/Corbis; **9.16c** David Muench/Corbis Images; **9.18a** Nick Cobbing/Alamy; **9.18b** Marc Steinmetz/Aurora Photos; **9.18c** Anthony Gow/U.S. Army Corps of Engineers, Cold Regions Research and Engineering Laboratory; **9.25a** ©Jean-Pierre Lescourre/Age Fotostock America, Inc.; **9.25b** Paulo Fridman/Latin Content/Getty Images, Inc.; **9.30** Intergovernmental Panel on Climate Change; **p. 218** Ria Novosti/Alamy; **p. 221 (bottom)** Jeff Schmaltz, MODIS Rapid Response Team, NASA/GSFC; **p. 225** PRISM, Oregon State University/NRCS.

CHAPTER 10

CO10 Philippe Bourseiller/Getty Images, Inc.; **10.1** Michio Hoshino/Minden Pictures, Inc.; **10.2a** ©B. Borrell Casals; Frank Lane Picture Agency/©Corbis; **10.2b** Joel Sartore/National Geographic Creative/Getty Images, Inc.; **10.2c** David M. Phillips/Photo Researchers; **10.2d** Novastock/Stock Connection/Aurora Photos Inc.; **10.2e** Duncan McEwan/Minden Pictures, Inc.; **10.2f** SOHO; **10.2** iStockphoto; **10.5** GSFC/NASA; **10.7b** Michael & Patricia Fogden/Corbis Images; **10.8** Robert Holmes/Corbis; **10.9** Alan Arbogast; **10.10** Geostock/PhotoDisc Green/Getty Images; **10.11** Rich Reid/Animals Animals/Earth Scenes; **10.12a** Alan Arbogast; **10.12b** Ken Biggs/Stone/Getty Images; **10.13** Chlaus Lotscher/Peter Arnold; **10.14** Alan Arbogast; **10.15b** Alan Arbogast; **10.15a** Gunter Ziesler/Peter Arnold, Inc.; **10.16** Malie Rich-Griffith/INFOCUS Photos/Alamy; **10.17** D. Robert & Lorri Franz/Corbis Images; **10.18a** Ron Sanford/Corbis; **10.18b** Alan Arbogast; **10.20a** Alan Arbogast; **10.20b** Alan Arbogast; **10.22** Scott Warren/Aurora Photos Inc.; **10.23** David Cavagnaro/Visuals Unlimited; **10.24** Alan Arbogast; **10.25** Alan Arbogast; **10.27** Stephanie Maze/Corbis; **10.28** Jacques Descloitres, MODIS Land Rapid Response Team at NASA/GSFC; **10.29** Alan Arbogast; **10.3** Gary Braasch/Corbis; **10.31** Ron Spomer/Visuals Unlimited; **p. 261** Paul A. Souders/Corbis; **p. 262 (left)** National Office of Fire and Aviation Bureau of Land Management, National Interagency Fire Center; **p. 262 (right)** Stephen J. Krasemann/Photo Rsearchers; **p. 265 (bottom)** National Geographic Society/Getty Images; **p. 274** USGS; **p. 275** Marbutt Collection/Soil Science Society of America.

CHAPTER 11

CO11 Jim Richardson/NG Image Collection; **11.1** Alan Arbogast; **11.7b** Prof. Randall Schaetzl; **11.8b** Alan Arbogast; **11.9** Alan Arbogast; **11.11a** Corbis Digital Stock; **11.11b** Alan Arbogast; **11.12a** Alan Arbogast; **11.12c** Alan Arbogast; **11.13a** NRCS/USDA; **11.13b** Visuals Unlimited;

11.19a Alan Arbogast; 11.19b Alan Arbogast; 11.23b Soils of the Great Plains Slide Set, University of Nebraska; 11.23c NRCS/USDA; 11.24b Alan Arbogast; 11.24c NRCS/USDA; 11.25b Marbutt Collection/Soil Science Society of America; 11.25d H. Curtis Monger, New Mexico; 11.25c Marbutt Collection/Soil Society of America; 11.26b Marbutt Collection/Soil Society of America; 11.26c Prof. Randall Schaetzl; 11.27c Prof. Randall Schaetzl; 11.27b Alan Arbogast; 11.28b Alan Arbogast; 11.28c Prof. Randall Schaetzl; 11.29b Marbut Collection/Soil Science of America; 11.29c University of Idaho, Courtesy of Paul McDaniel; 11.30b Alan Arbogast; 11.30c Mrbutt Collection/Soil Science Society of America; 11.31b Alan Arbogast; 11.31c H. Curtis Monger, New Mexico; 11.31d H. Curtis Monger, New Mexico; 11.33b Alan Arbogast; 11.33c NRCS; 11.34b Alan Arbogast; 11.34c Prof. Randall Schaetzl; 11.35b Marbutt Collection/Soil Science Society Collection; 11.35c NRCS/USDA; 11.36 iStockphoto; 11.37 USDA NRCS; 11.38a USDA; 11.38b USDA NRCS; p. 285 Brand X Pictures/Punchstock; p. 290 (top) Yann Layma/The Image Bank/Getty Images, Inc.; p. 310 Alan Arbogast; p. 318 Phyllis Picardi/Index Stock Imagery/Photolibrary.

CHAPTER 12

CO12 Tim Fitzharris/Minden Pictures, Inc.; 12.5 Roberto de Gugliemo/Photo Researchers, Inc.; 12.6b Photo by R.W. Decker, courtesy USGS Hawaiian Volcano Observatory; 12.7a Courtesy of Chip Fletcher, Dept. of Geology and Geophysics, University of Hawaii; 12.7b Courtesy of Chip Fletcher, Dept. of Geology and Geophysics, University of Hawaii; 12.7c Courtesy of Chip Fletcher, Dept. of Geology and Geophysics, University of Hawaii; 12.7d Courtesy of Chip Fletcher, Dept. of Geology and Geophysics, University of Hawaii; 12.7e Courtesy of Chip Fletcher, Dept. of Geology and Geophysics, University of Hawaii; 12.7f Courtesy of Chip Fletcher, Dept. of Geology and Geophysics, University of Hawaii; 12.8 Dough Martin/ Photo Researchers, Inc.; 12.10 Andrew J. Martinez/Photo Researchers, Inc.; 12.11 SuperStock/Alamy Images; 12.12a Michael Gadomski/Animals Animals/Earth Scenes; 12.12b Alan Arbogast; 12.13 Jacques Descloitres, MODIS Land Rapid Response Team, NASA/GSFC; 12.14 Courtesy of Alan Arbogast; 12.15b Charlie Knight/Alamy; 12.17a Joel Arem/Photo Researchers; 12.17b BRECK P.KENT/ Animals Animals; 12.20a PIXTAL/AGE; 12.22a Jacques Descloitres, MODIS Land Rapid Response Team, NASA/ GSFC; 12.22b Jim Wark/AirPhoto; 12.25 Time & Life Pictures/Getty Images, Inc.; 12.26 ©AP/Wide World Photos; 12.27 Bill Piere/Time & Life Pictures/Getty Images, Inc.; 12.30 TORU YAMANAKA/AFP/Getty Images, Inc.; p. 326 Digital Vision/Getty Images; p. 328 (top) Jim Wark/Peter Arnold, Inc.; p. 328 (bottom) Dynamic Graphic Group/

Creatas/Alamy Images; p. 338 (top) Jean-Yves Grospas/ Peter Arnold, Inc.; p. 349 USGS.

CHAPTER 13

CO13 Colin Monteath/Hedgehog House/Minden Pictures, Inc.; 13.4 Based on data from: Muller, R.D., Sdrolias, M., Gaina, C. and Rosets, W.R., 2008, Age, spreading rates and spreading asymmetry of the world's ocean crust, Geochemistry, Geophysics, Geosystems, 9 (4), doi:10.1029/2007/GC001743, Courtesy of Dietmar Muller; 13.7 Jacques Descloitres, MODIS Land Rapid Response Team, NASA/GSFC; 13.9b NASA; 13.9c Matt Fletcher/ Lonely Planet Images; 13.11b Photo by Dick Moore, 1992, USGS; 13.12b Martin Bond/Photo Researchers, Inc.; 13.12c Courtesy Prof. Allen F. Glazner, Dept. of Geological Sciences, University of North Carolina; 13.13a GSFC/NASA; 13.13b Dr. Marli Miller/Visuals Unlimited; 13.16 Getty Images, Inc.; 13.22 Alan Arbogast; 13.23 Alan Arbogast; 13.25 R.E. Wallace/USGS; 13.30a AP/Wide World Photos; 13.30b DigitalGlobe; 13.30c DigitalGlobe; 13.31a Steve Kaufman/Corbis Images; 13.31b G.E. Ulrich, Hawaiian Volcano Observatory, U.S. Geological Survey; 13.32 Courtesy National Park Service; 13.33b Akira Kaede/ Photodisc Green/Paste/Getty Images; 13.36a Jim Nieland, U.S. Forest Service, Mount St. Helens National Volcanic Monument; 13.36b Austin Post/USGS; 13.36c Digital Vision/Getty Images; 13.38b J.D. Griggs/USGS; 13.39 World Ocean Floor Panorama, Bruce C. Heezen and Marie Tharp, 1977 © by Marie Tharp 1977/2003; 13.41 Corbis Images; p. 368 Scott Drzyzga; p. 374 US Geological Survey/Photo Researchers, Inc.; p. 384 Lyn Topinka/USGS; p. 389 Lyn Topinka/USGS; p. 390 Dave Bunnell.

CHAPTER 14

CO14 Panoramic Stock Images/NG Image Collection; 14.2 Jules Cowan/Index Stock/Photolibrary; 14.3 Inga Spence/ Visuals Unlimited; 14.4 Susan Rayfield/Photo Researchers; 14.5 Dr. Marli Miller/Visuals Unlimited; 14.6 Tom Stack & Associates; 14.7 Dr. Marli Miller/Visuals Unlimited; 14.9a Gregory G. Dimijian/Photo Researchers; 14.9b Mark Newman/Visuals Unlimited; 14.10b Dr. Marlli Miller/Visuals Unlimited; 14.12 Michael Szoenyi/Photo Researchers; 14.13 Tom & Therisa Stack/Tom Stack & Associates, Inc.; 14.17a Dave Bunnell; 14.17b Marli Miller/Visuals Unlimited; 14.18 Ralph Lee Hopkins/Photo Researchers; 14.19 Courtesy of Gabriel Guttierez-Alonso; 14.2 USGS; 14.21b Alan Arbogast; 14.22 Bettman/Corbis; 14.23 USGS; 14.24a Dominique Favre/Reuters/Landov; 14.24b Photo Courtesy of Betsy Armstrong from the URL: http://nsidc.org/snow/ avalance/#ANATOMY; p. 396 (left) Yann Arthus-Bertrand/ Corbis; p. 396 (right) Alamy; p. 402 Alan Arbogast; p. 409 (bottom) Albert Copley/Visuals Unlimited; p. 410 China Tourism Press, The Image Bank/Getty.

A

Abiotic factors, in ecosystems, 250
Abrasion
    definition of, 510
    erosion by, 479–482, 510–512
Absolute location, definition of, 16
Absorbed radiation
    and global radiation budget, 81–85
    transfer of, 78–79
Absorption
    definition of, 76
    of solar radiation, 76–77
Acid rain, definition of, 401
Actual evapotranspiration (actual ET)
    definition of, 213
    in Thornthwaite system, 213–214
Additions, soil, characteristics of, 283–284
Adiabatic processes. *See also* Fronts
    characteristics of, 160–162
    definition of, 160
    GeoDiscoveries on, 162
    and precipitation, 169–175
Adret slope
    definition of, 266
    vegetation and, 266–267
Advanced Very High Resolution radiometer (AVHRR)
    characteristics of, 34–35
    and vegetation estimates, 254
Advection, definition of, 117
Advection fog, definition of, 164
Aerial photographs
    characteristics of, 33–34
    definition of, 33
Afforestation, process of, 270
Africa
    arid regions of, 217, 220, 521–523
    biome distribution, 257, 263
    desertification of the Sahel, 521–523
    forest cover changes in, 270–271
    GeoDiscoveries on, 305
    sandstorms in, 112–113
    soils of, 299–300, 304–305, 310–313
Age, geologic. *See also* Geologic time
    dating, 340–342
    timeline of, 338–340
Aggradation
    and outwash plains, 486
    in stream gradation, 446–449
Agriculture. *See also* Soil salinization
    arable land distribution, 572–573
    deforestation and, 270–273
    desertification and, 521–525
    farmland loss, 10–11, 565
    global loess deposits, 518–519
    greenhouse emissions and, 244
    groundwater depletion and, 419–422
    hail damage, 195
    impact on biomes, 244, 272–273

moisture measurements and, 158, 213–214
    population growth and, 562–563, 572–573
    soil science in, 298, 316–317
    wildlife habitat and, 584
A horizon, definition of, 296
Air. *See also* Winds
    fluid behavior of, 509–510
    in soils, 283
Airflow
    factors in, 119–125
    surface and upper-level interactions, 186–187
Air masses
    characteristics of, 180–181
    climate effects of, 212–213. *See also specific climate*
    definition of, 180
    polar front interactions, 184–187
Air pressure. *See also* Atmospheric pressure
    definition of, 114
Air temperature. *See also* Surface temperature
    atmospheric layers and, 92–95
    heat and wind chill indexes, 98–99
    humidity and, 153–156
    large-scale factors in, 99–101
    local factors in, 101–106
    and surface temperature, 95
Alaska
    biomes of, 256–257, 261
    forest ecosystem example, 250
    glaciers of, 473, 474–477
    Mount McKinley, 116
    oil reserves of, 346–348
    permafrost regions of, 496–497
    soils of, 299–300, 307–308, 496–497
    volcanoes of, 381–383, 386
Albedo
    definition of, 76
    in polar climates, 228
    solar radiation and, 76–77, 79–80
Albers equal-area projection, characteristics of, 26
Albers, H.C., and Albers projection, 26
Albic horizon, definition of, 305
Alfisols
    characteristics and distribution, 299–300, 304–305
    definition of, 304
Alluvial fans
    definition of, 508
    in deserts, 508–509
Alluvial terraces
    definition of, 455
    evolution of, 445–447
Alluvium
    definition of, 456
    in desert landscapes, 506–509
    in graded stream processes, 456–458
Alpine glaciers
    characteristics of, 474–476
    definition of, 474
    GeoDiscoveries on, 476

Alpine tundra, characteristics of, 265
Alps (European)
	avalanche hazards, 409
	formation of, 366–367
	glacial retreat, 493
	glaciers of, 474–475
	The Matterhorn, 482
	vertical zonation in, 267
Alps (New Zealand)
	avalanche hazards, 409
	glaciers of, 474–475
	tectonic origins of, 352–353
Altitude
	and air pressure, 114–115
	effects on temperatures, 106
Aluminum, as rock component, 326
American Clean Energy and Security Act of 2009, and
		greenhouse emissions, 244
American Samoa, tsunami (2009), 378–379
American Southwest. See Southwest United States
Andes Mountains
	annual temperature range, 106–109
	climate of, 223–224
	erosion and, 396
	glaciers of, 474–475
Andisols
	characteristics and distribution, 299–300, 315
	definition of, 315
Angle of incidence
	definition of, 79
	GeoDiscoveries on, 80
	global radiation budget and, 82–83
	solar radiation and, 79–80
Antarctica
	biome distribution, 257
	climates of, 227–228
	glaciers of, 478, 493–496
	ozone layer/hole, 71–73
	projections of, 25–26
Antarctic Ice Sheet
	characteristics of, 478
	retreat of, 493–496
Antarctic zones
	location of, 19
	seasons in, 53
Anthropogenic influences. See also Human/environmental
		interactions
	and climate change, 239–245
	definition of, 239
	future predictions, 242–245
Anticlinal valleys
	definition of, 365
	geomorphology and, 365–366
Anticlines
	definition of, 364
	geomorphology and, 363–368
Anticyclones
	characteristics of, 116–118
	definition of, 116
Aphelion, definition of, 48

Appalachian Mountains (United States)
	climate of, 223–224
	as drainage divide, 434–435
	evolution of, 364–368
Apparent motion, of Sun, 56
Aquicludes, definition of, 415
Aquifers. See also Water table
	Central Valley Aquifer, 577–580
	definition of, 417
	High Plains Aquifer, 417–422
Arabian Desert (Saudi Arabia), and Subtropical High (STH)
		Pressure System, 128
Arable land
	definition of, 572
	limitations of, 572–573
Arches National Park (Utah), rock outcrops of, 392–393
Arctic Ocean, geography of, 530
Arctic outbreak, characteristics of, 131–132
Arctic tundra, characteristics of, 264–265
Arctic zone
	location of, 19
	ozone layer/hole, 72–73
	seasons in, 53
	soils of, 299–300, 307–308
Arêtes, definition of, 482
Argentina, glaciers of, 493
Argillic horizon, definition of, 298
Aridisols
	characteristics and distribution, 299–300, 310–313
	definition of, 310
Arid landscapes
	characteristics and landforms, 504–509
	desertification in, 521–525
	eolian deposition and effects, 513–521
	eolian erosion and effects, 509–513
Arid regions
	soil salinization in, 572–583
	water issues in, 566–572
Arid and semi-arid (B) climates
	characteristics of, 215, 217, 219–222, 504
	soils of, 299–300, 308–313, 576–583
	water issues in, 566–572
Arizona
	Colorado River Compact and, 568–572
	Grand Canyon geography, 320–321, 342–343, 505, 507
	Phoenix sprawl, 568
Arroyos, definition of, 508
Artesian wells
	characteristics of, 419–420
	definition of, 419
Artificial levees
	definition of, 462
	purpose and issues of, 462–463
Ash .See Volcanic ash
Asia. See also Eurasia; Southeast Asia
	annual temperature range, 106–109
	arid regions of, 217, 220
	biome distribution, 257, 259, 263–265
	climates of, 217
	forest cover changes in, 270–271
	glaciers of, 473, 474–475

monsoons and, 133–134

soils of, 299–300, 308–313

Aspect, vegetation and, 266–267

Asteroid impacts, and climate change, 236

Asthenosphere. *See also* Plate tectonics

characteristics of, 322–324

continental drift and, 354–357

definition of, 324

isostatic adjustment and, 324–326

Atlanta (Georgia), as urban heat island, 105–106

Atlantic Ocean

circulation of, 139–143

geography of, 530

hurricane development in, 203–205

rifting in, 360–361

Atlas Mountains, erosion and, 396

Atmosphere. *See also* Atmospheric circulation; Atmospheric pressure

in coastline interactions, 531

data collection in, 8–9

definition of, 8

gas components of, 66–73

greenhouse effect and, 68–70

layered structure of, 92–95

ozone layer and hole, 70–73, 93

particulates in, 73–74

solar radiation flow in, 76–78

Atmospheric circulation. *See also* Oceanic circulation

climate effects of, 212–213

GeoDiscoveries on, 133, 134–135

global model, 125–126, 127

local systems of, 137–139

midlatitude, 129–132

polar, 132

seasonal migration patterns, 132–135

tropical, 126–129

wind energy production, 135–137

Atmospheric moisture. *See also* Precipitation

GeoDiscoveries on, 158

humidity, 152–156

and hydrologic cycle, 151–152

Atmospheric pressure. *See also* Atmospheric circulation; Winds

adiabatic processes and, 160–162

factors and measurement of, 114–116

the 500-mb pressure surface, 184–187

pressure systems, 116–118

Atolls, evolution of, 552–553

Auroras, characteristics of, 97

Australia

agricultural land use, 580–583

annual temperature range, 106–109

arid regions of, 217, 220, 505

biome distribution, 257, 263

climates of, 217, 580

geography of, 580

Great Barrier Reef, 552, 580

soil salinization in, 576, 580–583

soils of, 299–300, 304–305, 310–313

Autotrophs

definition of, 250

role in ecosystems, 250–251

Autumnal equinox. *See* Fall Equinox

Avalanches

definition of, 408

process and effects, 408–409

Average global temperatures, and climate change, 241–245

Axial tilt/orientation

characteristics of, 50–51

GeoDiscoveries on, 55

and insolation, 84–85

and the seasons, 51–55

Axis

definition of, 48

tilt of. *See* Axial tilt/orientation

B

Backshore, definition of, 545

Backslopes, definition of, 514

Backswamps

definition of, 453

in stream valleys, 453–454

Bahama Islands, rock of, 334, 338

Baltimore (Maryland), infrared images of, 12–13, 36–37

Banda Aceh (Indonesia), Sumatran tsunami, 377–378

Bankfull discharge, definition of, 440

Barchan dunes, characteristics of, 516–517

Barrier islands

definition of, 547

evolution of, 546–550

Barrier reefs, characteristics of, 552

Basal slip, and glacial movement, 473

Basalt

characteristics of, 329, 330

volcanic origin, 385

Base flows

definition of, 439

and flooding, 439–442

Base levels

definition of, 447

and stream gradation, 447–448

Basin and Range Province (United States)

formation and landforms of, 505–507

Las Vegas Valley, 566–567

Basins

in arid landscapes, 505–506

drainage. *See* Drainage basins

Bay of Fundy (Canada), tidal range, 534–535

Baymouth bars, definition of, 546

Bays, definition of, 531

Beach drift

definition of, 538

and deposition, 544–547

Beaches

formation and components of, 544–545

nourishment, 554

pocket, 540

Bedrock

folding of, 362–364

in soil formation, 285–286, 296–298

Berm scarps, definition of, 545

B horizon
    definition of, 296
    eluviation and, 296–298
Big Bang theory, definition of, 46
Biodiversity. *See also* Biomes
    deforestation and, 271–272
    wildfires and, 262
Biogeography, scope of, 250–251
Biomass
    definition of, 253
    global distribution of, 253–254
    and soil, 281
Biomes
    agricultural effects, 244, 272–273
    definition of, 255
    desert biomes, 264–265, 504
    forest biomes, 256–261, 267–268, 270–273, 401–402
    geographic distribution map, 257
    grassland biomes, 256–257, 261–264, 267–268, 299–300, 308–310
    major global, 256
    plant succession in, 268–269
    tundra, 264–265
Biosphere
    data collection in, 8–9
    definition of, 8
    GeoDiscoveries on, 275
Bioturbation, in soil formation, 287–288
Blowout dunes, characteristics of, 516–517, 547
Bluffs, evolution of, 540–541
Body waves, characteristics of, 370
Bolivia, slash and burn agriculture in, 271–272
Boreal forest biome, characteristics of, 256–257, 261
Borneo, forest cover changes in, 270–271
Boulder Dam. *See* Hoover Dam (United States)
Brackish water
    definition of, 531
    in lagoons, 546, 548–550
Braided streams
    definition of, 445
    GeoDiscoveries on, 445
Brazil
    Itaipu Dam, 465
    slash and burn agriculture in, 271–272
Brazos River (Texas), delta form of, 550–551
Breakers, definition of, 537
Breezes, characteristics of, 137–139
Bretz, J. Harlan, on the Channeled Scablands, 490–491
Brine, definition of, 531
Bristlecone pines, growth habits of, 231–232, 267
Bureau of Reclamation, in San Joaquin Valley, 577–579
Business, use of GIS in, 41
Buttes, definition of, 507

C

Calcic horizon, definition of, 309
Calcification
    definition of, 308
    GeoDiscoveries on, 310
    in soil formation, 308–313
Calcium, as rock component, 326

Calcium carbonate
    in coral reefs, 551
    in sedimentary rock formation, 333–335, 338
Calderas, Yellowstone National Park, 387–388
Caliche, characteristics of, 309
California
    bristlecone pines of, 231–232, 267
    coastal processes and, 540–543
    Death Valley sand dunes, 502–503
    earthquakes of, 374–375
    fault system of, 373–374
    GeoDiscoveries on, 544
    groundwater and subsidence, 423, 571–572, 577–578
    Imperial Valley satellite image, 38
    Los Angeles topography, 14–15
    mass wasting in, 407–408
    orographic patterns in, 171–172
    Santa Ana winds, 139
    soil salinization in, 576–580
Cambrian Period, characteristics of, 339–340
Canada
    annual temperature range, 106–109
    biomes of, 256–257, 261
    climates of, 217
    glaciers of, 474–475, 482
    permafrost regions of, 496–497
    soils of, 299–300, 304–305
    Vancouver (British Columbia), 105
Canyons
    of continental shelf, 532–533
    definition of, 507
    and landscape evolution, 507–508
Capillary action
    definition of, 148, 281
    in soils, 281–282, 414–415, 573–575
    in vegetation, 148
Caprock, definition of, 394
Cap and trade system, and greenhouse emissions, 244
Carbon–14 dating, process of, 340–341
Carbonation
    definition of, 400
    and karst topography, 425–426
    process and effects, 400
Carbon cycle, and greenhouse effect, 239–241
Carbon dioxide. *See also* Carbonation
    from permafrost thawing, 496
    and greenhouse effect, 68–70, 239–245, 565
    in photosynthesis, 251–254
    properties of, 68–70
    in soils, 283
    in weathering, 400
Caribbean Ocean
    circulation patterns of, 139–143
    hurricane activity in, 204
Carlsbad Caverns, size of, 425
Carotenoids, in photosynthesis, 253
Carrying capacity
    definition of, 562
    technology and, 562–563
Cartography. *See also* Maps
    definition of, 22

map projections, 22–27
Cascades Mountains (United States)
climate of, 223–224
GeoDiscoveries on, 476
glaciers of, 476–477
Cascade Volcanic Arc, characteristics of, 381–383
Case studies
giant panda conservation (China), 583–589
Las Vegas water issues, 566–572
soil salinization in arid lands, 572–583
Cation exchange capacity (CEC)
definition of, 294
and soil fertility, 294–295
Cations, definition of, 294
Cause-and-effect relationships, in physical geography, 10–11
Caverns, evolution of, 425–426
Caves
definition of, 425
evolution of, 425–426
Celestial dome
definition of, 56
GeoDiscoveries on, 58
and Sun arcs, 56–57
Celsius, Anders, temperature scale of, 95
Celsius temperature scale, characteristics of, 95–96, 98
Cenozoic Era, characteristics of, 339–340
Center-pivot irrigation, groundwater depletion and, 419–422, 525
Central America, soils of, 299–300, 304–305
Central Valley Project (CVP), actions of, 577–579
CFCs (chlorofluorocarbons), and ozone hole, 71–73
Chandeleur Islands, post-hurricane changes, 548
Channeled Scablands, origin of, 490–491
Chemical weathering
acid rain and, 401–402
climate and, 394
definition of, 398
processes and effects, 398–400
Chesapeake Bay (United States), characteristics of, 532–533
China
forest cover changes in, 270–271
giant panda conservation in, 583–589
karst topography of, 427
population growth in, 563–564, 584, 585
soils of, 299–300, 304–305
subsidence in, 423
terrace farming in, 290
Three Gorges Dam, 465
Chinook/foehn winds
characteristics of, 138–139
definition of, 138
Chlorophylls, in photosynthesis, 253
C horizon, definition of, 296
Cinder-cone volcanoes
characteristics of, 379–380
definition of, 379
Circle of illumination, definition of, 48
Circles
circle of illumination, 48
geographic grid and, 16–21

Cirques
definition of, 475
glaciers of, 475–477, 481–482
Cirrus clouds
characteristics of, 164, 166–167
definition of, 164
Clark County (Nevada), water issues of, 566–572
Clastic rock
characteristics and examples, 331–333
GeoDiscoveries on, 334
Clay
in eolian transport, 509–511
in soil textural classes, 291–292, 293
Climate. *See also* Köppen Climate Classification System
classification systems, 213–216
climate feedbacks, 243
definition of, 212
factors affecting, 212–213
GeoDiscoveries on, 219, 224, 225, 228–229, 305
glacier formation and, 470, 476–477, 478
Köppen climates, 214–229
microclimate and, 266
past climates. *See* Paleoclimatology
in soil formation, 286–287, 311–313
vegetation and, 255. *See also* Biomes
weathering and, 394
Climate change
anthropogenic factors, 10–11, 239–245
arid regions and, 571–572
coastline impacts, 556–557
GeoDiscoveries on, 245, 494
and glacial retreat, 493–496
Milankovitch theory and, 236–238
paleoclimatology and, 229–236
prediction of, 242–245
Climax vegetation
definition of, 268
in succession, 268–269
Climographs
definition of, 218
for Las Vegas (Nevada), 567
Clouds. *See also* Precipitation
formation and classification, 163–167, 198
in storms and fronts, 173–174, 182, 198
Coal. *See also* Fossil fuels
definition of, 335
Coastal dunes
formation and types, 546–547
succession in, 269
Coastal engineering, purposes and methods, 553–556
Coastal processes and landforms
climate change impacts, 556–557
depositional landforms, 544–553
erosional landforms, 539–544
GeoDiscoveries on, 537, 544, 549, 557
littoral processes in, 537–539
protection of, 553–556
tides and effects, 533–535
and water level changes, 531–533
wave formation and actions, 536–537

Cold fronts
  characteristics of, 182–183
  definition of, 182
Cold midlatitude desert climate (*BWk*)
  biomes of, 264–265
  characteristics of, 221
Cold midlatitude steppe climate (*BSk*)
  biomes of, 263
  characteristics of, 217, 220–221
Colloids
  definition of, 294
  GeoDiscoveries on, 295
  in soil fertility, 294–295
Colluvium, definition of, 403
Color, in soils, 289–291. *See also* Soil orders
Colorado, and Colorado River Compact, 568–572
Colorado River, recent drought and, 569–570
Colorado River Compact, contents and issues of, 568–572
Columns, formation of, 426
Composite volcanoes
  characteristics of, 380–381, 477
  definition of, 380
Condensation. *See also* Clouds
  adiabatic processes, 160–162, 169–175
  definition of, 150
Condensation nuclei
  in cloud formation, 162–163
  definition of, 163
Conduction
  definition of, 74
  principles of, 74–75
Cone of depression
  definition of, 420
  and water extraction, 420–422
Confluences, definition of, 434
Conformal projections
  characteristics of, 24–26
  definition of, 24
Conglomerate, definition of, 332
Coniferous forests, vertical zonation and, 267–268
Conservation
  of panda habitat, 583–589
  of soil, 290, 525
Constant gases
  definition of, 67
  role in atmosphere, 67
Continental Arctic (cA) air mass, characteristics of, 180–181
Continental crust
  characteristics of, 322, 324–326
  definition of, 324
  in subduction, 360–362
Continental divide, geography of, 434–435
Continental drift
  definition of, 355
  evidence and patterns, 354–358
  GeoDiscoveries on, 357
  and hotspots, 386–388
Continental glaciers
  characteristics of, 478
  definition of, 478

Continental locations
  definition of, 101
  temperatures and, 101–103
Continental Polar (cP) air mass, characteristics of, 180–181
Continental shelf
  barrier islands and, 547–548
  and eustatic change, 532–533
Continental Tropical (cT) air mass, characteristics of, 180–181
Continents. *See also* Continental drift
  continent-to-continent collisions, 362–364
  formation of, 354–355
  glaciers of, 478
Contours, definition of, 31
Convection. *See also* Thunderstorms
  airflow and, 119
  convectional uplift, 169, 173–175
  definition of, 75
  and plate movement, 355–357
  principles of, 74–75
Convectional uplift
  definition of, 169
  GeoDiscoveries on, 175
  and precipitation patterns, 173–175
Convection loops, and plate movement, 355–357
Convergent uplift
  definition of, 169
  and precipitation, 169–170
Copenhagen Climate Conference, and greenhouse emissions, 244
Coral reefs
  coral bleaching of, 557
  definition of, 551
  evolution and types, 551–553
Cordilleran Ice Sheet
  definition of, 489
  history and movement of, 488–492
Core (Earth), characteristics of, 322–323
Coriolis force
  airflow and, 122–125
  definition of, 122
  GeoDiscoveries on, 125
  and oceanic circulation, 139–140
Counterradiation
  definition of, 68
  and greenhouse effect, 68–70
Crater Lake (Oregon), volcanic history of, 384–385
Creep, in eolian transport, 510–511
Crests
  definition of, 514
  wave, 536–537
Cretaceous Period
  characteristics of, 339–340
  mountain formation in, 366–367
Crevasses
  definition of, 472
  formation of, 472–473
Cross dating, in dendrochronology, 232–233
Crust (Earth). *See also* Continental crust; Oceanic crust;
      Plate tectonics
  characteristics of, 322–324
  earthquakes and, 369–374

elemental composition, 326–327
isostatic adjustment and, 324–326
in subduction, 360–362
Cumulus clouds
characteristics of, 164, 166–167
and convectional precipitation, 173–174
definition of, 164
Currents, longshore, 537–539, 545–546, 549
Cyclogenesis
characteristics of, 188–190, 201–202
definition of, 184
Cyclones. *See also* Midlatitude cyclones
characteristics of, 116–118
definition of, 116
Cyclonic weather systems
midlatitude cyclones, 184–191
thunderstorms, 191–195
tornadoes, 195–201
tropical cyclones, 201–207

D

Dams
definition of, 463
purposes and issues of, 463–465
Data collection
in Earth spheres, 8–9
proxy data, 230–236
Day
GeoDiscoveries on, 58
length of and seasonal changes, 56–60, 82–85, 100–101
Dead Sea Fault (Middle East), characteristics of, 358–359
Debris flow
definition of, 408
process and effects, 408
Debris slides, definition of, 406
Deciduous forests
distribution of, 256–257
vertical zonation and, 267–268
Decomposers
definition of, 250
role in ecosystems, 250–251, 252–253
Decomposition, and soil formation, 280–281
Deep time. *See* Geologic time
Deflation, definition of, 510
Deflation hollows, definition of, 510
Deforestation
definition of, 270
effects of, 270–273, 584, 586–588
GeoDiscoveries on, 273
human role in, 10–11, 270–273
panda habitat and, 584, 586–588
satellite imagery of, 36, 274
soil fertility and, 302
suburbanization and, 565
Deformation
in glaciers, 472–474
of rock, 362–364, 369–374
Degradation
definition of, 446
in stream gradation, 446–449
Degrees, latitude measurement, 17–19

Delaware, beach nourishment in, 554
Deltas
definition of, 459
evolution of, 459–460, 550–551
Dendritic drainage pattern, characteristics of, 435–436
Dendrochronology
definition of, 231
in paleoclimatology, 231–233
Density, of air, 114–115
Depletions, soil, characteristics of, 283–284
Depositional processes and landforms
coastal, 544–553
definition of, 151
desert, 508–509
eolian, 513–521
fluvial. *See* Fluvial processes and landforms
GeoDiscoveries on, 445, 487, 549
glacial, 483–488
and sedimentary rock, 330–331
Deranged drainage pattern, characteristics of, 435–436
Desert biomes. *See also* Arid landscapes
characteristics and distribution of, 264–265, 504
Desertification
definition of, 521
process and effects, 521–525
prone global regions, 522
Desert pavement
characteristics of, 510, 512
definition of, 510
Deserts. *See also* Arid landscapes
and Subtropical High (STH) Pressure System, 128
water issues and, 566–572
Detroit (Michigan), suburbanization of, 565
Devil's Tower (Wyoming), exposed igneous intrusion, 328
Devonian Period, characteristics of, 339–340
Dew-point temperature
definition of, 156
and humidity levels, 156–158
Differential weathering, and landforms, 394, 483, 506
Digital technology, techniques of, 33–41
Dinosaurs, life and extinction of, 340
Direct radiation, definition of, 76
Direct tides, definition of, 534
Disappearing streams, definition of, 426
Discharge (stream)
fluctuations in, 438–442, 461–462
in graded stream model, 456–458
Disease, and global warming, 244
Diurnal cycle, definition of, 56
Dolomite, definition of, 335
Dome dunes, characteristics of, 516–517
Donner Pass (California), Donner Party story, 171
Doppler, Christian, and Doppler effect, 201
Doppler radar, principles and uses, 200–201
Downcutting. *See* Stream gradation
Downdrafts, definition of, 193
Downwelling currents, definition of, 141
Drainage basins
characteristics of, 433–435
Colorado River basin issues, 568–572
definition of, 433

Drainage basins *(cont.)*
    drainage patterns and density, 435–438
    flooding and, 440–442
    of United States, 434–435
Drainage density
    calculation of, 435–436
    definition of, 435
Drainage divides
    and basins, 433–438
    definition of, 433
Drainage patterns, characteristics of, 435–436
Drip curtains, formation of, 426
Drought
    and desertification, 523–525
    in Southwest United States, 569–572
    status and levels, 569–571
Drought-adapted plants, characteristics of, 264
Drumlins
    characteristics of, 484–486
    definition of, 485
Dry adiabatic lapse rate (DAR)
    characteristics of, 160–161
    definition of, 160
    in precipitation processes, 170–175
Dunes. *See* Sand dunes
Dust, atmospheric, 73, 523–525
The Dust Bowl, desertification and, 523–525

E

Earth
    age of, 338–342
    inner structure of, 322–326
    orbit around Sun, 47–49, 53, 236–238
    rotation and axial tilt, 48–51
    shape of, 46–47
    and solar radiation, 76–81
Earthflows, definition of, 407
Earthquakes
    definition of, 369
    fault types, 371–374
    mechanisms of, 369–371
    natural hazards of, 374–379
    seismic processes of, 369–371
Earth spheres. *See also specific sphere*
    components of, 8–9
    interactions in, 10–11, 278–279, 531
Earth Summit, and greenhouse emissions, 244
Earth-Sun geometry
    GeoDiscoveries on, 58, 238
    Milankovitch theory and, 236–238
    seasons and, 51–55
    solstices and equinoxes, 52–54, 56–58
    tides and, 533–535
East African Rift, characteristics of, 360–361
Easterly waves
    in cyclogenesis, 201–202
    definition of, 201
Eccentricity. *See* Orbital eccentricity
Ecosystems. *See also* Biomes
    biomes as, 255
    components of, 250–251

    definition of, 250
    manmade damage to, 347–348, 463, 465
Ecotones
    definition of, 255, 521
    the Sahel example, 521–523
E horizon, definition of, 296
Electromagnetic spectrum
    characteristics of, 64–66
    definition of, 64
    GeoDiscoveries on, 66
Elements, of Earth's crust, 326–327
Elevation, vegetation and, 267–268
El Niño
    characteristics of, 141–143
    coral bleaching and, 557
Eluviation
    definition of, 284
    in soil formation, 286–287, 296–298
Emissivity
    definition of, 35
    in satellite imagery, 35–37
Energy. *See* Fossil fuels; Solar radiation; *specific energy form*
Engineering, coastal, 553–557
Enhanced Fujita Scale, classification by, 196
Entisols
    characteristics and distribution, 299–300, 313
    definition of, 313
Entrenched meanders, process and effects, 455–456
Environment
    components of, 8
    legislation on, 401, 423, 584, 588
Environmental lapse rate
    in atmospheric layers, 92–95
    definition of, 93
Environmental management
    GIS in, 41
    ozone hole, 71–73
Eolian processes and landforms
    definition of, 509
    depositional, 513–521
    erosional, 509–513
    GeoDiscoveries on, 521
    human interactions with, 521–525
EPA. *See* United States Environmental Protection Agency (EPA)
Epicenter
    definition of, 369
    locating, 369–371
Epipedons
    definition of, 299
    in soil orders, 298–299
Epiphytes, in rainforests, 257–258
Equator
    definition of, 17
    geographic grid and, 17–21
    insolation at, 84–85
    and the seasons, 53–54, 56–58
Equatorial/tropical zones, location of, 19
Equatorial trough
    definition of, 126
    and tropical circulation, 126–129

Equilibrium line
    definition of, 471
    and glacial mass budget, 471–472
Equinoxes
    characteristics of, 52–54
    day length and, 56–58
Equivalent projections
    characteristics of, 26–27
    definition of, 26
Erosion and erosional landforms
    coastal, 553–556
    eolian, 509–515
    exposed igneous intrusions, 328
    fluvial. *See* Fluvial processes and landforms
    glacial, 479–483
    and mountain formation, 364–368
    rock outcrops and, 392–393
    and slope, 288, 290
Error
    in GPS systems, 40
    map distortion, 24–27
Eskers
    characteristics of, 484, 486–487
    definition of, 486
Eurasia
    acid rain in, 401
    biome distribution, 257, 259–260, 263
    climates of, 217
    forest cover changes in, 270–271
    soils of, 299–300, 304–305, 308–310
Eustatic change
    coastline evolution and, 541–543, 556–557
    definition of, 531
    Pleistocene Epoch example, 531–533
Evaporation. *See also* Evapotranspiration
    definition of, 79, 150
    factors affecting, 159–160
    in hurricane formation, 202
Evaporites
    in arid regions, 508
    definition of, 335
Evapotranspiration
    in climate classification, 213–214
    definition of, 159
    factors affecting, 159–160
    and soil water levels, 281–283, 414–415
Exfoliation
    definition of, 398
    process and effects, 398, 399
Explosive volcanoes, characteristics of, 379–385
Extinction
    in Cretaceous period, 340
    deforestation and, 271–272
Extrusive igneous rock
    definition of, 329
    formation and properties of, 327–330

F

Fahrenheit, Gabriel Daniel, temperature scale of, 95
Fahrenheit temperature scale, characteristics of, 95–96, 98

Fall Equinox
    day length and, 56–58
    definition of, 53
Farmland loss, human role in, 10–11, 565
Fault escarpment, definition of, 373
Faults. *See also* Earthquakes
    and Basin and Range Province, 505–507
    definition of, 369
    GeoDiscoveries on, 379
    types of, 371–374
Feldspars, weathering of, 399–400
Field capacity, definition of, 282, 414
Fire
    GeoDiscoveries on, 275
    wildfire, 262, 273
Firn, definition of, 470
Fjords
    characteristics of, 532
    definition of, 532
Flooding. *See also* Floodplains
    characteristics and examples, 440–442
    flashflooding, 461–462, 508
    mitigation and issues, 462–463
Floodplains
    abandonment of, 455–457
    characteristics of, 453–454
Flood stage, definition of, 440
Florida, beach nourishment in, 554
Flows (mass wasting), processes and effects, 407–408
Floyd (hurricane), satellite image of, 205
Fluid volcanoes, characteristics of, 385–386
Fluvial processes and landforms
    delta forms, 550–551
    erosion and deposition, 442–444
    floodplains, 453–461
    GeoDiscoveries on, 451, 453
    glaciofluvial processes, 486–488
    lower profile processes, 455–461
    stream gradation, 444–449
    stream valleys, 449–455
Fluvial systems
    drainage systems, 433–438
    hydraulic geometry and channel flow, 438–442, 456–458
    landforms. *See* Fluvial processes and landforms
    origins of, 432–433
Focus, of earthquake, 370
Fog, characteristics of, 164–165
Folding (rock)
    GeoDiscoveries on, 368
    process of, 362–364, 372–373
Food chains, characteristics of, 251
Food supply
    arable land available, 572–573
    soil salinization and, 573–580
Foredunes
    definition of, 546
    formation of, 546–547
Foreshore zones
    characteristics of, 544–545
    definition of, 544

Forest biomes
    acid rain and, 401–402
    characteristics and distribution, 256–261, 267–268
    deforestation and, 270–273
Fossil fuels
    definition of, 240
    economic geography of, 345–348
    and greenhouse effect, 68–70
Fossils, in sedimentary rock, 334–335, 367
France, mistral wind, 139
Freezing, definition of, 150
Freezing rain, definition of, 169
French Guyana (South America), vegetation of, 248–249
Freshwater, characteristics and issues, 412–413, 566–572
Frictional forces, airflow and, 123–125
Fringing reefs, characteristics of, 551–552
Frontal uplift
    characteristics of, 181–183
    definition of, 169
Fronts, types and characteristics of, 181–183
Frost heaving
    definition of, 396
    in periglacial regions, 497–498
    process and effects, 396–397
Frost wedging
    definition of, 395
    in periglacial regions, 497–498
    process of, 395
Fumaroles, definition of, 388

G

Gabbro, characteristics of, 329, 330
Ganges River (India), delta form of, 550–551
Gelisols
    characteristics and distribution, 299–300, 307–308
    definition of, 307
Geographic grid
    components of, 16–21
    GeoDiscoveries on, 21
Geographic Information Systems (GISs)
    characteristics and uses of, 40–41
    GeoDiscoveries on, 41
Geography. *See also* Physical geography
    scope of, 4–5
Geologic time
    definition of, 338
    GeoDiscoveries on, 341
    geomorphology and, 364–368
    glaciation and, 488–492
    in perspective, 341–342
    and the rock cycle, 342–345
    timescale, 338–340
Geomorphology. See also specific agent or process
    Appalachian Mountains example, 364–368
    definition of, 364
    GeoDiscoveries on, 453, 487
    water as primary agent, 430–432
Geostationary Operational Environmental Satellites
        (GOES)
    characteristics of, 35
    water vapor image, 68

Geostationary orbits
    definition of, 35
    satellite imagery from, 35
Geostrophic winds, definition of, 123
Geothermal features, at Yellowstone National Park, 388
Geysers, definition of, 388
Glacial abrasion
    definition of, 479
    process and effects, 479–482
Glacial drift, definition of, 483
Glacial erratics
    characteristics and examples, 479, 480
    definition of, 479
Glacial grooves, definition of, 479
Glacial ice
    deformation and movement, 472–474
    as sedimentary rock, 470
Glacial mass budget
    GeoDiscoveries on, 472
    as system, 471–472
Glacial outwash
    definition of, 486
    landforms of, 486–488
Glacial plucking
    definition of, 479
    process and effects, 479–480
Glacial processes and landforms
    GeoDiscoveries on, 487
    glacial deposition, 483–488
    glacial erosion, 479–483
    periglacial processes, 496–499
Glacial striations, definition of, 479
Glacial till
    definition of, 483
    landforms of, 483–486
Glacial troughs
    characteristics of, 481, 482–483
    definition of, 482
Glaciation
    and eustatic change, 531–533
    history of, 488–492
    Milankovitch theory and, 236–238
Glaciers
    definition of, 470
    development of, 470–474
    GeoDiscoveries on, 476, 494
    glacial mass budget, 471–472
    history of, 488–492
    human impact on, 493–496
    ice core analysis of, 233–235
    movement of, 472–474
    types of, 474–478
Global biomes. *See* Biomes
Global Positioning Systems (GPSs), principles and uses of,
        37–40
Global radiation budget
    characteristics of, 81–85
    climate feedbacks in, 243
    GeoDiscoveries on, 82
Global temperature patterns
    characteristics of, 106–109

GeoDiscoveries on, 108–109
Global warming. *See also* Climate change
    coastline impacts, 556–557
    evidence for, 241–242
    future predictions, 242–245
    and glacial retreat, 493–496
Gneiss, characteristics of, 336
Gobi Desert (Asia), characteristics of, 264
GOES. *See* Geostationary Operational Environmental
    Satellites (GOES)
Graben, definition of, 371
Graded streams
    definition of, 446
    evolution of stream valleys, 449–455
    holistic model of, 456–458
    processes of, 444–449
Grand Canyon (Arizona)
    characteristics of, 505, 507
    and geologic history, 320–321, 342–343
Granite
    characteristics of, 329, 330
    in earth layers, 323–324
    weathering of, 399–400
Grassland biomes
    characteristics and distribution, 256–257, 261–264, 267–268
    soils of, 299–300, 308–310
Great circles
    characteristics of, 16–17, 48
    definition of, 16
Great Lakes region (North America)
    coastal dune fields, 269, 547
    deforestation and, 272
    evolution of, 489–492
Great Plains (North America). *See also* Midwest United
    States
    characteristics of, 263–264, 269
    the Dust Bowl, 523–525
    High Plains Aquifer, 417–422
    soils of, 299–300, 308–313, 418–419
Great Smoky Mountains, history of, 364
Great Unconformity, erosion in, 342
Greenhouse effect
    definition of, 68
    GeoDiscoveries on, 245
    prediction of future, 241–245
    principles of, 68–70, 239–241
Greenland
    climates of, 227–228
    glaciers of, 478, 493–494, 496
    projections of, 25–26
Groins, definition of, 555
Ground ice
    definition of, 497
    landforms of, 497–499
Ground moraines, characteristics of, 484–485
Groundwater
    depletion and contamination of, 419–423, 571–572
    GeoDiscoveries on, 424
    importance of, 412–413
    and karst topography, 424–428

    movement and storage of, 414–417
    recharge, 414–417, 432–433
Gulf of Mexico
    barrier islands of, 547–548
    BP Gulf spill (2010), 347–348, 553
    GeoDiscoveries on, 557
    Hurricane Katrina effects, 205–207
    oil production in, 347–348
Gulfs
    characteristics of, 530–531
    definition of, 530
Gulf Stream, flow of, 139–141
Gullies
    definition of, 443
    formation of, 443–444
Gyres
    circulation in, 140–141
    definition of, 140
    pollution in, 563–564

H
Habitat, conservation issues, 583–589
Hadley cells
    definition of, 129
    in tropical circulation, 126–129
Hail
    definition of, 169
    formation of, 193, 195
Haiti, earthquake of 2010, 369, 375
Half-life, in radiometric dating, 340–341
Hanging valleys, definition of, 483
Hawaii
    igneous rock of, 327, 330
    soil of, 315
    as volcanic hotspot, 386–387
Headlands
    definition of, 539
    erosion of, 539–540, 542
Health hazards. *See also* Natural hazards
    from particulates, 74
Heat energy, and water phases, 148–151
Heat index
    calculation of, 98–99
    and dew-point temperatures, 158
Heat transfer
    airflow and, 119
    methods of, 74–75
    in water phase changes, 148–151
Heterotrophs
    definition of, 250
    ecosystem roles of, 250–251
Highland (*H*) climates, characteristics of, 228–229
High latitudes
    annual temperature range, 106–109
    biomes of, 256–257, 264–265
    definition of, 19
    periglacial processes and landforms, 496–499
    seasons in, 51–55
    soils of, 299–300, 307–308
High Plains Aquifer, formation and characteristics, 417–422

High-pressure ridges
  in cyclogenesis, 186–187
  definition of, 186
High-pressure systems
  characteristics of, 116–118
  definition of, 116
Hillslopes, erosional landforms, 442–443
Himalayan Mountains (Asia)
  climates of, 220, 228–229
  formation of, 367
  glaciers of, 473, 474–475
Histosols
  characteristics and distribution, 299–300, 306–307
  definition of, 306
Hogback ridges
  definition of, 365
  geomorphology and, 365–366
Holocene Climate Optimum, climate of, 235
Holocene Epoch
  characteristics of, 339–340, 562–563
  climate change in, 235, 562
Hook echoes, definition of, 201
Hoover Dam (United States), and Lake Mead, 463, 465, 568
Horns, definition of, 482
Horse Hollow Wind Energy Center (Texas), 136
Horsts
  definition of, 371
  and faults, 371–373
Hot and dry desert biome, characteristics of, 264
Hot low-latitude desert climate (*BWh*)
  biomes of, 264
  characteristics of, 220
Hot low-latitude steppe climate (*BSh*), characteristics of, 220
Hotspots
  characteristics and examples, 386–388
  definition of, 386
Human/environmental interactions
  acid rain, 401–402
  in climate change, 239–245
  coastlines and, 553–557
  critical issues in, 10–11
  deforestation, 270–272, 273
  desertification, 521–525
  earthquakes and natural hazards, 374–379
  glaciation and, 493–496
  groundwater depletion and contamination, 417–423
  heat and wind chill indexes, 98–99
  hurricane activity and impacts, 205–207
  large-scale agriculture, 272–273
  ozone hole, 71–73
  panda conservation case study, 583–589
  petroleum issues, 345–348
  physical geography and, 560–561
  population growth issues, 562–566, 572–573
  soil salinization case study, 572–583
  soils and, 316–317, 572–583
  solar power production, 86–87
  with streams, 461–465
  on vegetation, 269–275
  water issues case study, 566–572
  wildfires, 262

  wind energy production, 135–136
Humboldt Current
  flow of, 139–141
  maritime effect of, 106–108
Humid continental climates
  biomes of, 259
  characteristics of, 226
Humid continental hot-summer climates (*Dfa, Dwa*)
  biomes of, 259, 263
  characteristics of, 226
Humid continental mild-summer climates (*Dfb, Dwb*)
  biomes of, 259–260
  characteristics of, 226
Humidity. *See also* Precipitation
  definition of, 153
  dew-point temperature and, 156–158
  GeoDiscoveries on, 158
  types and measurement, 152–156
Humid subtropical hot-summer climates (*Cfa, Cwa*)
  biomes of, 259, 260
  characteristics of, 222–223
  GeoDiscoveries on, 224
  and giant panda habitat, 584–588
Humus
  definition of, 281
  in soil profiles, 296
Hurricanes
  activity and impacts, 203–207, 548
  Floyd, 205
  Katrina, 10–11, 178–179, 205–207, 548
  remote sensing of, 178–179, 205, 206
Hydraulics, of stream flow, 438–442, 456–458
Hydroelectric power, dams and, 463–465
Hydrogen bonding, in water, 148–151
Hydrographs, of dammed rivers, 463–465
Hydrologic cycle
  characteristics of, 151–152, 414
  definition of, 151
  GeoDiscoveries on, 424
Hydrology, stream, 438–442
Hydrolysis
  definition of, 399
  process and effects, 399–400
Hydrosphere. *See also* Coastal processes and landforms
  characteristics of, 151–152
  in coastline interactions, 531
  data collection in, 8–9
  definition of, 8, 151
Hygroscopic water, definition of, 414

I
Ice
  calving, 475
  in Earth's history, 468–469
  glacial. *See* Glacial ice
  ground ice, 497–499
  ice core analysis, 233–235, 242
  polar melting, 243, 244, 494–496
  in precipitation processes, 165–169
  properties of, 149–151
Icebergs, process of calving, 475

Ice-cap climate (*EF*), characteristics of, 228
Ice caps
    characteristics of, 474–475
    definition of, 474
Ice core analysis, in paleoclimatology, 233–235, 242
Ice fields
    characteristics of, 474–475
    definition of, 474
Iceland, glaciers of, 475
Ice sheets. *See* Continental glaciers
Ice wedges, formation of, 498
Igneous rock
    definition of, 329
    formation and properties of, 327–330
    GeoDiscoveries on, 329
    and rock cycle, 342–345
Illuviation
    definition of, 284
    in soil formation, 286–287, 296–298
Inceptisols
    characteristics and distribution, 299–300, 314–315
    definition of, 314
India
    biome distribution, 257, 258, 263
    monsoons in, 133–134
    population growth and, 563–564
    river deltas of, 460
Indian Ocean
    geography of, 530
    Sumatra-Andaman earthquake (2004), 375–378
Indirect radiation
    definition of, 77
    of solar radiation, 77
Indirect tides, definition of, 534
Indonesia
    coral bleaching in, 557
    forest cover changes in, 270–271
    Sumatra-Andaman earthquake (2004), 376–378
Industrialization, population issues of, 562–566
Infiltration, of water into soil, 281–283, 414
Infrared satellite images, uses of, 12–13, 38
Inner core (Earth)
    characteristics of, 322–323
    definition of, 323
Inner structure of Earth, layers of, 322–326
Inselbergs, Uluru (Ayer's Rock) example, 505
Insolation
    definition of, 74
    and global radiation budget, 81–85
Interfluves, definition of, 434
Intergovernmental Panel on Climate Change (IPCC), beliefs
    of, 242–244, 563–564
Interior of Earth. *See* Inner structure of Earth
International Date Line, definition of, 50
Intertropical Convergence Zone (ITCZ)
    circulation in, 126–129
    climate effects. See specific climate
    definition of, 126
    seasonal migration of, 132–134
Intrusive igneous rock
    definition of, 329

formation and properties of, 327–330
Iraq, soil salinization in, 574–576
Iron
    in earth layers, 323–324
    as rock component, 326
Irrigation
    and groundwater depletion, 419–422, 566–568
    and soil salinization, 573–575, 577–580, 582
Islands
    barrier, 546–550
    sea level changes and, 557
    urban heat, 102–106, 565
    volcanic, 362
Isobars
    airflow and, 120
    definition of, 31
Isohyets, definition of, 31
Isolines
    definition of, 28
    types and use of, 28–31
Isopachs, definition of, 31
Isostatic adjustment
    glacial, 478
    lithospheric, 324–326
Isotherms
    definition of, 31
    maritime vs. continental effect, 106–109
Isotopes. *See also* Oxygen isotope stages
    in radiometric dating, 340–341

J
Japan
    tsunamis in, 378
    volcanoes of, 380–381
Jet stream
    characteristics of, 110, 226
    polar front and, 130–132, 184–191
    unusual weather patterns and, 440–441
Jetties, definition of, 556
Joints, definition of, 395
Jurassic Period, characteristics of, 339–340

K
Kalahari Desert (Africa), characteristics of, 264
Kames
    characteristics of, 484, 486–487
    definition of, 486
Kansas River (United States), drainage basin of, 434–435
Karst topography
    caves and caverns, 425–426
    definition of, 424
    landforms, 424–428
Katabatic winds, definition of, 138
Katrina (Hurricane)
    as environmental issue, 10–11
    evolution and aftermath, 205–207, 548
    GeoDiscoveries on, 205
    satellite image of, 178–179, 206
Kelvin, Lord William Thomson, temperature scale of, 98
Kelvin temperature scale, characteristics of, 96

Kettle lakes
    characteristics of, 484, 487
    definition of, 487
Kinetic energy, definition of, 94
Knickpoints, of graded streams, 446
Köppen Climate Classification System
    arid and semi-arid (*B*) climates, 215, 217, 219–222
    characteristics of, 214–216
    GeoDiscoveries on, 219, 224, 225, 228
    highland (*H*) climates, 228–229
    mesothermal (*C*) climates, 216, 217, 222–225
    microthermal (*D*) climates, 216, 217, 226–227
    polar (*E*) climates, 216, 217, 226–227
    tropical (*A*) climates, 215–219
Köppen, Wladimir, climate classification and, 214
Krumholz zone, and tree lines, 268
Kyoto Protocol, and greenhouse emissions, 244

L
La Conchita (California) debris flow, process of, 408
Lagoons
    and barrier islands, 548–550
    definition of, 546
Lag time, and stream discharge, 461–462
Lake Mead (United States)
    drought and, 569–571
    origin of, 463, 465
Lake Michigan (United States)
    beach nourishment at, 554
    coastal dune fields, 269, 547
Lakes
    in evolution of stream valleys, 449–450
    manmade, 463–465, 568
    natural, 416, 487, 490–491
    oxbow, 452–453
    streams and, 432–433
    volcanic, 384–385
Lake Victoria (Africa), Nile River source, 432–433
Land. *See also* Surface temperature
    temperatures and, 101–103, 106–109
Land breezes, definition of, 138
Land cover, definition of, 586
Landforms
    coastal. *See* Coastal processes and landforms
    definition of, 364
    desert, 508–509
    of differential weathering, 394
    eolian, 510–513, 515–521
    fluvial. *See* Fluvial processes and landforms
    GeoDiscoveries on, 451, 453, 487, 521
    geomorphology and, 364
    glacial, 479–488
    karst, 424–428
    of mass wasting, 404–405
    periglacial, 496–499
Landsat satellite system
    characteristics of, 34
    glacial imagery, 493
Landscape evolution. *See also* Fluvial processes and
      landforms
    of desert environments, 506–509

glacial, 481–482
    process of, 392–393
Landscapes
    arid. *See* Arid landscapes
    karst, 426–428
    permafrost, 496–498
Land-sea breezes, characteristics of, 137–138
Landslides, types and effects, 406–407
Land-use practices, and forest cover changes, 270–271
Large-scale maps
    characteristics of, 28–29
    definition of, 28
Larsen B Ice Shelf, collapse of, 494–496
Las Vegas (Nevada)
    geography of, 566–567
    water issues of, 566–572
Latent heat. *See also* Cyclonic weather systems
    adiabatic processes and, 162
    definition of, 78
    of water, 148–151
Lateral moraines, characteristics of, 484–485
Laterization
    definition of, 301
    GeoDiscoveries on, 310
    in soil formation, 301–303
Latitude
    air temperature and, 99
    annual temperature range and, 106–109
    characteristics and measurement of, 17–19
    climate and, 212–213
    definition of, 17
    global radiation budget and, 81–85
    humidity and, 155
    seasons and. *See* Seasons and seasonality
Latitude/longitude coordinate system, characteristics of,
      17–21
Laurentide Ice Sheet
    definition of, 488
    history and movement of, 488–492
Lava, and igneous rock, 327, 329
Lava domes
    at Mt. St. Helens, 382–383
    definition of, 380
Layers of Earth, characteristics and properties of, 322–326
Leeward side
    definition of, 139
    in orographic processes, 170–172, 220–221, 223, 228–229
Length of day
    air temperature and, 100
    global radiation budget and, 82–85
Levees
    artificial, 462–463
    natural, 453–454
Level of condensation, definition of, 161
Lianas, in rainforests, 257–258
Lightning, formation of, 193–194
Limestone
    characteristics of, 334–335
    definition of, 335
    karst topography and, 424–428
    weathering of, 400, 506

Liquid outer core
    characteristics of, 322–323
    definition of, 323
Lithification
    definition of, 331
    and rock cycle, 342–345
    of sedimentary rock, 331
Lithosphere. *See also* Crust (Earth)
    characteristics of, 322–324
    in coastline interactions, 531
    data collection in, 8–9
    definition of, 8, 324
    isostatic adjustment and, 324–326
Lithospheric plates. *See also* Plate movements
    mechanism of movement, 354–357
    and volcanic activity, 362, 381–383
Little Ice Age, climate of, 235, 242, 493
Littoral drift
    characteristics of, 537–539
    definition of, 538
Littoral processes and drift, factors and effects, 537–539
Local winds, characteristics of, 137–139
Location
    continental and maritime, 101
    geographic grid and, 16–21
    and temperature, 106–109
Loess
    characteristics and distribution, 516–519
    definition of, 516
London (United Kingdom), Thames River, 430–431
Longitude
    characteristics and measurement of, 20–21
    definition of, 20
Longitudinal dunes, characteristics of, 516–517
Longitudinal profiles
    definition of, 446
    of graded streams, 446–447, 456–458
Longshore currents
    and coastal deposition, 545–546
    definition of, 537
    GeoDiscoveries on, 549
Longshore drift, definition of, 538
Longwave radiation
    definition of, 64
    global radiation budget and, 81–85
    greenhouse effect and, 68–70
    properties of, 64–66
    reflection of, 79–81, 81–85
Los Angeles (California), topography of, 14–15
Louisiana, *GeoDiscoveries* on, 557
Low latitudes
    atmospheric circulation in, 125–129
    biomes of, 256–257, 263
    climates of, 214–219, 220
    definition of, 19
Low-pressure systems, definition of, 116, 116–118
Low-pressure troughs
    in cyclogenesis, 186–187, 201–202
    definition of, 186

M
Madison slide (Montana), debris slide, 406–407
Magma. *See also* Volcanoes
    definition of, 327
    eruption types, 380, 385
    in igneous rock formation, 327–330
    and plate movement, 355–357
Magnesium
    in earth layers, 323–324
    as rock component, 326
Magnetic field/magnetosphere
    outer core role in, 323
    reversals, 357
Magnitude, of earthquakes, 371
Mammatus clouds, characteristics of, 198
Mantle
    characteristics of, 323–324
    convection in, 355–357
    definition of, 323
    in volcanic activity, 379–381
Map projections
    conformal and equivalent projections, 24–27
    definition of, 22
    principles of, 22–24
Maps
    GeoDiscoveries on, 32
    isolines, 28–31
    map projections, 22–27
    map scale, 28–29
Map scale
    definition of, 28
    large- and small-scale, 28–29
Marble
    characteristics of, 336
    weathering of, 400
Marine terraces, formation of, 541–543
Marine west-coast climates (*Cfb*, *Cfc*)
    characteristics of, 223–224
    GeoDiscoveries on, 225
Maritime locations
    definition of, 101
    temperatures and, 101–103
Maritime Polar (mP) air mass, characteristics of, 180–181
Maritime Tropical (mT) air mass, characteristics of, 180–181
Maritime vs. continental effect
    characteristics of, 101–103
    definition of, 101
    GeoDiscoveries on, 104
    global annual temperature range and, 106–109
    land-sea breezes and, 137–138
    monsoons and, 133–134
Maroon Bells (Colorado), climate regions of, 210–211
Mars, sand dunes of, 520
Marshall Islands (Pacific), sea level changes and, 557
Marshes, salt, 548–550
Mass wasting
    avalanches, 408–409
    definition of, 403
    factors in, 403
    flows, 407–408
    GeoDiscoveries on, 410

Mass wasting *(cont.)*
    landslides, 406–407
    rockfall, 403–405
    soil creep, 405–406
Matterhorn (Switzerland), glacial arête, 482
Mauna Loa (Hawaii)
    atmospheric carbon dioxide at, 69–70
    hotspot volcano, 386
Maximum humidity
    and atmospheric moisture, 153–158
    definition of, 153
Meandering streams
    definition of, 444
    GeoDiscoveries on, 451
    processes and landforms, 11, 444–445, 450–456
Mechanical weathering. *See also* Glacial processes and
    landforms
    climate and, 394
    definition of, 395
    processes and effects, 395–398
Medial moraines, characteristics of, 484–485
Medieval Warm Period, climate of, 235
Mediterranean dry-summer climate (*Csa, Csb*)
    biomes of, 259–260
    characteristics of, 222–223
Mediterranean woodland and shrub forest biome,
    characteristics of, 256–257, 259–260
Melting, of glaciers, 493–496
Meltwater. *See* Sea level changes
Mercator, Gerhardus, and Mercator projection, 24
Mercator projection, characteristics of, 24–26
Meridians
    characteristics of, 20–21
    definition of, 20
Meridional flow, definition of, 131
Mesas, definition of, 507
Mesa Verde National Park (United States), rock niches of,
    398
Mesocyclones, definition of, 195
Mesopause, definition of, 94
Mesosphere
    characteristics of, 94
    definition of, 94
Mesothermal (*C*) climates
    characteristics of, 216, 217, 222–225
    GeoDiscoveries on, 224, 225
Mesozoic Era, characteristics of, 339–340
Metamorphic rock
    characteristics and formation, 336
    definition of, 336
    and rock cycle, 342–345
Meteors/meteorites, in atmosphere, 94
Methane, and climate change, 242
Mexico
    and Colorado River Compact, 568–572
    soil salinization in, 576
Microclimate, definition of, 266
Microthermal (*D*) climates, characteristics of, 216, 217,
    226–227
Mid-Atlantic Ridge, rifting at, 360–361

Middle East
    desert areas of, 221
    oil reserves of, 346–348
    soils of, 299–300, 310–313
Midlatitude coniferous forest biome, characteristics of,
    256–257, 260
Midlatitude cyclones
    cyclogenesis, 188–191
    definition of, 184
    GeoDiscoveries on, 190–191
    migration of, 189–191
Midlatitude deciduous forest biome, characteristics of, 256–
    257, 259
Midlatitude grassland biome
    agriculture in, 272–273
    characteristics of, 256–257, 261, 262–264
Midlatitude jet stream, climate effects of, 226
Midlatitudes
    atmospheric circulation in, 121, 129–132
    climates of, 217, 220–221
    cyclone behavior in, 184–191
    definition of, 19
    GeoDiscoveries on, 269
    soils of, 299–300, 304–306, 308–313
Mid-oceanic ridge
    definition of, 360
    and plate divergence, 360–361
Midwest United States
    deforestation and, 272
    flooding in, 440–442
    soils of, 299–300, 304–305, 306–313
    tornadoes in, 198–200
Milankovich, Milutin, glacial cycles theory of, 236
Milankovitch theory
    components of, 236–238
    definition of, 236
    GeoDiscoveries on, 238
    and glaciation, 493
Milky Way Galaxy, characteristics of, 46
Minerals
    definition of, 326
    as rock component, 326–327
    silicate, 329–330, 331–332, 399–400
Minnesota, water table and lakes, 416, 487
Minutes, latitude measurement, 17–19
Mississippian Period, characteristics of, 339–340
Mississippi River (United States)
    delta of, 459–460, 548, 550–551
    drainage basin, 434–435
    meanders of, 452
    Mississippi Basin Floods (1993), 440–441, 462–463
    source of, 432
Mistral wind, characteristics of, 139
Mohorovicic discontinuity
    definition of, 324
    in isostatic adjustment, 324–326
Moisture, atmospheric . *See* Atmospheric moisture
Mojave Desert (United States), characteristics of, 264
Mollisols
    characteristics and distribution, 299–300, 308–310, 316, 418–419
    definition of, 308

Monoclines
characteristics of, 363–364
definition of, 364
Mono Lake (California), water issues and, 571–572
Monsoons
circulation of, 133–134
definition of, 133
GeoDiscoveries on, 134
monsoon climates, 217, 219
Montana, glacial Lake Missoula, 490–491
Montreal Protocol, and CFC decreases, 72–73
Monument Valley (Utah), landforms of, 507
Moon, tides and, 533–535
Moraines
definition of, 483
formation and types, 483–485
Moscow (Russia), climograph of, 218
Mountain breezes, definition of, 138
Mountains
alpine glaciers, 474–477
geomorphology and, 364–368
orogenies, 340, 364–365
orographic processes and, 170–172, 220–221, 223, 228–229
Mount Everest (Himalaya Mountains), characteristics of, 367
Mount Hood (Oregon), glaciers of, 476–477
Mount Kilimanjaro (Tanzania), glacial retreat, 493
Mount McKinley (Alaska), height of, 116
Mount Pinatubo (Philippines), 1991 eruption, 236
Mount Rainier (Washington), eruption hazards of, 383
Mt. Fuji (Japan), composite volcano, 380–381
Mt. St. Helens (Washington), recent activity of, 381–383
Muck, histosol soil, 307
Mud, in sedimentary rock formation, 332–333
Mudflats, evolution of, 548–549
Mudflows
definition of, 407
process and effects, 407–408
Mudpots, definition of, 388
Multipath error, definition of, 40
Municipal planning, GIS in, 41
Munsell color system, and soil color, 289–291
Murray-Darling River Basin (Australia), soil salinization and, 581–582

N

Namib Desert (Namibia), characteristics of, 515
National Weather Service. *See also* Weather monitoring and prediction
tornado monitoring, 200–201
Natural gas, definition of, 335
Natural hazards. *See also specific hazard*
coastal changes from, 547–548, 553–556
definition of, 375
examples of, 375
human role in, 10–11
protective measures, 553–556
Natural levees
definition of, 453
in stream valleys, 453–454
Natural Resources Conservation Service (NRCS), and soil maintenance, 316–317

Neap tides, characteristics of, 534–535
Nebraska Sand Hills (United States), characteristics of, 515, 518
Net radiation
definition of, 81
factors affecting, 82
spatial distribution, 59, 82–85
urban vs. rural, 103–105
Nevada, water issues of, 566–572
New Madrid Fault (United States), earthquakes along, 375
New Mexico
Carlsbad Caverns National Park, 425–426
Colorado River Compact and, 568–572
New Zealand
rias of, 532
sunrise in, 60
Niagara Falls (North America), fluvial processes of, 446, 448
Nickel, in earth layers, 323–324
Night, seasonal changes and length of, 56–60
Nile River (Egypt), delta of, 459
Nitrogen, as atmospheric component, 67
Noctilucent clouds, characteristics of, 165
Normal faults
characteristics of, 371–373
definition of, 371
North America
acid rain in, 401
air masses of, 180–181
biome distribution, 257, 259–260, 263–265
drainage basins of, 434–435
forest cover changes in, 270–271
Northern Hemisphere
annual temperature range, 106–109
atmospheric circulation patterns, 124
atmospheric pressure systems in, 116–118
climates of, 217
definition of, 17
insolation in, 84–85
ocean currents in, 139–141
ozone layer/hole, 72–73
periglacial regions, 496–497
predicted climate change, 243–244
seasons in, 51–55, 56–60
North Pole
geographic grid and, 16–21
insolation at, 84–85

O

Obliquity, and climate change, 237–238
Obsidian, characteristics of, 331
Occluded fronts, definition of, 189
Oceanic circulation
El Niño, 141–143
gyres and thermohaline circulation, 139–141
Oceanic crust
characteristics of, 322, 324–326
definition of, 324
in subduction, 360–362
Oceanic trenches, formation of, 362
Oceans. *See also specific ocean*
characteristics of, 530–531

Oceans. *See also* specific ocean *(cont.)*
  circulation patterns of, 139–143
  coral bleaching in, 557
  definition of, 530
  earthquakes and tsunamis, 375–379
Offshore bars
  characteristics of, 544–545
  definition of, 544
Offshore zones, definition of, 544
Ogallala Aquifer. *See* High Plains Aquifer
O horizon, definition of, 296
Oil
  economic geography of, 345–348
  global reserves, 348
Orbital eccentricity
  climate change and, 236–238
  tides and, 534–535
Orbital precession, climate change and, 237–238
Orbits
  Earth around Sun, 47–49, 53, 236–238
  satellite, 34, 35
Oregon
  Crater Lake, 384–385
  glaciers of, 476–477
Organic sedimentary rock
  characteristics of, 335
  petroleum issues, 345–348
Organisms, in soil formation, 287–288
Orogenies
  definition of, 364
  examples of, 364–368
Orographic uplift/processes
  definition of, 169
  GeoDiscoveries on, 172
  and precipitation patterns, 170–172, 220–221, 223, 228–229, 474
Outwash plains, definition of, 486
Overgrazing, vegetation and, 273
Overland flow, and drainage, 432–438
Overthrust faults, characteristics of, 372–373
Overthrust folds
  characteristics of, 363–364
  definition of, 364
Overturned folds
  characteristics of, 363–364
  definition of, 364
Oxbow lakes
  characteristics of, 452–453
  definition of, 453
Oxic horizon, definition of, 302
Oxidation, definition of, 400
Oxisols
  characteristics and distribution, 299–302
  definition of, 301
Oxygen
  as atmospheric component, 67, 94–95
  in ice core analysis, 233–235
  and photosynthesis, 251–254
  as rock component, 326
Oxygen isotope stages
  climate change and, 234–235
  definition of, 234

GeoDiscoveries on, 236
  in glaciation research, 489
Ozone hole
  definition of, 72
  properties and issues of, 70–73
Ozone layer
  definition of, 71
  properties and issues of, 70–73, 93

P
Pacific Northwest (United States)
  coastal processes of, 543
  deforestation and, 272–273
  glaciers of, 475–477
  volcanic activity, 381–385
Pacific Ocean
  circulation of, 139–143
  coral reefs of, 551–553
  drainage basins of, 434–435
  El Niño and, 141–143
  garbage patch, 563–564
  geography of, 530
  sea level changes and, 557
  tsunamis and, 377–379
  volcanic islands of, 362
Pacific Ring of Fire
  definition of, 381
  GeoDiscoveries on, 383
  volcanoes of, 381–383
Paleocene Epoch, characteristics of, 339–340
Paleoclimatology
  ice core analysis in, 233–235
  importance of, 229–230
  pollen records in, 230–231
  tree ring patterns in, 231–233
Paleozoic Era, characteristics of, 339–340
Palynology, and climate reconstruction, 230–231
Pandas, giant, conservation of, 583–589
Pangaea
  and continental drift, 354–355, 357
  definition of, 355
Parabolic dunes, characteristics of, 516–517, 547
Paraguay
  Itaipu Dam, 465
  slash and burn agriculture in, 271–272
Parallels, definition of, 18
Parent material
  definition of, 285
  in soil formation, 285–287, 303–304
Particulates, as atmospheric component, 73–74
Passive margins
  characteristics of, 357–359
  definition of, 357
Passive solar heating, design for, 86
Patterned ground, formation of, 499
Peak oil theory, principles of, 347
Peat, histosol soil, 307
Pedogenic processes
  calcification, 308–313
  definition of, 283
  GeoDiscoveries on, 310

interactions of, 283–284
laterization, 301–303
podzolization, 305–306
salinization, 310–313
Pennsylvanian Period, characteristics of, 339–340
Perched water table
definition of, 417
and salinization, 578
Percolation
of groundwater, 414–417
and karst topography, 424–426
in snow, 470
Periglacial processes and landforms, 496–499
definition of, 496
Perihelion, definition of, 48
Permafrost
definition of, 496
geography of, 217, 227, 307–308
processes and landforms, 496–498
Permian Period, characteristics of, 339–340
Petroleum
characteristics of, 335
definition of, 335
economic geography of, 345–348
PH
definition of, 293
GeoDiscoveries on, 295
of soils, 293–295
Phancrozoic Eon, characteristics of, 339–340
Phoenix (Arizona), urban sprawl in, 568
Photosynthesis
definition of, 251
GeoDiscoveries on, 254
process of, 251–254
Physical geography
climate change and, 244
definition of, 6, 202
elements and principles of, 5–9, 12–13
GeoDiscoveries on, 205
and human/environmental interactions, 560–561
panda conservation case study, 583–589
soil salinization case study, 572–583
water issues case study, 566–572
Physical weathering. *See* Mechanical weathering
Piedmont glaciers, characteristics of, 475, 477
Pingos, formation of, 498–499
Pinnacles, definition of, 507
Pioneer species, in succession, 268–269
Pixels, definition of, 34
Plane of the ecliptic
definition of, 47
Earth's orbit and, 47–48
Plant geography
ecosystems and, 250–251
of global biomes, 255–265. *See also* Biomes
photosynthesis, 251–254
Plant succession
definition of, 268
GeoDiscoveries on, 269
process of, 268–269

Plateaus
definition of, 507
and desert landscape evolution, 507–508
Plate convergence
characteristics of, 360–364
definition of, 360
Plate divergence, characteristics of, 360–361
Plate margins/boundaries. *See* Plate movements
Plate movements
earthquakes and, 369–374
passive margin movement, 357–360
plate convergence, 360–364
plate divergence, 360–361
subduction, 360–362
transform plate margins, 358–360
volcanoes at boundaries, 381–383
Plate tectonics
and continental drift theory, 354–357
definition of, 354
and earthquakes. *See* Earthquakes
GeoDiscoveries on, 368, 383
plate movements. *See* Plate movements
and volcanoes. *See* Volcanoes
Playas
characteristics of, 507–508
definition of, 507
Pleistocene Epoch
characteristics of, 339–340
glaciation in, 488–492
Pliocene Epoch, characteristics of, 339–340
Plutons
definition of, 329
formation of, 327–330
Pocket beaches, formation of, 540
Podzolization
definition of, 305
GeoDiscoveries on, 310
in soil formation, 305–306
Polar Easterlies, definition of, 132
Polar (*E*) climates, characteristics of, 216, 217, 226–227
Polar front
circulation of, 129–132
definition of, 129
and midlatitude cyclones, 184–191
Polar front jet stream
characteristics of, 130–132
definition of, 130
and midlatitude cyclones, 184–191
Polar High, definition of, 132
Polar ice, melting of, 243, 244, 494–496
Polar zones
atmospheric circulation in, 132
climates of, 227–229
Pollen records, in paleoclimatology, 230–231
Pollution
of groundwater, 423
population growth and, 563–566
Ponds
of sinkhole origin, 426
and streams, 432–433

Population growth (human)
    in China, 584, 585
    environmental impacts of, 562–566
    of Las Vegas (Nevada), 568–569
Pore spaces
    pore ice and, 498
    soil water levels and, 281–283, 414–415
Potassium, as rock component, 326
Potential evapotranspiration (potential ET)
    definition of, 213
    in Thornthwaite system, 213–214
Potential natural vegetation, definition of, 255
Precambrian time
    characteristics of, 339
    glaciation in, 488
    and Grand Canyon rock, 342–343
Precession, and climate change, 237–238
Precipitation. *See also* Drought; Orographic uplift/processes
    acid rain, 401–402
    adiabatic processes in, 160–162
    along fronts, 182–183
    biome classification and, 256
    climate classification and, 214–216
    flooding and, 439–442
    particulates and, 73–74
    processes of, 169–175
    in soil formation, 286–287
    soil water and, 281–283
    types of, 165–169
Precision agriculture, soil surveys in, 316–317
Pressure gradient force
    airflow and, 120, 124–125
    definition of, 120
    GeoDiscoveries on, 122
Pressure systems. *See* Atmospheric pressure, pressure
    systems
Primary succession, of vegetation, 268–269
Prime Meridian
    definition of, 20
    geographic grid and, 20–21
    time measurement by, 50
Processes, definition of, 6
Progradation, definition of, 544
Proxy data
    definition of, 230
    in paleoclimatology, 230–236
P waves, behavior of, 370
Pyroclastic material
    definition of, 380
    in eruptions, 380–384, 388

Q
Quartz
    properties of, 326–327
    in sedimentary rock, 331–332
Quaternary Period
    climate change in, 230–236
    epochs of, 339–340
    eustatic change during, 531–533

R
Radar
    Doppler, 200–201
    in remote sensing, 37
Radial dikes
    definition of, 344
    formation of, 344–345
Radial drainage pattern, characteristics of, 435–436
Radiation
    definition of, 74
    and global radiation budget, 81–85
    principles of, 64–66, 74–75
    solar. *See* Solar radiation
Radiation budget. *See also* Global radiation budget
    definition of, 81
Radiation fog, definition of, 164
Radioactive isotopes
    definition of, 340
    in radiometric dating, 340–341
Radiometric dating, principles of, 340–341
Rain
    definition of, 169
    formation of, 168–169
    stream hydrology and, 439–442
Rainbows, formation of, 70
Rainshadow deserts, characteristics of, 504
Rain shadows
    climate effects, 220–221
    definition of, 171
    rainshadow deserts, 504
    and San Joaquin valley climate, 576–577
Rajasthan (India), aerial photograph of, 33
Ranching, overgrazing effects, 273
Rayleigh scattering, and sky color, 77
Recessional moraines, characteristics of, 484–485
Rectangular drainage pattern, characteristics of, 435–436
Reflected radiation
    and Earth's surface, 79–80
    and global radiation budget, 81–85
Reflection
    definition of, 76
    of solar radiation, 76–78
Regolith
    definition of, 285
    in soil formation, 285–286, 296–298
Relative humidity
    and atmospheric moisture, 153–158
    definition of, 154
Relative location, definition of, 16
Relief
    of barrier islands, 547–548
    definition of, 288
    in soil formation, 288–289
Remote sensing
    at Wolong Nature Reserve, 586–587
    of continental drift, 356–357
    definition of, 33
    GeoDiscoveries on, 229, 275
    of hurricanes, 178–179, 205, 206
    methods and uses, 33–37, 356–357
    satellite. *See* Satellite remote sensing

Reservoirs. *See also* Aquifers
    environmental issues of, 465
    in hydrologic cycle, 151–152, 414
    purposes and effects, 463–465
Residual parent material
    definition of, 285
    in soil formation, 285–286
Respiration
    GeoDiscoveries on, 254
    and photosynthesis, 252–254
Retrogradation
    definition of, 540
    process of, 540–543
Reverse faults
    characteristics of, 372–373
    definition of, 373
Reversing dunes, characteristics of, 516–517
Revetments, erosion and, 554–555
Rhine River (Europe), *GeoDiscoveries* on, 437
R horizon, definition of, 296
Rhyolite, characteristics of, 329, 330
Rias
    characteristics of, 532–533
    definition of, 532
Richter scale, definition of, 371
Rifting
    definition of, 360
    examples of, 360–361
Rills
    definition of, 443
    formation of, 443–444
Riparian zones
    definition of, 269
    flooding along, 440–442
    vegetation distribution and, 269
Rivers. *See also* Fluvial systems; *specific river*; Streams
    spatial distribution in United States, 7
Robinson, Arthur H., and Robinson projection, 27
Robinson projection, characteristics of, 27
Roche moutonnée
    characteristics of, 479–480
    definition of, 479
Rock
    definition of, 326
    Earth layers, 322–326
    elemental composition of, 326–327
    folding of, 362–364, 372–373
    GeoDiscoveries on, 329, 334, 338, 368
    igneous, 327–330
    metamorphic, 336
    resistance and glaciation, 490–492
    rock cycle, 336–338, 342–345
    rockfall, 403–405
    sedimentary, 330–336, 345–348
    stress and faults, 369–374
    weathering of. *See* Weathering
Rock cycle
    characteristics and examples, 336–338, 342–345
    GeoDiscoveries on, 338
Rock deformation
    folding, 362–364

GeoDiscoveries on, 368
Rockfall
    definition of, 403
    process and effects of, 403–405
Rock outcrop, definition of, 327
Rock structure
    definition of, 364
    of deserts, 504–509
    folding and, 363–364
    GeoDiscoveries on, 368
Rocky Mountains (United States)
    avalanche hazards, 409
    climates of, 220, 228–229
    as drainage divide, 434–435
Root wedging, process and effects, 396–397
Rossby waves
    in cyclogenesis, 186–188
    definition of, 130
    midlatitude circulation and, 130–132
Rotation (Earth). *See also* Coriolis force
    characteristics of, 48–51
Runoff. *See* surface runoff
Rural areas. *See also* Agriculture
    net radiation in, 103–105
Russia
    biomes of, 256–257, 261, 263–264
    soils of, 299–300, 308–310

S
Saffir-Simpson scale, classification by, 203
Sahara Desert (Africa)
    characteristics of, 264, 515–516
    groundwater depletion, 422
    and Subtropical High (STH) Pressure System, 128
Sahel (Africa), desertification of, 521–523
Salic horizon, definition of, 311
Salinity. *See also* Soil salinization
    definition of, 531
Salinization
    of Australia, 581–583
    definition of, 310
    GeoDiscoveries on, 310
Saltation, in eolian transport, 510–511
Salt-crystal growth
    definition of, 397
    process and effects, 397–398
Salt marshes, evolution of, 548–550
Salts. *See* Soil salinization
Samoa, tsunami (2009), 378–379
San Andreas Fault (California)
    characteristics of, 358, 373–374
    imaging of, 14–15, 374
San Clemente Island (California), coastal evolution, 543
Sand. *See also* Depositional processes and landforms
    in eolian transport, 509–511
    in littoral processes, 537–539
    in sedimentation, 330–332
    in soil textural classes, 291–292, 293
Sand dunes
    classification of, 515–517
    coastal, 546–547

Sand dunes *(cont.)*
  of Death Valley (California), 502–503
  formation and characteristics of, 513–515
  of Mars, 520
  succession in, 269
Sandstone
  characteristics of, 332–333, 506
  definition of, 332
Sandstorms, wind/soil interaction, 112–113
San Francisco earthquake (1906), and San Andreas Fault, 374
Sangre de Cristo Mountains (United States), geologic history
    of, 344–345
San Joaquin Valley (California)
  irrigation and subsidence, 423, 577–578
  soil salinization in, 576–580
Santa Ana winds, characteristics of, 139
Satellite remote sensing
  characteristics of, 12–13, 34–37
  GeoDiscoveries on, 229, 275
  in GPS, 37–40
  of hurricanes, 178–179, 205, 206
  of Las Vegas growth, 569
  of Mars, 520
  of ozone layer/hole, 71–73
  of panda habitat, 586–587
  of Sumatra-Andaman earthquake (2004), 377–378
  of wildfires, 272
Saturated zones. *See also* Flooding
  definition of, 415
  and ground ice, 497–499
  and groundwater depletion, 420–422
  variations in, 418
Saturation (atmospheric)
  adiabatic processes and, 160–162
  humidity and, 153–158
Scales, temperature, 95–96, 98
Scarp. *See* Fault escarpment
Scattering
  definition of, 77
  of solar radiation, 77–78
Schist, characteristics of, 336
Scientific method, and physical geography, 8–9
Scotland, Kilt Rock cliff, 541
Sea arches, formation of, 541–542
Sea breezes, definition of, 137
Sea caves, formation of, 541–542
Seafloor
  spreading and rifting, 360
  and tsunami formation, 376
Sea fog
  characteristics of, 164–165
  definition of, 164
Sea level changes. *See also* Eustatic change
  from climate change, 243–244, 551–553, 556–557
Seas
  characteristics of, 530–531
  definition of, 530
Seasons and seasonality
  air temperature and, 100
  annual temperature range and, 106–109
  climate and, 212–213, 237–238

  day length, 56–60
  Earth-Sun geometry of, 51–55, 56–60
  GeoDiscoveries on, 55
  global radiation budget and, 83–85
  Milankovitch theory and, 236–238
  pressure system migration and, 132–134
Sea stacks, formation of, 542
Sea walls, erosion and, 554–555
Seawater, composition of, 531
Secondary succession, of vegetation, 268–269
Seconds, latitude measurement, 17–19
Sedimentary rock
  chemically precipitated, 333–335
  clastic rock, 331–333
  definition of, 330
  formation and properties of, 330–336
  GeoDiscoveries on, 334
  organic rock, 335, 345–348
  and rock cycle, 342–345
Sediment load. *See also* Stream gradation
  characteristics of, 444–445
  in coastal processes, 545–550
  GeoDiscoveries on, 445
Sediments. *See also* Beaches; Fluvial processes and landforms
  definition of, 331
  delta formation, 459–461
  glacial till and outwash, 483–488
  in ice core analysis, 233–235
  in littoral drift, 537–539
  mass wasting of, 403
  and rock cycle, 342–345
  sediment load, 444–445
Sediment transport
  eolian, 509–511
  littoral drift, 537–539
Seismic waves
  definition of, 322
  measurement of, 370–371
  tsunamis, 376–379
Seismograph, measurement with, 370
Semi-arid (*B*) climates, characteristics of, 215, 217, 219–222,
    504
Semi-arid and cold desert biome
  characteristics of, 264–265, 504
  soils of, 299–300
Sensible heat
  definition of, 78
  and latent heat, 148–149
Shale
  characteristics of, 332–333, 506
  definition of, 333
Shape, of Earth, 46–47
Shelf clouds, characteristics of, 198
Shenandoah National Park (United States), radar image of,
    37
Shield volcanoes
  characteristics of, 385–386
  definition of, 385
Shiprock (New Mexico), exposed igneous intrusion, 328
Shortwave radiation. *See also* Solar energy
  definition of, 64

global radiation budget and, 81–85

properties of, 64–66

Shuttle Radar Topography Mission (SRTM), characteristics of, 37

Sierra Nevada (California)

and desert regions, 504–509, 566–568

as fault escarpment, 373

Silicate minerals, characteristics of, 329–330, 331–332, 399–400

Silicon

in earth layers, 323–324

as rock component, 326, 329–330, 331–332

Silt

in eolian transport, 509–511

loess, 516–519

in soil textural classes, 291–292, 293

Sinkholes

definition of, 426

evolution and effects, 426–427

Sioux Quartzite (United States), glacial plucking and, 479, 480

Skjaldbrei[thorn]ur volcano (Iceland), shield volcano, 385–386

Skykomish River (Washington), flooding of, 440

Slate

characteristics of, 336

weathering of, 400

Sleet, definition of, 169

Slip faces, definition of, 514

Slope

definition of, 266

and mass wasting. *See* Mass wasting

in soil science, 288, 290

vegetation and, 266–267

Slumps, process and effects, 406–407

Small circles

characteristics of, 16–17

definition of, 17

Small-scale maps

characteristics of, 28–29

definition of, 28

Snow

definition of, 169

GeoDiscoveries on, 472

in glacier formation, 470–471

and river runoff, 569–570

Snowfall, and glacial mass budget, 471–472

Sodium, as rock component, 326

Soil. *See also* Erosion and erosional landforms; Mass wasting

basic properties of, 280–283

characteristics of, 289–292

chemistry of, 293–295

conservation measures, 290, 525

definition of, 280

fertility of, 294–295. *See also* Soil orders

GeoDiscoveries on, 295, 298, 305

human interactions with, 316–317

moisture characteristics of, 213–214

profiles and horizons, 295–298

salinization of. *See* Soil salinization

soil creep, 405–406

soil-forming factors, 285–288

soil-forming (pedogenic) processes, 283–288, 301–303, 305–306, 308–313

and urban heat islands, 102–106

Soil creep

definition of, 405

process and effects, 405–406

Soil-forming factors. *See also* Soil orders

characteristics of, 285–288

definition of, 285

Soil horizons

characteristics of, 295–296, 298, 302, 305, 311

definition of, 295

GeoDiscoveries on, 298

Soil orders

definition of, 298

geographic distribution, 300

organizational principles, 298–299

twelve soil orders, 299–315. *See also specific order*

Soil profiles

characteristics of, 295–296

definition of, 295

evolution of, 297–298

Soil salinization

in Australia, 580–583

definition of, 573

global extent of, 576

processes of, 573–575

in San Joaquin Valley (California), 576–580

Soil science

definition of, 298

relevance of, 298

in soil maintenance, 316–317

Soil structure, definition of, 292

Soil Survey, use of, 316–317

Soil taxonomy

definition of, 298

principles of, 298–299

Soil-water budget

definition of, 282

and soil texture, 292

Solar constant, definition of, 65

Solar energy. *See also* Solar radiation

power production, 86–87

and solar constant, 65

Solar noon, definition of, 56

Solar radiation

atmospheric flow, 76–78

clouds and, 163

Earth's surface and, 78–81

greenhouse effect and, 68–70

heat transfer and, 74–75

net radiation, 59

Solar system, components of, 46

Solar wind, and aurora formation, 97

Solifluction, definition of, 406

Solstices

characteristics of, 52–54

day length and, 56–58

Solum, definition of, 296

Sonoran Desert (United States), characteristics of, 264, 504–505
South America
annual temperature range, 106–109
biome distribution, 257, 258, 263
dams of, 465
forest cover changes in, 270–271
soils of, 299–300, 308–313
Southeast Asia
biome distribution, 257, 258
monsoons in, 133–134
Southeastern coniferous forest (United States), characteristics of, 260
Southern Hemisphere
annual temperature range, 106–109
atmospheric circulation patterns, 124
atmospheric pressure systems in, 116–118
climates of, 217
definition of, 17
ocean currents in, 139–141
ozone layer/hole, 71–73
seasons in, 51–55, 56–60
sunrise in, 60
Southern Nevada Water Authority (SNWA), actions of, 569–571
Southern Ocean, characteristics of, 530
South Pacific, coral reefs of, 551–553
South Pole, geographic grid and, 16–21
Southwest (United States)
desert regions of, 504–509
rock formations of, 392–393, 398, 400
water issues in, 566–572
Spanish Peaks (Colorado), geologic history of, 342–345
Spatial analysis, definition of, 5
Spatial resolution, definition of, 34
Species diversity, by biome. See specific biome
Specific humidity
atmospheric moisture and, 153–158
definition of, 154
Spheroidal weathering, process and effects, 399–400
Spits
definition of, 545
formation of, 545–546
Splash erosion, process of, 443
Spodic horizon, definition of, 305
Spodosols
characteristics and distribution, 299–300, 305–306
definition of, 305
Spring Equinox
day length and, 56–58
definition of, 53
Spring tides, characteristics of, 534–535
Sri Lanka, Sumatra-Andaman earthquake (2004), 376–378
Stable air
convectional precipitation and, 173–175
definition of, 173
Stalactites, formation of, 426
Stalagmites, formation of, 426
Star dunes, characteristics of, 516–517
Stationary fronts, definition of, 181

Steppe
characteristics of, 263–264
climates of, 217, 220–221
Stratopause, definition of, 93
Stratosphere
characteristics of, 93
definition of, 93
Stratovolcanoes. See Composite volcanoes
Stratus clouds
at frontal boundaries, 182
characteristics of, 164, 166–167
definition of, 164
Stream gradation
base level and, 447–448
evolution of streams and valleys, 449–453
floodplain evolution, 453–456
GeoDiscoveries on, 449
holistic model of, 456–458
processes and effects, 444–449
Stream hydrographs
of dammed rivers, 463–465
definition of, 439
urbanization example, 461
Stream hydrology, characteristics of, 438–442
Stream ordering
characteristics of, 436–438
in graded stream model, 456–458
Streams. See also Fluvial processes and landforms; Stream gradation
channel flow, 438–442
and drainage systems, 433–438
flooding of, 440–442
GeoDiscoveries on, 11, 437, 445, 451, 453
glaciofluvial processes, 486–488
human interactions with, 461–465
karst topography and, 425–428
meandering, 11, 444–445, 450–456
origin of, 432–433
rejuvenation of, 448, 455
stream ordering, 436–438
water tables and, 417, 419, 422
Stream valleys, evolution of, 449–455
Stress
faults and, 369–374
and joint formation, 395
Strike-slip faults
characteristics of, 372–374
definition of, 373
Subantarctic zones, location of, 19
Subarctic climates (Dfc, Dwc, Dwd)
biomes of, 256–257, 261, 264–265
characteristics of, 226–227
Subarctic zone
location of, 19
soils of, 299–300, 307–308
Subduction
definition of, 360
and plate convergence, 360–362
volcanic activity and, 381–383
Sublimation, definition of, 150

Subsidence
    definition of, 422
    groundwater depletion and, 422–423, 577–578
    isostatic adjustment and, 324–326
Subsolar point
    definition of, 47
    and ITCZ, 132–133
    and seasons, 51–55, 56–60
Subtropical High (STH) Pressure System
    arid landscapes and, 504
    circulation in, 128–129
    climate effects of, 222–223
    definition of, 128
    tropical savanna biome and, 263
Subtropical zones
    atmospheric circulation in, 128–129, 133–134
    biomes of, 256–257, 264
    location of, 19
    soils of, 299–300, 302–304
Sumatra-Andaman earthquake (2004), quake and tsunami,
        375–378
Sumeria, soil salinization in, 574–576
Summer Solstice
    day length and, 56–58
    definition of, 53
Sun. See also Earth-Sun geometry
    Earth's orbit around, 47–49, 53, 236–238
    GeoDiscoveries on, 58, 80, 238
    in photosynthesis, 251–254
    seasonal positional changes, 52–60
    solar energy production, 65, 86–87
    tides and, 533–535
Sun angle
    angle of incidence, 79–80
    definition of, 47
    GeoDiscoveries on, 58, 80
    and the seasons, 51–55, 56–60
Sun Belt, migration to, 568–569
Sun-synchronous orbit
    definition of, 34
    satellite imagery from, 34–35
Supercell thunderstorms, definition of, 195
Supercontinent theory, and continental drift, 354–355
Surf
    definition of, 537
    and drift, 537–538
Surface runoff
    flooding and, 440–442, 461–462
    stream water and, 432–433, 436, 440–442, 461–462
    suburbanization and, 565
Surface temperature
    air temperature and, 95
    GeoDiscoveries on, 101, 108–109
    global annual ranges, 106–109
    holistic assessment of, 106–109
    urbanization effects on, 102–106
Surface tension
    definition of, 282
    within soils, 282
Surface waves, in crust, 370
Surging, and glacial movement, 473

Sustainable development
    agriculture and. See Agriculture
    case studies in. See Case studies
    population growth and, 562–566
S waves, behavior of, 370
Synclinal valleys
    definition of, 365
    geomorphology and, 365–366
Synclines
    definition of, 364
    geomorphology and, 363–368
Systems. See also Ecosystems; Fluvial systems; Global
        radiation budget
    atmospheric pressure systems, 116–118
    definition of, 6
    glacial mass budget, 471–472
    in physical geography, 6–7, 40–41

T
Taiga, characteristics of, 261
Taliks, in permafrost landscapes, 496–498
Talus
    definition of, 404
    deposits of, 404–405
Tarns, definition of, 482
Tectonic plates. See Plate movements; Plate tectonics
Te Mata Peak (New Zealand), sunrise at, 60
Temperate rainforest, characteristics of, 260
Temperature. See also Atmospheric pressure; Climate
        change; Global warming
    in adiabatic processes, 160–162, 169–175
    of atmospheric layers, 92–95
    biome classification and, 256
    climate classification and, 214–216
    conversion formulas, 98
    dew-point, 156–158
    GeoDiscoveries on, 101, 108–109
    radiation and, 64–65
    scales, 95–96, 98
    in soil formation, 286–287
    surface. See Surface temperature
    thermohaline circulation and, 140–141
Temperature inversion, definition of, 164
Temporal lag, definition of, 100
Tephra, definition of, 380
Terminal moraines, characteristics of, 484–485
Tertiary Period
    characteristics of, 339–340
    mountain formation in, 367
Teton Mountains (Wyoming), internal processes of, 318
Texture, of soils, 291–292, 293
Thames River (United Kingdom), human uses of, 430–431
Thermodynamics, and heat transfer, 74–75
Thermohaline circulation
    characteristics of, 140–141
    definition of, 140
Thermosphere
    characteristics of, 94–95
    definition of, 94
Thornthwaite, C. Warren, climate classification and, 213–214
Thornthwaite system, and climate classification, 213–214

Three Gorges Dam (China), purposes and issues of, 465
Thunderstorms. *See also* Tornadoes
    definition of, 183
    evolution of, 173, 191–193
    GeoDiscoveries on, 196
    severe storms, 193–195
Tides
    characteristics of, 533–535
    and delta forms, 550–551
    GeoDiscoveries on, 537
Tidewater glaciers, characteristics of, 475, 477
Tilt obliquity, and climate change, 237–238
Timber industry, deforestation and, 270–272
Time
    geologic. *See* Geologic time
    geomorphology and, 364–368
    in soil formation, 288, 297–298
    and tides, 533–535
Time zones, and Earth's rotation, 48–50
Tombolos, definition of, 546
Tools of geographers
    geographic grid and, 16–21
    Geographic Information Systems (GISs), 40–41
    Global Positioning Systems (GPSs), 37–41
    maps, 22–32
    remote sensing. *See* Remote sensing
Topographic barriers, effects on temperatures, 106
Topographic maps
    conventions of, 31
    definition of, 31
Topographic winds, characteristics of, 138–139
Topography
    climate and, 212–213
    definition of, 31
    of drainage basins, 433–434
    geomorphology and, 364–368
    glaciation changes, 481
    soil salinization and, 573–574
    windflow and, 513–515
Tornado Alley, characteristics of, 198–200
Tornadoes
    evolution of, 195–197
    GeoDiscoveries on, 200
    geography of, 198–200
    monitoring of, 200–201
Tower karst, formations of, 427
Trade winds
    characteristics of, 126–129
    definition of, 126
Transformations, soil, characteristics of, 283–284
Transform plate margins
    characteristics of, 358–360
    definition of, 358
Transgressive dune fields, characteristics of, 547
Translocations, soil
    characteristics of, 283–284, 286–287, 296–298
    examples, 304–306
Transpiration. *See also* Evapotranspiration
    in climate classification, 213–214
    definition of, 159
    factors affecting, 159–160

Transported parent material
    definition of, 286
    in soil formation, 286
Trash, North Pacific garbage patch, 563–564
Tree lines
    characteristics of, 267–268
    definition of, 267
Tree ring patterns, in paleoclimatology, 231–233
Trellis drainage pattern, characteristics of, 435–436
Trenches, oceanic, 362
Triangulation
    of earthquake epicenter, 369–371
    and GPS, 38–39
Tributary glaciers, and glacial landforms, 481–483
Tributary streams. *See also* Streams
    definition of, 433
Trophic structure, of ecosystems, 250–251
Tropical (A) climates. *See also specific climate subtype*
    characteristics of, 215–219
    GeoDiscoveries on, 219
Tropical deciduous forest and scrub biome
    characteristics of, 256–257, 258–259, 267–268
    deforestation and, 271
Tropical depressions, definition of, 202
Tropical easterlies
    characteristics of, 126–129
    definition of, 126
Tropical monsoon climate (*Am*)
    biomes of, 256–258
    characteristics of, 217, 219
Tropical rainforest biome, characteristics of, 257–258
Tropical rainforest climate (*Af*)
    and biome, 256–258
    characteristics of, 214–217
Tropical savanna biome, characteristics of, 256–257, 261–263
Tropical savanna climate (*Aw*)
    biomes of, 256–259, 263
    characteristics of, 219
Tropical storms, definition of, 202
Tropical zones
    atmospheric circulation in, 126–129
    cyclones of, 201–207
    deforestation in, 270–272
    soils of, 299–300, 301–304
Tropic of Cancer
    definition of, 53
    and seasons, 53–54, 56–58
Tropic of Capricorn
    definition of, 53
    and seasons, 53–54, 56–58
Tropopause, definition of, 93
Troposphere
    characteristics of, 92–93
    definition of, 92
Troughs, wave, 536–537
Trunk streams
    definition of, 433
    in drainage basins, 433–438
Tsunamis
    definition of, 376
    Sumatra-Andaman, 376–378

Tundra biome, characteristics and distribution of, 264–265, 267–268, 406
Tundra climate (*ET*)
    biomes of, 264–265
    characteristics of, 227–228
Typhoons. *See* Hurricanes

U

Ubac slope
    definition of, 266
    vegetation and, 266–267
Ultisols
    characteristics and distribution, 299–300, 302–303
    definition of, 302
Ultraviolet (UV) radiation, and ozone formation, 70–73
United Nations Environmental Programme (UNEP), and climate change, 242
United States
    agricultural and grazing land-use, 272–273
    barrier islands of, 547–548
    biomes of, 256–257
    climate change issues, 243–244, 556–557
    deserts of, 504–509
    drainage basins of, 434–435
    glaciers of, 474–477
    life expectancy in, 562
    petroleum and, 345–348
    sea level changes, 556–557
    soil salinization in, 576
    soils of, 299–300, 303–313
    solar power in, 86–87
    time zones of, 48–50
    wind power in, 135–136
United States Department of Agriculture (USDA), and soil maintenance, 316–317
United States Environmental Protection Agency (EPA)
    on acid rain, 401
    on groundwater pollution, 423
    pollutant concerns, 74
United States Geological Survey (USGS)
    stream-gauging stations, 7, 438–439
    on Sumatra-Andaman earthquake, 376
    on tsunamis, 378
    volcano monitoring, 382–383
    and water management, 577–579
Universal Time Coordinated (UTC), measurement of, 50
Universe, nature of, 46
Unsaturated zones
    definition of, 414
    and groundwater depletion, 420–422
Unstable air
    convectional precipitation and, 173–175
    definition of, 173
    in thunderstorms, 191–193
Upcutting. *See* Stream gradation
Updrafts, definition of, 193
Upwelling currents, definition of, 141
Urban heat island
    characteristics of, 102–106
    definition of, 103
    suburbanization and, 565

Urbanization
    and stream behavior, 461–462
    temperature effects of, 102–106
    urban sprawl, 565, 568–572
Urban sprawl, environmental impacts of, 565, 568–572
Utah
    and Colorado River Compact, 568–572
    landforms of, 392–393, 507

V

Valley breezes, definition of, 138
Valley glaciers, characteristics of, 475, 477
Valleys
    flooded, 532–533
    geomorphology and, 365–366
    glacial troughs, 481, 482–483
    stream, 449–456
Vancouver (British Columbia), geography of, 105
Variable gases
    carbon dioxide, 68–70
    definition of, 67
    ozone, 70–73
    water vapor, 67–68
Vegetation. *See also specific region or biome*
    climate classification and, 213–216
    distribution factors, 255
    GeoDiscoveries on, 254, 269, 273, 305
    human influences on, 270–275
    karst topography and, 424
    local/regional factors in, 266–269
    proxy data and, 230–233
    sand dune formation and, 515
    soil formation and, 311–313
    Subtropical High (STH) Pressure System and, 128
    succession and, 268–269
    urban heat islands and, 102–106
Venezuela, oil reserves of, 346–348
Venice (Italy), subsidence of, 423
Ventifacts
    characteristics of, 511–512
    definition of, 511
Vernal equinox. *See* Spring Equinox
Vertical zonation
    characteristics of, 267–268
    definition of, 267
    in Qionglai Mountains, 585–586
Vertisols
    characteristics and distribution, 299–300, 303–304, 316
    definition of, 303
Volcanic arcs
    at plate boundaries, 362, 381–383
    definition of, 381
Volcanic ash, effects of, 388
Volcanoes
    ash plumes from, 3, 114, 388
    cinder-cone volcanoes, 379–380
    as climate change factor, 236
    composite volcanoes, 380–381, 477
    definition of, 379
    GeoDiscoveries on, 383, 389
    hotspots, 386–388

Volcanoes *(cont.)*
    and igneous rock, 327–330
    of Pacific Ring of Fire, 381–383
    shield volcanoes, 385–386
    and soil formation, 315
    types of, 379–380, 385–386
    underwater, 362

W

Wales, coastal landforms of, 542
Warm fronts
    characteristics of, 182
    definition of, 182
Warning systems. *See* Weather monitoring and prediction
Washington (United States)
    Channeled Scablands, 490–491
    glaciers of, 474–476, 482
    Skykomish River flooding, 440
    volcanic activity, 381–383
Water. *See also* Coastal processes and landforms ; Fluvial processes and landforms; Streams
    albedo of, 79–80
    freshwater, 412–413, 566–572
    GeoDiscoveries on, 424
    global geographic distribution of, 151
    groundwater. *See* Groundwater
    humidity, 152–159
    hydrosphere and hydrologic cycle, 151–152
    maritime effect of, 101–103
    phase changes of, 148–151
    photosynthesis and, 251–254
    physical properties of, 148–151
    as primary geomorphic agent, 430–432
    proximity to and deserts, 504
    resource management, 568–572, 577–580
    riparian zone effects, 269
    soil roles, 281–282
    urban vs. rural movement of, 104–105
    in weathering, 395–400
Waterfalls
    at hanging valleys, 483
    in evolution of stream valleys, 449–450
    Niagara Falls (North America), 446, 448
Water rights
    definition of, 568
    issues of, 568–572
Watershed. *See* Drainage basins
Water table. *See also* Streams
    cave formation and, 425–426
    characteristics of, 414–415, 417
    definition of, 415
    groundwater extraction and, 419–422
    soil salinization and, 573–574, 577–578
Water vapor
    evaporation and transpiration, 159–160
    humidity and, 152–158
    properties of, 67–68, 70, 148–151
Wave amplitude
    definition of, 64
    in electromagnetic spectrum, 64–66
Wave-cut platforms, formation of, 540–541

Wavelength
    definition of, 64
    and electromagnetic spectrum, 64–66
Wave refraction, definition of, 540
Waves (water). *See also* Coastal processes and landforms
    characteristics of, 536–537
    definition of, 536
    and delta forms, 550–551
    GeoDiscoveries on, 544
Weather, definition of, 212
Weathering. *See also* Arid landscapes
    chemical, 398–400
    climate and, 394
    definition of, 394
    factors affecting, 394
    GeoDiscoveries on, 402, 410
    mechanical, 395–398
    in soil formation, 285–286
Weather monitoring and prediction
    clouds in, 164, 166–167
    hurricane, 207
    tornado, 200–201
    tsunami warning systems, 377–379
Weather systems, cyclonic. *See* Cyclonic weather systems
Wegener, Alfred, on continental drift, 354–355
Wells, groundwater depletion and, 420–422
Westerlies
    definition of, 130
    in midlatitude circulation, 129–132
Wet adiabatic lapse rate (WAR)
    characteristics of, 161–162
    definition of, 162
    in precipitation processes, 170–175
Wheat, soil salinization and, 575–576, 580–583
Wildfires
    ecological effects of, 262
    overgrazing and, 273
Wilson Peak (Colorado), geography of, 2–3
Wilting points
    definition of, 282
    and soil water, 414
Windbreaks, and soil conservation, 525
Wind chill index, calculation of, 98
Wind energy, power production, 135–136
Windflow. *See also* Airflow
    and cyclone migration, 189–191
    direction, 117–118, 119–125
    in sand dune formation, 513–515
    speed, 120, 123, 159–160, 203
Windflow patterns, effects on temperatures, 106
Winds. *See also* Eolian processes and landforms; Thunderstorms; Tornadoes
    global, 126–134. *See also specific zone*
    local, 137–139
    oceanic circulation and, 139–140, 141–143
    topographic, 138–139
    wind energy, 135–136
Wind speed
    and evaporation rates, 159–160
    and storm surges, 203

Windward side
    definition of, 138
    in orographic processes, 170–172, 220–221, 223, 228–229
Winter Solstice
    day length and, 56–58
    definition of, 53
Wisconsin glaciation
    eustatic change in, 531–533
    history of, 489–492
Wolong Giant Panda Reserve Center, research of, 586–587
Wolong Nature Reserve (China), research work at, 586–588
World Meteorological Organization (WMO), and climate
        change, 242
World Wildlife Fund (WWF), actions of, 588
Wyoming, and Colorado River Compact, 568–572

X
Xerophytic plants, definition of, 264

Y
Yardangs
    characteristics of, 511–513
    definition of, 511

Year, calculation of, 47
Yellowstone National Park (Wyoming), as volcanic hotspot,
        387–388
Yosemite National Park (California), rock formation in, 398,
        399

Z
Zonal flow, definition of, 130
Zone of ablation
    definition of, 471
    and glacial mass budget, 471–472
Zone of accumulation
    definition of, 471
    and glacial mass budget, 471–472
Zones
    climate. *See specific zone*
    of glaciers, 471
    of latitude, 18–19
    time zones, 48–50